Nematology – Advances and Perspectives

Volume I: Nematode Morphology, Physiology, and Ecology

Editors: Z. X. Chen
S. Y. Chen
D. W. Dickson

Tsinghua University Press
CABI Publishing

CABI Publishing is a division of CAB International

CABI Publishing
CAB International
Wallingford
Oxfordshire OX10 8DE
UK

Tel: +44 (0)1491 832111
Fax: +44 (0)1491 833508
E-mail: info@cabi.org
Website: www.cabi-publishing.org

CABI Publishing
745 Atlantic Avenue
8th Floor
Boston, MA 02111
USA

Tel: +1 (617)682-9015

E-mail: cabi-nao@cabi.org

A catalogue record for this book is available from the British Library, London, UK.

The Library of Congress has cataloged the hardcover edition as follows:

Nematology : advances and perspectives / editors, Z.X. Chen, S.Y. Chen, D.W. Dickson.
p. cm.
Includes bibliographical references.
ISBN 0-85199-645-0 (alk. paper)
1. Nematoda. 2. Plant nematodes. 3. Soil nematodes. I. Chen, Z. X. II. Chen S. Y. III. Dickson, Donald W. (Donald Ward)

QL391.N4N43 2003
592'.57--dc21

2003048906

CABI ISBN 978 1 78639 929 8 (PB)

Tsinghua University Press ISBN 7-302-06312-5

Published in China by Tsinghua University Press, Xue Yan Building, Tsinghua University, Beijing, 1 00084, PR China.

Printed and bound by CPI Group (UK) Ltd, Croydon, CR0 4YY

Preface

Nematology, Advances and Perspectives

During the 20th century, science and technology have developed in an unparalleled manner. This is especially true for nematology. Since Cobb first recognized nematology as an independent discipline during the early part of the century, nematology has made unparalleled advances and become an integral part of the biological sciences. The development of nematology is largely attributed to the discovery of the importance of nematodes in agricultural ecosystems and their impact on society. Nematodes are the most abundant and diversified group in the animal kingdom, and four out of five animals on earth are nematodes. Marine nematology has become an independent discipline and it has been suggested that the secret of the natural history of our planet may lie in the nematodes dwelling deep in oceans. Animal-parasitic nematodes have had great impact on human heath and society throughout history. Soil nematodes play important roles in organic degradation, mineralization, and food webs in soil ecosystems. While most nematode species in soil are beneficial, some species are important pathogens of plants and cause severe damage to crops. Worldwide crop yield losses to nematodes have been estimated at approximately $ 78 billion annually. The advances in nematology and related disciplines, however, have drastically strengthened our ability to fight this unseen enemy. Thus, over the past century, billions of dollars of crop losses from nematode damages have been prevented.

One of the most exciting and important new fields of nematology includes recent advances made in the use of nematodes as model organisms for basic biological studies, especially in developmental biology, genetics, and cellular and molecular biology. Using *Caenorhabditis elegans* as a model, scientists have unraveled for the first time in history the complete genomic sequence of DNA from a multicellular organism. Also, they have documented cellular development from a single fertilized cell to a fully developed body.

Nematology is one of the most dynamic and exciting disciplines in the biological sciences. It is impossible to cover every aspect of nematology in this book. Furthermore, several topics in nematology reached their climax during the past century, but interest has now ebbed. The major focus of this book is on topics that have made remarkable advances in the latter part of the century and currently are of primary research emphasis. These topics will relate mainly to free-living, plant-parasitic, and entomopathogenic nematodes.

The following lists of chapters evolved after lengthy discussion and consultation with fellow nematologists. We are pleased that so many renowned nematologists with international reputations and experiences have responded positively to the book proposal and agreed to make authoritative contributions. It is overwhelming to have nematologists around the world making a collaborative effort to produce a quality book. The book is to be published in 2004. Therefore, we shall consider it a memorial to the achievements in nematology in the 20th century and as a beacon for future developments in nematology during the new century. For many reasons, we believe that nematology will become a primary focus in the agricultural sciences during the new century. With the rapid growth of the world's population, food and fiber demands in the next century will increase tremendously. But the task for nematologists is daunting in that funding, personnel in nematology, and graduate education has been sharply curtailed. Also, with the suspensions and phase-outs of many soil fumigants and nonfumigants and the threat for even more suspensions in the foreseeable future, the challenges will become even greater. Increased endeavors in nematode research will be essential to achieve the sustained growth of agriculture in the 21st century.

The objective of this book is to summarize advances in nematology that have been made during the 20th century and to provide perspectives for the development of nematology in the next century. It is aimed at researcher and graduate student communities. The international representation of the science will be a key component of this book. It is our hope that the book will provide a road map to the most important aspects of the science at this time and that it will provide critical thoughts and ideas for researchers and graduate students to carry forward into the next millennium.

The editors are indebted to the authors of the respective chapters for their excellent contributions. We are grateful to the large number of scientists who have provided valuable comments and suggestions during book proposal discussion and reviews of each individual chapter manuscripts. Finally, we express our gratitude to the Tsinghua University Press and CAB International for fulfilling this endeavor.

Editors

Zhongxiao Chen
National Food Safety and Toxicology Center
Michigan State University
East Lansing, Michigan

Senyu Chen
Southern Research and Outreach Center
University of Minnesota
Waseca, Minnesota

Donald W. Dickson
Entomology and Nematology Department
University of Florida
Gainesville, Florida

This Book Is Dedicated To

Alan Francis Bird (1928 – 1999)

Alan Bird was born on 11 February 1928 in Seremban, Malaysia. He died suddenly in Adelaide, Australia on 13 December 1999. He began his education in Malaysia (then the Federated Malay States) but went to boarding school in Northern Ireland in 1937. Apart from his final year at a school in Dublin, he completed his primary and secondary education there. Towards the end of 1946, Alan went to Perth, Western Australia to join his parents whom he had not seen for almost ten years, due to the intervention of World War II. He was awarded a B. Sc. (Hons) from the University of Western Australia in 1952.

Soon after graduation, Alan went to Adelaide, South Australia as a Demonstrator in Zoology. It was there that he developed a passion for nematodes which was to last for the rest of his life. He graduated from the University of Adelaide (M. Sc. in Zoology) in 1955.

At the end of 1954, newly married, Alan went to Scotland, where, as an Assistant Lecturer in Zoology he taught helminthology and nematode physiology. He also managed to complete, in what must be close to record time, a Ph. D. while he was in Edinburgh. He graduated towards the end of 1956 and returned to Australia early in 1957. He was awarded a D. Sc. from the same university in 1975. The title of his thesis ("The nematode cuticle") reflected his continuing interest in nematodes.

During his illustrious career, Alan published two editions of "The structure of nematodes" and 120 papers, in most of which he was the sole or principal author. His interests were wide ranging and, as well as classical plant nematology, encompassed nematode ecology, ultrastructure, behavior and nematode-microbe interactions. He was particularly interested in relating structure to function. Alan was principally a zoologist and believed strongly that the distinctions between plant and animal parasitic nematodes were artificial and that nematodes should be considered as a whole, distinct group.

In his distinguished career, Alan was made a Fellow of the Society of Nematologists (1983), a Fellow of the Australian Society for Parasitology (1993), an Honorary Member of the Helminthological Society of Washington

(1997), and an Honorary Fellow of the Royal Society of South Australia (1999). He had been awarded the Sir Joseph Verco Medal for Scientific Research of that Society in 1991.

Alan Bird was a warm, kind, gentle and unassuming man with a keen sense of humour. He listed his hobbies as microscopy and history but was also very fond of sport, particularly rugby union football which he had played with distinction as a young man. When he watched a rugby match, either in person or on television, Alan became quite different from his usual quiet self. He shouted and told the players what to do! He watched the telecast of every match in the 1999 World Cup Series, regardless of the fact that most of the games were played in the early hours of the morning Australian time. It was a source of great joy to him that Australia won the World Cup for the second time and he would have been very proud of Australia's rugby achievements in 2000, had he lived to see them.

Alan was an outstanding and inspirational nematologist who was still actively engaged in research at the time of his death. His enthusiasm for research never waned and he unfailingly encouraged and helped young scientists. He is sadly missed not only by his family and friends but also by his many scientific colleagues.

Contents

List of Contributors

Kenneth R. Barker	Professor. Emeritus, Plant Pathology Department, North Carolina State University. Raleigh, NC 27695, USA Email: krbpp@unity. ncsu. edu
John M. Webster	Professor. Centre for Environmental Biology, Simon Fraser University. Burnaby, Vancouver, BC, Canada Email: jwebster@sfu. ca
Marie-Anne Félix	Institut Jacques Monod, C. N. R. S., Universités Paris 6 and 7, Tour 43, 2 place Jussieu, 75251 Paris cedex 05, France Email: felix@ijm. jussieu. fr
James G. Baldwin	Professor. Department of Nematology University of California. Riverside, CA 92521, USA Email: james. baldwin@ucr. edu
Roland N. Perry	Professor. Department of Entomology and Nematology, IACR Rothamsted. Harpenden, Herts, AL5 2JQ, UK Email: roland. perry@bbsrc. ac. uk
Richard S. Hussey	Department of Plant Pathology, University of Georgia. Athens, GA 30602 – 7274, USA Email: hussey@uga. cc. uga. edu
Eric L. Davis	Department of Plant Pathology. Box 7616, N. C. State University. Raleigh, NC 27695-7616, USA Email: eric_davis@ncsu. edu
Alan F. Bird B. Sc., MSc, Ph. D., D. Sc.	2 Playford Rd, Mitcham, South Australia 5062, Australia Email: jeanbird@ozemail. com. au(Contact person: Dr. Jean Bird)

A. Forest Robinson	USDA, ARS. College Station, TX 77845, USA Email: frobinson@cpru. usda. gov
Gregor W. Yeates	Landcare Research Private Bag 11052, Palmerston North, New Zealand Email: YeatesG@Landcare. cri. nz
Brian Boag	Scottish Crop Research Institute, Invergowrie, Dundee DD2 5DA, Scotland Email: b. boag@scri. sari. ac. uk
P. John D. Lambshead	The Natural History Museum, London SW75BD, United Kingdom Email: pjdl@nhm. ac. uk
Robert McSorley	Professor. Department of Entomology and Nematology, University of Florida, Gainesville, FL 32611, USA Email: rmcs@gnv. ifas. ufl. edu
Larry W. Duncan	Professor. Citrus Research and Education Center, University of Florida, Lake Alfred, FL 33850, USA Email: lwdn@icon. lal. ufl. edu
Robin M. Giblin-Davis	Professor. University of Florida, 3205 College Avenue Davie, FL 33314, USA Email: giblin@ufl. edu
Kerrie A. Davies and Gary S. Taylor	Visiting Research Scientist and Research and Publications Manager, Department of Applied & Molecular Ecology, Waite Campus, University of Adelaide. PMB 1 Glen Osmond, South Australia 5064, Australia
W. Kelley Thomas	Codirector and Associate Professor. Hubbard Center for Genome Studies, University of New Hampshire. 35 Colovos Road, Durham, NH 03820, USA
Paul De Ley and Manuel Mundo-Ocampo	Department of Nematology, University of California. Riverside, Riverside, CA 92507, USA Email: pdeley@citrus. ucr. edu manuel. mundo@ucr. edu

1 A Century of Plant Nematology

Kenneth R. Barker

1.1 Introduction

Although most nematologists consider our science to be very young, Chitwood and Chitwood (1974) concluded that nematology has a history of its own that is quite as ancient as that of any other field of zoology. They further deduced that the history of this science, in general, parallels the political and social development of Europe. During the last four decades, nematology has developed as an independent, though often limited discipline in many countries throughout the world. Historically, nematodes were important in the development of general zoology (Chitwood and Chitwood, 1974) and in production agriculture (Wilson, 1982). These organisms, however, often have been and continue to be overlooked. The "home" for the study of nematodes was a major issue for N. A. Cobb who proposed the word *Nematology* for the discipline and was largely responsible for the creation of the "Division of Nematology", United States Department of Agriculture (USDA) in 1918 (Campbell, et al., 1999). Even then, scientists continued to be uncertain as to who should study the often severe nematode problems in agriculture as well as fundamental research. In 1938 the Dean of the School of Agriculture at the University of Florida, H. H. Hume, succinctly described the limited awareness of nematodes and their roles in agriculture as follows: "If there were no nematodes in the South (USA) and they should suddenly appear in their present numbers, they would be seen as the pestilence they are. Funds for their control would be supplied quickly and in large amount. Such was the situation which allowed complete eradication of the Mediterranean fruit fly. But the nematode problem arouses no particular interest. Nematodes are always working havoc, taking their toll on crops, sometimes causing complete destruction. We blind ourselves by accepting them as a matter of course. The problem is here; we live with it; but there is no solution. There must be a general awakening all along the line to the magnitude of this situation" (Wilson, 1982).

Many advancements related to the biology, ecology, diversity, and function of nematodes were recorded prior to the 1940s. During the 1940s and 1950s, however, the development of nematode-resistant cultivars, the discovery of effective and inexpensive nematicides, and the characterization of the role of nematodes and disease complexes contributed much to an increased awareness of the role of plant-parasitic nematodes. Even today, with readily available and extensive crop-loss data related to plant-parasitic nematodes (Koenning, et al., 1999; Sasser and Freckman, 1987), however, a great need for increased awareness and fundamental research on the role of nematodes in the diverse niches of our biosphere is still encountered. Undoubtedly, the recently completed characterization of the genome of the microbivorous nematode *Caenorhabditis elegans* (Esnard, 1999; Hodgkin, et al., 1998; Riddle, et al., 1997; The *C. elegans* Sequencing Consortium, 1998), and the ecological research on the role of beneficial nematodes in soil and plant health (Barker and Koenning, 1998; Blaxter, 1998) will have a major impact on the general awareness and evolving history of nematology.

Since the innumerable recent developments in nematology will be discussed in detail in subsequent chapters, this treatment will focus primarily on the period of 1870 through 1970 – the first Century of Plant Nematology. Thus, this treatment will not address the many advances on animal-parasitic nematodes nor marine species. Even with this restricted emphasis, some of the very significant early contributions to this discipline likely have been excluded due to oversight and/or information being lost.

1.2 Early Records and References to Nematodes

Excellent accounts of the early history of nematology, with emphasis on developments in Europe and North America, are given by Chitwood and Chitwood (1950, 1974), Thorne (1961), Poinar (1988), and Viglierchio (1991). Coomans (1992) summarized the extensive and diverse contributions by nematologists in the European "Low Land Countries". More recently, de Moura (1998, 2000) presented the historical account of major advancements in Brazil, including contributions by Goeldi and Lordello. Other treatments of recent developments in nematology, including those in Europe, are discussed by Evans (1995) and Webster (1998). Somewhat selective reviews of nematology in Canada (Potter, 1998) and the USA (Barker, 1998a, 1998b) also are available. Other historical treatments focus on the discovery and development of information on given nematode species or groups. For example, Noel (1992) presented a fascinating story of how soybeans were brought into the USA and the early observations on the long association of damage on this crop by the soybean cyst nematode, even before 235 B. C. in China. In addition to a historical treatment, Poinar (1983) provided a synopsis of the rather sparse but intriguing fossilized resin records of nematodes such as *Oligaphelenchoides atrebora* that date back between 20 and 120 million years. Finally, Sayre (1994) has given an incisive perspective of the role of art in nematology and science with a focus on the work of N. A. Cobb and associates.

Thorne (1961) suggested that knowledge of animal-parasitic nematodes is probably almost as ancient as the history of man. Ancient scholars and leaders who addressed the issues of worms and nematodes included Moses, Hippocrates, Agatharchides, Aristotle, Pliny, Celsus, and Vegetius (Chitwood and Chitwood, 1974). Two of our first records of parasitic nematodes (Table 1.1) appear in passages in the Chinese Yellow Emperor's classic (2700 B.C.) and in the "Papyrus Ebers" records of the early Egyptian

Table 1.1 Early Development of Nematology[a].

2700 B. C.	China Yellow Emperor's Classic of Internal Medicine (Huang di Nei Ching) discussed symptoms of the intestinal worm, *Ascaris*.
1550 B. C.	Egyptian "Papyrus Ebers" referred to the intestinal worm, *Ascaris*; Guinea worm, *Dracunculus*; and possibly a hookworm, *Ancylostoma*.
1250 B. C.	Middle East Bible 21: 6 – 9. Moses referred to the Guinea worm as the "Fiery serpent of the Israelites". The serpent of brass constructed by Moses may be the origin for the symbol of the physicians = caduceus.
430 B. C.	Hippocrates in "Aphorisms" discussed nematodes (probably pinworm, *Enterobius*).
384 – 322 B. C.	Aristotle in "Historia Animalium" discussed nematodes (*Ascaris* and others).
181 – 146 B. C.	Agatharchides credited by Plutarch for describing the "Guinea worm" and so naming it.
53 B. C. – 7 A. D.	A. C. Celsus distinguished between roundworms (nematodes) and flatworms (cestodes).
100 A. D.	Columella noted an ascarid (*Neoascaris vitulorum*) from a calf.
400 A. D.	Vegetius noted the horse ascarid (*Parascaris equorum*).
980 – 1037 A. D.	Avicenna referred to ascarids and dracunculids in a medical text.
1200 – 1280	A. Magnus first noted nematodes from birds in the falcon.

Continued

1519 – 1630	Caesalpinus discovered *Dioctophyma renale* (kidney worm) in the dog kidney.
1547	Vinegia described a filarid and two intestinal nematodes from the falcon.
1650	O. Leeuwenhoek constructed the microscope.
1656	P. Borellus discovered the vinegar eelworm (*Turbatrix aceti*).
1667	Robert Hooke reported an eelworm in paste (*Panagrellus redivivus*).
1683	E. Tyson conducted early attempts at nemic anatomy by describing morphological structures.
1684	F. Redi compiled reports of worms in vertebrates and new hosts, e. g., lion, fish.
1707 – 1778	Linnaeus established modern taxonomic nomenclature.
1743	Discovery of first plant-parasitic nematode *Anguina tritici* by Needham.
1780	A prize offered for the best assay on intestinal worms by the Copenhagen Academy of Sciences indicated the widening interest of biologists in nematodes and other worms. Incidentally, the winner's theory that the entire life history of these worms took place within the host tissue was in fact largely erroneous, whereas the one loser's opposing view was not. "The triumph of form over substance apparently has been a confounding factor in science for some time."
1835	R. Owen discovered *Trichinella spiralis* in human muscle.
1855	M. J. Berkeley discovered the root-knot nematode.
1857	Kühn described the bulb and stem nematode, *Ditylenchus dipsaci*.
1859	H. Schacht discovered the sugarbeet nematode, which effected the collapse of the German economy of the time.
1865	R. Leuckart and I. I. Metchnikoff established the alternation of generation in the life history of animal-parasitic nematodes and the obligate development of nematodes in an intermediate host. H. C. Bastian published the description of over 100 new species of free-living nematodes.
1871	Kühn first to use soil fumigation to control nematodes (*H. schachtii*).
1878	P. Manson established the mosquito as the intermediate host in transmission of *Wuchereria bancrofti* (elephantiasis).
1879	Cornu characterized the root-knot nematode, *A. marioni*.
1883	Van Benden's description of meiosis (in *Parascaris equorum*).

Continued

1892	Atkinson and Goeldi independently discovered disease complexes associated with root-knot nematodes.
1934	K. I. Skrjabin published a book "Veterinary Parasitology and Invasion Diseases of Domestic Animals" and became the nematode authority in the U. S. S. R. A. A. Paramanov, a student of Skrjabin, advocated the evolutionary approach to nematological study.

[a]In large part, after Viglierchio (1991).

physicians (1550 B. C.). The China Yellow Emperor's classic of internal medicine included discussions of the symptoms associated with the intestinal worm, *Ascaris* (Viglierchio, 1991). The Egyptian "Papyrus Ebers" also referred to the intestinal worm *Ascaris lumbricoides* and the guinea worm, *Dracunculus medinensis* (Fig. 1.1), and possibly the hookworm, *Ancylostoma* (Viglierchio, 1991). According to Chitwood and Chitwood (1974), the term "fiery serpents", as recorded in the Bible (Numbers 21: 6 – 9), may have referred to guinea worms. Chitwood and Chitwood (1974) discussed the Guinea-worm story at length, including the following passage attributed to

Figure 1.1 Extraction of the Guinea worm (*Dracunculus medinensis*) (after Chitwood and Chitwood, 1974 (originally after Velschius, 1674)).

Agatharchides of Cnidus (181 – 146 B. C.), teacher of Ptolmaeus Alexander: "That the people taking ill on the Red Sea suffered from many strange and unheard of attacks, amongst others, worms, upon them, which gnawed away their legs and arms, and when touched, retracted themselves up in the muscles and there gave rise to the most unsupportable pains; but their evil has only been found then, neither before nor since amongst any other people."

During the years between 450 B. C. and 200 A. D., general knowledge of the people, especially that of physicians of the day, apparently was reduced to writing. The "aphorisms" of Hippocrates (ca. 430 B. C.) included statements on the existence of worms. For example, he states: "To persons somewhat older, afflictions of the tonsils, and curvature of the spine at the vertebra next to the occiput, asthma, calculus, roundworms, ascarids . . ." (Chitwood and Chitwood, 1974; Viglierchio, 1991). Aristotle (384 – 322 B. C.), often considered the "father of zoology", mentions nematodes on numerous occasions in his "Hitoria Animalium".

Nematodes were differentiated from cestodes (flat worms) by Celsus (53 B. C. –7 A. D.). Excluding a few advancements (Table 1. 1), Chitwood and Chitwood (1974) suggested that the next ensuing several hundred years could be considered the dark ages of nematology, as was true for all of science. The renewal of scientific investigation was delayed until the 16th century. Upon the development and application of the microscope, however, advances in knowledge of nematodes came rapidly. For a discussion of the discoveries of this period, including many of those listed in Table 1. 1, see Chitwood and Chitwood(1974).

1.3 The Early Foundations of the Science of Nematology

Although the origin of nematology was gradual, 1870 has been fixed as an approximate date as the "beginning of recent nematology" (Chitwood and Chitwood, 1974; Thorne, 1961). The description of *Anguina tritici* by T. Needham (1743), Carolus Linnaeus (1768) and J. A. Scopoli (1777) and others (see Campbell, et al., 1999; Thorne, 1961) was a somewhat isolated but crucial landmark event in the initial characterization of a plant-parasitic nematode. Needham dissected distorted wheat plants and observed many minute fibers. Further, he placed these fiber-like structures in water, and observed their coming to life. While now recognized as the first documentation of anhydrobiosis, this observation was used by many scientists of the day to support the ongoing concept of spontaneous generation of organisms.

Figure 1.2 Henry Charlton Bastian, 1837 – 1915.

Figure 1.3 Otto Bütschli, 1848 – 1920.

Figure 1.4 Emilio Goeldi.

Figure 1.5 Heinrich Micoletzky, 1883 – 1929.

Figures 1.2 – 1.5 Early pioneers in nematology (Figs 1.2, 1.3, and 1.5 after Thorne, 1961; Fig. 1.4 courtesy of Jaime Maia dos Santos).

Dujardin (1845), Diesing (1851), and Schneider (1866) advanced the field of taxonomy by adding numerous species and compiling the information into a more useable form. Bastian (Fig. 1.2) in 1865 contributed much to the field of nematology through the development of descriptions of 100 new species of free-living nematodes in a single paper. Thorne (1961) concluded that Bastian's "monograph of the Anguilullida" (1866) marked the beginning of nematology as a science. However, he further indicated that perhaps the credit for founding the science of nematology should belong to Bütschli (1873) (Fig. 1.3), cited in Thorne (1961) rather than Bastian. This statement was based on Bütschli's providing the first detailed descriptions of the morphology of free-living nematodes, including a number of characters that are still in use in differentiating genera and species of nematodes today. The classic taxonomic monograph of Johannes G. de Man (1876), cited in Thorne (1961) closely followed the contributions of Bütschli (Table 1.2; Thorne, 1961). Heinrich Micoletzky (Fig. 1.5) in Austria published his first paper on nematodes in 1914, and in 1922 provided an extensive key to species and discussion of diverse groups encountered in different habitats. The title of the latter paper was "Die freilebende Erd-Nematoden" (Thorne, 1961).

Table 1.2 Some of the important books and monographs published during the first 100 years of nematology as an emerging science[a].

Author	Year	Title	Reference
Henry Charlton Bastian	1866	"Monograph of the Anguillulidae"	Thorne, 1961
A. Schneider	1866	"Monographie der, Nematoden"	Nicholas, 1975
Otto Bütschli	1873	"Beiträge zur Kenntnis der freilebenden Nematoden" [detailed morphology of microbivorous nematodes]	Thorne, 1961
Johannes Govertus de Man	1876	"Die einheimschen, frei in der reinen Erde und im süssen-Wasser Iebenden Nematoden der Neiderlandischen Fauna" [taxonomic monograph]	Thorne, 1961
E. Maupas	1900	"Modes et formes de reproduction des nématodes"	Nicholas, 1975
Ernst A. Bessey	1911	"Root knot and its Control"	Thorne, 1961
Nathanial A. Cobb	1913	"Contributions to the Science of Nematology"	Thorne, 1961

Continued

Author	Year	Title	Reference
N. A. Cobb	1918	"Estimating the Nema Populations of Soil"	Thorne, 1961
Heinrich Micoletsky	1922	"Die freilebende Erd-Nematoden" [keys/discussion of nematodes in different habitats]	Thorne, 1961
Tom Goodey	1933	"Plant Parasitic Nematodes and the Diseases They Cause"	Thorne, 1961
I. N. Filipjev	1934	"Nematodes That Are of Importance to Agriculture"	Thorne, 1961
G. Thorne and H. H. Swanger	1936	"A Monograph ..." (see reference)	
Benjamin G. Chitwood & Mabel B. Chitwood	1937	"An Introduction to Nematology"	
W. Schneider	1939	"Freilebende und pftanzenparasitische nematoden"	
G. Thorne	1939	[Compilation of well known species] "A Monograph..."(see reference)	
I. N. Filipjev & J. H. Schuurmans-Stekhoven, Jr.	1941	"Manual of Agricultural Helminthology"	Thorne, 1961
C. Overgaard Nielsen	1949	"Studies on the Soil Microfauna"	Nicholas, 1975
Benjamin G. Chitwood & Mabel B. Chitwood	1950	"An Introduction to Nematology"	
Michiel Oostenbrink	1950	"The Potato Nematode (*Heterodera rostochiensis* Wollenweber) a Dangerous Parasite to Potato Monoculture" (in Dutch)	Coomans, 1992
Hans Goffart	1951	"Nematoden der Kulturpflanzen Europas" (summary of plant parasites in Europe)	Thorne, 1961
Tom Goodey	1951	"Soil and Fresh Water Nematodes"	Thorne, 1961
M. T. Franklin	1951	"The Cyst-forming Species of *Heterodera*"	
C. L. Duddington	1957	"The Friendly Fungi"	
J. R. Christie	1959	"Plant Nematodes – Their Bionomics and Control"	

Continued

Author	Year	Title	Reference
J. F. Southey	1959	"Plant Nematology"	
Joseph N. Sasser and W. R. Jenkins (eds.)	1960	"Nematology – Fundamentals and Recent Advances with Emphasis on Plant Parasitic and Soil Forms"	
William F. Mai and H. L. Lyon	1960	"Plant-Parasitic Nematodes – A Pictorial Key to Genera" (with five subsequent editions)	
Gerald Thorne	1961	"Principles of Nematology"	
A. M. Shepherd	1962	"The Emergence of Larvae from Cysts in the Genus *Heterodera*"	
A. D. Baker	1962	"Check Lists of the Nematode Superfamilies Dorylaimoidea, Rhabditoidea, Tylenchoidea, and Aphelenchoidea"	
A. A. Paramonov	1962, 1964	"Principles of Plant Nematology" Vol. I & II[b]	Nicholas, 1975
Harry R. Wallace	1963	"The Biology of Plant Parasitic Nematodes"	
J. B. Goodey	1963	"Soil and Fresh Water Nematodes"	
J. B. Goodey, M. T. Franklin, and D. J. Hooper	1965	T. Goodey's "The Nematode Parasites of Plants Catalogued Under Their Hosts" (3rd Revision)	
D. L. Lee	1965	"The Physiology of Nematodes"	
J. E. Peachy	1969	"Nematodes of Tropical Crops"	
Neil A. Croll	1970	"The Behaviour of Nematodes"	
A. F. Bird	1971	"The Structure of Nematodes"	
Zuckerman, B. M., W. F. Mai, and R. A. Rohde (eds.)	1971	"Plant Parasitic Nematodes" 2 Vols. (with 3rd volume in 1981)	
J. M. Webster	1972	"Economic Nematology"	
L. G. E. Lordello	1973	"Nematoides Das Plantas Cultivadas"	
Harry R. Wallace	1973	"Nematode Ecology and Plant Disease"	

[a] The many nematological books published after 1973 are not included here (see literature citations).

[b] See paramanov literature citations for English translations.

More than 100 years elapsed after Needham's report on *A. tritici* before additional plant-parasitic nematodes were described. After Berkeley's (1855) observation of "Vibrios" from galls or root knot on the roots of cucumber in England, a number of important plant-parasitic nematodes were described during a short period (Thorne, 1961). Although overlooked until the classic work by Chitwood in 1949, Goeldi (Fig. 1.4) in 1887 first used the genus *Meloidogyne* for root-knot nematodes that were attacking Brazilian coffee. He also indicated that these nematodes were predisposing coffee to attack by root-infecting fungi. Of interest, de Moura (1998) recently provided a tribute to Goeldi through a special publication of that 1887 classic on root-knot nematodes. Julius Kühn (1857) found *Anguillula dipsaci* (*Ditylenchus dipsaci*) infecting the heads of teasel *Dipsacus fullonun* (Thorne, 1961). In 1859, H. Schacht described a major disease caused by a nematode that had been inflicting severe losses in sugarbeet fields of Germany for many years (see Campbell, et al., 1999; Thorne, 1961). This nematode problem had developed as the commercial sugarbeet industry had grown rapidly in Germany in the early 1800s. This pest was named *Heterodera schachtii* by A. Schmidt (1871), and Strubell (1888) published a detailed morphological characterization of the species. Kühn in 1881 also found cyst nematodes attacking potato, but apparently concluded that they were *H. schachtii* (Thorne, 1961). This cyst problem on potato was not resolved until 1914 when Zimmermann concluded that *H. schachtii* was a pathogen of potato. Wollenweber in 1923 deduced that the cysts on potato were of a different shape and named that parasite *H. rostochiensis* (Thorne, 1961).

The first major efforts to control a plant-parasitic nematode focused on the sugarbeet cyst nematode in Europe from 1870 to 1910 (Thorne, 1961). Kühn was the primary leader in this campaign and undoubtedly established several of the strategies and tactics still used for nematode management. He and others directed efforts toward characterizing host plants, life history, habitats, distribution, etiology, and methods of control. His application of carbon disulfide in 1871 in sugarbeet fields was apparently the first soil fumigation for nematode control. Unfortunately, this chemical proved to be too difficult and expensive to apply as a control method (Campbell, et al., 1999). Kühn and his associates also invested extensive efforts on the use of "trap crops" or *Fangpflanzen*, including the planting of rape and other rapidly growing hosts in an attempt to induce nematode juveniles to hatch and emerge from the brown cyst. The goal was the apparent removal and destruction of infected plants such as rape (canola) and thereby prevention of the completion of the nematode's life cycle. Repeating this procedure several times during a growing season was not highly effective, whereas crop rotation proved to be the most effective and economical control tactic (Campbell, et al., 1999; Thorne, 1961). Surprisingly, the principal methods of nematode management for-mulated by Kühn and his associates are still employed in both Europe and the United States today.

The research on animal-parasitic nematodes during the 1800s greatly advanced our understanding of the nature of parasitism as well as the biology of these parasites (Chitwood and Chitwood, 1974). Research in the late 1800s resulted in striking advancements of our knowledge on the life history, anatomy, and systematics of various nematode groups. After experimental failures by a number of earlier workers, several researchers including Grassi (1887, 1888; cited in Chitwood and Chitwood, 1974) and others developed experimental evidence on the direct infection of man by *Ascaris* eggs. The discovery that mosquitoes act as intermediate hosts in the transmission of *Wuchereria bancofti* by Manson (1878) received much attention as it was the first case involving a biting insect as an intermediate host (Chitwood and Chitwood, 1974). Subsequent discoveries of interest include: insects also serve as intermediate hosts for *Physocephalus* and *Ascarops* (Seurat, 1913); the discovery of migration of *Ascaris lumbricoides* through the lungs by Stewart (1916); and the addition of a new type of an intermediate host (earthworms) for lungworms, by Hobmaier (1929) (Chitwood and Chitwood, 1974). The contributions by Chitwood and Chitwood (1974) on nematode structure, embryology and post-embryonic development clearly are classics in their own right.

Joseph Leidy, in 1851, provided the first record of a free-living nematode in America (cited in Thorne, 1961). His taxon for this nematode, *Anguillula longa* n. s., was later redescribed as *Trilobus longus* (Leidy) by Cobb in 1914. George Francis Atkinson in Alabama and J. C. Neal in Florida in 1888 made independent studies on the morphology, host range, and crop damage caused by root-knot nematodes and published their results in 1889 (Atkinson, 1889; Campbell, et al., 1999). Before initiating his research at the Agricultural and Mechanical College of Alabama Polytechnic Institute at Auburn, Atkinson had earned his doctorate (with training in botany and zoology) from Cornell University in 1885 and held posts for short terms as professor of zoology at the University of North Carolina and professor of botany and zoology at the University of South Carolina (Campbell, et al., 1999). In addition to expanding Kühn's strategies and tactics for nematode management, Atkinson (1892) also concluded that these nematodes predispose cotton to Fusarium wilt.

As already indicated, the 1890s was a very productive decade for nematology. Atkinson's first paper "Nematode Root-Galls", published in

1889, superseded and corrected related research done in Florida by J. C. Neal. The 1889 paper by Atkinson created much interest in the American plant pathology community with many understanding the work even though they were not familiar with nematodes (Campbell, et al., 1999). Statements of appreciation included: Pammel of Iowa – "one of the notable bulletins of the year '89, and is the most comprehensive document upon nematodes that has been published in this country", and Galloway stated "such publications as this raises a station greatly in ... estimation" (Campbell, et al., 1999). The year 1889 also was significant in the USA in that Frank Lamson-Scribner had just initiated his research on the lesion nematode on potato in Tennessee.

An immense leap in the management of plant-parasitic nematodes occurred very early when scientists commenced selecting and breeding cultivars for host resistance. The first early suggestion of potential root-knot nematode resistance appeared in Neal's 1889 entomology bulletin in which he recommended selected orange-tree cultivars to control these parasites (Campbell, et al., 1999). In 1897, Albrecht Zimmerman showed *Coffea arabica* to be more susceptible to root-knot than *C. liberica*; he, in turn, recommended the grafting of the more desirable *C. arabica* onto the resistant rootstock of *C. liberica* (see Campbell, et al., 1999; Thorne, 1961). The earliest practical research in the United States occurred in 1902 with H. J. Webber and W. A. Orton discovering that Iron cowpea was resistant to root knot as well as wilt diseases (Campbell, et al., 1999). Webber concluded their work to be "a genuine out and out discovery... which... opens up important new fields" since "so far as my knowledge of this class of diseases goes no resistant strain of any effected plants [*sic*] have ever been known" (Campbell, et al., 1999).

Additional leads on crop resistance followed during the next few years, including evidence offered by A. D. Shamel and W. W. Cobey (1907) for root-knot resistance in some tobacco strains (Campbell, et al., 1999). Charles Bessey found certain grape cultivars, particularly American lines with resistance to phylloxera, to have nematode resistance. Orton, in his continuing search for wilt resistance, noted that wilt-resistant watermelon was free of root knot. He recommended that "simple selection can be and ought to be practiced by everyone who grows his own seed; more complicated breeding work, unless performed by men who devote considerable time to it, hardly pays of the time and expense required" (Campbell, et al., 1999).

Even after these rather notable advances in nematode management, fundamental research on nematode biology remained stalled during the early 1900s (Campbell, et al., 1999). The USDA provided additional but limited resources, and in 1900 assigned Ernst A. Bessey (Fig. 1.16) to research plant-pathogenic nematodes. While not an expert on nematodes, he worked on nematodes and many other projects. Still, Bessey finally published an excellent summary bulletin on root-knot nematodes in 1911. He attempted to clarify the confusing root-knot nematode taxonomy, and incorrectly grouped all previously used names under one species, *Heterodera radicicola*. Campbell et al. (1999) concluded that Bessey's efforts to resolve the taxonomy morass still were justified, showing the need for more coordination among scientists in the field. Correcting the lack of understanding of the magnitude of existing nematode diversity and almost non-existent knowledge on nematode physiology and morphology was paramount. Unfortunately, major research advancements on the root-knot nematodes were delayed until the work of Christie and Albin (1944) on host races of these pathogens and Chitwood's 1949 milestone taxonomic treatment on *Meloidogyne*. These topics, while receiving much study in the 1950s, continue to be researched today.

A singular important event for the development of nematology in the United States occurred in 1907 – Nathan Augustus Cobb (Figs 1.6 – 1.7) joined the USDA. Cobb, an eccentric but brilliant individual, was born on June 30, 1859 in Spencer, Massachusetts. He had earned his doctorate at the University of Jena in Germany in 1889 (in only 10 months). He then worked at the Zoological Research Station, Naples, Italy where he collected nematodes and mounted them in balsam, apparently representing some of the first permanent slides of nematodes. While in Naples, Cobb named his first genus, the marine nematode *Tricoma*; over his career, he described well over 1000 species of nematodes (Campbell, et al., 1999). Cobb's relationship with a noted landscape painter, Joseph Greenwood, apparently helped him hone his own unique artistic skills (Campbell, et al., 1999). Thorne (1961) indicated that Cobb then proceeded to Australia where he sold watches for some time until an opening for a plant pathologist developed in the New South Wales' Department of Agriculture in 1890. While training himself as a plant pathologist, Cobb's first paper on plant-parasitic nematodes, "*Tylenchus* and root galls", was published in the Agricultural Gazette of New South Wales in 1890. With the limited means of communication in that day, he apparently was unaware that the names *Heterodera* and *Meloidogyne* had been proposed previously for gall-forming nematodes (Campbell, et al., 1999).

In addition to research on taxonomy and morphology, Cobb had keen interests in nematode ecology, their effects as parasites of farm animals, and how they impacted the economics of crop production. For example, he prepared capsules containing eggs of *Oxyuris* spp., swallowed these materials and proved that these nematodes could survive in humans (Campbell, et al., 1999). A key breakthrough made in 1906 involved his observations on *Dorylaimus bulbiferus* that nematodes have parasites later known as *Pasteuria* spp. He also coined a number of new nematological terms, including *amphid*,

Figure 1.6

Figure 1.7

Figures 1.6 – 1.7 Nathan Augustus Cobb, 1859 – 1932 (after Thorne, 1961) and N. A. Cobb at age 14 (after Sayre, 1994).

deirid, *phasmid*, and *nema*, a shortened version of *nematode*. Impressed by Cobb's research in Australia, E. F. Smith recommended him for a USDA position in 1905. For his first two years with the USDA, Cobb worked as the first chief of the Hawaiian Sugar Planters' Association Experiment Station and studied sugarcane diseases. He finally arrived in Washington, DC in 1907, but it was not until 1915 that he was able to study nematodes exclusively (Campbell, et al., 1999).

Prior to turning to nematology full time, Cobb developed USDA cotton grades for the standardization of cotton quality. This system and the scale for cereal-rust assessment, which he devised while in Australia, were widely used for decades. After his work on cotton, his attention was redirected back to nematodes. The first paper "New nematode genera found inhabiting fresh-water and non-brackish soils" was published in 1913. Cobb and his associates made amazing contributions that provided a framework with exceptionally high standards for the burgeoning science of nematology. Thorne (1961) and Campbell et al. (1999) concluded that Cobb's microscopic abilities and unusual ingenuity, coupled with the artistry of his associate W. E. Chambers (Fig. 1.8), established a new era in nematology on which our present science is largely founded, not only in America, but in other parts of the world. In

1914, Cobb initiated a publication, "Contributions to a Science of Nematology", a periodical which he paid for himself. During the next year, he published "Nematodes and Their Relationships" as a feature article in the *Yearbook of the Department of Agriculture* which he viewed as the unveiling of a new field of science with the goal of convincing readers of its potential (Campbell, et al., 1999). That article and specific titles such as "Estimating the Nema Populations of Soil" (1918) provided many of the concepts, methods, and apparati that are still in use in the study of nematodes today. For example, his suggested methods for sampling nematode populations and procedures for their extractions or modifications thereof, are employed throughout the world.

Although the glamour of handsome instruments was unimportant to Cobb, he continually made changes in his laboratory instrumentation to improve their functionality (Sayre, 1994). Cobb's most famous invention was a revolving table on which as many as eleven microscopes were placed. This system facilitated the successive examination of a number of nematode specimens, enabling Cobb to compare the taxonomic features of one after the other in rapid sequence by rotating the table. Clearly, N. A. Cobb had a unique interest and ability in the development of complex equipment and apparati. To ensure vibration-free viewing of microscopic specimens, he used steel pillows embedded in concrete as mountings for his microscope (Sayre, 1994). This system with a steel pedestal showing a chain-driven mechanism for adjusting the heights of the microscope and a drawing of his apparatus for utilizing sunlight as a source of light for the microscope are depicted by Sayre (1994). Thorne (Fig. 1.10), who was trained in Cobb's laboratory, offered this impression: "Often these devices resulted in weird and complicated assemblies, for he (Cobb) lacked the ability to simplify either apparati or procedures." His laboratories were always a showplace of revolving tables with several microscopes on each and devices to perform laboratory techniques. In addition to this techniques such as the Cobb sieving extraction procedure and his invention of the metal slide have been widely used. Another important contribution was the use of a simple beam splitter and large mirror to replace the old type camera lucida.

J. R. Christie (Fig. 1.9) (1966) stated in a Society of Nematologists News Letter article entitled "Days with N. A. Cobb", that he neither had the language nor the ability to write a character sketch of this unusual, picturesque, and likable person. He further indicated that to all who knew Cobb well, none bore him enmity. At that time Cobb was called "the Skipper", a name used with much respect and affection. After contributing much during a transition period with great research accomplishments, N. A. Cobb died suddenly at the age of 73. Christie concluded that to break through

Figure 1.8 W. E. Chambers, 1866 – 1920.

Figure 1.9 Jesse Roy Christie, 1889 – 1978.

Figure 1.10 Gerald Thorne, 1890 – 1975.

Figure 1.11 A. D. Baker, 1894 – 1974.

Figures 1.8 – 1.11 Selected early North American nematologists (Figs 1.8 – 1.9 after Thorne, 1961; Fig. 1.11 after Nematology News Letter 12(4), 1966).

Figure 1.12 Gotthold Steiner, 1866 – 1961.

Figure 1.13 Benjamin G. Chitwood, 1907 – 1972.

Figure 1.14 George H. Godfrey.

Figure 1.15 Albert L. Taylor, 1901 – 1988.

Figures 1.12 – 1.15 Early United States nematologists (continued) (after Thorne, 1961).

the wall of skepticism that hemmed N. A. Cobb in, was more than could be accomplished by the efforts of one man. Although Cobb unfortunately did not live to see the transition in plant nematology (Christie, 1966), his contributions to nematology, including fundamental and applied research, training of the first generation of true nematologists in the USA, and his efforts to enhance awareness of nematodes and nematology, definitively established his niche as one of the most important individuals in plant pathology in the twentieth century (Campbell, et al., 1999).

The first generation of true nematologists included many exceptionally able scientists in the USA and around the world. Thanks, in large part, to the efforts and contributions by N. A. Cobb, the U. S. Department of Agriculture was on the cutting edge of organized research on nematodes. Gotthold Steiner (Fig. 1.12) succeeded Cobb as the leader of the Nematology unit. Other leading nematologists and members of the USDA nematology division included B. G. Chitwood (Fig. 1.13), M. B. Chitwood, G. H. Godfrey (Fig. 1.14), A. L. Taylor (Fig. 1.15), Edna M. Buhrer (Fig. 1.18), Grace S. Cobb (Fig. 1.19), Helen H. Swanger, Louise Grossman, Josephine Danforth, Florence E. Albin, Margaret N. Corder, and Sevilla Brooks (Thorne, 1961).

A dedicated nematologist of true genius in that second generation was Benjamin G. Chitwood (Fig. 1.13). The 1937 classic "An Introduction to Nematology" (revised in 1950 and 1974) by B. G. Chitwood and M. B. Chitwood clearly stands alone as a superb early contribution to nematology in the USA. Their detailed descriptions of the anatomy and structure of nematodes provided much of the foundation on which nematode systematics is based. This includes later publications such as his 1949 paper in the Proceedings of the Helminthological Society of Washington in which he resolved the confusion in the diversity and taxonomy of root-knot nematodes.

Several interesting Memorium views of Benjamin Goodwin Chitwood's life (1907 – 1972) were presented by a number of his friends and associates in the Nematology News Letter (1973). He received a Ph. D. degree from George Washington University in 1929, but commenced working as a junior nematologist under the direction of N. A. Cobb in the USDA in 1928. Among his several positions, he was appointed as the first Chief Nematologist of the Florida State Plant Board in 1955 where he established the Nematology Section. Although sometimes offering challenges for his associates, Chitwood had a profound impact on nematology over a 40-year career. His exceptionally informative books and his almost 200 publications still serve as invaluable resources today. While working on nematodes parasitizing vertebrates, invertebrates and plants, free-living and marine forms, Chitwood's update and

Figure 1.16 Ernst Bessey, 1877 – 1957.

Figure 1.17 Walter Carter, 1897 – 1977.

Figure 1.18 Edna M. Buhrer, 1898 – 1988.

Figure 1.19 Grace S. Cobb, 1905 – 1993.

Figures 1.16 – 1.19 Selected early United States nematologists (continued) (after Thorne, 1961).

revision of the genus *Meloidogyne,* the establishment of the two nematode classes, and his detailed nematology books were among his most important contributions.

In Canada, nematologists generally worked in one-person or small units, as is still true for much of our discipline (Potter, 1998). The first research program in that country was initiated by W. Newton. He and his associates characterized the biology and management for various nematodes on a number of crops. Alex D. Baker (Fig. 1.11), a contemporary of Steiner and Thorne, contributed much to the taxonomy of *Heterodera* and other nematodes (Baker, 1962). In the 1950s, W. B. Mountain modified Koch's Postulates for proving the pathogenicity of nematodes in plant diseases (Mountain, 1960). The establishment of the Nematode Taxonomic Unit in the Agriculture Canada Biosystematics Research Center in Ottawa in the 1950s and its success through the 1980s (Potter, 1998) reflect the importance and international focus of nematology during that period.

The very limited number of exceptional scientists working on nematodes during the first five decades of the 20th century made many significant advances in our discipline. J. G. de Man's (Fig. 1.20) contributions to nematology in Europe are comparable to those of Cobb in the USA and focused largely in the area of taxonomy of marine, freshwater, and soil nematodes; he also contributed much in the area of nematode anatomy (Coomans, 1992). Jan Ritzema bos, also around the turn of the century, described damage caused by plant-parasitic nematodes including instances especially involving *Aphelenchoides* and *Ditylenchus* species (Coomans, 1992). As indicated under subsequent topics, Tom Goodey and others provided crucial early contributions to nematology.

Nematologists in the Netherlands have long collaborated with associates in many other countries. For example, Thorne and de Man frequently corresponded and benefitted mutually from their interactions. J. H. Schuurmans Stekhoven (Fig. 1.21), a Dutch nematologist with a keen interest in plant-parasitic as well as free-living nematodes, provided taxonomic training for L. De Coninck (Fig. 1.22) from Belgium. According to Coomans (1992), this was the first connection between nematology in the Netherlands and Belgium. These scientists also published a number of joint papers on marine nematodes. Coomans (1992) also provides a most insightful review of the many significant contributions to nematology by the European "Low Land" scientists. In 1948 Van den Brande (Fig. 1.23) organized a yearly "International Symposium on Phytopharmacy and Phytiatry" in which many nematologists participated. The contributions by Oostenbrink (Fig. 1.24) to nematode management, including population dynamics, and those sentinel crop-damage and population models developed by Seinhorst (Fig. 1.25) provided a much needed nudge to nematologists worldwide. For equally important contributions by Low Land scientists to nematode systematics, population assessment methods, and biology, the reader is referred to Cooman's 1992 treatment.

Figure 1.20 J. G. de Man, 1850 – 1930.

Figure 1.21 J. H. Schuurmans Stekhoven, 1892 – 1958.

Figure 1.22 L. de Coninck, 1909 – 1988.

Figure 1.23 J. Van den Brande, 1899 – 1989.

Figure 1.24 M. Oostenbrink, 1922 – 1979.

Figure 1.25 J. W. Seinhorst, 1918 – 1997.

Figures 1.20 – 1.25 Pioneer Nematologists in the European "low land countries" (after Coomans, 1992).

1.4 Nematology Textbooks and Education

While the outstanding monographs and other key publications by Bastian, Schneider, Bütschli, de Man, Bessey, Maupas, Cobb, and Micoletsky served as a framework for the development of the science of nematology, the first of many textbooks on plant nematodes was by Tom Goodey (Fig. 1.27). Goodey's 1933 book entitled "Plant-parasitic Nematodes and the Diseases They Cause" clearly represents a milestone in nematology. A careful examination of the major nematology books published in the 1930s (Table 1.2) tends to dispel the notion that nematology developed as a science only after the discovery of D-D by Carter (Fig. 1.17) in 1943. For example, Tom Goodey's 1933 nematology textbook encompassing nine chapters included treatments on most groups of damaging plant-parasitic nematodes. More than 50 species of plant-parasitic nematodes such as *A. tritici*, *D. dipsaci*, *A. angusta* (rice nematode), *Tylenchulus semipenetrans*, *H. schachtii*, *H. punctata*, *H. marioni*, *Aphelenchoides fragariae*, *A. cocopholus*, *A. ritzema-bosi*, *Hoplolaimus coronatus*, *Pratylenchus bukowinensis*, and *Criconema rusticum* were discussed in detail. Chapter 9 by Tom Goodey was entitled "Biological Races in Plant-parasitic Nematodes with Extensive Discussion on the Host Race Specialization with *D. dipsaci*". He also had a chapter that included parasitic nematodes of doubtful pathogenicity such as some of the *Aphelenchoides* sp.,

Aphelenchus avenae, *Hoplolaimus cornatus* and others as well as extensive treatments on the bacterivorous and predaceous nematodes known at that time. Tom Goodey and associates, including his son J. B. Goodey (Fig. 1.38), published a wealth of nematology books (Table 1.2).

In 1934, the original publication by I. N. Filipjev (Fig. 1.26) entitled "Nematodes That Are of Importance to Agriculture" was published in Russia (this book was later translated to English as discussed later). Chitwood and Chitwood's nematology classic "An Introduction to Nematology" appeared 3 years later. The rarely matched quality of this book by the Chitwoods and subsequent additions clearly has served as a cornerstone for the development of nematology, especially in relation to anatomy, structure, function, and even physiology. Another milestone contribution came through the collaboration of I. N. Filipjev and J. H. Schuurmans-Stekhoven (1941) which resulted in the widely used nematology reference book "A Manual of Agricultural Helminthology". In fact, this invaluable early textbook, focusing on general morphology, biology, systematics, techniques, plant-parasitic as well as insect-parasitic forms, served as a model for W. R. Nickle's 1991 "Manual of Agricultural Nematology". A number of additional Russian nematology books have been translated to English, including the two volumes on "Plant Parasitic Nematodes", by Paramonov (Fig. 1.30; Table 1.2).

With the severe setback of nematology and related biological sciences caused by World War II in the late 1930s through 1945, the increasing numbers of aspiring nematologists had major problems in locating up-to-date textbooks on this subject. The Chitwoods' 1950 revision of their introductory nematology

Figure 1.26 I. N. Filipjev, 1889–1940.

Figure 1.27 Tom Goodey, 1885–1953.

Figure 1.28 Hans Goffart, 1900 – 1965.

Figure 1.29 Mary T. Franklin, 1911 – 1982.

Figure 1.30 A. A. Paramonov, 1891 – 1970.

Figure 1.31 Minora Ichinohe

Figures 1.26 – 1.31 Selected international nematology pioneers (Figs 1.26 – 1.29 after Thorne, 1961; Fig. 1.30 after Nematology News Letter 16(3), 1970; Fig. 1.31 courtesy of N. Ishibashi and M. Ichinohe).

textbook and Tom Goodey's 1951 "Soil and Freshwater Nematodes" were excellent reference books. Michiel Oostenbrink's 1950 study on the potato cyst nematode is documented in a 230-page book (in Dutch) that served as a basis for the development of legislative measures for combating this nematode that included obligate or required crop rotation, field inspection, and quarantine in the Netherlands (Coomans, 1992). In 1951, Hans Goffart (Fig. 1.28) summarized the plant-parasitic nematodes in Europe. The "Cyst-forming species of *Heterodera*" by Mary Franklin (Fig. 1.29) was released during that same year. The 1957 book "The Friendly Fungi" by Duddington enhanced the interest of many in studying the antagonists and parasites of nematodes as a means of their biological control. The series of books edited by Southey, especially the 1959 and 1978 editions of "Plant Nematology", J. B. Goodey's 1963 revision of T. Goodey's "Soil and Fresh Water Nematodes", and the various editions of T. Goodey's "Host Index" for nematodes are still used worldwide. Thorne with his numerous monographs and the 1961 textbook, Christie with his 1959 sorely needed book "Plant Nematodes – Their Bionomics and Control", Sasser with a number of key books and especially the 1960 "Nematology..." co-edited by W. R. Jenkins, Lee's treatments on the physiology of nematodes, and Neil Croll's series of books ("The Behaviour of Nematodes", "The Organization of Nematodes", "Biology of Nematodes" – with B. E. Matthews) contributed much to the advancement of the discipline. Subsequent textbooks developed by Bird (Fig. 1.39) and Bird (1971, 1991), Dropkin (Fig. 1.36) (1980, 1989), Lamberti and Taylor (1979), Lordello (Fig. 1.40) (1973), Norton (1978), Poinar (1983), Sasser (Fig. 1.33) and Carter (1985), Southey (1978), Wallace (Fig. 1.42) (1963, 1973), Zuckerman and other co-editors (1971, 1980, 1981), and others added greatly to the development of nematology (see Table 1.2, list of recent books at end of chapter text, and literature cited).

Figure 1.32 Merlin W. Allen, 1912 – 1974.

Figure 1.33 Joseph N. Sasser.

Figure 1.34 Vernon G. Perry.

Figure 1.35 William (Bill) B. Mountain.

Figure 1.36 Victor H. Dropkin, 1916 – 1995.

Figures 1.32 – 1.36 First five presidents of the Society of Nematologists (in part, after Nematology News Letter 12(3), 1966; photo of M. W. Allen courtesy of Mrs. Bonnie Allen and Ms. Mae Noffsinger, and photo of J. N. Sasser courtesy of Marvin Williams).

Figure 1.37 F. G. W. (Freddie) Jones.

Figure 1.38 J. Basil Goodey, 1914 – 1965.

Figure 1.39 Alan F. Bird, 1928 – 1999.

Figure 1.40 Luiz Gonzaga E. Lordello, 1926 – 2002.

Figure 1.41 William (Bill) F. Mai.

Figure 1.42 Harry R. Wallace.

Figures 1.37 – 1.42 Additional "second-generation" international nematologists (Figs 1.37 – 1.38 courtesy of R. N. Perry; Fig. 1.39 courtesy of David Bird; Fig. 1.41 courtesy of George Abawi; Fig. 1.42 courtesy of David Bird and H. R. Wallace).

As indicated by the number of new books (Table 1.2), the period of the 1950s through the early 1970s brought much exciting growth to the field of nematology. This included the development of formal as well as specialized short courses in nematology, the organizations of nematology departments and other centers for research and graduate education. The formation of nematology societies, and specialized nematology journals reflected the growth of the discipline. Most nematology programs in the USA encompassed research and teaching as well as outreach/extension, reflecting the "Land-Grant University tradition" (Westerdahl and Caswell-Chen, 1998).

With an explosive demand for courses in nematology, a wide range of short courses as well as formal university courses were organized in the USA and in other countries. The first graduate-level course was taught in 1948 by Merlin Allen (Fig. 1.32) at the University of California at Berkeley where he rapidly developed an international graduate program in nematology. Allen, who trained in nematology with G. Thorne, moved to the University of California campus at Davis in 1958. In the Nematology Department there, Allen and Dewey Raski and others continued and expanded superb graduate and research programs. In 1955 – 1956, Joseph Sasser and Hedwig Hirschmann taught their first year-long series of two nematology courses at North Carolina State University. During the following year (1956 – 1957), W. F. Mai (Fig. 1.41) taught the first nematology course at Cornell University. The academic program in nematology at the University of Florida also was initiated in the 1950s with George Weber teaching the first course. J. R. Christie was recruited to lead nematology research, and V. G. Perry, after completing a leave and doctorate with G. Thorne at the University of Wisconsin, led the teaching of nematology. Additional early faculty members joining the University of Florida group included A. J. Overman, G. C. Smart, Jr., and A. C. Tarjan. A second Department of Nematology with graduate and research programs was established at the University of California at Riverside with a faculty that included R. G. Baines, I. J. Thomason, S. D. Van Gundy, and S. A. Sher.

During the 1950s, many plant pathologists, entomologists, and other agricultural scientists trained themselves to work in the field of nematology. This endeavor was augmented through the establishment of a number of short courses, including those sponsored by Shell Chemical Company, and more intensive courses offered at a number of universities. For example, J. N. Sasser and W. R. Jenkins organized an intensive course at North Carolina State University that involved leading scientists from around the world as lecturers (Barker, 1998a). Those lectures resulted in an important textbook entitled "Nematology, Fundamentals and Recent Advances with Emphasis on Plant-parasitic and Soil Forms" edited by Sasser and Jenkins (1960). A similar course was also offered shortly thereafter at Cornell University.

Gotthold Steiner, leader of the USDA Nematology Division after Cobb, developed a successful collaborative educational program with the University of Maryland, through which a number of third generation nematologists, including J. N. Sasser and A. C. Tarjan, earned their doctorates in the early 1950s. After considerable downsizing, the USDA nematologists still continue as the largest group in the discipline in the USA, but the graduate program has been very limited in recent years. Even with a relatively early focus on nematological research in the USA, only three academic nematology departments have been organized, these being at the University of California, one at Davis campus and one at Riverside campus, and the Department of Entomology and Nematology at the University of Florida, Gainesville. Although housed in Plant Pathology or Entomology Departments, the nematologists at Purdue University (G. Bergeson, V. Ferris, and J. M. Ferris), Cornell University (led by W. F. Mai), and N. C. State University (J. N. Sasser, H. Hirschmann Triantaphyllou, C. J. Nusbaum, N. T.

Powell, A. C. Triantaphyllou, and others) developed and have continued major education/research centers for nematology.

A number of departments devoted solely to nematology have been established in other countries. Some of the major departments or groups in Europe include the one at Rothamsted in the UK; the Landbougheschool, Gent, Belgium; the Institut voor Pflantezeiktenkundig, Wageningen; and the Division of Nematology, IARI, New Delhi. As in the USA, additional research and academic groups of nematologists are housed in other disciplinary units such as plant pathology, entomology, and plant science.

The development of formal courses dedicated to nematology was very slow in Europe (Webster, 1998). One or two scattered courses were offered at different institutions, including the London School of Hygiene and Tropical Medicine offered by B. G. Peters and D. Fenwick and at the School of Agriculture, Cambridge University, offered by Eddie Don and F. G. W. (Freddie) Jones from 1949 to 1950 (Webster, 1998). The first short course including lectures and labs, offered by F. G. W. Jones (Fig. 1.37), was in 1955 at the Botany School, Cambridge University (Webster, 1998). The first comprehensive educational program in nematology in the UK involved a one-year diploma course at the Silwood Park Field Station of Imperial College in 1958. That course resulted from the efforts of B. G. Peters who was the first professor of parasitology at Imperial College, University of London, and also was the former head of the nematology department at Rothamsted (Webster, 1998).

In addition to the courses developed in Europe and the USA, interest in nematodes in other countries soon increased. In 1961, F. G. W. Jones joined with local scientists and developed a course on plant nematodes in the Department of Botany, Aligarh Moslim University in India (Webster, 1998). Upon the subsequent detection of the potato cyst nematodes in South India, endeavors directed towards these organisms were increased further. A second nematology course, developed by Freddie Jones in concert with J. Basil Goodey from Rothamsted and Dewey Raski from the University of California at Davis, was offered at the Agriculture Research Institute, Delhi (Webster, 1998).

Many international, collaborative educational/research programs have and continue to contribute much to nematology. These efforts range from support for graduate stipends by industry-, private-, and government-agency foundations to major projects such as the USAID-North Carolina State University-sponsored International *Meloidogyne* Project (IMP) that involved more than 700 scientists in more than 60 countries. That program and the subsequent "Crop Nematode and Control Project" were directed by J. N. Sasser. Both programs, with the numerous regional conferences, intensive courses, books and other specialized publications, crop-loss data for nematodes, and the *International Nematology Network Newsletter* increased the worldwide awareness of nematodes. Volumes I and II of "An Advanced Treatise on *Meloidogyne*" (Barker, et al., 1985; Sasser and Carter, 1985) and "Biology, Identification and Control of Root-knot Nematodes (*Meloidogyne* species)" (Taylor and Sasser, 1978) were distributed worldwide.

Many other individual scientists, including D. J. Raski, A. M. Golden, J. W. Seinhorst, M. Oostenbrink, P. A. A. Loof, A. D. Baker, R. H. Mulvey, F. G. W. Jones, J. B. Goodey, H. R. Wallace, A. F. Bird, R. S. Pitcher, J. F. Southey, M. T. Franklin, A. Shepherd, C. E. Taylor, M. Luc, F. E. Caveness, I. J. Thomason, S. D. Van Gundy, M. Golden, G. Steiner, M. W. Allen, G. Thorne, J. R. Christie, W. F. Mai, and M. Ichinohe contributed much to the international nematology community. Universities and other research organizations in the UK, France, Italy, Germany, The Netherlands, Japan, the Americas, and other countries have played important roles in advancing nematology worldwide. Foundations such as the International Foundation of Science and various agencies continue to support international projects, including nematology.

1.5 Professional Societies and International Nematology Meetings

Professional nematological organizations developed in parallel or shortly after academic programs in nematology were established. For example, the First International Nematology Symposium was held in 1951 at Harpenden, UK. According to Webster (1998), Tom Goodey was responsible for the development of this symposium with B. G. Peters organizing the meeting. This group had a second meeting in Copenhagen in 1953 in association with the Zoological Congress. In 1955, P. A. van der Lan, M. Oostenbrink and J. W. Seinhorst organized the third and larger meeting of this group at which time the Society of European Nematologists (now European Society of Nematologists) was organized. The first journal of nematology (*Nematologica* (now *Nematology*)) commenced publication one year later (1956) with a volume containing 358 pages (Webster, 1998).

Throughout the 1950s, nematology as a distinct discipline was still under debate. In fact, nematodes continued to receive minimal study throughout much of the world. For example, when the first nematology library was established by Orstrom in the Ivory Coast in 1955, only nine published references on plant-parasitic nematodes were available in the whole of West Africa and Zaire (Luc et al., 1990). Luc, et al. (1990) estimated that only 400 scientists worked full- or part-time on nematode problems in subtropical and tropical agriculture. While most major crop problems caused by plant-parasitic nematodes have been characterized, new disease problems induced by these pathogens continue to be discovered in subtropical and tropical agriculture (Luc, et al., 1990).

In North America, most nematologists during the 1950s and even the early 1960s were affiliated largely with the societies of plant pathology such as the American Phytopathological Society. The *Nematology News Letter*, initially edited by Quentin Holdeman (1956 – 1959), as a member of the Southern Nematology Research Project (S-19) in the USA, served a critical role in the communication among nematologists in North America. This news letter was a primary vehicle for interactions among nematologists in the continent prior to the establishment of the *Journal of Nematology* in 1969. Although the Society of Nematologists had been organized in 1961, much discussion-debate ensued before S. D. van Gundy was finally appointed as the first editor-in-chief of that journal. This breakthrough – the establishment of the *Journal of Nematology* – required continuous work that extended beyond the tenure of the first five Presidents (Figs 1.32 – 1.36) of the young Society. Since that time, more than 30 societies of nematology have been organized around the world. The Organization of Nematologists of Tropical America was founded in 1968, and the "Chinese Society of Plant Nematologists" was organized as recently as 1996.

While the European Society of Nematologists clearly had a major international focus, the first worldwide Congress of Nematology was sponsored by The European Society of Nematologists, Society of Nematologists, and The Organization of Nematologists of Tropical America in Guelph, Canada, in 1984 (First International Congress of Nematology, 1984). The second Congress was organized by these same three societies in the Netherlands in 1990 (Second International Nematology Congress, 1990) with the third Congress also sponsored by this group being held in Guadalupe, in 1996 (Third International Nematology Congress, 1996). At the third Congress, an

International Federation of Nematology Societies was organized. The details of this umbrella organization are available on its web page (www.ifns.org).

1.6 Milestones and Trends in Nematology Research

Many of the major advancements in nematology over the last 150 years are summarized in Table 1.3. Rather than address each item, some of the highlights are discussed here.

Table 1.3 Chronology of major advancements in the study of plant and soil nematodes with emphasis on developments in USA during the last 100 years.

Advancement	Dates	Early contributors	Impact[a] (Immediate and/or delayed effects)
Multiple tactics for nematode control	1880s – 1911	Atkinson; Bessey; Kühn	Set stage for integrated management
Early classical systematics/taxonomy	1890s – 1930s	Bastian; Cobb; de Man; Filipjev; Goeldi; T. Goodey; Micoletzky; Steiner; S. Stekhoven; Thorne	Established initial criteria for nematode types and taxa
Early host resistance to nematodes	1880s – 1950s	Barrons; Bessey; Bingefors; Burkhart; Clayton; Ellenby; Mai and Peterson; Neal; Orton; Ross and Brim; Smith; Weber; Zimmerman	Provided an ideal means of control
Early integrated nematode controls	1911 – 1930s	Bessey; Tyler	Early integrated nematode management
Concept of biological control	1930s	Duddington; Linford; Tyler	Natural management tactic
Nematode anatomy and physiology	1930s	Chitwood	Basis for much of current understanding/knowledge of nematodes
Discovery of the potato cyst nematode in the USA	1941	Betz; Buhrer; Chitwood; Feldmesser; Mai	Effective regulatory measures to limit spread
Discovery of nematicidal activity of "D-D"	1943	Carter	Definitive method to show and assess dramatic crop damage
Taxonomy of *Meloidogyne*	1930s – 1940s	Chitwood; Christie	Provided sound criteria for species and host races
Discovery of the role of ectoparasites in plant disease	1940s – 1950s	Christie and Perry	Broadened concepts of parasitism-pathogenesis

Continued

Advancement	Dates	Early contributors	Impact[a] (Immediate and/or delayed effects)
Agronomic importance of disease complexes	1950s	Hewitt; Mountain; Nusbaum; Pitcher; N. Powell; Weischer	Documented role of nematodes in predisposing plants to other parasites, including "breaking" of resistance
Burrowing nematode on citrus	1920s – 1950s	Feldmesser; Rebois and A. Taylor; Suit and DuCharme	Showed extreme damage potential of nematodes on major fruit trees
Regulatory programs	1940s – 1990s	Esser; Hart; Mai; O'Bannon; Spears; A. Taylor	Contributed to containment of two or more major nematodes (golden nematode of potato, burrowing nematode of citrus)
Discovery of soybean cyst nematode in USA	1954	Winstead and Sasser	Contributed much to study of nematodes as a science
Development of Federal-State research projects and workshops in USA	1950s	Allen; Cairns; Mai; Norton; Raski; Sasser; Thorne	Established funding and means of collaborative research and graduate education, and training of first large group of nematologists in USA
"Axenic and monoxenic" culturing of nematodes	1914 – 1950s	Byars; Daugherty; Krusberg; Mountain; Nicholas; Tyler[b]	Proved pathogenicity and enhanced study of physiology
Methods of extraction and population assessment	1917 – 1960s	Baerman; Christie and Perry; Cobb[b]	Provided basis for diverse studies on population biology and related crop-loss assessments
Development of educational centers for nematology	1950s	Allen; Christie; Fenwick; T. and J. B. Goodey; Jones; Mai; Oostenbrink; Peters; V. Perry; Raski; Sasser; Seinhorst; Sher; Steiner; Thomason; Thorne; Van Gundy[b]	Advanced specialized education in nematology
Discovery of nematodes vectoring plant viruses	1956	Hewitt, Raski and Goheen; C. E. Taylor; Weischer	Further broadened role of nematodes in disease etiology
Concepts of population dynamics and ecology	1950s – 1980s	Godfrey; Jones; Norton; Oostenbrink; Seinhorst; Wallace[b]	Developed systems for ecology-based control/management
Development of nonfumigant nematicides	1950s – 1960s	Largely industry-oriented	Highly effective, short-term control
General and integrated management	1930s – 1990s	Baines; Griffin; Jensen; Jones; McBeth; L. I. Miller; N. A. Minton; Oostenbrink; Overman; Reynolds; Tyler; Thomason[b]	Interfacing management options on an ecologically sound basis
Biological control of plant-parasitic nematodes	1930s – 1990s	Barron; Duddington; Mankau; Linford; Sayre; Zuckerman[b]	Enhanced understanding of antagonists and related biology enhancing potential for practical biocontrol

Continued

Advancement	Dates	Early contributors	Impact[a] (Immediate and/or delayed effects)
Extensive systematics of many nematode groups	1940s – 1990s	Allen; Andrássy; Baker; Coomans; DeConinck; Franklin; Geraert; Golden; B. Goodey; Hirschmann; Ichinohe; Jairajpuri; Khan; Kivjanova; Loof; Lordello; Luc; Maggenti; Meyl; Mulvey; Paramonov; Raski; Sher; Siddiqi; Steiner; Stone; Tarjan; Thorne[b]	Provided concept and insight into diversity and enormity of nematode communities
Cytogenetics	1950s – 1990s	Mulvey; Triantaphyllou[b]	Enhanced understanding of evolution and taxa interrelationships
Modes of reproduction/sexuality	1950s – 1990s	Triantaphyllou; *C. elegans* group[b]	Provided information database for genetics/molecular research
Survival mechanisms	1940s – 1990s	Van Gundy; *C. elegans* group[b]	Fundamental knowledge and facilitated practical control
Genetic and physiological diversity in major species	1960s – 1990s	Dropkin; Jones; Ludders; L. I. Miller; Ross; Sasser; Triantaphyllou[b]	Facilitated development of durable resistant cultivars and recent molecular nematology
Damage functions/advisory programs	1960s – 1990s	Oostenbrink; Seinhorst; Jones[b]	Objective basis for nematode management
Population simulators	1960s – 1990s	Jones[b]	Facilitated effectiveness of research and management of nematodes
Tropical and international programs such as "International *Meloidogyne* Project"	1950s – 1990s	Caveness; Jones; Oostenbrink; Sasser; Seinhorst[b]	Expanded educational base of nematologists worldwide and provided ecological-taxonomic data base
Physiology, behavior, and host/parasite relationships, including nature of feeding cells	1960s – 1990s	A. Bird; Bergeson; Cohn; Croll; Dropkin; Krusberg; Shepherd; Van Gundy; Veech; Viglierchio; *C. elegans* group[b]	Database essential for molecular research and genetics engineering
Nematode ultrastructure	1960s – 1990s	A. Bird; Endo; Hirschmann; Wright; *C. elegans* group[b]	Contributed to understanding of structure and function
Biological control of insects with nematodes	1960s – 1990s	Dutky; Glaser; Nickle; Poinar; Steiner[b]	Decades of research now resulting in effective control of some insects

Continued

Advancement	Dates	Early contributors	Impact[a] (Immediate and/or delayed effects)
Caenorhabditis elegans developmental biology and genetics model system	1960s – 1990s	Brenner; Dougherty; Nicholas; Sulston; *C. elegans* group[b]	Providing fundamental information on cell lineage, behavior, gene function, aging, and overall genome for this model biological system
Molecular systematics/taxonomy	1960s – 1990s	- b -	Provided new tools for identification and qualitative population assessments
Nutrient cycling (soil-plant system)	1980s – 1990s	- b -	Information being developed providing possible means of enhancing "soil health" via practices that favor beneficial nematodes
Molecular markers for resistance genes	1990s	- b -	Increasing efficiency of breeding for resistance
Molecular analysis of host-parasite interactions	1990s	- b -	Providing fundamental knowledge on mechanisms of pathogenesis
Cloning of resistance genes	1990s	- b -	Elucidation of the molecular mechanisms of resistance
Transgenic host resistance to plant-parasitic nematodes	1990s	- b -	Great potential, but limited to model systems to date
Precision agriculture	1990s	- b -	"GPS" and "GIS" have much potential in nematode management
Genomics	1990s	- b -	Entire genome of *C. elegans* sequenced. Study of whole organism genomes commences

[a] This compilation is somewhat subjective; but the kind reviews and inputs by many nematologists hopefully have minimized this problem (in large part, after Barker, 1998b).

[b] Active and recently retired scientists are not listed under their research activities (due to large numbers and risks of failing to include key individuals).

Many discoveries and advancements in nematology commencing in the 1940s and 1950s added much impetus to the study of plant-parasitic nematodes. The documentation of the distribution and genetic diversity of key, highly damaging nematodes, including the potato and soybean cyst nematodes (Brodie, 1993; Marks and Brodie, 1998; Riggs and Wrather, 1992; Sharma, 1998), root-knot species (Jepson, 1987; Nickle, 1991; Sasser and Carter, 1985; Triantaphyllou, 1985), *Radopholus* species (Luc, et al., 1990; Nickle, 1991), *Ditylenchus* species (Barker, et al., 1998; Luc, et al., 1990; Nickle, 1991) and other taxa set the stage for much applied and fundamental nematological research. The 1952 description of *Heterodera glycines* by

Ichinohe (Fig. 1.31) was most timely for nematologists in the USA as this pathogen was detected in North Carolina in 1954 (see Riggs and Wrather, 1992). The discovery of given nematodes, such as the potato cyst *Globodera rostochiensis* in the USA, also led to regulatory measures that often prevented extensive spread (Brodie, 1993).

Building upon the very early work of Kühn and Atkinson, Bessey (1911) and Tyler (1933) clearly established the fundamental strategies and tactics for integrated nematode management. New and improved strategies and tactics continue to be adapted for intensive yet sustainable as well as subsistence agriculture (Barker and Koenning, 1998; Bridge, 1996). Management approaches being integrated include biological, cultural, chemical, and physical tactics, including genetically improved crop cultivars. New technologies and resources becoming available for sustainable nematode management systems include molecular diagnostics for nematode identification (Curran and Robinson, 1993), genetic engineering for host resistance (Atkinson, et al., 1998), precision agriculture with options for site-specific treatments (Ellsbury, et al., 2000), and the application of emerging principles of soil biology/ecology for general cropping systems (Barker and Koenning, 1998; Bridge, 1996). Remaining challenges include greater integration of tactics directed at the many, highly diverse crop pests (Kennedy and Sutton, 2000; Whitten, 1992). In this regard, nematology has much to offer, including the increasingly attractive use of entomopathogenic nematodes for biological control of a number of insects (Webster, 1998; Whitten, 1992).

In the 1930s through the 1950s, the concept and potential of biocontrol of plant-parasitic nematodes was delineated by Duddington (1957), Linford (1930s), and others. This approach to nematode management was researched increasingly, as discussed by Kerry and Evans (1996) and Stirling (1991). Entomopathogenic nematodes are now being used to control insects (Webster, 1998), while bacterial and fungal parasites/antagonists have much potential for suppressing plant-parasitic nematodes as well as other pathogens (Barker and Koenning, 1998; Chen and Dickson, 1998; Kerry and Evans, 1996; Stirling, 1991). Certain rhizobacteria may induce acquired systemic resistance in crop plants to nematodes and other pathogens (see Barker and Koenning, 1998).

In addition to providing an effective, low-cost means of control, the discovery of the fumigant D-D (1,3-dichloropropene-1,2-dichloropropane) by Carter in 1943 and the subsequent development of additional, highly effective nematicides provided a reliable procedure for determining the economic losses caused by these pathogens. Unfortunately, a number of the more effective nematicides have been removed from the market, and the prospects for environmentally acceptable compounds are cloudy (Thomason, 1987).

Shortly after the introduction of D-D, agronomically acceptable cultivars of crops with nematode resistance (*Meloidogyne* spp. – tomato, tobacco, etc.; *Ditylenchus dipsaci* – alfalfa; *Heterodera glycines* – soybean; *Globodera rostochiensis* – potato; etc.) were deployed successfully (see Table 1.3). The increasing resource of crop resistance to nematodes (Young, 1998) and greater understanding of the related genetics (Roberts, et al., 1998) hold great promise for contributing to sustainable nematode management systems. Chitwood's 1949 delineation of the major species of *Meloidogyne* and the ongoing discoveries and descriptions of many new nematode species greatly facilitated progress in the development of management tactics, especially host resistance.

The documentation that ectoparasitic nematodes are often serious pathogens by Christie and Perry (Fig. 1.34) (1951) was soon followed by the discoveries that a number of ectoparasites in the Dorylaimoida also serve as vectors of important plant viruses (Abawi and Chen, 1998; Hewitt, et al., 1958; see Table 1.3). Elegant research has elucidated the nature of virus retention and transmission by nematodes (Taylor and Brown, 1997). In parallel, the elucidation and extension of the early conclusions by Goeldi (1887) and Atkinson (1892) documented the magnitude of economic losses resulting from disease complexes induced by plant-parasitic nematodes (Abawi and Chen, 1998; Pitcher, 1978).

Early ecological studies and the pioneering work by Seinhorst (1970), Oostenbrink (1966), and Jones et al. (1967) in modeling nematode population dynamics and developing crop-response models provided models for objective determinations of crop losses caused by nematodes (McSorley, 1998). Developments in precision agriculture, including "global positioning systems" and "geographical information systems", have only recently been applied to improve nematode-associated crop losses and integrated pest management (Copeland, et al., 1996; S. M. Schneider, pers. comm.). This new technology, combined with rapidly expanding information on nematode and soil ecology (including that of beneficial nematodes and bacteria), especially when interfaced with nematode-resistant crops and cultivars, is bringing innovative approaches to integrated nematode/pest/soil management and sustainable production systems (Barker and Koenning, 1998; Bridge, 1996; Duncan and Noling, 1998; Niles and Freckman, 1998; Yeates, 1996). Decades of ecological research (Norton, 1978; Wallace, 1963, 1973) are now providing a foundation for this ecology-based nematode management.

The early work on monoxenic culture of nematodes by Byars (1914), the intensive research by Dougherty and associates on axenic culture of microbivorous nematodes (Nicholas, 1975), the related monoxenic culture of *Pratylenchus penetrans* by Mountain (Fig. 1.35) (1960), and similar research by others opened avenues for more definitive studies on the nature of nematode-induced damage to plants, and the nature of nematode-host interactions, as well as investigations of nematode biology and genetics. The long-term contributions in characterizing nematode cytogenetics and related work on plant-parasitic nematodes by Triantaphyllou (1985) and others contributed much to the latter.

An area that is undergoing great change is that of nematode taxonomy. As Cobb, de Man, Chitwood, Filipjev, T. Goodey, Thorne and many others built upon the taxonomic/systematics concepts initiated by Bastian and Bütschli, nematologists over the last five decades developed a phenomenal resource in nematode taxonomy/systematics (Table 1.3). While classical taxonomy still continues to be crucial to nematode management, few experts are active in this endeavor. Fortunately, molecular research has confirmed the validity of a number of classical nematode species, and molecular systematics, combined with related work in genetics, has moved to the forefront (Curran and Robinson, 1993; Barker and Davis, 1996). The superb progress in nematode anatomy, ultrastructure, biochemistry, and basic physiology contributed much to systematics (Bird and Bird, 1991; Chitwood and Chitwood, 1974; Hussey and Williamson, 1998; Lee and Atkinson, 1977; Perry and Wright, 1998). This enhanced understanding of the nematode structure and function, combined with improved molecular probes for identification (Curran and Robinson, 1993) and the new information on nematode genomics (Blaxter, 1998), should result in renewed interest in systematics.

Molecular research approaches have resulted in an ongoing revolution in nematological research and other areas of biology (Blaxter, 1998; Fenoll, et al., 1997). In addition to new systematics, markers for host resistance have become available, and resistance genes have been cloned (Cai, et al., 1997). Transgenic resistance to nematodes is on the horizon (Atkinson, et al., 1998; Opperman and Conkling, 1998). Molecular responses and manipulations in nematode-host interactions are being documented (Bird, 1996; Bird and Koltai, 2000; Bird and Wilson, 1994). Genes likely involved in parasitism of plants have been isolated and cloned, and present evidence suggests that certain nematode genes may have a bacterial origin. For example, Smant et al. (1998) cloned cellulase genes from two species of cyst nematodes. Biocontrol organisms can be genetically engineered, including the use of entomopathogenic nematodes for insect control (Webster, 1998).

The entire genome of *Caenorhabditis elegans* has been characterized

(Blaxter, 1998; Esnard, 1999; The *C. elegans* Sequencing Consortium, 1998), and functional genomics has opened an entirely new field of research. In December, 1998, a special issue of *Science* highlighted a landmark in biology – the complete DNA sequence of *C. elegans* (Blaxter, 1998; Hodgkin, et al., 1998; The *C. elegans* Sequencing Consortium, 1998). This accomplishment is the culmination of the vision and efforts of Sydney Brenner who first used *C. briggsae* as a model system for his genetics/molecular research before selecting *C. elegans* as the preferred organism (http://elegans.swmed.edu/Sydney.html). Although this bacterivorous nematode is very different from plant-parasitic species, the technology and wealth of information available through the *C. elegans* project are already having an enormous positive impact on the study of plant and other nematodes. For example, projects are being initiated to clone the genome of selected plant parasitic species. Having the whole genome sequence of *C. elegans* should expedite the isolation and cloning of genes from plant parasites as well as those from other trophic groups (Blaxter, 1998; Esnard, 1999). Of interest, the sequence for this 959-cell worm consists of some 97 million base pairs of DNA which is about one-thirtieth that of the human genome (Hodgkin, et al., 1998). Blaxter et al. (1998) interfaced the *C. elegans* molecular information with that available for other organisms to assess the phylogeny of various groups of animals as well as orders and trophic groups of nematodes.

In addition to augmenting fundamental research on plant-parasitic nematodes, the information on the basic biology of *C. elegans* could bring renewed research on the behavior of plant forms. The earlier work on plant-nematode movement/behavior and survival (Croll, 1970; Van Gundy, 1965; Wallace, 1963) could be revisited in light of the wealth of related data on *C. elegans* (Riddle, et al., 1997). The genetic regulation of the life cycle of *C. elegans* may provide new clues as to how a host metabolite such as glycinoeclepin A at concentrations in parts per billion induces egg hatch by *Heterodera glycines* (Masamune, et al., 1982) or similar responses by *Globodera* spp. to potato root diffusates (Shepherd, 1962). This system and other host-parasite interactions may well be exploited for developing novel nematode management approaches.

In concert with the hundreds of advancements in nematology, much synergism emerged via the education and regional research programs discussed earlier. Graduate education and research were central to most of the major research centers during the 1950s through the 1980s. For many institutions, however, much of that graduate student research was replaced by post-doctorate contributions during the 1990s. In recent years, more and more of the nematological research on molecular genetics, ecology, and integrated management has become heavily interdisciplinary oriented. These developments offer great promise for future nematological studies, but maintaining nematology as a distinct discipline is becoming a greater challenge. As was true for the 1890s, the 1990s were good years for nematology. In addition to the announcement of the sequencing of the *C. elegans* genome in 1998, at least 20 new books were published during the decade (Barker, et al., 1998; Bird and Bird, 1991; Evans, et al., 1993; Fenoll, et al., 1997; Gommers and Maas, 1992; Hunt, 1993; Jairajpuri, 1990; Lamberti, et al., 1994; Luc, et al., 1990; Mai and Mullin, 1996; Marks and Brodie, 1998; Nickle, 1991; Perry and Wright, 1998; Riddle, et al., 1997; Riggs and Wrather, 1992; Sayre, 1994; Sharma, 1998; Stirling, 1991; Taylor and Brown, 1997; Viglierchio, 1991).

With this dramatic development and flow of new nematological information, the next 100 years have even greater promise than the past century. Nevertheless, many challenges as well as opportunities face the discipline today. As in N. A. Cobb's career and as emphasized by Hume (see Wilson, 1982), nematodes and nematology continue to suffer from unawareness by the public and scientific communities. Although the modern, often molecular-based research programs are well supported and are making exciting advancements, they must be interfaced effectively with applied research programs to resolve nematode-induced problems facing the grower and homeowner. This includes the need to develop integrated, ecologically and precision agriculture-based nematode management programs. And those programs urgently demand innovative, nematode resistance in a wide range of crops and ornamentals, environmentally acceptable nematicides, biology-based integrated cropping system, and highly efficient and reliable identification and diagnostic services. For all of these goals to come to fruition and be documented in the history of the second century of nematology, graduate education for the discipline must be given a very high priority, as was done in the 1950s through the 1980s.

References

Abawi, G. S., and J. Chen. 1998. Concomitant pathogen and pest interactions. Pp. 135–158. In K. R. Barker, G. A. Pederson, and G. L. Windham, eds. Plant and nematode interactions. Madison, WI: American Society of Agronomy, Crop Science Society of America, and Soil Science Society of America.

Atkinson, G. F. 1889. Nematode root-galls. Report of Agriculture Experiment Station, Alabama Polytechnic Institute, Auburn, Alabama. Pp. 176–226.

Atkinson, G. F. 1892. Some diseases of cotton. Alabama Agricultural Experiment Station,

Auburn University, Bulletin 41: 5 –65.

Atkinson, H. J., C. J. Lilley, P. E. Urwin, and J. J. McPherson. 1998. Engineering resistance to plant-parasitic nematodes. Pp. 381 – 413. In R. N. Perry and D. J. Wright, eds. The physiology and Physiology and Biochemistry of Free-living and Plant-parasitic Nematodes. Wallingford, UK: CAB Int.

Baker, A. D. 1962. Check Lists of the Nematode Superfamilies Dorylaimoidea, Rhabditoidea, Tylenchoidea, and Aphelenchoidea. Leiden: E. J. Brill.

Baldwin, J. G., and M. Mundo-Ocampo. 1991. Heteroderinae, cyst- and non-cyst forming nematodes. Pp. 275 – 362. In W. R. Nickle, ed. Manual of Agricultural Nematology. New York: Marcel Dekker.

Barker, K. R. 1998a. Nematology: Then, then and now. Nematology News Letter 44(1): 24 –31.

Barker, K. R. 1998b. Introduction and synopsis of advancements in nematology. Pp. 1 – 20. In K. R. Barker, G. A. Pederson, and G. L. Windham, eds. Plant and Nematode Interactions. Madison, WI: American Society of Agronomy, Crop Science Society of America, and Soil Science Society of America.

Barker, K. R., and E. L. Davis. 1996. Assessing plant nematode infestations and infections. Pp. 103 – 136. In S. H. De Boer, ed. Advances in Botanical Research. Vol. 23: Pathogen Indexing Technologies. London: Academic Press.

Barker, K. R., C. C. Carter, and J. N. Sasser. eds. 1985. An Advanced Treatise on *Meloidogyne* – Vol. II. Methodology. Raleigh, NC: North Carolina State University Press.

Barker, K. R., and S. R. Koenning. 1998. Developing sustainable systems for nematode management. Annual Review of Phytopathology 36: 165 –205.

Barker, K. R., G. A. Pederson, and G. A. Windham. eds. 1998. Plant and Nematode Interactions. Madison, WI: American Society of Agronomy, Crop Science Society of America, and Soil Science Society of America.

Bastian, H. C. 1865. Monograph on the Anguillulidae. Transactions of the Linnaean Society of London 25: 73 – 184.

Berkeley, M. J. 1855. Vibrio forming excrescences on the roots of cucumber plants. Gardener's Chronicle (14): 220.

Bessey, E. A. 1911. Root-knot and its control. USDA Bureau Plant Industry Bulletin 217.

Bird, A. F. 1971. The Structure of Nematodes. New York: Academic Press.

Bird, A. F., and J. Bird. 1991. The Structure of Nematodes, 2 nd edn. San Diego, CA: Academic Press.

Bird, D. M. 1996. Manipulation of host gene expression by root-knot nematodes. Journal of Parasitology 82: 881 –888.

Bird, D. M., and H. Koltai. 2000. Plant parasitic nematodes: Habitats, hormones and horizontally-acquired genes. Journal of Plant Growth Regulation 19: 183 – 194.

Bird, D. M., and M. A. Wilson. 1994. DNA sequence and expression analysis of root-knot nematode-elicited giant cell transcripts. Molecular Plant-Microbe Interactions 7: 410 –424.

Blaxter, M. 1998. *Caenorhabditis elegans* is a nematode. Science (Special Edition – December) 282: 2041 –2046.

Blaxter, M. L., P. Delay, G. R. Garey, L. X. Liu, P. Scheldeman, A. Vierstraete,

J. R. Vanfleteren, L. Y. Machey, D. Dorris, L. M. Frisse, J. T. Vida, and W. K. Thomas. 1998. A molecular evolutionary framework for the phylum Nematoda. Nature 392: 71 –75.

Bridge, J. 1996. Nematode management in sustainable and subsistence agriculture. Annual Review of Phytopathology 34: 201 –225.

Brodie, B. B. 1993. Probability of *Globodera rostochiensis* spread on equipment and potato tubers. Journal of Nematology 25:291 –296.

Byars, L. P. 1914. Preliminary notes on the cultivation of the plant parasitic nematode, *Heterodera radicicola*. Phytopathology 4: 223 –226.

Cai, D., M. Kleine, S. Kifle, H. J. Harloff, N. N. Sandal, K. A. Marcker, R. M. Klein-Lankhorst, E. M. J. Salentijn, W. Lange, W. J. Stiekema, U. Wyss, F. M. W. Grundler, and C. Jung. 1997. Positional cloning of a gene for nematode resistance in sugarbeet. Science (Washington, DC) 275:32 –34.

Campbell, C. L., P. D. Peterson, and C. S. Griffith. 1999. The Formative Years in Plant Pathology in the United States. St. Paul, MN: APS Press.

Carter, W. 1943. A promising new soil amendment and disinfectant. Science 97: 383 –384.

Chen, Z. X., and D. W. Dickson. 1998. Review of *Pasteuria penetrans*: Biology, ecology, and biological control potential. Journal of Nematology 30: 313 –340.

Chitwood, B. G. 1949. Root-knot nematodes. I. A revision of the genus *Meloidogyne* Goeldi, 1887. Proceedings of the Helminthological Society of Washington 16: 90 – 104.

Chitwood, B. G., and M. B. Chitwood. 1937. An Introduction to Nematology. Baltimore, MD: Monumental Printing.

Chitwood, B. G., and M. B. Chitwood. 1950. An Introduction to nematology. Baltimore, MD: Monumental Printing.

Chitwood, B. G., and M. B. Chitwood (Consolidated Edition). 1974. Introduction to Nematology. Baltimore, MD: University Park Press.

Christie, J. R. 1959. Plant Nematodes – their bionomics and control. Agricultural Experiment Stations, University of Florida, Gainesville, FL.

Christie, J. R. 1966. Days with N. A. Cobb. Nematology News Letter 12(2): 17 –19.

Christie, J. R., and F. E. Albin. 1944. Host-parasite relationships of the root-knot nematode, *Heterodera marioni*. 1. The question of races. Proceedings of the Helminthological Society of Washington 11: 31 –37.

Christie, J. R., and V. G. Perry. 1951. A root disease of plants caused by a nematode of the genus *Trichodorus*. Science 113: 491 –493.

Cobb, N. A. 1914. Nematodes and their relationships. USDA Yearbook, 1914. Pp. 457 –490.

Cobb, N. A. 1918. Estimating the nema populations of soil. United States Department of Agriculture, Bureau of Plant Industry, Agriculture Technical Circular 1: 1 –48.

Coomans, A. V. 1992. Development of nematology in the low land countries. Pp. 11 –20. In F. J. Gommers and P. W. Th. Maas, eds. Nematology from Molecule to Ecosystem. Invergowrie, Dundee, Scotland: European Society of Nematologists.

Copeland, P. J., G. L. Malzer, J. G. Davis, J. A. Lamb, P. C. Robert, and T. W. Bruulsema. 1996. Using harvest index to locate environmental stress. Pp. 531 –540. In P. C. Robert, R. H. Rust, and W. E. Larson, eds. Precision Agriculture. Madison,

WI: American Society of Agronomy, Crop Science Society of America, and Soil Science Society of America.

Croll, N. A. 1970. The Behaviour of Nematodes: Their Activity, Senses, and Responses. New York: St. Martin's Press.

Croll, N. A. 1976. The Organization of Nematodes. London: Academic Press.

Croll, N. A., and B. E. Matthew. 1977. Biology of Nematodes. New York: Wiley.

Curran, J., and M. P. Robinson. 1993. Molecular aids to nematode diagnosis. Pp. 545–564. In K. Evans, D. L. Trudgill, and J. M. Webster, eds. Plant Parasitic Nematodes in Temperate Agriculture. Wallingford, UK: CAB International.

Diesing, C. M. 1851. Systema Helminthum, Vol. 1. Vindobonae.

Dropkin, V. H. 1980. Introduction to Plant Nematology. New York: Wiley.

Dropkin, V. H. 1989. Introduction to Plant Nematology, 2 nd edn. New York: John Wiley & Sons.

Duddington, C. L. 1957. The Friendly Fungi. London: Faber and Faber.

Dujardin, F. 1845. Histoire Naturelle des Helminths ou Vers Intestinaux. Paris.

Duncan, L. W., and J. W. Noling. 1998. Agricultural and sustainability and nematode integrated pest management. Pp. 251–287. In K. R. Barker, G. A. Pederson, and G. L. Windhamn, eds. Plant and Nematode Interactions. Madison, WI: American Society of Agronomy, Crop Science Society of America, and Soil Science Society of America.

Ellsbury, M. M., S. A. Clay, S. J. Fleischer, L. D. Chandler, and S. M. Schneider. 2000. Use of GIS/GPS systems in IPM: Progress and reality. Pp. 419–438. In G. G. Kennedy, and T. B. Sutton, eds. Emerging Technologies for Integrated Pest Management – Concepts, Research and Implementation. St. Paul, MN: APS Press.

Esnard, J. 1999. Implications of sequencing the nematode *Caenorhabditis elegans* genome for plant pathology. APSnet http://www. scisoc. org/feature/Celegans/top/html.

Evans, K. 1995. Closing the gap between molecular biologists and traditional nematologists. Nematologica 41:385–394.

Evans, K., D. L. Trudgill, and J. M. Webster. 1993. Plant Parasitic Nematodes in Temperate Agriculture. Wallingford, UK: CAB International.

Fenoll, C., F. M. W. Grundler, and S. A. Ohl. 1997. Cellular and Molecular Aspects of Plant-nematode Infections. Dordrecht, Netherlands: Kluwer Academic Publishers.

Filipjev, I. N., and J. H. Schuurmans-Stekhoven. 1941. A Manual of Agricultural Helminthology. Leiden: Brill.

First International Congress of Nematology. 1984. Program and Abstracts. European Society of Nematology, Organization of Tropical American Nematologists, and the Society of Nematologists. Guelph, Canada.

Franklin, M. T. 1951. The Cyst-forming Species of *Heterodera*. Farnham Royal, UK: CAB.

Goeldi, E. A. 1887. Relatório sobre a moléstia do cafeeiro na Provincia do Rio de Janeiro. Archivos do Museu Nacional Rio de Janeiro 8:7–123 (publ. 1892).

Goffart, H. 1951. Nematoden der Kulturpflanzen Europas. Paul Parey, Verllag für Landwirtschaftr, Gartenbau und Forstwesen. Berlin.

Gommers, F. J., and P. W. Th. Maas. eds. 1992. Nematology from Molecule to Ecosystem. Invergowrie, Dundee, Scotland: European Society of Nematologists.

Goodey, J. B. 1963. Soil and Fresh Water Nematodes (2 nd Edition/Revision of T. Goodey's 1951 edition). London: Methuen.

Goodey, J. B., M. T. Franklin, and D. J. Hooper. 1965. T. Goodey's The Nematode Parasites of Plants Catalogued under their Hosts (3 rd edn.). Farnham Royal, UK: CAB.

Goodey, T. 1933. Plant-parasitic Nematodes and the Diseases they Cause. New York: E. P. Dutton and Company.

Goodey, T. 1951. Soil and Fresh Water Nematodes. London: Methuen.

Hewitt, W. B., D. J. Raski, and A. C. Goheen. 1958. Nematode vector of soil-borne fanleaf virus of grapevines. Phytopathology 48: 586 – 595.

Hobmaier, M. 1929. Über die Entwicklung der Lungenwürmer des Schweines *Metastrongylus elongatus*, und *Choerostrongylus brevivaginatus*. Eesti Loomaarstlik Ringvaade 5(3): 67 – 71.

Hodgkin, J., H. R. Horvitz, B. R. Jasny, and J. K. Kimble. 1998. *C. elegans*: Sequence to biology. Science (Special Edition – December) 282: 2011.

Hunt, D. J. 1993. Aphelenchida, Longidoridae and Trichodoridae: The systematics and bionomics. Wallingford, UK: CAB International.

Hussey, R. S., and V. M. Williamson. 1998. Physiological and molecular aspects of nematode parasitism. Pp. 87 – 108. In K. R. Barker, G. A. Pederson, and G. L. Windham, eds. Plant and Nematode Interactions. Madison, WI: American Society of Agronomy, Crop Science Society of America, and Soil Science Society of America.

Ichinohe, M. 1952. On the soybean nematode, *Heterodera glycines* n. sp. from Japan. Oyo-Do-but-sogaku-Zasshi (Magazine of Applied Zoology) 17: 1 – 4.

Jairajpuri, M. S. 1990. Nematode Biocontrol. Delhi, India: CBS Publishers.

Jenkins, W. R., and D. P. Taylor. 1967. Plant Nematology. New York: Reinhold Publishing.

Jepson, S. B. 1987. Identification of the Root-knot Nematodes (*Meloidogyne* species). Farnham Royal, UK: CAB.

Jones, F. G. W., D. M. Parrott, and G. J. S. Ross. 1967. The population genetics of the potato cyst-nematode *Heterodera rostochiensis*: Mathematical models to simulate the effects of growing eelworm-resistant potatoes bred from *solanum tuberosum* spp. *andigena*. Annals of Applied Biology 60: 151 – 171.

Kennedy, G. G., and T. B. Sutton. eds. 2000. Emerging Technologies for Integrated Pest Management – Concepts, research, and implementation. St. Paul, MN: APS Press.

Kerry, B. R., and K. Evans. 1996. New strategies for the management of plant-parasitic nematodes. Pp. 134 – 152. In R. Hall, ed. Principles and Practices of Managing Soilborne Plant Pathogens. St. Paul, MN: APS Press.

Koenning, S. R., C. Overstreet, J. W. Noling, P. A. Donald, J. O. Becker, and B. A. Fortnum. 1999. Survey of crop losses in response to phytoparasitic nematodes in the United States for 1994. Supplement to Journal of Nematology 31: 587 – 618.

Kühn, J. 1857. Über das Vorkommen von *Anguillula* an erkrankten Blühtenköpfen von *Dipsacus fullonum* L. Zeitschrift fuer Wissenschaftliche Zoologie 9: 129 – 137.

Kühn, J. 1881. Die Ergebnisse der Versuche zur Ermittenlung der Ursache der Rübenmüdigkeit und zur Erforschung der natur der Nematoden. Berichte aus dem. Physiologischen Laboratorium Univ. Halle 3: 1 – 153.

Lamberti, F., and C. E. Taylor. eds. 1979. Root-knot Nematodes (*Meloidogyne* species) – Systematics, biology, and control. London: Academic Press.

Lamberti, F., C. De Giorgi, and D. McK. Bird. eds. 1994. Advances in Molecular Plant Nematology. New York: Plenum Press.

Lee, D. L. 1965. The Physiology of Nematodes. San Francisco, CA: W. H. Freeman & Company.

Lee, D. L., and H. J. Atkinson. 1977. Physiology of Nematodes. 2 nd edn. London: MacMillan Press.

Linnaeus, C. 1768. Systema Naturae, Vol. 1. Holmiae. (10 th edn.).

Lordello, L. G. E. 1973. Nematòides das Plantas Cultivadas (2 nd edn.). São Paulo, Brazil: Livraria Nobel.

Luc, M., J. Bridge, and R. A. Sikora. eds. 1990. Plant-parasitic Nematodes in Subtropical and Tropical Agriculture. Wallingford, UK: CAB International.

Mai, W. F., and A. H. Lyon. 1960. Pictorial Key to Genera of Plant Parasitic Nematodes. Ithaca, NY: Cornell University.

Mai, W. F., and P. G. Mullin. 1996. Plant-parasitic Nematodes – A pictorial key to genera (5 th edn.). Ithaca, NY: Cornell University Press.

Manson, P. 1878. On the development of *Filaria sanguinis hominis,* and on the mosquito considered as a nurse. Journal of the Linnean Society of London Zoology (75), v. 14: 304 – 311.

Marks, R. J., and B. B. Brodie. eds. 1998. Potato Cyst Nematodes. Biology, distribution, and control. Wallingford, UK: CAB International.

Masamune, T., M. Anetai, M. Takasugi, and N. Katsui. 1982. Isolation of natural stimulus, glycinoeclepin A for the soybean cyst nematode. Nature 297: 495 – 496.

McSorley, R. 1998. Population dynamics. Pp. 109 – 133. In K. R. Barker, G. A. Pederson, and K. R. Barker, eds. Plant and Nematode Interactions. Madison, WI: American Society of Agronomy, Crop Science Society of America, and Soil Science Society of America.

Mountain, W. B. 1960. (a) Theoretical considerations of plant-nematode relationships. (b) Acceptable standards of proof and approaches for evaluating plant-nematode relationships. (c) Mechanisms involved in plant-nematode relationships. Pp. 419 – 431. In J. N. Sasser and W. R. Jenkins, eds. Nematology: Fundamentals and Recent Advances with Emphasis on Plant-parasitic and Soil Forms. Chapel Hill, NC: University of North Carolina Press.

Moura de, R. M. 1998. Reedição de: E. A. Goeldi, 1887 "Relatório sobre a" Moléstia do Cafeeiro na Província do Rio de Janeiro. FADURPE/UFRPE, Recife, Pp. 121.

Moura de, R. M. 2000. Fundamentos históricos e eavolutivos da nematologia de interesse agrícola: Uma visão no ano 2000. Nematologia Brasileira 24: 1 – 21.

Neal, J. C. 1889. The Root-knot disease in peach, orange, and other plants in Florida, due to the work of *Anguillula.* USDA Division of Entomology Bulletin 20. Washington, DC.

Needham, T. 1743. A letter concerning certain chalky tubulous concretions called malm; with some microscopical observations on the farina of the red lily, and worms discovered in smutty corn. Philosophical Transactions of the Royal Society 42: 173, 174, 634 – 641.

Nematology News Letter. 1973. Benjamin Goodwin Chitwood – In memorium. 19 (1): 1–4.

Nicholas, W. L. 1975. The Biology of Free-living Nematodes. Oxford, UK: Clarenden Press.

Nickle, W. R. ed. 1991. Manual of Agricultural Nematology. New York: Marcel Dekker.

Niles, R. K., and D. W. Freckman. 1998. Nematode ecology in bioassessment and ecosystem health. Pp. 65 – 85. In K. R. Barker, G. A. Pederson, and G. L. Windham, eds. Plant and Nematode Interactions. Madison, WI: American Society of Agronomy, Crop Science Society of America, and Soil Science Society of America.

Noel, G. R. 1992. History, distribution, and economics. Pp. 1–13. In R. D. Riggs and J. A. Wrather, eds. Biology and Management of the Soybean Cyst Nematode. St. Paul, MN: APS Press.

Norton, D. C. 1978. Ecology of Plant-parasitic Nematodes. New York: John Wiley & Sons.

Oostenbrink, M. 1950. Het Aardappelaaltje (*Heterodera rostochiensis* Wollenweber). (Ph. D. Dissertation) Verslagen Mededelingen Plantenziekten Dienst. Wageningen, No. 115 Pp. 230.

Oostenbrink, M. 1966. Major characteristics of the relation between nematodes and plants. Mededelingen Landbouwhogeschool Wageningen 66(4): 1–46.

Opperman, C. H., and M. A. Conkling. 1998. Bioengineering resistance to plant-parasitic nematodes. Pp. 239–250. In K. R. Barker, G. A. Pederson, and G. L. Windham, eds. Plant and Nematode Interactions. Madison, WI: American Society of Agronomy, Crop Science Society of America, and Soil Science Society of America.

Paramonov, A. A. 1968. Plant-parasitic Nematodes Vol. 1. Translation of 1962 Russian Edition by the Israel Program for Scientific Translation and Published for the USDA and NSF, Washington, DC.

Paramonov, A. A. 1976. Plant Parasitic Nematodes Vol. 2. Translated from the 1964 Russian Edition and Published for the USDA/NSF (Washington, DC.) by the Indian National Scientific Documentation Center, Delhi, India.

Peachey, J. E. ed. 1969. Nematodes of Tropical Crops. Technical Communication No. 40. St. Albans, England: Commonwealth Bureau of Helminthology.

Perry, R. N., and D. J. Wright. 1998. The Physiology and Biochemistry of Free-living and Plant-parasitic Nematodes. Wallingford, UK: CAB International.

Pitcher, R. S. 1978. Interactions of nematodes with other pathogens. Pp. 63–77. In Plant Nematology. London: Her Majesty's Stationery Office.

Poinar, G. O., Jr. 1983. The Natural History of Nematodes. Englewood, NJ: Prentice Hall.

Poinar, G. O., Jr. 1988. The impact of nematodes on mankind – A historical perspective. Nematologica 34: 249–251.

Potter, J. W. 1998. A selective history of nematology in Canada. Nematology News Letter 44(1): 14–18.

Riddle, D. L., T. Blumenthal, B. J. Meyer, and J. R. Priess. 1997. *C. elegans* II. New York: Cold Springs Harbor Laboratory Press.

Riggs, R. D., and J. A. Wrather. eds. 1992. Biology and Management of the Soybean Cyst Nematode. St. Paul, MN: APS Press.

Roberts, P. A., W. C. Matthews, and J. C. Veremis. 1998. Genetic mutations of host-plant resistance to nematodes. Pp. 209 – 238. In K. R. Barker, G. A. Pederson, and G. L. Windham, eds. Plant and Nematode Interactions. Madison, WI: American Society of Agronomy, Crop Science Society of America, and Soil Science Society of America.

Sasser, J. N., and C. C. Carter. 1985. An Advanced Treatise on *Meloidogyne*. Vol. I. Biology and Control. Raleigh, NC: North Carolina State University Graphics.

Sasser, J. N., and D. W. Freckman. 1987. A world perspective of nematology: The role of the Society. Pp. 7 – 14. In J. A. Veech, and D. W. Dickson, eds. Vistas on Nematology. Hyattsville, MD: Society of Nematologists.

Sasser, J. N., and W. R. Jenkins. eds. 1960. Nematology – Fundamentals and Recent Advances with Emphasis on Plant-parasitic and Soil Forms. Chapel Hill: The University of North Carolina Press.

Sayre, R. M. 1994. Art in Phytopathology – Portfolios of Nathan A. Cobb, Nematologist. St. Paul, MN: APS Press.

Schacht, H. 1859. Über einige Feinde der Rubenfelder. Zeitschrift fuer die Rübenzuckerindustrie im Zollverein 9: 175 – 179.

Schmidt, A. 1871. Über den Rüben-Nematoden (*Heterodera schachtii*). Zeitschrift fuer die Rübenzuckerindustraie Zollverein 21: 1 – 19.

Schneider, A. 1866. Monographie der Nematoden. Berlin.

Schneider, W. 1939. Freilebende und Pflanzenparasitische Nematoden. Tierwelt Deutschlands 36: 1 – 260.

Scopoli, J. A. 1777. Introductio ad historiam naturalem sistens genera lapidum, plantarum, et animalium. Prague.

Second International Nematology Congress. 1990. Program and Abstracts. Department of Nematology, Agricultural University, Wageningen, The Netherlands.

Seinhorst, J. W. 1970. Dynamics of populations of plant parasitic nematodes. Annual Review of Phytopathology 8: 131 – 156.

Seurat, L. O. G. 1913. Sur l'évolution du *Physocephalus sexalatus* (Molin). Comptes Rendus Societe de Biologie 79: 517 – 520.

Shamel, A. D., and W. W. Cobey. 1907. Tobacco breeding. United States Department of Agriculture, Bureau of Plant Industry Bulletin 86. Washington, DC.

Sharma, S. B. ed. 1998. The Cyst Nematodes. Dordrecht: Kluwer Academic Publishers.

Shepherd, A. M. 1962. The Emergence of Larvae from Cysts in the Genus *Heterodera*. Technical Communication No. 32. St Albans, England: Commonwealth Bureau of Helminthology.

Smant, G., J. Stokkermans, Y. Yan, J. M. deBoer, T. Baum, X. Wang, R. S. Hussey, E. L. Davis, F. J. Gommers, B. Henrissat, J. Helder, A. Schots, and J. Bakker. 1998. Endogenous cellulases in animals: Cloning of expressed β-1, 4-endoglucanase genes from two species of plant-parasitic cyst nematodes. Proceedings of the National Academy of Science 95: 4906 – 4911.

Southey, J. F. ed. 1959. Plant Nematology. Technical bulletin No. 7. London: Her Majesty's Stationery Office.

Southey, J. F. ed. 1978. Plant Nematology (3rd edn.) London: Ministry of Agriculture, Fisheries and Food, Her Majesty's Stationery Office.

Stewart, F. H. 1916. On the life history of *Ascaris lumbricoides*. British Medical Journal 2: 6 – 7, 486 – 488, 753 – 754.

Stirling, G. R. 1991. Biological Control of Plant-parasitic Nematodes. Wallingford, UK: CAB International.

Strubell, A. 1888. Untersunchungen über den Bau und die Entwicklung des Rüben-Nematoden, *Heterodera schachtii* Schmidt. Bibliotheca Zoologica 2: 1 – 52.

Taylor, A. L., and J. N. Sasser. 1978. Biology, Identification and Control of Root-knot Nematodes (*Meloidogyne* species). Raleigh, NC: North Carolina State University Graphics.

Taylor, C. E., and D. J. F. Brown. 1997. Nematode Vectors of Plant Viruses. Wallingford, UK: CAB International.

The *C. elegans* Sequencing Consortium. 1998. Genome sequence of the nematode *C. elegans*: A platform for investigating biology. Science (Special Edition – December) 282: 2012 – 2018. [For list of authors see: <genome. wustl. edu. gsc/C-elegans/> and www. sanger. ac. uk/projects/C-elegans/>].

Third International Nematology Congress. 1996. Programs and Abstracts. Organization of Nematologists of Tropical America, European Society of Nematologists and the Society of Nematologists. Gosier, Guadeloupe.

Thomason, I. J. 1987. Challenges facing nematology: environmental risks with nematicides and the need for new approaches. Pp. 469 – 476. In J. A. Veech, and D. W. Dickson, eds. Vistas in Nematology. Hyattsville, MD: Society of Nematologists.

Thorne, G. 1939. A monograph on the nematodes of the superfamily Dorylaimoidea. Capita Zoologica 8(5): 1 – 190.

Thorne, G. 1961. Principles of Nematology. New York: McGraw-Hill Book Company.

Thorne, G., and H. H. Swanger. 1936. A monograph of the nematode genera *Dorylaimus* Dujardin, *Aporcelaimus* n. g., *Dorylaimoides* n. g. and *Pungentus* n. g. Capita Zoologica 6(4): 1 – 223.

Triantaphyllou, A. C. 1985. Cytogenetics, cytotaxonomy and phylogeny of root-knot nematodes. Pp. 113 – 126. In J. N. Sasser, and C. C. Carter, eds. An Advanced Treatise on *Meloidogyne*, Vol. I. Biology and Control. Raleigh, NC: North Carolina State University Graphics.

Tyler, J. 1933. The root-knot nematode. Circular 330. College of Agriculture, Agriculture Experiment Station, University of California, Berkeley, CA.

Van Gundy, S. D. 1965. Factors in surviving of nematodes. Annual Review of Phytopathology 3: 43 – 68.

Viglierchio, D. R. 1991. The World of Nematodes. Davis, CA: David R. Viglierchio.

Wallace, H. R. 1963. The Biology of Plant Parasitic Nematodes. London: Edward Arnold.

Wallace, H. R. 1973. Nematode Ecology and Plant Disease. New York: Edward Arnold.

Weber, H. J., and W. A. Orton. 1902. A cowpea resistant to root-knot (*Heterodera radicicola*). United States Department of Agriculture, Bureau of Plant Industry Bulletin 17(2): 23 – 38.

Webster, J. M. 1972. Economic Nematology. London: Academic Press.

Webster, J. M. 1998. Nematology: From curiosity to space science in fifty years. Annals of Applied Biology 132: 3 – 11.

Westerdahl, B. B., and E. P. Caswell-Chen. 1998. Information transfer. Pp. 289 – 315. In K. R. Barker, G. A. Pederson, and G. L. Windham, eds. Plant and Nematode Interactions. Madison, WI: American Society of Agronomy, Crop Science Society of America, and Soil Science Society of America.

Whitten, M. J. 1992. Pest management in 2000: What we might learn from the twentieth century. Pp. 9 – 44. In A. A. S. A. Kadir, and H. S. Barlow, eds. Pest Management and the Environment in 2000. Wallingford, UK: CAB International.

Wilson, C. T. 1982. Development of an expanded program of teaching and research in nematology in the Southern region. Pp. 1 – 7. In R. D. Riggs (Chair of Editorial Committee). Nematology in the Southern United States. Southern Cooperative Series Bulletin 276. Arkansas Agricultural Experiment Station. University of Arkansas, Fayetteville, AR.

Yeates, G. W. 1996. Nematode ecology. Russian Journal of Nematology 4: 71 – 75.

Young, L. D. 1998. Breeding for nematode resistance and tolerance. Pp. 187 – 207. In K. R. Barker, G. A. Pederson, and G. L. Windham, eds. Plant and Nematode Interactions. Madison, WI: American Society of Agronomy, Crop Science Society of America, and Soil Science Society of America.

Zuckerman, B. M. ed. 1980. Nematodes as Biological Models. Vols. I & II. New York: Academic Press.

Zuckerman, B. M., and R. A. Rohde. eds. 1981. Plant Parasitic Nematodes Vol. III. New York: Academic Press.

Zuckerman, B. M., W. F. Mai, and R. A. Rohde. eds. 1971. Plant Parasitic Nematodes. Vols. I & II. New York: Academic Press.

Additional websites on *C. elegans:* WormBase (http://www. wormbse. org/) and *Caenorhabditis elegans* www Server (http://elegans. swmed. edu/).

2 Perspectives on Nematology for the 21st Century

John M. Webster

2.1 The Way Ahead

Nematology has never had it so good, as we consider what the future holds for the study of nematodes. This young discipline has been extraordinarily successful in achieving many of its objectives in the closing decades of the twentieth century. In so doing, it has laid a foundation upon which the subject can blossom during the century ahead. What next?

"You can never plan the future by the past"

Edmund Burke

Predicting the future is a fool's game, but based on current trends and bearing in mind the inevitability and enormity of change, some contributions that nematology could make to science and society can be considered.

The impact of an increasing human population will become increasingly obvious during the first half of this century and will surely test the carrying capacity of the planet in a multitude of ways. In particular, the per capita increase in food consumption and quality worldwide will continue to significantly impact crop production and processing and, therefore, influence agriculture and its related industries. Associated with this, will be the challenge of land availability for agriculture in light of growing urbanization and soil perturbation. Water availability for agriculture and other purposes will become a precious commodity. Inter-related with these pervasive factors will be the challenge of managing our diminishing forests, be they old-growth or managed plantations, and other agroecosystems. Concern for the natural environment will be reflected globally in the changing practices of the world's population and in legislation.

Political and scientific pundits have predicted impending global food shortage many times in the past, and once again there is a prevailing belief that current agronomic practices and crop cultivars are close to producing maximum yields. Of particular note are the statements by Ehrlich (1995) that although global grain production tripled between 1950 and 1993 it remained virtually the same on a per capita basis during this period, and by Pimentel and Lehman (1993) that worldwide crop losses due to pests have increased from 32% to 37% over the last 50 years despite increased use of pesticides. What are the prospects for continuing to increase food production and what are the limitations? What are the choices and trade offs as we surge toward seven billion humans on a planet with finite resources? On the other hand, how would the challenges change if and when the global human population begins to decline?

Gro Haarlem Brundtland focused our attention on a putative route of achievement with regards to food production and the environment in the following statement: "Ensure that [development] meets the needs of the present without compromising the ability of future generations to meet their own needs" (World Commission on Environment and Development, 1987). This highlighting of sustainable development is viewed by many as a maxim that should strongly influence society's goals for the foreseeable future. In view of the prevailing human concern for the global environment, as reflected in our changing use of chemical pesticides and fertilizers, and of the globalization of transport and communication, the challenges of providing adequate food in this century will be very different from those of the last. The adage of Cook and Veseth (1991) regarding the production of optimum wheat crops of actual, affordable, attainable and absolute yield will take on a somewhat different perspective when we consider the full implications of gene technology in the intensive growth of healthier and more productive crop plants. Concurrent with this continued manipulation of organismic productivity for the benefit of the human race we can expect the progressive development of more productive but less intrusive agriculture in climatic regions where soil conditions allow.

It can be reasonably presumed that, despite or with the assistance of changing technologies, mechanisms will continue to be found to alleviate the worst of the environmental challenges and so help the world community to improve its lot. Technical innovations will continue to modify the daily pattern of our lives, and the speed of change will continue to increase as the computerization of processes encompasses our lives. Nevertheless, one of society's greatest challenges will be that the diverse environmental stresses of the over-populated world will result in photosynthate, in all its forms, becoming a very precious commodity.

The capacity to develop and grow disease-resistant crops will expand as will the utilization of plant products for non-food purposes. Technology will enable us to increase controlled-environment crop production and also the shelf life of food products. The trends in globalization will continue as will the increased speed of all forms of communication, thus ensuring more rapid, global distribution of food and migration of people. Consequently, the impacts on increased plant quarantine-related activities will be substantial. The ultimate breakthrough in synthetic photosynthesis will help solve the perennial natural-resource-driven energy crisis. This will lead also to satisfying some of our food requirements through more widespread production of synthetic foods. Mankind's concern for improved health and longer life will continue to provide a powerful technological, demographic and sociological influence of our global practices that will impact the agro-forest industries and the terrestrial and marine environment through the requirement for higher quality of food and habitation.

2.2 Healthy Soil and Crops

There is a general recognition that crops are not producing yields at their genetic potential because of biotic and abiotic stress (Browning, 1998). These stresses are diverse, multicultural, and interactive, and we need to better recognize/diagnose the causes. We need to understand how these stresses diminish crop yield and how their negative effects can be ameliorated. Rather than continuing to focus on growing high yields of crops in soils using a succession of fertilizers and other high cost inputs to plant growth there will be a move toward growing crops in a more sustainable way. Traditionally, the short goal has been to prevent significant loss of crop yield and quality by controlling plant-parasitic nematodes, and the long-term one has been to keep nematode population levels low. These objectives have been achieved by using a combination of methods that included chemical nematicides in the control strategy (Whitehead, 1998). In future, the use of such chemicals will probably be prohibited completely unless their effects can be restricted selectively to plant-parasitic nematodes and they can be proven to be non-toxic to vertebrates. Some nematode species (e.g., *Bursaphelenchus xylophilus*, *Rhadinaphelenchus* (sensu lato) *cocophilus*, *Xiphinema* spp., *Longidorus* spp., *Helicotylenchus* spp.) are difficult to control without nematicides, but a substantial range of options are available for integrated nematode management of most species of plant-parasitic nematodes (Table 2.1). Integrated pest management (IPM) practices will become more sophisticated and effective as information for management models builds on our increased understanding of the components of plant nematode management. The intensification of agronomic practices, such as excessive soil disturbance and application of chemical fertilizers and pesticides in order to meet our food requirements, has a negative impact on soil communities that results in a depletion of assimilable carbon. There is justification for changing the research emphasis from increasing the efficacy of nematicides to determining the factors that comprise a healthy, plant-nematode suppressive soil.

Table 2.1 Methods for controlling common species of plant-parasitic nematodes without applying chemical nematicides to the soil or field crop (based on A. G. Whitehead, 1998).

Nematode species	* Heat S. P.	Rotation	Resistance	Seed treatment	Transplant dip
Anguina tritici		+	–	+	
Aphelenchoides besseyi	+		+	+	+
A. fragariae	+				+
Bursaphelenchus xylophilus		–	+		
B. cocophilus		–	–		
Ditylenchus angustus		+		+	
D. destructor	+	+		+	
D. dipsaci	+ +	+/–	+/–	+	
Belonolaimus longicaudatus	? +	–	–		
Criconemella xenoplax		–	–		
Hoplolaimus columbus		? +	–		
Longidorus elongatus		–	–		
Longidorus spp.		–	–		
Paratrichodorus minor		? +	–		
Other *trichodorids*		–	–		
Tylenchorhynchus spp.		–	–		
Xiphinema diversicaudatum		–	–		
X. americanum		+/–	–		
X. index		? +	–		
Helicotylenchus dihystera		–	–		
H. multicinctus		–	–		+
Rotylenchulus reniformis	+	+	+		+
Tylenchulus semipenetrans	+	–	+		+
Hirschmanniella oryzae		? +	? +		+
Pratylenchus brachyurus			–		+

Continued

Nematode species	* Heat S. P.	Rotation	Resistance	Seed treatment	Transplant dip
P. coffeae	+	+	+		+
P. loosi		−	+		
P. penetrans	+		−	+	
P. thornei	+	? +	−	? +	
P. vulnus	+		−		+
P. zeae		+	−		
Radopholus similis	+	+/−	? +		+
R. citrophilus	+		+		+
Scutellonema bradys	+	+	−		+
Globodera pallida		+	+		
G. rostochiensis	+/−	+	+		
Heterodera avenae		+	+	+	
H. cajani			−	+	
H. carotae	? +	? +	−		
H. ciceri	+	+	−		
H. cruciferae		+	−		
H. glycines		+	+		
H. goettingiana		+	−		
H. schachtii		+	?		
H. zeae		? +	−		
Meloidogyne arenaria	+	+	+	+	
M. chitwoodi		? +			
M. exigua		+	+		
M. graminicola		+	+		
M. hapla		+	+	+	+
M. incognita	+ +	+	+	+	+
M. javanica	+ +	+	+	+	+
Nacobbus aberrans		? +		+	+

"+" = actually or potentially effective; "?" = uncertain; " − " = inappropriate, ineffective or unavailable. * Heat therapy; S. (of soil); P. (of plant).

The withdrawal from use, through legislation, of many effective nematicides is concurrent with the increased emphasis on the use of crop rotation, plant resistance, and biological control methods in IPM practices against nematodes and other pests. For centuries mankind has recognized the value of crop rotation for maintaining crop yields, and for a century we have experimented and quantified the benefits of rotation (Lawes, 1899). Our increasing focus on the benefits of crop rotation and on other agronomic practices such as conservation tillage requires us to examine more closely the biotic and abiotic interactions in the soil environment. The soil microflora and associated free-living nematode population vary greatly with cropping systems and with the age and development of plants, due to the constantly changing nature of plant root secretions. Conservation tillage has the advantage of conserving moisture and maintaining a higher level of organic matter in the soil and thereby enhancing microbial activity (Stinner and House, 1990). Products of the microflora and excretions of nematodes, such as *Mesodiplogaster*, significantly influence the type of nitrogen available to plant roots (Anderson, et al., 1983). The soil microflora and the nematodes that selectively graze in and on it are key components in the decomposition of organic matter and release of nutrients (Yeates, 1998). Such microbivorous nematodes play a substantial role in distributing bacteria and fungi in the soil and in protecting them from adverse conditions (Wasilewska and Webster, 1975). In order to achieve sustained, maximum crop yields on limited land areas our knowledge of the biology of soil organisms, including free-living and parasitic nematodes and their complex interactions, must increase greatly. Soil health is the foundation for growing healthy plants, and organismal diversity is an important component of soil health. Therefore, nematode population management could become the backbone of a comprehensive IPM program for high intensity, agricultural soils.

It will be essential to have a better understanding of the ecology of plant-parasitic nematodes in the rhizosphere, and of the associated etiology in the root systems. As well as focusing on the physiology of host – parasite relationships in isolation within experimental, potted plants, additional emphasis will be placed on understanding the complex interactions among free-living nematode populations, the rhizospheric flora and other microorganisms associated with plant roots and a healthy ecosystem (Neipp and Becker, 1999). By so doing, we also shall more completely understand the role of free-living nematodes in sustaining soil fertility in suppressive soils, and in the cycling and partitioning of carbon, the core of our organic life. The beneficial interactions of ectomycorrhizal fungi and nematodes will be exploited more fully, especially in perennial crops. Unless we put more emphasis into understanding these complex edaphic interactions that contribute to healthy, suppressive soils, our goal to optimize the value of rotation, intercropping, solarization and biological control will be thwarted.

The search for effective biological control agents against plant-parasitic nematodes will continue unabated. However, as shown by Chen et al. (1997) using *Pasteuria penetrans* against *Meloidogyne arenaria* and by Hallman and Sikora (1996) using secondary metabolites of endophytic fungi, a fundamental breakthrough is needed in our understanding and approach in order to optimize the effectiveness of natural edaphic agents in the control of plant-parasitic nematodes. In order to effectively exploit biological control, we need a greater understanding of the array of interactions between the various nematode antagonists and saprophytes within and adjacent to the rhizosphere, in the presence and absence of plant stress.

The removal of many of the conventional, chemical insecticides from common practice despite the need to increase crop yields or protect human health continues to be a prime motivator behind research into alternative forms of insect pest control. Successes in the application of entomopathogenic nematodes (EPNs), e. g., *Steinernema* spp. and *Heterorhabditis* spp. in the biocontrol of insect pests, such as black vine weevil and citrus weevil, have been muted by our recognition that we do not fully understand the biological relationships of EPNs with other soil organisms, including plant roots (Kaya and Gaugler, 1993). The intensive research needed to overcome uncertainties associated with the applications of steinernematids and heterorhabditids as biological control agents of insects will focus on the critical determinants of pathogenicity and epizootic development, such as EPN survival in the soil and their interactions with populations of other soil organisms. The population dynamics of EPNs and other soil biota following an EPN field application is virtually a "black box", and a new research focus is necessary for this to be overcome (Anderson and May, 1986; Hominick and Reid, 1990).

2.3 Agronomic Alternatives

In order to achieve larger quantities of good quality food, we can expect larger percentages of some of our cash crops and vegetables to be grown on artificial substrates in controlled environment glasshouses in cool, cold, and hot climates. Application of conservation technologies and the use of non-carbon substrates and energy from novel sources will optimize plant-growth conditions, help to conserve water and minimize soilborne pests, including plant-parasitic nematodes. These controlled growth conditions could successfully overcome many problems such as those caused by *Meloidogyne* spp., but may lead to a resurgence of *Aphelenchoides ritzemabosi* and *Hirschmanniella oryzae* as pests if conditions are not carefully managed.

Undoubtedly, traditional human diets will change due to an overall increased quality of life and to increased travel, and this will change not only the volume of so-called specialty crops produced in a geographic area but will lead to production shifts in some of the more staple food crops. As well, increased globalization and relatively lower distribution costs will lead to more crops being grown in their optimum climates rather than within the sociopolitical boundaries of individual consumer countries. It is reasonable to expect the development of some new agricultural crops as only a small fraction of the existing quarter of a million plant species are used currently as food crops. We will supplement our use of plants as a major food source with much more intensive cultivation of diverse types of fungi in non-soil environments. The emergence of new pest species from among the mycophagous nematodes could ensue and add to the losses caused by *Ditylenchus myceliophagus* and *Aphelenchoides composticola*. As well, rapidly reproducing bacteriophagous forms may become a problem in such high intensity agri-units if rigorous management practices are not pursued. Similarly, the development of mariculture on a worldwide rather than regional basis will help ensure the adequate production of food and other natural products. This, inevitably, will be associated with algal parasitic and microbivorous nematode pest problems, and will lead to the emergence of a whole new field of economic aquatic nematology.

For the large-volume, staple, carbohydrate crops (e.g., grain or root crops), oil legumes (e.g., soybean), sugarbeet, or fiber crops (e.g., cotton), the challenge will continue to be to manage successive monocultures against nematodes and other pests while retaining optimum soil structure. The utilization of rotation, non-tillage, and low fertilizer applications within an IPM program will help minimize nematode-induced loss of yield. Nevertheless, the development of cultivars with durable nematode resistance will be required in most areas in order to sustain crop yields.

2.4 Resistance in Plants

Most plants have evolved highly effective defense mechanisms against most diseases and pests, including nematodes. However, when genetically selected varieties are grown intensively, these conditions over time lead to relative imbalances in the phenotypic expressions of the crop plants and an increase in pests and disease with a resulting loss of yield. The use of molecular markers has enabled us to begin to track, map, and clone nematode resistance genes in crop plants, and we now have the technological ability to develop durable resistance in some plant species that will survive the dynamic genetic pressures of aggressive populations of nematode pests. With the increased knowledge of the genetic complement and expression of many crop plant species and the availability of new molecular techniques, we can expect that the development of genetically based resistance to nematode diseases will become a major means of controlling plant-parasitic nematodes. The elucidation of genome sequences in our crop plants will soon become commonplace and should provide a large number of genes or clusters of genes to manipulate for nematode resistance and crop improvement. It should soon be possible, using transgenic techniques, to develop durable resistance to nematode pests in several major crop species (e. g., in wheat against *Heterodera avenae*, in potato against *Globodera pallida*, and in cotton against *Meloidogyne* spp.). For some crops it will be more challenging, as in soybean where the complexity of the genes for resistance is matched by a very large number of resistance-breaking races of *H. glycines*. Durable resistance lines of a range of crop species should be particularly effective using gene pyramiding techniques (Burrows and DeWaele, 1997; Roberts, 1995) especially if these genes differ significantly in their mode of action and the plants' resistance to other diseases is not compromised.

In a classical piece of genetic engineering, a cyst-resistant gene against *H. schachtii* was introduced from wild *Beta* species into sugarbeet (Cai, et al., 1997). To do the equivalent in a wide range of host plants against the economically important *M. incognita* is a goal that is keenly sought. However, there are particular challenges to overcome with the rapidly changing variations in pathogenicity among populations of the polyploid, mitotic pathenogenic *Meloidogyne* spp. The *Mi* gene for resistance in tomatoes against some species of *Meloidogyne* (e. g., *M. javanica*, *M. anenaria*, *M. incognita*, and some isolates of *M. chitwoodi*) is very effective, though not at high temperatures, and there are already reports of resistance breaking in field populations of *M. incognita* and *M. javanica* (Kaloshian, et al., 1995; Tzortzakakis and Gowan, 1996; Williamson, et al., 1994). Other resistance genes to *Meloidogyne* have been identified in tomato, and it is probable that one or more of these (e. g., *Mi-3*) could be introduced into tomato cultivars (Yaghoobi, et al., 1995). Roberts (1995) considers the concept of developing a shield of resistance, combining direct action and residual impact on soil population densities to control a range of *Meloidogyne* species and subspecific forms. Despite the emergence of new technologies the goal of developing nematode-resistant cultivars will not be easily achieved in all crop plant species.

In a recent review, Vrain (1999) lucidly described the potential of various forms of genetic engineering for the control of plant-parasitic nematodes. It seems possible that proteinase inhibitors, lectins, plantibodies, small molecule compounds or the equivalent of Bt proteins could be incorporated into the genome of plants in order to kill or modify the behavior of plant-parasitic nematodes (Burrows, et al., 1998; Hu, et al., 1999; Steikima, et al., 1997; Urwin, et al., 1998; Vrain, 1999). The aim for sometime to come must be to develop desirable, novel, nematode resistance mechanisms for field use through a combination of distinct defense strategies (Atkinson, et al., 1995). Perhaps, as suggested by Tanksley and McCouch (1997), we shall seek multiple gene resistance in crop plants that have yield-increasing synergistic effects. By so doing, the breakdown of genetically controlled yield increases would be less likely because the complex of genes involves several types of interacting organisms. Genetic manipulation of the regulatory mechanisms of resistance and of the signaling processes that control and coordinate activation of multiple defense responses is a route that may provide effective resistance to nematodes in some crop species. It will require knowledge not only of the signaling pathways but also of the subsequent metabolic pathways that they trigger. It also may be possible to induce defense-related protein responses in a plant to particular nematode species parasitic on the roots by spraying elicitor compounds on the leaves (Hammond-Kosack, et al., 1989; Rahimi, et al., 1993). Inclusion of transgenic crops in a rotation may become a control method of choice.

Exploitation of the genetic approach to parasitism, as advocated by Bird et al. (1999), could be a powerful route to novel methods of controlling plant-parasitic nematodes. As obligate nematode parasites need to locate, invade, select, and establish a feeding site before being able to reproduce it should be possible, by drawing behavioral parallels to gene function in *Caenorhabditis elegans*, to design effective controls based on perturbation of nematode behaviour. As well, an understanding of the genetic basis and associated biochemical pathways for the hatching, molting, diapause, and dauer phenomena in nematodes should lead to new control methods for plant-parasitic forms. The prevailing fear among some about the growth of transgenic plants and consumption of their produce will change due to the need for adequate food from crops. Concurrent with this will be a substantial increase in legislative action globally to guide the commercial application of new and emerging technologies. A parallel development must be a much more effective communication between scientists, seed producers, growers, teachers, and consumers. This will help in the optimum and appropriate development and marketing of food and fiber enhancing technologies together with the management of plant-parasitic nematodes and other crop pests.

2.5 Technological Applications

Through the help of molecular technology we shall gain an understanding of the molecular basis of the physiological and structural response of plants to the sedentary root-knot and cyst nematodes within the root and to the migratory and ectoparasitic forms around the root tips. In some situations, resistance may take the form of plant secretions that deter feeding by ectoparasitic nematodes or modifies the rhizosphere flora such that metabolites from rhizospheric bacteria repel plant-parasitic nematodes.

Nematode parasitism is based on a series of signaling events that lead to various changes in the nematode and complementary ones in the host plant. Undoubtedly, an avenue that will be fruitful is the molecular exploitation of the nematode's sensory system and associated behavior to prevent nematode feeding on plant roots. Whole arrays of control strategies are possible that involve disruption of the nematode's sensory perception and behavior. Some have been identified already (Perry, 1996). Through information arising from the *C. elegans* model and from studies of the sensory behavior of nematodes in the presence of plant roots, bacteria, fungi, and insects (Hui and Webster, 2000), semiochemicals will be developed to control plant-parasitic nematodes. The integration of such semiochemicals into IPM programs, as advocated for insect control, based on the "push-pull" or stimulate-deterrent strategy (Pickett, et al., 1999) would be a logical development. Thus it should be possible to repel nematodes from a food plant species and attract them to an adjacent, genetically modified, non-food intercrop that contains a resistant mechanism or allelochemical. Despite some recent advances (Ferris and Zheng, 1999; Halbrendt, 1996; Perry and Aumann, 1998) relatively little attention has been given to the use of allelochemicals. However, by cloning the gene(s) that produce certain chemical inhibitors into a transgenic crop it may be possible, for example, to inhibit cyst nematode egg hatch. Now is the time to also vigorously exploit sex pheromones as another means of nematode control, though until recently (Jonz, et al., 2001) little progress has been made in this field since it was first proposed in nematology by Greet (1964).

Our developing knowledge of gene sequences and of the use of DNA technologies could lead to the development of methods to breakdown the integrity or disrupt the function of key structures such as the amphids or cuticle of nematodes. The cuticular surface and its molecular contact with the host plant is a prime research target for nematode control. Is it the secretions of the amphids, excretory pore, anus, or the Demanian organ that flow along the lateral lines and bathe the cuticle with an epicuticular, monomolecular layer to prevent plant host recognition of a nematode parasite or, conversely, to prevent parasite attachment to or a predator recognition of the nematode? What are the species-specific signals that attract some nematode species to the roots of some plant species and not to others? The answers to these questions should prove exciting and rewarding.

Rapid and accurate nematode diagnosis to the subspecific level will soon be possible and this will be a necessary complement to developing durable resistance. Molecular diagnosis requires identification of stable, reproducible, polymorphic differences between closely related species, and this is proving to be challenging in the major, economically important *Meloidogyne* species, especially at the subspecific level (Roberts, 1995). The possibility of developing nematode diagnostic biochips for identifying single specimens of nematodes in the field by plant quarantine authorities and by the grower is virtually upon us (Vrain, pers. comm.). Such rapid diagnosis and the use of DNA technology to facilitate more precise population estimates of multiple nematode infestations will enable on-the-spot population models to be developed so that growers and advisors can predict nematode populations and then use the most appropriate IPM program (Barker and Davis, 1996). Plant quarantine regulations concerning the movement of soil and plant products are likely to become more exacting and more important as global trade and travel continues to increase. Rapid molecular differentiation of *B. xylophilus* and *B. mucronatus* for example (Lawler and Harmey, 1993), will assist in managing wood product importations by legislative authorities (Dwinell, 1997).

Molecular diagnostics will benefit, in particular, the situation of *Longidorus* sp. and *Trichodorus* sp., where each population is ecologically distinct and often has different virus transmission rates to plants. Molecular discrimination of the *Xiphinema americanum* group would be a first step to relating the molecular basis of the 40 putative species to that of the recognition phenomenon of the transmitted virus particles (Taylor and Brown, 1997). Effective management strategies against these difficult-to-control virus vectors also should become possible, and this will be enhanced by a greater understanding of the genetic determinants involved in the nematode-virus recognition process (Brown, et al., 1995). Molecular techniques will prove to be the "life-saver" of nematode taxonomy and, indeed, of all those researchers in diverse disciplines that need to know the identity of a nematode. Persistent, successive demands from researchers in agriculture, environmental sciences, and medicine will require the availability worldwide of scientists with training in molecular and morphological taxonomy of nematodes. It will be essential too to combine these skills with knowledge of nematode behaviour, ecology, and embryological development in the overall quest to identify nematodes and to determine their phylogenetic relationships (Coomans, 2000; de Ley, 2000).

2.6 Environmental Quality

Human attention to environmental quality and the survival of organisms is reflected also in the current debate on the "greenhouse effect" and on its consequences for plants and animals through perceived changes in the global climate. Although anthropogenic sources of CO_2 comprise only a small fraction of a plant's carbon budget, these sources are potentially significant destabilizing components of the atmospheric CO_2 concentration. The ancient trees in our forests have been exposed to substantial differences in CO_2 concentrations in their lifetime, e. g., 28% increase over the last 200 years (Bowes, 1993), and have adapted to this. However, the rapidity of change of the $CO_2 : O_2$ ratio that is now occurring may well have measurable adverse effects on the photosynthetic potential of our forests, seaweeds, and crops thereby decreasing plant tolerance to pests and pathogens (Amthor and Cumming, 1988). The ozone component of urban pollution should diminish with the replacement of gasoline by alternative forms of fuel (e. g., hydrogen and natural gas) for automobiles, thereby decreasing the ozone which itself has the effect of decreasing plant tolerance to pests and pathogens (Amthor and Cumming, 1988).

Persistent global warming will influence the geographic distribution of some nematode species, e. g., *Longidorus caespiticola* (Boag and Neilson, 1996), and some *Meloidogyne* species (Barker and Koenning, 1998) so that they thrive in locations where they do not now exist. The detailed studies of Griffith et al. (1997) on population growth patterns of *Ditylenchus dipsaci* in white clover suggest that global warming could lead to greater crop damage due to differential nematode parasite and plant growth. Global warming would change also the area where pine trees grow, not only by directly influencing plant growth but by changing the distribution of pine wilt disease through its influence on the trees' response to *B. xylophilus* (Rutherford, et al., 1990). Other studies (Sohlenius and Bostrom, 1999) have suggested that warmer temperatures will have no rapid, direct influence on the composition of the soil nematode fauna but rather an indirect one. Above certain temperature thresholds the rate of soil mineralization increases, and this would lead to a greater abundance of microbivorous nematodes.

Other less desirable effects of urbanization could be monitored, using nematodes as "environmental canaries", by measuring nematode response to pollution in the air (Samoiloff, 1987), oceans (Tietjen and Lee, 1977), or in the soil (Wasilewska, 1998). In fact, the use of nematodes to monitor the consequences of spatial, temporal, and environmental variability and change will abound. The recent discovery of pine wilt disease and the causative pinewood nematode, *B. xylophilus*, in Korea (Yi, et al., 1989) and Portugal (Mota, et al., 1999) serves to emphasize not only the human role in terms of trade and in spreading disease but also the critical effect of temperature. The increasing frequency of such pathogenic outbreaks will lead to greater international collaboration to establish high profile, plant quarantine legislation together with the associated technical, diagnostic capability to slow the spread of nematode pests.

2.7 Human Health

The completion of the *C. elegans* Genome Project and the Human Genome Project is the beginning of a whole new era of opportunity as the ability to understand and relate the precise function and mode of operation of specific genes in the life of the nematode and of its hosts becomes possible. There will be a burgeoning of research into human genetic malfunctions as well as the genetic control of plant diseases, caused by nematodes. The completion of the nematode genome sequence provides the opportunity, for instance, to link various developmental and behavioural patterns and neurological and mobility malfunctions with their respective genes in nematodes and to extrapolate this to the human condition.

The exploitation of some bioactivities of the secondary metabolites from the bacterial symbionts, *Xenorhabdus* spp. and *Photorhabdus* spp. of EPNs, may prove rewarding in the years ahead. Since the early 1980s (Paul, et al., 1981) there has been a growing interest in the antimicrobial properties of these substances, and a large number of antibiotic and antimycotic substances have been identified (Webster, et al., 1998). Although these substances may not all have a role in the biology of their respective symbionts (Boemare, et al., 1997), the range and strength of activity of these compounds or their derivatives has stimulated research into their potential as naturally-derived pharmaceutical drugs and agroforestry pesticides (Chen, et al., 1996). Many of these secondary metabolites have chemical structures that are atypical for conventional antibiotics (Li, et al., 1997). Their diverse effects, such as nematicidal (Hu, et al., 1999) and antineoplastic (Webster, et al., 1998) from *Photorhabdus* sp. and *Xenorhabdus* sp., respectively, lead one to speculate at the range of potential benefits to society of the products from these inconspicuous edaphic symbionts. The early decades of this century will reveal the level to which such speculation is realized.

Whether or not the human race will try to establish itself on another planet during this century is a moot point, but life in space will continue to be explored as a vacation to escape from the crowd or for a variety of medical treatments. We know from experiments with *C. elegans* that nematodes feed and reproduce normally at zero gravity (Nelson, et al., 1994), which gives us the confidence to explore this high tech health spa in comfort as long as we do not expose ourselves to cosmic radiation. On earth, we will continue to test the boundaries of our cultural, physical, and nutritional limits. The mass culture of nematodes (e. g., *C. elegans*, *Steinernema* spp., and *B. xylophilus*), could lead to the development of a nematode-based, low fat or low protein diet or a marine nematode based, isotonic beverage for humans. We can speculate further that the pheromone: food ratios that trigger, through the *daf* genes, entry into and exit from the Dauer development pathway in *C. elegans*, are a form of suspension of the aging process in response to environmental conditions. Recognizing that genes are highly conserved across the phyla, does this Dauer pathway hold the key to the human ageing process that mankind has sought to resolve over the centuries?

Within this strangled encapsulation of the multitude of interacting and sometimes conflicting paradigms for the future, one can see that opportunities and challenges for nematology abound. Nematologists should develop their own prioritized list of these and, as and when the opportunity arises, focus on and develop an answer to the nematological challenge of the day. Nematology, whether focussed on plant, animal, insect parasitic, or on free-living forms, is in a position to influence some of the fundamental issues that will continue to challenge society, namely food production, environmental management, and the human concern for ageing and disease.

References

Amthor, J. S., and J. R. Cumming. 1988. Low levels of ozone increase bean leaf maintenance respiration. Canadian Journal of Botany 66: 724 – 726.

Anderson, R. M., and R. M. May. 1986. The invasion, persistence and spread of infectious diseases within animal and plant communities. Philosophical Transactions of the Royal Society of London B314: 533 – 546.

Anderson, R. V., W. D. Gould, L. E. Woods, C. Cambardella, R. E. Ingham, and D. C. Coleman. 1983. Organic and inorganic nitrogenous losses by microbivorous nematodes in soil. Oikos 40: 75 – 80.

Atkinson, H. J., P. E. Urwin, E. Hansen, and M. J. McPherson. 1995. Designs for engineered resistance to root-parasitic nematodes. Tibtech 13: 1 – 16.

Barker, K. R., and E. L. Davis. 1996. Assessing plant-nematode infestations and infections. Advances in Botanical Research 23: 103 – 136.

Barker, K. R., and S. R. Koenning. 1998. Developing sustainable systems for nematode management. Annual Review of Phytopathology 36: 165 – 205.

Bednarek, A., and R. Gaugler. 1997. Compatibility of soil amendments with entomopathogenic nematodes. Journal of Nematology 29: 220 – 227.

Bird, D. D. Mck, C. H. Opperman, S. J. M. Jones, and D. L. Baillie. 1999. The *Caenorhabditis elegans* genome: a guide in the post genomics age. Annual Review of Phytopathology 37: 247 – 265.

Boag, B., and R. Neilson. 1996. Effects of potential climatic changes on plant-parasitic nematodes. Annals of Applied Biology 45: 331 – 334.

Boemare, N. E., A. Givaudan, M. Brehelin, and C. Laumond. 1997. Symbiosis and pathogenicity of nematode – bacterium complexes. Symbiosis 22: 21 – 45.

Bowes, G. 1993. Facing the inevitable: Plants and increasing atmospheric CO_2. Annual Review of Plant Physiology and Plant Molecular Biology 44: 309 – 332.

Brown, D. J. F., W. M. Robertson, and D. L. Trudgill. 1995. Transmission of viruses by plant nematodes. Annual Review of Phytopathology 33: 223 – 249.

Browning, J. A. 1998. One phytopathologist's growth through IPM to holistic plant health: The key to approaching genetic yield potential. Annual Review of Phytopathology 36: 1 – 24.

Burrows, P. R., and D. De Waele. 1997. Engineering resistance against plant-parasitic nematodes using anti-nematode genes. Pp. 217 – 236. In C. Fenoll, F. M. W. Grundler, and S. A. Ohl, eds. Cellular and Molecular Aspects of Plant-nematode Interactions. The Netherlands: Kluwer Academic Publishers.

Burrows, P. R., A. D. Barker, C. A. Newell, and W. D. Hamilton. 1998. Plant-derived enzyme inhibitors and lectins for resistance against plant-parasitic nematodes in transgenic crops. Pesticide Science 52: 176 – 183.

Cai, D., M. Kleine, S. Kifle, H.-J. Harloff, N. N. Sandal, K. A. Marcker, R. M. Klein-Lankhorst, E. M. Salentijn, W. Lange, W. J. Stiekma, U. Wyss, M. W. Grundler, and C. Jung. 1997. Positional cloning of a gene for nematode resistance in sugar beet. Science 275: 832 – 834.

Chen, G., Y. Zhang, J. Li, G. B. Dunphy, Z. K. Punja, and J. M. Webster. 1996. Chitinase activity of *Xenorhabdus* and *Photorhabdus* species, bacterial associates of entomophogenic nematodes. Journal of Invertebrate Pathology 68: 101 – 108.

Chen, Z. X., D. W. Dickson, D. J. Mitchell, R. McSorley, and T. E. Hewlett. 1997. Suppression mechanisms of *Meloidogyne arenaria* race 1 by *Pasteuria penetrans*. Journal of Nematology 29: 1 – 8.

Cook, R. J., and R. J. Veseth. 1991. Wheat Health Management. St. Paul, Minnesota USA: American Phytopathological Society.

Coomans, A. 2000. Nematode systematics: Past, present and future. Nematology 2: 3 – 7.

De Ley, P. 2000. Lost in worm space: Phylogeny and morphology as road maps to nematode diversity. Journal of Nematology 2: 9 – 16.

Dwinell, L. D. 1997. The pinewood nematode: Regulation and mitigation. Annual Review of Phytopathology 35: 153 – 166.

Ehrlich, A. H. 1995. Implications of population pressure on agriculture and ecosystems. Advances in Botanical Research 21: 79 – 104.

Ferris, H., and L. Zheng. 1999. Plant sources of Chinese herbal remedies: Effects on *Pratylenchus vulnus* and *Meloidogyne javanica*. Journal of Nematology 31: 241 – 263.

Greet, D. N. 1964. Observations on sexual attraction and copulation in the nematode *Panagrolaimus rigidus* (Schneider). Nature 204: 96 – 97.

Griffith, G. S., R. Cook, and K. A. Mizen. 1997. *Ditylenchus dipsaci* infestation of *Trifolium repens*. II. Dynamics of infestation development. Journal of Nematology 29: 356 – 369.

Halbrendt, J. M. 1996. Allelopathy in the management of plant-parasitic nematodes. Journal of Nematology 28: 8 – 14.

Hallman, J., and R. A. Sikora. 1996. Toxicity of fungal endophyte secondary metabolites to plant-parasitic nematodes and soil-borne plant pathogenic fungi. European Journal of Plant Pathology 102: 155 – 162.

Hammond-Kosack, K. E., H. J. Atkinson, and D. J. Bowles. 1989. Systemic changes in the composition of leaf apoplast following root infection with cyst nematode *Globodera rostochiensis*. Physiological and Molecular Plant Pathology 35: 495 – 506.

Hominick, W. M., and A. P. Reid. 1990. Perspectives on entomopathogenic nematology. Pp. 327 – 345. In R. Gaugler, and H. K. Kaya, eds. Entomopathogenic Nematodes in Biological Control. Boca Raton, FL: CRC Press.

Hu, K., J. Li, and J. M. Webster. 1999. Nematicidal metabolites produced by *Photorhabdus luminescens* (Enterobacteriaceae), bacterial symbiont of entomopathogenic nematodes. Nematology 1: 457 – 469.

Hui, E., and J. M. Webster. 2000. Influence of insect larvae and seedling roots on the host-finding ability of *Steinernema feltiae* (Nematoda: Steinernematidae). Journal of Invertebrate Pathology 75: 152 – 162.

Jonz, M. G., E. Riga, A. J. Mercier, and J. W. Potter. 2001. Partial isolation of a water soluble pheromone from the sugar beet cyst nematode, *Heterodera schachtii*, using a novel bioassay. Nematology 3: 65 – 74.

Kaloshian, I., W. H. Lang, and V. M. Williamson. 1995. An aphid-resistance locus is tightly linked to the nematode-resistance gene, *Mi*, in tomato. Proceedings of the National Academy of Science, USA 92: 622 – 625.

Kaya, H., and R. Gaugler. 1993. Entomopathogenic nematodes. Annual Review of Entomology 38: 181 – 206.

Lawes, J. B. 1899. The wheat crop of 1899. Agricultural Gazette 1899, No. 2: 248.

Lawler, C., and M. A. Harmey. 1993. Immunological detection of nematode antigens on the surface of a wood section. Fundamental and Applied Nematology 16: 521 – 523.

Li, J., G. Chen, and J. M. Webster. 1997. Nematophin, a novel antimicrobial substance produced by *Xenrohabdus nematophilus* (Enterobacteriaceae). Canadian Journal of Microbiology 43: 770 – 773.

Mota, M., H. Braasch, M. A. Bravo, A. C. Penas, W. Burgermeister, K. Metge, and E. Sousa. 1999. First report of *Bursaphelenchus xylophilus* in Portugal and in Europe. Nematology 1: 727 – 734.

Neipp, P. W., and J. O. Becker. 1999. Evolution of biocontrol activity of rhizobacteria from *Beta vulgaris* against *Heterodera schachtii*. Journal of Nematology 31: 54 – 61.

Nelson, G. A., W. W. Schubert, G. A. Kazarians, and G. F. Richards. 1994. Development and chromosome mechanics in nematodes: Results from IML-1. Advanced Space Research 14: 209 – 214.

Paul, V. J., S. Frautschy, W. Fenical, and K. H. Nelson. 1981. Antibiotics in microbial ecology: Isolation and structure assignment of several new antibacterial compounds from the insect-symbiotic bacteria *Xenorhabdus* spp. Journal of Chemical Ecology 7: 589 – 597.

Perry, R. N. 1996. Chemoreception in plant-parasitic nematodes. Annual Review of Phytopathology 34: 181 – 199.

Perry, R. N., and J. Aumann. 1998. Behaviour and sensory responses. Pp. 75 – 102. In R. N. Perry, and D. J. Wright, eds. The Physiology and Biochemistry of Free-living and Plant-parasitic Nematodes. Wallingford, UK: CAB International.

Pickett, J. A., D. W. M. Smiley, and C. M. Woodcock. 1999. Secondary metabolites in plant insect interactions: Dynamic systems of induced and adaptive responses. Advances in Botanical Research 30: 91 – 113.

Pimentel, D., and H. Lehman. 1993. The Pesticide Question. New York: Chapman and Hall.

Rahimi, S., R. N. Perry, and D. J. Wright. 1993. Induction and detection of pathogenesis related proteins in leaves and roots of potato plants infected with pathotypes of *Globodera pallida*. Fundamental and Applied Nematology 16: 549 – 556.

Roberts, P. A. 1995. Conceptual and practised aspects of variability in root-knot nematodes related to host plant resistance. Annual Review of Phytopathology 33: 199 – 221.

Rutherford, T. A., Y. Mamiya, and J. M. Webster. 1990. Nematode-induced pine wilt disease: Factors influencing its occurrence and distribution. Forest Science 36: 145 – 155.

Samoiloff, M. R. 1987. Nematodes as indicators of toxic environmental contaminants. Pp. 433 – 439. In J. A. Veech, and D. W. Dickson, eds. Vistas on Nematology. Hyattsville, Maryland, USA: Society of Nematologists INC.

Sohlenius, B., and S. Bostrom. 1999. Effects of global warming on nematode diversity in a Swedish tundra soil – a soil transplantation experiment. Nematology 1: 695 – 709.

Steikema, W. J., D. Bosch, A. Wilmink, J. M. DeBoer, A. Schouten, J. Roosien, A. Goverse, G. Smart, J. Stokkermans, F. J. Gommers, A. Schots, and J. Bakker. 1997. Towards plantebody-mediated resistance against nematodes. Pp. 262 – 271. In C. Fenoll, F. M. Grundler, and S. A. Ohl, eds. Cellular and Molecular Aspects of Plant-Nematode Interactions. The Netherlands: Kluwer Academic Publishers.

Stinner, B. R., and G. J. House. 1990. Arthropods and other invertebrates in conservation – tillage agriculture. Annual Review of Entomology 35: 299 – 318.

Tanksley, S. D., and S. R. McCouch. 1997. Seed banks and molecular maps: Unlocking genetic potential from the wild. Science 277: 1063 – 1066.

Taylor, C. E., and D. J. F. Brown. 1997. Nematode Vectors of Plant Viruses. Wallingford, UK: CAB International.

Tietjen, J. H., and J. J. Lee. 1977. Life histories of marine nematodes. Influence of temperature and salinity on the reproductive potential of *Chromadorina germanica*,

Butschli. Mikrofauna Meeresboden 61:263 –270.

Tzortzakakis, E. A., and S. R. Gowan. 1996. Occurrence of resistance-breaking pathotypes of *Meloidogyne javanica* on tomatoes in Crete, Greece. Fundamental and Applied Nematology 19:960 –967.

Urwin, P. E., C. J. Lilley, M. J. McPherson, and H. J. Atkinson. 1998. Resistance to both cyst and root-knot nematodes conferred by transgenic *Arabidopsis* expressing a modified plant cystatin. Plant Journal 12:455 –461.

Vrain, T. C. 1999. Engineering natural and synthetic resistance for nematode management. Journal of Nematology 31: 424 –436.

Wasilewska, L. 1998. Changes in the proportion of groups of bacterivorous soil nematodes with different life strategies in relation to environmental conditions. Applied Soil Ecology 9:215 –220.

Wasilewska, L., and J. M. Webster. 1975. Free living nematodes as disease factors of man and his crops. International Journal of Environmental Studies 7:201 –204.

Webster, J. M., G. Chen, and J. Li. 1998. Parasitic worms: An ally in the war against the superbugs. Parasitology Today 14:161 –163.

Whitehead, A. G. 1998. Plant Nematode Control. Wallingford, UK: CAB International.

Williamson, V. M., K. N. Lambert, and I. Kaloshian. 1994. Molecular biology of nematode resistance in tomato. Pp. 211 –219. In F. Lambert, C. De Giorgi, and D. M. Bird, eds. Advances in Molecular Plant Nematology. London: Plenum Press.

World Commission on Environment and Development. 1987. Our Common Future. London: Oxford University Press.

Yaghoobi, J., I. Kaloshian, Y. Wen, and V. M. Williamson. 1995. Mapping a new nematode resistance locus in *Lycopesicon peruvianum*. Theoretical Applied Genetics 91: 457 –464.

Yeates, G. 1998. Feeding in free-living soil nematodes: A fundamental approach. Pp. 245 – 269. In R. N. Perry, and D. J. Wright, eds. The Physiology and Biochemistry of Free-living and Plant-parasitic Nematodes. Wallingford, UK: CAB International.

Yi, C. K., B. H. Byon, J. D. Park, S. I. Yang, and K. H. Chang. 1989. First finding of the pine wood nematode *Bursaphelenchus xylophilus* (Stiener and Buhrer) Nickle and its insect vector in Korea. Research Report of the Forest Research Institute of Korea 38: 141 –149.

3 Developmental Biology of Nematodes – What We Learn from *Caenorhabditis elegans*

Marie-Anne Félix

3.1 Introduction: the *Caenorhabditis elegans* Cell Lineage

3.1.1 Historical Perspective on Nematode Developmental Studies

Nematodes have been the subject of classical developmental studies at the end of the 19th century and helped to unravel some universal aspects of reproduction and development, for example meiosis (Van Beneden, 1883). The distinction between soma and germ line could be visualized in *Ascaris* through the specific process of chromatin diminution (Boveri, 1892) (note however that chromatin diminution does not occur in most other nematodes).

As knowledge accumulated for many phyla, distinctions could be drawn between them, and the nematodes became a prime example of mosaic (autonomous) development, whereby the different cells of the embryo cannot replace each other and diverge in fates through successive asymmetric divisions (Laufer, et al., 1980; zur Strassen, 1959). This mosaic development was contrasted with the regulative development of mammals or sea urchins, in which each cell of a 2-cell or 4-cell embryo can give rise to an entire organism (twins). The equivalence of some early blastomeres during nematode development and the role of cell interactions were, however, advocated by some (e.g., Bonfig, 1925).

Another historical aspect of nematode development was the study of variations between nematodes in their early embryonic development (Chitwood and Chitwood, 1974; Malakhov, 1994; Nigon, 1965). More recently, with *Caenorhabditis elegans* becoming a major model system in developmental biology, the techniques developed for cell lineaging, cell ablations, genetics, and molecular biology have been extended to other free-living soil nematodes. We will cover some of the evolutionary variations within nematodes in relevant sections. It is now clear that the development of *Drosophila melanogaster* is special among insects and that, likewise, *C. elegans* is not representative of all nematodes, even if often familiarly called "the nematode" or "the worm". It offers, however, an excellent starting point for the study of other nematodes.

3.1.2 *C. elegans* as a Model System

Caenorhabditis elegans was first described by Maupas, who was interested in its hermaphroditic mode of reproduction (Maupas, 1900). Its early embryonic development was described by Nigon and his collaborators (Nigon, et al., 1960). In the 1960s, Sydney Brenner chose *C. elegans* as a new model system for studies in development and neurobiology. His goal was to study these processes genetically in a very simple metazoan. With his collaborators in Cambridge, they developed a set of tools during the following 20 years: a genetic map (Brenner, 1974), the cell lineage (Kimble and Hirsh, 1979; Sulston and Horvitz, 1977; Sulston, et al., 1983), electron microscopic reconstructions of the pharynx (Albertson and Thomson, 1976) and of the nervous system (White, et al., 1986). These thorough studies founded the consecutive expansion of the field in the 1980s and 1990s. With new molecular techniques and the genome entirely sequenced (The *C. elegans* Sequencing Consortium, 1998), *C. elegans* is now a favored model to study gene function in a multicellular context, like yeast for subcellular events.

Reference books on *C. elegans* are: The Nematode *Caenorhabditis elegans* by Wood (1988) and *C. elegans* II by Riddle, et al. (1997). For methods, see Epstein and Shakes (1995).

3.1.3 Invariance of the *C. elegans* Cell Lineage

3.1.3.1 Lineaging

In *C. elegans*, the continuous observation of development at the cellular level is possible because the animals are transparent and the nuclei appear contrasted under Nomarski optics (the cell contour cannot always be determined) (Fig. 3.1). The animals can feed and develop for hours between a slide and a coverslip, on a thin agar layer in the presence of bacteria.

The total life cycle lasts 3.5 days. Embryonic development is completed in 15 hours and is followed by hatching and four larval stages separated by lethargic periods and molts (Fig. 3.2; note that the *C. elegans* nomenclature uses larval stages – called L1 to L4 – instead of juvenile stages). The postembryonic cell lineage was first determined (Kimble and Hirsh, 1979; Sulston and Horvitz, 1977). The few blast cells that divide during larval development mostly give rise to the reproductive system (Fig. 3.3).

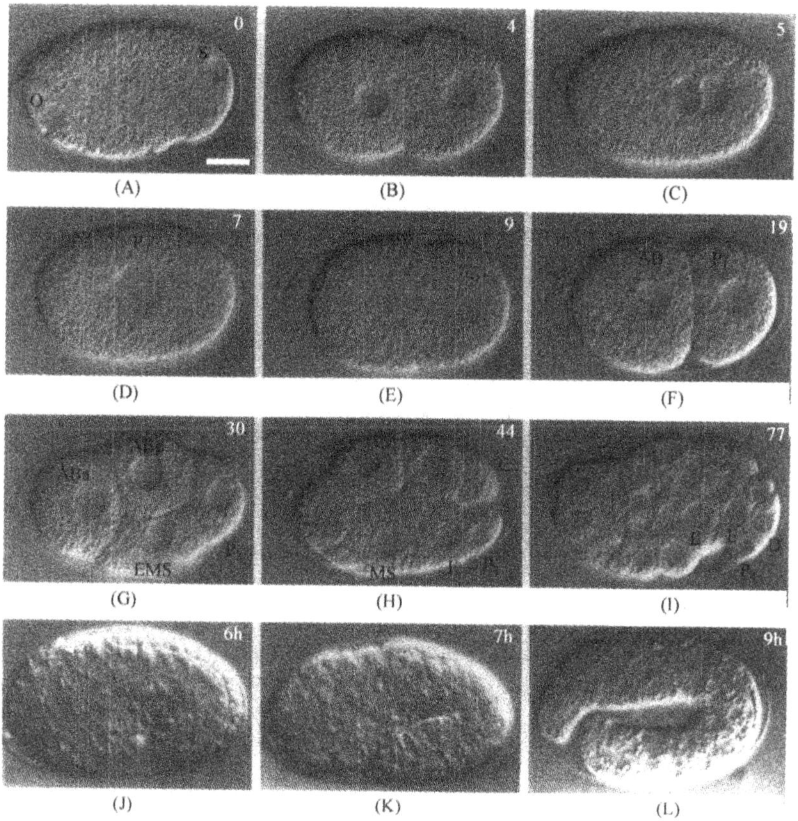

Figure 3.1 Nomarski micrographs of embryonic stages of *Caenorhabditis elegans*. (A) About 45 min after fertilization, the sperm nucleus (s) becomes visible at the future posterior side of the egg, and the oocyte nucleus (o) at the anterior side. (B) Cortical contractions of the egg result in a "pseudocleavage" furrow; the oocyte nucleus migrates towards the sperm nucleus and its associated microtubule aster, and the sperm nucleus/aster migrates towards the center. (C) The two pronuclei meet. (D) The sperm centrosomes separate and migrate to either side of the nuclei. (E) The mitotic spindle forms along the long axis of the egg, slightly displaced towards the future posterior side. (F) 2-cell stage. The anterior AB blastomere is slightly larger than its posterior sister P1. (G) 4-cell stage. AB divides symmetrically, P1 asymmetrically into EMS and P2. (H) 8-cell stage. EMS and P2 have divided asymmetrically. (I) 26-cell stage. Gastrulation starts by the invagination of the E daughters on the ventral side. P3 has divided asymmetrically into P4 and D. (J) Comma-stage embryo. Most embryonic divisions have taken place. The embryo starts elongating. (K) 1½-fold embryo. (L) 2-fold embryo. In all pictures, the anterior pole is to the left and the ventral side down. The number of minutes (A) – (I) or hours (J) – (L) elapsed since the first picture is shown at the top right corner of each picture. Bar = 10μm. Reprinted with permission from Schierenberg (1988).

Undoubtedly, the masterpiece was the determination of the embryonic lineage from egg to hatching (Sulston, et al., 1983). Embryogenesis can be divided in a first period of rapid cell divisions, and a second period of morphogenesis with few divisions. Determining the entire lineage required following many simultaneous cell divisions in the tridimensional embryo, as well as cell migrations and fates throughout the second period of embryogenesis, during which the embryo elongates into a worm and rotates within the egg shell. Automatic recording of images of embryonic development is now possible (Schnabel, et al., 1997; Thomas, et al., 1996).

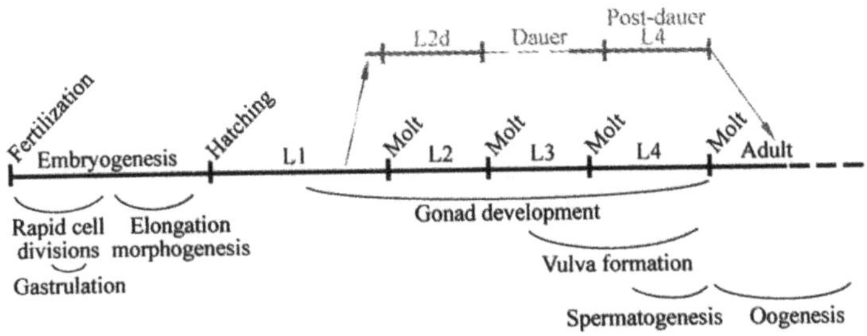

Figure 3.2 Developmental stages of *Caenorhabditis elegans*. The life cycle lasts 3.5 days at 20℃. The adult lives for about 2 weeks. The non-mated hermaphrodite produces progeny during the first 4 days of adulthood in laboratory culture conditions. The alternative life cycle through a Dauer stage is shown in grey.

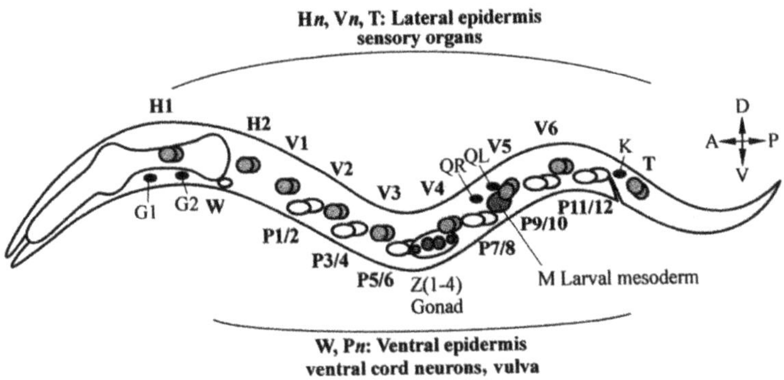

Figure 3.3 Larval blast cells of *Caenorhabditis elegans*. L1 larva at hatching, showing the nuclei of hermaphrodite larval blast cells (except the intestinal cells). Q, G, and K cells are neuroblasts. M is located on the right side of the animal, K on the left side. QL/R, P*n*, V*n*, H*n*, and T cells are left – right pairs indicated by double circles. After Sulston and Horvitz (1977).

Determination of the cell lineage was followed by the electron microscopic reconstruction of the entire nervous system of the hermaphrodite, including axon morphology and neuronal connections (White, et al., 1986). Ultrastructural studies were also undertaken for the pharynx (Albertson and Thomson, 1976), buccal capsule (Wright and Thomson, 1981), male tail (Sulston, et al., 1980), uterus (Newman, et al., 1996), and vulva (Sharma-Kishore, et al., 1999). The goal was a complete description of *C. elegans* at the cellular level.

The embryonic cell lineage of *Turbatrix aceti* (Sulston, et al., 1983) and the entire larval cell lineage of *Panagrellus redivivus* (Panagrolaimidae) (Sternberg and Horvitz, 1981, 1982) were also followed. The larval blast cells are almost identical to those of *C. elegans*; cell homologies can be easily drawn and development thus compared at the cellular level. Cell lineaging in different nematodes is also a way to define homologous cells in less obvious instances. For example, cell homologies in the buccal capsule of *Zeldia punctata* (Cephalobidae) were redefined by cell lineaging and comparisons with *C. elegans* (Dolinski, et al., 1998).

3.1.3.2 Invariance of the Cell Lineage

Obviously, the cell lineage could only be determined because of its quasi-invariance from individual to individual in the species. The final number of somatic cells is 959 in the adult hermaphrodite, 1031 in the male. Each of these cells has an identical position, ancestry, and fate in each worm (germ line divisions, however, are variable).

Lineage invariance and small cell number constitute great experimental advantages to study development, compared to flies or mice. Each cell that is formed during development is given a name. The lineage name corresponds to cell ancestry; for instance, ABpr is the right daughter of the posterior daughter of AB. Terminal cells are renamed to designate their fate; for instance, ABprpappaaa is called RIH as one of the nerve ring interneurons. Larval blast cells are also renamed; for example, MSpppaap is called Z1.

The few exceptions to somatic lineage invariance concern pairs of cells that can adopt either of two fates, either in a fully random fashion as for anchor cell determination (Section 3.5.4), or in a biased fashion for instance for P11/P12 orientation of migration (Delattre and Félix, 2001a). Some division patterns of precursor cells are also variable (Delattre and Félix, 2001b; Sulston and Horvitz, 1997). The timing of divisions and cell positions also vary somewhat from embryo to embryo (Sulston, et al., 1983); interestingly, these variations appear greater at some developmental stages than others, reflecting perhaps the level of constraints on the pattern of cell contacts (Schnabel, et al., 1997).

In other nematodes, the division patterns of a given blast cell are different from those of *C. elegans* (see, for example, Sommer and Sternberg, 1994; Sternberg and Horvitz, 1981, 1982). Some polymorphisms exist within a species and correspond to lineages that show inter-specific variations (Delattre and Félix, 2001b). At least in some cases, evolutionary divergence in cell lineage could thus be the result of an intra-specific polymorphism.

In more distant nematodes, the early embryonic lineage is not invariant and alternate cleavage patterns may be used in a given species (Malakhov, 1994). Moreover, in the marine nematode *Enoplus brevis*, individual blastomeres labeled at the 8-cell stage do not give rise to a definite set of tissues (except for one endodermal precursor) (Voronov and Panchin, 1998). These striking variations suggest a great variety in developmental mechanisms among nematodes.

3.1.3.3 Cell lineage and Developmental Mechanisms

The cell lineage provides information concerning the link between body parts and cell ancestry, and symmetry relationships in the body. For example, the pharynx develops from descendants of both ABa and MS and its 3-fold final symmetry results from recruitment of pairs of bilaterally symmetric cells plus the piecemeal recruitment of a third cell by symmetry breakage of two bilateral homologs in a distinct lineage. Most surprisingly, the bilateral symmetry of the larva is built in a complicated fashion from a highly left-right asymmetric cell lineage (Sulston, et al., 1983).

The lineage shows repetitive patterns (see Section 3.7.1 and Fig. 3.25). These sublineages suggest that an autonomously determined program of asymmetric divisions is activated in the precursor cell and unfolds independently of external cues. The asymmetric divisions in the sublineage are likely polarized by cues from the preceding division or from another cell in the same sublineage. This sublineage is modified repeatedly, "en bloc", during evolution (Sternberg and Horvitz, 1982) or in homeotic mutants that transform the fate of the precursor cell. Some mutants, on the other hand, transform each instance of a repeated sublineage, for example, by disrupting the asymmetry of a given division (Chalfie, et al., 1981; Horvitz, 1988).

The invariance of the cell lineage and the analysis of repeated sublineages supported the view that *C. elegans* development was "determined by its lineage", i.e., by a succession of asymmetric cell divisions. Despite words of caution (Sulston, et al., 1983), the cell lineage was almost seen as an explanation of the developmental process. This view was reinforced by the fascinating drawing of the whole lineage as a genealogical tree, which, however, loses the information on cell positions. Neighbor relationships that were not represented by cell lineage drawings were subsequently found to be playing an important role.

3.1.4 Cell Interactions and Asymmetric Divisions

The lineage-centered view of *C. elegans* development was seriously revised in the following years by the discovery of extensive intercellular signaling.

3.1.4.1 Laser Ablations

The alignment of a laser microbeam in the microscope (a setup devised by John White in Cambridge) made it possible to eliminate specific cells during development. In some instances, ablation of a cell affects the fate of its neighbors. For example, ablation of the cells that normally divide and form the vulva results in their neighbors changing fate and replacing them (Sulston and White, 1980) (Section 3.6.2). This phenomenon is called replacement regulation and defines an equivalence group of cells that are competent to adopt a given cell fate. A different type of cell interaction can also be demonstrated by laser ablation (Fig. 3.4). For example, if the uterine anchor cell is ablated, all cells of the vulval equivalence group adopt a non-vulval fate, demonstrating an inductive signaling (direct or indirect) by the anchor cell (Kimble, 1981).

3.1.4.2 Blastomere Manipulations

Several cell interactions were unraveled in the early embryo by mechanical manipulation of blastomeres, either by inverting the orientation of a division or by physically separating and recombining blastomeres (Fig. 3.4). Thus, in *C. elegans*, as in any other animals, cell fates diverge during development through both cell intrinsic (asymmetric divisions) and extrinsic (intercellular signaling) mechanisms. Both types of mechanisms can occur concomitantly, with an intercellular signal polarizing cell division (see Section 3.2.3). What characterizes *C. elegans* is the small number of cells, which allows cell interactions to be reproducible. Combined with a tight regulation of the cell cycle, the outcome is an invariant lineage.

3.1.4.3 Molecular Pathways

These developmental processes can now be analyzed at the molecular level using classical and molecular genetics. Mutants defective in a given intercellular signaling pathway or in an asymmetric division are isolated. The genes affected can be isolated by genetic mapping of the mutation followed by transformation rescue with the corresponding genomic region, narrowing it

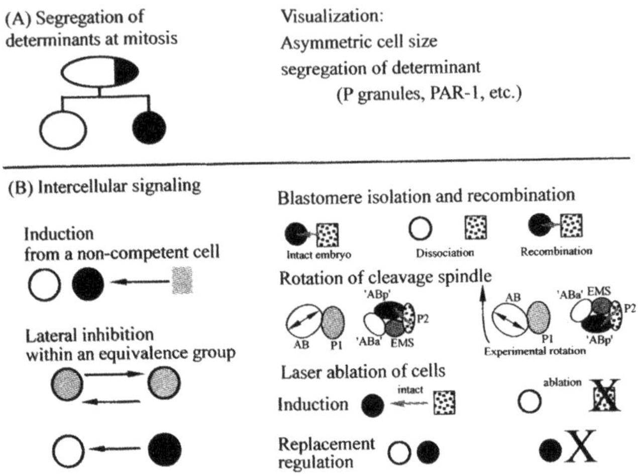

Figure 3.4 Cell specification mechanisms in development. Cells acquire specific fates (black versus white) either by asymmetric division of their mother (A) or by intercellular signaling (B). On the right are methods used to demonstrate these mechanisms. In the early embryo, cell interactions can be demonstrated by blastomere isolations and manipulations. For example, by separating and recombining blastomeres at the 4-cell stage, the P2 blastomere was shown to polarize the EMS division and induce MS formation (Goldstein, 1992) (see Section 3.2.3). Another type of experiment in the embryo is to modify the orientation of a division by mechanical pressure. Inversion of the position of the AB daughters results in normal embryos, demonstrating that they are equivalent in potential and that their fate is determined by their position (Priess and Thomson, 1987) (Section 3.2.3). Similarly, the left-right skew of the next AB divisions can be inverted, resulting in left-right reversed, but otherwise normal animals (Wood, 1991) (Section 3.2.4). Finally, changes of cell fates observed after ablation of cells with a laser indicate that cellular interactions are required for normal cell fates, either because the ablated cell normally sends an inductive signal, or because replacement occurs within a group of equivalent cells (Sections 3.6.2 and 3.5.4).

down to a single rescuing gene. This process is facilitated by the fine correspondance between the genetic map (the genetic linkage between mutated genes) and the physical map (the linear DNA sequence). DNA-mediated transformation is achieved easily in *C. elegans* by injection of DNA into the syncytial hermaphrodite germ line, under any form (circular plasmids and cosmids as well as linear cloning vectors like phages or yeast artificial chromosomes).

The striking result is that a small number of pathways are used over and over again during development of a given animal and appear extremely conserved between animals. Some of these pathways, like the vulval EGF/Ras/MAP kinase pathway, were previously known from other animals and new regulators are now found by genetic screens in *C. elegans*. Other pathways, like that regulating cell death, have been uncovered in *C. elegans* and later found to be conserved. Sequencing of the genome gives an overview of all developmental pathway components encoded in the *C. elegans* genome (Ruvkun and Hobert, 1998). Once the molecules are known, their expression pattern can be determined by the use of a reporter transgene or by *in situ* hybridization. The site of action of a gene can be determined by isolating genetic mosaics in which some cells are mutant and some not. These mosaic animals are obtained by making use of the fact that partial chromosome duplication or extra-chromosomal arrays obtained by DNA injection are lost randomly at some frequency during mitosis. The loss of the duplication or transgene in a given animal is followed by a cell-autonomous marker, for example, GFP (green fluorescent protein) on a transgene driven by a ubiquitous promoter (see example of use in Section 3.6.5). Since molecular pathways are conserved throughout the animal kingdom and reused in different developmental contexts, it becomes interesting to understand how their recruitment evolves in other nematodes. We will see some examples of variations in the use of intercellular signals and molecular pathways.

3.2 Early Development

3.2.1 Fertilization and Antero-posterior Axis Specification

Mature oocytes are elongated, and their long axis corresponds to that of the ovary. Sperm entry takes place at the leading end when the oocyte enters the spermatheca (Hirsh, et al., 1976) (Section 3.5.6). Goldstein and Hird (1996) demonstrated that the sperm entry site determines the posterior side of the embryo by obtaining animals in which sperm entry occurs at random around the oocyte. The sperm nucleus is not required for polarization of the egg (Sadler and Shakes, 2000), but the sperm centrosome or aster appears to be (O'Connell, et al., 2000; Wallenfang and Seydoux, 2000).

In the family Rhabditidae, some variations occur between species as to the site of sperm entry, but it always correlates with the future posterior side. In more distant species in the suborder Cephalobina, however, the posterior of the embryo does not correlate with the site of sperm entry, but with the distal end of the oocyte in the ovary (Goldstein, et al., 1998).

3.2.2 Asymmetry of the First Division and in the Posterior Lineage

3.2.2.1 Asymmetric Division of the Egg

After fertilization, the sperm centrosome nucleates an aster of microtubules at the posterior pole, while the female pronucleus undergoes its two meiotic divisions at the anterior pole (Fig. 3.5). The egg undergoes major cytoskeletal rearrangements, which lead to the asymmetric distribution of determinants along the antero-posterior axis and their differential segregation in the two daughter cells. Cortical actin contractions result in a "pseudocleavage" furrow

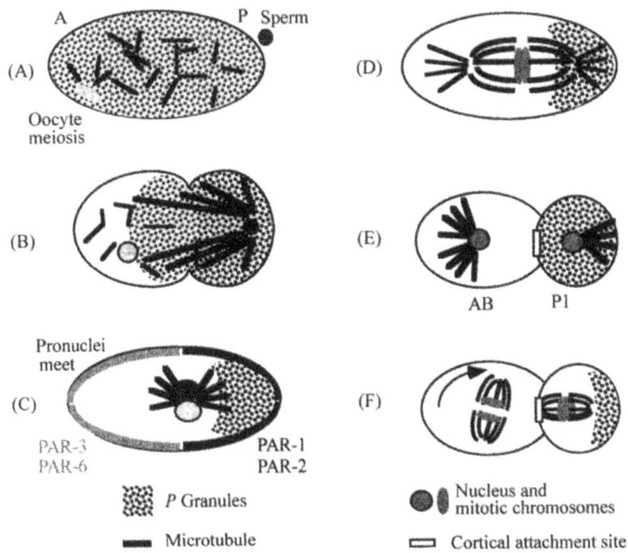

Figure 3.5 Fertilization and the establishment of egg polarity. (A) Fertilization. The sperm entry site determines the posterior pole. (B) Pseudocleavage and cortical actin contractions. (C) PAR protein and P granule segregation. (D) First division. The spindle is slightly posterior. (E) 2-cell stage. The anterior AB is slightly larger than the posterior P1. (F) Second division. AB divides slightly before P1. The AB spindle is initially orthogonal to the antero-posterior axis and undergoes a late rotation (arrow) under physical pressure inside the vitelline membrane. In P1, one of the centrosomes is attached to the anterior cortex (corresponding to the previous division site).

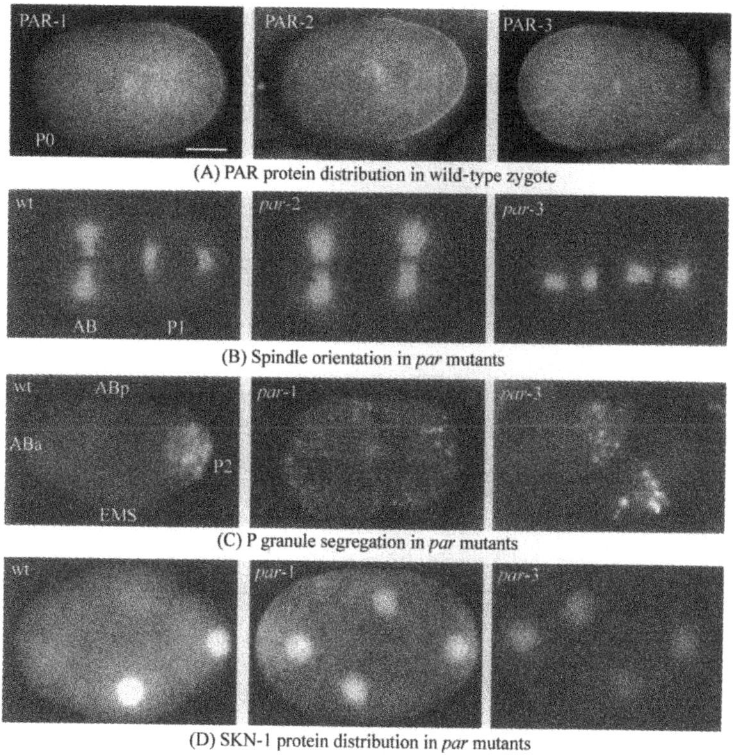

Figure 3.6 Localization and role of the PAR proteins in *Caenorhabditis elegans*. (A) Immunofluorescence staining using antibodies directed against the PAR proteins. PAR-1 and PAR-2 are localized to the posterior cortex of wild type 1-cell stage embryos, PAR-3 to the anterior cortex. The chromosomes are labeled with DAPI and visible in the center of the embryo. Reprinted with permission from Kemphues and Strome (1997). (B) Tubulin immunostaining at the 2-cell stage. In the wild type (wt), the AB spindle is orthogonal and the P1 spindle parallel to the long axis of the egg, and AB divides before P1. In *par* mutants, the divisions are synchronous. In *par-2* mutants, both spindles at the 2-cell stage orient like in wild type AB. In *par-3* mutants, both spindles orient like in P1. Reprinted with permission from Kemphues et al. (1988). (C) P granule immunostaining. In the wild type, the P granules are localized successively to the P1 cell and then to the P2 cell at the 4-cell stage. This asymmetric segregation is defective in *par* mutants. Reprinted with permission from Kemphues et al. (1988). (D) SKN-1 immunostaining. The SKN-1 transcription factor is localized in the wild type to the nuclei of the posterior blastomeres EMS and P2. In *par* mutants, it is found in all nuclei at the 4-cell stage. Reprinted with permission from Bowerman et al. (1997). In all pictures, the anterior pole is to the left and the ventral side down. Bar = 10 μm.

that almost cuts through the egg and then retracts (Fig. 3.1(b)). The female and male pronuclei meet (Fig. 3.1(c)). A mitotic spindle forms along the long axis of the egg, slightly displaced towards the posterior pole, and the first division is subsequently asymmetric (Hird and White, 1993; Strome, 1983) (Fig. 3.1(d) – (f)). The anterior blastomere is called AB (S1 in earlier nematode embryological studies; Chitwood and Chitwood, 1974), the posterior one P1. AB is slightly larger than P1, and most importantly, they contain a different repertoire of determinants (Kemphues and Strome, 1997). For example, the germ line P granules are segregated into the posterior blastomere (Figs 3.5, 3.6(C)).

Isolation of AB or P1 results in a partial embryo that cannot develop to hatching. The two first blastomeres have therefore already lost the potential to form all tissues (Laufer, et al., 1980). This feature defines a mosaic-type development. In *Acrobeloides nanus* (Cephalobidae), however, although the first division is asymmetric, P1 can replace an eliminated AB and produce a larva. Thus an inhibitory interaction from AB must normally prevent P1 or its progeny from producing some tissues. Further inhibitory cell interactions prevent neighboring blastomeres from adopting similar fates (Wiegner and Schierenberg, 1998, 1999). Thus, in *A. nanus*, although symmetry between the early blastomeres is broken by asymmetric divisions, it is reinforced by subsequent lateral interactions, resulting in a regulative-type development.

3.2.2.2 Asymmetric Divisions in the P Lineage

A series of asymmetric divisions follows at the posterior end and results in the segregation of the P granules and other germ line determinants successively to P2, P3 and P4. The latter gives rise to the germ line. The P granules contain several RNA binding proteins, including proteins similar to *Drosophila* vasa, itself located to the germ granules (Gruidl, et al., 1996). In this posterior lineage, spindles of successive divisions keep the same orientation, parallel to the antero-posterior axis of the egg (Hyman and White, 1987). One of the two centrosomes is attracted by an actin cortical site located at the previous division site (Hyman, 1989; Skop and White, 1998; Waddle, et al., 1994). By contrast, in the AB lineage, divisions are symmetric and successive spindles are oriented orthogonal to each other.

3.2.2.3 The *par* Genes

The asymmetric division of the egg is an excellent genetic model to study general cell biological events like mitosis (Gönczy, et al., 1999, 2000), and more specifically cell polarization and asymmetric division (Kemphues and Strome, 1997). Mutations in these processes affect the progeny of the mutant mother (maternal effect), because the proteins of the fertilized egg are expressed in the oocytes from the maternal genome. Such mutations in early development most often result in embryonic lethality. Kemphues et al. (1988) screened for maternal-effect lethal mutations with defects in the polarity of the egg and the first division, which they called *par*, for *par*titioning-defective. In the *par-2* class, both blastomeres behave like the AB blastomere (for instance for spindle positioning), whereas in the *par-3* class, both behave like the P1 blastomere (Fig. 3.6). (Note that in the *C. elegans* nomenclature, both a gene and a mutant in this gene are noted *par-3*, whereas the corresponding protein is called PAR-3.)

PAR-2 and the PAR-1 serine/threonine kinase are localized at the posterior cortex prior to the first division (Boyd, et al., 1996; Guo and Kemphues, 1995), whereas the PDZ-domain proteins PAR-3 and PAR-6 are localized at the anterior cortex (Etemad-Moghadam, et al., 1995; Hung and Kemphues, 1999) (Fig. 3.6(A)). Interest in the *C. elegans* egg polarity greatly increased when this molecular system was shown to be at least partly conserved. The homolog of PAR-1 was found to be localized at the basolateral side of a mammalian epithelial cell line (Böhm, et al., 1997), whereas a protein similar to PAR-3 was found at the apical tight junctions (Izumi, et al., 1998). Other proteins with PDZ domains like PAR-3 and PAR-6 are involved in epithelial (Section 3.6.4) and neuronal polarity (Section 3.7.3) in *C. elegans*. This family is now also found to be required in *Drosophila* for polarity of neuroblasts (Kuchinke, et al., 1998) and of the oocyte (Schulman, et al., 2000; Tomancak, et al., 2000).

The asymmetric localizations of PAR-2 and PAR-3 are not affected by mutations in *par-1*, but PAR-1 localization is uniform in *par-2* and *par-3* mutants; thus, the posterior localization of PAR-1 requires PAR-2 and PAR-3. Localization of PAR-3 and PAR-6 is only slightly asymmetric in *par-2* mutants, whereas that of PAR-2 is uniform in *par-3* mutants. Localization of PAR-3 and PAR-6 is strongly affected in *par-5* mutants. Thus the polarity cue generated by sperm entry may be interpreted somehow through *par-5*, followed by posterior PAR-2 localization and mutual exclusion of the PAR-2 and PAR-3/PAR-6 domains. PAR-1 is subsequently localized to the posterior cortex (Hung and Kemphues, 1999; Kemphues and Strome, 1997).

Downstream, the asymmetry built by the PAR proteins results in the asymmetric segregation of cell fate determinants (Bowerman, et al., 1997). For example, the Notch-like transmembrane receptor GLP-1 is restricted to the anterior blastomeres (Fig. 3.8) by post-transcriptional regulation acting through the 3' untranslated region of its mRNA (Evans, et al., 1994). The MEX-5 protein is also restricted to anterior blastomeres and acts by restricting the putative transcription factors SKN-1, PIE-1 and PAL-1 to the posterior blastomeres (Bowerman, et al., 1993; Mello, et al., 1996; Schubert, et al., 2000; Seydoux and Fire, 1994) (Fig. 3.5(D)). PIE-1 is eventually segregated into P4 and is required for germ cell development. SKN-1 is required for EMS development.

The PAR system is reused at successive asymmetric divisions in the posterior lineage. The polarity of P2 and P3 divisions is reversed compared to that of the egg and P1. P3 and P4 are the anterior daughters (Schierenberg, 1987). During P2 division into P3 and C, the P granules do not move to the posterior cortex but congregate around the nucleus, which moves towards EMS, concomitantly with centrosome rotation. This movement fails in *mes-1* mutants (Hird, et al., 1996). Inversion of the polarity of P2 division is not seen in *Bradynema rigidum* (zur Strassen, 1959) and *Cephalobus* sp. (Skiba and Schierenberg, 1992), but the same final blastomere configuration is achieved by cell repositioning after the division (Sternberg, 1991).

3.2.3 Dorsoventral Specification and Cell Interactions at the 4-Cell Stage

3.2.3.1 Induction of ABp by P2

The second round of cleavage starts with the symmetric transverse division of AB into ABa and ABp. The asymmetric division of P1 into EMS and P2 occurs shortly thereafter, in the same orientation as the first egg division (Figs 3.6(A), 3.7). Although the resulting arrangement of the blastomeres should be in a "T-shape" configuration, the vitelline membrane physically constrains the cells to adopt a rhombomeric configuration (Fig. 3.1(G)) (Schierenberg and Junkersdorf, 1992), thereby breaking symmetry in a second direction, the dorsoventral axis. The AB spindle becomes tilted at the end of the division towards the antero-posterior axis and the posterior daughter (ABp) defines the dorsal side. The anterior daughter of P1 (EMS) subsequently occupies the ventral side (Fig. 3.7).

Priess and Thomson (1987) interchanged with a microneedle the position of the AB daughters immediately after the division (thereby inverting the dorso-ventral axis; Fig. 3.4). The resulting embryo developed normally; both AB daughters are thus initially equivalent and their distinct fates are dictated by their position. The direction of AB spindle skewing may be random around the egg axis.

This rhombomeric blastomere configuration has an important consequence. As a result, ABp contacts P2, but ABa does not, and an inductive signal from P2 distinguishes ABp from ABa. The P2 signal is the APX-1/Delta protein,

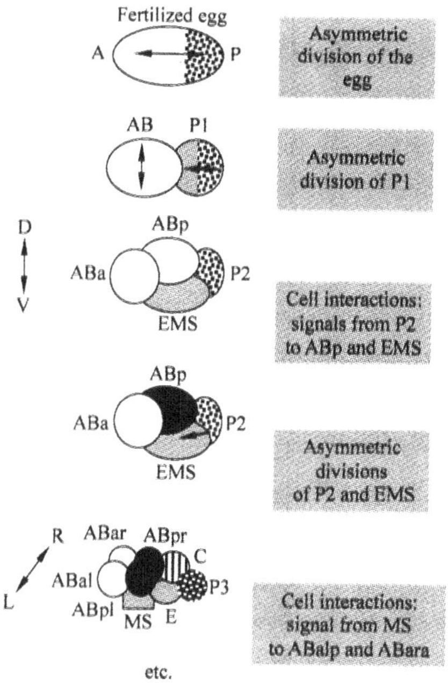

Figure 3.7 Determination of embryonic axes and early blastomere specification in *Caenorhabditis elegans*. The fates of the early blastomeres are determined by a combination of asymmetric divisions (egg, P1, P2, EMS) and cell interactions (arrows from P2). Black double-headed arrows indicate the spindle axis. Asymmetries are determined successively along the antero-posterior, dorso-ventral, and left-right axes. The vitelline membrane (not represented) constrains the positions of blastomeres at the 4-cell stage.

which binds the GLP-1/Notch receptor localized to the anterior blastomeres (Fig. 3.8) (Mango, et al., 1994; Mello, et al., 1994).

Other configurations of blastomeres at the 4-cell stage have been described in different nematode species (Malakhov, 1994). The tetrahedral configuration (in which each blastomere contacts every other one) does not allow for a specific interaction between P2 and one AB daughter. The T-shape and linear configurations do not allow for contact between P2 and any of the AB progeny (zur Strassen, 1959). Changes in the timing of divisions also result in a different pattern of blastomere contacts; for example, in *Acrobeloides/Cephalobus nanus*, P2 divides before AB (Skiba and Schierenberg, 1992).

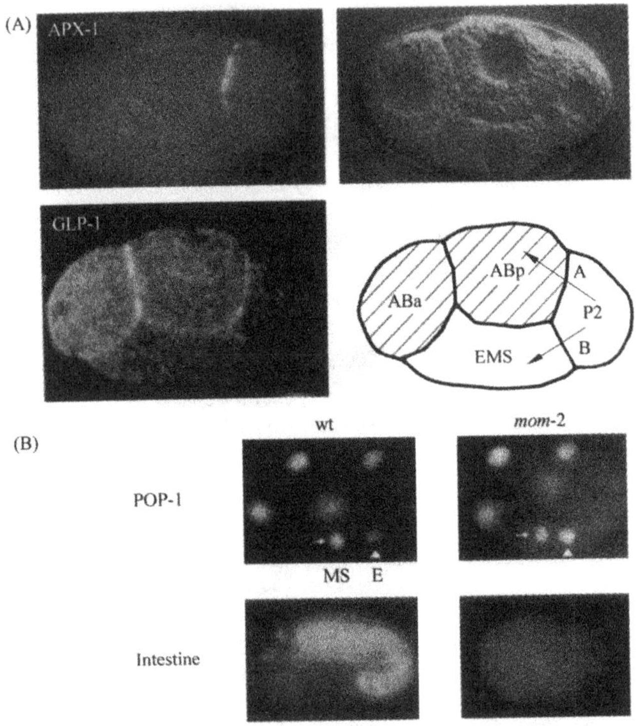

Figure 3.8 Two cell interactions occur at the 4-cell stage in *Caenorhabditis elegans*. (A) P2 contacts and induces ABp through APX-1 and GLP-1. Left panels: Antibody stainings. The APX-1 transmembrane ligand is expressed and localized specifically at the P2 plasma membrane. Reprinted with permission from Mickey et al. (1996). The GLP-1 transmembrane receptor is expressed in the AB daughters. Reprinted with permission from Evans et al. (1994). Right panels: Nomarski micrograph and drawing of the corresponding embryos at the 4-cell stage (the arrows correspond to the two inductions described in A and B). (B) P2 induces the polarization of the asymmetric EMS division. The POP-1 transcription factor is expressed in MS (arrow) and down-regulated in E (arrowhead) in wild-type embryos, but not in embryos mutant for the *mom*-2 Wnt signal. In such mutant embryos, no intestine is produced. Stainings with anti-POP-1 and anti-intestinal marker antibodies. Reprinted with permission from Thorpe et al. (1997). In all pictures, anterior is to the left and ventral down.

3.2.3.2 Induction of EMS Asymmetry by P2

Another cell interaction takes place at the 4-cell stage and polarizes the division of EMS. The EMS spindle orients with one aster close to P2 and the daughter on this side becomes E (endoderm), whereas its sister becomes MS (mesoderm mostly) (Fig. 3.7). Goldstein (1992, 1995) dissociated and recombined blastomeres at the 4-cell stage and showed that the interaction of EMS with P2 was necessary and sufficient for correct spindle orientation and asymmetry of fates between the EMS daughters. In the absence of P2, both daughters adopt the MS fate.

Mutants defective in endoderm specification (*mom* mutants, for more of mesoderm) were isolated. The reverse phenotype (excess of endoderm at the expense of mesoderm) was also found (*pop* mutants, for posterior pharynx-defective). Molecular cloning and genetic epistasis experiments have demonstrated that a Wnt pathway is involved in this cellular interaction (Rocheleau, et al., 1997; Thorpe, et al., 1997). P2 produces the MOM-2/Wnt signal, which binds to the MOM-5/Frizzled receptor in the EMS cell and activates a branching signal transduction pathway that on one hand acts in spindle orientation and on the other hand leads to the down-regulation of the POP-1 transcription factor in the closest daughter (Fig. 3.8) (Thorpe, et al., 2000).

In summary, P2 sends two distinct signals to its two neighbors at the 4-cell stage. The first, through Delta-Notch signaling, specifies ABp versus ABa after their division. The second, through Wnt signaling, results in the asymmetric division of EMS into E and MS and specifies the E fate.

3.2.4 Left-Right Symmetry and Asymmetry

The second division in the AB lineage is mostly oriented left-right, but the left daughters are slightly anterior to their right sisters (Fig. 3.7). The mechanism for this reproducible bias is unknown so far, but it constitutes the first apparent symmetry breakage along the left-right axis of the embryo, and the resulting left-right sisters (ABal and ABar on one hand, and ABpl and ABpr on the other) have very distinct fates.

The larva, however, has a better bilateral symmetry than the cell lineage. Left-right symmetrical cells in the larva often come from cells that are not left-right lineage analogs in the embryonic lineage (Sulston, et al., 1983). For instance, ABara and ABalp have more symmetric fates that ABara and ABala, or ABalp and ABarp. Some left-right asymmetries, however, are left in the worm, such as the position of the gonad relative to the intestine, or of the ventral cord relative to the vulva. They serve as markers of left-right asymmetry.

Wood physically inverted the skewing of the ABa and ABp cleavages by pressing on the dividing cells with a microneedle. Development proceeded, and the resulting animals had an inverted left-right axis (Wood, 1991). In normal development, the MS blastomere contacts ABara and ABalp and induces their specific fates through the GLP-1 receptor (Hutter and Schnabel, 1995a). In the inverted embryos, the pattern of cell contacts is reversed and MS contacts ABala and ABarp instead. As a consequence, the handedness of the whole animal is reversed. Cuticle chirality, however, is determined independently and is not affected (Bergmann, et al., 1998).

Further left-right asymmetries in the ABp lineage (Fig. 3.9) involve other cell interactions through Delta-Notch signaling around the 50-cell stage, at a time when cell positions of left and right blastomeres are very asymmetric in

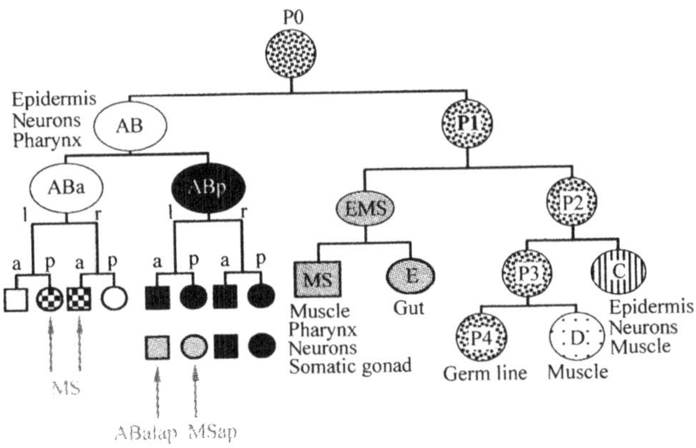

Figure 3.9 Early embryonic lineage in *Caenorhabditis elegans*. The contribution of each blastomere to different tissues is indicated. The germ line P granules are segregated into P4 by a series of asymmetric divisions. Following an inductive polarization of EMS by P2, the division of EMS is asymmetric. E gives rise to the intestine, MS mostly to mesoderm. The divisions in the AB lineage are apparently symmetric, except that the anterior great-granddaughters of AB, like MS, express POP-1 (squares), whereas the posterior ones and E do not (circles). Other fate differences between the AB progeny arise by inductive signaling of P2 to ABp, of MS to ABalp and ABara, and later of ABalap and MSap (or their progeny) to ABpla and ABplp progeny respectively. These interactions result in each AB great-granddaughter acquiring a specific fate and lineage (color/shape code). For instance, following MS induction, ABalp and ABara produce pharyngeal precursors (and other cells). Except for E and P4, each blastomere gives rise to a variety of cell types.

the embryo (Hutter and Schnabel, 1995a; Moskowitz and Rothman, 1996). Later, when bilateral symmetry has been practically reconstructed, a few left-right asymmetries of the ABp lineage require signaling within a left-right pair of cells that are both competent to adopt a given fate, for example the G2/W blast cells (Sulston et al., 1983) or the intestinal valve/rectal epithelial precursors (Bowerman et al., 1992). Some of these late interactions again involve *lin*-12/Notch (Bowerman et al., 1992; Greenwald et al., 1983).

Most nematodes are dextral (zur Strassen, 1951), i. e., their anterior gonadal arm is located on the right of the intestine and their ventral nerve cord passes the vulva on the right. An exception is the sinistral *Acrobeloides bodenheimeri*: in this species, handedness of the ABa and ABp divisions is reversed, thereby inverting the whole body plan handedness (Félix et al., 1996).

3.2.5 Asymmetries after the Third AB Division

The third division in the AB lineage is oriented along the antero-posterior axis. Like for the EMS division and other antero-posterior divisions in the embryo (for instance that of E, MS or C), nuclear staining of the POP-1 transcription factor is higher in the anterior sister (Lin, et al., 1998). In the absence of POP-1, the fate of the anterior cell is often transformed to that of the posterior sister. Whereas asymmetry in the EMS lineage is the result of the initial P2 induction, that in the AB lineage has an unclear origin. It is possible that it is induced by P1 early in development (Hutter and Schnabel, 1995b).

In summary, the combination of asymmetric divisions and cell interactions between the early blastomeres results in a combinatorial variety of cell fates. Asymmetries can be then propagated during later embryogenesis and larval development, giving rise to a large reservoir of specific cell environments. How are these cell specifications recruited in later embryonic development into the coherent formation of organs and tissues?

3.3 Specification of Body Regions and Cell Types

3.3.1 Phases in Embryonic Development

After the period of founder cell generation (Section 3.2), gastrulation starts by the invagination of the daughters of the E (endoderm) blastomere at the ventral posterior side (100 minutes after fertilization) (Fig. 3.1(I)). The site of gastrulation extends into a cleft, first posteriorly with the entrance of P4 (germ line) and of the progeny of MS (mesoderm), then anteriorly with that of the precursors of the anterior pharynx (derived from ABa) (Sulston, et al.,

1983). The ventral side is then occupied by neuroblasts. The closure of this ventral cleft starts by 230 minutes, followed by spreading of epithelial cells over the neuroblasts and ventral enclosure by 340 min. The mouth forms independently at the anterior end. Most embryonic divisions occur during the first 400 minutes, as well as a good part of cell differentiation.

In the second half of embryogenesis, the embryo elongates about 4-fold. Successive stages are called comma-shape (400 min), 1½-fold, 2-fold, and 3-fold (550 min) (Fig. 3.1(J) – (L)). Embryo elongation is likely driven by microfilaments in the epidermis squeezing circumferentially, which results in cell shape changes and overall elongation into a worm shape (Priess and Hirsh, 1986). Cell shape changes during elongation (as well as during ventral enclosure) depend on an actin-catenin-cadherin system in the epidermis (Costa, et al., 1998). Cell intercalation, nuclear migration, and cell fusion take place in the dorsal epidermis (Podbilewicz and White, 1994) and build a large epidermal syncytium called hyp7. At hatching, hyp7 extends between the pharynx and the rectum, mostly on the dorsal side. During larval development, it will extend and surround the animal throughout the central region through the fusion of progeny of the lateral V*n* and ventral P*n* postembryonic blast cells (Fig. 3.3).

Most cells are born close to their final position. A few undergo long-range longitudinal migrations, for example, the larval mesodermal precursor M (see Section 3.3.3), the gonad precursors Z1 and Z4, and the hermaphrodite-specific neurons HSN (which innervate the vulval muscles; Section 3.7.1).

Organ morphogenesis can be described by observation of cell behavior, but much needs to be learned about the underlying mechanisms. For example, the buccal capsule and the pharynx are formed by a combination of ABa and MS progeny, with no simple relation between the lineal origin and the fate within the organ. These pharyngeal precursor cells meet in the anterior end after their gastrulation and form a cylinder. They become linked by desmosomes and secrete a basal membrane. This cylinder gradually elongates and occupies the anterior half of the embryo at the comma stage. The two bulbs and isthmus become apparent. The internal cavity then forms and is lined with the L1 cuticle. The gland cell bodies occupy most of the terminal bulb and become active before hatching. Specific cell fusions occur in the pharynx around hatching (Portereiko and Mango, 2001; Sulston, et al., 1983). The intestine is a simpler epithelial tube that forms from the 20 progeny of the E blastomere. Part of its morphogenesis can be reconstituted *in vitro* after isolation of the E blastomere, for example the polarization of cells, showing that cell polarization does not require interactions with surrounding tissues (Leung, et al., 1999), whereas the left-right twist does (Hermann, et al., 2000). After cytoplasmic polarization and nuclear migration, a cavity is formed within the primordium and the cells form a polarized epithelium with adherent junctions (Leung, et al., 1999).

The excretory cell is a simple one-cell "organ" that extends a long cellular process, the excretory canal, on either side of the animal. Tubulogenesis of the canal can also be reconstituted in isolated cell culture. Mutants in canal formation have been isolated, particularly mutants that cause canal enlargements/cysts (Buechner, et al., 1999).

The cuticle starts forming towards the end of embryogenesis, by 600 min, and the embryo hatches by 800 min. Little is known about how the onset of hatching is regulated in *C. elegans*. It is delayed in mutants for the *hch*-1 gene, which encodes a TOLLOID/BMP-1-like metalloprotease (Hishida, et al., 1996).

3.3.2 Tissue and Organ Specification

From the set of early blastomeres, how are the different tissues (muscle, epidermis, etc.) and organs (intestine, pharynx, etc.) specified?

In some cases, a simple relation can be drawn between an early blastomere and the final organ. For example, the E blastomere gives rise to all and only the intestinal cells (Sulston, et al., 1983). The P4 blastomere gives rise to all the germ line and only germ line (Fig. 3.9).

In most other cases, no simple relations exist between the lineage and the structure. For example, the epidermis, the nervous system or the muscles have a complex lineal origin, with cells of a given tissue coming from completely different and asymmetrical positions in the lineage. Such complex and bizarre relationships between cell lineage and cell fate were called "perverse assignments" by Sulston et al. (1983). Neither do body regions map to early blastomeres: the correspondance between the body axes defined early and those of the hatched larva is only partial. For example, the anterior AB blastomere contributes cells to the larval tail, and the posterior P1 blastomere contributes cells to the head. Moreover, the conventional germ layers do not easily map to the early blastomeres. MS produces neurons as well as mesodermal derivatives, and muscles are derived from MS, AB, C, and D (Fig. 3.9). In fact, some neurons are sisters to muscle cells.

Some of these groups of cells nevertheless express the same set of transcription factors, which define their identity, such as the MyoD homolog for muscle cells (Krause, et al., 1990, 1994), PHA-4 in the pharynx, ELT-1 in the epidermis, and END-1 in the intestine (Labouesse and Mango, 1999). Downstream, PHA-4 directly activates most genes that are expressed specifically in the pharynx. PHA-4 levels increase during development and appear to activate early targets during embryogenesis through high-affinity binding sites in the target gene regulatory sequences, and later targets through low-affinity binding sites (Gaudet and Mango, 2002). The same transcription factor (with cofactors) may thus control both early and late expressed genes required for different steps in the formation and physiology of a given organ.

Upstream of these genes, since the groups of cells that express one of them do not have a common lineal origin (except for the intestine), it has been proposed that their common fate is the result of clustering of their precursors around the 80-cell stage (Labouesse and Mango, 1999). Intercellular signaling could be responsible for their common fate, or they could be specified cell by cell by independent mechanisms. This alternative is a major puzzle for the understanding of *C. elegans* development.

In some cases, it seems that cells with common fates (for example, muscles, neurons, or cells expressing a given transcription factor) are not only scattered in the lineage but also geographically in the embryo, and that they are recruited one by one by specific mechanisms independently of their position (Ahringer, 1997; Cowing and Kenyon, 1996; Sulston, et al., 1983; see next section). The left-right skewing of the embryo and the rising number of blastomeres provide a specific environment for each blastomere, allowing for the specific regulation of every one of them. Convergence of expression of some genes in independently specified cells might require complex gene promoters. Such specification by piecemeal recruitment is probably only possible for an organism like *C. elegans*, in which cell specification occurs with less than a thousand cells. In other organisms, it might occur at the level of cell populations (compartments).

Observing this baroque relation of cell fates with lineage and positions makes one wonder how it evolved. Which of these mechanisms of tissue specification is ancestral: simple fate/lineage relationship, neighboring positions of the precursors in the early embryo, or piecemeal recuitment of each precursor? Is there a general trend? Evolution of a new cell fate and symmetry breakage between equivalent cells can evolve by the addition of a specific cell interaction, because each cell has a specific cellular environment in the embryo, resulting from small asymmetries propagating through development. Do cell interactions tend to be later replaced by autonomous programs? More generally, is the invariant lineage primitive or derived among nematodes?

3.3.3 Antero-posterior Patterning and the Homeotic Gene Cluster

In animal development, some mechanisms specify cells according to their position in the embryo rather than by the tissue or organ they will form. The best-known example is specification by the homeotic cluster or HOM-C group of genes, which helps define cell identities along the antero-posterior axis of all metazoa. For example, expression of a given HOM-C gene specifies identity of a segment in the fly *Drosophila melanogaster* (but is not required for the formation of segments) (Lawrence, 1992). These genes encode transcription factors with a DNA-binding homeodomain. They are typically aligned along the chromosome in the same order as they are expressed along the animal antero-posterior axis.

Only six HOM-C genes are found in *C. elegans*: a labial-like gene called *ceh-13*, a deformed-like gene called *lin-39*, an Antp/Ubx/AbdA-like gene called *mab-5*, and an Abd-B-like gene called *egl-5* (Bürglin et al., 1991), as well as two additional Abd-B-like genes, *nob-1* and *php-3* (Van Auken, et al., 2000) (Fig. 3.10). The "cluster" is actually spread over several megabases, with many intervening genes. Moreover, the order of the genes does not correspond to that in other animals: *ceh-13* is found in between *lin-39* and *mab-5* on the chromosome, but corresponds to a more anterior gene by its sequence (Kenyon, et al., 1997).

Unlike the *Drosophila* HOM-C genes, *lin-39*, *mab-5*, and *egl-5* are not essential for embryogenesis, and loss-of-function mutations are not lethal. These genes are not required for embryonic organ specification. Their expression pattern specifies different antero-posterior domains in the larva. Although nematodes are non-segmented, a repetition along the antero-posterior axis of six units (plus some in the head and tail) is found at hatching in the ectoderm (Sulston and Horvitz, 1977). Each unit comprises four larval blast cells: one left-right pair of V*n* cells and one pair of P*n* cells (Fig. 3.3), which adopt a fate and a lineage characteristic of their position along the antero-posterior axis. Expression of a specific HOM-C gene (or of two of them) confers them a specific cell fate: *lin-39* is expressed and specifies cells in the midbody region, *mab-5* then *egl-5* more posteriorly (Clark, et al., 1993; Wang, et al., 1993). For instance, expression of LIN-39 in P(3-8) and their daughters is required for competence of the vulval precursor cells P(3-8).p (Fig. 3.10). HOM-C genes are also expressed in some non-repeated cells, for instance *mab-5* in the larval mesodermal precursor M or the QL neuroblast.

How are their domains of expression specified? The cells expressing *mab-5* are found at a similar position along the antero-posterior axis, but are not related by lineage. Among cells that express *mab-5* are the mesodermal precursor M and the epidermal blast cell pair V6(L/R). The M cell is born in the anterior part of the embryo and migrates posteriorly into the region of *mab-5* expression. Blocking its migration with actin or microtubule drugs does not prevent its *mab-5* expression (Cowing and Kenyon, 1996). Thus *mab-5* expression in M is not the consequence of its final position. The same holds true for *mab-5* expression in V6. Expression of *mab-5* in different cells in the

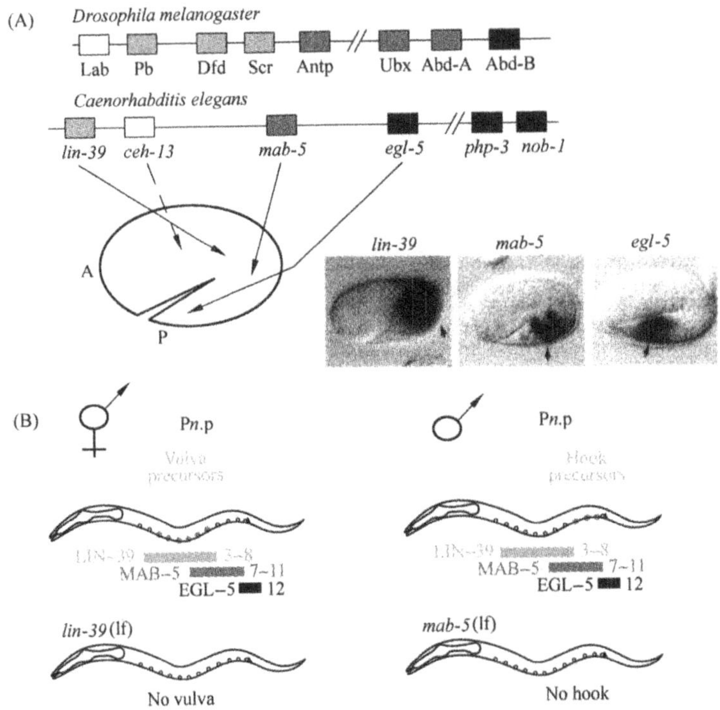

Figure 3.10 The *C. elegans* HOM-C cluster. (A) Six HOM-C genes are found in *C. elegans*. Paralogy can be established with the *Drosophila* HOM-C genes (color code). Mouse has four such clusters. The genes are expressed along the antero-posterior axis of the embryo and of the larva and determine cell fates along this axis. *ceh-13* has a broad domain of expression, which starts more anteriorly than the other HOM-C cluster genes. The expression patterns shown are β-galactosidase stainings of worms carrying a transgene with the promoter of *lin-39*, *mab-5* or *egl-5* fused to the β-galactosidase reporter gene. Reprinted with permission from Kenyon et al. (1997). (B) *lin-39*, *mab-5* and *egl-5* are expressed at different (though overlapping) positions along the antero-posterior axis of the larva. Expression of *lin-39* in P(3-8).p renders them competent to form the vulva in hermaphrodites. Expression of *mab-5* in P(9-11).p in males renders them competent to form some posterior male-specific structures, such as the hook.

same body region is not a consequence of their final position, but of their precise origin in the cell lineage.

Moreover, during larval development, expression of *mab-5* comes on and off several times in the V6 lineage, and this specific spatio-temporal pattern of expression is required for correct fate specification (Salser and Kenyon, 1996) (Fig. 3.11). The regulation of HOM-C genes is thus very fine-grained. In some cases, cell interactions are required for correct HOM-C gene expression (Hunter, et al., 1999; Jiang and Sternberg, 1998; Maloof and Kenyon, 1998). Analysis of HOM-C gene promoters will be interesting.

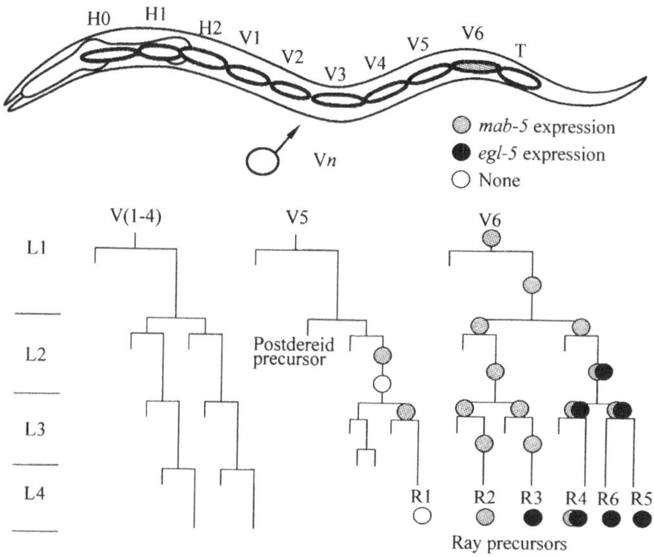

Figure 3.11 Fine-grained regulation of *Caenorhabditis elegans* HOM-C gene expression: the V*n* example. The V*n* lateral epidermal seam cells are located symmetrically along the left and right sides of the larva. They divide after each molt and give rise to one (or two) seam blast cells and sensory organ precursors (postdereid and ray precursors in the male). Other cells fuse with the epidermal syncytium hyp7. HOM-C gene expression is finely regulated temporally and spatially. For example, *mab-5* expression comes on and off several times within the V5 and V6 lineages. This precise temporal pattern in V5 and its level of expression relative to that of *egl-5* in V6 are required for proper specification of male ray identity. From Salser and Kenyon (1996); Emmons and Sternberg (1997); Ferreira et al. (1999).

In contrast with *lin-39*, *mab-5*, and *egl-5*, the knockout of *ceh-13* causes embryonic lethality, with specific defects in the lateral epidermis of the head

(H1, H2 and V1) (Brunschwig, et al., 1999). However, its expression is not fully restricted to the anterior of the embryo, and interestingly, it is expressed in the posterior daughter of many embryonic cell divisions (in contrast to the anterior POP-1 localization, Section 3.2.5) (Wittman, et al., 1997). The posterior Abd-B-like genes *nob-1* (no back end) and *php-3* cause abnormalities in the posterior region and early larval lethality (Van Auken, et al., 2000).

In summary, the HOM-C cluster of *C. elegans* contains fewer genes than that of other animals, and this reduction could be a derived situation (Ruvkun and Hobert, 1998). They participate in defining cell identities along the antero-posterior axis of the animal, but at least three of them do not play an essential role during embryonic development. It will be interesting to understand the evolution in structure and regulation of the HOM-C cluster genes in nematodes. It is already clear that their downstream function has evolved (Jungblut and Sommer, 1998; Sommer, et al., 1998).

3.4 Cell Death

3.4.1 Programmed Cell Death in Development

One major contribution of *C. elegans* as a genetic model system concerns the molecular pathway of programmed cell death, which is conserved between nematodes and vertebrates.

During the development of the wild type *C. elegans* hermaphrodite, 131 somatic cells reproducibly undergo programmed cell death (apoptosis), which is thus one of the most common cell fates. Many cell deaths occur during the middle part of embryogenesis. As we shall see, cells die after entering a specific molecular and cellular death program. In most cases, they die shortly after their birth and do not appear to play any role. The division of their mother could have been suppressed and the fate of the mother transformed to that of its surviving daughter. Programmed cell death is nevertheless used many times rather than the suppression of division, especially in sublineages that are repeated and modified several times (see Section 3.7.1), or for sexually dimorphic structures. Cell death is often recruited as a source of variation in cell fate during the evolution of nematodes. For example, cell death is absent in the endoderm lineage in *C. elegans*, but present in *Turbatrix aceti* (Sulston, et al., 1983). It plays a role during morphological evolution of the gonad (Section 3.5.7).

In a few cases, the cell dies long after birth, after having performed its function; death may then be important to get the cell out of the way. One example is the linker cell of the male gonad. The linker cell is the most posterior cell in the male gonad and leads its posterior elongation (Kimble and Hirsh, 1979) (Section 3.8.2). The linker cell death only takes place once it has reached the proctodeum, and it is actually induced by two proctodeum cells (Sulston, et al., 1980). This regulated cell death ensures the right connection between the proctodeum and the vas deferens of the gonad. Therefore, in contrast to the other "suicidal" cell deaths, linker cell death is a murder.

The cells undergoing programmed cell death have a characteristic refractile appearance by Nomarski optics (Fig. 3.30). At the electron microscopic level, the chromatin condenses, the cell volume decreases, and autophagic vacuoles appear. The cell then disappears altogether, being engulfed and digested by neighboring cells. This type of apoptotic cell death differs from the degenerative cell deaths that can be induced in *C. elegans* in some experimental conditions. Degenerative cell deaths do not share most of the molecular program of cell death determination of apoptotic cells (Hedgecock, et al., 1983).

3.4.2 The *ced* Mutants Define Distinct Steps in Cell Death

The reproducibility of programmed cell death in *C. elegans* facilitates the genetic analysis of the process. Because dying cells appear refractile for only approximately half an hour, they are, however, difficult to visualize and count. This problem was alleviated by the isolation of the *ced-1* (cell death) mutant, in which the dead cells are not engulfed and persist as corpses for several hours (Hedgecock, et al., 1983). Screens for mutations that prevent cell death were easier in the *ced-1* mutant background, as the mutants appear as animals with no cell corpses. Mutants without cell deaths, surprisingly, are viable and do not display any gross phenotype (Ellis, 1985). When cell death is suppressed, the surviving cell does not divide further and sometimes takes on the fate of its sister or aunt, and can replace it in some instances (Avery and Horvitz, 1987; White, et al., 1991). Many *ced* mutants were isolated. Analysis of their phenotype at the cellular level defines different steps in cell death (Fig. 3.12).

First, some mutations affect only a subset of cell deaths. For example, *ces-1* and *ces-2* prevent cell death of the sisters of the NSM pharyngeal motoneurons. They encode transcription factors that are presumably involved in determining these specific cells, and not cell death *per se* (Metzstein and Horvitz, 1999). Again, cells with the same final fate (here cell death) are determined individually by distinct mechanisms and finally express a common set of molecules (*egl-1*, see below) specifying their common fate. Second, some mutants, like *ced-3*, *ced-4*, and gain-of-function (gf) alleles of *ced-9*,

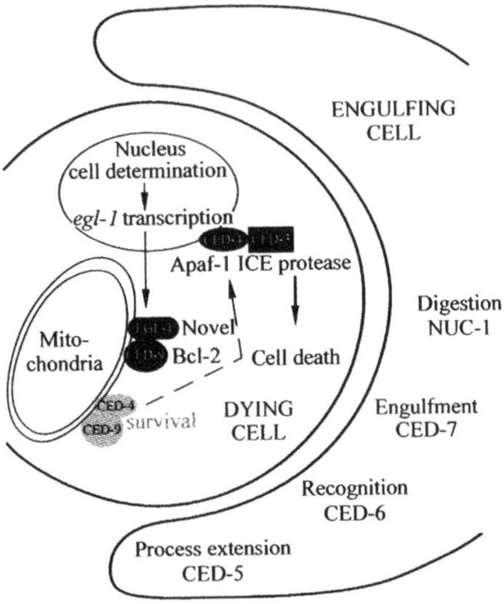

Figure 3.12 Programmed cell death in *Caenorhabditis elegans*. The *egl-1* transcription is under the control of cell-specific transcription factors. The protein EGL-1 binds CED-9 and thereby releases CED-4, which can activate the CED-3 protease that triggers cell death. The dying cell is then engulfed by neighboring cells. Some of the genes required in the engulfing cell have now been determined: CED-5 is similar to human DOCK180 and may be involved in extension of cell surfaces during engulfment as well as cell migration. The *ced-5* mutant is also defective in the migration of the gonadal distal tip cells (Wu and Horvitz, 1998b). CED-6 contains a phosphotyrosine binding domain and may be involved in signal transduction for cell recognition before engulfment (Liu and Hengartner, 1998). CED-7 is a member of a family of ABC transporters that are also involved in the engulfment of cell corpses in mammals (Wu and Horvitz, 1998a).

prevent all cell deaths (Ellis and Horvitz, 1986). By mosaic analysis, they are required in the dying cell, and define therefore a suicide program (Yuan and Horvitz, 1990). The *ced-9* loss-of-function (lf) mutants have the inverse phenotype: too many cells die, which is not viable for the animal (Hengartner et al., 1992). Wild-type *ced-3* and *ced-4* are required for the ectopic cell deaths caused by a *ced-9*(lf) mutation. Thus, *ced-9* appears to prevent cell death by inhibiting the activities of *ced-3* and *ced-4*. The *egl-1* gene appears to act further upstream of *ced-9*. The EGL-1 protein is expressed in all cells that are programmed to die and its expression is sufficient to induce cell death

(Conradt and Horvitz, 1998, 1999). Third, some mutants, like *ced-1*, are involved in the engulfment process of a dead cell by neighboring cells. Finally, *nuc-1* is involved in the degradation of the dead cell by the engulfing cell; it more specifically affects DNA degradation (Hedgecock et al., 1983).

3.4.3 The Molecular Pathway of Cell Death

The genes affected by these mutations were isolated and sequenced. CED-3 is a putative cysteine protease of the ICE family (interleukin-1β-converting enzyme) and may be involved directly in the proteolytic degradation of cellular substrates leading to cell death (Yuan, et al., 1993). It is synthesized as a proenzyme, and must be proteolytically cleaved, probably autocatalytically (Xue, et al., 1996). A homolog of CED-4 was later found in mammals, where it was called Apaf-1. The CED-4 gene is expressed ubiquitously (Yuan and Horvitz, 1992). The CED-9 gene shows some similarity to and can be replaced by the mammalian proto-oncogene Bcl-2, the function of which was previously unknown (Hengartner and Horvitz, 1994; Vaux, et al., 1992). CED-9 is localized to mitochondrial membranes. Its overexpression prevents cell death. The *egl-1* gene encodes a novel protein (Conradt and Horvitz, 1998). Its binding to CED-9 disrupts the association between CED-4 and CED-9 and results in the translocation of CED-4 from mitochondrial to nuclear membranes. CED-4 then activates the CED-3 protease (Fig. 3.12) (Chen, et al., 2000; del Peso, et al., 1998).

The homologs of CED-3, 4 and 9 have now been shown to play a similar role in mammalian cell death induction. The molecular pathway and its cellular function are thus conserved. *Caenorhabditis elegans* as a model system helped greatly to dissect this cell death pathway.

3.5 Development of the Gonad

3.5.1 Overview of Gonad Development

At hatching, the gonad primordium consists of four cells: two somatic precursors called Z1 and Z4 (anterior right and posterior left, respectively) and two germ line precursors called Z2 and Z3 (the daughters of embryonic P4) (Fig. 3.13). These cells are born at different positions in the embryo and subsequently migrate and meet in the midbody region (Sulston et al., 1983). The germ line divides slowly throughout larval development; its differentiation into sperm cells and then oocytes begins at the end of the L4 stage (Section 3.5.6).

Z1 and Z4 divide in a mirror-image pattern in the L1 stage and give rise each

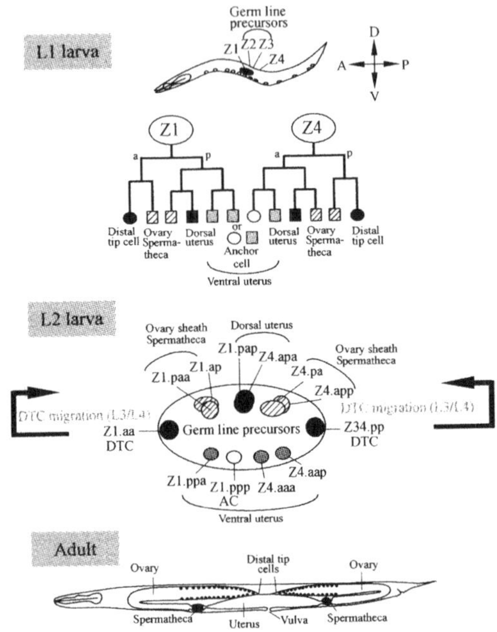

Figure 3.13 Development of the hermaphrodite gonad in *Caenorhabditis elegans*. The somatic gonad precursors Z1 and Z4 divide in the L1 stage and give rise to 12 cells, some of which further divide in the L3 stage and form the ovary sheath, spermatheca wall, and uterus. The two distal tip cells (DTC) are required for gonad arm elongation and germ line proliferation. The single anchor cell (AC) (either Z1. ppp or Z4. aaa) is involved in induction of the vulva in the underlying epidermis; it then fuses with a uterine syncytium. From Kimble and Hirsh (1979).

to six cells (Kimble and Hirsh, 1979). Among these six cells, the three distal ones are ovary and spermatheca precursors, the three central ones are uterine precursors (Fig. 3.13). The three ovary/spermatheca precursors in the Z1 lineage give rise to the anterior arm, and those in the Z4 lineage to the posterior arm. The most distal cell in each arm is a specialized cell that does not divide further. The six uterine precursors from Z1 and Z4 intermingle in the center and form a common uterus structure. Among them, the most central cell becomes the anchor cell. The ovary/spermatheca and uterine precursors further divide in the L3 and L4 stages. We will review the development of the gonad from distal to proximal.

3.5.2 The Distal Tip Cells

One distal tip cell (DTC) is found at the tip of each ovary. It is required for arm growth and for germ line proliferation (Fig. 3.14) (Kimble and White, 1981).

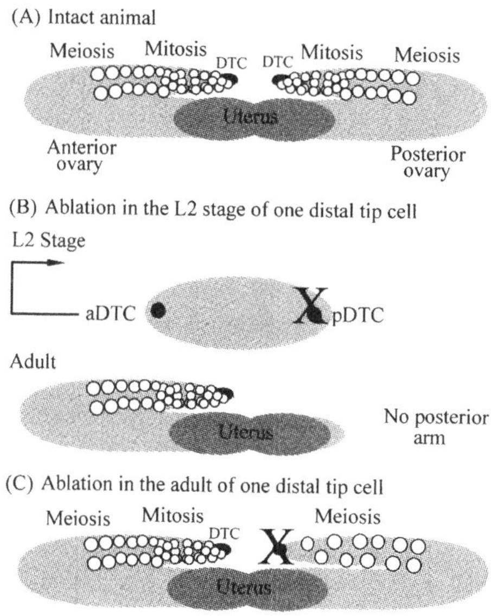

Figure 3.14 Role of the distal tip cells in gonad arm elongation and germ line proliferation in *Caenorhabditis elegans*. (A) Female gonads with two intact ovaries. (B) If a distal tip cell (DTC) is ablated with a laser microbeam in the L2 stage, the corresponding arm does not elongate and germ line does not proliferate in this arm (the few germ cells that remain differentiate as sperm). (C) If a distal tip cell is ablated in the adult, the germ line enters meiosis precociously in the corresponding ovary. Based on Kimble and White (1981).

The gonadal arms grow during the L3 and L4 stages along a defined path (Fig. 3.13). Mutants defective in the dorsal turn of the distal tip cell were isolated (Hedgecock, et al., 1990). The UNC-6 gene encodes a netrin molecule (Ishii, et al., 1992), whereas UNC-5 and UNC-40 encode netrin receptors expressed on the motile cells (Chan, et al., 1996; Leung-Hagesteijn, et al., 1992). By itself, the UNC-40 receptor (homolog of DCC-deleted in colorectal cancer) guides cells toward the ventral UNC-6 sources. When coexpressed with UNC-5 (for instance in the distal tip cells), it orients the cells away from the UNC-6 source (dorsalwards). The DTCs start expressing UNC-5 in the L3 stage in response to the DAF-12 temporal control system (Section 3.9.1), and subsequently initiate their dorsal turn (Su, et al., 2000). The UNC-5, UNC-6 and UNC-40 mutants are generally defective in dorso-ventral guidance, and also affect axon pathfinding (Section 3.7.2). The DTCs secrete a metalloprotease called GON-1, which allows gonadal arm extension, probably by digesting away the extracellular matrix (Blelloch and Kimble, 1999). The role of the DTC in germ line proliferation will be detailed below (Section 3.5.6).

3.5.3 Ovary and Spermatheca

The two precursors of each gonadal arm give rise to a few ovary sheath cells that partially surround the female germ line (Hall, et al., 1999), and to more proximal cells that form the spermatheca (Kimble and Hirsh, 1979).

3.5.4 Anchor Cell Specification: Lateral Inhibition Through Notch

The anchor cell (AC) is a specialized cell of the ventral uterus. It does not divide further and plays a role in inducing the vulva in the underlying epidermis (Section 3.6.2), as well as in patterning different cell fates within the ventral uterus (Section 3.5.5).

The anchor cell origin in the lineage is one of the rare cases of variation in *C. elegans* cell lineage. Two symmetric great-granddaughters of Z1 and Z4 are equally competent to become the anchor cell: in half of the individuals, Z1.ppp becomes the anchor cell and Z4.aaa a ventral uterine precursor (VU), whereas in the other half, the lineage is reversed (Fig. 3.13) (Kimble and Hirsh, 1979). After laser ablation of one of them, the remaining one always adopts the anchor cell fate (Kimble, 1981). Z1.ppp and Z4.aaa thus form an equivalence group, with the anchor cell fate as the 1° fate, the ventral uterine fate as 2°. Cell fate determination occurs by cell-cell signaling between Z1.ppp and Z4.aaa. This lateral (i.e., within the equivalence group) inhibition is similar to the choice of sensory organ precursor cells in *Drosophila* (Simpson, 1990). Strikingly, the same Delta/Notch pathway is used in most or all cases of lateral inhibition (Artavanis-Tsakonas, et al., 1995; Kimble and Simpson, 1997). In the case of AC/VU signaling, the Delta-like ligand is LAG-2 and its receptor is LIN-12 (Greenwald, et al., 1983; Henderson, et al., 1994; Lambie and Kimble, 1991; Tax, et al., 1994). At their birth, the two cells express both the ligand and the receptor. The activated receptor activates its own transcription and inhibits that of the ligand in a feedback loop. The system is therefore metastable: a small asymmetry in expression is amplified, and finally in the late L2 stage, one cell expresses LAG-2 only and the other LIN-12 (Wilkinson, et al., 1994). The signaling cell becomes the anchor cell, the receiving cell a ventral uterine precursor (Fig. 3.15(A)) (Seydoux and Greenwald, 1989).

Genetic screens have identified additional components of the signal

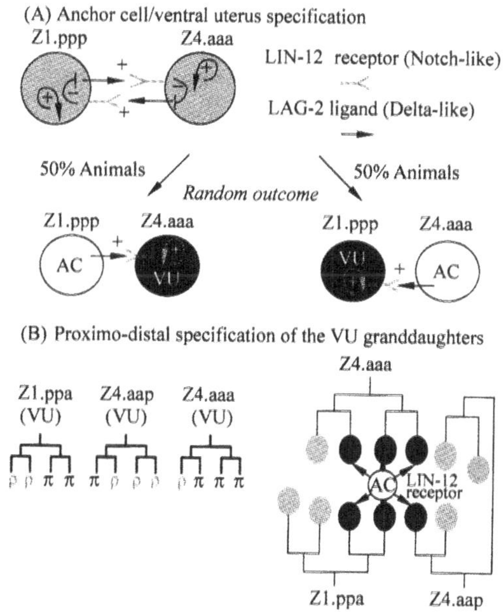

Figure 3.15 Anchor cell determination and uterus patterning in *Caenorhabditis elegans*. (A) Anchor cell (AC) specification by lateral inhibition between Z1.ppp and Z4.aaa. Both Z1.ppp and Z4.aaa are competent to become the anchor cell. One of them becomes the anchor cell and the other one a ventral uterine precursor (VU). At birth, they both express the LIN-12/Notch receptor (grey) and its LAG-2/Delta ligand (black). A positive feedback of the activated receptor on its own transcription and a negative feedback on LAG-2 expression render the system metastable: a small variation in LIN-12/LAG-2 activity between the two cells is amplified. (B) Induction of the uterine π cells by the anchor cell. The twelve granddaughters of the three ventral uterine precursors adopt either of two fates: π (black, central uterus) or ρ (grey, distal uterine branches). The lineage tree by itself does not give any information about how these fates may be specified. Spatially, the granddaughters adopting the π fate are those closest to the anchor cell. A single signal from the anchor cell creates the pattern of π versus ρ fates (Newman, et al., 1995). The lineage and spatial conformation are drawn for the case where Z1.ppp is the anchor cell.

transduction pathway, such as the downstream transcription factor LAG-1 (similar to human CBF1 and *Drosophila* Su(H); Christensen, et al., 1996), or SEL-12, a homolog of presenilin (affected in early-onset Alzheimer's disease), as a suppressor of a gain-of-function mutation in *lin-12* (Levitan and Greenwald, 1995).

3.5.5 Uterus Patterning

Among the six uterine precursors, one becomes the anchor cell, two are dorsal uterine precursors (DU), and three are ventral uterine precursors (Fig. 3.13). These three ventral uterine precursors cells divide twice in the L3 stage and give rise to 12 cells located around the anchor cell (Kimble and Hirsh, 1979). Among these 12 "intermediate precursor" cells, the 6 central ones divide once more and adopt the "π" fate, whereas the six distal ones divide twice and take on a "ρ" fate (Newman, et al., 1995) (Fig. 3.15(B)). The π cells give rise to the central uterine structures building the connection with the vulva. The ρ cells give rise to the cells that form the walls of the two distal branches of the uterus (Newman, et al., 1996).

If the anchor cell is ablated, all uterine intermediate precursors adopt a ρ fate, demonstrating that the π fate is induced by the anchor cell. This induction requires the LIN-12 receptor (Newman, et al., 1995). LIN-12 is thus first used for anchor cell determination in the symmetric lateral inhibition between Z1.ppp and Z4.aaa, and reused after two divisions of the VUs in the inductive interaction between the anchor cell and the VU granddaughters (Fig. 3.15).

3.5.6 Germ Line Development

3.5.6.1 Mitotic Versus Meiotic Divisions

The germ line divides slowly during larval development, and begins differentiation at the end of the L4 stage. Mitotic divisions continue to occur throughout adulthood in the distal part of the gonad. The undifferentiated germ line is syncytial, with a central cytoplasm (the rachis). The mitotic divisions require the distal tip cell; its ablation results in the germ line entering meiosis precociously (Fig. 3.14) (Kimble and White, 1981). The germ line cells close to the distal tip cell divide mitotically, those far away from the distal tip cell enter the meiotic pathway by default. This effect of the distal tip cell also involves a Delta-Notch system; the *glp-1* gene (defective in germ line proliferation) encodes a Notch-type receptor expressed in the germ line (Austin and Kimble, 1987, 1989; Crittenden, et al., 1994; Yochem and Greenwald, 1989) and the distal tip cell expresses the Delta-signal LAG-2 (Henderson, et al., 1994; Tax, et al., 1994).

We now have seen several examples of Delta-Notch signaling. There are two Notch-like receptors in *C. elegans*: *glp*-1, which is involved in germ line proliferation (with Delta-ligand LAG-2) and early embryonic inductions (with Delta-ligand APX-1, Section 3.2.3); and LIN-12, which plays a role in the anchor cell lateral inhibition (with ligand LAG-2), in patterning of the uterus (Section 3.5.5), in vulva development (Section 3.6.4), as well as in several other cell fate specifications (Greenwald, et al., 1983). These two genes lie close to each other on the chromosome and are thus likely the product of a recent duplication event. Their coding regions are interchangeable (Austin and Kimble, 1989; Fitzgerald, et al., 1993; Lambie and Kimble, 1991; Yochem and Greenwald, 1989). Four Delta-like ligands are found in the genome (Ruvkun and Hobert, 1998). Two of them, APX-1 and LAG-2, have been analyzed genetically, and their coding sequences also appear interchangeable (Fitzgerald and Greenwald, 1995; Gao and Kimble, 1995). Duplication of these genes might have allowed for a divergence in expression pattern through promoter evolution.

3.5.6.2 Spermatogenesis

In the hermaphrodite, the first germ cells to differentiate (those furthest away from the distal tip cell) undergo spermatogenesis and form approximately 300 spermatozoa. A switch to oogenesis then occurs (Section 3.8.1). In this self-fertilizing hermaphroditic species, the number of self-progeny is limited by the sperm produced in early adulthood (Hirsh, et al., 1976). More progeny can only be produced after male mating. There is probably a tradeoff between the number of self-progeny and generation time, as an increase in sperm production results in a longer life cycle (Hodgkin and Barnes, 1991).

The spermatocyte buds off the mitotic germ line syncytium when it initiates prophase of meiosis I. Its two meiotic divisions produce four small haploid spermatids and additional cytoplasm is cast away (residual body) (Ward, et al., 1981; Wolf, et al., 1979). The chromatin remains condensed after the second meiotic division and the nuclear envelope does not reform. The centriole pair remains close to the chromatin. During early spermatogenesis, a specialized structure is built, the fibrous body – membranous organelle (FB-MO), which ensures asymmetric partitioning of components during separation of the spermatocyte from the mitotic syncytium and from the residual body. The membranous organelle derives from the Golgi apparatus. It will eventually fuse with the plasma membrane and secrete its contents before fertilization. The main component of the fibrous body is the major sperm protein (MSP), the cytoskeletal protein used by the motile spermatozoa (whereas most actin and tubulin are removed into the residual body). The spermatids are activated when they enter the spermatheca in young adults. They become asymmetrical and motile, with a single pseudopod (Ward and Carrel, 1979). Like the spermatozoa of other nematodes, they lack a flagellum and move by crawling on surfaces. This pseudopode motility relies on the multigenic family of major sperm proteins, which form filaments pushing the plasma membrane forward (Roberts and Stewart, 1997).

Mutants affecting many of the steps of spermatogenesis have been isolated. For example, the *spe-4* mutant is defective in the association between the fibrous body and the membranous organelle. Like SEL-12 (Section 3.5.4), the *spe-4* gene encodes a transmembrane protein of the presenilin family; it is localized to the FB-MO membranes (Arduengo, et al., 1998).

3.5.6.3 Oogenesis

Oogenesis occurs continuously in the adult gonad. The germ cells away from the distal tip cell enter meiosis, and are blocked in pachytene of meiosis I, until release by the activation of the Ras-MAP kinase pathway (Fig. 3.16) (Church, et al., 1995). The nature of the signal activating this pathway is unknown, but it likely originates from the distal ovary sheath cells, since their ablation affects this step (McCarter, et al., 1997). The oocyte enlarges and takes up yolk from the intestine through pores in the ovary sheath cells (Hall, et al., 1999; Kimble and Sharrock, 1983); the nucleolus disappears and the nucleus migrates to the distal part of the oocyte (McCarter, et al., 1999). Some of the germ line nuclei undergo programmed cell death at this stage, which may act as a quality control step, or as a way to eliminate supernumerary nuclei after they have functioned as nurse cells. These germ cell deaths are specific to oogenesis, require Ras/MAP kinase activation, and involve the cell death genes *ced-9*, *ced-4*, and *ced-3*, but not *egl-1* (Gumienny, et al., 1999). In the absence of sperm in the spermatheca, oocyte development arrests in diakinesis of meiosis I. In the presence of sperm, the nuclear envelope of the most proximal oocyte breaks down. The oocyte becomes more spherical (cortical rearrangements) (Fig. 3.16).

Ovulation occurs through the contraction of the proximal ovary sheath cells and the dilation of the distal spermatheca valve, which pushes the oocyte into the spermatheca (McCarter, et al., 1999). These movements are regulated by a signal that probably originates from the maturing oocyte, involves the *lin-3* and *let-23* genes (EGF and EGF receptor homologs, see Section 3.6.4) and is

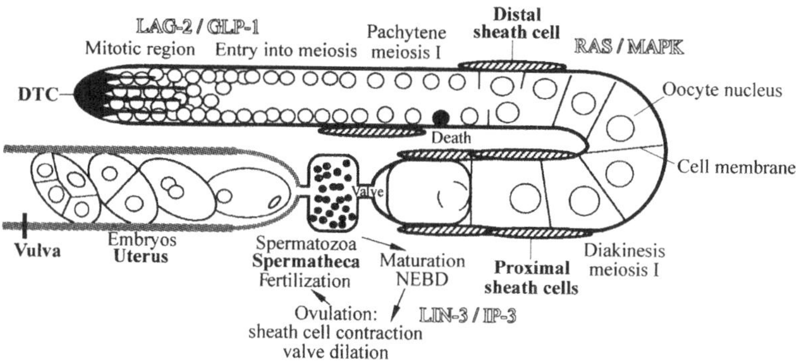

Figure 3.16 Oogenesis and fertilization in *Caenorhabditis elegans*. The different steps of hermaphrodite adult germ line development are depicted with the associated somatic structures (bold) and the molecular pathways involved (outlined). At the distal end of the gonad, the germ line nuclei divide in a syncytium; the distal tip cell prevents them from entering meiosis through LAG-2/GLP-1 signaling. More proximally, the germ line enters meiosis and is blocked in pachytene of meiosis I until release through the Ras/MAP kinase (RAS/MAPK) pathway, probably activated by signaling from distal ovary sheath cells. Maturation is induced in the presence of sperm in the spermatheca. The maturing oocyte sends a signal (LIN-3/IP3 pathway) which results in proximal sheath cell contractions, spermatheca valve dilation, ovulation and entry into the spermatheca, where fertilization takes place. The fertilized egg enters the uterus and is expelled through the vulva. (DTC = distal tip cell. NEBD = nuclear envelope breakdown.)

transduced through the inositol triphosphate pathway, probably triggering calcium release and regulating myocontractile activity (Clandinin, et al., 1998; McCarter, et al., 1999). A cycle of ovulation takes place approximately every 25 minutes.

3.5.6.4 Fertilization and Embryo-Laying

Fertilization occurs at the leading end of the oocyte when it enters the spermatheca. The oocyte then exits the spermatheca through the spermatheca-uterine valve. The two meiotic divisions take place and the egg shell is formed. In laboratory conditions in *C. elegans*, the first embryonic cleavages take place within the mother, and embryos are laid at approximately the 40-cell stage. Different nematode species lay progeny at different stages, from 1-cell eggs to L1 larvae.

3.5.7 Evolution of Gonad Development in Nematodes

Many nematode species do not have two symmetric gonadal arms (didelphy), but a single one (monodelphy) (Chitwood and Chitwood, 1974). In species with a single anterior arm, the posterior arm is reduced to a post-vulval sac or totally eliminated.

Whereas the development of Z1 and Z4 is symmetric in *C. elegans*, it is asymmetric in monodelphic species (Fig. 3.17). Z1 divides before Z4 (Félix and Sternberg, 1996; Sternberg and Horvitz, 1981). The putative posterior distal tip cell (Z4. pp) dies by programmed cell death. Depending on the

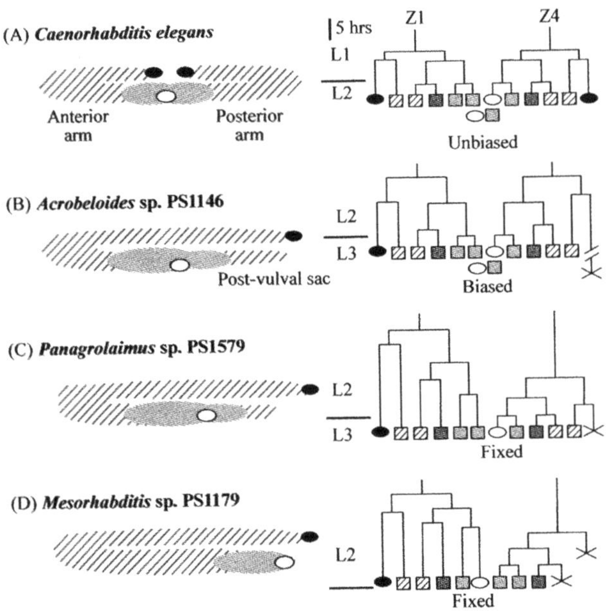

Figure 3.17 Evolution of gonad development in nematodes. Schematic morphology of female/hermaphrodite gonads of four nematode species are shown on the left, and the corresponding early larval lineage on the right. Cell fates are represented by the same code as in Fig. 3.13. X = cell death. The posterior distal tip cell dies in species with a single anterior gonadal arm. All posterior ovary precursors die in *Mesorhabditis* sp. The anchor cell is Z1. ppp in half of the individuals in *C. elegans*, Z4. aaa in the other half. It is more often Z4. aaa in *Acrobeloides*. It is always Z4. aaa in *Panagrolaimus* and always a Z1 descendant in *Mesorhabditis*. Based on Félix and Sternberg (1996).

species, the posterior ovary precursors either die or do not divide at all or divide less than their anterior counterparts and give rise to the post-vulval sac. Thus asymmetry of the organ structure correlates with asymmetry of cell fates between Z1 and Z4.

In the symmetrically developing *C. elegans* uterus, Z1. ppp and Z4. aaa are equally likely to become the anchor cell, and symmetry between them has to be broken *de novo* in each individual by the metastable lateral inhibition between them. In monodelphic species, the symmetry is broken between Z1 and Z4, and consequently downstream between Z1. ppp and Z4. aaa. Monodelphic species show a partial or total bias towards one of them becoming the anchor cell (Fig. 3.17). In some of those species, replacement can occur after anchor cell ablation; in some others, it cannot (Félix and Sternberg, 1996). In this simple 2-cell system, large evolutionary variations occur in the developmental mechanisms that specify the two cells: symmetric lateral inhibition, biased lateral signaling, unilateral induction, and autonomous specification. Biased lateral signaling seems derived from symmetric signaling in some evolutionary lines (Félix and Sternberg, 1996; Sternberg and Félix, 1997).

3.6 Development of the Vulva

3.6.1 Formation of the Vulva

The vulva is a specialized opening in the ventral epidermis of hermaphrodites, which is used for egg-laying and copulation with males. Vulva development consists of the formation of an epithelial tube in the ventral epidermis and its connection with the uterus lumen. It develops from a set of precursor cells in the ventral epidermis, the P*n*. p cells (posterior daughters of the P*n* cells). The determination of these vulva precursor cells has been particularly well studied, and we will present it as an example of genetic analysis of *C. elegans* development.

At hatching in *C. elegans*, there are six left-right pairs of P*n* cells on either side of the young larva (Fig. 3.3). During the L1 stage, their nuclei migrate ventrally and the cells rotate and become aligned along the ventral midline of the animal; they are then numbered $n = 1$ to 12 from anterior to posterior. They divide and give rise to a P*n*. a neuroblast (anterior daughter) and a P*n*. p ectoblast (posterior daughter). The vulva forms from P(5 – 7). p, which undergo a characteristic lineage in the L3 stage (Fig. 3.18). P6. p has a specific central fate, whereas P5. p and P7. p form the lateral sides of the vulva

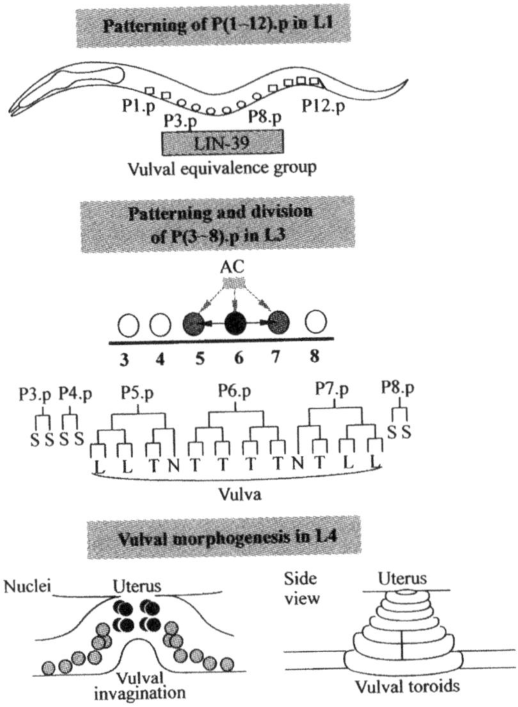

Figure 3.18 Vulva formation in *Caenorhabditis elegans*. The different stages of P*n*.p cell development and vulva formation are depicted from top to bottom. The P(4–8).p cells are competent to form vulval tissue as a consequence of *lin-39* expression (circles). P3.p is competent in half of the animals. The other P*n*.p cells (except P12.p) fuse to the hyp7 epidermal syncytium and are not competent (squares). In wild-type animals, P(3, 4, 8).p progeny fuse to hyp7 (S) in the late L3 stage and the vulva is formed from the progeny of P(5–7).p. P6.p adopts a central vulval fate (black), P5.p and P7.p a lateral vulval fate (grey). One way to determine P*n*.p fate is to follow the cell lineage. The letters correspond to the orientation of the last/third division for P(5–7).p: L = longitudinal (antero-posterior); T = transverse (left-right); N = no division. Their progeny form seven concentric toroidal rings with an internal lumen in the L4 stage. Connection to the uterus lumen creates a passageway for the embryos between the uterus and the outside. Vulval opening in the adult is regulated by specific neurons and muscles. AC = anchor cell.

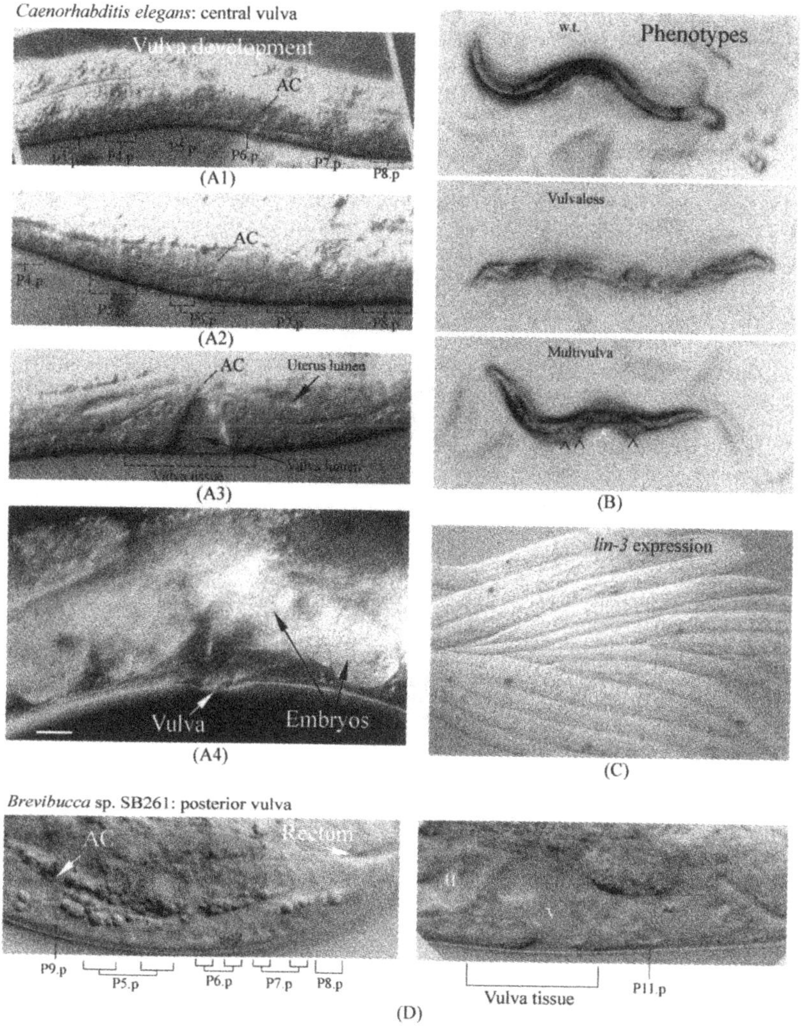

Figure 3.19 Vulva development. (A) Nomarski micrographs of wild-type vulva development in *C. elegans*. (A1) Mid-L3 stage. The cells of the vulval equivalence group are aligned in the ventral cord. P3.p, P4.p and P8.p have already divided once. The gonad is separated from the ventral cord by a basal lamina, visible as a black line. The anchor cell (AC) is dorsal dorsal to P6.p. Anterior is to the left, ventral down. (A2): Late L3 stage. Division of the VPCs. Each cell has divided once, except P6.p which has divided twice. The AC starts

to invade the basal lamina and contacts P6. p progeny. (A3): Mid-L4 stage. Formation of the vulval invagination and uterine lumen. The anchor cell is at the junction. (A4): Adult. The vulval invagination has collapsed and forms a slit. Embryos undergo their first cleavage divisions inside the adult uterus before being expelled through the vulva. Bar = 10μm. (B) Phenotypes of the wild type and of vulva mutants. Wild-type vulva (white arrowhead) allows egg-laying. In Vulvaless mutants, the progeny hatches inside the mother. In Multivulva mutants, ventral bumps (black arrowheads) are visible in addition to the vulva. Courtesy of Paul Sternberg. (C) The *lin-3* gene, which encodes the anchor cell inductive signal, is expressed specifically in the anchor cell. The worms carry a transgene with the *lin-3* promoter fused to the reporter β-galactosidase gene and have been stained for β-galactosidase activity (dot). Courtesy of Russell Hill. (D) Development of a posterior vulva in *Brevibucca* sp. SB261 (Brevibuccidae). Left panel: Late L3 stage. P(5 – 8). p have migrated posteriorly and divided. The nuclei of the granddaughters of P(5 – 7). p are visible in the ventral cord. The P8. p daughters are out of focus. Stars mark P6. ppa and P6. ppp, which are the first cells to invaginate in this species. The anchor cell is at the posterior edge of the elongating uterus and will connect with the vulval cells P6. ppa and P6. ppp. The rectum is visible on the right. Right panel: Mid-L4 stage. The vulval invagination ("v") is tilted towards the uterus ("u"). Anterior is to the left, ventral down.

and have mirror-image symmetric lineages (Sulston and Horvitz, 1977). In the L4 stage, the central vulval cells detach from the cuticle, whereas pairs of symmetric cells on either side extend processes toward each other, thus pushing the central cells dorsally. An invagination forms, surrounded by seven toroidal rings of cells (Figs 3. 18, 3. 19). The symmetric cells then undergo cell-to-cell fusion (Sharma-Kishore, et al., 1999). This epithelial tube is connected with the uterus through the anchor cell, which eventually fuses to a uterine syncytium (Newman, et al., 1996). The anchor cell has a central role in establishing vulva precursor cell fates and more generally in coordinating vulval and gonadal tissues.

3.6.2 The Vulval Equivalence Group and Vulval Fate Induction by the Anchor Cell

P3. p, P4. p and P8. p do not participate to the vulva in wild-type development; they divide once in the L3 stage and fuse with hyp7, the large epidermal syncytium that surrounds most of the animal (Section 3. 3. 1). However, they are competent to replace P(5 – 7). p if these cells are ablated (Sulston and White, 1980). P(3 – 8). p thus form the vulval equivalence group group, which is defined by the expression of the HOM-C gene *lin-39* (Section 3. 3. 3). The study of *C. elegans* vulva development centers on how cell fates are specified within the equivalence group.

The three distinct cell fates of P(3 – 8).p are called 1°, 2°, and 3° by reference to their hierarchy of priority upon replacement regulation: a 3° cell (P3.p, P4.p or P8.p) can replace an ablated 2° or 1° cell; a 2° cell (P5.p or P7.p) can replace a 1° cell (P6.p), but the converse does not occur (Fig. 3.20) (Sternberg and Horvitz, 1986). The normal pattern of cell fates of P3.p to P8.p is 3°/3°/2°/1°/2°/3° (Fig. 3.20).

P6.p is located immediately ventrally to the anchor cell of the gonad. If the

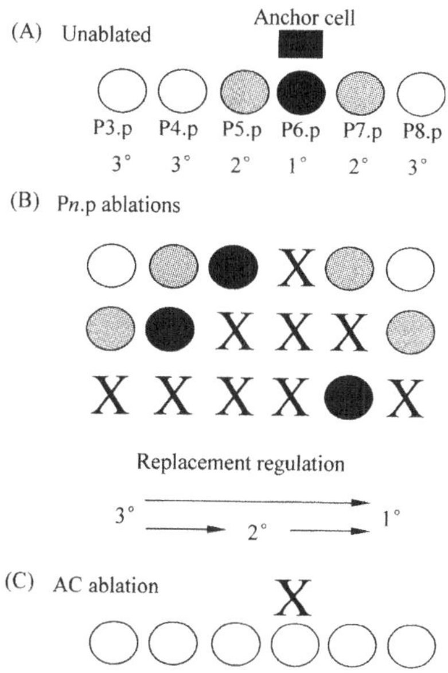

Figure 3.20 Cell ablation experiments and *Pn*.p cell determination in *Caenorhabditis elegans*. (A) In intact animals, P6.p adopts a central vulval fate (black), P5.p and P7.p a lateral vulval fate (grey) and P(3, 4, 8).p progeny fuse to hyp7 (white). (B) Ablated P6.p can be replaced by P5.p (or P7.p) and P5.p by P4.p. If P(5 – 7).p are ablated, P(3, 4, 8).p can replace them. If more cells are ablated, P(1, 2, 9 – 12).p cannot replace them and are therefore not part of the equivalence group. Thus, the vulval equivalence group contains P(3 – 8).p. A single isolated cell of the equivalence group adopts a central fate. Replacement regulation occurs within the equivalence group, the central vulval fate being the 1° fate, the lateral vulval fate 2° fate, and the non-vulval fate 3°. (C) Ablation of the uterine anchor cell results in all cells of the equivalence group adopting a 3° fate: the anchor cell normally sends a signal that induces vulval fates in P(5 – 7).p.

anchor cell is ablated, the vulva does not form, and P(5 –7). p adopt the non-vulval 3° fate (Sulston and White, 1980; Kimble, 1981) (Fig. 3.20). Thus, the anchor cell normally sends a signal that induces vulval fates in P(5 –7). p.

3.6.3 The Vulva Mutants

Mutants in vulva development were isolated based on their egg-laying defective (Egl) phenotype (Ferguson and Horvitz, 1985; Trent, et al., 1983). Vulvaless mothers cannot lay eggs; their progeny hatch inside, producing a "bag of worms" easily scored under a dissecting microscope. These strains are not sterile and can be propagated easily. A second class of vulva mutants was found, with an excess of vulval differentiation of P(3,4,8). p, which produces bumps on the adult ventral cuticle (Multivulva) (Fig. 3.19).

The phenotype of the egg-laying mutants was analyzed at the cellular level. Some of them affect the vulva *per se*; others affect the vulval muscles or the neurons. Mutants defective in vulval development affect different steps (Ferguson, et al., 1987). Some mutants are defective in the initial migration of the P*n* nuclei along the ventral midline (Malone, et al., 1999; Starr, et al., 2001), some in their asymmetric divisions into P*n*. a and P*n*. p cells (Labouesse, et al., 1994), and others in the acquisition of vulval competence (like *lin-39*) (Clark, et al., 1993; Wang, et al., 1993), in the specification of 1°, 2° and 3° fates, or in the execution of vulval cell lineages.

Let us focus on the mutations affecting the specification of 1°, 2°, and 3° vulval fates. Some are Vulvaless because P(5 –7). p adopt a 3° fate (Fig. 3.21(B)). These mutants define genes required for induction of vulval fates by the anchor cell (for instance the signal itself, its receptor or components of the signal transduction pathway within the P*n*. p cells). Another class contains the Multivulva mutations in which P (3 –8). p all adopt a vulval fate. These mutants define genes that ***inhibit*** the vulval induction pathway. Upon hyperactivation of the pathway, the cells do not all become 1°; rather, 1° and 2° fates alternate (Fig. 3.21 (B)), suggesting the presence of a lateral inhibition of the 1° cells on their neighbors. The last mutant category affects the 2° fate specifically. The *lin-12* loss-of-function (lf) mutants have no 2° cells, whereas in gain-of-function (gf) *lin-12* alleles, P (3 –8). p adopt a 2° fate (Fig. 3.21 (C)) (Ferguson, et al., 1987). The combination of a Multivulva mutation and a *lin-12*(lf) allele abolishes the alternation of 1° and 2° fates: all cells adopt a 1° fate. Thus, alternation of 1° and 2° fates in a Multivulva mutant requires a wild-type *lin-12* gene, suggesting that 1° cells normally send a lateral inhibitory signal through *lin-12* preventing their neighbors from adopting a 1° fate (Sternberg, 1988). This analysis of mutant phenotypes thus sheds light on developmental mechanisms; the three cell fates

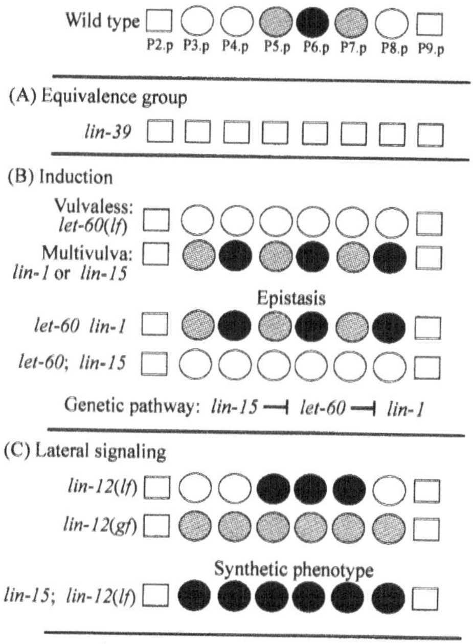

Figure 3.21 Vulva mutants and P*n*.p cell determination in *Caenorhabditis elegans*. The cellular phenotype of several categories of vulva development mutants is shown. Black = central vulval (1°) fate; grey = lateral vulval (2°) fate; white = non-vulval (3°) fate. Square = non-competent cell. (A) *lin-39* mutations abolish the vulval equivalence group: none of the P*n*.p cells is competent to respond to anchor cell signaling. (B) Mutants in the anchor cell induction pathway may decrease the amount of induction and result in a Vulvaless phenotype (3° fates), or increase induction and result in a Multivulva phenotype (alternation of 1° and 2° fates). These two distinct phenotypes of mutants acting in the same pathway allow for genetic epistasis studies. A double mutant combination between *lin-1* and *let-60* has the same Multivulva phenotype as *lin-1*; thus, *lin-1* likely acts downstream of *let-60*. The converse is true for *lin-15*. (C) The 2° fate is specifically affected in *lin-12* mutants. In *lin-12* loss-of-function (lf) mutants, the 2° fate is absent; in *lin-12* gain-of-function (gf) mutants, all competent cells adopt a 2° fate. The alternation of 1° and 2° fates in a Multivulva mutant (for example *lin-15*) is abolished in a double mutant with *lin-12* (lf); thus lateral signaling between the P*n*.p cells relies on wild type *lin-12*. After Ferguson et al. (1987) and Sternberg and Horvitz (1989).

of P(3 – 8). p may be determined by the inductive signal from the anchor cell and by a lateral inhibitory signal, which prevents P6. p neighbors from adopting a 1° fate (Sternberg and Horvitz, 1989).

The phenomenon of the two opposite phenotypes observed in the induction pathway (too little or too much induction) makes vulva development favorable for genetic epistasis studies (Ferguson, et al., 1987). For instance, the double mutant combination of the Multivulva *lin-1* mutant and the Vulvaless *let-60* mutant has a Multivulva phenotype, suggesting that wild type *lin-1* is required for the action of the *let-60* gene and that it may act downstream of *let-60* in a pathway (Fig. 3.21). Using gain-of-function alleles available for some of the genes (Han, et al., 1990; Katz, et al., 1996; Lackner and Kim, 1998), a linear pathway of gene action was defined for vulval induction. Anchor cell ablation and genetic mosaic experiments help determine in which cells a given gene function is required (Ferguson, et al., 1987; Koga and Ohshima, 1995; Simske and Kim, 1995).

3.6.4 The Molecular Pathways: EGF-Ras and Notch Signaling

The genes defined by these mutations were isolated and sequenced, based on genetic mapping of the mutation and transformation rescue of the vulval defect. The most upstream gene, *lin-3*, encodes a molecule similar to the mammalian growth factors EGF and TGF-α and is expressed specifically in the anchor cell (Fig. 3.19) (Hill and Sternberg, 1992). The next gene in the pathway, *let-23*, encodes a homolog of the EGF receptor (Aroian, et al., 1990). It is required and expressed in the P*n*. p cells, and most likely acts as the receptor for the LIN-3 signal (Koga and Ohshima, 1995; Simske and Kim, 1995; Simske, et al., 1996). A conserved signal transduction pathway that activates the Ras GTPase encoded by *let- 60* is found downstream of the receptor (Fig. 3.22) (Han and Sternberg, 1990; Sundaram and Han, 1996). Ras/LET-60 can bind GTP and hydrolyzes it into GDP. When bound to GTP it activates downstream effectors, in the case of vulval induction the Raf/LIN-45 serine/threonine kinase (Han, et al., 1993). A kinase pathway leads to the phosphorylation and inhibition of downstream transcription factors, like LIN-1 (Beitel, et al., 1995) and LIN-31 (Jacobs, et al., 1998; Miller, et al., 1993; Tan, et al., 1998).

This molecular pathway, or parts of it, are used several times during *C. elegans* development (Fig. 3.22). The same pathway is used in an induction in the male B lineage (Section 3.8.2) (Chamberlin and Sternberg, 1994) and for P12 fate specification (Jiang and Sternberg, 1998), but with different downstream targets. The cellular context confers specificity to its downstream

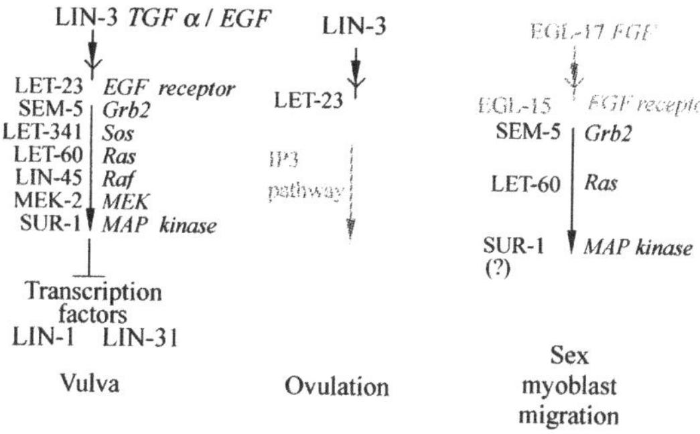

Figure 3.22 The molecular pathway of *Caenorhabditis elegans* vulval induction: use and reuse. The anchor cell induces the vulva precursor cells through the LIN-3/EGF signal. LIN-3 binding to the LET-23 receptor in the P*n*.p cells activates a pathway that leads to inhibition of some transcription factors that inhibit vulval fates (hence leading to activation of vulval fates). The protein names are indicated on the left, and the corresponding names in mammals on the right. Distinct parts of this pathway are used in other developmental contexts. For example, in the activation of ovulation by the oocyte (Section 3.5.6), a different signal transduction cascade through the IP3 receptor is found downstream of *lin-3* and *let-23* (Clandinin, et al., 1998). Conversely, for precise positioning of the sex myoblasts on either side of the vulva, another growth factor, EGL-17 (FGF homolog) and its tyrosine kinase receptor, EGL-15 (FGF receptor homolog) (Burdine, et al., 1997; Devore, et al., 1995) activate the same Ras-MAP kinase pathway (Chen and Stern, 1998; Sundaram, et al., 1996). The role of *sur-1* in sex myoblast migration is unclear (Lackner and Kim, 1998).

effectors. This same pathway is conserved in evolution. For instance, it is involved in inductive signaling during *Drosophila* eye development and in the control of mammalian cell proliferation by growth factors. The Ras-MAP kinase portion of this pathway is also found in yeast, where one of its roles is in the mating response.

As described above, the *lin-12* gene involved in lateral signaling between the P*n*.p cells encodes a Notch-type receptor. Its ligand in the vulva is unknown. Among other targets, LIN-12 activation in P5.p and P7.p results in transcriptional upregulation of the MAP kinase phosphatase LIP-1, which can downregulate the Ras pathway and prevent 1° fate specification in these cells (Berset, et al., 2001).

Some of the genes affecting vulval fate induction do not belong directly to the linear EGF receptor/Ras/MAP kinase pathway. For instance, *lin-2*, *lin-7* and *lin-10* are involved in the localization of the LET-23 receptor at the basolateral (dorsal) plasma membrane of the P*n*. p cells. Mislocalization of LET-23 in the mutants results in a reduced level of vulval induction. LIN-2, LIN-7, and LIN-10 are PDZ-domain containing proteins (like PAR-3 and PAR-6; Section 3. 2. 2) and are localized at the tight junctions (Kaech, et al., 1998; Simske, et al., 1996). In the embryonic epidermis, the establishment of epithelial polarity also requires a PDZ-domain protein called LET-413, which is localized at the basolateral side and may provide a link to a small GTPase pathway (Legouis, et al., 2000).

Screens for suppressor mutations of some of the known mutations have defined genes that were not found in the first screens because they do not display a phenotype when mutated in a wild-type background. These new mutations will help to understand the regulation of this pathway. Some of the genes are novel, and some others are similar to proto-oncogenes, whose site of action was previously unknown. For instance, *sli-1* was isolated as a suppressor of a *let-23* mutation and encodes a homolog of the c-Cbl proto-oncogene (Yoon et al., 1995). The easy use of genetics in *C. elegans* makes it a good system to explore further the regulation of this pathway, with possible applications in the understanding of mammalian cell proliferation and cancer.

3.6.5 Patterning of Vulval Cell Fates

Let us come back to the initial developmental problem of how the pattern of P*n*. p cell fates is specified. The ablation experiments and mutant analysis indicate that the LIN-3 inductive signal from the anchor cell activates vulval fates in the P*n*. p cells through the Ras/MAP kinase pathway. The 1° cell sends a lateral inhibitory signal to its neighbors, received by the LIN-12/ receptor, which prevents them from adopting a 1° fate. What determines the 2° fate in P5. p and P7. p?

Two series of experiments give two apparently different answers (Fig. 3. 23). The first one shows that LIN-3 can act as a morphogen, i. e., a low LIN-3 dose induces the 2° fate, whereas a high dose induces the 1° fate. The second series of experiments, using genetic mosaics, shows that the LET-23 receptor is not required in P5. p and P7. p for 2° fate specification, but that the 2° fate always appears in the neighbor of a 1° cell. In the proposed model, P6. p is first induced by the anchor cell and then not only inhibits the 1° fate in P5. p and P7. p, but also induces the 2° fate through the LIN-12 receptor. These two modes of formation of the centered pattern of vulval precursor cell fates (morphogen and sequential induction) are in fact not incompatible and might act redundantly to specify the pattern reproducibly.

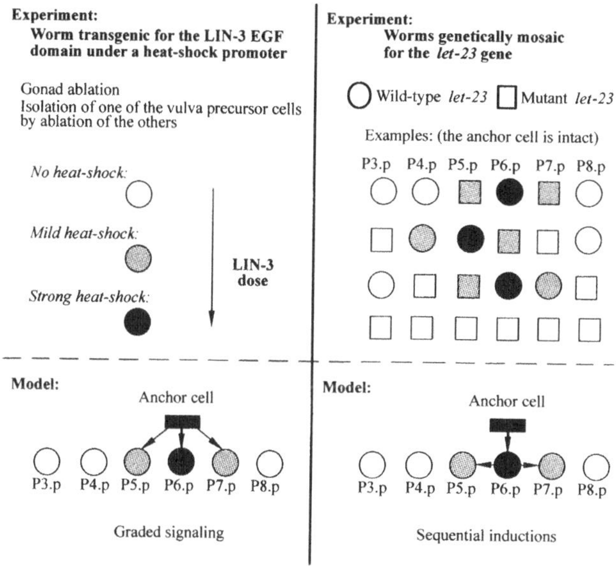

Figure 3.23 Two mechanisms of vulval precursor cell fate patterning in *Caenorhabditis elegans*. (A) Graded LIN-3 signaling. Different doses of LIN-3 are applied on a P*n*. p cell, in worms in which the anchor cell and the other P*n*. p cells were ablated. In the absence of LIN-3, an isolated P*n*. p cell adopts a 3° fate (white); at low doses, it usually adopts a 2° fate (grey), and at high doses a 1° fate (black). LIN-3 can thus directly activate the 2° fate in P5. p and P7. p (Katz et al., 1995). (B) Sequential inductions. Another way of addressing the problem of how the 2° fate is specified is by asking whether the LET-23 receptor is required for 2° fate specification. Among worms which are genetic mosaics for *let-23*, some have a subset of P*n*. p cells with a functional receptor. Analysis of many individual mosaic worms suggests that the LET-23 receptor is required for 1° but nor for 2° fate specification, and that the 2° fate always appears in the immediate neighborhood of a 1° cell (Koga and Ohshima, 1995; Simske and Kim, 1995). The 1° cells thus not only inhibit the 1° fate in their neighbors, but also activate the 2° fate through LIN-12. The pattern can be formed by sequential inductions: the anchor cell induces the 1° fate in P6. p; P6. p then induces the 2° fate in its neighbors P5. p and P7. p.

3.6.6 Evolution of Vulva Development in Nematodes

A similar centered pattern of vulva precursor cell fates is found in other nematode species, but the network of cell interactions that specifies it varies highly (Fig. 3. 24) (Félix, 1999; Sommer, 1996). For example, *Oscheius* sp. 1 CEW1 (Rhabditidae) uses a 2-step induction by the anchor cell: the anchor cell first induces P5. p, P6. p and P7. p to adopt a vulval (vs. non-vulval) fate, and then induces the two daughters of P6. p to adopt an inner vs. outer vulval fate (1° vs. 2°) (Félix and Sternberg, 1997).

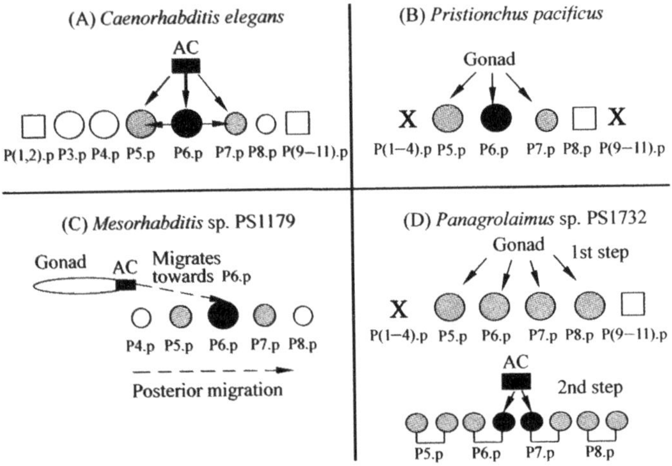

Figure 3. 24 Evolution of vulva patterning mechanisms in nematodes. (A) In *C. elegans*, the pattern of P*n*. p cell fates is the result of several mechanisms: graded induction by the anchor cell (AC), lateral signaling between the P*n*. p cells. Moreover, the P*n*. p cells are not equally competent. P7. p and P8. p are less sensitive to anchor cell signaling due to their expressing the HOM-C gene *mab-5* in addition to *lin-39*; this difference in competence might ensure centering on P6. p (Clandinin, et al., 1997). (B) In *Pristionchus pacificus* (Diplogastridae), only P(5 – 7). p are competent and they are induced by a gonadal signal before the anchor cell is born. P7. p is not competent to adopt the 1° fate. Whether lateral signaling is involved in P5. p specification is not known. (C) In *Mesorhabditis* sp. PS1179 (Rhabditidae), P(4 – 8). p are competent and migrate posteriorly. P(5 – 7). p form the vulva, even in the absence of the gonad. The anchor cell then drives elongation of the gonad towards the posterior vulva. (D) In *Panagrolaimus* sp. PS1732 (Panagrolaimidae), the vulva is formed from P(5 – 8). p and the pattern of vulval cell fates is centered between P6. p and P7. p. Fate specification is achieved by two successive nested inductions of the P*n*. p cells and their daughters. Black = central vulval (1°) fate; grey = lateral vulval (2°) fate; white = non-vulval (3°) fate. Square = non-competent cell. X = cell death. The size of the circles represents the level of competence of the P*n*. p cells.

Species with a single anterior gonadal arm often have a posterior vulva. The vulva is still formed by the central P*n*. p cells (rather than by P(9 – 11). p), which migrate posteriorly in the L2 stage (Sommer and Sternberg, 1994). In some of these species, like *Mesorhabditis sp.* (Rhabditidae) (Sommer and Sternberg, 1994) or *Brevibucca* sp. (Félix, et al., 2000), the anchor cell is not positioned close to the P*n*. p cells when they start dividing in the L3 stage and it is not required for vulval fate patterning.

In some species, the vulval equivalence group is restricted to the cells that actually form the vulva; consequently spatial restriction of vulval fates is achieved by the restriction of vulval competence, and not by localized induction (even if the induction by the gonad or the anchor cell is required for vulval fates). For example, in *Pristionchus pacificus* (Diplogastridae), only P(5 – 7). p are competent to form the vulva. P(1 – 4). p and P(9 – 11). p die by programmed cell death in the L1 stage, and P8. p loses its competence early in the first larval stage (Sommer and Sternberg, 1996). The gonad is required for vulval fate induction in P(5 – 7). p before the anchor cell is born (Sigrist and Sommer, 1999).

In the families Panagrolaimidae and Cephalobidae, the vulva is formed from four precursor cells (P(5 – 8). p) instead of three, and the pattern is centered in between P6. p and P7. p (Félix and Sternberg, 1997; Sternberg and Horvitz, 1982). A two-step induction is used in *Panagrolaimus* sp. PS1732: the gonad induces P(5 – 8). p in the L2 stage to adopt a vulval fate and the anchor cell then induces P6. pp and P7. pa to adopt a central vulval fate (Félix and Sternberg, 1997). Large variations in vulval fate patterning mechanisms are also found within this family (Félix, et al., 2000).

In summary, the mechanisms described above for *C. elegans* vulva development are far from general among nematodes. It will be interesting to pinpoint the genetic changes involved in these variations in cell interaction networks. Genetic screens have been performed in other nematode species. It is already clear that the role of the *lin-39* and *mab-5* genes differs between *C. elegans* and *P. pacificus* (Eizinger and Sommer, 1997; Jungblut and Sommer, 1998; Sommer, et al., 1998).

3.7 Development of the Nervous System

The adult hermaphrodite nervous system contains 302 neurons and 56 glial/support cells (381 neurons and 92 support cells in the male). Neurons can be classified according to whether they are sensory neurons, interneurons, or motoneurons, and more finely divided in 118 types according to their morphology, connectivity, and type of neurotransmitter. The morphology of neuronal processes and the diagram of connectivity between neurons and to the muscles are mostly constant, although the exact placement and number of gap junctions and synapses do vary between individuals (Albertson and Thomson, 1976; White, et al., 1976, 1986).

The development of the nervous system can be divided into two successive steps: (1) cell determination; and (2) axon growth and synapse formation. The first step involves developmental mechanisms that are common with other cell types. The second step is specific to the nervous system.

3.7.1 Determination of Neural Cells

The nervous system is formed during two main waves of neural differentiation, namely during embryogenesis and at the end of the L1 stage (Sulston and Horvitz, 1977; Sulston, et al., 1983). In the male, a later wave of neural development occurs during the L3/L4 stages and builds the male-specific posterior structures used in mating (Sulston, et al., 1980).

Neurons originate from several of the early embryonic blastomeres: AB, MS and C. As for other cell types (Section 3.1.3), a number of repeated structures in the nervous system are produced by repeated sublineages, for example, the sensilla around the head, the motoneurons along the ventral cord and the male tail sensory structures/rays (Fig. 3.25) (Sulston and Horvitz, 1977; Sulston, et al., 1983). Sensilla contain a specialized sensory neuron with a ciliated ending, a sheath cell (a glial cell that surrounds the neuron ciliated ending), and a socket cell (an epidermal support cell that ensures the connection with the rest of the epidermis). For many sensilla, a small sublineage gives rise to the three cells. The mother of such a sublineage is a neuroblast or sensillum precursor cell. Strikingly, similar sublineages are used in the development of the nervous system in *Drosophila*. There are, however, exceptions to the use of sublineages in sensory organ formation in *C. elegans*. For example, the phasmid sensory organ in the tail is built from cells that are distant in the lineage. The amphid in the head also contains several sensory cell types that come from different parts of the lineage.

From laser ablation experiments, only two neurons are absolutely required for survival. Moreover, paralyzed mutant strains that are impaired in egg-laying can still be propagated through internal fertilization and hatching in the hermaphrodite. Mutants in the development of many neurons could therefore be easily isolated. Many general nervous system mutants were first isolated by S. Brenner as locomotion mutants and are called *unc* for <u>unc</u>oordinated (Brenner, 1974). Screens based on other behaviors or sensory responses such as mechanosensation (*mec* mutants; Chalfie and Sulston, 1981) yielded more

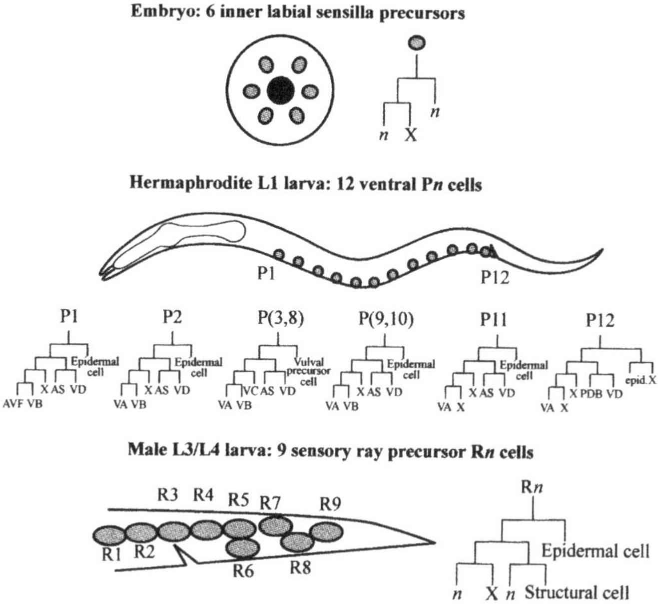

Figure 3.25 Examples of repeated neuroblast sublineages in *Caenorhabditis elegans*. Examples of sublineages of the inner labial sensilla precursors located around the mouth in the embryo, the P*n* cells aligned in the ventral cord along the antero-posterior axis in the larva and the ray R*n* precursors in the male tail. The repetitions are imperfect as the sublineage is modified according to the position along the antero-posterior axis, as shown for the P*n* lineages, and according to the sex of the animal. The lateral epidermal V*n* blast cells also undergo repeated sublineages in a temporal as well as spatial fashion (see Fig. 3.11). *n* = neuron. X = cell death. AVF, VB, AS, etc. are neurons of differnet types. After Sulston and Horvitz (1977); Sulston et al. (1983).

specific mutants. Such screens based on behaviors yield mutants of the development as well as the function of the nervous system. Detailed cellular phenotypic analysis is then required to distinguish between these two classes of defects.

No general determinant of neural cell fate has been found. However, some subsets of neuroblasts are determined by similar genetic systems. Neuroblasts are often sisters of epidermoblasts (for example in the P*n* or V*n* lineages), and their fate choice is between these two cell types. In the P*n* lineages as well as in early embryogenesis, the epidermal cell fate requires expression of the

LIN-26 zinc-finger transcription factor. In *lin-26* mutants, both daughters of P*n* cells are neuroblasts. LIN-26 thus represses neuroblast fate (Labouesse, et al., 1994). In the V5, Q or male R*n* cells, the neuroblast fate requires the *lin-32* gene, a member of the achaete-scute family of basic helix-loop-helix transcription factors (Zhao and Emmons, 1995). These "proneural genes" are required for neuroblast determination in the central as well as in the peripheral sensory nervous systems in *Drosophila*. As in *Drosophila*, the expression of *lin-32* is in some cases repressed by the *hairy* homolog *lin-22* (Wrischnik and Kenyon, 1997). In contrast to *Drosophila*, however, the Notch pathway does not seem to be recruited for lateral inhibition in neuroblast determination in *C. elegans*, nor later in cell specification within sublineages.

After neuroblast determination, the second step is the execution of neuroblast sublineages. As discussed above (Section 3. 1. 3), these sublineages involve asymmetric divisions, but may also involve cell interactions. A division can also be rendered asymmetric by the action of an asymmetric extracellular signal, as described for the Wnt signal polarizing the EMS blastomere (Section 3. 2. 3). Whereas Wnt pathway mutations affect a large array of asymmetric cell divisions, the *unc-86* mutation appears to be specific for neuroblast lineages and affects many of them. In *unc-86* mutants, the fate of one daughter is changed to that of its mother, which results in a repetition of sublineages (Chalfie, et al., 1981). The *unc-86* gene encodes a putative POU-domain transcription factor, which is asymmetrically segregated to the daughter that is affected by the mutation (Finney and Ruvkun, 1990). It is required in many neuroblast sublineages, but its expression is not regulated by the same upstream factors in all cases. The *unc-86* promoter is modular: different parts of the promoter are responsive to different upstream cues in different cells (Baumeister, et al., 1996). In the lineage leading to the HSNs (hermaphrodite-specific neurons), one upstream component is HAM-1, an asymmetrically segregated protein that appears to act as a tether for cell fate determinants during neuroblast division (Guenther and Garriga, 1996).

The next step is the specification and differentiation of different neural cell types in the terminal progeny of the neuroblast sublineages. For example, anterior daughters of a subset of *unc-86* expressing cells differentiate as touch receptors. Touch receptor specification requires the *mec-3* gene, which encodes a founding member of the LIM-domain family of transcription factors (Way and Chalfie, 1988). The MEC-3 protein acts with *unc-86* in a complex to activate transcription of target genes, among them the *mec-3* gene itself as well as downstream genes such as *mec-7*, which encodes a touch-cell specific tubulin isoform required for formation of 15-protofilament microtubules (Duggan, et al., 1998). Thus, further restriction of cell fate potential within the sublineages of *unc-86* expressing cells appears to operate through the combinatorial control of UNC-86 and MEC-3 transcriptional regulation. Specific LIM-domain proteins are also required for the determination of the AFD, AIY, and AIZ neurons (*ceh-14*, *ttx-3*, and *lin-11*, respectively), which are required for thermosensation (Cassata, et al., 2000).

Other types of transcription factors are involved in the determination of other neural cell types. In the P*n*.a lineage, the P*n*.aaaa cells differentiate as VA motoneurons required for backward movement, and their sisters differentiate as VB motoneurons required for forward movement (Fig. 3.25). In *unc-4* mutants, the P*n*.aaaa cells acquire characteristics of the VB fate of their sisters, such as their presynaptic interneuron partner, and the animals are defective in backing (White, et al., 1992). The *unc-4* gene encodes a homeodomain-containing protein (Miller, et al., 1992) and the *groucho* homolog UNC-37 acts as a corepressor of the VB fate in the Pn.aaaa cells (Winnier, et al., 1999). The P*n*.app cells differentiate as VD neurons, which are inhibitory GABAergic motoneurons (Fig. 3.25). In *unc-30* mutants, inhibitory motoneurons are defective and the animals shrink upon touch because of simultaneous contraction of both dorsal and ventral muscles. The *unc-30* gene encodes another homeodomain-containing transcription factor, which is required for specification of the VD (and DD) GABAergic motoneuron fate. In particular, *unc-30* directly activates the expression of the glutamate acid decarboxylase gene *unc-25*, which is required for production of the GABA inhibitory neurotransmitter, and of the *unc-47* gene that codes for the GABA vesicular transporter (Eastman, et al., 1999; Jin, et al., 1994). Thus, as in vertebrates, neuronal specificity is in part controlled by a variety of homeodomain-containing transcription factors.

Terminal cell types sometimes differ between repeated sublineages. For example, the different P*n*.a cells give rise to different neuron types along the antero-posterior axis, or the R*n* cells to different types of rays (Fig. 3.25; Section 3.8.2). As for other cell types, antero-posterior differentiation is accomplished in part by the differential expression of genes of the HOM-C complex (Section 3.3.3). For example, the survival and differentiation of the VC motoneurons (that connect to vulval muscles) require *lin-39* expression in the central body region (Clark, et al., 1993). Note that the differences between neural sublineages often involve removal of cells by cell death (Section 3.4).

Some neuroblast and neuronal cells undergo long-range cell migrations. For example, the QL and QR neuroblasts are located at hatching on the left and right sides, respectively. QL then migrates posteriorly and gives rise, among other neurons, to the PVM touch receptor cell. QR migrates anteriorly and gives rise to the anterior AVM touch cell. The orientation of migration of these two cells is regulated by the HOM-C gene *mab-5*, which is expressed in QL, but not QR (Salser and Kenyon, 1992). The HSN neurons (a pair of hermaphrodite-specific neurons that regulate egg-laying) migrate towards the end of embryogenesis from a posterior position to the central body region, before differentiating and connecting to the vulval muscles in the L4 stage. Their migration requires expression of the *egl-5* HOM-C gene, whereas their final differentiation requires the *unc-86* gene, among others (Desai, et al., 1988).

In summary, neural cell types are determined by the specific expression of transcription factors, for example of the LIM-domain or homeodomain containing families. These factors are not specific to the nervous system and also contribute in combination to the specification of other cell types.

3.7.2 Axon Outgrowth and Pathfinding

Developmental processes that concern the nervous system specifically are the extension, branching, and migration of axons and the formation of synapses with other neurons or with muscles. This field of study is very active, in part through the design of new mutant screens based on transgenic lines expressing GFP (or different color derivatives) in specific axons (Fig. 3.26). It is, for example, possible to screen for the aberrant migration of a specific neuronal process.

In *C. elegans*, most neurons have only one or two processes, with little branching. These neurites/axons extend between the basal side of the epidermal cells and the basement membrane, led by a dynamic growth cone. Extension of the VD motoneurons can be directly visualized with a specific GFP reporter (Fig. 3.26). Growth cone collapse occurs when the growth cone stalls in front of an obstacle (such as body wall muscle attachments to the epidermis), sends fingers out and resumes a new growth cone at the tip of one finger (Knobel, et al., 1999). Axon elongation of the motoneurons usually stops when the neuron contacts a motoneuron of the same class and establishes a gap junction with it (White, et al., 1986). Mutant screens and reverse genetic experiments have uncovered parts of the cellular machinery required for axon extension, particularly proteins that link the extracellular matrix to the actin cytoskeleton, for example integrin, ankyrin, or spectrin.

How is the direction of migration specified? During embryogenesis, the first neurons to send axons, called pioneers, extend their process on the basal side of the forming epidermis. The others then follow along the pioneers, which results in a reproducible placement of axons within a bundle (White, et al., 1986). Growth cone migration occurs along either of two orientations: circumferential (dorso-ventrally around the animal) and longitudinally

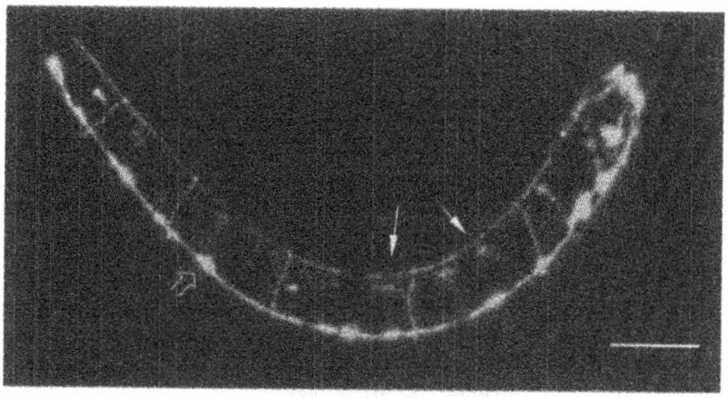

(A)

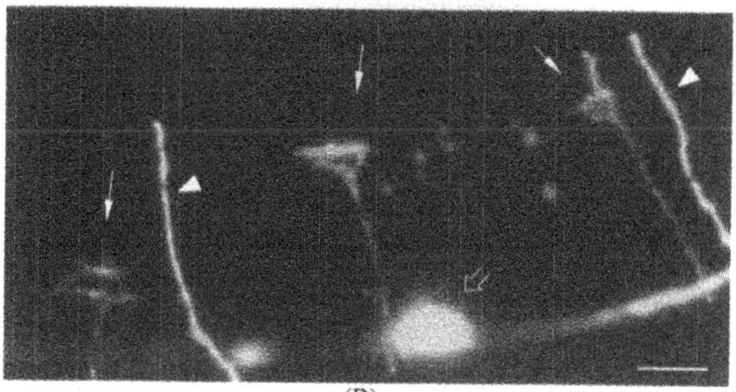

(B)

Figure 3.26 Dorsal migration of growth cones of the VD motoneurons in *Caenorhabditis elegans*. (A) Visualization of the DD and VD motoneurons in a L2 larva by confocal fluorescence microscopy in an *unc-47*: : GFP transgenic line. The *unc-47* gene encodes the vesicular transporter for the GABA neurotransmitter, which is expressed in the DD and VD inhibitory motoneurons. The DD motoneurons have already formed ventro-dorsal commissures during embryogenesis (arrowheads). The VD motoneurons are born during the L1 stage from the P*n*.a lineage (Fig. 3.25) and will synapse to both ventral and dorsal muscles. In the L2 stage, they extend a ventro-dorsal process from the ventral cord (arrows). The open arrow designates a cell body in the ventral cord. Anterior is to the right; dorsal is up. (B) Higher magnification. The shape and dynamics of the migrating growth cone can be observed by time-lapse confocal microscopy. The growth cone on the right extends a finger towards the dorsal cord. Reprinted with permission from Knobel et al. (1999).

(antero-posteriorly). In the nerve ring, migration is mostly circumferential around the pharynx. Along the body, most longitudinal migration tracts run in the ventral nerve cord, but some axons extend laterally or dorsally along the body. The ventral nerve cord is divided in two tracts. In *C. elegans*, the right tract contains most axons (whereas the two tracts are much more symmetric in other nematodes). The right tract is pioneered by the AVG neuron and the left tract by the PVPR neuron. Ablation of either of them results in subsequent axon migration defects in the tract (Antebi, et al., 1997; Durbin, 1987; White, et al., 1986). Pioneer and later growth cone migrations along the left tract can also be distinguished genetically, opening the way to the characterization of the specific molecules involved in each process (Wightman, et al., 1997). A specific system involving the *sax-3* gene is required for keeping the axons along their tract in the ventral cord without crossing the midline. In the *sax-3* mutant, axons cross the midline between the left and right tracts. The transmembrane protein SAX-3 has a *Drosophila* counterpart called Roundabout (Robo), which serves a similar function (Zallen, et al., 1998).

Direction changes occur reproducibly during axon growth. For example, the axons of the DA/DB/DD and VD motoneurons leave (defasciculate) the ventral tract (where their cell nucleus is found), migrate circumferentially and resume an antero-posterior migration in the dorsal "nerve cord", where they stop upon encountering another neuron of their type (Fig. 3.26).

Circumferential migrations were first studied because the same mutations affect guidance of whole cell migration and of axon elongation. Defects in the migration of the gonadal distal tip cell are easily visible and formed the basis of the earlier mutant screens. Analysis of these mutations defined the UNC-6 netrin and UNC-5/UNC-40 netrin receptor interaction (Section 3.5.2). The *unc-5* mutants show specific defects in pathfinding along the ventral-to-dorsal route, whereas in *unc-6* and *unc-40* mutants, axons that normally run along the ventral cord also display aberrant routes (Hedgecock, et al., 1990). The UNC-6 netrin is expressed embryonically in the underlying ventral epidermal cells (P1/2 to P11/12) and in the pioneer ventral axons like AVG, and is therefore highest at the ventral midline (Wadsworth, et al., 1996). The UNC-40 receptor in the absence of UNC-5 specifies ventralwards migration. The pioneer motoneurons express UNC-40 (Chan, et al., 1996). Coexpression of the UNC-5 receptor specifies dorsalwards migration. Specific axons express the UNC-40 and/or UNC-5 receptors and, like the distal tip cell, are thereby either attracted or repulsed (sometimes successively) by the UNC-6 ventral cue, guiding them towards a longitudinal ventral tract or sending them along a ventral-to-dorsal route away from the ventral cue source. The same molecular system is used in dorso-ventral migration in *Drosophila* or in the mammalian neural tube. In addition, the TGF-β UNC-129 is expressed dorsally and is involved in the dorsal migration of the DA and DB axons (Colavita, et al., 1998).

Posteriorwards longitudinal cell and axon migrations are disrupted specifically in *vab-8* mutants (Wightman, et al., 1996). The VAB-8 protein is expressed in neurons that display posterior axon elongation and encodes a protein with a kinesin (microtubule motor) domain (Wolf, et al., 1998).

One signal transduction system for guidance cues to the cytoskeleton appears to use small G-proteins. The *unc-73* mutants are defective in the direction of migration of a number of axons and cells and in the polarity of cell lineages. The *unc-73* gene encodes a small G-protein nucleotide exchange factor required in the migrating cells and axons. Injection of a part of the molecule into mammalian cells activates the G-protein Rac, acts on the actin cytoskeleton and promotes membrane ruffling (Steven, et al., 1998). One possible substrate could be the Rho family G-protein MIG-2 (Zipkin, et al., 1997).

3.7.3 Synapse Formation

Synapses between neurons in the *C. elegans* nervous system are made "en passant" along the length of axons contacting neighboring axons. Synapses between neurons and muscles (neuromuscular junctions) are generally formed by the extension of muscle processes past the basal membrane towards neurons (unlike in other phyla). An exception is the branching of the HSN and VC motoneurons towards the vulval muscles. The same neuronal processes may contain both pre- and post-synaptic sites, which renders the formal distinction between axons and dendrites often arbitrary.

As described above for *unc-4* regulating the presynaptic partner of the VA motoneurons (Section 3.7.1), synaptic specificity is precisely regulated for each neuron class. Some plasticity exists, however, in the wiring of the *C. elegans* nervous system. The birth of new motoneurons in the ventral cord in the L1 stage is accompanied by a rewiring of the DD embryonic motoneurons with changes of their pre- and post-synaptic partners (White, et al., 1978), and additional synapses are added during larval growth. Moreover, the maintenance of sensory neuron structure depends on some aspects of neuronal activity (Peckol, et al., 1999).

Mechanisms of protein targeting within the neuron, for example the localization of the glutamate receptor GLR-1 at post-synaptic sites, employ a conserved molecular system with PDZ-domain-containing proteins such as LIN-10 (see Section 3.6.4) (Rongo, et al., 1998). Mechanisms regulating synapse formation are now studied using mutant screens based on transgenic lines expressing GFP in synaptic vesicles (Zhen and Jin, 1999).

3.8 Sex Determination and Male Development

3.8.1 Sex Determination

Caenorhabditis elegans reproduces through self-fertilization in hermaphrodites or through cross-fertilization of the hermaphrodites with males. The hermaphrodite is a modified female that produces sperm as well as oocytes. *Caenorhabditis elegans* is thus an excellent model for studying the mechanisms specifying the sex of the developing animal (somatic sex determination) and those specifying the sex of the gametes (germ line sex determination).

3.8.1.1 Chromosomal Signals for Sex Determination

Hermaphrodites of *C. elegans* have six pairs of chromosomes (5 pairs of autosomes and one pair of X chromosomes, or XX), while males have only one X chromosome (XO). Males result from the accidental non-disjunction of the X chromosome during meiosis, which produces an X-less gamete. Fertilization of this X-less gamete with an X-containing gamete gives rise to an XO embryo that develops as a male (Nigon, 1949). Subsequent mating of males with hermaphrodites produces cross-progeny, half of which are hermaphroditic, half male. X chromosome non-disjunction and male production occur more frequently under stress conditions (high temperature, starvation). Mutations that increase X non-disjunction, called *him* (high incidence of males), have been isolated (Hodgkin, et al., 1979).

Sex determination first requires sensing the number of X chromosomes in the embryo. This counting then results in a set of sex-specific events: development of sex-specific somatic structures, regulation of germ line maturation, and dosage compensation of X expression. We will first review how the X chromosomes are counted.

The number of X chromosomes is not counted as an absolute number, but relative to the dose of autosomes, with 0.5 X per autosome in the male and 1 X per autosome in the hermaphrodite. The occurrence of several genes on the X chromosome contributes to the counting (Akerib and Meyer, 1994; Hodgkin, et al., 1994), among which *fox-1* encodes an RNA-binding protein (Nicoll, et al., 1997) and *sex-1* encodes a nuclear hormone receptor (Carmi, et al., 1998). These two genes inhibit the expression of the same downstream target gene, *xol-1*, SEX-1 transcriptionally, and FOX-1 post-transcriptionally. In XX animals, these genes are present at a high dose and *xol-1* is repressed (Fig. 3.27) (Miller, et al., 1988). The *xol-1* gene then acts by inhibiting *sdc-2* (sex and dosage compensation). XOL-1 is absent in hermaphrodites and present in males; SDC-2 is thus present in hermaphrodites and absent in males. Both are “novel” proteins with no similarity to others in databases. They both control dosage compensation as well as sex determination *per se* (Nusbaum and Meyer, 1989; Rhind et al., 1995). Downstream of *sdc-2*, however, distinct genes control either dosage compensation or sex determination (Fig. 3.27).

3.8.1.2 Dosage Compensation

A 2-fold difference in expression of an entire chromosome matters for the

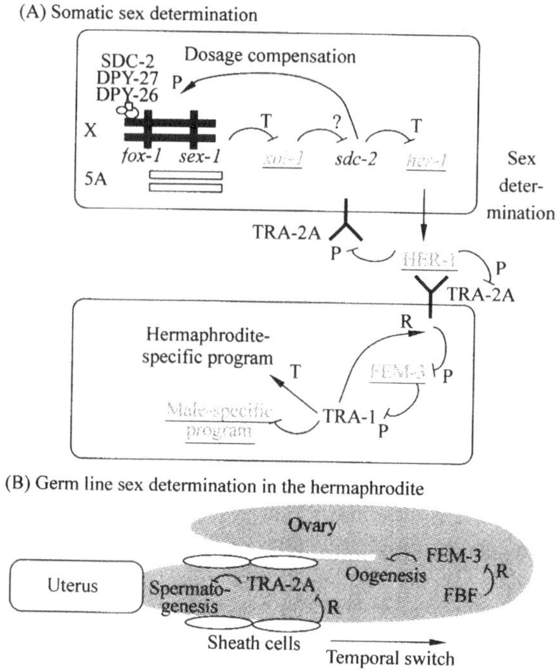

Figure 3.27 Sex determination in *Caenorhabditis elegans*. (A) Somatic sex determination. The counting of the ratio of X chromosomes to sets of autosomes (5 pairs in wild type) activates a pathway that leads (i) to assembly of a dosage compensation complex that halves transcription on the X chromosomes (the signal elements on the X chromosome like *fox-1* and *sex-1* act early, and not concurrently with dosage compensation) and (ii) to production of the HER-1 intercellular signal in males. HER-1 acts through another negative intercellular pathway, which ultimately leads to the sex-specific activation of TRA-1 in hermaphrodites. (B) Germ line sex determination. In the hermaphrodite, *tra-2* is inhibited through its 3'UTR in the proximal gonad. This inhibition results in male germ line production. A temporal switch to oogenesis through FBF (*fem-3* binding factor) then occurs by the regulation of the *fem-3* 3'UTR. Nonunderlined indicates genes activated in hermaphrodites, underlined in males. T = transcriptional regulation; R = post-transcriptional regulation through the 3'UTR of the mRNA; P = regulation of protein activity.

animal. Many genes on the X chromosome are involved in non-sex related processes and their expression level needs to be equalized between males and hermaphrodites to compensate for the different number of X chromosomes. In mammals, dosage compensation is achieved by silencing of one of the two X chromosomes in females. In contrast, in *C. elegans*, the transcription on both X chromosomes is halved in hermaphrodites (Meyer and Casson, 1986). A defect in dosage-compensation results in hermaphrodite-specific lethality; the rare XX dosage-compensation defective survivors are Dumpy (Dpy: short and fat) and Egg-laying defective (Egl) (Hodgkin, 1983b).

In the hermaphrodite, SDC-2 triggers assembly of a dosage compensation complex (Fig. 3.27), which assembles specifically on the X chromosomes of XX embryos from the 30-cell stage on (Dawes, et al., 1999). This complex contains the DPY-26 and DPY-27 proteins (Chuang, et al., 1996). DPY-27 is a member of the SMC (structural maintenance of chromosomes) family. It may increase the degree of chromatin condensation, thereby diminishing gene expression levels. Males do not activate this dosage compensation program.

3.8.1.3 Somatic Sex Determination

Apart from activating dosage compensation in the early embryo, XOL-1 and SDC-2 are required for the activation of sex-specific developmental programs, most of which concern larval development of the reproductive apparatus.

Downstream of SDC-2, a negative regulatory cascade with four steps results in the activation of TRA-1 in hermaphrodites (Fig. 3.27). The genes involved in this pathway are called *tra* for transformer of XX into males (Hodgkin and Brenner, 1977), *her* for hermaphrodization, and *fem* for feminization (when it also affects the male germ line). Note that, as for most genes, these names describe the phenotypes of reduction-of-function mutants, and hence the reverse of what is promoted by the wild-type gene product.

In the first step, SDC-2 inhibits *her-1* transcription by direct binding to its promoter (Dawes, et al., 1999). HER-1 is probably a secreted protein. Genetic mosaics for the *her-1* gene show that it can act in other cells than the ones in which it is produced (cell-nonautonomously), thus acting like a "masculinizing hormone" (Hunter and Wood, 1992; Perry, et al., 1993). Its receptor is likely the transmembrane TRA-2A receptor (similar to *Drosophila* patched) (Kuwabara, et al., 1992; Okkema and Kimble, 1991). Production of HER-1 in males inhibits TRA-2A activity (Fig. 3.27).

On the cytoplasmic side, TRA-2A interacts with the FEM-3 protein and thereby prevents the inactivation of TRA-1 (Mehra, et al., 1999). This signal transduction pathway requires FEM-1 and FEM-2 (a type 2C-phosphatase). As a result, the TRA-1 protein is active in hermaphrodites and inactive in males. In contrast to *her-1*, *tra-1* acts cell-autonomously to confer sex-specificity to each cell (Hunter and Wood, 1990). Whereas sensing of X chromosome number results in regulating *sdc-2* transcription, the downstream pathway between *sdc-2* and *tra-1* regulates *tra-1* transcription in all cells through an intercellular signal (HER-1). This may ensure that a possible defect in *sdc-2* regulation in some cells is averaged out and does not result in unharmonious development of hermaphrodite and male structures in the same animal. This linear pathway is complemented by a positive feedback loop of *tra-1* on the activity of the intermediate *tra-2* gene, which might ensure a switch-like behavior of the system.

TRA-1 is a transcription factor of the Gli/Ci family (Zarkower and Hodgkin, 1992), whose targets are beginning to be identified. It acts on *egl-1* transcription (Conradt and Horvitz, 1999), preventing death of the hermaphrodite-specific neurons (HSN, involved in egg-laying) in the hermaphrodite (Section 3.4.3). TRA-1 also seems to be able to act post-transcriptionally; its feedback action on *tra-2* acts by its binding the 3'UTR of the *tra-2* mRNA, thereby regulating its export from the nucleus (Graves, et al., 1999). Other putative transcriptional targets of TRA-1 are the *fog-3* and *mab-3* genes, which contain TRA-1 binding sites in their promoters (Clarke and Berg, 1998). The *fog-3* gene is required for spermatogenesis. The *mab-3* gene controls the development of some sex-specific neurons, as well as yolk production by binding to the vitellogenin promoters (Raymond, et al., 1998).

3.8.1.4 Germ Line Sex Determination

In the male, the germ line produces exclusively sperm, whereas in the hermaphrodite, both sperm and oocytes are produced successively (Section 3.5.6). Production of the male germ line involves the down-regulation of *tra-2* activity. In males, *tra-2* inhibition operates through the production of HER-1. In hermaphrodites, HER-1 is not produced and the *tra-2* gene is down-regulated by the binding of other factors to the 3' UTR of its mRNA (Goodwin, et al., 1993), probably as a result of a signal from surrounding somatic gonadal cells (Fig. 3.27(B)) (McCarter, et al., 1997). After about 300 sperm are produced, the germ line switches to the female mode: the downstream gene *fem-3* is itself down-regulated through its 3'UTR (Ahringer and Kimble, 1991) by binding of a factor called FBF (similar to *Drosophila* Pumilio, which binds Hunchback 3'UTR during antero-posterior patterning of the *Drosophila* embryo) (Kraemer, et al., 1999; Zhang, et al., 1997).

3.8.1.5 Evolution of Sex Determination

Several modes of reproduction exist in nematodes: amphimixis (males and females), hermaphroditism (self-fertilizing hermaphrodites with facultative males), and parthenogenesis. Several mechanisms for sex determination have been found, including environmental sex determination. Chromosomal sex determination often involves XX females (or hermaphrodites) and XO males, although the presence of Y chromosomes has been reported (Chitwood and Chitwood, 1974; Nigon, 1949).

Transitions between modes of reproduction are easy to perform experimentally in *C. elegans* with some mutant combinations (Hodgkin, 1983a, 1992). It seems that hermaphroditism is derived within the genus *Caenorhabditis* from the amphimictic mode. Candidate genes for this variation are those involved in the temporal regulation of *tra-2* activity in the germ line of hermaphrodites.

Genes and pathways involved in sex determination appear to evolve rapidly, even within the genus *Caenorhabditis* (de Bono and Hodgkin, 1996; Kuwabara, 1996), and are unrelated to the genes involved in mammalian or *Drosophila* sex determination. The exception is the most downstream gene *mab-3*, which is related by sequence to the *Drosophila* gene *doublesex*, which is itself also involved in controlling yolk production. Although specific genes and molecular mechanisms diverge, it is remarkable that the *C. elegans* and *Drosophila* sex determination pathways share a long cascade of regulative interactions (Hodgkin, 1992). It has been proposed that such a pathway could have evolved from downstream to upstream (Wilkins, 1995).

3.8.2 Development of the Male Reproductive System

At hatching, males and hermaphrodites appear very similar, except for a few details such as the absence of the hermaphrodite-specific neurons (HSN) in the male (via a sex-specific cell death) (Sulston and Horvitz, 1977). Most differences between the two sexes arise during larval development of the reproductive system.

3.8.2.1 Male Gonad

The male gonad develops from the same precursors as the hermaphrodite gonad (Kimble and Hirsh, 1979). Asymmetric migrations between the progeny of Z1 and Z4 result in an asymmetrical gonad primordium with two distal tip cells at the posterior end and the linker cell at the anterior end (the lineal equivalent to the hermaphrodite anchor cell, which is also determined by LIN-12 signaling; Greenwald, et al., 1983) (Fig. 3.28). The linker cell leads the elongation of the gonad, turns dorsally and posteriorly, links the gonad to the proctodeum, and finally dies (Section 3.4.1). As a result, a single testis forms, instead of the two gonadal arms in the hermaphrodite. Mitotic divisions occur at the distal end of the testis and spermatogenesis occurs throughout adulthood at the proximal end. The spermatids are stored in the seminal vesicle, and pass through the vas deferens during mating. They are activated by contact with the seminal fluid during sperm transfer to the hermaphrodite.

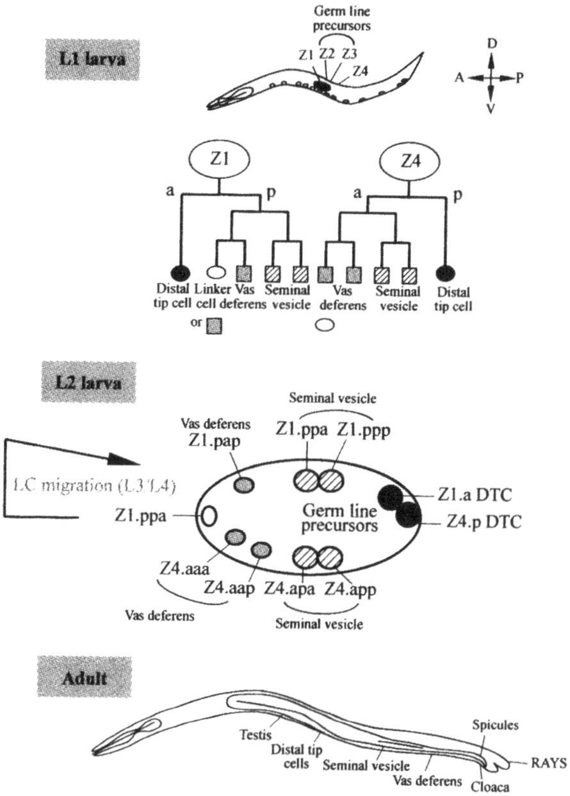

Figure 3.28 Development of the male gonad of *Caenorhabditis elegans*. The somatic gonad precursors Z1 and Z4 divide in the L1 stage and give rise to 10 cells, some of which will further divide in the L3 stage and form the seminal vesicle and vas deferens. The distal testis does not have any sheath cells and only contains germ line. Due to asymmetric migrations of Z1 and Z4 progeny in the gonad primordium, the two distal tip cells are posterior in the primordium; they are required for germ line proliferation, but not for gonad elongation. The single linker cell (either Z1.paa or Z4.aaa) is anterior in the primordium; it drives arm elongation, connects to the proctodeum and then dies (Section 3.4.1). After Kimble and Hirsh (1979).

3.8.2.2 Sensory Rays in the Male Tail

Location of the hermaphrodite vulva requires sensory perception by the male. The male posterior region contains several specialized sensory structures, which account for the greater number of somatic cells of the adult male (Fig. 3.29). Male-specific blast cells in the posterior region give rise to this copulatory apparatus (Fig. 3.30).

The mating behavior can be divided in several motor steps, each controlled by specific posterior sensory structures in the male (Liu and Sternberg, 1995). Contact of the hermaphrodite by the male tail results in a specific response of the male. He positions the ventral side of his tail towards the hermaphrodite

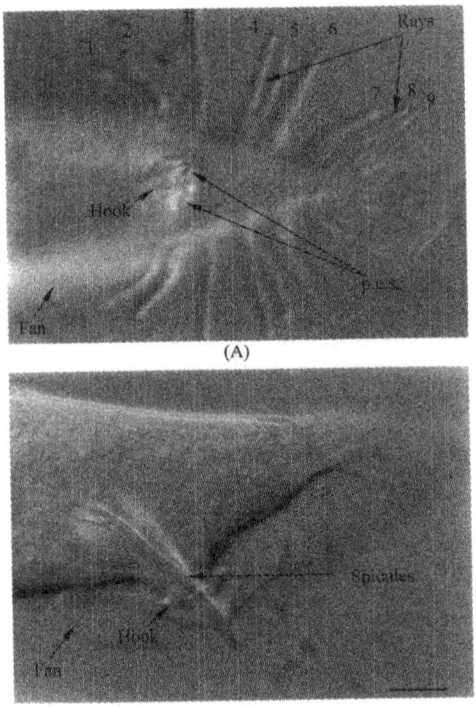

(B)

Figure 3.29 Adult mating structures in the male tail of *Caenorhabditis elegans*. (A) Ventral view. Anterior is to the left. (B) Lateral view. Ventral is down. Reprinted with permission from Liu and Sternberg (1995). Nomarski micrographs. Bar = 10 μm.

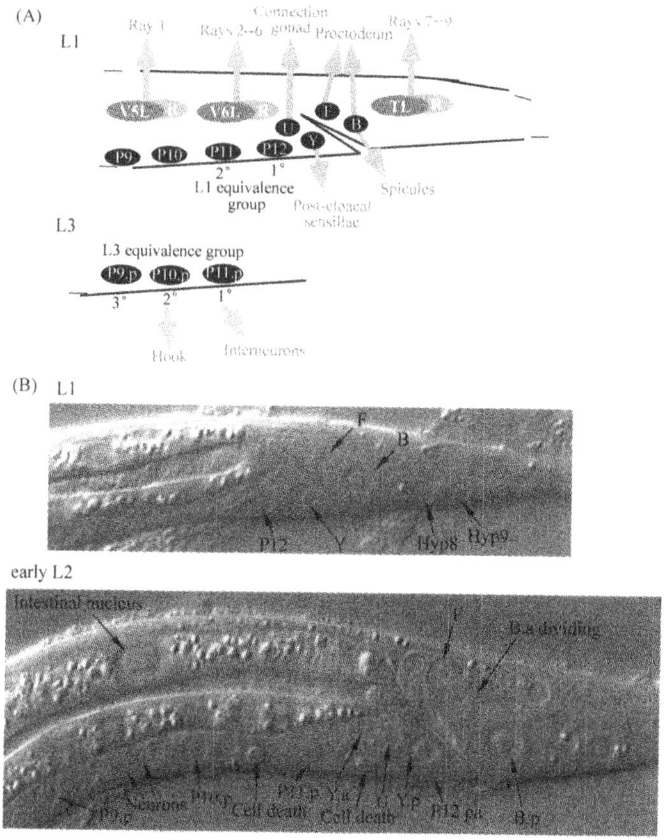

Figure 3.30 Development of the male tail of *Caenorhabditis elegans*. (A) Larval blast cells in the male posterior region. The adult structures formed from each blast cell are indicated. V*n* and T cells are left-right pairs. As in the hermaphrodite, P12 has a specific fate in the L1 stage, and P11 can replace it; they form an equivalence group. In the L3 stage, P(9 – 11). p form another equivalence group. (B) Nomarski micrographs. In the L1 stage, the B, F, and Y blast cell nuclei are visible on either side of the rectum. The P12 nucleus has just migrated from its right lateral position. U and P11 are out of the plane of focus on the left. In the early L2 stage, B and Y have divided once asymmetrically. The large anterior daughter of B is dividing again (hence its nucleus is not visible). P(9 – 11). p large nuclei are visible in the ventral cord. P12. p has divided once more and P12. pp died. Some cell deaths occur in the P11. a and P12. a lineages.

and starts backing along her body until he reaches the vulva, or the end of the body in which case he turns and continues to back on the other side of the hermaphrodite body. This response to contact and turning requires the male tail sensory rays (caudal papillae). The different rays have specific roles in sensing dorsal versus ventral contact and in turning before reaching the end of the hermaphrodite body (Liu and Sternberg, 1995). Shape, position, neurotransmitter use, and neuronal connections are specific for each of the nine rays.

The rays are generated by the posterior lateral epidermal blast cells V5, V6, and T (left-right pairs, Fig. 3.30). Like their most anterior counterparts H(1, 2) (in the head) and V(1 –4), they undergo repeated divisions after each molt, each time generating a new blast cell (Fig. 3.11) (Sulston and Horvitz, 1977). In both sexes, the T lineage gives rise in the L1 stage to specific neurons and neural cells (some associated with the phasmid, a posterior sensory organ) and the V5 lineage gives rise in the L2 stage to the postdereid sensory organs. In the male, the three posterior cells V5, V6, and T generate by mid-L3 nine pairs of "ray precursor cells" called R*n* cells. V5 gives rise to R1, V6 to R(2 –6), and T to R(7 –9). The R*n* cells then divide during late L3 and each gives rise to an R*n*. a neuroblast and an Rn. p epidermal cell. R*n*. a divides with a characteristic sublineage and generates three ray cells: one structural cell and two neurons (Sulston, et al., 1980). The ray sublineage is a typical example of a repeated sublineage (Fig. 3.25).

The production of ray precursors within the lateral epidermal lineages is the outcome of successive cell specifications at the level of the V*n* cells and of their descendants. Many cell interactions are involved. The V*n* cells send early on an inhibitory signal to their anterior neighbors and prevent them from adopting the same fate (Waring and Kenyon, 1990). This regulation occurs through the control of HOM-C gene expression. The HOM-C gene *mab-5* is not expressed in V(1 –4). In V5, it is initially off and is then switched on and off in the posterior part of V5 lineage (Fig. 3.11). This specific expression pattern is required for development of a ray from this lineage (Salser and Kenyon, 1996). In V6, it is initially on, but is later extinguished in posterior parts of the lineage, where *egl-5*, the more posterior HOM-C gene, becomes expressed. The HOM-C genes probably act in part by activating expression of *lin-32*, a member of the basic helix-loop-helix *achaete-scute* family of transcription factors, involved in neuroblast precursor determination in *Drosophila* (Section 3. 7. 1; Zhao and Emmons, 1995) and here in ray precursor determination. *lin-32* expression is repressed in the anterior V cells by *lin-22*, a homolog of *Drosophila hairy*, indicating that the regulatory interaction between these two classes of genes is conserved between *C. elegans* and *Drosophila* (Wrischnik and Kenyon, 1997).

The determination of ray identity appears to occur at the level of the R*n* cell. Rays can be considered as serially homologous structures, and mutants with homeotic transformations in their identity can be isolated. In such mutants, two rays with the same identity undergo fusion, probably because ray identity recruits specific cell recognition and adhesion molecules in the structural cell. The ratio of expression of *mab-5* and *egl-5* in the R*n* cell specifies ray identity (Chow and Emmons, 1994). Other mutations that result in ray transformations define different genes that participate in defining the nine identities. For instance, mutation of *mab-18*/Pax (homeodomain portion of the Pax6 homolog) causes transformation of ray 6 to 4 (Zhang and Emmons, 1995). This locus also encodes a longer transcript required for development of neurons in the head (Chisholm and Horvitz, 1995).

Fan (bursa) morphogenesis occurs during the final molt through retraction of the posterior cells and of the inner layer of the cuticle away from the outer cuticular layer (Sulston, et al., 1980). The rays stay extended.

Number and position of rays as well as tail morphology appear to evolve rapidly and are used extensively in nematode taxonomy. Variations are found even within the *Caenorhabditis* genus, for example with the occasional fusion of rays 3 and 4 in *C. briggsae*, probably through loss of distinct identities. The number and position of rays vary and can be interpreted as changes in cell position, shape and adhesive properties. For example, ray 8 appears to be lost in *Rhabditis blumi* (Fitch, 1997; Fitch and Emmons, 1995).

A major evolutionary change in overall male tail morphology occurs between the blunt peloderan and the pointy leptoderan tails. The tail tip is formed in males like in hermaphrodites by the posterior hyp8 to hyp11 cells. The peloderan male tail of *C. elegans* is formed by fusion of hyp8 to hyp11 followed by their anterior migration and retraction of the tail tip. The development of the pointy leptoderan male tail of *Oscheius myriophila* resembles more that of the hermaphrodite, with no fusion and no retraction of hyp8 to hyp11 (the retractions involved in fan formation still occur) (Nguyen, et al., 1999). The *lep-1* mutant in *C. elegans* has a defect in fusion of hyp8 to hyp11, which results in a leptoderan-like tail (Nguyen, et al., 1999).

3.8.2.3 Hook and Preanal Ganglion

After the male has sensed the hermaphrodite body, the next step in mating requires finding the vulva. The male sensory structure used for approximate vulva location is the hook and its associated neurons. The hook is located immediately anterior to the cloaca (Fig. 3.29).

The hook develops from P(9−11).p, which form an equivalence group during the L3 stage, like P(3 –8).p for the hermaphrodite vulva (Fig. 3.30) (Sulston and White, 1980). Within this group, P11.p adopts the 1° fate (male-specific interneurons), P10.p the 2° fate (hook) and P9.p the 3° fate (fusion with the epidermal syncytium hyp7). Like the hermaphrodite P(3 – 8).p, male P(9 –11).p patterning involves components of the LET-23 and LIN-12 signaling pathways, which promote the 1° and 2° fates respectively (Emmons and Sternberg, 1997; Greenwald, et al., 1983). Surprisingly, the male-specific response of P(9 –11).p to the activation of the LET-23 pathway is mediated by the expression of a HOM-C gene rather than a sex-determination gene, since replacement of *mab-5* expression in the posterior P*n*.p cells by the more anterior HOM-C gene *lin-39* makes these cells adopt hermaphrodite-specific (vulval) fates, and conversely (Maloof and Kenyon, 1998). Thus, antero-posterior specification by the HOM-C genes at least partially obviates the need for sex-determination pathways in these cells. The source and regulation of putative LIN-3 signaling in the male is unclear.

3.8.2.4 Spicules and Postcloacal Sensillae

After approximate vulva location with the hook, the male uses its postcloacal sensillae for precise vulva location, as well as direct sensing by its extended spicules. The sclerotized spicules then penetrate the hermaphrodite vulva. Sensing of a uterine cue by spicule-associated neurons results in sperm transfer (Liu and Sternberg, 1995).

Four male-specific blast cells are located around the rectum at hatching: B, Y, U, and F (Fig. 3.30). They give rise to the spicules and postcloacal sensillae, as well as male-specific interneurons, and play a role in remodeling of the proctodeum during the last molt. The spicules and their associated neurons are formed from the progeny of the B cell, the postcloacal sensillae by that of the Y cell. In addition, the sex myoblast precursors give rise to the spicule protactors and retractors among other muscles. During the last molt, the anal muscles undergo male-specific rearrangements (Sulston, et al., 1980).

The B cell lineage is complex. The B cell first undergoes an asymmetric division along the posterior wall of the rectum, giving rise to a large anterior-dorsal daughter and a smaller posterior-ventral daughter. This division is a well-studied example of an asymmetric division. In *lin-44* mutants, it is often reversed (Herman and Horvitz, 1994; Herman, et al., 1995). In *lin-17* mutants, asymmetry is lost and both daughters adopt the anterior daughter fate (Sawa, et al., 1996). The *lin-44* gene encodes a Wnt signal secreted by more posterior cells (Herman, et al., 1995), and *lin-17* encodes a Wnt receptor (Sawa, et al., 1996). The B cell division is thus polarized by an intercellular

Wnt signaling that may act on one side of the cell, like the asymmetric division of EMS in the early embryo (Section 3.2.3).

Later in the B cell lineage, several cell interactions specify different fates among the eight great-granddaughters of B.a (Chamberlin and Sternberg, 1993). An induction from the anterior F and U cells promotes anterior fates and requires the LIN-3/LET-23/Ras/MAP kinase pathway (Chamberlin and Sternberg, 1994). The LIN-12/Notch receptor on the other hand is required for some posterior fates, but like in vulva development, it can be overriden by activation of LET-23/EGFR. Some left-right equivalents among the B.a great-granddaughters can adopt either of two fates, like during AC/VU determination in the hermaphrodite (Section 3.5.4).

The sclerotized spicules and gubernaculum are secreted during the end of the L4 stage from spicule socket cells, whereas the sheath cells and spicule-associated neurons are not involved in spicule morphology (Jiang and Sternberg, 1999).

3.9 Identity of Larval Stages: the Heterochronic Genes

3.9.1 The Heterochronic Mutants

Like most nematodes, *C. elegans* develops through four larval (or "juvenile") stages, separated by a lethargic period and a molt. This temporal repetitive pattern of events is manifest in the lineage of the V*n* cells, the lateral epidermal blast cells. At each molt, the V*n* cells undergo one (or two) divisions, giving rise to a new stem cell for the next molt and a daughter that fuses to the hyp7 epidermal syncytium that surrounds the animal (Fig. 3.13) (Sulston and Horvitz, 1977).

Apart from these repetitive patterns, each larval stage is filled with characteristic events (Sulston and Horvitz, 1977). In the L1 stage, the P*n* cells become aligned along the ventral midline and undergo characteristic divisions. The somatic gonad precursors Z1 and Z4 and the mesodermal precursor M also divide. In the L2 stage, few divisions happen; the sex myoblast precursors SM migrate to their final position (central in the hermaphrodite, posterior in the male); in the gonad primordium, cells rearrange and the anchor cell is determined; entry into the Dauer program occurs at the end of the L2 stage. In the L3 stage, the P*n*.p vulva precursor cells and the SM sex myoblast precursors divide, and gonadal divisions resume. In the L4 stage, vulval morphogenesis and further gonadal divisions take place. In the adult, the lateral epidermal cells stop dividing and form the characteristic lateral alae of the adult cuticle.

Some mutations identified in screens for lineage mutants affect the identity of larval stages (Ambros and Horvitz, 1984; Chalfie, et al., 1981). These mutations have been called heterochronic, because timing of developmental events is altered (Ambros and Horvitz, 1984). One class of mutants is "precocious". For example, in animals homozygous for loss-of-function (lf) alleles of *lin-14*, the events of the L1 stage are skipped and those of the L2 stage occur precociously. Only three larval stages take place, corresponding to L2, L3 and L4; the latter leads to an adult stage without further molt (Fig. 3.31). Another class of mutants is "retarded". For example, in animals with loss-of-function (lf) alleles of *lin-4* or gain-of-function (gf) alleles of *lin-14*, the L1 stage is repeated several times, and the number of molts increases (Ambros and Horvitz, 1987). Other mutants skip the L2 stage, including *lin-28* (lf) and some *lin-14* (lf) mutants (Ambros and Horvitz, 1984, 1987). Some *daf-12* mutants (which are also abnormal in entry into Dauer; Section 3.10.2) reiterate the L2 stage (Antebi, et al., 1998). Finally, other mutations specifically affect entry into the adult stage, like *let-7* or *lin-29*, and result in a repetition of the L4 stage. The *lin-41* and *lin-42* mutants conversely show adult features in L4, without affecting cell lineage patterns at early stages (Jeon, et al., 1999; Slack and Ruvkun, 1997; Slack, et al., 2000).

The recurrence and timing of molts are not affected in these mutants, but their number is. Not much is known on the regulation of molting, although it is of interest in view of the proposed relationship of nematodes with arthropods as an Ecdysozoa clade sharing molting as a specific character (Aguinaldo, et al., 1997). Several nuclear receptors (including the homolog of *Drosophila* Ftz-F1) are required for molting in *C. elegans*, suggesting a possible role for a steroid hormone ligand, as in insects (Asahina, et al., 2000; Gissendanner and Sluder, 2000; Kostrouchova, et al., 1998). Sterol-starved animals or mutants in a member of the LDL receptor family also have problems with molting (Yochem, et al., 1999).

Remarkably, most heterochronic genes seem to affect developmental events other than those of gonadal development. Both classes (precocious and retarded) result in fertile animals (the temporal defects in vulva development lead, however, to impairment in egg-laying). Thus, mutations in heterochronic genes dissociate the temporal control of gonad development from that of the rest of the animal. The only mutant that also affects gonadal development is *daf-12*, which encodes another putative nuclear hormone receptor (Antebi, et al., 1998; Yeh, 1991). Sterile mutants that specifically affect gonadal divisions have also been isolated (Sun and Lambie, 1997). These mutants do not affect the repetition and timing of molts.

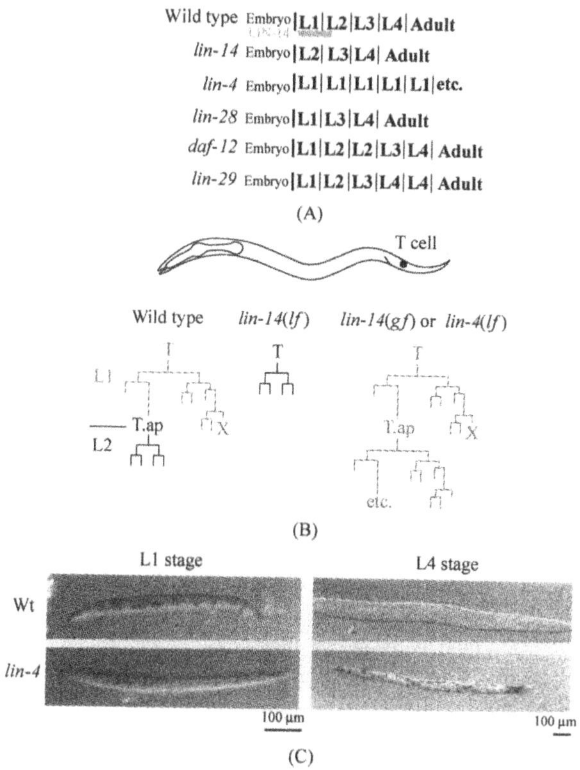

Figure 3.31 Heterochronic mutations in *Caenorhabditis elegans*. (A) Heterochronic mutations affect the identity and number of larval stages. Some delete a specific stage, some repeat it (once or several times). (B) As an example, the modification of the cell lineage of the T blast cell is shown for two such mutants. After Ambros and Horvitz (1984). (C) *lin-14* gene expression is high in the L1 stage and low in the L4 stage (top panels). This temporal down-regulation does not occur in *lin-4* mutants (bottom panels). The worms carry a transgene with a collagen promoter fused with the β-galactosidase reporter gene fused to the 3' untranslated region (3'UTR) of the *lin-14* gene and stained for β-galactosidase activity (originally in blue). The *lin-14* 3' UTR region confers stage-specific regulation of the transgene. Reprinted with permission from Wightman et al. (1993).

The heterochronic mutants affect a number of downstream cell-specific events, such as the regulation of cell division, cell migration or the rewiring of neurons. For instance, *lin-14* and *lin-28* affect the cell cycle of the vulva precursors (Euling and Ambros, 1996), as well as many other developmental events. One mutant that affects specifically the timing of vulval divisions has been described. This *n300* mutant is retarded exclusively for the vulva precursor cell divisions, and is genetically downstream of *lin-28* and thus a possible candidate for a cell-specific downstream regulator of developmental timing (Ambros, 1997).

3.9.2 Molecular Regulation Through 3' UTRs and Small RNAs

The sequences of the heterochronic genes indicate an interesting molecular mechanism for the temporal regulation of larval stage identity. The *lin-4* gene turns out to produce a small RNA molecule (22 nucleotides), which is not translated and is complementary to a portion of the 3' untranslated region (3' UTR) of the *lin-14* and *lin-28* RNAs (Lee, et al., 1993; Moss, et al., 1997; Wightman, et al., 1993). The *lin-14* gene encodes a novel nuclear protein (Ruvkun and Giusto, 1989) and the *lin-28* gene encodes a protein with a cold-shock domain (Moss, et al., 1997).

The proposed mechanism of regulation of these genes is the following. In the L1 stage, *lin-4* is absent, *lin-14* is transcribed and translated, and a high level of LIN-14 protein induces L1-specific events. Towards the end of the L1 stage, *lin-4* transcription starts and progressively inhibits LIN-14 translation by binding to its 3' UTR. An intermediate level of LIN-14, together with LIN-28, specifies L2-specific events. Finally, by the L3 stage, LIN-14 and LIN-28 translations are totally inhibited by *lin-4* RNA, and L3-specific events occur (Arasu, et al., 1991; Ruvkun and Giusto, 1989; Wightman, et al., 1991, 1993). The mechanism of transcriptional activation of *lin-4* is unknown, but is likely coordinated with some L1-specific events and requires feeding.

Among the heterochronic genes which act later in development, the *let-7* gene produces a 21-nucleotide RNA, which, like *lin-4*, is complementary to the 3' UTR of some other heterochronic genes, such as *lin-41* or *lin-42* (Reinhart, et al., 2000). The *let-7* small RNA and its complementary sequence in the 3' UTR of the *lin-41* homolog are conserved in other animals (Bilateria), where *let-7* is also expressed at late developmental times (Pasquinelli, et al., 2000). In *C. elegans*, *let-7* transcription starts from the end of the L3 stage and inhibits LIN-41 translation. The abundance of LIN-41 protein (an RBCC-family member) declines during the L4 stage, thereby relieving the inhibition of downstream targets like *lin-29* transcription (Slack, et al., 2000). The *lin-29* gene encodes a zinc-finger protein of the C2H2 class of transcription factors (Rougvie and Ambros, 1995); its expression starts at the L4 stage (Bettinger, et al., 1996) and may directly activate the transcription of stage-specific cuticular collagens (Liu, et al., 1995). Like *lin-41* mutants, *lin-42* mutants show precocious adult-stage specification. Interestingly, *lin-42* shows a weak similarity to the *Drosophila period* gene, a gene involved in circadian rhythm. Like *period*, the *lin-42* gene is transcribed in a cyclical fashion, with a peak in the middle of each larval stage, whereas the *lin-42* mutants only show defects in late stages (Jeon, et al., 1999). Its function and mode of regulation are still unclear.

The *lin-4* and *let-7* loci both actually encode longer RNAs (about 70 nucleotides) with double-stranded hairpins, that are then processed to 21 or 22-nt RNAs by DCR-1, an RNAse that was first described for its role in double-stranded RNA interference (Grishok, et al., 2001). Many such small RNAs that are putatively processed have been found in *C. elegans* and other animal genomes, but their function remains to be studied (Lau, et al., 2001; Lee and Ambros, 2001).

In summary, regulation of larval stage identity recruits successively two short RNAs, *lin-4* and *let-7*, which are transcribed at distinct stages and bind the 3' UTRs of specific genes and thereby inhibit their translation. The regulation of *lin-4* or *let-7* transcription touches an important point. Clearly, heterochronic genes are central to the mechanism of larval stage specification, but we do not yet understand the mechanistic basis for temporal change in their activities. Mutations in the genes specifying the early larval stages result in defects in the late stages (Ambros, 1989). This dependence may stem from the fact that the gene activities that specify the late stages are under the control of some earlier event, possibly very undirectly downstream of *lin-14*. Passage through Dauer actually suppresses the late effects of the early-acting genes like *lin-14* and *lin-28* (Euling and Ambros, 1996; Liu and Ambros, 1991), maybe because it influences the basic regulation of larval stage identity. One interesting issue, like for other temporally regulated events like the cell cycle (Murray and Kirschner, 1989), is how much the temporal regulation of the heterochronic genes is regulated by each other in a cascade (and if so, how the long time constants are generated) and how much the downstream events that they control feed back on them.

3.9.3 Developmental Heterochronies between Nematodes

Heterochrony has a long history as a proposed mechanism of evolutionary change, although it is now recognized that it likely plays only a minor, if spectacular, contribution to evolutionary variations between species (Gould, 1977; Raff, 1996). The finding of "heterochronic" mutants in *C. elegans* has been very instrumental in understanding the link between molting cycle and larval stage identity. Their naming, however, should not imply that they are necessarily the source of evolutionary change in developmental timing between nematodes. To date, nothing is known about their presence and regulation in nematodes outside the genus *Caenorhabditis*. Moreover, they only concern the timing of larval development.

The timing of developmental events actually varies among nematode species, e. g., the relative timing of early embryonic divisions (Skiba and Schierenberg, 1992). Also, during larval development, gonadal divisions start in *Panagrellus redivivus* later in the L1 stage than in *C. elegans* (Sternberg and Horvitz, 1981), and only in the L2 stage in *Panagrolaimus* species (Family Panagrolaimidae) (Félix and Sternberg, 1996). The rapid evolution of gonadal division timing might reflect the developmental dissociation between gonadal and non-gonadal timing. A more dramatic change in larval development timing is found in the Diplogastridae. Only three larval stages are found in this family and the events of postembryonic development are shifted relative to each other and to the molts (Fig. 3.32) (Félix, et al., 1999). Some other nematodes have a delayed hatching and an internal molt in the embryo.

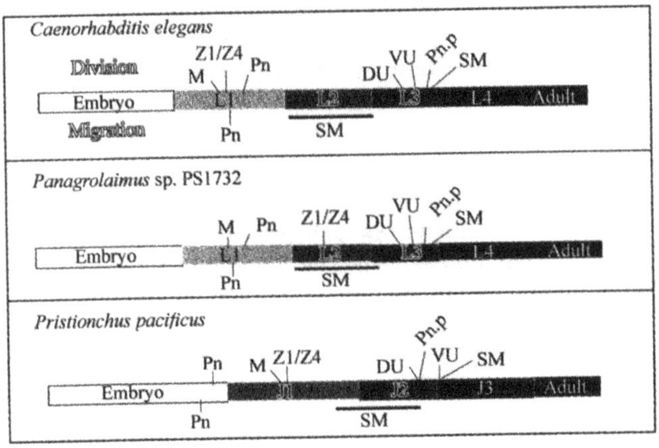

Figure 3.32 Heterochronies between nematodes. Some landmark developmental events are shown for three free-living soil nematodes. The timing of division (top) and migration (bottom) of several cells is indicated relative to hatching (the transition from embryo to L1 or J1) and the molts. The gonadal divisions occur later in *Panagrolaimus* sp. PS1732 (Panagrolaimidae) than in *C. elegans*. *Pristionchus pacificus* (Diplogastridae) has only three larval stages and developmental events are shifted relative to hatching, the molts, and each other.

3.10 Dauer and Longevity

3.10.1 The *C. elegans* Dauer

Two alternative life cycles are possible for *C. elegans*: the dauer larva is an alternative L3 stage favored in some external conditions (Cassada and Russell, 1975). The sensory cues for dauer entry are high population density (through a pheromone), absence of food and high temperature (Golden and Riddle, 1984). The decision is made in the L1 stage to enter a pre-dauer L2 stage (L2d; Fig. 3.2). The L2d stage is characterized by a greater fat accumulation, but advancement into dauer development is still reversible. The L2d enters the dauer stage at the next lethargic period. Dauer larvae can survive several months (Klass and Hirsh, 1976). They do not feed (having a closed mouth and no pharyngeal pumping) and are characterized by a thin dark intestine, a specific cuticle with lateral alae, a modified metabolism (decrease of the tricarboxylic acid pathway relative to the glyoxylate pathway), and a higher resistance to oxidative stress. A dauer larva usually does not move. However, if it is touched it is able to move rapidly. It also displays a specific behavior of standing on its tail and moving its body, which is thought to help with finding a bigger animal for dispersal (Riddle, 1988; Riddle and Albert, 1997). Exit from dauer is induced by the set of cues opposite to that inducing dauer entry. The animal starts feeding after 3 hours, and after 10 hours, it molts into the post-dauer L4 (Fig. 3.2).

The dauer larva corresponds to the parasitic stage of many nematodes, which themselves may have alternative life cycles depending on the generation. Even if the primary cues for entry/exit are different in the specific ecological context of each species, the downstream effectors might be similar in different nematodes.

3.10.2 Insights into the Neuronal, Genetic and Molecular Pathways Regulating Dauer Entry

Like other aspects of development, dauer entry has been studied through a combination of mutant analysis and cell ablation experiments. Mutants in dauer formation (*daf*) have one of two opposite phenotypes: dauer-defective (Dafd) or dauer-constitutive (Daf-c). They have been isolated through visual screens or by selection using the resistance of dauer larvae to detergents. Their phenotype was analyzed more precisely at the cellular level (e.g., for defects in the structure of certain neurons). The two opposite phenotypes allow for genetic epistasis analysis between the many genes, defining a complex pathway, which has recently begun to be well dissected at the molecular level

(Fig. 3.33).

The first gene in the pathway, *daf-22*, is involved in pheromone production, as its mutation can be suppressed by pheromone addition (Kuwabara, et al., 1992). The pheromone has been semi-purified. It is hydrophobic and heat-resistant, but its exact chemical nature and its receptor are unknown.

With laser ablation experiments, it has been shown that the regulation of

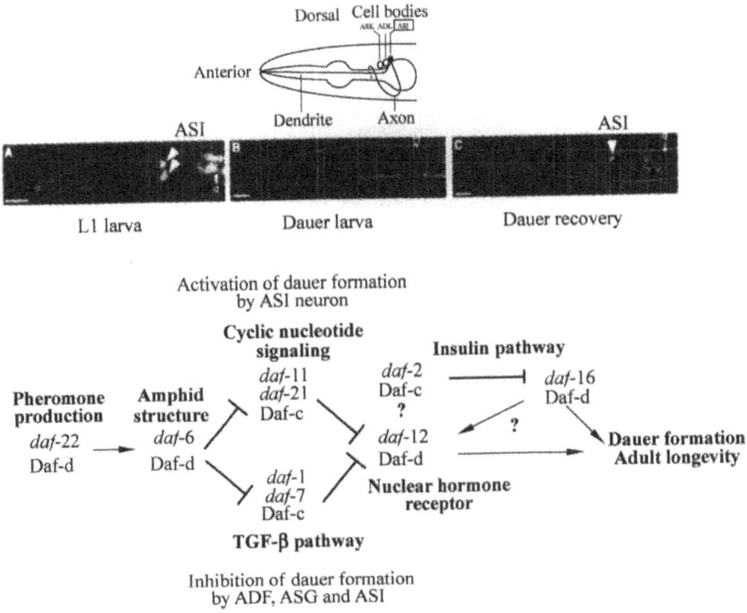

Figure 3.33 Regulation of entry into the dauer stage of *Caenorhabditis elegans*. *Top*: Localization of ASI neurons near the basal bulb (one side is shown). *Center*: *daf-7* expression. The worms carry a transgene with the *daf-7* promoter fused to the green fluorescent protein (GFP). Only the anterior part of the worm is shown. (A) *daf-7* is expressed in ASI neurons in a L1 larva. Fluorescence is visible in the cell bodies (arrowheads) and axons (anteriorly). g = unspecific gut autofluorescence. (B) *daf-7* is repressed when entry into dauer is induced by pheromone (the arrowhead indicates the position of ASI). (C) Recovering larva 8 hours after addition of food. From Ren et al. (1996). *Bottom*: The genetic pathway represented covers, from left to right: pheromone production, sensation of external cues by amphid neurons, processing in a neuronal network and putative signaling throughout the animal. The downstream genes also regulate adult lifespan. Only a few genes are represented.

dauer entry requires some amphid neurons (Bargmann and Horvitz, 1991). Mutants generally defective in amphid structure or sensory cilia are defective in dauer entry. For instance, the *daf-6* mutant shows a defect in the amphid sheath cell that results in abnormal closure of the amphid channel (Albert, et al., 1981).

From cell ablations, several amphid neurons were found to be involved in dauer entry/recovery (Bargmann and Horvitz, 1991). Among the amphid neurons, ADF, ASG, and ASI inhibit dauer formation in good growth conditions, and ASJ promotes dauer recovery. The different neurons might be involved in sensing different sets of cues (e. g., crowding, food, and temperature).

Downstream of *daf-6*, the mutants can be divided into two more or less redundant pathways, which we call the *daf-1* and *daf-11* branches (Fig. 3.33). Single mutants tend to enter dauer easily (for instance, at a high temperature), but double mutants between the two branches always enter the dauer stage at all temperatures (Schackwitz, et al., 1996; Thomas, et al., 1993). The *daf-1* branch is required for the inhibition of dauer formation by ADF, ASG and ASI neurons. ASJ plays a role in responding to pheromone and promoting dauer formation through the *daf-11* branch (Schackwitz, et al., 1996).

The *daf-1* branch encodes components of TGF-β signaling. The *daf-1* and *daf-4* genes encode two subunits of a receptor (Estevez, et al., 1993; Georgi, et al., 1990), and *daf-7* encodes a secreted TGF-β ligand, which is transcribed in favorable growth conditions in the ASI neuron (Fig. 3.33) (Ren, et al., 1996). The DAF-1 receptor is expressed in the ASI neuron but also in interneurons; thus, surprisingly, TGF-β signaling might also be involved in neurotransmission. The receptors probably phosphorylate the Smad proteins encoded by *daf-3*, *daf-8*, and *daf-14*, which then enter the nucleus and bind DNA. Similar TGF-β pathways are also used in body size regulation, male tail morphology, spicule formation and axon guidance in *C. elegans*. These developmental events share the DAF-4 type II-receptor, but specificity is found in the second receptor subunit that uses different downstream Smads (Krishna, et al., 1999; Padgett, et al., 1998). In summary, in good growth conditions, the sensory neuron ASI (and possibly others) expresses the *daf-7* TGF-β secreted protein, which probably acts on other neurons and inhibits dauer formation.

The *daf-11* branch encodes components of cyclic nucleotide signaling. The *daf-11* gene itself encodes a putative transmembrane guanylyl cyclase that produces cyclic GMP (Riddle and Albert, 1997). The *gpa-2* and *gpa-3* genes encode subunits of heterotrimeric G proteins, which are usually found downstream of seven transmembrane receptors (Zwaal, et al., 1997). The *gpa-3* gene is expressed in amphid neurons and could act directly downstream of the unknown pheromone receptor. This *daf-11* branch might function in the ASJ neuron to promote dauer entry (Schackwitz, et al., 1996). The detailed understanding of the neuronal network involved is still unclear. The *daf-1* and *daf-11* branches of the pathway subsequently meet downstream on the *daf-12* gene (Fig. 3.33).

The *daf-2* gene and downstream genes define a third branch that interacts with the *daf-1* and *daf-11* branches. The *daf-2* gene prevents dauer formation and promotes L2/L3/L4 development, and encodes a protein with similarity to mammalian insulin receptor (Kimura, et al., 1997). The genome contains several insulin genes that are potential ligands for DAF-2. The intracellular pathway downstream of DAF-2 has been recently unraveled. DAF-2 activates the AGE-1 phosphatidylinositol-3-OH (PI3) kinase (Morris, et al., 1996), which converts PI(4,5)P2 into PI(3,4,5)P3. The reverse reaction is catalyzed by the DAF-18 phosphatase (Ogg and Ruvkun, 1998). PIP3 acts through other factors (AKT) and finally inhibits the DAF-16 forkhead transcription factor, which itself promotes dauer formation (Lin, et al., 1997; Ogg, et al., 1997). By mosaic analysis, *daf-2* can act in other cells than the cells that express it. Thus its activity probably controls the production of another downstream intercellular signal, which could be the ligand for *daf-12* (Apfeld and Kenyon, 1998).

The TGF-β, cyclic nucleotide and insulin receptor-like pathways converge on the regulation of the Daf-d gene *daf-12*. The *daf-12* gene is required for dauer entry. It encodes a putative nuclear hormone receptor and is also required for temporal progression to the L3 stage (Section 3.9.1) (Antebi, et al., 1998; Yeh, 1991). Its ligand and targets are unknown. However, the Daf-c gene *daf-9* that is genetically just upstream of *daf-12* encodes a cytochrome P450 related to vertebrate steroidogenic enzymes. *daf-9* is expressed in sensory neurons, the epidermis and the somatic gonad and probably activates the production of a steroid hormone by these tissues (Gerisch, et al., 2001).

In summary, sensory cues like pheromone, food and temperature sensation act on sensory amphid neurons. The output is probably decided by a small neuronal network. Good growth conditions result in the activation of the *daf-2*, *daf-7*, and *daf-11* groups of Daf-c genes and the inhibition of the nuclear hormone receptor *daf-12*. The *daf-2* and *daf-12* genes appear to mutually repress each other by unknown means. In unfavorable conditions, *daf-12* and *daf-16* are activated and finally promote a cell-specific dauer program.

3.10.3 Regulation of Adult Lifespan

The *daf-2* branch of the dauer pathway has been recently implicated in the regulation of *C. elegans* adult lifespan. In the wild type in defined culture conditions, adult hermaphrodites live for about 2 weeks. Some Daf-c mutations that promote dauer formation in the larva lengthen lifespan of the adult (Kenyon, et al., 1993). Like in the dauer pathway, *daf-16* is required downstream of *daf-2* for extended lifespan, and *daf-12* affects this regulation (Larsen, et al., 1995).

Adults with extended lifespans, for instance *age-1* and *daf-2* mutants, have a higher stress resistance (like dauers), especially to oxidative damage (Larsen, 1993; Vanfleteren, 1993). Lifespan is influenced by temperature: the animals live longer at low temperatures (for a review, see Kenyon, 1997). Endogenous signals from the gonad (Hsin and Kenyon, 1999) and signals from the sensory system (Apfeld and Kenyon, 1999) also appear to regulate lifespan. Multiple processes are involved in lifespan determination. It is unclear, however, how much of an influence each has. For instance, metabolism may have a large influence on lifespan. The generally accepted view on why a life-shortening allele like *daf-2* is found in wild populations is that the aging rate is not driving selection but rather optimal growth and reproduction.

3.11 Conclusions

Development of *C. elegans* is studied at the cellular level using cell lineaging, ultrastructural studies, and laser ablations. Mutants obtained by genetic screens reveal underlying mechanisms. Molecular cloning defines the molecular pathways involved and provides new tools such as transformation, expression patterns, or mosaic analysis. These same complementary approaches are used in all fields of study in *C. elegans*, from studies of meiosis, cuticle formation, or animal behavior (Riddle, et al., 1997).

Many molecules and entire pathways are used several times during development and are very conserved in other phyla. We saw for instance the Notch/Delta, Wnt, EGF/RAS/MAP kinase, TGF-β and insulin signaling pathways, and the PAR proteins in asymmetric division. What is remarkable about the developmental genes in the *C. elegans* genome is the absence of parts of the hedgehog and JAK/STAT signaling pathways, the high number of nuclear hormone receptor genes whose regulation and function are unknown, as well as the presence of many genes with no obvious homologs in other species

(Ruvkun and Hobert, 1998). The conserved pathways sometimes have similar functions in different organisms (for instance the cell death pathway), or operate in similar cellular contexts (Notch/Delta in lateral inhibition), but they are mostly used and reused in different non-homologous developmental contexts, associated with changes in morphology.

As in other animals, the specification of cell fates during *C. elegans* development combines two cellular processes: asymmetric division and intercellular signaling. The low number of cells and reproducibility of important cell contacts result in invariant fate specification in *C. elegans*. It will be interesting to understand how the cell cycle is so tightly regulated.

The overall organization of *C. elegans* development appears at first rather chaotic, with "perverse assignments" in the lineage and mechanistic piecemeal recruitment of cells to form a given tissue. We saw, however, cases of modularities in the cell lineage (E, P4) and in the use of one common intercellular signal to specify a given tissue (central uterus). It will be interesting to see how much the latter mechanism is used, especially with long-range diffusible signals, like FGF or TGF-β (used as cell migration cues), Wnt (polarity cue), and peptidic or non-peptidic hormones that may function throughout the animal for molt induction, dauer entry, developmental timing, etc.

The good understanding of developmental mechanisms achieved in *C. elegans* will help to study the development of other nematodes. For example, alternative life cycles may be regulated by homologous neurons or molecular pathways; germ-line proliferation by the same distal tip cell induction. The development of larger nematodes may involve mechanisms that act at the level of cell populations, although it is as yet not clear whether cell proliferation in such species occurs before or after cells are specified. The pattern of conserved and divergent features at the molecular and cellular levels will shed light on evolutionary mechanisms that produce developmental variations.

References

Aguinaldo, A. M. A., J. M. Turbeville, L. S. Linford, M. C. Rivera, J. R. Garey, R. A. Raff, and J. A. Lake. 1997. Evidence for a clade of nematodes, arthropods and other moulting animals. Nature 387:489–493.

Ahringer, J. 1997. Maternal control of a zygotic patterning gene in *Caenorhabditis elegans*. Development 124:3865–3869.

Ahringer, J., and J. Kimble. 1991. Control of the sperm-oocyte switch in *Caenorhabditis elegans* hermaphrodites by the *fem-3* 3' untranslated region. Nature 349:346–348.

Akerib, C. C., and B. J. Meyer. 1994. Identification of X chromosome regions in *C.*

elegans that contain sex-determination signal elements. Genetics 138: 1105 – 1125.

Albert, P. S., S. J. Brown, and D. L. Riddle. 1981. Sensory control of dauer larva formation in *Caenorhabditis elegans*. Journal of Comparative Neurology 198: 435 – 451.

Albertson, D. G., and J. N. Thomson. 1976. The pharynx of *Caenorhabditis elegans*. Philosophical Transactions of the Royal Society (London) B 274: 299 – 325.

Ambros, V. 1989. A hierarchy of regulatory genes controls a larval to adult developmental switch in *C. elegans*. Cell 57: 49 – 57.

Ambros, V. 1997. Heterochronic genes. Pp. 501 – 518. In D. L. Riddle, T. Blumenthal, B. J. Meyer, and J. R. Priess, eds. *C. elegans* II. New York: Cold Spring Harbor Laboratory Press.

Ambros, V., and H. R. Horvitz. 1984. Heterochronic mutants of the nematode *Caenorhabditis elegans*. Science 226: 409 – 416.

Ambros, V., and H. R. Horvitz. 1987. The *lin-14* locus of *C. elegans* controls the time of expression of specific postembryonic developmental events. Genes and Development 1: 398 – 414.

Antebi, A., C. R. Norris, E. M. Hedgecock, and G. Garriga. 1997. Cell and growth cone migrations. Pp. 583 – 609. In D. L. Riddle, T. Blumenthal, B. J. Meyer and J. R. Priess, eds. *C. elegans* II. New York: Cold Spring Harbor Laboratory Press.

Antebi, A., J. G. Culotti, and E. M. Hedgecock. 1998. *daf-12* regulates developmental age and the dauer alternative in *Caenorhabditis elegans*. Development 125: 1191 – 1205.

Apfeld, J., and C. Kenyon. 1998. Cell nonautonomy of *C. elegans daf-2* function in the regulation of diapause and life span. Cell 95: 199 – 210.

Apfeld, J., and C. Kenyon. 1999. Regulation of lifespan by sensory perception in *Caenorhabditis elegans*. Nature 402: 804 – 809.

Arasu, P., B. Wightman, and G. Ruvkun. 1991. Temporal regulation of *lin-14* by the antagonistic action of two other heterochronic genes, *lin-4* and *lin-28*. Genes and Development 5: 1825 – 1833.

Arduengo, P. M., O. K. Appleberry, P. Chuang, and S. W. L'Hernault. 1998. The presenilin protein family member SPE-4 localizes to an ER/Golgi derived organelle and is required for proper cytoplasmic partitioning during *Caenorhabditis elegans* spermatogenesis. Journal of Cell Science 111: 3645 – 3654.

Aroian, R. V., M. Koga, J. E. Mendel, Y. Ohshima, and P. W. Sternberg. 1990. The *let-23* gene necessary for *Caenorhabditis elegans* vulval induction encodes a tyrosine kinase of the EGF receptor subfamily. Nature 348: 693 – 699.

Artavanis-Tsakonas, S., K. Matsuno, and M. E. Fortini. 1995. Notch signaling. Science 268: 225 – 232.

Asahina, M., T. Ishihara, M. Jindra, Y. Kohara, I. Katsura, and S. Hirose. 2000. The conserved nuclear receptor Ftz-Fl is required for embryogenesis, moulting and reproduction in *Caenorhabditis elegans*. Genes to Cells 5: 711 – 723.

Austin, J., and J. Kimble. 1987. *glp-1* is required in the germ line for regulation of the decision between mitosis and meiosis in *C. elegans*. Cell 51: 589 – 599.

Austin, J., and J. Kimble. 1989. Transcript analysis of *glp-1* and *lin-12*, homologous genes required for cell interactions during development of *C. elegans*. Cell 58: 565 – 571.

Avery, L., and H. R. Horvitz. 1987. A cell that dies during wild-type *C. elegans* development can function as a neuron in a *ced-3* mutant. Cell 51: 1071 – 1078.

Bargmann, C. I., and H. R. Horvitz. 1991. Control of larval development by chemosensory neurons in *Caenorhabditis elegans*. Science 251: 1243 – 1246.

Baumeister, R., Y. Liu, and G. Ruvkun. 1996. Lineage-specific regulators couple cell lineage asymmetry to the transcription of the *Caenorhabditis elegans* POU gene *unc-86* during neurogenesis. Genes and Development 10: 1395 – 1410.

Beitel, G. J., S. Tuck, I. Greenwald, and H. R. Horvitz. 1995. The *Caenorhabditis elegans* gene *lin-1* encodes an ETS-domain protein and defines a branch of the vulval induction pathway. Genes and Development 9: 3149 – 3162.

Bergmann, D. C., J. R. Crew, J. M. Kramer, and W. B. Wood. 1998. Cuticle chirality and body handedness in *Caenorhabditis elegans*. Developmental Genetics 23: 164 – 174.

Berset, T., E. F. Hoier, G. Battu, S. Canevascini, and A. Hajnal. 2001. Notch inhibition of RAS signaling through MAP kinase phosphatase LIP-1 during *C. elegans* vulval development. Science 291: 1055 – 1058.

Bettinger, J. C., K. Lee, and A. E. Rougvie. 1996. Stage-specific accumulation of the terminal differentiation factor LIN-29 during *Caenorhabditis elegans* development. Development 122: 2517 – 2527.

Blelloch, R., and J. Kimble. 1999. Control of organ shape by a secreted metalloprotease in the nematode *Caenorhabditis elegans*. Nature 399: 586 – 590.

Böhm, H., V. Brinkmann, M. Drab, A. Henske, and T. V. Kurzchalia. 1997. Mammalian homologues of *C. elegans* PAR-1 are asymmetrically localized in epithelial cells and may influence their polarity. Current Biology 7: R603 – R606.

Bonfig, R. 1925. Die Determination der Hauptrichtungen des Embryos von Ascaris megalocephala. Zeitschrift für Wissenschaftliche Zoologie 124: 407 – 456.

Boveri, T. 1892. Über die Entstehung des Gegensatzes zwischen den Geschlechtszellen und den somatischen Zellen bei *Ascaris megalocephala*. Sitzungsberichte der Gesellschaft für Morphologie und Physiologie in München.

Bowerman, B., B. W. Draper, C. C. Mello, and J. R. Priess. 1993. The maternal gene *skn-1* encodes a protein that is distributed unequally in early *C. elegans* embryos. Cell 74: 443 – 452.

Bowerman, B., F. E. Tax, J. H. Thomas, and J. R. Priess. 1992. Cell interactions involved in development of bilaterally symmetrical intestinal valve cells during embryogenesis in *Caenorhabditis elegans*. Development 116: 1113 – 1122.

Bowerman, B., M. K. Ingram, and C. P. Hunter. 1997. The maternal *par* genes and the segregation of cell fate specification activities in early *Caenorhabditis elegans* embryos. Development 124: 3815 – 3826.

Boyd, L., S. Guo, D. Levitan, D. T. Stinchcomb, and K. J. Kemphues. 1996. PAR-2 is asymmetrically distributed and promotes association of P granules and PAR-1 with the cortex in *C. elegans* embryos. Development 122: 3075 – 3084.

Brenner, S. 1974. The genetics of *Caenorhabditis elegans*. Genetics 77: 71 – 94.

Brunschwig, K., C. Wittman, R. Schnabel, T. R. Bürglin, H. Tobler, and F. Müller. 1999. Anterior organization of the *Caenorhabditis elegans* embryo by the labial-like gene *ceh-13*. Development 126: 1537 – 1546.

Buechner, M., D. H. Hall, H. Bhatt, and E. M. Hedgecock. 1999. Cystic canal mutants in *Caenorhabditis elegans* are defective in the apical membrane domain of the renal (excretory) cell. Developmental Biology 214: 227 – 241.

Burdine, R. D., E. B. Chen, S. F. Kwok, and M. J. Stern. 1997. *egl-17* encodes an invertebrate fibroblast growth factor family member required specifically for sex myoblast migration in *Caenorhabditis elegans*. Proceedings of the National Academy of Sciences (USA) 94: 2433 – 2437.

Bürglin, T. R., G. Ruvkun, A. Coulson, N. C. Hawkins, J. D. McGhee, D. Schaller, C. Wittman, F. Müller, and R. H. Waterston. 1991. Nematode homeobox cluster. Nature 351: 703.

Carmi, I., J. B. Kopczynski, and B. J. Meyer. 1998. The nuclear hormone receptor SEX-1 is an X-chromosome signal that determines nematode sex. Nature 396: 168 – 173.

Cassada, R. C., and R. L. Russell. 1975. The dauer larva, a post-embryonic developmental variant of the nematode *Caenorhabditis elegans*. Developmental Biology 46: 326 – 342.

Cassata, G., H. Kagoshima, Y. Andachi, Y. Kohara, M. B. Durrenberger, D. H. Hall, and T. R. Bürglin. 2000. The LIM homeobox gene *ceh-14* confers thermosensory function to the AFD neurons in *Caenorhabditis elegans*. Neuron 25: 287 – 297.

Chalfie, M., and J. Sulston. 1981. Developmental genetics of the mechanosensory neurons of *Caenorhabditis elegans*. Developmental Biology 82: 358 – 370.

Chalfie, M., H. R. Horvitz, and J. E. Sulston. 1981. Mutations that lead to reiterations in the cell lineages of *Caenorhabditis elegans*. Cell 24: 59 – 69.

Chamberlin, H. M., and P. W. Sternberg. 1993. Multiple cell interactions are required for fate specification during male spicule development in *Caenorhabditis elegans*. Development 118: 297 – 323.

Chamberlin, H., and P. W. Sternberg. 1994. The *lin-3/let-23* pathway mediates inductive signalling during male spicule development in *Caenorhabditis elegans*. Development 120: 2713 – 2721.

Chan, S.-Y., H. Zheng, M.-W. Su, R. Wilk, M. T. Killeen, E. M. Hedgecock, and J. G. Culotti. 1996. UNC-40, a *C. elegans* homolog of DCC (deleted in colorectal cancer), is required in motile cells responding to UNC-6 netrin cues. Cell 87: 187 – 195.

Chen, E. B., and M. J. Stern. 1998. Understanding cell migration guidance: Lessons from sex myoblast migration in *C. elegans*. Trends in Genetics 14: 322 – 327.

Chen, F., B. M. Hersh, B. Conradt, Z. Zhou, D. Riemer, Y. Gruenbaum, and H. R. Horvitz. 2000. Translocation of *C. elegans* CED-4 to nuclear membranes during programmed cell death. Science 287: 1485 – 1489.

Chisholm, A. D., and H. R. Horvitz. 1995. Patterning of the *Caenorhabditis elegans* head region by the Pax-6 family member *vab-3*. Nature 377: 52 – 55.

Chitwood, B. G., and M. B. Chitwood. 1974. Introduction to Nematology, 2nd edition. Baltimore, MD: University Park Press.

Chow, K. L., and S. W. Emmons. 1994. HOM-C/Hox genes and four interacting loci determine the morphogenetic properties of single cells in the nematode male tail. Development 120: 2579 – 2593.

Christensen, S., V. Kodoyianni, M. Bosenberg, L. Friedman, and J. Kimble. 1996. *lag-1*, a gene required for *lin-12* and *glp-1* signaling in *Caenorhabditis elegans*, is

homologous to human CBF1 and *Drosophila* Su(H). Development 122: 1373 – 1383.
Chuang, P.-T., J. D. Lieb, and B. J. Meyer. 1996. Sex-specific assembly of a dosage compensation complex on the nematode X chromosome. Science 274: 1736 – 1739.
Church, D., K.-L. Guan, and E. J. Lambie. 1995. Three genes of the MAP kinase cascade, *mek-2*, *mpk-1/sur-1* and *let-60 ras*, are required for meiotic cell cycle progression in *Caenorhabditis elegans*. Development 121: 2525 – 2535.
Clandinin, T. R., J. Demodena, and P. W. Sternberg. 1998. Inositol triphosphate mediates a RAS-independent response to LET-23 receptor tyrosine kinase activation in *C. elegans*. Cell 92: 523 – 533.
Clandinin, T. R., W. S. Katz, and P. W. Sternberg. 1997. *Caenorhabditis elegans* HOM-C genes regulate the response of vulval precursor cells to inductive signal. Developmental Biology 182: 150 – 161.
Clark, S. G., A. D. Chisholm, and H. R. Horvitz. 1993. Control of cell fates in the central body region of *C. elegans* by the homeobox gene *lin-39*. Cell 74: 43 – 55.
Clarke, N. D., and J. M. Berg. 1998. Zinc fingers in *Caenorhabditis elegans*: Finding families and probing pathways. Science 282: 2018 – 2022.
Colavita, A., S. Krishna, H. Zheng, R. W. Padgett, and J. G. Culotti. 1998. Pioneer axon guidance by UNC-129, a *C. elegans* TGF-β. Science 281: 706 – 709.
Conradt, B., and H. R. Horvitz. 1998. The *C. elegans* protein EGL-1 is required for programmed cell death and interacts with the Bcl-2-like protein CED-9. Cell 93: 519 – 529.
Conradt, B., and H. R. Horvitz. 1999. The TRA-1A sex determination protein of *C. elegans* regulates sexually dimorphic cell deaths by repressing the *egl-1* cell death activator gene. Cell 98: 317 – 327.
Costa, M., W. Raich, C. Agbunag, B. Leung, J. Hardin, and J. R. Priess. 1998. A putative catenin-cadherin system mediates morphogenesis of the *Caenorhabditis elegans* embryo. Journal of Cell Biology 141: 297 – 308.
Cowing, D., and C. Kenyon. 1996. Correct *Hox* gene expression established independently of position in *Caenorhabditis elegans*. Nature 382: 353 – 356.
Crittenden, S. L., E. R. Troemel, T. C. Evans, and J. Kimble. 1994. GLP-1 is localized to the mitotic region of the *C. elegans* germ line. Development 120: 2901 – 2911.
Dawes, H. E., D. S. Berlin, D. M. Lapidus, C. Nusbaum, T. L. Davis, and B. J. Meyer. 1999. Dosage compensation proteins targeted to X chromosomes by a determinant of hermaphrodite fate. Science 284: 1800 – 1804.
de Bono, M., and J. Hodgkin. 1996. Evolution of sex determination in *Caenorhabditis*: Unusually high divergence of *tra-1* and its functional consequences. Genetics 144: 587 – 595.
Delattre, M., and M.-A. Félix. 2001a. Evolution and development of a variable left-right asymmetry in nematodes: the handedness of P11/P12 migration. Developmental Biology 232: 362 – 371.
Delattre, M., and M.-A. Félix. 2001b. Polymorphism and evolution of vulval precursor cell lineages within two nematode genera, *Caenorhabditis* and *Oscheius*. Current Biology 11: 631 – 643.
del Peso, L., V. M. Gonzalez, and G. Nunez. 1998. *Caenorhabditis elegans* EGL-1

disrupts the interaction of CED-9 with CED-4 and promotes CED-3 activation. Journal of Biological Chemistry 273: 13495 – 13500.

Desai, C., G. Garriga, S. L. McIntire, and H. R. Horvitz. 1988. A genetic pathway for the development of the *Caenorhabditis elegans* HSN motor neurons. Nature 336: 638 – 646.

Devore, D. L., H. R. Horvitz, and M. J. Stern. 1995. An FGF receptor signaling pathway is required for the normal cell-migrations of the sex myoblasts in *C. elegans* hermaphrodites. Cell 83: 611 – 620.

Dolinski, C., G. Borgonie, R. Schnabel, and J. G. Baldwin. 1998. Buccal capsule development as a consideration for phylogenetic analysis of Rhabditida (Nemata). Development, Genes and Evolution 208: 495 – 503.

Duggan, A., C. Ma, and M. Chalfie. 1998. Regulation of touch receptor differentiation by the *Caenorhabditis elegans mec-3* and *unc-86* genes. Development 125: 4107 – 4119.

Durbin, R. 1987. Studies on the development and organisation of the nervous system of *Caenorhabditis elegans*. Ph. D. thesis, University of Cambridge, England.

Eastman, C., H. R. Horvitz, and Y. Jin. 1999. Coordinated transcriptional regulation of the *unc-25* glutamic acid decarboxylase and the *unc-47* GABA vesicular transporter by the *Caenorhabditis elegans* UNC-30 homeodomain protein. Journal of Neuroscience 19: 6225 – 6234.

Eizinger, A., and Sommer, R. J. 1997. The homeotic gene *lin-39* and the evolution of nematode epidermal cell fates. Science 278: 452 – 455.

Ellis, H. M. 1985. Genetic control of programmed cell death in the nematode *Caenorhabditis elegans*. Ph. D. thesis, Massachusetts Institute of Technology, Massachusets, USA.

Ellis, H. M., and H. R. Horvitz. 1986. Genetic control of programmed cell death in the nematode *Caenorhabditis elegans*. Cell 44: 817 – 829.

Emmons, S. W., and P. W. Sternberg. 1997. Male development and mating behavior. Pp. 295 – 334. In D. L. Riddle, T. Blumenthal, B. J. Meyer, and J. R. Priess, eds. *C. elegans* II. New York: Cold Spring Harbor Laboratory Press.

Epstein, H. F., and D. C. Shakes. 1995. *Caenorhabditis elegans*: Modern Biological Analysis of an Organism. San Diego, CA: Academic Press.

Estevez, M., L. Attisano, J. L. Wrana, P. S. Albert, J. Massagué, and D. L. Riddle. 1993. The *daf-4* gene encodes a bone morphogenetic protein receptor controling *C. elegans* dauer larva development. Nature 365: 644 – 649.

Etemad-Moghadam, B., S. Guo, and K. J. Kemphues. 1995. Asymmetrically distributed PAR-3 protein contributes to cell polarity and spindle alignment in early *C. elegans* embryos. Cell 83: 743 – 752.

Euling, S., and V. Ambros. 1996. Reversal of cell fate determination in *Caenorhabditis elegans* vulval development. Development 122: 2507 – 2515.

Evans, T. C., S. L. Crittenden, V. Kodoyianni, and J. Kimble. 1994. Translational control of maternal *glp-1* mRNA establishes an asymmetry in the *C. elegans* embryo. Cell 77: 183 – 194.

Félix, M.-A. 1999. Evolution of developmental mechanisms in nematodes. Journal of Experimental Zoology (Molecular and Developmental Evolution) 28: 3 – 18.

Félix, M.-A., and P. W. Sternberg. 1996. Symmetry breakage in the development of one-armed gonads in nematodes. Development 122: 2129 – 2142.

Félix, M.-A., and P. W. Sternberg. 1997. Two nested gonadal inductions of the vulva in nematodes. Development 124: 253 – 259.

Félix, M.-A., P. W. Sternberg, and P. De Ley. 1996. Sinistral nematode population. Nature 381: 122.

Félix, M.-A., R. J. Hill, H. Schwarz, P. W. Sternberg, W. Sudhaus, and R. J. Sommer. 1999. *Pristionchus pacificus*, a nematode with only three juvenile stages, displays major heterochronic changes relative to *C. elegans*. Philosophical Transactions of the Royal Society (London) B 266: 1617 – 1621.

Félix, M.-A., P. De Ley, R. J. Sommer, L. Frisse, S. A. Nadler, W. K. Thomas, J. Vanfleteren, and P. W. Sternberg. 2000. Evolution of vulva development in the Cephalobina (Nematoda). Developmental Biology 221: 68 – 86.

Ferguson, E., and H. R. Horvitz. 1985. Identification and characterization of 22 genes that affect the vulval cell lineages of *Caenorhabditis elegans*. Genetics 110: 17 – 72.

Ferguson, E. L., P. W. Sternberg, and H. R. Horvitz. 1987. A genetic pathway for the specification of the vulval cell lineages of *Caenorhabditis elegans*. Nature 326: 259 – 267.

Ferreira, H. B., Y. Zhang, C. Zhao, and S. W. Emmons. 1999. Patterning of *Caenorhabditis elegans* posterior structures by the *Abdominal-B* homolog, *egl-5*. Developmental Biology 207: 215 – 228.

Finney, M., and G. Ruvkun. 1990. The *unc-86* gene product couples cell lineage and cell identity in *Caenorhabditis elegans*. Cell 63: 895 – 905.

Fitch, D. H. A. 1997. Evolution of male tail development in rhabditid nematodes related to *Caenorhabditis elegans*. Systematic Biology 46: 145 – 179.

Fitch, D. H. A., and S. W. Emmons. 1995. Variable cell positions and cell contacts underlie morphological evolution of the rays in the male tails of nematodes related to *Caenorhabditis elegans*. Developmental Biology 170: 564 – 582.

Fitzgerald, K., and I. Greenwald. 1995. Interchangeability of *Caenorhabditis elegans* DSL proteins and intrinsic signalling activity of their extracellular domains *in vivo*. Development 121: 4275 – 4282.

Fitzgerald, K., H. Wilkinson, and I. Greenwald. 1993. *glp-1* can substitute for *lin-12* in specifying cell fate decisions in *C. elegans*. Development 119: 1019 – 1027.

Gao, D., and J. Kimble. 1995. APX-1 can substitute for its homolog LAG-2 to direct cell interactions throughout *Caenorhabditis elegans* development. Proceedings of the National Academy of Sciences (USA) 92: 9839 – 9842.

Gaudet, J., and S. E. Mango. 2002. Regulation of organogenesis by the *Caenorhabditis elegans* FoxA protein PHA-4. Science 295: 821 – 825.

Georgi, L. L., P. S. Albert, and D. R. Riddle. 1990. *daf-1*, a *C. elegans* gene controlling dauer larva development, encodes a novel receptor protein kinase. Cell 61: 635 – 645.

Gerisch, B., C. Weitzel, C. Kober-Eisermann, V. Rottiers, and A. Antebi. 2001. A hormonal signaling pathway influencing *C. elegans* metabolism, reproductive development, and lifespan. Developmental Cell 1: 841 – 851.

Gissendanner, C. R., and A. E. Sluder. 2000. *nhr-25*, the *Caenorhabditis elegans* ortholog of *ftz-f1*, is required for epidermal and somatic gonad development.

Developmental Biology 221: 259 – 272.

Golden, J. W., and D. L. Riddle. 1984. The *Caenorhabditis elegans* dauer larva: Developmental effects of pheromone, food, and temperature. Developmental Biology 102: 368 – 378.

Goldstein, B. 1992. Induction of gut in *Caenorhabditis elegans* embryos. Nature 357: 255 – 257.

Goldstein, B. 1995. Cell contacts orient some cell division axes in the *Caenorhabditis elegans* embryo. Journal of Cell Biology 129: 1071 – 1080.

Goldstein, B., and S. Hird. 1996. Specification of the antero-posterior axis in *Caenorhabditis elegans*. Development 122: 1467 – 1474.

Goldstein, B., L. M. Frisse, and W. K. Thomas. 1998. Embryonic axis specification in nematodes: Evolution of the first step in development. Current Biology 8: 157 – 160.

Gönczy, P., C. Echeverri, K. Oegema, A. Coulson, S. J. M. Jones, R. R. Copley, J. Duperon, J. Oegama, M. Brehm, E. Cassin, E. Hannak, M. Kirkham, S. Pichler, K. Flohrs, A. Goessen, S. Leidel, A. -M. Alleaume, C. Martin, N. Özlü, P. Bork, and A. A. Hyman. 2000. Functional genomic analysis of cell division in *C. elegans* using RNAi of genes on chromosome III. Nature 408: 331 – 336.

Gönczy, P., H. Schnabel, T. Kaletta, A. Duran Amores, T. Hyman, and R. Schnabel. 1999. Dissection of cell division processes in the one cell stage *Caenorhabditis elegans* embryo by mutational analysis. Journal of Cell Biology 144: 927 – 946.

Goodwin, E. B., P. G. Okkema, T. C. Evans, and J. Kimble. 1993. Translational regulation of *tra-2* by its 3' untranslated region controls sexual identity in *Caenorhabditis elegans*. Cell 75: 329 – 339.

Gould, S. J. 1977. Ontogeny and Phylogeny. Cambridge, MA: Harvard University Press.

Graves, L. E., S. Segal, and E. B. Goodwin. 1999. TRA-1 regulates the cellular distribution of the *tra-2* mRNA in *C. elegans*. Nature 399: 802 – 805.

Greenwald, I. S., P. W. Sternberg, and H. R. Horvitz. 1983. The *lin-12* locus specifies cell fates in *Caenorhabditis elegans*. Cell 34: 435 – 444.

Grishok, A., A. E. Pasquinelli, D. Conte, N. Li, S. Parrish, I. Ha, D. L. Baillie, A. Fire, G. Ruvkun, and C. C. Mello. 2001. Genes and mechanisms related to RNA interference regulate expression of the small temporal RNAs that control *C. elegans* developmental timing. Cell 106: 23 – 34.

Gruidl, M. E., P. A. Smith, K. A. Kuznicki, J. S. Mc Crone, J. Kirchner, D. L. Roussell, S. Strome, and K. L. Bennett. 1996. Multiple potential germ-line helicases are components of the germ-line-specific P granules of *Caenorhabditis elegans*. Proceedings of the National Academy of Sciences (USA) 93: 13837 – 13842.

Guenther, S., and G. Garriga. 1996. Asymmetric distribution of the *C. elegans* HAM-1 protein in neuroblasts enables daughter cells to adopt distinct fates. Development 122: 3509 – 3518.

Gumienny, T. L., E. Lambie, E. Hartwieg, H. R. Horvitz, and M. O. Hengartner. 1999. Genetic control of programmed cell death in the *Caenorhabditis elegans* hermaphrodite germline. Development 126: 1011 – 1022.

Guo, S., and K. J. Kemphues. 1995. *par-1*, a gene required for establishing polarity in *C. elegans* embryos, encodes a putative Ser/Thr kinase that is asymmetrically distributed. Cell 81: 611 – 620.

Hall, D. H., V. P. Winfrey, G. Blaeuer, L. H. Hoffman, T. Furuta, K. L. Rose, O. Hobert, and D. Greenstein. 1999. Ultrastructural features of the adult hermaphrodite gonad of *Caenorhabditis elegans*: Relations between the germ line and soma. Developmental Biology 212: 101 – 123.

Han, M., and P. W. Sternberg. 1990. *let-60*, a gene that specifies cell fates during *C. elegans* vulval induction, encodes a ras protein. Cell 63: 921 – 931.

Han, M., R. V. Aroian, and P. W. Sternberg. 1990. The *let-60* locus controls the switch between vulval and non-vulval cell types in *C. elegans*. Genetics 126: 899 – 913.

Han, M., A. Golden, Y. Han, and P. W. Sternberg. 1993. *C. elegans lin-45 raf* gene participates in *let-60 ras* stimulated vulval differentiation. Nature 363: 133 – 140.

Hedgecock, E. M., J. E. Sulston, and J. N. Thomson. 1983. Mutations affecting programmed cell deaths in the nematode *Caenorhabditis elegans*. Science 220: 1277 – 1279.

Hedgecock, E. M., J. G. Culotti, and D. H. Hall. 1990. The *unc-5, unc-6,* and *unc-40* genes guide circumferential migrations of pioneer axons and mesodermal cells on the epidermis in *C. elegans*. Neuron 4: 61 – 85.

Henderson, S. T., D. Gao, E. J. Lambie, and J. Kimble. 1994. *lag-2* may encode a signaling ligand for the GLP-1 and LIN-12 receptors of *C. elegans*. Development 120: 2913 – 2924.

Hengartner, M. O., and H. R. Horvitz. 1994. *C. elegans* cell survival gene *ced-9* encodes a functional homolog of the mammalian proto-oncogene *bcl-2*. Cell 76: 665 – 676.

Hengartner, M. O., R. E. Ellis, and H. R. Horvitz. 1992. *Caenorhabditis elegans* gene *ced-9* protects cells from programmed cell death. Nature 356: 494 – 499.

Herman, M. A., and H. R. Horvitz. 1994. The *Caenorhabditis elegans* gene *lin-44* controls the polarity of asymmetric cell divisions. Development 120: 1035 – 1047.

Herman, M. A., L. L. Vassillieva, H. R. Horvitz, J. E. Shaw, and R. K. Herman. 1995. The *C. elegans* gene *lin-44,* which controls the polarity of certain asymmetric cell divisions, encodes a Wnt protein and acts cell nonautonomously. Cell 83: 101 – 110.

Hermann, G. J., B. Leung, and J. R. Priess. 2000. Left-right asymmetry in *C. elegans* intestine organogenesis involves a LIN-12/Notch signaling pathway. Development 127: 3429 – 3440.

Hill, R. J., and P. W. Sternberg. 1992. The *lin-3* gene encodes an inductive signal for vulval development in *C. elegans*. Nature 358: 470 – 476.

Hird, S., and J. G. White. 1993. Cortical and cytoplasmic flow polarity in early embryonic cells of *Caenorhabditis elegans*. Journal of Cell Biology 121: 1343 – 1355.

Hird, S. N., J. E. Paulsen, and S. Strom. 1996. Segregation of germ granules in living *Caenorhabditis elegans* embryos: Cell-type-specific mechanisms for cytoplasmic localization. Development 122: 1303 – 1312.

Hirsh, D., D. Oppenheim, and M. Klass. 1976. Development of the reproductive system of *Caenorhabditis elegans*. Developmental Biology 49: 200 – 219.

Hishida, R., T. Ishihara, K. Kondo, and I. Katsura. 1996. *hch-1*, a gene required for normal hatching and normal migration of a neuroblast in *C. elegans*, encodes a protein related to TOLLOID and BMP-1. EMBO Journal 15: 4111 – 4122.

Hodgkin, J. 1983a. Two types of sex determination in a nematode. Nature 304: 267 – 268.

Hodgkin, J. 1983b. X chromosome dosage and gene expression in *Caenorhabditis elegans*. Two unusual dumpy genes. Molecular and General Genetics 192: 452 – 458.

Hodgkin, J. 1992. Genetic sex determination mechanisms and evolution. Bioessays 14: 253 – 261.

Hodgkin, J., and S. Brenner. 1977. Mutations causing transformation of sexual phenotype in the nematode *Caenorhabditis elegans*. Genetics 86: 275 – 287.

Hodgkin, J., and T. Barnes. 1991. More is not better: Brood size and population growth in a self-fertilizing nematode. Philosophical Transactions of the Royal Society (London) B 246: 19 – 24.

Hodgkin, J., H. R. Horvitz, and S. Brenner. 1979. Nondisjunction mutants of the nematode *Caenorhabditis elegans*. Genetics 91: 67 – 94.

Hodgkin, J., J. D. Zellan, and D. G. Albertson. 1994. Identification of a candidate primary sex determination locus, *fox-1*, on the X chromosome of *Caenorhabditis elegans*. Development 120: 3681 – 3689.

Horvitz, H. R. 1988. Genetics of cell lineage. Pp. 157 – 190. In W. B. Wood, ed. The Nematode *Caenorhabditis elegans*. New York: Cold Spring Harbor Laboratory Press.

Hsin, H., and C. Kenyon. 1999. Signals from the reproductive system regulate the lifespan of *C. elegans*. Nature 399: 362 – 366.

Hung, T. J., and K. J. Kemphues. 1999. PAR-6 is a conserved PDZ domain-containing protein that colocalizes with PAR-3 in *Caenorhabditis elegans* embryos. Development 126: 127 – 135.

Hunter, C. P., and W. B. Wood. 1990. The *tra-1* gene determines sexual phenotype cell-autonomously in *C. elegans*. Cell 63: 1193 – 1204.

Hunter, C. P., and W. B. Wood. 1992. Evidence from mosaic analysis of the masculinizing gene *her-1* for cell interactions in *C. elegans* sex determination. Nature 355: 551 – 555.

Hunter, C. P., J. M. Harris, J. N. Maloof, and C. Kenyon. 1999. Hox gene expression in a single *Caenorhabditis elegans* cell is regulated by a *caudal* homolog and intercellular signals that inhibit *Wnt* signaling. Development 126: 805 – 814.

Hutter, H., and R. Schnabel. 1995a. Establishment of left-right asymmetry in the *Caenorhabditis elegans* embryo: A multistep process involving a series of inductive events. Development 121: 3417 – 3424.

Hutter, H., and R. Schnabel. 1995b. Specification of antero-posterior differences within the AB lineage in the *C. elegans* embryo. Development 121: 1559 – 1568.

Hyman, A. A. 1989. Centrosome movement in the early divisions of *Caenorhabditis elegans*: A cortical site determining centrosome position. Journal of Cell Biology 109: 1185 – 1193.

Hyman, A. A., and J. G. White. 1987. Determination of cell division axes in the early embryogenesis of *Caenorhabditis elegans*. Journal of Cell Biology 105: 2123 – 2135.

Ishii, N., W. G. Wadsworth, B. D. Stern, J. G. Culotti, and E. M. Hedgecock. 1992. UNC-6, a laminin-related protein, guides cell and pioneer axon migrations in *C. elegans*. Neuron 9: 873 – 881.

Izumi, Y., T. Hirose, Y. Tamai, S.-I. Hirai, Y. Nagashima, T. Fujimoto, Y. Tabuse, K. J. Kemphues, and S. Ohno. 1998. An atypical PKC directly associates and colocalizes at the epithelial tight junction with ASIP, a mammalian homologue of

Caenorhabditis elegans polarity protein PAR-3. Journal of Cell Biology 143: 95 – 106.

Jacobs, D., G. J. Beitel, S. G. Clark, H. R. Horvitz, and K. Kornfeld. 1998. Gain-of-function mutations in the *Caenorhabditis elegans lin-1* ETS gene identify a C-terminal regulatory domain phosphorylated by ERK MAP kinase. Genetics 149: 1809 – 1822.

Jeon, M., H. F. Gardner, E. A. Miller, J. Deshler, and A. E. Rougvie. 1999. Similarity of the *C. elegans* developmental timing protein LIN-42 to circadian rhythm proteins. Science 286: 1141 – 1146.

Jiang, L. I., and P. W. Sternberg. 1998. Interactions of EGF, Wnt and HOM-C genes specify the P12 neuroectoblast fate in *C. elegans*. Development 125: 2337 – 2347.

Jiang, L. I., and P. W. Sternberg. 1999. Socket cells mediate spicule morphogenesis in *Caenorhabditis elegans* males. Developmental Biology 211: 88 – 99.

Jin, Y., R. Hoskins, and H. R. Horvitz. 1994. Control of type-D GABAergic neuron differentiation by *C. elegans* UNC-30 homeodomain protein. Nature 372: 780 – 783.

Jungblut, B., and R. J. Sommer. 1998. The *Pristionchus pacificus mab-5* gene is involved in the regulation of ventral epidermal cell fates. Current Biology 8: 775 – 778.

Kaech, S. M., C. W. Whitfield, and S. K. Kim. 1998. The LIN-2/LIN-7/LIN-10 complex mediates basolateral membrane localization of the *C. elegans* EGF receptor LET-23 in vulval epithelial cells. Cell 94: 761 – 771.

Katz, W. S., G. M. Lesa, D. Yannoukakos, T. R. Clandinin, J. Schlessinger, and P. W. Sternberg. 1996. A point mutation in the extracellular domain activates LET-23, the *C. elegans* EGF receptor homolog. Molecular and Cellular Biology 16: 529 – 537.

Katz, W. S., R. J. Hill, T. R. Clandinin, and P. W. Sternberg. 1995. Different levels of the *C. elegans* growth factor LIN-3 promote distinct vulval precursor fates. Cell 82: 297 – 307.

Kemphues, K. J., and S. Strome. 1997. Fertilization and establishment of polarity in the embryo. Pp. 335 – 359. In D. L. Riddle, T. Blumenthal, B. J. Meyer, and J. R. Priess, eds. *C. elegans* II. New York: Cold Spring Harbor Laboratory Press.

Kemphues, K. J., J. R. Priess, D. G. Morton, and N. Cheng. 1988. Identification of genes required for cytoplasmic localization in early *C. elegans* embryos. Cell 52: 311 – 320.

Kenyon, C. 1997. Environmental factors and gene activities that influence lifespan. Pp. 791 – 813. In D. L. Riddle, T. Blumenthal, B. J. Meyer, and J. R. Priess, eds. *C. elegans* II. New York: Cold Spring Harbor Laboratory Press.

Kenyon, C., J. Chang, E. Gensch, A. Rudner, and R. Tabtiang. 1993. A *C. elegans* mutant that lives twice as long as wild type. Nature 366: 461 – 464.

Kenyon, C. J., J. Austin, M. Costa, D. W. Cowing, J. M. Harris, L. Honigberg, C. P. Hunter, J. N. Maloof, M. M. Muller-Immerglück, S. J. Salser, D. A. Waring, B. B. Wang, and L. A. Wrischnik. 1997. The dance of the *Hox* genes: Patterning the anteroposterior body axis of *Caenorhabditis elegans*. Cold Spring Harbor Symposia on Quantitative Biology 62: 293 – 305.

Kimble, J. 1981. Alterations in cell lineage following laser ablation of cells in the somatic gonad of *Caenorhabditis elegans*. Developmental Biology 87: 286 – 300.

Kimble, J., and D. Hirsh. 1979. Post-embryonic cell lineages of the hermaphrodite and male gonads in *Caenorhabditis elegans*. Developmental Biology 70: 396 – 417.

Kimble, J., and W. J. Sharrock. 1983. Tissue-specific synthesis of yolk proteins in

Caenorhabditis elegans. Developmental Biology 96: 189 – 196.

Kimble, J., and P. Simpson. 1997. The LIN-12/Notch signaling pathway and its regulation. Annual Reviews in Cell and Developmental Biology 13: 333 – 361.

Kimble, J. E., and J. G. White. 1981. On the control of germ cell development in *Caenorhabditis elegans*. Developmental Biology 81: 208 – 219.

Kimura, K. D., H. A. Tissenbaum, Y. Liu, and G. Ruvkun. 1997. *daf-2*, an insulin receptor-like gene that regulates longevity and diapause in *Caenorhabditis elegans*. Science 277: 942 – 946.

Klass, M., and D. Hirsh. 1976. Nonaging developmental variant of *Caenorhabditis elegans*. Nature 260: 523 – 525.

Knobel, K. M., E. M. Jorgensen, and M. J. Bastiani. 1999. Growth cones stall and collapse during axon outgrowth in *Caenorhabditis elegans*. Development 126: 4489 – 4498.

Koga, M., and Y. Ohshima. 1995. Mosaic analysis of the *let-23* gene function in vulval induction of *Caenorhabditis elegans*. Development 121: 2655 – 2666.

Kostrouchova, M., M. Krause, Z. Kostrouch, and J. E. Rall. 1998. CHR3: A *Caenorhabditis elegans* orphan nuclear hormone receptor required for proper epidermal development and molting. Development 125: 1617 – 1626.

Kraemer, B., S. Crittenden, M. Gallegos, G. Moulder, R. Barstead, J. Kimble, and M. Wickens. 1999. NANOS-3 and FBF proteins physically interact to control the sperm-oocyte switch in *Caenorhabditis elegans*. Current Biology 9: 1009 – 1018.

Krause, M., A. Fire, S. White Harrison, J. Priess, and H. Weintraub. 1990. CeMyoD accumulation defines the body wall muscle cell fate during *C. elegans* embryogenesis. Cell 63: 907 – 919.

Krause, M., S. White-Harrison, S.-X. Xu, L. Chen, and A. Fire. 1994. Elements regulating cell- and stage-specific expression of the *C. elegans* MyoD family homolog *hlh-1*. Developmental Biology 166: 133 – 148.

Krishna, S., L. L. Maduzia, and R. W. Padgett. 1999. Specificity of TGF-β signaling is conferred by distinct type I receptors and their associated SMAD proteins in *Caenorhabditis elegans*. Development 126: 251 – 260.

Kuchinke, U., F. Grawe, and E. Knust. 1998. Control of spindle orientation in *Drosophila* by the Par-3-related PDZ-domain protein Bazooka. Current Biology 8: 1357 – 1365.

Kuwabara, P. E. 1996. Interspecies comparison reveals evolution of control regions in the nematode sex-determining gene *tra-2*. Genetics 144: 597 – 607.

Kuwabara, P. E., P. G. Okkema, and J. Kimble. 1992. *tra-2* encodes a membrane protein and may mediate cell communication in the *Caenorhabditis elegans* sex determination pathway. Molecular Biology of the Cell 3: 461 – 473.

Labouesse, M., and S. E. Mango. 1999. Patterning the *C. elegans* embryo: Moving beyond the cell lineage. Trends in Genetics 15: 307 – 313.

Labouesse, M., S. Sookhareea, and H. R. Horvitz. 1994. The *Caenorhabditis elegans* gene *lin-26* is required to specify the fates of hypodermal cells and encodes a presumptive zinc-finger transcription factor. Development 120: 2359 – 2368.

Lackner, M. R., and S. K. Kim. 1998. Genetic analysis of the *Caenorhabditis elegans* MAP kinase gene *mpk-1*. Genetics 150: 103 – 117.

Lambie, E. J., and J. Kimble. 1991. Two homologous regulatory genes, *lin-12* and *glp-1*, have overlapping functions. Development 112: 231 – 240.

Larsen, P. L. 1993. Aging and resistance to oxidative damage in *Caenorhabditis elegans*. Proceedings of the National Academy of Sciences (USA) 90: 8905 – 8909.

Larsen, P. L., P. S. Albert, and D. L. Riddle. 1995. Genes that regulate both development and longevity in *Caenorhabditis elegans*. Genetics 139: 1567 – 1583.

Lau, N. C., L. P. Lim, E. G. Weinstein, and D. P. Bartel. 2001. An abundant class of tiny RNAs with probable regulatory roles in *Caenorhabditis elegans*. Science 294: 858 – 862.

Laufer, J. S., P. Bazzicalupo, and W. B. Wood. 1980. Segregation of developmental potential in early embryos of *Caenorhabditis elegans*. Cell 19: 569 – 577.

Lawrence, P. A. 1992. The Making of a Fly: The Genetics of Animal Design. Oxford: Blackwell Scientific Publications.

Lee, R. C., and V. Ambros. 2001. An extensive class of small RNAs in *Caenorhabditis elegans*. Science 294: 862 – 864.

Lee, R. C., R. L. Feinbaum, and V. Ambros. 1993. The *C. elegans* heterochronic gene *lin-4* encodes small RNAs with antisense complementarity to *lin-14*. Cell 75: 843 – 854.

Legouis, R., A. Gansmuller, S. Sookhareea, J. M. Bosher, D. L. Baillie, and M. Labouesse. 2000. LET-413 is a basolateral protein required for the assembly of adherens junctions in *Caenorhabditis elegans*. Nature Cell Biology 2: 415 – 422.

Leung, B., G. J. Hermann, and J. R. Priess. 1999. Organogenesis of the *Caenorhabditis elegans* intestine. Developmental Biology 216: 114 – 134.

Leung-Hagesteijn, C., A. M. Spence, B. D. Stern, Y. Zhou, M. Su, E. M. Hedgecock, and J. G. Culotti. 1992. UNC-5, a transmembrane protein with immunoglobulin and thrombospondin type 1 domains, guides cell and pioneer axon migration in *C. elegans*. Cell 71: 289 – 299.

Levitan, D., and I. Greenwald. 1995. Facilitation of *lin-12*-mediated signalling by *sel-12*, a *Caenorhabditis elegans* S182 Alzheimer's disease gene. Nature 377: 351 – 354.

Lin, K., J. B. Dorman, A. Rodan, and C. Kenyon. 1997. *daf-16*: An HFN-3/forkhead family member that can function to double the life-span of *Caenorhabditis elegans*. Science 278: 1319 – 1322.

Lin, R., R. Hill, and J. R. Priess. 1998. POP-1 and antero-posterior fate decisions in *C. elegans* embryos. Cell 92: 229 – 239.

Liu, K., and P. W. Sternberg. 1995. Sensory regulation of male mating behavior in *Caenorhabditis elegans*. Neuron 14: 79 – 89.

Liu, Q. A., and M. O. Hengartner. 1998. Candidate adaptor protein CED-6 promotes the engulfment of apoptotic cells in *C. elegans*. Cell 93: 961 – 972.

Liu, Z., S. Kirch, and V. Ambros. 1995. The *Caenorhabditis elegans* heterochronic gene pathway controls stage-specific transcription of collagen genes. Development 121: 2471 – 2478.

Liu, Z. C., and V. Ambros. 1991. Alternative temporal control systems for hypodermal cell differentiation in *Caenorhabditis elegans*. Nature 350: 162 – 165.

Malakhov, V. V. 1994. Nematodes: Structure, Development, Classification, and Phylogeny. Washington DC: Smithsonian Institute Press.

Malone, C. J., W. D. Fixsen, H. R. Horvitz, and M. Han. 1999. UNC-84 localizes to the nuclear envelope and is required for nuclear migration and anchoring during *C. elegans development. Development* 126:3171 –3181.

Maloof, J. N., and C. Kenyon. 1998. The Hox gene *lin-39* is required during *C. elegans* vulval induction to select the outcome of Ras signaling. Development 125: 181 – 190.

Mango, S. E., C. J. Thorpe, P. R. Martin, S. H. Chamberlain, and B. Bowerman. 1994. Two maternal genes, *apx-1* and *pie-1*, are required to distinguish the fates of equivalent blastomeres in the early *Caenorhabditis elegans* embryo. Development 120: 2305 – 2315.

Maupas, E. 1900. Modes et formes de reproduction des nématodes. Archives de Zoologie Expérimentale et Générale 8: 463 – 624.

McCarter, J., B. Bartlett, T. Dang, and T. Schedl. 1997. Soma-germ cell-interactions in *Caenorhabditis elegans* – multiple events of hermaphroditic germline development require the somatic sheath and spermathecal lineages. Developmental Biology 181: 121 – 143.

McCarter, J., B. Bartlett, T. Dang, and T. Schedl. 1999. On the control of oocyte meiotic maturation and ovulation in *Caenorhabditis elegans*. Developmental Biology 205: 111 – 128.

Mehra, A., J. Gaudet, L. Heck, P. E. Kuwabara, and A. M. Spence. 1999. Negative regulation of male development in *Caenorhabditis elegans* by a protein-protein interaction between TRA-2A and FEM-3. Genes and Development 13: 1453 – 1463.

Mello, C. C., B. W. Draper, and J. R. Priess. 1994. The maternal genes *apx-1* and *glp-1* and establishment of dorsal-ventral polarity in the early *C. elegans* embryo. Cell 77: 95 – 106.

Mello, C. C., C. Schubert, B. Draper, W. Zhang, R. Lobel, and J. R. Priess. 1996. The PIE-1 protein and germline specification in *C. elegans* embryos. Nature 382: 710 – 712.

Metzstein, M. M., and H. R. Horvitz. 1999. The *C. elegans* cell death specification gene *ces-1* encodes a snail family zinc finger protein. Molecular Cell 4: 309 – 319.

Meyer, B. J., and L. P. Casson. 1986. *Caenorhabditis elegans* compensates for the difference in X chromosome dosage between the sexes by regulating transcript levels. Cell 47: 871 – 881.

Mickey, K. M., C. C. Mello, M. K. Montgomery, A. Fire, and J. R. Priess. 1996. An inductive interaction in 4-cell stage *C. elegans* embryos involves APX-1 expression in the signaling cell. Development 122: 1791 – 1798.

Miller, D. M., M. M. Shen, C. E. Shamu, T. R. Bürglin, G. Ruvkun, M. L. Dubois, M. Ghee, and L. Wilson. 1992. *C. elegans unc-4* gene encodes a homeodomain protein that determines the pattern of synaptic input to specific motor neurons. Nature 355: 841 – 845.

Miller, L. M., J. D. Plenefisch, L. P. Casson, and B. J. Meyer. 1988. *xol-1:* A gene that controls the male modes of both sex determination and X chromosome dosage compensation in *C. elegans*. Cell 55: 167 – 183.

Miller, L. M., M. E. Gallegos, B. A. Morisseau, and S. A. Kim. 1993. *lin-31*, a *Caenorhabditis elegans* HNF-3/*fork head* transcription factor homolog, specifies three alternative cell fates in vulval development. Genes and Development 7: 933 – 947.

Morris, J. Z., H. A. Tissenbaum, and G. Ruvkun. 1996. A phosphatidylinositol-3-OH kinase family member regulating longevity and diapause in *Caenorhabditis elegans*.

Nature 382: 536 – 539.

Moskowitz, I. P. G., and J. H. Rothman. 1996. *lin-12* and *glp-1* are required zygotically for early embryonic cellular interactions and are regulated by maternal GLP-1 signaling in *Caenorhabditis elegans*. Development 122: 4105 – 4117.

Moss, E. G., R. C. Lee, and V. Ambros. 1997. The cold-shock domain protein LIN-28 controls developmental timing in *C. elegans* and is regulated by the *lin-4* RNA. Cell 88: 637 – 646.

Murray, A. W., and M. W. Kirschner. 1989. Dominoes and clocks: The union of two views of cell cycle regulation. Science 246: 614 – 621.

Newman, A. P., J. G. White, and P. W. Sternberg. 1995. The *C. elegans lin-12* gene mediates induction of ventral uterine specialization by the anchor cell. Development 121: 263 – 271.

Newman, A. P., J. G. White, and P. W. Sternberg. 1996. Morphogenesis of the *C. elegans* hermaphrodite uterus. Development 122: 3617 – 3626.

Nguyen, C. Q., D. H. Hall, Y. Yang, and D. H. A. Fitch. 1999. Morphogenesis of the *Caenorhabditis elegans* male tail tip. Developmental Biology 207: 86 – 106.

Nicoll, M., C. C. Akerib, and B. J. Meyer. 1997. X-chromosome counting mechanisms that determine nematode sex. Nature 388: 200 – 204.

Nigon, V. 1949. Modalités de la reproduction et déterminisme du sexe chez quelques nématodes libres. Annales des Sciences Naturelles, Zoologie 11: 1 – 132.

Nigon, V. 1965. Développement et reproduction des nématodes. In Traité de Zoologie. Paris: Masson.

Nigon, V., P. Guerrier, and H. Monin. 1960. L'architecture polaire de l'oeuf et les mouvements des constituants cellulaires au cours des premières étapes du développement chez quelques Nématodes. Bulletin biologique de la France et de la Belgique 93: 131 – 202.

Nusbaum, C., and B. J. Meyer. 1989. The *Caenorhabditis elegans* gene *sdc-2* controls sex determination and dosage compensation in XX animals. Genetics 122: 579 – 593.

O'Connell, K. F., K. N. Maxwell, and J. G. White. 2000. The *spd-2* gene is required for polarization of the anteroposterior axis and formation of the sperm asters in the *Caenorhabditis elegans* zygotes. Developmental Biology 222: 55 – 70.

Ogg, S., and G. Ruvkun. 1998. The *C. elegans* PTEN homolog, DAF-18, acts in the insulin receptor-like metabolic signaling pathway. Molecular Cell 2: 887 – 893.

Ogg, S., S. Paradis, S. Gottlieb, G. I. Patterson, L. Lee, H. A. Tissenbaum, and G. Ruvkun. 1997. The Fork head transcription factor DAF-16 transduces insulin-like metabolic and longevity signals in *C. elegans*. Nature 389: 994 – 999.

Okkema, P. G., and J. Kimble. 1991. Molecular analysis of *tra-2*, a sex determining gene in *C. elegans*. EMBO Journal 10: 171 – 176.

Padgett, R. W., P. Das, and S. Krishna. 1998. TGF-β signaling, Smads and tumor suppressors. Bioessays 20: 382 – 391.

Pasquinelli, A. E., B. J. Reinhart, F. Slack, M. Q. Martindale, M. I. Kuroda, B. Maller, D. C. Hayward, E. E. Ball, B. Degnan, P. Müller, J. Spring, A. Srinivasan, M. Fishman, J. Finnerty, J. Corbo, M. Levine, P. Leahy, E. Davidson, and G. Ruvkun. 2000. Conservation of the sequence and temporal expression of *let-7* heterochronic regulatory RNA. Nature 408: 86 – 89.

Peckol, E. L., J. A. Zallen, J. C. Yarrow, and C. I. Bargmann. 1999. Sensory activity affects sensory axon development in *C. elegans*. Development 126: 1891 – 1902.

Perry, M. D., W. Li, C. Trent, B. Robertson, A. Fire, J. M. Hageman, and W. B. Wood. 1993. Molecular characterization of the *her-1* gene suggests a direct role in cell signaling during *Caenorhabditis elegans* sex determination. Genes and Development 7: 216 – 278.

Podbilewicz, B., and J. G. White. 1994. Cell fusions in the developing epithelia of *C. elegans*. Developmental Biology 161: 408 – 424.

Portereiko, M. F., and S. E. Mango. 2001. Early morphogenesis of the *Caenorhabditis elegans* pharynx. Developmental Biology 233: 482 – 494.

Priess, J. R., and D. I. Hirsh. 1986. *Caenorhabditis elegans* morphogenesis: The role of the cytoskeleton in elongation of the embryo. Developmental Biology 117: 156 – 173.

Priess, J. R., and J. N. Thomson. 1987. Cellular interactions in early *Caenorhabditis elegans* embryos. Cell 48: 241 – 250.

Raff, R. A. 1996. The Shape of Life. Chicago: Chicago University Press.

Raymond, C. S., C. E. Shamu, M. M. Shen, K. J. Seifert, B. Hirsch, J. Hodgkin, and D. Zarkower. 1998. Evidence for evolutionary conservation of sex-determining genes. Nature 391: 691 – 695.

Reinhart, B. J., F. J. Slack, M. Basson, A. E. Pasquinelli, J. C. Bettinger, A. E. Rougvie, H. R. Horvitz, and G. Ruvkun. 2000. The 21-nucleotide *let-7* RNA regulates developmental timing in *Caenorhabditis elegans*. Nature 403: 901 – 906.

Ren, P., C.-S. Lim, R. Johnson, P. S. Albert, D. Pilgrim, and D. L. Riddle. 1996. Control of *C. elegans* larval development by neuronal expression of a TGF-β homolog. Science 274: 1389 – 1392.

Rhind, N. R., L. M. Miller, J. B. Kopczynski, and B. J. Meyer. 1995. *xol-1* acts as an early switch in the *C. elegans* male/hermaphrodite decision. Cell 80: 71 – 82.

Riddle, D. L. 1988. The dauer larva. Pp. 393 – 412. In W. B. Wood, ed. The Nematode *Caenorhabditis elegans*. New York: Cold Spring Harbor Laboratory Press.

Riddle, D. L., and P. S. Albert. 1997. Genetic and environmental regulation of dauer larva development. Pp. 739 – 768. In D. L. Riddle, T. Blumenthal, B. J. Meyer and J. R. Priess, eds. *C. elegans* II. New York: Cold Spring Harbor Laboratory Press.

Riddle, D. L., T. Blumenthal, B. J. Meyer, and J. R. Priess, eds. 1997. *C. elegans* II. New York: Cold Spring Harbor Laboratory Press.

Roberts, T. M., and M. Stewart. 1997. Nematode sperm: Amoeboid movement without actin. Trends in Cell Biology 7: 368 – 373.

Rocheleau, C. E., W. D. Downs, R. Lin, C. Wittman, Y. Bei, Y.-H. Cha, M. Ali, J. R. Priess, and C. C. Mello. 1997. Wnt signaling and an APC-related gene specify endoderm in early *C. elegans* embryos. Cell 90: 707 – 716.

Rongo, C., C. W. Whitfield, A. Rodal, S. K. Kim, and J. M. Kaplan. 1998. LIN-10 is a shared component of the polarized protein localization pathways in neurons and epithelia. Cell 94: 751 – 759.

Rougvie, A. E., and V. Ambros. 1995. The heterochronic gene *lin-29* encodes a zinc finger protein that controls a terminal differentiation event in *Caenorhabditis elegans*. Development 121: 2491 – 2500.

Ruvkun, G., and J. Giusto. 1989. The *Caenorhabditis elegans* heterochronic gene *lin-14* encodes a nuclear protein that forms a temporal developmental switch. Nature 338: 313 – 319.

Ruvkun, G., and O. Hobert. 1998. The taxonomy of developmental control in *Caenorhabditis elegans*. Science 282: 2033 – 2041.

Sadler, P. L., and D. C. Shakes. 2000. Anucleate *Caenorhabditis elegans* sperm can crawl, fertilize oocytes and direct antero-posterior polarization of the 1-cell embryo. Development 127: 355 – 366.

Salser, S. J., and C. Kenyon. 1992. Activation of a *C. elegans Antennapedia* homologue in migrating cells controls their direction of migration. Nature 355: 255 – 258.

Salser, S. J., and C. Kenyon. 1996. A *C. elegans* Hox gene switches on, off, on and off again to regulate proliferation, differentiation and morphogenesis. Development 122: 1651 – 1661.

Sawa, H., L. Lobel, and H. R. Horvitz. 1996. The *Caenorhabditis elegans* gene *lin-17*, which is required for certain asymmetric cell divisions, encodes a putative seven-transmembrane protein similar to the *Drosophila* Frizzled protein. Genes and Development 10: 2189 – 2197.

Schackwitz, W. S., T. Inoue, and J. H. Thomas. 1996. Chemosensory neurons function in parallel to mediate a pheromone response in *C. elegans*. Neuron 17: 719 – 728.

Schierenberg, E. 1987. Reversal of cellular polarity and early cell-cell interactions in the embryo of *Caenorhabditis elegans*. Developmental Biology 122: 452 – 463.

Schierenberg, E. 1988. Mikroskopische Untersuchungen an freilebenden Nematoden. Mikrokosmos 77: 377 – 380.

Schierenberg, E., and B. Junkersdorf. 1992. The role of the eggshell and underlying vitelline membrane for normal pattern formation in the early *C. elegans* embryo. Roux's Archives of Developmental Biology 202: 10 – 16.

Schnabel, R., H. Hutter, D. G. Moerman, and H. Schnabel. 1997. Assessing normal embryogenesis in *C. elegans* using a 4D microscope: Variability of development and regional specification. Developmental Biology 184: 234 – 265.

Schubert, C. M., R. Lin, C. J. de Vries, R. H. A. Plasterk, and J. R. Priess. 2000. MEX-5 and MEX-6 function to establish soma/germline asymmetry in early *C. elegans* embryos. Molecular Cell 5: 671 – 682.

Schulman, J. M., R. Benton, and D. St Johnston. 2000. The *Drosophila* homolog of *C. elegans* PAR-1 organizes the oocyte cytoskeleton and directs oskar mRNA localization to the posterior pole. Cell 101: 377 – 388.

Seydoux, G., and A. Fire. 1994. Soma-germline asymmetry in the distributions of embryonic RNAs in *Caenorhabditis elegans*. Development 120: 2823 – 2834.

Seydoux, G., and I. Greenwald. 1989. Cell autonomy of *lin-12* function in a cell fate decision in *C. elegans*. Cell 57: 1237 – 1245.

Sharma-Kishore, R., J. G. White, E. Southgate, and B. Podbilewicz. 1999. Formation of the vulva in *Caenorhabditis elegans*: A paradigm for organogenesis. Development 126: 691 – 699.

Sigrist, C. B., and R. J. Sommer. 1999. Vulva formation in *Pristionchus pacificus* relies on continuous gonadal induction. Development, Genes and Evolution 209: 451 – 459.

Simpson, P. 1990. Lateral inhibition and the development of the sensory bristles of the adult peripheral nervous system of *Drosophila*. Development 109: 509 – 519.

Simske, J. S., and S. K. Kim. 1995. Sequential signalling during *Caenorhabditis elegans* vulval induction. Nature 375: 142 – 146.

Simske, J. S., S. M. Kaech, S. A. Harp, and S. K. Kim. 1996. LET-23 receptor localization by the cell junction protein LIN-7 during *C. elegans* vulval induction. Cell 85: 195 – 204.

Skiba, F., and E. Schierenberg. 1992. Cell lineages, developmental timing and spatial pattern formation in embryos of free-living soil nematodes. Developmental Biology 151: 597 – 610.

Skop, A. R., and J. G. White. 1998. The dynactin complex is required for cleavage plane specification in early *Caenorhabditis elegans* embryos. Current Biology 8: 1110 – 1116.

Slack, F., and G. Ruvkun. 1997. Temporal pattern formation by heterochronic genes. Annual Review of Genetics 31: 611 – 634.

Slack, F. J., M. Basson, Z. Liu, V. Ambros, H. R. Horvitz, and G. Ruvkun. 2000. The *lin-41* RBCC gene acts in the *C. elegans* heterochronic pathway between the *let-7* regulatory RNA and the LIN-29 transcription factor. Molecular Cell 5: 659 – 669.

Sommer, R., A. Eizinger, K.-Z. Lee, B. Jungblut, A. Bubeck, and I. Schlak. 1998. The *Pristionchus* HOX gene *Ppa-lin-39* inhibits programmed cell death to specify the vulva equivalence group and is not required during vulval induction. Development 125: 3865 – 3873.

Sommer, R. J. 1996. Evolution and development – the nematode vulva as a case study. Bioessays 19: 225 – 231.

Sommer, R. J., and P. W. Sternberg. 1994. Changes of induction and competence during the evolution of vulva development in nematodes. Science 265: 114 – 118.

Sommer, R. J., and P. W. Sternberg. 1996. Apoptosis and change of competence limit the size of the vulva equivalence group in *Pristionchus pacificus*: A genetic analysis. Current Biology 6: 52 – 59.

Starr, D. A., G. J. Hermann, C. J. Malone, W. Fixsen, J. R. Priess, H. R. Horvitz, and M. Han. 2001. *unc-83* encodes a novel component of the nuclear envelope and is essential for proper nuclear migration. Development 128: 5039 – 5050.

Sternberg, P. W. 1988. Lateral inhibition during vulval induction in *Caenorhabditis elegans*. Nature 335: 551 – 554.

Sternberg, P. W. 1991. Control of cell lineage and cell fate during nematode development. Current Topics in Developmental Biology 25: 177 – 224.

Sternberg, P. W., and M.-A. Félix. 1997. Evolution of cell lineage. Current Opinion in Genetics and Development 7: 543 – 550.

Sternberg, P. W., and H. R. Horvitz. 1981. Gonadal cell lineages of the nematode *Panagrellus redivivus* and implications for evolution by the modification of cell lineage. Developmental Biology 88: 147 – 166.

Sternberg, P. W., and H. R. Horvitz. 1982. Postembryonic nongonadal cell lineages of the nematode *Panagrellus redivivus*: Description and comparison with those of *Caenorhabditis elegans*. Developmental Biology 93: 181 – 205.

Sternberg, P. W., and H. R. Horvitz. 1986. Pattern formation during vulval development in *Caenorhabditis elegans*. Cell 44: 761 – 772.

Sternberg, P. W., and H. R. Horvitz. 1989. The combined action of two intercellular signalling pathways specifies three cell fates during vulval induction in *C. elegans*. Cell

58:679 – 693.

Steven, R., T. J. Kubiseski, H. Zheng, S. Kulkarni, J. Mancillas, A. R. Morales, C. W. V. Hogue, T. Pawson, and J. Culotti. 1998. UNC-73 activates the Rac GTPase and is required for cell and growth cone migrations in *C. elegans*. Cell 92:785 – 795.

Strome, S. 1983. Generation of asymmetry and segregation of germ-line granules in early *C. elegans* embryos. Cell 35:15 – 25.

Su, M. W., D. C. Merz, M. T. Killeen, Y. W. Zhou, H. Zheng, J. M. Kramer, E. M. Hedgecock, and J. G. Culotti. 2000. Regulation of the UNC-5 netrin receptor initiates the first reorientation of migrating distal tip cells in *Caenorhabditis elegans*. Development 127:585 – 594.

Sulston, J., and H. R. Horvitz. 1977. Postembryonic cell lineages of the nematode *Caenorhabditis elegans*. Developmental Biology 56:110 – 156.

Sulston, J. E., and J. G. White. 1980. Regulation and cell autonomy during postembryonic development of *Caenorhabditis elegans*. Developmental Biology 78: 577 – 597.

Sulston, J. E., D. G. Albertson, and J. N. Thomson. 1980. The *Caenorhabditis elegans* male: Postembryonic development of nongonadal structures. Developmental Biology 78: 542 – 576.

Sulston, J. E., E. Schierenberg, J. G. White, and J. N. Thomson. 1983. The embryonic cell lineage of the nematode *Caenorhabditis elegans*. Developmental Biology 100:64 – 119.

Sun, A. Y., and E. J. Lambie. 1997. *gon-2*, a gene required for gonadogenesis in *Caenorhabditis elegans*. Genetics 147:1077 – 1089.

Sundaram, M., and M. Han. 1996. Control and integration of cell signaling pathways during *C. elegans* vulval development. Bioessays 18:473 – 480.

Sundaram, M., J. Yochem, and M. Han. 1996. A Ras-mediated signal transduction pathway is involved in the control of sex myoblast migration in *Caenorhabditis elegans*. Development 122:2823 – 2833.

Tan, P. B., M. R. Lackner, and S. K. Kim. 1998. MAP kinase signaling specificity mediated by the LIN-1 Ets/LIN-31 WH transcription factor complex during *C. elegans* vulval induction. Cell 93:569 – 580.

Tax, F. E., J. J. Yeargers, and J. H. Thomas. 1994. Sequence of *C. elegans lag-2* reveals a cell-signalling domain shared with *Delta* and *Serrate* of *Drosophila*. Nature 368:150 – 154.

The *C. elegans* Sequencing Consortium. 1998. Genome sequence of the nematode *C. elegans*: A platform for investigating biology. Science 282:2012 – 2018.

Thomas, C., P. DeVries, J. Hardin, and J. White. 1996. Four-dimensional imaging: Computer visualization of 3D movements in living specimens. Science 273:603 – 607.

Thomas, J. H., D. A. Birnby, and J. J. Vowels. 1993. Evidence for parallel processing of sensory information controlling dauer formation in *Caenorhabditis elegans*. Genetics 134:1105 – 1117.

Thorpe, C. J., A. Schlesinger, J. C. Carter, and B. Bowerman. 1997. Wnt signaling polarizes an early *C. elegans* blastomere to distinguish endoderm from mesoderm. Cell 90:695 – 705.

Thorpe, C. J., A. Schlesinger, and B. Bowerman. 2000. Wnt signalling in

Caenorhabditis elegans: Regulating repressors and polarizing the cytoskeleton. Trends in Cell Biology 10: 10 – 17.

Tomancak, P., F. Piano, V. Riechmann, K. C. Gunsalus, K. J. Kemphues, and A. Ephrussi. 2000. A *Drosophila melanogaster* homologue of *Caenorhabditis elegans par-1* acts at an early step in embryonic axis formation. Nature Cell Biology 2: 458 – 460.

Trent, C., N. Tsung, and H. R. Horvitz. 1983. Egg-laying defective mutants of the nematode *Caenorhabditis elegans*. Genetics 104: 619 – 647.

Van Auken, K., D. C. Weaver, L. G. Edgar, and W. B. Wood. 2000. *Caenorhabditis elegans* embryonic axial patterning requires two recently discovered posterior-group Hox genes. Proceedings of the National Academy of Sciences (USA) 97: 4499 – 4503.

Van Beneden, E. 1883. Recherches sur la fécondation et maturation. Archives de Biologie 4: 265 – 640.

Vanfleteren, J. R. 1993. Oxidative stress and ageing in *Caenorhabditis elegans*. Biochemical Journal 292: 605 – 608.

Vaux, D. L., I. L. Weissman, and S. K. Kim. 1992. Prevention of programmed cell death in *Caenorhabditis elegans* by human *bcl-2*. Science 258: 1955 – 1957.

Voronov, D. A., and Y. V. Panchin. 1998. Cell lineage in marine nematode *Enoplus brevis*. Development 125: 143 – 150.

Waddle, J. A., J. A. Cooper, and R. H. Waterston. 1994. Transient localized accumulation of actin in *Caenorhabditis elegans* blastomeres with oriented asymmetric divisions. Development 120: 2317 – 2328.

Wadsworth, W. G., H. Bhatt, and E. M. Hedgecock. 1996. Neuroglia and pioneer neurons express UNC-6 to provide global and local netrin cues for guiding migrations in *C. elegans*. Neuron 16: 35 – 46.

Wallenfang, M. R., and G. Seydoux. 2000. Polarization of the anterior-posterior axis of *C. elegans* is a microtubule-directed process. Nature 408: 89 – 92.

Wang, B. B., M. M. Muller-Immergluck, J. Austin, N. T. Robinson, A. Chisholm, and C. Kenyon. 1993. A homeotic gene cluster patterns the anteroposterior body axis of *C. elegans*. Cell 74: 29 – 42.

Ward, S., and J. S. Carrel. 1979. Fertilization and sperm competition in the nematode *Caenorhabditis elegans*. Developmental Biology 73: 304 – 321.

Ward, S., Y. Argon, and G. A. Nelson. 1981. Sperm morphogenesis in wild-type and fertilization-defective mutants of *Caenorhabditis elegans*. Journal of Cell Biology 91: 26 – 44.

Waring, D., and C. Kenyon. 1990. Selective silencing of cell communication influences anteroposterior pattern formation in *C. elegans*. Cell 60: 123 – 131.

Way, J., and M. Chalfie. 1988. *mec-3*, a homeobox-containing gene that specifies differentiation of the touch receptor neurons in *C. elegans*. Cell 54: 5 – 16.

White, J. G., E. Southgate, J. N. Thomson, and S. Brenner. 1976. The structure of the ventral nerve cord of *Caenorhabditis elegans*. Philosophical Transactions of the Royal Society (London) B 275: 327 – 348.

White, J. G., D. G. Albertson, and M. A. R. Anness. 1978. Connectivity changes in a class of motoneurone during the development of a nematode. Nature 271: 764 – 766.

White, J. G., E. Southgate, J. N. Thomson, and S. Brenner. 1986. The structure of the nervous system of the nematode *Caenorhabditis elegans*. Philosophical Transactions of the Royal Society (London) B 314: 1 –340.

White, J. G., E. Southgate, and J. N. Thomson. 1991. On the nature of undead cells in the nematode *Caenorhabditis elegans*. Philosophical Transactions of the Royal Society (London) B 331: 263 –271.

White, J. G., E. Southgate, and J. N. Thomson. 1992. Mutations in the *Caenorhabditis elegans unc-4* gene alter the synaptic input to ventral cord motor neurons. Nature 355: 838 –841.

Wiegner, O., and E. Schierenberg. 1998. Specification of gut cell fate differs significantly between the nematodes *Acrobeloides nanus* and *Caenorhabditis elegans*. Developmental Biology 204: 3 – 14.

Wiegner, O., and E. Schierenberg. 1999. Regulative development in a nematode embryo: A hierarchy of cell fate transformations. Developmental Biology 215: 1 – 12.

Wightman, B., I. Ha, and G. Ruvkun. 1993. Posttranscriptional regulation of the heterochronic gene *lin-14* by *lin-4* mediates temporal pattern formation in *C. elegans*. Cell 75: 855 –862.

Wightman, B., R. Baran, and G. Garriga. 1997. Genes that guide growth cones along the *C. elegans* ventral nerve cord. Development 124: 2571 –2580.

Wightman, B., S. G. Clark, A. M. Taskar, W. C. Forrester, A. V. Maricq, C. I. Bargmann, and G. Garriga. 1996. The *C. elegans* gene *vab-8* guides posteriorly directed axon outgrowth and cell migration. Development 122: 671 – 682.

Wightman, B., T. R. Bürglin, J. Gatto, P. Arasu, and G. Ruvkun. 1991. Negative regulatory sequences in the *lin-14* 3'-untranslated region are necessary to generate a temporal switch during *Caenorhabditis elegans* development. Genes and Development 5: 1813 –1824.

Wilkins, A. S. 1995. Moving up the hierarchy: A hypothesis on the evolution of a genetic sex determination pathway. Bioessays 17: 71 –77.

Wilkinson, H. A., K. Fitzgerald, and I. Greenwald. 1994. Reciprocal changes in expression of the receptor *lin-12* and its ligand *lag-2* prior to commitment in a *C. elegans* cell fate decision. Cell 79: 1187 –1198.

Winnier, A. R., J. Y. J. Meir, J. M. Ross, N. Tavernakis, M. Driscoll, T. Ishihara, I. Katsura, and D. M. Miller. 1999. UNC-4/UNC-37-dependent repression of motor neuron-specific genes control synaptic choice in *Caenorhabditis elegans*. Genes and Development 13: 2774 –2786.

Wittman, C., O. Bossinger, B. Goldstein, M. Fleischmann, R. Kohler, K. Brunschwig, H. Tobler, and F. Müller. 1997. The expression of the *C. elegans labial*-like *Hox* gene *ceh-13* during early embryogenesis relies on cell fate and on anteroposterior cell polarity. Development 124: 4193 – 4200.

Wolf, N., D. Hirsh, and J. R. McIntosh. 1979. Spermatogenesis in males of the free-living nematode *Caenorhabditis elegans*. Journal of Ultrastructural Research 63: 155 – 169.

Wolf, F. W., M. Hung, B. Wightman, J. Way, and G. Garriga. 1998. *vab-8* is a key regulator of posteriorly directed migrations in *C. elegans* and encodes a novel protein with kinesin motor similarity. Neuron 20: 655 – 666.

Wood, W. B., ed. 1988. The Nematode *Caenorhabditis elegans*. New York: Cold Spring Harbor Laboratory Press.

Wood, W. B. 1991. Evidence from reversal of handedness in *C. elegans* for early cell interactions determining cell fates. Nature 349: 536 – 538.

Wright, K. A., and J. N. Thomson. 1981. The buccal capsule of *Caenorhabditis elegans* (Nematoda: Rhabditoidea): An ultrastructural study. Canadian Journal of Zoology 59: 1952 – 1961.

Wrischnik, L. A., and C. J. Kenyon. 1997. The role of *lin-22*, a hairy/Enhancer of split homolog, in patterning the peripheral nervous system of *C. elegans*. Development 124: 2875 – 2888.

Wu, Y., and H. R. Horvitz. 1998a. The *C. elegans* cell corpse engulfment gene *ced-7* encodes a protein similar to ABC transporters. Cell 93: 951 – 960.

Wu, Y. C., and H. R. Horvitz. 1998b. *C. elegans* phagocytosis and cell migration protein CED-5 is similar to human DOCK180. Nature 392: 501 – 504.

Xue, D., S. Shaham, and H. R. Horvitz. 1996. The *Caenorhabditis elegans* cell-death protein CED-3 is a cysteine protease with substrate specificities similar to those of the human CPP32 protease. Genes and Development 10: 1073 – 1083.

Yeh, W.-H. 1991. Genes acting late in the signalling pathway for *Caenorhabditis elegans* dauer larval development. Ph. D. thesis, University of Missouri, Columbia, Missouri.

Yochem, J., and I. Greenwald. 1989. *glp-1* and *lin-12*, genes implicated in distinct cell-cell interactions in *C. elegans*, encode similar transmembrane proteins. Cell 58: 553 – 563.

Yochem, J., S. Tuck, I. Greenwald, and M. Han. 1999. A gp330/megalin-related protein is required in the major epidermis of *Caenorhabditis elegans* for completion of molting. Development 126: 597 – 606.

Yoon, C. H., J. Lee, G. D. Jongeward, and P. W. Sternberg. 1995. Similarity of *sli-1*, a regulator of vulval development in *Caenorhabditis elegans*, to the mammalian proto-oncogene, *c-cbl*. Science 269: 1102 – 1105.

Yuan, J., and H. R. Horvitz. 1990. The *Caenorhabditis elegans* genes *ced-3* and *ced-4* act cell autonomously to cause programmed cell death. Developmental Biology 138: 33 – 41.

Yuan, J., and H. R. Horvitz. 1992. The *Caenorhabditis elegans* cell death gene *ced-4* encodes a novel protein and is expressed during the period of extensive programmed cell death. Development 116: 309 – 320.

Yuan, J., S. Shaham, S. Ledoux, H. M. Ellis, and H. R. Horvitz. 1993. The *C. elegans* cell death gene *ced-3* encodes a protein similar to mammalian interleukin-1b converting enzyme. Cell 75: 641 – 652.

Zallen, J. A., B. A. Yi, and C. I. Bargmann. 1998. The conserved immunoglobulin superfamily member SAX-3/Robo directs multiple aspects of axon guidance in *C. elegans*. Cell 92: 217 – 227.

Zarkower, D., and J. Hodgkin. 1992. Molecular analysis of the *C. elegans* sex-determining gene *tra-1*: A gene encoding two zinc finger proteins. Cell 70: 237 – 249.

Zhang, B., M. Gallegos, A. Puoti, E. Durkin, S. Fields, J. Kimble, and M. P. Wicken. 1997. A conserved RNA-binding protein that regulates sexual fates in the *C. elegans* hermaphrodite germ line. Nature 390: 477 – 484.

Zhang, Y., and S. W. Emmons. 1995. Specification of sense-organ identity by a *Caenorhabditis elegans* Pax-6 homologue. Nature 377: 55 – 59.

Zhao, C., and S. W. Emmons. 1995. A transcription factor controlling development of peripheral sense organs in *C. elegans*. Nature 373: 74 – 78.

Zhen, M., and Y. Jin. 1999. The liprin protein SYD-2 regulates the differentiation of presynaptic termini in *C. elegans*. Nature 401: 371 – 375.

Zipkin, I. D., R. M. Kindt, and C. J. Kenyon. 1997. Role of a new Rho family member in cell migration and axon guidance in *C. elegans*. Cell 90: 883 – 894.

zur Strassen, O. 1951. Der Erbgang der Nematoden-Asymmetrie. Verhandlungen – Deutsche Zoologische Gesellschaft 77 – 81.

zur Strassen, O. 1959. Neue Beiträge zur Entwicklungsmechanik der Nematoden. Zoologica 107: 1 – 142.

Zwaal, R. R., J. E. Mendel, P. W. Sternberg, and R. H. A. Plasterk. 1997. Two neuronal G proteins are involved in chemosensation of the *Caenorhabditis elegans* dauer inducing pheromone. Genetics 145: 717 – 727.

4 Nematode Morphology, Sensory Structure and Function

James G. Baldwin and Roland N. Perry

4.1 Introduction

What is the role of morphology in 21st century nematology? Classically, it is often the basis for predicting function, and typically, for most taxa, morphology is the initial point for formulating testable hypotheses of feeding biology, ecosystem roles, pathogenicity, and evolution. The biodiversity of Nematoda is first discerned from comparative morphology, and the taxonomic underpinnings of the phylum are first rooted in comparative morphology.

Throughout biology, morphology is gaining new power from integration with new technologies, and this is particularly so for nematology. In many cases, these involve familiar instruments of differential interference contrast (DIC) light microscopy, scanning electron microscopy (SEM) and transmission electron microscopy (TEM). The change is in the marrying of these instruments to powerful computers for 4D microscopy (recording spatial dimensions over time) (Dolinski, et al., 1998), confocal microscopy, telemicroscopy (i. e., Web access and control of remote microscopes), and TEM tomography (Perkins, et al., 1997), as well as in new applications in immunocytochemistry, labeling and morphology of protein complexes, morphological analysis of transgenics, pathophysiological consequences of altered gene expression, symbiotic and pathogenic relationships, and comparative character homologies (e. g., Curtis, et al., 1996; Fire, et al., 1998; Lopez de Mendoza, et al., 1999; Zhang and Baldwin, 1999, 2000, 2001). New technologies will advance our understanding of comparative morphology of nematodes, and this new understanding will help the underlying genetic, ontogenetic, and phylogenetic basis of nematode structural diversity.

Caenorhabditis elegans has proven to be a useful model for its closest evolutionary relatives; the exciting morphological, developmental, genetic, and functional insights that have been gained into this organism are being extended to a wide range of taxa including nematode species of agricultural and medical significance. New character sets, including those from molecular biology, provide an independent source against which to compare, test, and refine our classical morphology-based phylogenetic taxonomy, thereby providing a more meaningful evolutionary framework and extending the utility of the *C. elegans* model (Blaxter, et al., 1998).

The starting point for our discussion of nematode morphology, emphasizing sensory systems and function, is *C. elegans*. Structurally, this microscopic bacterial-feeding worm appears deceptively simple; like all nematodes, it includes body wall, digestive, reproductive, secretory/excretory, and nervous systems, while lacking systems recognizable as circulatory, respiratory, and endocrine (Fig. 4.1).

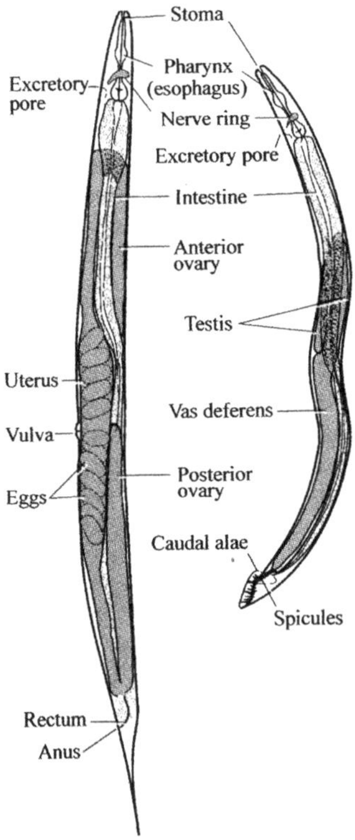

Figure 4.1 Diagrammatic representation of a hermaphrodite and male of *Caenorhabditis elegans* (redrawn from Sulston and Horvitz, 1997).

4.2 Body Wall

The body wall of nematodes is conserved in basic pattern as hypodermis (=epidermis) sandwiched between cuticle to the exterior and somatic muscles to the inside (Fig. 4.2(E)). This pattern is undoubtedly associated with the basic nematode-defining characteristics of elongate shape and body-form integrity sustained by internal hydrostatic pressure. The radially inflexible outer layers of the cuticle, together with the tone of somatic muscles, provide the resistance for this pressure (Cox, et al., 1981a). The tone of the somatic muscles is regulated in part by the hypodermis, and all the layers of the body wall thus contribute to the hydrostatic skeleton (Bird and Bird, 1991; Chalfie and White, 1988). The importance of this internal pressure means that most nematodes, when punctured, cannot move, and generally collapse or contract.

4.2.1 Hypodermis

The external epithelium, or hypodermis, is the outermost "living" layer of the nematode and the progenitor of the extracellular secretion, which is the cuticle or exoskeleton. In gross structure it consists of a thin plasmalemma-bound layer interrupted by thick cords that protrude into the body cavity and run nearly the full length of the worm (Fig. 4.2(E)). Although most nematodes have four cords (dorsal, ventral, and two lateral), exceptions include mermithids with eight cords. The cords comprise specialized seam cells/syncytia that play a critical role in postembryonic development and particularly in development of the longitudinal cuticular alae (White, 1988). Particularly between molts, most organelles, including nuclei, lipid globules, glycogen granules, Golgi bodies, and mitochondria, are concentrated in the cords; the cords also are notable for their close association with longitudinal nerve process bundles, and, where present, the lateral canals of the excretory system (Figs 4.12(A) – (C), 13 (A)). The thin sheet-like intercordal hypodermis is characterized by regularly spaced clumps of cytoplasmic tonofilaments that anchor somatic muscles to the body wall cuticle. The anchor points for these tonofilaments are on hemidesmosomes adjacent to the basal lamina that underlies the somatic muscle and the body wall cuticle (Baldwin and Hirschmann, 1975) (Fig. 4.13(C)).

The cytology of the hypodermis varies among taxa and developmental stages and is most clearly understood in *C. elegans* (Rhabditinae) in which the hypodermis is primarily a large syncytium; however, smaller syncytia and individual cells occur, for example, at extremities and in the region of the

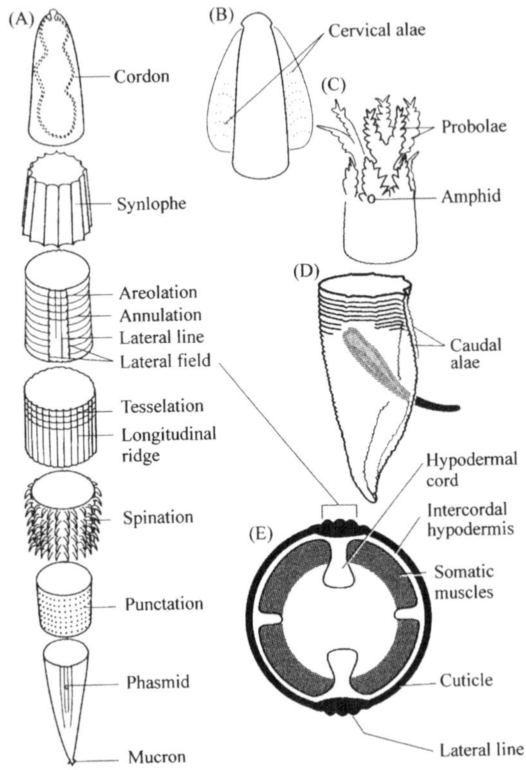

Figure 4. 2 Cuticular structures of nematodes. (A) Variations and terminology of surface patterns (redrawn from Freckman and Baldwin, 1990). (B) The anterior end of a nematode showing lateral cervical alae. (C) The lip region of *Acrobeles* sp. showing probolae. (D) The posterior end of a male including caudal alae. (E) A cross-section of a nematode showing the layers of the body wall.

vulva (White, 1988). In the Monhysterida, in several other orders of non-parasitic Adenophorea, and particularly in the juvenile stages of certain animal parasites, the hypodermis apparently is entirely cellular (Bird and Bird, 1991; Malakhov, 1994). The structure is known in so few taxa that it is unclear if in adults syncytial versus cellular structure is simply a function of overall size of the nematode. Alternately, syncytial versus cellular structure may be congruent with major evolutionary categories (Malakhov, 1994).

The hypodermis is established early in embryonic development and to a large extent regulates the overall architecture of the nematode. Specific models of regulation include vulval formation, which often is taxon-specific throughout the Secernentea that have been examined (Eizinger and Sommer, 1997; Sommer, et al., 1998). Similarly, diversity of male tail formation, including positions of sensory organs, is regulated by positions and adhesions among hypodermal cells (Fitch and Emmons, 1995). Other roles attributed to the hypodermis include exchange of water and solutes and storage of lipid and glycogen. An unusual adaptation occurs in certain insect parasites; for example in the parasitic stages of *Bradynema* (Tylenchida) and *Gastromermis* (Mermithida), nutrients are directly absorbed into the hypodermis and this absorption is facilitated by a surface of hypodermal microvilli and loss of cuticle (Batson, 1979; Riding, 1970, Subbotin, et al., 1994). Perhaps the most critical role of the hypodermis is in cuticle formation, not only in embryogenesis but also during its dissolution and regeneration at each molt. It is often the plasmalemma of the most peripheral part of the hypodermis that serves as a template for cuticle formation, including the folds that determine alae and annulation (Cox, et al., 1981a, 1981b; Singh and Sulston, 1978).

4.2.2 Cuticle

Although the body wall cuticle is a "non-living" layer secreted by the hypodermis, it is remarkably adaptable and complex in composition and function. The cuticle typically covers the entire nematode surface and a thinner extension of the cuticle lines each body opening, including the stoma, pharynx, rectum, vulva, and cloaca, as well as ducts of the excretory system and certain sensory organs. Between molts, the cuticle retains elasticity and the potential to grow; for example, after the final molt *Ascaris* may increase in size 30-fold. In addition to providing the primary antagonistic exoskeletal structure against which somatic muscles act, the cuticle has important physiological and biochemical roles, including protection from the environment, selective permeability, nutrition, and excretion (Blaxter and Robertson, 1998). Kramer (1997) reviewed the extensive literature on the structure, function, genetics, and molecular biology of the *C. elegans* cuticle. The biochemical components of the cuticle were divided by Blaxter and Robertson (1998) into three groups: (1) the structural components; (2) the soluble proteins, which may have enzymatic functions; and (3) low molecular weight components, including lipids.

4.2.2.1 Chemical Composition

The nematode body wall cuticle lacks the polysaccharide, chitin. Collagen is the principal structural component and in *C. elegans*, for example, the genome contains 50 – 150 collagen genes, most of which are believed to encode cuticle components (Johnstone, 1994). Collagen fibers are inextensible and resist longitudinal strain. The polypeptide chains are folded back on themselves to form three-stranded hydrogen-bonded superhelices, which are held together by sulphydryl bridges. Johnson and Barry (1996) defined the temporal expression patterns of six genes encoding different cuticle collagens of *C. elegans*. The abundance of mRNA for each of the six genes was found to oscillate, with each gene having a peak of expression at each of the juvenile stages. However, at each molting cycle the same ordered temporal sequence of expression occurred with the different genes being expressed at different times, resulting in differences in the timing of the peak abundance of each mRNA. Other workers have found that some genes are active only at certain life-cycle stages (Cox and Hirsh, 1985), which is likely to result in stage-specific cuticle characteristics. The other structural proteins present in the cuticle are termed cuticulins, which have a biased amino acid composition (rich in prolines and glycines, like collagens) with di-, tri- and isotri-tyrosine linkages (Fujimoto and Kanaya, 1973). The hypodermis synthesizes and secretes the collagens and cuticulins. Cuticle components are extensively glycosylated (Kiefer, et al., 1989; Rudin, 1990) but the molecular nature and protein association of these glycans is not yet known.

The soluble components of the cuticle include proteins, glycoproteins that are not crosslinked to the matrix, lipids, and glycolipids. The proteins and glycoproteins are likely to include enzymes with, for example, a role in molting. However, little is known about these enzymes. Much more research has been done on the proteins and glycoproteins associated with the outer cortical zone (= epicuticle) and surface coat (see below for structural description). In free-living nematodes, mucin-like proteins may help protect the nematodes from pathogen invasion (Davies, et al., 1988); in animal-parasitic nematodes, they have a role in resistance to the host immune response (Blaxter and Robertson, 1998). Proteins may also play a similar protective role in plant-parasitic nematodes. For example, thioredoxin peroxidase is present on the surface of second-stage juveniles (J2) of *Globodera rostochiensis* and specifically breaks down hydrogen peroxide (Jones, et al., 2000). Waetzig et al. (1999) have demonstrated that hydrogen peroxide is produced by plants as part of the defense response against nematode attack. The source of the surface coat may be the secretory-excretory systems or pharynx in secernentean nematodes and the epidermal gland cells, cuticular pores or caudal glands in adenophorean nematodes (Bird and Bird, 1991, 1998). Variations in the complement of surface proteins may reflect differences in the requirements of the phases of the life cycle during, for example, the transition from a free-living to a parasitic phase.

The outer cortical layer or epicuticle, described below in relation to other layers, is made up of lipids. Information on these lipids is sparse, although it is likely that they have an important role in the resistance of host attack. For example, the surface lipid of *Brugia malayi* includes labile, possibly antioxidant, components (Smith, et al., 1996). Molted cuticles, retained as protective sheaths, and resistant cuticles aid protection from environmental extremes, such as desiccation (Perry, 1999). An important function in this respect is to reduce the rate of water loss, and lipids are thought to be involved in this control of cuticular permeability. The surface of the sheath of a desiccation-tolerant mutant line of the entomopathogenic nematode *Heterorhabditis megidis* is more negatively charged than that of the wild-type; removal of the outer layer, possibly the outer cortical layer, resulted in loss of the mutant phenotype (O'Leary, et al., 1998). A strongly negative charge on the outer cortical layer has been found in juveniles of *Strongyloides ratti* and related to desiccation tolerance (Murrell, et al., 1983). O'Leary et al. (1998) suggested that the presence of a strongly ionized or polar coat on the surface of nematodes could facilitate the maintenance of a film of water over the cuticle.

4.2.2.2 Cuticle Layers

Although the structure of nematode cuticle is highly variable among taxa and stages, a basic pattern that is widely expressed throughout the phylum is a cortical, medial, and basal layer (Fig. 4.3(A)). These layers were first recognized by light microscopy of large nematodes but have been redefined and characterized for a larger range of nematodes with TEM (Anya, 1965; Bird and Bird, 1991; Inglis 1964a, 1964b). The cortex usually includes a thin, electron-dense outer zone and a homogeneous thicker electron-lucent inner zone. The outer zone somewhat resembles a membrane that often can be further resolved into three or more fine films, and it typically has a surface coat of carbohydrates sometimes referred to as a glycocalyx (Figs 4.4(A), 4.15 (B)). In Heteroderinae a specialized surface coat, referred to as a subcrystalline layer, forms on certain taxa, but its origin is controversial (Brown, et al., 1971; Zunke, 1986). The cortical layer is the underlying basis for specialized surface profiles (see below), such as furrows, that develop from templates of actin filaments in the hypodermis (Costa et al., 1997).

Beneath the cortical layer is a medial layer of intermediate density, often with periodic electron-dense globules or columns, and usually highly variable

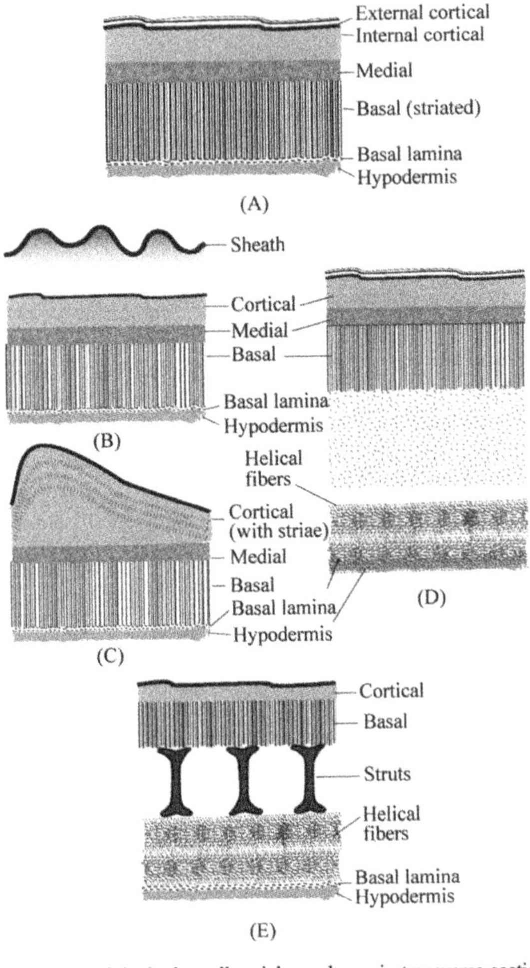

Figure 4.3 Layering of the body wall cuticle as shown in transverse section. (A) Basic cuticle pattern of layers. (B) Modification of the basic layering by addition of layers external to the cortex. (C) Modification of the basic layer by thickening and subdivision of the cortical layer into several zones. (D) Modification of the basic pattern by addition of layers internal to the basal striated layer. (E) Modification of the basic pattern by addition of struts and additional layers internal to the basal striated layer.

in thickness, even within different regions of an individual (Figs 4.3, 4.4 (A)). The basal layer is typically striated and appears to be a crystalline form of collagen. The basal layer is separated from the underlying hypodermis by basal lamina.

In a given taxon, the basic cortex, medial and basal cuticle pattern may be conserved in certain juvenile stages with more complex layering in other juvenile and adult stages. In addition to structure, the cuticle of juveniles, dauer or other specialized stages may differ from adults and with aging, in permeability, biochemical makeup, and surface carbohydrates (Cordero and Baldwin, 1990; Cox, et al., 1981a, 1981b; Searcy, et al., 1976; Zuckerman and Kahane, 1983). Often variations can be characterized as: (1) additional zones or sublayers within one or more of the basic layers; (2) additional layers external to the cortex; and (3) additional layers internal to the basal layer (Fig. 4.3). Examples of taxa with variation of the basic pattern include members of the Criconematoidea (Tylenchida), which have a typical medial and basal layer but the cortical layer is modified by additional zones, and they thicken to provide the basic structural component of heavy annuli or spines (Fig. 4.3(C)). Similarly, the highly convoluted surface of adult *Brugia* (Filarioidea) is associated with additional zones of the cortex and medial layers (Rogers, et al., 1974; Vincent, et al., 1975). Additional layers are added external to the cortex, for example, in *Hemicycliophora* (Tylenchida), in which the adult has an external sheath previously incorrectly interpreted as a persistent cuticle of the fourth-stage juvenile (Johnson, et al., 1970) (Fig. 4.3(B)). Similarly, the highly specialized cuticle of *Bunonema* includes an elaborate surface network interpreted as layers external to the cortex (Dolinski and Baldwin, unpublished; Raski and Jones, 1973). In still other taxa, including *Hoplolaimus*, Heteroderinae (Tylenchida), *Steineria* (Monhysterida), *Gastromermis* (Mermithida), and some species of Enoplida and Chromodorida, layers of at least some stages occur internal to a striated basal layer (Baldwin, 1983; Baldwin and Eddleman, 1992; Cliff and Baldwin, 1985; Shepherd, et al., 1972; Smith and Stephenson, 1970; Wright and Hope, 1968; Yushin and Malakhov, 1994b). These inner layers often include stacks of helically arranged fibers, a character shared with the sister phylum Nematomorpha that probably reflects similar structural regulation in large worms, including shortening, elongation, bending, and coiling, or relative rigidity in sedentary parasites (Lee, 1970a; Seymour, 1983; Shepherd, et al., 1972) (Figs 4.3(D), (E)). Many marine nematodes also have elaborate struts or columns internal to the basal striated layer (Wright and Hope, 1968; Yushin and Malakhov, 1994a) (Fig. 4.3(E)). Many Adenophorea appear to have layers additional to the basic pattern and next to

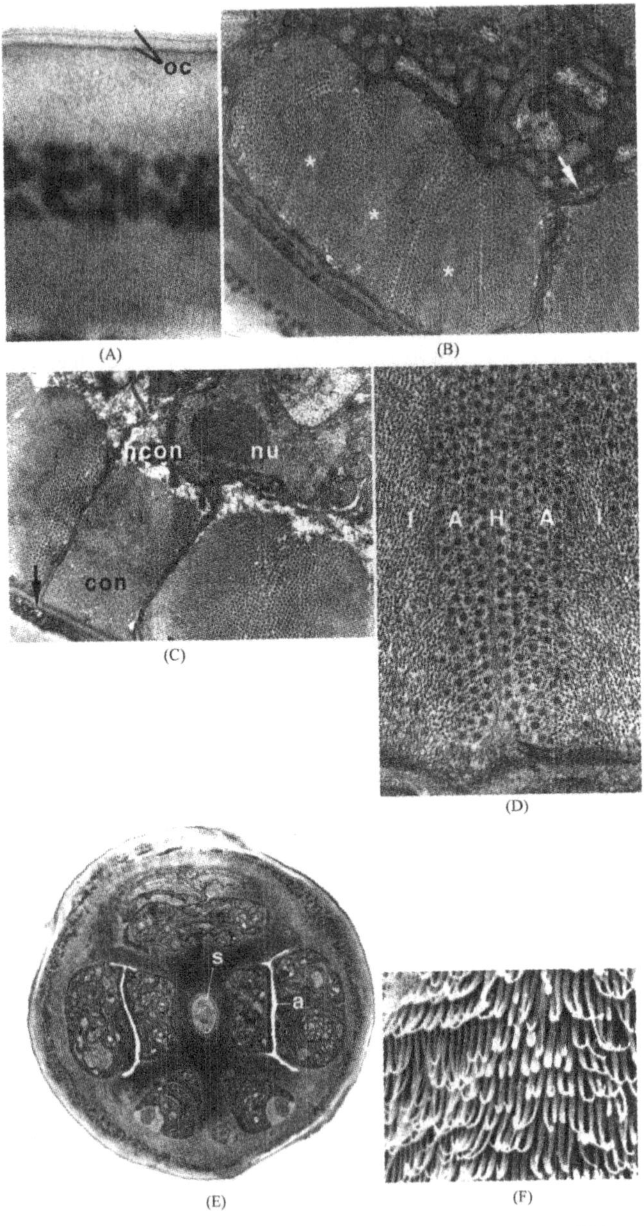

Figure 4.4 Electron micrographs of aspects of the body wall. (A) – (E) Transverse

transmission electron micrograph sections of *Meloidogyne incognita* male. (A) Basic cuticle pattern, including outer cortical layers (oc). (B) Platymyarian muscle cell including four sarcomeres delimited by three I-bands (asterisks). Arrow indicates innervation process. (C) Platymyarian muscle cell including the contractile (con) region and the noncontractile (ncon) region, with a large nucleus (nu). Arrow indicates a peripheral neuron within the intercordal hypodermis. (D) Sarcomere of a platymyarian muscle cell showing I, A, H-bands. A membrane-like (M-line) bisects the H-band. (E) The cephalic region showing the I-shaped amphid openings (a) and the stylet (s) that moves through the vestibule of the framework. (F) Scanning electron micrograph of the cuticle surface of *Medinema alii* (Diplogasterida) (courtesy Lien Luong).

the hypodermis, but combined with further modifications of the cortex or basal (i.e., striated) regions (Grootaert and Lippens, 1974; Platzer and Platzer, 1988). Some nematodes, including a range of large vertebrate parasites and certain marine nematodes, have complex cuticle structures where apparent modification from the basic pattern is so extensive that it is difficult to recognize specific layers (Kulikov, et al., 1998; Lee, 1965; Maggenti, 1979; Yushin and Malakhov, 1994a). Indeed, in many cases cuticle structure alone may be so divergent from the basic pattern, and specialized in adaptation to environment, that it is impossible to establish homologies by pattern and position alone. Instead, this variation must be investigated in a phylogenetic context, including aspects of comparative development, genetics, and molecular biology. Although the phylogenetic significance of cuticle layering has not been explored for deep phylogenetic relationships, it has proven to be of value in particular taxa such as Heteroderinae (Baldwin, 1992).

4.2.2.3 Surface Features

Diversity in surface features may have adaptive and phylogenetic significance. Transverse grooves covering most of the body, longitudinal grooves on each lateral side, and triradiate lip patterns appear to be common in the phylum and may represent a basic pattern upon which further adaptations are imposed to provide an amazing array of diversity in surface topography.

Transverse markings might be essential structures because they are associated with undulatory locomotion, accommodating deep reflexes in the moving nematode, without radial collapse of the body. The reason for diversity in size and depth of transverse grooves is not known, but it could be in part an adaptation to the coarseness of the nematodes' substrate. Certain migratory endo-plant-parasites, including *Pratylenchus*, soil microbivores such as *Caenorhabditis*, and vertebrate parasites like *Ascaris*, may have relatively fine transverse striations, whereas certain migratory ectoparasites, such as *Hoplolaimus*, may have heavy transverse annuli and others, e. g., *Criconema*, may have large modified transverse scales. Some marine taxa such as *Draconema* and *Desmoscolex* have especially heavy transverse annuli that may facilitate locomotion in the rough surfaces or wet sand in which they often live, although some species of *Desmoscolex* are also found in deep-sea ooze or mud. In some cases transverse lines such as "lip annuli" or modifications of transverse lines in the perineal region (e. g., *Meloidogyne* spp.), may be highly diagnostic for species. In other cases, such as plant-parasitic cyst nematodes (e. g., *Heterodera* spp., *Globodera* spp.), transverse annulations appear to be important adaptations for mobile juveniles and males, but they are lost in adult sedentary females where cuticle patterns may be highly modified into taxon-specific irregular ridges and grooves.

Although lateral longitudinal grooves are common in members of Nematoda, and have been referred to as "tread like", suggesting a general function in locomotion, they are diverse. The grooves, often called lateral lines or incisures, are separated by lateral ridges or alae, and the number of lines per lateral field often is species and/or age specific (Figs 4. 2(A), (E)). For example, *C. elegans* adults have four lateral lines per field, but they are absent in juveniles. Where the lateral longitudinal lines are interrupted by transverse grooves, the pattern is called areolation.

In some taxa, including certain members of Dorylaimida, Diplogasterida, Rhabditida, and Strongylida, longitudinal grooves and ridges cover the body. In a few plant-parasitic nematodes in Tylenchida, and certain members of Cephalobina, for example, longitudinal ridges encircling the body may be combined with transverse markings, and this combined "checkered" pattern is called tesselation (Fig. 4. 2(A)). Quite commonly, tesselation or areolation may be limited to the anterior or posterior end of the body.

In certain members of Strongylida, especially pronounced ridges are called synlophe, and these have considerable taxonomic importance and may function in holding the parasitic nematode to the gut wall (Commonwealth Institute of Helminthology, 1974) (Fig. 4. 2 (A)). Complex longitudinal ridges, cordons, extend about one third of the body length from the head of some species of Spirurida, and these, with further adaptations of secondary ridges and spines, may be associated with feeding (Kennedy, et al., 1973). Simpler cervical alae, or lateral projections from the anterior end, also characterize certain members of Ascarida such as *Toxocara* (Fig. 4. 2 (B)). More common, particularly in Secernentea, are caudal alae; they occur in lateral pairs on the male tail, may be highly innervated, and assist in clasping the female or hermaphrodite during copulation (Figs 4. 1, 4. 2(D), 4. 10(A),

4.11(A)).

The cuticle of nematodes throughout a wide range of diverse taxa is described from LM as having so-called punctations, typically referring to rows of refractive dots, but closer examination reveals a diversity of structures (Fig. 4.2(A)). For example, in Chromodorida, an order mostly of marine nematodes, regular rows of punctations may be canals that traverse the cuticle layers, and/or they may represent regular patterns of dense columns limited to specific cuticle layers as is the case for punctations in Cephalobina or Diplogasterida (Baldwin, unpublished; Mounport, et al., 2000; Siddiqui and Viglierchio, 1977; Wright and Hope, 1968; Yushin and Malakhov, 1994a). In fact, periodic struts or columns of the medial layer (Fig. 4.3(E)) are common, but only in certain taxa do they refract with light microscopy to suggest punctations. Punctations in the perineal pattern of the plant parasite *Meloidogyne hapla* are considered diagnostic to the species, but scanning electron microscopy reveals that they are not on the cuticle surface but occur beneath it (Sasser and Carter, 1985).

The cuticle of some nematodes, including certain members of Tylenchida (e.g., *Criconema*), Diplogastrida (e.g., *Mehdinema*), Spirurida (e.g., *Echinuria*), and oxyurids, has striking non-innervated cuticular projections or spines (Kennedy, et al., 1973; Luong, et al., 1999) (Figs 4.2(A), 4.4(F)). In some, the most elaborate cuticular projections are restricted to the head region, where the triradiate lips may be expressed as elongate projections; internal to the lips, elaborate unbranched or highly branched and fringed probolae may extend from the underlying cuticular layers associated with the stoma opening (Fig. 4.2(C)). Speculation is that these variable cuticular structures of the lip region may be associated with specific feeding habits; clearly they have arisen independently throughout divergent taxa including Cephalobina (e.g., *Acrobeles*), Rhabditina (e.g., *Diploscapter*), and Teratocephalina (e.g., *Metateratocephalus*), and somewhat similar convergent elaborations occur in the head region of certain members of Araeolaimida (e.g., *Tylocephalus, Wilsonema*) (Fig. 4.15(C)). While not as prominent as lips and probolae, the tail region also may include unusual elaborations, mucrons, of taxonomic value but unknown function (Fig. 4.2(A)). These structures may range from simple to star-shaped and multi-rayed protrusions.

4.2.3 Muscle

The internal side of the body wall cuticle is effectively attached by tonofilaments that extend from hemidesmosomes on each side of the hypodermis and are complemented by corresponding hemidesmosomes on the muscle plasmalemma (= sarcolemma) (Fig. 4.13(C)). The connection is complete to the extent that contraction of somatic muscles is fully transferred through the body wall (Francis and Waterston, 1985; Moerman and Fire, 1997). Typically, the somatic muscles are spindle-shaped longitudinally-oriented cells that intersect to form a layer separated into four fields by the hypodermal cords (Figs 4.2(E), 4.5(A), (D)). Details of organization of individual cells are highly variable among taxa, but certain features are conserved in the phylum. A given somatic muscle cell includes a contractile region with highly organized myofilaments (i.e., actin, myosin), a noncontractile region that includes more general organelles such as the cell nucleus, mitochondria, and glycogen, and one or more innervation processes that extend as part of the muscle cell to the nerve cord (Figs 4.4(A), (C), 4.5(B), (E)). Innervation processes originating from muscle (versus neuron), while unusual, also occur in certain other invertebrates, including some Gastrotrichs (Trechert, 1977).

The contractile region of somatic muscles is composed of abundant proteins assembled into a highly specific lattice consisting of bundles of thick (myosin) filaments flanked and interdigitated with bundles of thin (actin) filaments. Interestingly, both the actin and myosin filaments of somatic muscles are much longer than the filaments of vertebrates (Moerman and Fire, 1997), and they vary further in structure among nematode taxa, stages of nematode development, and between somatic muscles and muscles associated with organs (e.g., the pharynx). As in other organisms, muscle contraction is associated with ATP-dependent movement of myosin across actin and thus increased overlap of thin and thick filaments. The basic unit of contraction and pattern of filament arrangement is the sarcomere, which is a sequence of IAHAI patterns where I is a bundle of thin filaments, A is a lattice of mixed thin and thick, and H is thick filaments (Figs 4.4(B) – (D), 4.5). The I-band may be divided by an additional electron-dense zone, and a membrane-like M-line bisects the H-band; both the dense zone and M-line are part of the system that integrates the contractile system with the overlying exoskeleton (Fig. 4.4 (D)). In transverse section, the sarcomeres including IAHAI bands are readily recognizable, and they can also be viewed in radial section where the IAHAI pattern is offset at a 6% oblique angle relative to the longitudinal axis of the nematode (Figs 4.4(B) – (D), 4.5). This oblique striation, which occurs in many phyla, is critical to the smooth bending of the nematode body (Burr and Gans, 1998). A given muscle cell, depending on its diameter, may have only a few, or it may have dozens of sarcomeres (Figs 4.4(B), (C), 4.5(B), (D)).

The arrangement of the contractile region and noncontractile region of somatic muscles varies among nematode taxa and stages of development, and

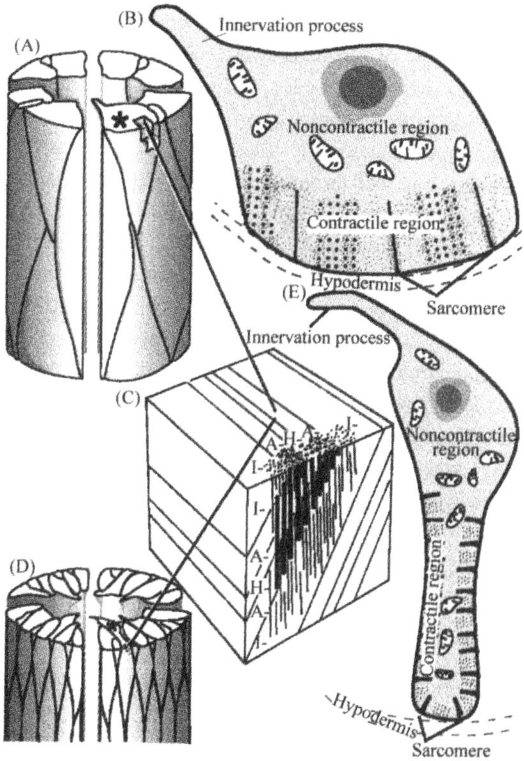

Figure 4.5 Diagrammatic representation of somatic musculature in nematodes. (A) A thick cross section of a nematode, in which only the somatic musculature remains. The four quadrants of musculature are each comprised of spindle-shaped platymyarian muscle cells. Asterisk indicates the cell enlarged in (B). (B) Transverse section of platymyarian muscle cell from (A). Cell consists of contractile region, noncontractile region and innervation process. Contractile region is comprised of sarcomeres. (C) Three-dimensional representation of a sarcomere showing I, A, H bands in transverse section as well as their oblique striation in longitudinal section. (D) A thick cross-section of a nematode in which only the somatic musculature remains. The four quadrants of musculature are each comprised of spindle-shaped coelomyarian muscle cells. Asterisk indicates the cell enlarged in (E). (E) Transverse section of coelomyarian muscle cell from (D). Cell consists of a contractile region, noncontractile region, and innervation process.

may, to some extent, be a function of nematode size. The simplest pattern, platymyarian, occurs in smaller nematodes, and in this case the contractile region consists of a row of sarcomeres that closely follows the contour of the adjacent hypodermis (Fig. 4. 5 (B)). In contrast, larger nematodes have coelomyarian muscle cells in which the single layer of many sarcomeres, as viewed in transverse section, is folded, with noncontractile cytoplasm sandwiched between, as well as ballooning into, the body cavity (Fig. 4. 5 (E)). In a coelomyarian muscle cell, only a few sarcomeres are adjacent to the hypodermis, and the remainder are folded against adjacent muscle cells. We suggest that there may be a functional size limit to sarcomeres considering their need for energy and metabolites from the noncontractile part of the cytoplasm. Whereas the simple platymyarian system is suitable for small nematodes, coelomyarian muscles exhibit a remarkable approach to increasing contact with cytoplasmic organelles for a maximum amount of contractile power in many sarcomeres. For example, J2s of *Ascaris* have 83 platymyarian somatic muscles, whereas adults may have 50,000 coelomyarian muscle cells. Nematodes of intermediate size typically have cells intermediate in structure, sometimes referred to as shallow coelomyarian (Hirumi, et al., 1971). Other rare variations include circomyarian somatic muscles of certain large aquatic nematodes, where the contractile region completely encloses the noncontractile region (Maggenti, 1981).

Although it is not part of the body wall, the nematode body cavity is lined externally by the somatic muscles and internally by the digestive system. The considerable evolutionary significance given to the embryonic development and incomplete mesodermal lining of this cavity is reflected in the name "pseudocoelom", contrasting to the coelom, for example, of vertebrates (Barnes, 1987; Hyman, 1951). Although the name pseudocoloem persists, there is reduced emphasis on the distinctiveness of this cavity, in the light of new understanding of the developmental ambiguity of basic embryonic tissues, including mesoderm (Bird and Bird, 1991), together with new insight into the artificial nature of the "Pseudocoelomates" as a taxonomic unit of invertebrates (Aguinaldo, et al., 1997; Winnepenninckx, et al., 1995). In fact, a distinct body cavity is most recognizable in large parasites; in some nematodes, including most plant-parasitic members of Tylenchida, there is typically no fluid-filled space between the body wall and the digestive tract (Baldwin and Hirschmann, 1975). Coelomocytes (= pseudocoelomocytes) are large, often branched, irregularly shaped cells that occur in the body cavity of most nematodes, but their number varies according to species (Bird and Bird, 1991; Chitwood and Chitwood, 1950). In *C. elegans*, for example, there are six such cells in the hermaphrodite and five in the male. The function of these cells is unknown; laser ablation in *C. elegans* indicates that they are not essential for survival, but other observations, including their remarkable affinity for stains, indicate they might function as scavengers (Chitwood and Chitwood, 1950; White, 1988).

4.3 Digestive System

Diversity of the nematode digestive system or alimentary canal reflects feeding adaptations so that its structure is a plausible predictor of specific feeding behavior. Fundamentally, the digestive system is a conduit through the body's longitudinal axis, and both ends of the conduit are an invagination of the body wall cuticle (Figs 4.1, 4.6(A)). The anterior cuticle-lined region is the stomodeum (= foregut), which includes the stoma (= buccal capsule enclosing the buccal cavity) and its transition from the lip region. The stoma opens into the pharynx (= esophagus) and the stomodeum terminates as a cuticle-lined pharyngeal-intestinal junction. Posterior to the junction, the conduit is the mesenteron (= midgut) or intestine, which is not lined by cuticle, and is the major organ of nutrient absorption and storage. The mesenteron terminates posteriorly in a cuticle-lined proctodeum (= hindgut), including the rectum and anus in females and cloaca in males (Figs 4.1, 4.6 (A), 4.9(F)).

The basic pattern of the stomodeum is triradiate or hexaradiate (Figs 4.7 (K) – (L), 4.8(A)). The region surrounding the oral opening typically includes three or six lips and is surrounded by a hexaradiate pattern of sensory structures (De Coninck, 1942) (Figs 4.8(F), 4.16(A)). Most of the stomodeum lumen has two subdorsal and one ventral radii and most of the cells surrounding the lumen are repeated in multiples of three (Figs 4.7(K), (L), 4.8(A)). This basic pattern of radial symmetry, however, may be modified: as in Spirurida with two lips (pseudolabia), or certain members of Diplogasterida or Strongylida, Mononchida, Enoplida and others that may have asymmetrical cuticular specializations including teeth (Figs 4.6(D), 4.8(E) – (H)). Unlike the stomodeum, the mesenteron or proctodeum is generally asymmetrical.

4.3.1 Stoma

The stoma is the cavity transition between the oral opening and the narrower lumen of the pharynx (Fig. 4.6). Anteriorly, the stoma is lined by an extension of the body wall cuticle, and this region, highly conserved the phylum,

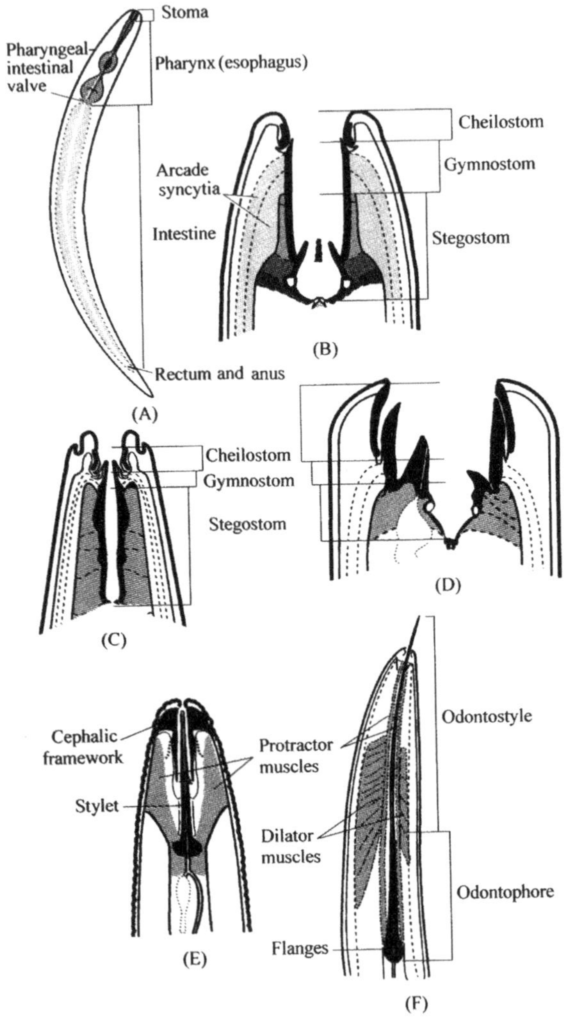

Figure 4.6 The stoma. (A) The stoma in relationship to the pharynx and intestine. (B) *Caenorhabditis elegans* (Rhabditina) (dorsal). (C) *Zeldia punctata* (Cephalobina) (dorsal). (D) *Aduncospiculum halicti* (Diplogasterida) (left side dorsal, right side subventral). (E) *Heterodera glycines* male (Tylenchida) (lateral). (F) *Xiphinema* sp. (lateral). (B – D redrawn from Baldwin, et al., 1997b; F redrawn from Coomans, 1963.)

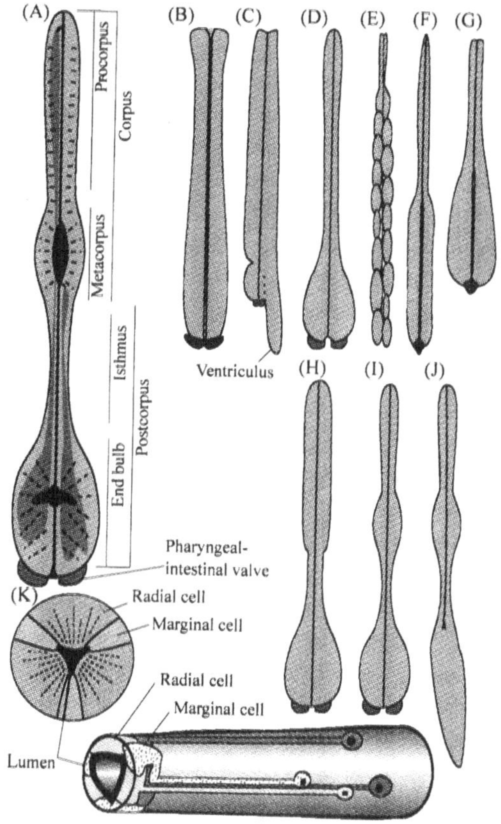

Figure 4.7 The pharynx. (A) General representation showing the regions of a three-part pharynx. (B) One-part pharynx. (C) Two-part pharynx with ventriculus. (D) Two-part pharynx with basal bulb. (E) Two-part pharynx with posterior part specialized as a stichostome. (F) Two-part pharynx with posterior part cylindrical. (G) Two-part pyriform-shaped pharynx. (H) Three-part pharynx lacking a metacorpus. (I) Three-part pharynx with a metacorpus and basal bulb. (J) Three-part pharynx with a basal glandular lobe. (K) Transverse view of the triradiate pharynx showing the relative position of marginal cells in relation to radial cells (redrawn after Albertson and Thomson, 1976). (L) A three-dimensional view of the marginal and radial cells in the stegostom of the stoma showing their posterior elongate processes and positional relationship to the cell nuclei.

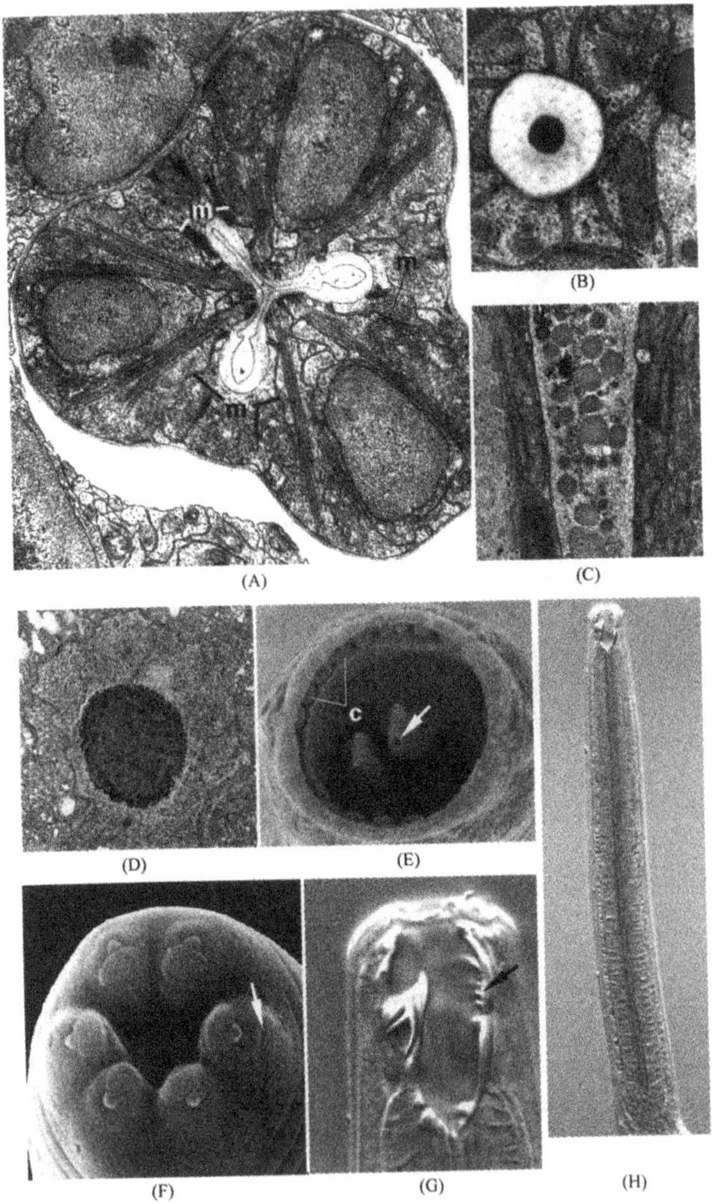

Figure 4.8 Light (LM), scanning electron (SEM), and transmission electron (TEM)

microscopy of the stomodeum. (A) TEM transverse section of the pharynx of *Zeldia punctata* (Cephalobina). Surrounding the triradiate lumen, the boundaries of three marginal cells (m) are delimited by distinct membranes. Between the marginal cells are the three muscular radial cells which each include a large nucleus. (B) TEM transverse section of the round lumen of the nonmuscular procorpus of *Meloidogyne incognita* (Tylenchida) male. (C) TEM longitudinal section of the procorpus showing the dorsal gland process (*M. incognita* male). (D) TEM of the dorsal gland nucleus (*N. aberrans*). (E) SEM en face view of the oral opening of *Diplenteron* sp. (Diplogasterida) showing the dorsal gland orifice (arrow) on the dorsal tooth. The stoma opening is surrounded by a large number of cheilorhabions (c). (F) SEM en face view of *Caenorhabditis elegans* showing the hexaradiate pattern of lips, each with a distinct papillum. Arrow indicates an amphid opening. (G) LM lateral view of a mononchid showing a large dorsal tooth and ventrally rows of denticles (arrow). (H) LM lateral view of a mononchid showing the elongate muscular cylindrical pharynx. (D is by courtesy of R. Souza.)

is the cheilostom (Figs 4.6(B) – (D)). The basic pattern of cuticular thickenings lining the cheilostom is three; however, in Tylenchida, for example, they are expressed as a hexaradiate cuticular cephalic framework that underlies the head region, and in most Diplogasterida, and apparently Strongylida, the cheilostom lining may be repeated to more than 30 units (i.e., cheilorhabdions) (Figs 4.6, 4.8(E)).

Posteriorly from the cheilostom, the thin cuticle lining of the stoma is disjunct from the body wall cuticle and lined by a stacked pair of arcade syncytia. This region, while highly variable in form, appears to be common in nematodes examined with TEM (Baldwin and Eddleman, 1995; De Ley, et al., 1995; Endo and Nickle, 1994, 1995; Nicholas and Stewart, 1997; Stewart and Nicholas, 1994; Van de Velde and Coomans, 1991; Van de Velde, et al., 1994), and is aptly named gymnostom. Further posteriorly, the stoma of most nematodes is enclosed by specialized marginal and radial anterior cells of the pharynx; this region is the stegostom (Figs 4.6(B) – (D)). The basic pattern in the stegostom, and in the pharynx in general, includes a triradiate lumen with an elongate marginal cell at the apex of each radius, and with a stack of two or more sets of radial cells between each of the apices (Figs 4.7(K), (L), 4.8(A)). Additional regions including prostom, mesostom, metastom, and telostom, respectively from anterior to posterior, have been described for subregions of the stegostom defined by thickenings of the cuticle lining. However, fine structure and comparative embryonic cell lineages have falsified a previous assumption that the position of these thickenings is a reliable indicator of homology; therefore, the consistent application of these terms, particularly outside Rhabditina, is usually implausible (Baldwin, et al., 1997b; De Ley, et al., 1995; Dolinski, et al., 1998). Use of prostom in particular has been confusing because, perhaps due to limitations of light microscopy, it has been applied to regions both in the gymnostom and stegostom (De Ley, et al., 1995; Steiner, 1933). Maggenti (1981), who also noted that in many nematodes the posterior portion of the stoma is closely linked to the pharynx, proposed esophastom for this region. Considering that in some nematodes the stoma is almost exclusively associated with the body wall whereas in others this region is more closely linked to the pharynx, Wright (1976) proposed stomatous and astomatous, respectively, to describe such nematodes. Many of the taxa considered by Wright (1976) to be astomatous appear to have a greatly reduced gymnostom.

Previous assumptions of homology of stoma structures and evolution of the stoma (= stomatostylet) in Secernentea (Andrássy, 1976) from the simple cylindrical stoma of Cephalobina and Rhabditina through more complex stomas of Diplogasterida and finally the stylet of Tylenchida, are now considered doubtful (Fig. 4.6). This falsification is in part due to incongruence with molecular phylogenies and new morphological character analyses in representatives of Cephalobina, Rhabditina, Diplogasterida (Baldwin and Eddleman, 1995; Baldwin, et al., 1997b; De Ley, et al., 1993, 1995; Endo and Nickle, 1994, 1995; Van de Velde, et al., 1994; Wright and Thomson, 1981). Homologies of stoma characters, however, remain only partly resolved for the complex stoma of Tylenchida, adapted for plant parasitism. Nevertheless, from TEM evidence it is clear that the tylenchid stylet is a modified stoma, or stomatostylet, that the stylet moves through a chamber lined by cheilostom (specifically, the cephalic framework) and gymnostom, and that the stomatostylet is an anterior extension of the stegostom lining, and the three primary and three secondary protractor cells are homologs of stegostom radial muscle cells (Baldwin and Hirschmann, 1976; Byers and Anderson, 1972; Chen and Wen, 1972, 1980; Endo, 1984, 1985) (Fig. 4.4(E)).

The fine structure of the stoma of Secernentean animal parasites, often with highly specialized feeding adaptations, is known for a number of taxa. Although cheilostom, gymnostom, and stegostom have not been directly described in these taxa, homologs of these basic regions are usually recognizable in micrographs (Dick and Wright, 1973; Talluri, et al., 1986; Weise, 1977; Wright, 1976). The stoma of few Adenophorean animal parasites have been examined with TEM; however, in Adenophorea, the cheilostom, gymnostom, and stomatostom are clearly resolved in Monhysterida (Van de Velde and Coomans, 1991) and appear to be present from micrographs of Enoplida (Hope, 1988) and Trichocephalida (Wright, 1974).

Homologies of gymnostom and stegostom are more difficult to resolve from the complex odontostylets of Dorylaimida and Triplonchida (Coomans and Van der Heiden, 1971; Grootaert and Coomans, 1980; Grootaert and Wyss, 1978; Hirumi, et al., 1968; Jacobs, et al., 1987; Lippens, et al., 1974; Taylor, et al., 1970; Taylor and Robertson, 1971; Wright, 1975, 1976) (Fig. 4.6 (F)). Although the components of the odontostyle, putatively derived from an axial tooth, and odontophore develop separately in postembryogenesis, both appear to be of pharyngeal origin and thereby are analogous to the stegostom. Wright et al. (1983) speculated that the protractor and other muscles of the odontostyle/odontophore in *Xiphinema* may be derived from arcade syncytia homologous with those of *C. elegans*; these arcade syncytia define the gymnostom.

In addition to stylets, regions of the cuticle lining of the stoma may be highly specific in their adaptations for feeding, including one or more teeth (onchia), rasps, and small denticles (Fig. 4.8(E), (G), (H)). This diversity is reviewed in a number of classical studies including Chitwood and Chitwood (1950), but, in general, the fine structure of such cuticular elaborations is only known for a few taxa and homologies of these structures among taxa are mainly unresolved.

4.3.2 Pharynx

Since the pharynx encloses the posterior end of the stoma (the stegostom), these two parts of the stomodeum may seem not to be demarcated clearly. However, an abrupt narrowing of the lumen of the stomodeum marks a change to that of the main (non-stoma) part of the pharynx. The pharynx, which is covered by basal lamina, may be variously modified in overall shape from a single cylinder as in Mononchida, Enoplida, and some Monhysterida, to a two-part pharynx with a narrow anterior region (corpus) and broader posterior cylinder or bulb (postcorpus) as in Dorylaimida, Triplonchida, Tricocephalida, Mermithida, Chromodorida, and Araeolaimida (Figs 4.7(B) – (G), 4.8 (H)). In general, Secernentea, including Rhabditida, Diplogasterida, and Tylenchida, are characterized by a pharynx with a distinct corpus and by division of the postcorpus into a narrow isthmus and a basal bulb; this results in a three-part pharynx (Fig. 4.7(H) – (J)). In addition to these three parts, the corpus also may be subdivided to include anteriorly a narrow nonmuscular region (procorpus) and posteriorly a broader muscular pump (metacorpus). Some Secernentea, including certain animal parasites of Strongylida and some members of Ascarida, undergo a postembryonic change from a three-part pharynx to a more nearly cylindrical muscular pharynx in adults, and in some plant parasites (e.g., Tylenchida) the basal bulb may be specialized by glandular lobes that overlap the intestine.

The primary components of the pharynx are cuticle as well as muscle, epithelial, gland, and nerve cells; diverse feeding strategies are reflected in part by the relative proportions of these components, and specialization of pharyngeal regions to break, grind, condense, move, or digest food. However, feeding for some nematodes emphasizes mechanical activity, and a highly muscular pharynx throughout reflects this, as is the case for Mononchida (Fig. 4. 8 (H)). Others, such as Tylenchida, are adapted for minimal mechanical breaking of food and the most obvious muscles are limited to radial cells that operate the metacorpus pump and isthmus for ingestion and usually constrictor muscles that occur just anterior and posterior to the metacorpus that may help to regulate flow of gland contents. These constrictor muscles are combined with large highly developed pharyngeal glands for chemical digestion, or perhaps for highly specialized aspects of feeding biology that in some cases include the establishment of host nurse cells (Anderson and Byers, 1975; Baldwin, 1982; Endo, 1984; Hussey, 1989; Hussey and Grundler, 1998; Shepherd and Clark, 1983a; Shepherd, et al., 1980). However, there are some rare species in Tylenchida that lack pharyngeal muscles, and feed only passively (Shepherd and Clark, 1976a).

As in the stegostom of the stoma, the cuticle-lined lumen throughout the pharynx is typically triradiate and is dilated by radial muscle cells that occur between the ventral and subdorsal radii; the system is anchored by tonofilaments that extend through marginal epithelial cells to the periphery of the pharynx (Figs 4. 7(K), 4. 8(A)). In some cases, radial cells are epithelial rather than muscular, as is the case for the anteriormost cells of the stegostom of Rhabditina and Diplogasterida (Baldwin, et al., 1997a, 1997b; Wright and Thomson, 1981). Radial cells also are expressed as epithelial cells throughout the corpus of Dorylaimida and the procorpus of Tylenchida, and in these regions the lumen is round and cannot be dilated (Fig. 4. 8(B)). Within those Secernentea examined, the numbers and positions of marginal and radial cells seem to be surprisingly conserved; typically, at a given level the radial cells occur in pairs between radii, but often a taxon-specific change is that these pairs fuse in adults to form a binucleate cell, and in Ascarida the radial cells are described as syncytial (Reger, 1966). However, reconstructing the radial and marginal cells of the pharynx is difficult, because the cells are highly polarized, with the nucleus to a given cell often connected by a narrow cytoplasmic process to the main mass of the cell body (Fig. 4. 7 (L)). Consequently, nuclei for a large number of muscle cells may all occur at about the same region, and their number and morphology can be assessed by specific nuclear stains such as Dapi (Dolinski, et al., 1998).

Pharyngeal glands may be relatively small single cells adjacent to short ducts that open into the pharyngeal lumen, or they may be the largest cells of the stomodeum, with terminal ducts and ampullae far from the main body of the cell (Figs 4.7(A), 4.8(C), (D)). For example, in most Secernentea the dorsal gland duct opening is in the stegostom or far anterior within the pharynx; in some Diplogasterida and Strongylida the orifice of the dorsal gland duct may be on the dorsal tooth with a gland cell process typically terminating as an ampulla at the base of the tooth (Baldwin, et al., 1997b; McLaren, et al., 1974; Weise, 1977; Wilfred and Lee, 1981; Zhang and Baldwin, 1999) (Fig. 4.8(E)). The position of duct openings may define higher taxa; for example, in Mononchida five gland openings occur below the nerve ring, whereas in Enoplida two of five gland openings are above the nerve ring. The basic number of gland cells in Nematoda appears to be five, including one dorsal and four subventral, i. e., an anterior and posterior pair, but in many groups there appears to be a loss of an anterior pair of subventral glands. Rhabditina and Cephalobina have five glands (Albertson and Thomson, 1976; Zhang and Baldwin, 2000) whereas Tylenchida have three; but depending on the species, Diplogasterida (Zhang and Baldwin, 1999, 2000, 2001), Dorylaimida, Triplonchida (Decraemer and De Waele, 1981), and some of Chromodorida and Araeolaimida may have three or five (Maggenti, 1981). Often it is difficult to determine the number of gland cells without careful reconstruction. For example, in *C. elegans* and *Zeldia punctata*, TEM reveals, in addition to a pair of larger subventral glands, a small pair of anterior subventral glands not visible with light microscopy. The ducts for these glands are easily obscured within the complex cuticle of the grinder, and certain gland cells may partially fuse with other glands. The glandular region of the parasites of Trichocephalida and Mermithida are further modified by a glandular stichostome; here the glands are given the special name, stichocytes, and each has a duct opening into the pharyngeal lumen (Bruce, 1970; Wright, 1972; Wright, et al., 1985) (Fig. 4.7(E)). Maggenti (1981) interprets the stichostome as a proliferation of subventral glands. Some other taxa, including Spirurida, may have the typical three or five gland openings, but each gland is multinucleate (Chitwood and Chitwood, 1950). In Diplogasterida and Tylenchida, the glands of the basal bulb are particularly large, and homologs of the obvious radial muscles associated with the grinder of other Secernentea are not observed with light microscopy. Recent TEM, however, reveals that in addition to the large glands a full complement of four sets of radial muscle cells may be present in the basal bulb of Tylenchida (Baldwin, et al., 2001), and that in Diplogasterida only one homolog of the four sets of radial muscle cells is absent; this is the set that in other nematodes is associated with the grinder

(Zhang and Baldwin, 1999). In some other species of Tylenchida, including those with a posterior lobe, the complement of radial muscle cells otherwise associated with the bulb may instead be confined to a shorter region beginning in the isthmus and ending at the terminus of the cuticle-lined lumen; the lobes are entirely comprised of the gland cells (Baldwin, et al., 1977; Endo, 1984; Endo, et al., 1997; Souza and Baldwin, 1999). In certain ascaridoids, a highly specialized, simple or complex, sac-like glandular lobe (= ventriculus) extends posteriorly from the base of the pharynx and its junction with the intestine (Fig. 4.7(C)). Unlike the gland lobe in certain Tylenchida, the ventriculus is penetrated by an extension of pharyngeal lumen associated with diminutive radial muscle cells; however, the dorsal gland and a portion of the subventral glands are its predominant components (Buzzell and Sommerville, 1985; Race, et al., 1968). Innervation of the pharynx is closely integrated with glands and radial muscles and is discussed in conjunction with the nervous system.

The posteriormost structure of the stomodeum is the pharyngeal-intestinal junction or valve (Fig. 4.7(A)). Previously, this has been called the cardia because of apparent pumping action; however, in those cases where the structure has been examined, TEM revealed it is not muscular and its apparent dilation and contraction must be the result of changing pressure from more anterior muscular parts of the pharynx, or from rapid changes in hydrostatic pressure of the specialized cells of this region. In the presence of this movement, the cellular connections between the pharynx, pharyngeal-intestinal junction and intestine are reinforced by an abundance of desmosomes (White, 1988). Chitwood and Chitwood (1950) describe Phasmidia (= Secernentea) as having five pharyngeal-intestinal nuclei, two of which occur internally and three externally. The two pharyngeal-intestinal cells with particularly large nuclei has been confirmed with TEM of this region in a number of nematodes in Secernentea (Endo and Wergin, 1988; Seymour and Shepherd, 1974; Shepherd and Clark, 1976a, 1983a; Shepherd, et al., 1980; White, 1988), but there appears to be considerable variation among taxa. For example, in *C. elegans* the two inner cells are surrounded by six cells that, by the presence of microvilli, seem to be more closely associated with the intestine; in contrast, in a representative of Diplogasterida there are five similar large epithelial cells surrounding a triradiate lumen (Zhang and Baldwin, 1999). More typically, the pharyngeal lumen in this region is flattened, the cuticle lining is thin, and it extends between the two internal cells. The pharyngeal-intestinal junction in Adenophorea is often large, well developed, may or may not be surrounded by intestinal tissue, and is otherwise variable at a level to be of systematic value, although for many taxa, details of its structure are not known from TEM

(Clark, 1960; Mapes, 1965; Yeates, 1972).

4.3.3 Intestine

The mesenteron or intestine is the largest organ of the digestive system and begins anteriorly at the point at which the cuticle lining of the stomodeum ends and terminates at the protodeum (Figs 4.1, 4.6(A)). Relative to the diversity of the stomodeum, it is highly conserved among taxa in gross morphology (Chitwood and Chitwood, 1950). Although basically tube-like it may vary from anterior to posterior end with respect to the shape of cells and their contents. It also varies in overall shape to accommodate other organs including the adjacent reproductive system. In many taxa, a portion of the intestine wall extends anteriorly adjacent to the stomodeum, and this extension is a caecum, a structure particularly well developed in certain species Ascarida. The walls of the intestine are composed of a single layer of epithelial cells or syncytia of endodermal origin that surround a relatively flattened lumen. Some classical classification systems gave particular importance to the number of such cells in the intestine, classifying nematodes as oligocytous (<128), polycytous (128 – 8224), and myricytous (> 8224), but these are now considered as primarily a reflection of the size of the nematode, and are highly convergent, in the sense that similar numbers of cells have evolved independently.

The intestine is primarily an organ for absorption and storage, and most likely secretion of digestive enzymes. Typically, the surface adjacent to the lumen is modified for absorption with microvilli, first recognized by distinctive staining as a bacillary layer (Chitwood and Chitwood, 1950). However, in some nematodes, such as adults of *Meloidogyne*, there does not appear to be distinct lumen or microvilli, but rather the intestine is derived as a large storage organ. Contents of the intestinal wall typically include lipid globules, glycogen, and a range of materials that often have a crystalline-like refraction. While these "crystals" have previously been considered "waste products" (Chitwood and Chitwood, 1950), this clearly is not always the case since such autoflorescent granules appear early in embryogenesis in *C. elegans* and are used as markers to follow development of intestinal cells (Laufer, et al., 1980; White, 1988). An unanticipated function of the intestine, as demonstrated in *C. elegans*, is the formation of yolk proteins that are transmitted to the oocytes (Kimble and Sharrock, 1983; White, 1988).

4.3.4 Rectum

Posteriorly, the intestine terminates in an intestino-rectal valve, and the lumen is continuous with that of the cuticle-lined protodeum. This valve includes epithelial cells (usually two) similar to those of the pharyngeal-intestinal junction. The valve is followed posteriorly by the rectum, which is enclosed by endothelial cells (three pairs in *C. elegans*) covered by a flattened muscular tube (sphincter). The tube opens ventrally into the rectum and anus that is continuous with body wall cuticle (Figs 4.1, 4.6(A)).

The endothelial cells lining the rectum are closely linked with the hypodermis and most serve as blast cells in postembryonic development of the male (Sulston and Horvitz, 1977; White, 1988). In addition to the sphincter muscle, the rectum opens by a highly conserved unicellular H-shaped anal depressor muscle, and in *C. elegans* two additional intestinal muscles; the muscles are linked by gap junctions and innervated by a single neuron (Chitwood and Chitwood, 1950; White, 1988). The rectum typically is associated with glands, and in a few cases these are known to have specialized functions. For example, the enlarged sedentary females of *Meloidogyne* have six very large rectal glands that produce a gelatinous matrix into which eggs are deposited. In males, the rectum is replaced by a cloaca that is a common duct of the digestive and reproductive systems (Figs 4.1, 4.9(F)). The dorsal wall of the cloaca has pouches that accommodate cuticular structures, including spicules. Spicules and their associated musculature will be further considered in the following section (Fig. 4.10).

4.4 Reproductive System

Nematodes typically occur as males and females and usually can be readily distinguished by secondary sexual structures such as copulatory spicules in males. In some species sexual dimorphism is pronounced, as in many insect parasites of Allantonematidae (Chitwood and Chitwood, 1950; Siddiqi, 1986) or the plant parasites, e.g., *Meloidogyne*, where males remain vermiform and migratory but females are enlarged, globose, and sedentary. In other taxa, including many Criconematoidea, males, unlike females, apparently do not feed and have only a rudimentary digestive tract. Certain marine taxa, e.g., *Calyptronema* and *Symplocostoma*, have males with buccal capsule cuticularization that is small compared to the heavy cuticularization and armature of females (Platt and Warwick, 1983). Often males are considerably smaller than females and this is particularly true in some vertebrate parasites, including *Philometra* and *Trichosomoides*, where the male is so small that it lives within the uterus of the female (Chitwood and Chitwood, 1950; Hirschmann, 1960).

4.4.1 General Reproductive Tract

The female, hermaphrodite, or male genital tract, including the gonad, is a thin tube comprising sheet-like epithelial cells that enclose a column of germ cells. An individual nematode may have one or two such genital tubes (rarely multiple) (Fig. 4.9), that originate postembryonically from a genital primordium consisting of two germ cells and two epithelial cells. Typically the gonad is telegonic, with germ cells forming only at the distal end, or more rarely hologonic, with synchronous development of germ cells throughout the length. Trichurids, for example, have hologonic development of spermatids. Laser ablation from the most distal epithelial (cap) cell demonstrates that telegony is controlled by signals from the cap cell that prevent meiosis in the nearest germ cells (Figs 4.9(E), (F)). Thus, the region of meiosis of germ cells typically marks a transition between germinal and growth zones; the areas are further distinct because the former is a region of rapid germ cell division and the latter an area of gradual increase in germ cell size. Ovary, testes, or ovotestes (in the case of hermaphrodism) are terms commonly used to describe the entire genital tube, but apply more specifically to the germinal and growth zones that comprise the gonad (Hirschmann, 1960) (Figs 4.9(E), (F)). Proximal to the growth zone of the gonad, the genital tube is a specialized conduit including glands and muscles that are involved with final packaging and delivery of the gametes. This tube also is the site of fertilization in hermaphrodites and amphimictic females.

Whilst sharing the above basic structural characteristics, nematode reproductive systems are highly variable among taxa and between sexes in number, length, overall shape, gamete productivity, adaptation for particular modes of reproduction, and accessory structures. Females may be didelphic, having two gonads, or monodelphic, having one gonad, and there is broad evidence that one ovary has evolved from two in many groups of nematodes (Figs 4.9(A) – (D)). From the two germ and two cap cells of the primordia, monodelphy can develop from postembryogenic migration of both germ cells together, versus one anterior and one posterior as in didelphy. An alternative mechanism for monodelphy is apoptosis of a germ cell or cap cell that controls further development of one branch (Felix and Sternberg, 1996; Hirschmann and Triantaphyllou, 1967; Roman and Hirschmann, 1969; Sternberg and Horvitz, 1981). It is the anterior genital branch that is expressed in monodelphy, with very few exceptions discussed by Chitwood and Chitwood (1950). Often didelphic gonads are amphidelphic, i.e., one genital branch is directed anteriorly and one posteriorly (Fig. 4.9(C)). One or more anteriorly directed female gonads are called prodelphic as in *Ditylenchus* (Figs 4.9(A)).

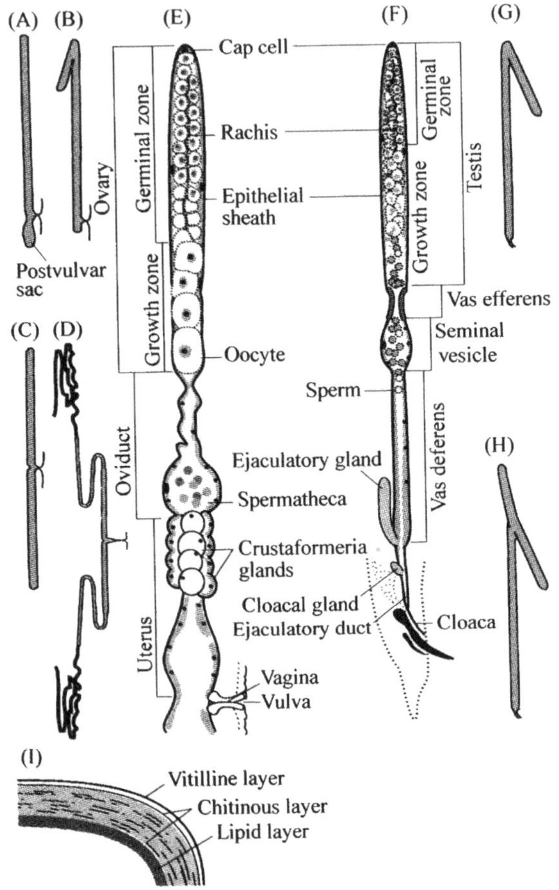

Figure 4.9 Reproductive system. (A) – (D) Basic orientation of female and hermaphrodite reproductive systems. (A) Monodelphic and prodelphic with a postvulvar sac. (B) Monodelphic, prodelphic, and reflexed. (C) Amphidelphic and outstretched. (D) Amphidelphic and coiled. (E) General representation of a female or hermaphrodite reproductive branch. (F) General representation of a male reproductive system. (G) –(H) Basic orientation of male reproductive system. (G) Monorchic. (H) Diorchic. (I) A portion of an egg showing basic layers.

The distal terminus of the gonad may be outstretched as in *Ditylenchus*, or the distal ends may be reflexed as in *C. elegans* (Figs 4.9(B), (D)). In some females, the genital tube(s) may be posteriorly directed, or opisthodelphic as in *Ascaris* and *Dirofilaria*. In some nematodes, such as *Ascaris* and *Meloidogyne*, the ovary is particularly long and coiled (Fig. 4.9(D)). The length of the ovary is congruent with egg productivity to some extent; for example, *Ascaris* may produce 200,000 eggs per day, whereas the shorter pair of ovotestes in *C. elegans* may produce only a few hundred eggs in its lifetime of a few days. Vertebrate parasites are usually adapted for production of large numbers of gametes (Chitwood and Chitwood, 1950). These large numbers help ensure adequate dispersal so that a few offspring will contact a suitable host.

Depending on the taxon, an individual male may be monorchic as in *C. elegans*, having a single testis, or diorchic, with two testes, as in *Tobrilus* (Figs 4.9(G), (H)). Testes usually are reflexed or may be coiled, but there are no special terms to describe their anterior or posterior spatial orientation. Unlike the two completely separate reproductive tracts of didelphic females, the two testes of diorchic males join at their proximal end from where they share a single tract that ultimately connects the testes posteriorly to the accessory structures and external opening (Fig. 4.9(H)). Typically, males of Secernentea have only a single testis, whereas those of Adenophorea may have one or two testes.

4.4.2 Modes of Reproduction

Morphology of the reproductive tracts of a given taxon reflects the mode of reproduction for that taxon. Amphimixis or cross fertilization is common in nematodes (Evans, 1998; Triantaphyllou, 1983). Also widely occurring is parthenogenesis where development occurs without fertilization, and hermaphrodism, or self fertilization, where sperm and egg production, and fertilization take place within a single individual. Amphimixis may be inferred where males and females are present in about the same numbers, but amphimixis is often a secondary option in taxa where parthenogenesis or hermaphrodism are predominant, and sometimes it is difficult to recognize cross fertilization with certainty. Hermaphrodism in nematodes generally is syngonic, where a single gonad (ovotestis) produces both sperm and eggs, and it is usually protandric in that sperm are produced first and stored followed by egg production. However, in some taxa, after producing sperm and then eggs, the gonad can revert to sperm production if needed. Parthenogenesis, or development of an individual without fertilization, is common in nematodes, and it may be meiotic or mitotic. Less common is sperm-dependent parthenogenesis or pseudofertilization (Beukeboom and Vrijenhoek, 1998). In meiotic parthenogenesis following meiosis, the 2N condition may be re-established by fusion with a polar body, or by simple doubling of the chromosomes, but in mitotic parthenogenesis meiosis never occurs. Whereas mitotic parthenogenesis is always obligate, as in certain species of *Meloidogyne*, meiotic parthenogenesis (as occurs in other *Meloidogyne* and some *Xiphinema* species) is associated with facultative amphimixis. Marine species of *Chromadorina* and *Viscosia* also may be parthenogenetic with facultative parthenogenesis (Hopper and Meyers, 1966). In some cases amphimixis, and thus recombination, and an increased ratio of males, are promoted by environmental stress. Similarly, taxa that are usually hermaphroditic may also have the potential to produce males, and in hermaphrodites sperm introduced from males out-competes sperm from the hermaphrodite, thus favoring recombination (Ward and Carrel, 1979). In some taxa, including plant-parasitic *Meloidogyne* and some of insect-parasitic Mermithida, sex expression may thus be determined postembryonically by environment (Triantaphyllou, 1960, 1983). Haplodiploidy, in which fertilized eggs mature into females and unfertilized eggs develop into males, has been reported in the oxyuroid *Desmicola* (Adamson, 1989). Genotypic determination of sex is most commonly by an XX/XO system, but some taxa have an XX/XY or multiple X female determination, and in some cases these genotypes may be overridden by postembryonic phenotypic determinants.

4.4.3 Females and Hermaphrodites

4.4.3.1 Ovary and Ovotestes

The epithelial covering of the ovary or ovotestes may be remarkably simple; in *C. elegans* it consists of the cap and two sheath-like cells (Fig. 4.9(E)), but in *Ascaris*, for example, it may consist of thousands of elongate multinucleate high columnar epithelial cells (Chitwood and Chitwood, 1950; Musso, 1930). This epithelial sheath encloses germ cells of the germinal and growth zone, which may occur in a single row, as is the case for some Secernentea, including *Ditylenchus*, and many free-living Adenophorea. In contrast, the germ cells may be arranged in multiple rows around a central narrow strand of cytoplasm, the rachis (Fig. 4.9(E)). In vertebrate parasites, including *Ascaris*, the growth zone surrounds thousands of oogonia and occupies most of the length of the coiled ovary. Chitwood and Chitwood (1950) remark that in some nematodes germ cell boundaries are difficult to see in the germinal zone. Although boundaries between germ cells have been demonstrated for many nematodes, in *C. elegans* the germ "cell" nuclei of the distal end are syncytial, making this model species remarkably amenable to introduction of sequences and transgenic offspring.

4.4.3.2 Oviduct

The oviduct is a narrow tube between the ovary and uterus and is lined by epithelium continuous with that of the ovary (Fig. 4.9(E)); it may or may not be distinctly set off from the growth zone of the ovary, but sometimes the transition is marked by a sphincter muscle. The oviduct varies greatly in length and specialization; indeed, in some taxa including *Chromadora* and *Theristus* it is apparently absent (Chitwood and Chitwood, 1950).

Geraert (1983) suggests that the number of cells and rows of cells comprising the oviduct may be specific and correspond with major taxa, but the number of cells in the oviduct has only been confirmed by TEM for a few species. Whereas only four thin myoepithelial cells comprise the region in *C. elegans* (excluding the spermatheca as discussed below), in *Ascaris* the walls are thick and comprised of multiple columnar epithelial cells.

Near its proximal end, the oviduct may include a spermatheca (= spermatotheca), or a broad region specialized for sperm storage, and usually the site of fertilization (Fig. 4.9(E)). In *C. elegans*, the spermatheca wall is composed of 22 sheath-like cells, whereas in *Meloidogyne javanica* it consists of 14 rounded cells (McClure and Bird, 1976). The morphology of this structure may be highly specific in shape and spatial arrangement, with utility for the definition of higher taxonomic units. For example, Cephalobina is defined in part by an offset spermatheca that typically occurs at the point at which the gonad reflexes; furthermore, certain species of *Pratylenchus* can be distinguished partly by the overall shape of the spermatheca. In some parthenogenetic species, a rudimentary and apparently nonfunctional spermatheca is retained. The Z-differentiation, a muscular organ on the reproductive tract occurring in some Dorylaimida, includes a zigzag lumen that could store sperm like a spermatheca, but this function is not yet clearly established (Van de Velde, et al., 1990). In some cases, structures specialized for sperm storage clearly are out-pouches of the uterus rather than the oviduct; indeed, Chitwood and Chitwood (1950) considered the spermatheca typically to be a specialization of the uterus rather than the adjacent region of the oviduct. However, the spermatheca is defined by function and its divergent position, and its origin may indicate that it is not homologous across taxa.

4.4.3.3 Uterus

The oviduct terminates at the uterus, which is an epithelial tube enclosed by an outer muscular layer (Fig. 4.9(E)). In *C. elegans*, the uterus is comprised of

50 epithelial and eight muscle cells, but the number of cells and other specializations undoubtedly varies among taxa. Quite commonly, the uterus is associated with regions specialized for secretion and variously named to suggest function or the number of gland cells included: crustaformeria, tricolumella, tetracollumella, quadricollumella (Hirschmann, 1971; Maggenti, 1981). Clearly, the uterus is muscular, but in some taxa it includes notable specializations for egg ejection, as is the case for the ovejector of certain members of Dorylaimida (Van de Velde, et al., 1990); other taxa have been described with ovejectors, but it has been noted that this is defined by function and not necessarily applied to equivalent structures. In fact "ovejector" often describes a specialized muscular transition of the uterus to the vagina (Chitwood and Chitwood, 1950). Other structures, such as the Z-differentiation of some species of *Xiphinema*, may retard reflux of sperm from the spermatheca during passage of eggs (Van de Velde, et al., 1990). In didelphic species, the two genital tracts join at the combined uterus, but in monodelphic species a small postvulvar sac, or even a rudimentary nonfunctional ovary, may branch opposite the genital tract (Fig. 4.9(A)). The structure and length of the postvulvar sac may be taxon specific and in some cases it may function for sperm storage (Hirschmann, 1960).

4.4.3.4 Vagina and Vulva

The uterus opens into the vagina that, in turn, leads to the external ventral opening, the vulva (Fig. 4.9(E)). The vagina is surrounded by a few large epithelial cells (eight or less) and its lumen is lined by an extension of the body wall cuticle. The epithelial cells are enclosed by dilator muscles closely associated with the musculature of the uterus, and in some taxa a constricting sphincter muscle also is present. In some cases the true cuticle-lined vagina is distinguished as *vagina vera* from an unlined continuation, the *vagina uterina*, that makes the connection to the uterus (Chitwood and Chitwood, 1950; Maggenti, 1981). Although vulval glands are reported for some taxa, Chitwood and Chitwood (1950) warn that "bladder-like structures", sometimes misinterpreted as glands, are actually the cell bodies and nuclei of dilator muscles. According to Maggenti (1981), the dilator and constrictor muscles of the vulva are independent from those of the vagina. The vulva is usually a transverse slit, but rarely the slit may be oriented parallel to the longitudinal axis of the nematode (Decraemer, 1995). It further varies in shape and profile as well as associated structures including epiptyma, anterior and posterior cuticular flaps of tissue that overlap the vulva opening. The vulva also varies in position. Typically in didelphic/amphidelphic species, the vulva is located near the midbody, whereas in monodelphy it may occur further posteriorly.

The vagina and vulva, which do not appear until the final molt, are essentially derived separately from the reproductive tract, from cells of the hypodermis, but the specific triggers for development are variable and may be taxon specific (Sommer and Sternberg, 1994; Sommer, et al., 2000).

4.4.3.5 Eggs

Species of some nematode genera, such as *Panagrellus*, *Trichinella*, and *Turbatrix*, are ovoviviparous, with juveniles hatching within the female uterus and subsequently emerging. However, these are exceptions, as in the majority of species juveniles hatch from eggs laid by the adult female. Each egg contains a single juvenile, and in many species of nematodes the first-stage juvenile hatches, whereas in most species of plant-parasitic nematodes the juvenile molts within the egg and it is the second-stage juvenile that hatches. In some animal parasites it is the third stage that hatches. Despite the vast difference in size between adults of various species of nematodes, the majority of eggs are of similar size and basic morphology. However, there are numerous variations in detailed structure that may reflect specializations for survival, dispersal, and resistance to adverse conditions. The "shell" of the egg includes an outer vitelline layer, a middle layer consisting of the polysaccharide chitin and protein microfibrils, as well as an inner lipid layer (McClure and Bird, 1976) (Fig. 4.9(I)). The outer layer may be absent in some nematodes, including most Tylenchida. Elaborate structures on the surface of some nematode eggs, such as the mammillation of *Ascaris*, spines of *Anoplectus*, or elaborate pairs of branched byssi of mermithids, are probably secreted from the uterus, but the basis for the highly specific organization is not known. The middle chitinous layer provides the eggshell with its structural rigidity and is the only nematode structure in which chitin has been demonstrated conclusively. The inner lipid layer may be critical to the remarkable impermeability of eggs of some species of nematodes. In certain nematodes, such as *G. rostochiensis* and *Ascaris*, host-mediated alteration of the permeability characteristics of the lipid layer is a necessary prerequisite for juvenile activation and hatching (Perry, 2002). In *M. incognita*, hatching was positively correlated with lipase activity in the hatching fluid, and proteinase, including collagenase, and chitinase activity also were detected (Perry, et al., 1992). In *M. incognita*, and probably the majority of species, such enzymes are likely to be associated with increased flexibility of the eggshell prior to eclosion. However, the source of the enzymes is uncertain. In a few taxa, hatching is facilitated by structural specializations of the egg, including the single operculum of *Aspiculuris tetraptera* (Wharton, 1979) and the two opercula, one at each pole, of *Trichurus* spp. (Hirschmann, 1960). The opercula of *Trichurus suis* consist of a fine network of chitin-protein microfibrils that differs in organization from the rest of the eggshell and may make the opercula more susceptible to enzymic degradation (Wharton and Jenkins, 1978).

4.4.4 Males

4.4.4.1 Testes

Like the ovary, the testis is a single epithelial sac, and, in all but rare hologonic forms (e. g., Trichuroidea), the germinal zone is confined to the distal end (the end furthest from the cloacal opening). The growth zone, where sperm mature, is immediately proximal to the germinal zone (Fig. 4.9 (F)). In *C. elegans*, as is also the case for the ovotestes, distally in the testes the germ nuclei are syncytial. Similarly, the testes of *Dipletalonema* have syncytial germ nuclei, but this may be unusual, relative to distinct germ cells (McLaren, 1973; Wolf, et al., 1978). Testes commonly are reflexed and often have a rachis in Secernentea but not in Adenophorea (Chitwood and Chitwood, 1950) (Fig. 4.9(F)).

4.4.4.2 Seminal Vesicle, Vas Efferens, and Vas Deferens

Posterior to the testis the duct in some nematodes, such as *Dirofilaria*, is dilated as a seminal vesicle for storage of sperm (Delves, et al., 1986). Rarely, a vas efferens, or narrow duct, connects the growth zone of the testis to the seminal vesicle, and there may also be a pronounced constriction between the seminal vesicle and vas deferens (Chitwood and Chitwood, 1950).

The vas deferens is the prominent part of the male reproductive tract and the degree to which it is muscular and/or glandular varies greatly among taxa, although often it is more glandular anteriorly and muscular posteriorly. Where the muscular region is particularly developed near the cloaca, it may be called an ejaculatory duct as in Enoplida; in contrast, for example, there is little musculature in Chromadorida (Fig. 4. 9(F)). Similarly regions may be specialized for secretion, as is the case for the ejaculatory glands or cloacal glands (Fig. 4.9(F)), as well as male "rectal glands" in Dorylaimida and Mononchida, or a pair of glandular pouches thought to form adhesive cement at copulation in *Rhabditis strongyloides* (Chitwood and Chitwood, 1950; Coomans and Lima, 1965; Coomans and Loof, 1986). Secretions may also provide changes in pH or other factors that promote activation of sperm at copulation (Abbas and Cain, 1979; Nelson and Ward, 1980; Ward, et al., 1983). Posteriorly, the vas deferens creates a common duct with the digestive system, and thus develops a cuticle-lined cloaca (Fig. 4.9(F)) surrounded by a large sphincter muscle that constricts the gut during copulation.

4.4.4.3 Spicules and Accessory Structures

The dorsal wall of the cloaca has a cuticle-lined out-pouch that typically includes a pair of the male cuticular copulatory organs, spicules (Figs 4.9(F), 4.10, 4.11(A),(B)). Each spicule includes a head (manubrium), shaft and blade (lamina), and sometimes the blade has membranous winglike extensions called the vellum (Maggenti, 1981) (Figs 4.10, 4.11(B)). Spicules are mobilized by protractor and retractor muscles; in *C. elegans*, these 41 muscle cells are added during the fourth stage (White, 1988). Spicule innervation is discussed in the section on the sympathetic nervous system. Spicules do not include a lumen for transfer of sperm but rather usually the two cup around one another so that sperm is transferred between the pair (Figs 4.10(B), 4.11 (A), (B)); the vellum, when present, may fold to further complete the conduit (Högger and Bird, 1974). Details of spicules often define taxa; modifications may include distal fusion as is the case for many rhabditids, or one spicule may be larger than the other, or more rarely, one spicule may be absent as in some Mermithida. In Tylenchida, spicule length, often used in taxonomy, generally is a function of the overall length of the male (Geraert and De Grisse, 1981). The lining of the duct in which spicules move may have further specializations. Often the floor of the spicular pouch is lined with

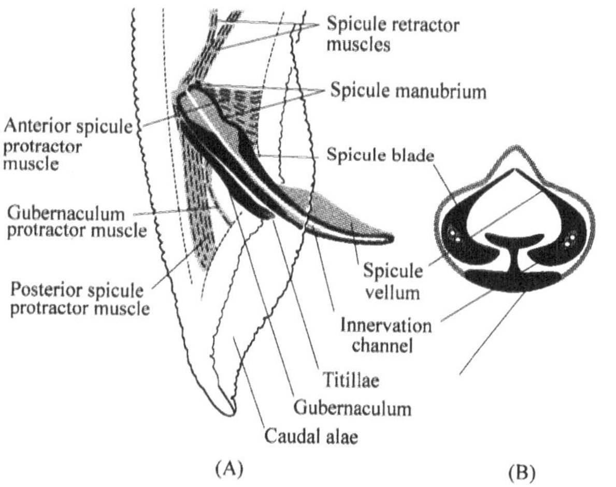

Figure 4.10 Male copulatory structures. (A) Lateral view. (B) Ventral cross section showing the relationship of spicules to the gubernaculum. (A is redrawn from Coomans, 1962.)

thickened cuticle, a gubernaculum (Fig. 4.10), and less commonly a smaller thickening, the capitulum lines the roof of the pouch. The gubernaculum may be stationary, or regulated by specialized muscles as in *C. elegans* (White, 1988), and distinguished by various taxon-specific thickenings or projections including a pair of small projections, titillae, on the ventral surface (Högger and Bird, 1974) (Fig. 4.11(A)). Although often confused with the gubernaculum, Maggenti (1981) argues that the telamon, specific to strongylids, is distinctive as an immobile thickening of the cloacal wall. Unlike the gonad, which often diverges according to sex early in postembryonic development, investigations of *C. elegans* demonstrate that the secondary sexual apparatus of males, including spicules, specialized muscles, innervation, and caudal alae, occur late in development as an overlay on the existing female hermaphrodite (Sulston, et al., 1980). The complex of intercellular signals leading to spicule development in *C. elegans* has been described (Chamberlin and Sternberg, 1993).

Male accessory structures include a wide range of specializations for clasping females, and many of these are taxon specific. Cloacal glands that may secrete an adhesive cement occur in many taxa. In *Syphacia*, fields of fine cuticular spines comprise the surface of mamelons that may aid in keeping the male's position during copulation (Dick and Wright, 1974a, 1974b).

Often the male tail is highly modified relative to that of the female, including a range of sensory receptors as well as the pair of latero-ventral cuticular caudal alae (= bursae) that sometimes are adapted for clasping the female during copulation (Figs 4.10, 4.11(A), 4.13(D),(E)). Caudal alae are common in Secernentea and mostly absent in Adenophorea, but some Trichodoridae (Decraemer, 1995) and Oncholaimidae (Chitwood and Chitwood, 1950) and others are exceptions with narrow caudal alae. In Rhabditina, the male tail may be completely surrounded by the caudal alae (peloderan) or the tail may extend beyond the caudal alae (leptoderan), or caudal alae may be absent. The caudal alae also vary in shape and width and by patterns of innervation which may include sensory rays. The often large and elaborate caudal alae of Strongylida differ from those of Rhabditina primarily by the presence of a dorsal and two lateral lobes (although these are greatly reduced in Metastrongyloidea), and, according to Chitwood and Chitwood (1950), by the presence of muscles associated with the caudal papillae or rays. In Ascarida, Spirurida, and Tylenchida caudal alae are reduced compared to Strongylida and genital papillae are not associated with prominent rays. Whereas members of Cephalobina and most species of Diplogasterida lack caudal alae, the pattern of distribution of their male caudal sensilla is similar to the pattern in Rhabditina. Sensory organs, including those of male

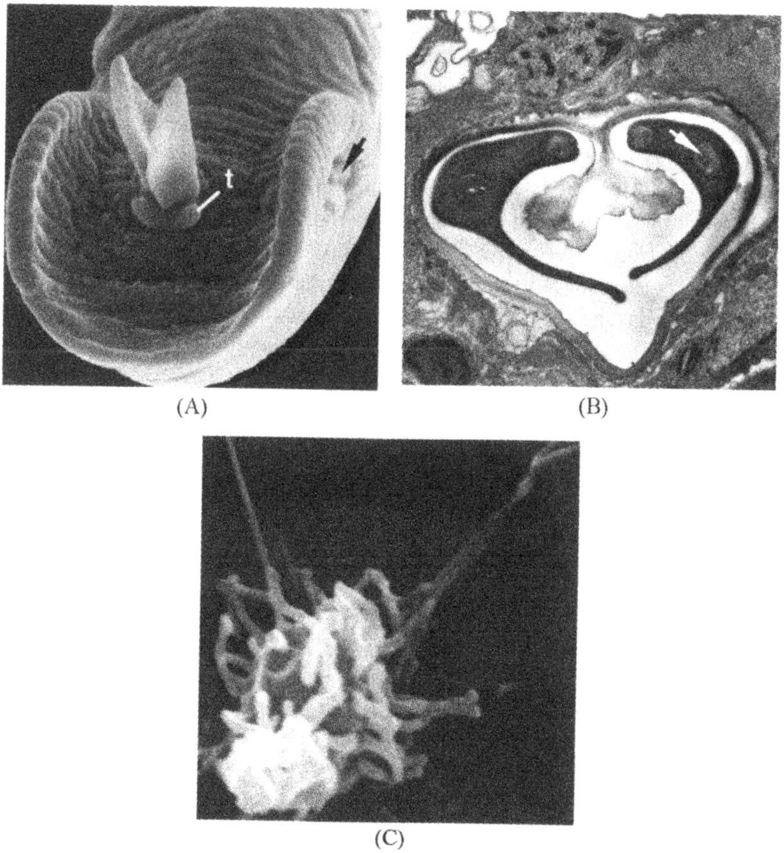

Figure 4.11 Scanning (SEM) and transmission (TEM) electron micrographs showing male reproductive structures. (A) SEM of ventral/posterior view of tail of *Hoplolaimus* sp. with caudal alae and phasmid opening (arrow). A portion of the gubernaculum with titillae (t) protrudes at the base of the pair of spicules. (B) Transverse TEM section through spicules of *Meloidodera* sp. surrounding the gubernaculum (ventral is toward the bottom of the plate). Note that the spicules are each penetrated by sensory neurons (courtesy Lynn Carta). (C) SEM of a sperm from *Meloidodera* sp. (Cares and Baldwin, 1994a).

Adenophorea, are discussed under "nervous system".

4.4.4.4 Sperm

Nematode sperm share unusual characteristics including lack of a flagellum and acrosome, as well as, with rare exception, loss of the nuclear envelope during spermatogenesis and presence of a special major sperm protein (Baccetti and Afzelius, 1976; Baccetti, et al., 1983; Foor, 1970; Klass and Hirsh, 1981; Nelson and Ward, 1980; Turpeenniemi, 1998; Ward and Klass, 1982; Ward, et al., 1982) (Fig. 4.11(C)). Major sperm proteins in nematodes are cytoskeletal and are essential to crawling mobility via pseudopodia (Nelson, et al., 1982; Nourysrairi, et al., 1993; Scott; 1996). Unlike other crawling organisms, nematode sperm apparently are unique in not utilizing actin or myosin for this process (Scott, 1996). Generally, nematode sperm have organelles of crystalline-like fibers; these are believed to be protein, stored early in spermatogenesis, for utilization in activation (Scott, 1996), but the persistence of fibrous organelles into maturity varies among taxa (Cares and Baldwin, 1994a, 1994b).

Regardless of shared characteristics unique to nematodes, nematode sperm vary considerably among taxa, and in a few cases mature sperm are reportedly polymorphic within a given species or even an individual male (Riemann, 1986; Turpeenniemi, 1998). Among nematode taxa, sperm vary from 2 μm to 50μm diameter and in shape from elongate with a cone-like tail extension to amoeboid or rounded (Jamuar, 1966; Lee and Anya, 1967; Turpeenniemi, 1998; Wright and Sommerville, 1985; Yushin and Malakhov, 1994b). In most Tylenchida, the sperm are rounded and the surface is covered with thin pseudopodia, i.e., filopodia (Cares and Baldwin, 1995; Endo, et al., 1998; Goldstein and Triantaphyllou, 1980; Shepherd and Clark, 1976b, 1983b) (Fig. 4.11(C)). Sperm of some Enoplida, Rhabditida, Strongylida, Ascarida, Spirurida, and Trichinellida include a system of membranous organelles (for synonyms, see Bird and Bird, 1991) (Baccetti and Afzelius, 1976; Baccetti, et al., 1983; Foor, 1970; Hess and Poinar, 1989; Klass and Hirsh, 1981; Lee and Anya, 1967; Mansir and Justine, 1995; McLaren, 1973; Turpeenniemi, 1998; Ward, et al., 1982; Ward and Klass, 1982) that, with a few exceptions, are absent in Tylenchida, Dorylaimida, and Chromadoridae (Cares and Baldwin, 1995; Kruger, 1991; Yushin and Coomans, 2000). Where present, membranous organelles apparently are involved in sperm activation through a process that involves their fusion with the plasma membrane (Scott, 1996), but it is not known how this process differs from taxa that lack membranous organelles. Sperm have been shown to vary among taxa in the role they take in axis specification during early cleavages of the fertilized eggs (Goldstein, et al., 1998).

4.5 Secretory-Excretory System

Early descriptions in large ascarids of the system now known as secretory-excretory (S-E) (Bird and Bird, 1991) assumed that associated canals, typical of these and other Secernentea, functioned as part of a circulatory system (Chitwood and Chitwood, 1950). More recent evidence indicates that specific function of this system varies, but in many taxa secretion is indisputable, and osmoregulation, first proposed in the 1860s by Schneider and Bastian (Chitwood and Chitwood, 1950), is plausible but not yet well-substantiated (Bird and Bird, 1991; Wright, 1998; Wright and Newall, 1976). Chitwood and Chitwood (1950) noted two basic types of S-E systems that defined the two classes of Nematoda (Fig. 4.12). Where an S-E system is present at all in the Class Adenophorea (= Aphasmidia sensu Chitwood and Chitwood, 1950), it is expressed as one or two excretory gland cells with a ventral opening to the exterior via a cuticle-lined duct (Bird and Bird, 1991) (Fig. 4.12(D)). The S-E system in the Class Secernentea (= Phasmidia sensu Chitwood and Chitwood, 1950) also is glandular with a cuticle-lined ventral opening but, unlike that of Adenophorea, it consistently has tubular fluid-filled canals in lateral hypodermal chords. In both Adenophorea and Secernentea the cuticle-lined duct most often opens in the cervical region near the nerve ring (Figs 4.1, 4.12).

The S-E system is diverse among taxa, and sometimes it is further specialized between postembryonic stages of a particular species. However, the complex S-E system of *C. elegans* has been most completely reconstructed and serves as a point for comparison; the system consists of four cells: gland, duct, pore, and canal (Nelson, et al., 1983) (Fig. 4.12(A)). The gland cell is binucleate and A-shaped; near the apex it joins with a duct cell that encloses the distal two-thirds of the cuticle-lined duct. The proximal end of the duct joins the body wall cuticle at the ventral opening in the cervical region and is lined by a pore cell. A single large H-shaped excretory cell comprises the pairs of anterior and posterior canals that extend nearly the entire length of the nematode and are embedded in the lateral chords of the hypodermis; the lumens of the four lateral canals converge as small channels at a sinus in the transverse or "bridge" region, at the distal end of the duct. The secretory function of the S-E system of *C. elegans* appears to be active through all life

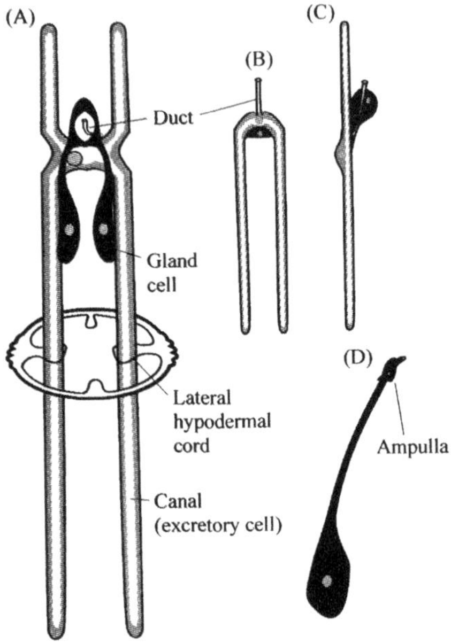

Figure 4.12 Excretory system. (A) Rhabditoid (ventral view). (B) Cephaloboid (ventral view). (C) Tylenchoid (lateral view). (D) Adenophorean (lateral view).

stages except the dauer (Nelson, et al., 1983).

Other nematodes differ from *C. elegans* by the number of excretory glands. In Adenophorea, the taxa examined by TEM have one gland cell (Narang, 1970; Turpeenniemi and Hyvarinen, 1996; Van de Velde and Coomans, 1987). However, in some Secernentea including Strongylida, there are two large excretory glands (McLaren, et al., 1974; Waddell, 1968). The ultrastructure and organelles of S-E glands are similar among taxa. However, the secretory products clearly vary; they may include specific enzymes likely to be involved in parasitic phases of penetration or digestion, protectants from a hostile host environment, or products essential to molting or egg protection (Davey and Kan, 1968; Maggenti, 1962; McLaren, 1974; Rhoads and Romanowski, 1974; Romanowski, et al., 1973). Often the position of the duct, described below, provides additional evidence of secretory function.

Unlike *C. elegans*, particular cells associated with the cuticle-lined duct and pore, each with distinct types of cuticle, have not been recognized in most nematodes. Endo and Wergin (1988), however, noted a transition in cuticle lining between the pore and distal duct region in infective juveniles of *Meloidogyne*. In the monhysterid *Sphaerolaimus gracilis*, a separate ampulla cell and other supporting cells were observed near the duct opening (Turpeenniemi and Hyvarinen, 1996). In some taxa, the duct opening varies from its typical position near the nerve ring, as in some of Ascarida, Strongylida, and insect-parasitic Aphelenchoididae where it opens far anterior near the lip region (Sprent, et al., 1983; Giblin-Davis, personal communication), a position more suited to a possible role in parasitic penetration or extracorporeal digestion (Lee, 1970b). Conversely, in *Tylenchulus*, where the S-E gland produces a gelatinous matrix for eggs, the duct opening occurs posteriorly near the tail (Maggenti, 1962). In the Adenophoreans, *Enoplus brevis* and *Sphaerolaimus gracilis*, the ventral opening is in the cervical region, but in *Monhystera disjuncta* the opening is near the lip region (Narang, 1970; Turpeenniemi and Hyvarinen, 1996; Van de Velde and Coomans, 1987). In some taxa, specialized ampullae or valves are associated with the duct (Dick and Wright, 1973; Evans and Fisher, 1970).

The S-E canals are variable within Secernentea and, relative to the H-shaped canal system of *C. elegans* or *Necator americanus*, often reduced in length or, as in Tylenchida, to only a single lateral canal (Allen, 1960; Chitwood and Chitwood, 1950; McLaren, 1974; Narang, 1972; Nelson, et al., 1983) (Fig. 4.12(C)). It is the tubular region, with its complex inner surface area of canaliculi, that is a possible site of osmoregulation or excretion, and indeed in a few nematodes or stages the canals and/or ducts have been observed to pulsate in response to changes in solutes in the environment (Atkinson and Onwuliri, 1981; Croll, et al., 1972; Narang, 1972; Nelson and Riddle, 1984; White, 1988; Wright, 1976).

4.6 Nervous System

The nervous system is organized into a complex array of sensilla as well as central, peripheral, and sympathetic systems (Figs 4.13 – 4.15). Early insight into the structure of the nematode nervous system was derived from remarkably detailed mapping of neurons, first in *Ascaris* (Butschli, 1874; Goldschmidt, 1908; Otto, 1816) and other large nematodes of both Secernentea and Adenophorea (Chitwood and Chitwood, 1950; Meissner, 1853), and subsequently in several small free-living forms including *Rhabditis* (Chitwood, 1930; Chitwood and Chitwood, 1950; Chitwood and Wehr, 1934). TEM work on *C. elegans* and other microscopic nematodes (Hall and Russell, 1991; Ward, et al., 1975; White, et al., 1976) has shown, as

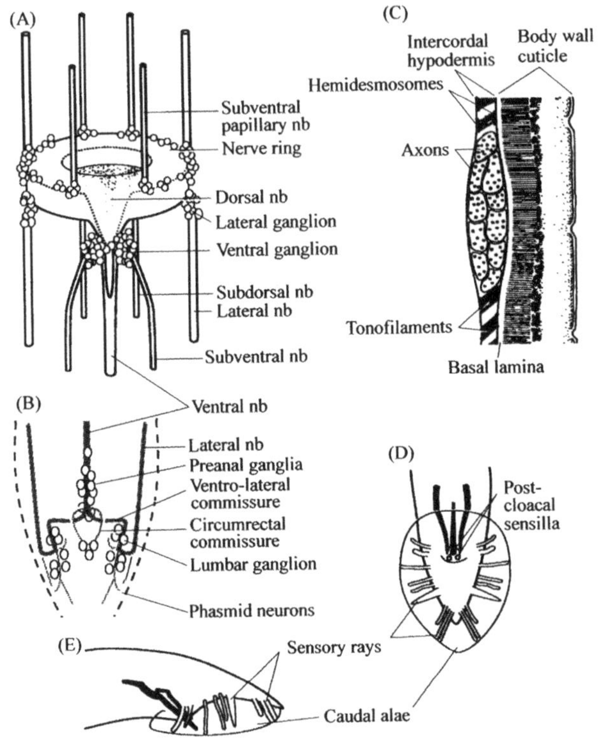

Figure 4.13 Nervous system. (A) The nerve ring and associated ganglia and major nerve bundles (nb). (B) Ganglia and major nerve bundles of the female tail region. (C) Representation of a longitudinal section of a hemizonian from transmission electron micrographs showing peripheral nerve bundles penetrating the intercordal hypodermis. Note that the regular pattern of tonofilments that attaches the cuticle to underlying layers is interrupted by these neurons. (D) Ventral view of the male tail in *Caenorhabditis elegans*. (E) Lateral view of the male tail in *C. elegans*. ((D) and (E) are redrawn from Fitch and Emmons, 1995.)

predicted by Chitwood and Chitwood (1950), that the basic pattern of neurons is remarkably consistent throughout the phylum. In some cases, the function of individual neurons can be inferred from structure and position. In addition, the role of particular neurons has been unraveled through TEM reconstruction or neuron-specific (e. g., monoclonal) staining that associates particular morphological neuron aberrations with specific behavior mutants in *C. elegans*

(Lewis and Hodgkin, 1977; Perkins, et al., 1986). Alternatively, laser ablation of particular neurons can allow evaluation of particular behavioral attributes in a single nematode. Presently, knowledge of the electrophysiological or biochemical properties of individual neurons is limited to large nematodes such as *Ascaris* (Chalfie and White, 1988). Testing a hypothetical endocrine-like role of some neurons in regulating molting or sex determination has proven more elusive (Davey, 1966, 1976).

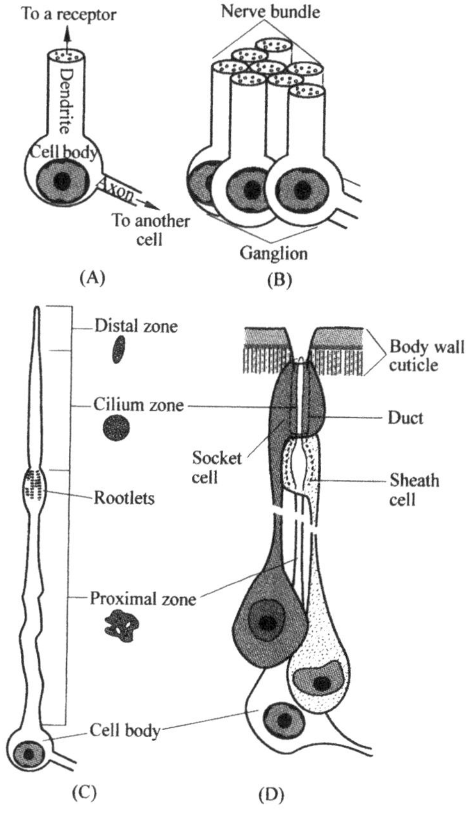

Figure 4.14 Basic neuron structure and associations. (A) Bipolar neuron with truncated nerve processes. (B) A ganglion comprised of the cell bodies of a group of neurons. (C) Differentiation of regions of a typical receptor (dendrite) process of a neuron. (D) Basic components of a typical sensory organ in relation to the body wall cuticle. ((D) is redrawn from Wright, 1980.)

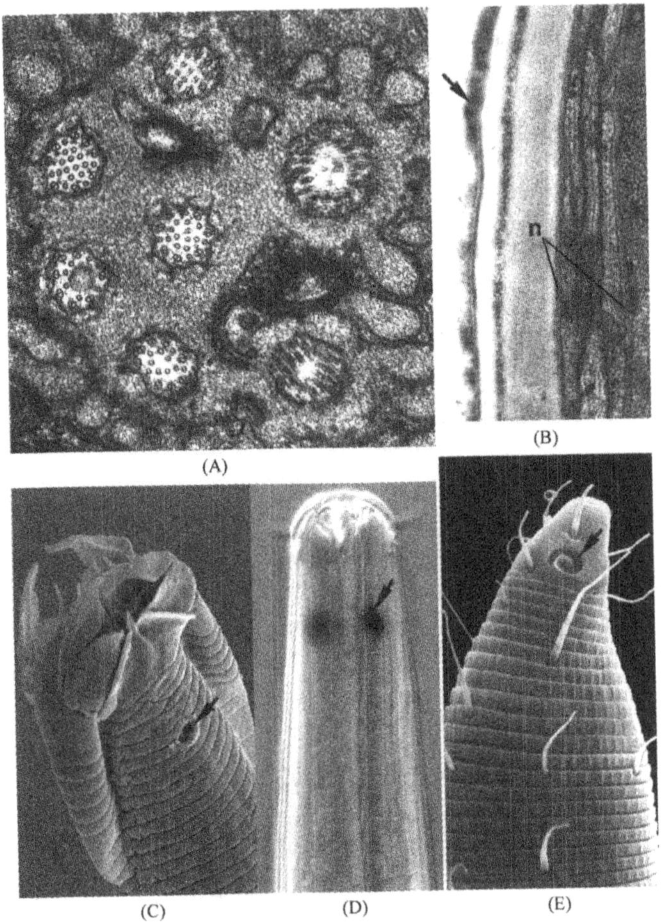

Figure 4.15 Scanning (SEM), transmission (TEM) electron and light (LM) micrographs of neuron and sensory structures. (A) Transverse TEM of the ciliary zone of the amphid of a *Meloidogyne incognita* male. (B) Transverse TEM of peripheral neurons (n) underlying the cuticle and within the intercordal hypodermis of *M. incognita*. Arrow indicates deposits on the cuticle surface. (C) SEM of *Tylocephalus* sp. (Araeolaimida). Arrow indicates amphid opening. (D) LM showing dark pigment associated with ocelli of *Phanoderma* sp. (nearly ventral view). (E) SEM of head region of *Epsilonema* sp. Note large numbers of setae and large spiral amphid opening (arrow).

4.6.1 Basic Nerve Elements

Essential to the discussion of the basic patterns of nematodes is a description of the fundamental nerve elements that comprise such systems. The most basic element, of course, is the neuron, which is a single nerve cell with typically one (monopolar), two (bipolar), or more elongate processes (Fig. 4.14 (A)). The nerve cell including nucleus, often distinctive from other nuclei by its extensive peripheral chromatin and excluding the processes, is often called the "nerve cell body". A given nerve process may be either an axon or dendrite. An axon is specialized to conduct nerve impulses over a distance and usually forms a synapse with another type of cell such as muscle, or with another neuron (interneuron). A dendrite receptor in nematodes typically terminates within a sensory organ. However, Chalfie and White (1988) note that an individual neuron may have multiple synapses combining, for example, interneuron and dendrite and/or axon functions. Slightly proximal to the terminus of a given dendrite, the cytoplasm is usually specialized with a regular arrangement of microtubules, thus resembling cilia (Figs 4.14(C), (D), 4.15(A)). The presence of such sensory "cilia", or a ciliary region, now known to be highly conserved in metazoa, was previously considered significant in defining deep phylogenetic relationships among pseudocoelomates (Hyman, 1951; Roggen, et al., 1966), but may also have certain features specific to particular phyla (Tyler, 1979). Dendrites are usually either not branched at all, or they may split into just two cilia-like tips. In nematodes, unlike vertebrates, highly branched dendrites are relatively rare. Typically in nematodes, neuron cells are organized in clusters. Collections of nerve cell bodies are ganglia (Figs 4.13(A), (B), 4.14(B)). Large cords containing a number of axons or dendrites may be called "nerve bundles" (Figs 4.13(A), (B), 4.14(B), 4.16(B)). Groups of neurons are organized into a conspicuous central nervous system as well as the less apparent peripheral and sympathetic systems.

4.6.2 Central Nervous System

The central nervous system includes a nerve ring where most sensory integration takes place (White, 1988) (Figs 4.1, 4.13(A)). The nerve ring, also called the circumpharyngeal commissure, encircles the isthmus, or equivalent region, of the pharynx. In addition to the nerve ring, the central nervous system primarily comprises interneurons and connected ganglia, as well as the ventral nerve bundle. Anterior to the nerve ring, clusters of nuclei comprise six ganglia (sometimes called papillary ganglia); the largest of such ganglia are lateral and include cell bodies associated with innervation of amphids and most of the neurons of the head (Figs 4.13(A), 4.15(A), 4.16(A),(B)). In some nematodes, including *C. elegans*, the papillary ganglia are not readily demarcated and are simply referred to as the anterior ganglia (Chalfie and White, 1988). Posterior to the nerve ring, the ventral nerve bundle, extending to near the anus, is comprised of most of the motor neurons and interneurons, whereas the smaller dorsal nerve bundle is comprised mostly of motor axons (Bird and Bird, 1991) (Figs 4.13(A),(B)); however, this division, while well-established for *C. elegans*, may not be true for large animal parasites where the ventral nerve bundle is described as also innervating motor neurons (Chitwood and Chitwood, 1950). In larger nematodes, the ventral nerve bundle is described as a chain of ganglia, and connections to the nerve ring are noted as highly variable among taxa (Chitwood and Chitwood, 1950). Typically, however, the ventral nerve cord is divided into two subventral branches near the nerve ring and associated posterior ganglia; fusion into a single ventral branch most often occurs just posterior to the region of the excretory pore. Similarly, in some taxa the dorsal cord may be anteriorly divided into two subdorsal cords. Additional nerve bundles extending posteriorly from the nerve ring may include up to four submedial nerve and one or more pairs of lateral nerve bundles; posterior to the nerve ring, ganglia of the lateral nerve bundles may be further subdivided in some large parasites (Chitwood and Chitwood, 1950; Goldschmidt, 1908). All major longitudinal nerve bundles tend to occur adjacent to, or even partly enclosed by, the hypodermal cords.

4.6.3 Peripheral Nervous System

The demarcation between the central and peripheral nervous system is not explicit. Chitwood and Chitwood (1950) considered all as peripheral except the nerve ring and its ganglia, ventral nerve bundle, and sympathetic systems. Although they included innervation of sensory organs, it is clear that the cell bodies and ganglia for most such organs are part of the central nervous system. With increased understanding from TEM, including recognition of a network of neurons in the intercordal hypodermis, the peripheral nervous system can now be understood in a more restricted sense to include neurons and small commissures just beneath the cuticle surface (Figs 4.4(C), 4.13(C), 4.15(B)). These small neurons typically have cell bodies in the hypodermal cords, and processes, visible only with TEM, occur within the intercordal hypodermis. Whereas these processes generally occur singly rather than in bundles, at a few regions they are connected by small commissures of bundles of interneurons (Baldwin and Hirschmann, 1975; Smith, 1974; Timm, 1960), and these bundles often are readily visible with light microscopy.

Where they are visible they are described, depending on location, as hemizonid (anterior to the excretory pore), hemizonion (posterior to the excretory pore), caudalids, and cephalids (Hirschmann, 1971) (Figs 4.13 (C), 4.15(B)). In some cases peripheral commissures are more apparent in fixed than in living specimens. This is because the commissure interrupts the pattern of tonofilaments traversing the intercordal hypodermis that connect the body wall cuticle to underlying tissues; the result is that in fixed material the cuticle pulls away in these regions, leaving what appears in whole mounts, to be a lens-like structure beneath the cuticle surface. Peripheral neurons, including commissures, have been demonstrated with TEM, but attempts to elucidate a peripheral nervous system with specific optical stains have been difficult to repeat and controversial, and the structures are easily confused with components of cell membranes or junctional complexes (Croll and Maggenti, 1968; Smith and Stephenson, 1970).

4.6.4 Sympathetic Nervous System

Chitwood and Chitwood (1950) recognized that innervation of the nematode pharynx and rectal/cloacal region are semi-independent of the central nervous system and thus described these as two sympathetic nervous systems. In *C. elegans*, the pharyngeal-sympathetic system comprises 20 neurons concentrated in a ring in the metacorpus and a secondary ring in the terminal bulb, with three bundles of nerve processes extending most of the length of the pharynx. Several of the cells are interneurons; others synapse with muscles, and two may be neurosecretory. The pharyngeal neurons in *C. elegans*, including cell bodies, are completely contained within the pharynx, except for two interneurons that connect with the central nervous system; this pattern appears to be highly conserved as it also has been described for *Ascaris* and *Parascaris* and presumed for *Spironeoura* by Chitwood and Chitwood (1950). Recent comparisons of *C. elegans* with representative diplogastrids and cephalobids suggest that the pharyngeal-sympathetic system is generally conserved but with minor variation in numbers of neurons among taxa of Secernentea; these differences can be largely attributed to variation in musculature associated with the presence/absence of a posterior grinder (Zhang and Baldwin, 1999, 2000, 2001). Innervation of more specialized pharynxes, including those of plant parasitic Tylenchida with reduced musculature, has been partly described (Baldwin, 1982; Endo, 1984; Endo, et al., 1997; Shepherd and Clark, 1983a; Shepherd, et al., 1980), but homologies with neurons in *C. elegans*, for example, have not been elucidated.

The rectal/cloacal-sympathetic nervous system was first described in *Ascaris* (Chitwood and Chitwood, 1950; Rohde, 1885) but now has been elucidated in much finer detail for *C. elegans* (Hall and Russell, 1991; White, 1988) (Fig. 4. 13(B)). Unlike the pharyngeal-sympathetic nervous system, the rectal-sympathetic nervous system morphologically appears to be less autonomous from the central nervous system, and indeed Chitwood and Chitwood (1950) only suggested that it "might" be considered a sympathetic nervous system. The complex of rectal innervation extends from the posterior terminus of the ventral cord as a preanal-dorsal-rectal ganglion, and posteriorly from this ganglion one or more commissures associated with this encircle the rectum. Additional commissures extend from a pair of (lumbar) ganglia, associated with the lateral nerve bundles, and connect the ventral and dorsal nerve bundles (Bird and Bird, 1991; Chitwood and Chitwood, 1950) (Fig. 4.13(B)). Innervation of the paired caudal sensory organs, the phasmids (see below), extend from the lumbar ganglia. In males of *C. elegans*, more than one-fifth of the 381 neurons are sexually specialized including 18 neurons, with one innervating each sensillum of the nine pairs of sensory rays, as well as additional neurons innervating the postcloacal sensilla, hook sensilla, spicules, and spicule musculature as well as interneurons (Bird and Bird, 1991) (Figs 4. 10(B), 4. 11(B), 4. 13(D), (E)). In addition, each spicule possesses one or more nerves that end at a pore or pores in the tip and are likely to have a chemosensory or mechanosensory function before or during copulation. Laser ablation of these spicule neurons can interfere with locating a mate (Samoiloff, 1980). Caudal sensory rays or papillae vary among taxa; even within Rhabditina they vary in morphology, developmental patterns, arrangement, and fusion (Fitch, 1997).

4.6.5 Sense Organs

4.6.5.1 Basic Sense Organ Structure

Nematodes interact with their external environment by a wide range of sensory organs/sensilla (Figs 4.14 – 4.16); a sense organ typically consists of one or more specialized receptor nerve endings together with functionally associated non-nervous tissue (Fig. 4. 14(D)). Specialization of the receptor nerve ending originates proximally as a basal-body-like (Chalfie and White, 1988) transition region with highly ordered arrangement of single and typically nine peripheral doublet microtubules that are anchored in rootlets; this basal-body region is characterized by specialized complexes of the surrounding cell membrane with adjacent supporting cells. Distal to the basal body, the receptor approximates to a classic cilium with respect to arrangement of doublet microtubules, and is thus called the ciliary region; often the ciliary region occurs within a duct (Figs 4. 14(C), (D), 4. 15(A)). Further approaching the distal end, the receptor tip typically is reduced in size, irregular in shape,

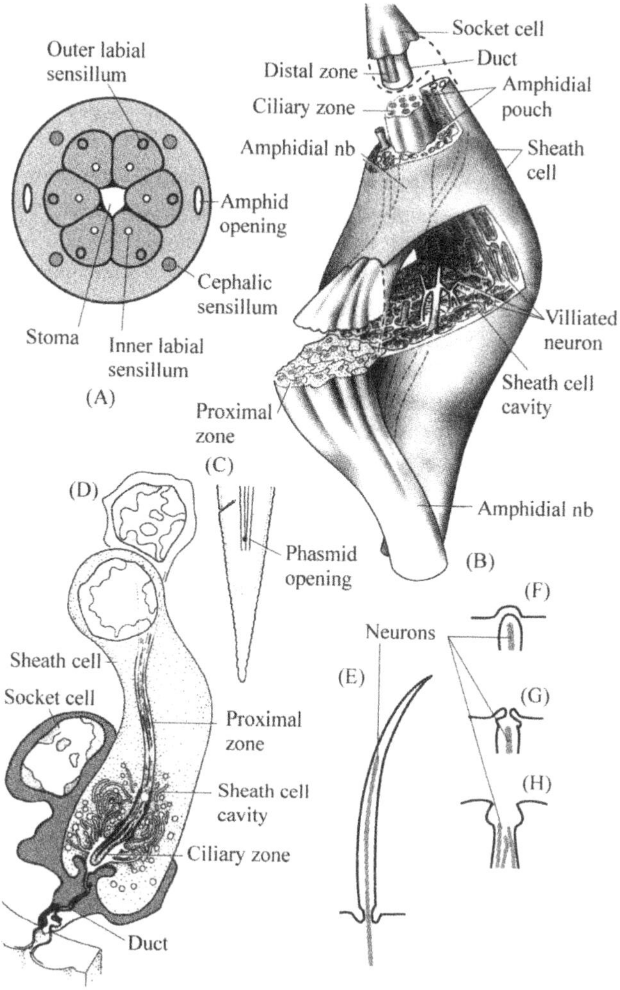

Figure 4.16 Sensory organs. (A) Typical en face view showing basic hexaradiate lip pattern and associated sensilla. (B) Amphid of *Meloidogyne incognita* male. nb = nerve bundle. (C) Tail of second-stage juvenile of *Heterodera schachtii* showing phasmid opening. (D) Phasmid of *H. schachtii*. (E) Setae. (F) Mechanoreceptive papillum. (G) Chemoreceptive papillum. (H) Chemoreceptor opening with multiple neuron receptors. ((B) modified from Baldwin and Hirschmann, 1973; (C), (D) modified from Baldwin, 1985.)

and microtubules are single and appear to be distributed randomly (Figs 4.14 (C), (D)). However, in some cases the receptor terminates with the ciliary region, or it may include a more specialized ending such as a sheet, bulb, or complex of microvilli-like extensions (Baldwin and Hirschmann, 1973; Perkins, et al., 1986; Ward, et al., 1975; Wergin and Endo, 1976) (Figs 4.16(B), (D)).

In addition to the sensory receptor, usually a sensory organ also includes non-nervous supporting cells that function with the neuron; most commonly, this non-nervous tissue is a socket and a sheath cell (Figs 4.14(D), 4.16(B), (D)). The sheath cell encloses the area of the rootlets of the ciliary region of the receptor, whereas the socket cell surrounds the tip of the neuron near to where it enters the body wall cuticle. Although the function of the sheath cell is not precisely known, it is described as glia-like and may regulate the external chemical environment of the ciliary region (Chalfie and White, 1988). Where a sensory organ involves a cuticle-lined duct, it is the socket cell that forms this duct. Primarily in chemoreceptors, characterized by direct contact with the external environment, the socket cell also may secrete substances, such as mucoproteins that coat the duct and receptor tip. There is some experimental support for the assumption that those sensilla that open to the environment, including amphids (see below), have a chemosensory function. For example, mutants of *C. elegans* with impaired response to chemical stimulants had structural defects in the sensory neurons or in the sheath cells (Lewis and Hodgkin, 1977).

4.6.5.2 Chemoreception and Mechanoreception

Chemoreception, which may involve several anterior sensory organs (see below), plays an important role in the oriented movement (taxis) of nematodes (Bargmann and Horvitz, 1991; Dusenbery, 1974; Jansson, et al., 1984; Perry, 1996; Perry and Aumann, 1998; Ward, 1973). In theory, orientation by an animal in a stimulus field can be achieved with one sensor by sequentially sensing and comparing the stimulus on either side of the path of movement (klinotaxis) or with two or more sensors on different parts of the body by simultaneously sensing different points within in the stimulus field (tropotaxis). See Robinson (chapter 7 in this volume) for further discussion of behavior terminology. Bargmann and Horvitz (1991) demonstrated that killing neurons in either the left or the right chemosensory sensilla of *C. elegans* did not prevent chemotaxis, indicating that the nematode does not perform tropotaxis. Repeated side-to-side movements of the head during sinusoidal body movement may indicate that an alternating sampling in space from either side of the head occurs and that klinotaxis is involved in nematode orientation.

Mechanoreceptors differ from chemoreceptors because they may terminate beneath the cuticle surface without ducts opening to the exterior; nevertheless, socket cells of mechanoreceptors may also form a fluid-filled chamber near the surface that might aid in mechanical transmission. Unlike chemoreceptors, among other unique morphologies, the microtubules of the distal regions of mechanoreceptors are often embedded in electron dense material (Perkins, et al., 1986). Rarely, as in the case of touch receptors, or proprioceptors, sensory receptors may occur as "free-nerve endings" without apparent supporting cells (Chalfie, et al., 1985; Hall and Russell, 1991). For example, there are sensory, or touch cell, processes that end near the cuticle and which, in *C. elegans*, extend along the lateral and mid-ventral lines. Also, in *C. elegans* hermaphrodites have an elongate tail that has been observed to include sensory neurons lacking in sheath and socket cells (Bird and Bird, 1991; Hall and Russell, 1991).

Mechanoreception may be expressed in a range of sensory organs on the cuticle surface, as well as in structures less readily recognized as sensory organs. Extensive research on the functioning of mechanosensilla of *C. elegans* (summarized by Driscoll and Kaplan, 1997) is not matched by information from other species of nematodes. Mechanosensory perception is involved in locomotion and responses to external environmental stimuli as well as internal responses associated, for example, with egg-laying and pharyngeal pumping. Four behavioral classes have been used to study sensory responses to mechanical stimuli: nose touch, gentle body touch, tap, and harsh body touch. The nose touch response, associated with hitting an object "head-on", results in backward movement. It is controlled by three classes of mechanosensory neurons, one of which, the ASH neurons, is located in the amphids (see below). Touch also regulates continuous, exploratory head movements, called foraging behavior. The neurons that mediate "nose" (anterior end) touch avoidance and foraging behavior have sensory endings that contain a single cilium (Albert, et al., 1981; Perkins, et al., 1986; Ward, et al., 1975). Touch cell processes, unciliated but filled with microtubules, respond to gentle body touch, usually administered experimentally with an eyelash (Chalfie, et al., 1985). The touch cell processes synapse onto many other neurons, including interneurons that control locomotion. Stimulation of the anterior touch receptors results in the nematode moving backward, whereas stimulation of the posterior ones results in the nematode moving forward. Response to the tap stimulus (a tap to the side of the dish containing the nematodes) is manifest either by accelerated forward movement or by backward movement (Rankin, 1991). Harsh touch is achieved by prodding with a platinum wire and usually elicits backward movement. The responses to mechanical stimuli are complex and mutants defective in one type of response are not necessarily non-responsive to other mechanical stimulation. For example, mutants defective in response to gentle body touch, respond normally to mechanostimuli acting on the anterior sensilla. The neuronal circuitry for the transduction of mechanical stimuli into each behavioral class has been described (Driscoll and Kaplan, 1997).

Of all the mechanosensilla of *C. elegans*, the six touch receptor cells have been studied most extensively, and a model for their functioning has been detailed by Driscoll and Kaplan (1997). Several aspects have been determined at the molecular level. Hong and Driscoll (1994) characterized a subunit domain of a putative movement-sensitive ion channel. Huang et al. (1995) then found a stomatin-like protein that may link the channel with microtubules of the mechanosensory dendritic endings. Gu et al. (1996) proposed a model of *C. elegans* mechanotransduction where opening of an ion channel is stimulated by a touch-induced displacement of microtubules that are bound to the channel proteins by stomatin.

4.6.5.3 Papillae

Frequently, in nematodes, sensory organs are expressed at the cuticle surface as slight elevations, termed papillae (Figs 4.8(F), 4.16(A), (F) – (H)). Papillae are often found at the anterior and posterior ends of a nematode, although, depending on the taxon, they may occur anywhere on the body. Throughout Nematoda, the basic pattern on the head region is six inner, six outer and four cephalic sensilla (Coomans, 1979), and in Secernentea and many species of Adenophorea these sensilla are expressed as papillae (Fig. 4.16(A)). The pattern, however, may vary slightly among taxa by altered positions, or loss of particular sensilla (Endo and Wergin, 1977). In Secernentea, these anterior papillae, and specifically the outer labial and cephalic papillae, are the main anterior concentration of mechanosensilla. However, in Tylenchida and perhaps others in Secernentea, the inner labial papillae, surrounding the oral opening, are chemoreceptors (Baldwin and Hirschmann, 1973, 1975).

In the cervical region located laterally near the region of the nerve ring of some Secernentea, deirids are paired papillae structurally similar to cephalic papillae (Coomans and De Grisse, 1981); lacking an opening through the cuticle, they appear to be mechanoreceptors (Sulston, et al., 1975; Ward, et al., 1975). Analogous structures have been reported in the midbody region of some nematodes and have been termed postdeirids or centrids (Bird and Bird, 1991). Many nematodes, particularly in Adenophorea, have additional pores on the body that have been shown to be sensory, but often with external openings and highly secretory socket cells (McLaren, 1976a, 1976b; Wright and Carter, 1980; Wright and Chan, 1974).

Frequently sensilla in the tail region, and particularly on males, are also expressed as papillae. These papillae often are tactile receptors involved in copulation; they may be absent altogether or vary in number from few to complexes of ten or more pairs (Figs 4.13(D), (E)). Often caudal papillae are simply raised areas of thin cuticle, but in males of Rhabditida and Strongylida, most are specialized on rays associated with caudal alae. In *C. elegans*, there are nine pairs of caudal papillae (sensory rays) embedded in the fan-shaped caudal alae. In this species, the sensory rays help in the orientation of the male tail to the hermaphrodite female. The male then locates the approximate position of the vulva using the hook sensillum (adjacent to the male cloaca) and then uses the spicules and post-cloacal sensilla to locate the vulva precisely (Emmons and Sternberg, 1997). In *C. elegans*, laser ablation studies of caudal papillae have shown that these sensory organs are tactile and essential in mating behavior and in locating the vulva (White, 1988).

4.6.5.4 Supplementary Organs

Supplementary organs sometimes are not readily distinguished from papillae, except that they are located anterior to the cloaca and ventrally on the male of many Adenophorea. However, in some cases they are large and "wartlike" or cuticularized tuboid structures in which any sensory function is also associated with obvious secretion (Hirschmann, 1971); it is not known if this secretion is a specialization of the socket cell, or if additional glands are present, but regardless of source, it is likely to function as a cement in copulation.

4.6.5.5 Setae

Setae are most obviously differentiated from papillae by their elongation; they are mechanoreceptors penetrated by one or more receptor ends and sometimes they are narrowed and "hinge-like" at their base for easy deflection (Croll and Smith, 1974) (Fig. 4.16(E)). In some Adenophorea, outer labial and/or cephalic sensilla (Fig. 4.16(A)) are expressed as setae; in some taxa setae also may occur cervically, caudally, and elsewhere. Often it is difficult with light microscopy to determine if a given cuticular protrusion is indeed an innervated seta or if superficially similar structures may be specialized for other nonsensory functions, such as adhesion; adhesion appears to be the primary function, for example, of some caudal bristle-like extensions in Desmodorida.

4.6.5.6 Amphids

4.6.5.6.1 Structure of amphids

Amphids apparently are universally present in nematodes as complex paired organs, positioned laterally and with cephalic or cervical external openings. Although each amphid generally is large, often highly specialized, and includes several sensory receptors (the number varying among taxa), the basic structure, including a socket and sheath cell, is fundamentally the same as other smaller sensory organs (Ward, et al., 1975) (Figs 4.14(D), 4.16(A), (B)). The socket cell surrounds the amphidial duct, which encloses the distal regions of the receptors. In some cases the socket cell includes bundles of fibers believed to provide support (McLaren, 1976a), and often the socket cell secretes copious material (discussed below) into the duct, which may exude from the amphid openings (Baldwin and Hirschmann, 1973). The ciliary region of the receptors is enclosed by the sheath cell (= amphidial gland, sensu Chitwood and Chitwood, 1950), and within the sheath cell, and near the base of the duct, gap junctions occur between adjacent receptors. In at least some nematodes, the base of the duct appears to be sealed from the sheath cell by junctional complexes between receptors entering the duct as well as between the receptors and the sheath cell (Fig. 4.16(A)). Typically, the amphid sheath cell is deeply folded or otherwise modified resulting in a very large surface area. In *Meloidogyne* males and to a lesser extent in infective juveniles, for example, it resembles Swiss cheese, with many extracellular, fluid-filled caverns continuous with a larger sensilla pouch surrounding the receptors (Baldwin and Hirschmann, 1973; Wergin and Endo, 1976) (Fig. 4.16(B)). Although the sheath cell appears to be secretory, and the secretions may be specialized for parasitism (Jones, 1975; McLaren, 1974; Riga, et al., 1995), it appears that the receptor cavity of the sheath cell is closed, and sheath cell secretory products are not in direct contact with the amphidial duct. An unusual feature of the amphid sheath cell is that in many Secernentea examined with TEM, it is penetrated by one or two specialized neurons considered the "finger cell" in *C. elegans* (Chalfie and White, 1988; Ward, et al., 1975) with distal tips that radiate multiple villiated endings, and these endings are embedded in the extracellular caverns defined by the sheath cell (Fig. 4.16(B)). In most Secernentea, the amphid openings are simple cephalic pores, slits, clefts (Geraert, 1965), and the duct is relatively narrow; in Adenophorea, the duct generally is much broader than in Secernentea (Wright, 1980) (Figs 4.4(E), 4.8(F), 4.15(C), (D), 4.16(A), (B)). In Dorylaimida, Mononchida, and Enoplida, openings may be slits or round, but characteristically the opening leads to a pocket-like duct ampulla and thus the opening is considered cyanthiform. Other Adenophorea, including Chromodorida, have large circular openings, hoops, or elaborate spiral "trenches" that lead into the duct (Fig. 4.15(E)).

4.6.5.6.2 Function of the amphids

In contrast to the detailed information on the ultrastructure of sensilla of many species of nematodes, research on the functional physiology and biochemistry of nematode sensilla derives primarily from research on *C. elegans*. Sensilla that open to the environment have long been assumed to have a chemosensory function, and the amphids are considered to be the primary chemosensilla. There is experimental support for a chemosensory role. Mutants of *C. elegans* with impaired response to chemical stimulants had structural defects in the sensory neurons or in the sheath cells (Lewis and Hodgkin, 1977). Laser ablation of individual amphidial neurons of *C. elegans* (Bargmann, et al., 1990) and direct electrophysiological recordings from amphids of the animal-parasitic nematode *Syngamus trachea* (Riga, et al., 1995) also demonstrated the chemosensory role of the amphids. In addition to structure, laser-ablated neurons and behavioral mutants indicate the amphid is chemoreceptive (Perry and Aumann, 1998), but the complex structure of the sheath cell and the diverse opening morphology of some taxa suggest multiple functions and specialization among taxa (Ashton and Schad, 1996).

The twelve sensory neurons which enter the amphid of the *C. elegans* hermaphrodite have been divided into three classes, based on morphology and function (Ward, et al., 1975; Ware, et al., 1975). The first class comprises eight sensory neurons whose dendritic processes end in the amphidial duct and, thus, are exposed directly to the external environment through the amphidial pore. These neurons detect water-soluble odorants including salts, amino acids, and the dauer pheromone. Volatile odorants are detected by the three neurons, called wing cells, of the second class that have flattened, branched dendritic processes embedded within the amphidial sheath cell. The third class of neuron, comprising a single neuron called the finger cell, has a sensory ending bearing microvillar projections that are embedded in the sheath cell; this neuron is thought to detect thermal cues (Mori and Ohshima, 1995).

Different neurons may detect the same odorant and individual neurons can detect multiple odorants. For example, the neurons designated "ASH", which have sensory endings exposed to the environment, are polymodal sensory neurons with two chemosensory functions, mediating avoidance of osmotic and volatile repellents (Bargmann, et al., 1990; Troemel, et al., 1995), and a mechanosensory function involving an avoidance response when the nematode touches an object "head-on" (Kaplan and Horvitz, 1993). Thus, the amphids are not exclusively chemosensory organs but have chemosensory, thermosensory, and mechanosensory roles.

The neurons appear to be "hard wired", and stimulation of a particular chemosensory neuron results in a distinct behavioral response (attraction or avoidance). Thus, it is the neuron that determines the response, not the nature of the receptor molecule being stimulated (Bargmann and Horvitz, 1991). This was demonstrated experimentally by expressing the receptor for diacetyl, an attractant for *C. elegans*, in a sensory neuron that normally detects repellent chemicals; the transgenic nematodes showed a reversal of the wild-type behavior by avoiding diacetyl (Troemel, et al., 1997).

The ODR-10 protein is the receptor for diacetyl. The *odr-10* mutant phenotype responds normally to all odorants tested except for diacetyl, which elicited no response. Transforming the mutant nematodes with the cloned *odr-10* gene restored the normal attraction to diacetyl. When expressed in mammalian cells, the ODR-10 receptor is capable of detecting diacetyl (Zhang, et al., 1998). Thus, there is genetic and biochemical evidence that ODR-10 is the diacetyl receptor. The *odr-10* gene is one of a large and diverse family of genes thought to detect odorant receptor molecules (Troemel, et al., 1995), but functional data are needed for the other genes.

Some receptors are expressed in a single sensory neuron while others are present in up to five sensory neurons. The odorant receptors are members of the seven-transmembrane domain superfamily of receptors, that are linked to G-proteins and often referred to as G-protein coupled receptors. Interactions between ligands and seven-transmembrane receptors usually lead to the activation of secondary messenger cascades through activation of G-proteins. After binding of signal molecules to their receptors, G-proteins activate enzymes that catalyze the formation of secondary messengers from inactive precursors. For example, adenylate cyclase catalyses the formation of cyclic AMP (cAMP) and phospholipase C catalyses the formation of inositol triphosphate (IP3). These secondary messengers bind to ion channels, causing them to open, and the resulting ion influx causes voltage-gated ion channels to open thus generating electrical activity in neurons. No data are available for the secondary messenger systems in nematodes. However, in *C. elegans* the G-protein G_o has been shown to play a role in different forms of behavior (Mendel, et al., 1995; Ségalat, et al., 1995), and the identification of G-protein coupled seven-transmembrane receptor proteins as olfactory receptors (Bargmann, et al., 1993; Sengupta, et al., 1996; Tromel, et al., 1995) implies a role for the secondary messenger systems in transduction of olfactory signals. Full details of the mechanisms of nematode neurotransmission are given by Wright and Perry (1998).

4.6.5.6.3 Role of the amphid sheath cell secretions

The supporting (socket and sheath) cells of amphids both produce secretions that appear to have multiple roles. Often the secretions from the duct are primarily attributed to the sheath cell, but for many nematodes, this needs to be more clearly resolved. TEM studies of some nematodes have indicated the sheath cell is sealed from the duct. Furthermore, continuity of secretory vesicles within the socket cell and with material in the duct has been shown (e.g., Baldwin and Hirschmann 1973, 1975). Regardless of source, secretions in the duct may serve to maintain electrical continuity between the bases and tips of the dendritic processes (Trett and Perry, 1985a), a function thought to be important in the generation of receptor currents in some insect sensilla (Zacharuk, 1980). They may also protect the dendritic endings of sensory nerve cells against desiccation and microbial attack (Aumann, 1993). It has been suggested (Endo, 1978) that the feeding plug of *H. glycines* is secreted by the amphids, although alternative origins for the feeding plug have been reported (Jones, et al., 1993). Several animal-parasitic nematodes that feed on blood secrete anticoagulants from the amphids (Jones, 1975; McLaren, 1976a, 1976b), and the amphids become very enlarged at the feeding stages. Stanssens et al. (1996) have described a family of small proteins from the hookworm *Ancylostoma caninum* that inhibit coagulation of host blood. The amphidial glands (= supporting cells, i.e., socket and/or sheath cells) of many animal-parasitic nematodes contain acetylcholinesterase. Various functions have been suggested for secretory cholinesterases (see Pritchard, 1993), including a possible role as a biochemical holdfast, enabling some species to maintain position in the host's gut (Ogilvie and Jones, 1971), and altering the permeability of the host membranes to facilitate leakage of nutrients (Lee, 1970a). The fact that cholinesterase has also been detected in plant parasitic nematodes (Rohde, 1960) and *C. elegans* (Pertel, et al., 1976), and has been detected in amphids, phasmids, and papillary glands of the animal-parasitic nematode *Dirofilaria immitis* (McLaren 1976a, 1976b), indicates that this enzyme may have a more general role in chemoreception in addition to specialized functions in some species.

Molecules of chemical substances initially come into contact with the amphidial secretions. In insects, secretions of sense organs contain olfactory binding proteins (OBPs) which bind hydrophobic odorant molecules and transport them through the aqueous lymph to their receptors (Stengl, et al., 1992). Different OBPs bind different odorant molecules (Vogt, et al., 1991a, 1991b). The secretions of insect sense organs also contain a range of odorant degrading enzymes (ODEs), which destroy odorant molecules released from the receptors at the nerve processes and thus prevent repeated restimulation of the nerve by a single odorant molecule (Vogt and Riddiford, 1981). The possibility that OBPs are present in nematode sense organs has been investigated by Jones et al. (1992), but convincing evidence that the secretions of nematode sensilla contain the same suite of molecules as found in insect sense organs is still lacking.

Premachandran et al. (1988) found that the protein-specific dye, Coomassie Brilliant Blue R-250, bound to the amphidial secretions of a number of nematodes including J2s of *M. incognita*, indicating that the secretions contain protein. Lectins have been used to demonstrate the presence of carbohydrate residues in amphidial secretions of *H. schachtii* (Aumann and Wyss, 1987, 1989) and species of *Meloidogyne* (McClure and Stynes, 1988) and *Globodera* (Forrest and Robertson, 1986). Alterations in binding patterns observed after digestion with certain proteases suggested that at least some of these residues were constituents of glycoproteins (Aumann, 1989; Forrest, 1985), but sialic acids, common glycoproteins of vertebrates and higher invertebrate taxa, have not been found in nematodes (Bacic, et al., 1990; Reuter, et al., 1991). Components of amphidial secretions are thought to include *N*-acetylgalactosamine and fucose in J2s of *M. incognita* (McClure and Stynes, 1988; Spiegel and McClure, 1991). The amphidial exudates of J2s of *H. schachtii* are composed of O-glycans with mannose or glucose, N-acetylglucosamine and galactose and/or N-acetylgalactosamine forming the oligosaccharide chains (Aumann, 1994). Information from behavioral studies using lectins is conflicting and inconclusive (Zuckerman and Esnard, 1994) perhaps because, in some instances, the lectins did not penetrate the amphidial secretions. Although certain lectins inhibited the attraction of *C. elegans* to its food source, possibly because the lectins bound to membrane receptors of dendritic nerve extensions, Aumann et al. (1990) found that lectins did not inhibit sex pheromone reception by males of *H. schachtii*. Treating males of *H. schachtii* with the sulphydryl reagent, mersalyl acid, inhibited their ability to move to a pheromone-emitting source; including dithiothreitol (a sulphydryl group protectant) in the treatment blocked the effect, indicating that sulphydryl groups may be involved in pheromone reception (Aumann, 1991).

Low concentrations (<1 ppm) of nematicides impair chemotaxis, but not motility, of plant-parasitic nematodes. Ultrastructural examination of the amphids of adult female *Pratylenchus penetrans* after exposure to 5 and 10 ppm of the nematicide aldicarb, equivalent to field rate, showed hypertrophy of the internal dendrite terminals within the sheath cell and large, electron-lucent granules in the sheath cell cytoplasm, indicating that the sheath cell metabolism is adversely affected (Trett and Perry, 1985b). The effect that these changes may have on signal reception and/or transduction is unknown.

Differences in the composition of amphidial secretions between second-stage juveniles and females of *M. incognita* have been demonstrated by Davis et al. (1992), who used a monoclonal antibody that reacted with the amphids of adult females but not with the amphids of second-stage juveniles. Stewart et al. (1993a) demonstrated the presence of a 32 kDa glycoprotein in the amphidial duct secretions and the sheath cell of J2s of six species of *Meloidogyne*, but it was not found in representatives from eight other genera including *Globodera* and *Heterodera*, indicating a more specialized function for this protein in *Meloidogyne*. The protein was found in all stages of the *Meloidogyne* life cycle, including males of *M. javanica*, but not in the sedentary adult female, where the amphids appear to be non-functional. In addition, electron microscopy indicated a difference in the morphology of amphidial secretions in the second-stage juveniles and the adult female (Stewart, et al., 1993b). Incubation of infective *M. javanica* second-stage juveniles in the antiserum against the protein significantly retarded chemotaxis (Stewart, et al., 1993b). Thus, there are indications that at separate stages of the life cycle the amphids of some plant-parasitic nematodes may have a different function or a different combination of functions; they may even be non-functional, as in unstimulated, unhatched *G. rostochiensis* second-stage juveniles (Jones, et al., 1994). This is further underscored in the vertebrate parasite, *Haemonchus contortus*, where amphids differ between the first stage and the infective third stage (Li, et al., 2000, 2001).

4.6.5.7 Ocelli

Ocelli are a specialized pair of red/brownish pigmented sensory organs that occur on the lateral or dorsolateral sides of the pharynx (Fig. 4.15(D)). The distribution of ocelli in several orders of Adenophorea might suggest convergent evolution, and that the term "ocelli" may not always be applied to homologous or even morphologically similar structures. Van de Velde and Coomans (1988) demonstrated that in *Diplolaimiella* (Monhysterida) ocelli include a lens-like structure slightly anterior to a lamellae stack receptor; the receptor is almost completely embedded in pigment granules apparently produced by a modified sheath cell. In other taxa examined, a lens was absent (Croll, et al., 1972, 1975; Siddiqui and Viglierchio, 1970, 1971), in some cases the pigment was located within, rather than adjacent to the pharynx (Bollerup and Burr, 1979; Burr and Webster, 1971), and in one case the ocelli apparently are structurally linked to the amphids (Burr and Burr, 1975). Ocelli appear to have evolved independently in taxa that are primarily marine, as well as in Mermithida that includes parasites of insects. In the latter group, the mechanisms of refraction, the location of the receptors and the composition of the pigment have been investigated (Burr and Babinski, 1990; Burr, et al., 1989; Robinson, et al., 1990). Some nematodes lacking ocelli are also believed to respond to light by a dermal light sense, or otherwise unidentified receptors.

4.6.5.8 Phasmids

Phasmids, specialized pairs of sensory organs in the posterior lateral field, are restricted to Secernentea (Figs 4. 11 (A), 4. 13 (B), 4. 16 (C), (D)). Typically the structure is that of a simple chemoreceptor, including one or two neuron receptors, each with a typical ciliary region surrounded by a sheath cell, and the receptors terminate in a duct surrounded by one or two socket cells. Consistent with a chemoreceptor, the duct opens to the exterior (Baldwin, 1985; Carta and Baldwin, 1990a, 1990b). Similar to the amphid, the socket cell is highly secretory, apparently producing a mucoprotein that often exudes from the duct. The duct may be variously modified to include an ampulla or large cup near the opening, as is the case for the scutellum of *Scutellonema*, or the lens-like phasmid of certain Heteroderinae (Carta and Baldwin, 1990b).

Although there is much information on the ultrastructure of the phasmids of many species of nematodes, there is only speculation about their function. Wang and Chen (1985) suggested that one function of the phasmids of *Scutellonema brachyurum* may be as receptors for female sex pheromones. If true, this could be a specialized function in a limited number of species, as phasmids are present on females of many species of nematodes and Riga et al. (1996, 1997) obtained changes in electrical activity from the anterior sense organs of males of *G. rostochiensis* and *G. pallida* in response to female sex pheromones. Clearly, in species of the Adenophorea, where phasmids are usually absent, other sensilla must assume the sensory role of phasmids of the Secernentea. Baldwin (1985) found two types of phasmids (termed A and B) in second-stage juveniles of *H. schachtii*; in A, the larger type, the sheath cell was deeply invaginated to form a receptor cavity in which secretions accumulated, whereas in B no receptor cavity or secretions were present. The functional significance of these differences is unknown.

References

Abbas, M., and G. D. Cain. 1979. *In vitro* activation and behavior of the ameboid sperm of *Ascaris suum* (Nematoda). Cell and Tissue Research 200: 273 – 284.

Adamson, M. L. 1989. Evolutionary biology of the Oxyurida (Nematoda): Biofacies of a haplodiploid taxon. Advances in Parasitology 28: 175 – 228.

Aguinaldo, A. M. A., J. M. Turbeville, L. S. Linford, M. C. Rivera, J. R. Garey,

R. A. Raff, and J. A. Lake. 1997. Evidence for a clade of nematodes, arthropods and other moulting animals. Nature 387: 489 – 493.

Albert, P. S., S. J. Brown, and D. L. Riddle. 1981. Sensory control of dauer larvae formation in *Caenorhabditis elegans*. Journal of Comparative Neurology 198: 435 – 451.

Albertson, D. G., and J. N. Thomson. 1976. The pharynx of *Caenorhabditis elegans*. Philosophical Transactions of the Royal Society (London) B 275: 299 – 325.

Allen, M. W. 1960. Alimentary canal, excretory and nervous systems. Pp. 136 – 139. In J. N. Sasser, and W. R. Jenkins, eds. Nematology fundamentals and recent advances with emphasis on plant parasitic and soil forms. Chapel Hill, NC: The University of North Carolina Press.

Anderson, R. V., and J. R. Byers. 1975. Ultrastructure of the esophageal procorpus in the plant parasitic nematode, *Tylenchorhynchus dubius*, and functional aspects in relation to feeding. Canadian Journal of Zoology 53: 1581 – 1595.

Andrássy, I. 1976. Evolution as a basis for the systematization of nematodes. London: Pitman Publishing.

Anya, A. O. 1965. The structure and chemical composition of the nematode cuticle. Observations on some oxyurids and *Ascaris*. Parasitology 56: 179 – 198.

Ashton, F. T., and G. A. Schad. 1996. Amphids in *Strongyloides stercoralis* and other parasitic nematodes. Parasitology Today 12: 187 – 194.

Atkinson, H. J., and C. O. E. Onwuliri. 1981. *Nipplostrongylus brasiliensis* and *Haemonchus contortus*: Function of the excretory ampulla of the third-stage larva. Experimental Parasitology 52: 191 – 198.

Aumann, J. 1989. Enzymatic effects on lectin binding to *Heterodera schachtii* (Nematoda: Heteroderidae) males. Nematologica 35: 461 – 468.

Aumann, J. 1991. Effect of the sulphydryl reagent, mersalyl acid, on sex pheromone reception by *Heterodera schachtii* (Nematoda: Heteroderidae) males. Parasitology 102: 299 – 301.

Aumann, J. 1993. Permeability of chemosensillum-associated exudates for lectins in a plant-parasitic and in a free-living nematode. Fundamental and Applied Nematology 16: 381 – 382.

Aumann, J. 1994. The chemical nature of the amphidial and "excretory" system secretions of *Heterodera schachtii* (Nematoda: Heteroderidae) males. Fundamental and Applied Nematology 17: 186 – 189.

Aumann, J., and U. Wyss. 1987. Lectin binding sites on mobile stages of *Heterodera schachtii* Schmidt (Nematoda: Heteroderidae). Nematologica 33: 410 – 418.

Aumann, J., and U. Wyss. 1989. Histochemical studies on exudates of *Heterodera schachtii* (Nematoda: Heteroderidae) males. Revue de Nématologie 12: 309 – 315.

Aumann, J., C. D. Clemens, and U. Wyss. 1990. Influence of lectins on female sex pheromone reception by *Heterodera schachtii* (Nematoda: Heteroderidae) males. Journal of Chemical Ecology 16: 2371 – 2380.

Baccetti, B., and B. A. Afzelius. 1976. The biology of the sperm cell. Monographs in Developmental Biology 54: 1 – 254.

Baccetti, B., R. Dallai, S. G. De Zio, and A. Marinari. 1983. The evolution of the nematode spermatozoon. Gamete Research 8: 309 – 323.

Bacic, A., I. Kahane, and B. M. Zuckerman. 1990. *Panagrellus redivivus* and

Caenorhabditis elegans: evidence for the absence of sialic acids. Experimental Parasitology 71: 483 – 488.

Baldwin, J. G. 1982. Fine structure of the esophagus of males of *Sarisodera hydrophila* (Heteroderoidea). Journal of Nematology 14: 279 – 291.

Baldwin, J. G. 1983. Fine structure of body wall cuticle of females of *Meloidodera charis, Atalodera lonicerae* and *Sarisodera hydrophila* (Heteroderidae). Journal of Nematology 15: 370 – 381.

Baldwin, J. G. 1985. Fine structure of the phasmid of second-stage juveniles of *Heterodera schachtii* (Tylenchida: Nematoda). Canadian Journal of Zoology 63: 534 – 542.

Baldwin, J. G. 1992. Evolution of cyst and noncyst-forming Heteroderinae. Annual Review of Phytopathology 30: 271 – 290.

Baldwin, J. G., and C. D. Eddleman. 1992. Fine structure of body wall cuticle of females of *Bellodera utahensis* and *Ekphymatodera thomasoni* (Sarisoderini, Heteroderinae). Canadian Journal of Zoology 70: 268 – 273.

Baldwin, J. G., and C. D. Eddleman. 1995. Buccal capsule of *Zeldia punctata* (Nemata: Cephalobidae). Canadian Journal of Zoology 73: 648 – 656.

Baldwin, J. G., R. M. Souza, and C. M. Dolinski. 2001. Fine stracture and phylogenetic significance of a muscular basal bolbin *Basiria gracilis* (Tylenchida). Nematology 3: 681 – 688.

Baldwin, J. G., and H. Hirschmann. 1973. Fine structure of cephalic sense organs in *Meloidogyne incognita* males. Journal of Nematology 5: 285 – 302.

Baldwin, J. G., and H. Hirschmann. 1975. Body wall fine structure of the anterior region of *Meloidogyne incognita* and *Heterodera glycines* males. Journal of Nematology 7: 175 – 193.

Baldwin, J. G., and H. Hirschmann. 1976. Comparative fine structure of the stomatal region of males of *Meloidogyne incognita* and *Heterodera glycines*. Journal of Nematology 8: 1 – 17.

Baldwin, J. G., H. Hirschmann, and A. C. Triantaphyllou. 1977. Comparative fine structure of the esophagus of males of *Heterodera glycines* and *Meloidogyne incognita*. Nematologica 23: 239 – 252.

Baldwin, J. G., L. M. Frisse, J. T. Vida, C. D. Eddleman, and W. K. Thomas. 1997a. An evolutionary framework for the study of developmental evolution in a set of nematodes related to *Caenorhabditis elegans*. Molecular Phylogenetics and Evolution 8: 249 – 259.

Baldwin, J. G., R. M. Giblin-Davis, C. D. Eddleman, D. S. Williams, J. T. Vida, and W. K. Thomas. 1997b. The buccal capsule of *Aduncospiculum halicti* (Nemata: Diplogasterina): an ultrastructural and molecular phylogenetic study. Canadian Journal of Zoology 75: 407 – 423.

Bargmann, C. I., and H. R. Horvitz. 1991. Chemosensory neurons with overlapping functions direct chemotaxis to multiple chemicals in *C. elegans*. Neuron 7: 729 – 742.

Bargmann, C. I., J. H. Thomas, and H. R. Horvitz. 1990. Chemosensory cell function in the behaviour and development of *Caenorhabditis elegans*. Cold Spring Harbor Symposium on Quantitative Biology 55: 529 – 538.

Bargmann, C. I., E. Hartwieg, and H. R. Horvitz. 1993. Odorant-selective genes and neurons mediate olfaction in *C. elegans*. Cell 74: 515 – 527.

Barnes, R. B. 1987. Invertebrate Zoology. New York: Saunders College Publishing.

Batson, B. S. 1979. Body wall of juvenile and adult *Gastromermis boophthorae* (Nematoda: Mermithidae): Ultrastructure and nutritional role. International Journal for Parasitology 9: 495 – 503.

Beukeboom, L. W., and R. C. Vrijenhoek. 1998. Evolutionary genetics and ecology of sperm-dependent parthenogenesis. Journal of Evolutionary Biology 11: 755 – 782.

Bird, A. F., and J. Bird. 1991. The Structure of Nematodes. New York: Academic Press.

Bird, A. F., and J. Bird. 1998. Introduction to functional organization. Pp. 1 – 24. In R. N. Perry and D. J. Wright, eds. The Physiology and Biochemistry of Free-living and Plant-parasitic Nematodes. Wallingford, UK: CAB International.

Blaxter, M. L., and W. M. Robertson. 1998. The cuticle. Pp. 25 – 48. In R. N. Perry and D. J. Wright, eds. The Physiology and Biochemistry of Free-living and Plant-parasitic Nematodes. Wallingford, UK: CAB International.

Blaxter, M. L., P. De Ley, J. R. Garey, L. X. Liu, P. Scheldeman, A. Vierstraete, J. R. Vanfleteren, L. Y. Mackey, M. Dorris, L. M. Frisse, J. T. Vida, and W. K. Thomas. 1998. A molecular evolutionary framework for the phylum Nematoda. Nature (London) 392: 71 – 75.

Bollerup, G., and A. H. Burr. 1979. Eyespot and other pigments in nematode esophageal muscle cells. Canadian Journal of Zoology 57: 1057 – 1069.

Brown, G., R. K. Callow, C. D. Green, G. W. Jones, J. H. Rayner, A. M. Shepherd, and T. D. Williams. 1971. The structure, composition and origin of the sub-crystalline layer in some species of the genus *Heterodera*. Nematologica 17: 591 – 599.

Bruce, R. G. 1970. Structure of the esophagus of the infective juvenile and adult *Trichinella spiralis*. Journal of Parasitology 56: 540 – 549.

Burr, A. H. J., and C. P. F. Babinski. 1990. Scanning motion, ocellar morphology and orientation mechanisms in the phototaxis of the nematode *Mermis nigrescens*. Journal of Comparative Physiology A 167: 257 – 268.

Burr, A. H. J., and C. Burr, 1975. The amphid of the nematode *Oncholaimus vesicarius*: Ultrastructural evidence for a dual function as chemoreceptor and photoreceptor. Journal of Ultrastructure Research 51: 1 – 15.

Burr, A. H. J., and C. Gans, 1998. Mechanical significance of obliquely striated architecture in nematode muscle. Biological Bulletin 194: 1 – 6.

Burr, A. H. J., and J. M. Webster. 1971. Morphology of the eyespot and description of two pigment granules in the esophageal muscle of a marine nematode *Oncolaimus vesicarius*. Journal of Ultrastructure Research 36: 621 – 632.

Burr, A. H. J., D. K. Eggleton, R. Patterson, and J. T. Leutscher-Hazelhoff. 1989. The role of hemoglobin in the phototaxis of the nematode *Mermis nigrescens*. Photochemistry and Photobiology 49: 89 – 96.

Butschli, O. 1874. Beitrage zur Kenntniss des nervensystems der nematoden. Archiv fuer Mikroskopische Anatomie 10: 74 – 100.

Buzzell, G. R., and R. I. Sommerville. 1985. The structure of the esophagus in the third-stage infective larva of *Anisakis* sp. (Nematoda: Anisakidae). Transactions of the American Microscopical Society 104: 86 – 94.

Byers, J. R., and R. V. Anderson. 1972. Ultrastructural morphology of the body wall, stoma, and stomatostyle of the nematode, *Tylenchorhynchus dubius* (Butschli, 1873) Filipjev. Canadian Journal of Zoology 50: 457 – 465.

Byers, J. R., and R. V. Anderson. 1973. Morphology and ultrastructure of the intestine in a plant-parasitic nematode, *Tylenchorhynchus dubius*. Journal of Nematology 5: 28 – 37.

Cares, J. E., and J. G. Baldwin. 1994a. Comparative fine structure of sperm of *Verutus volvingentis* and *Meloidodera floridensis* (Heteroderinae, Nematoda). Canadian Journal of Zoology 72: 1481 – 1491.

Cares, J. E., and J. G. Baldwin. 1994b. Fine structure of sperm of *Ekphymatodera thomasoni* (Heteroderinae, Nemata). Journal of Nematology 26: 375 – 383.

Cares, J. E., and J. G. Baldwin. 1995. Comparative fine structure of sperm of *Heterodera schachtii* and *Punctodera chalcoensis* with phylogenetic implications for Heteroderinae (Nemata: Heteroderidae). Canadian Journal of Zoology 73: 309 – 320.

Carta, L. K., and J. G. Baldwin. 1990a. Ultrastructure of phasmid development in *Meloidodera floridensis* and *M. charis* (Heteroderinae). Journal of Nematology 22: 362 – 385.

Carta, L. K., and J. G. Baldwin. 1990b. Phylogenetic implications of phasmid absence in males of three genera in Heteroderinae. Journal of Nematology 22: 386 – 394.

Chalfie, M., and J. White. 1988. The nervous system. Pp. 337 – 391. In W. B. Wood, ed. The Nematode *Caenorhabditis elegans*. New York: Cold Spring Harbor Laboratory Press.

Chalfie, M., J. E. Sulston, J. G. White, E. Southgate, J. N. Thomson, and S. Brenner. 1985. The neural circuit for touch sensitivity in *Caenorhabditis elegans*. Journal of Neurosciences 5: 956 – 964.

Chamberlin, H. M., and P. W. Sternberg. 1993. Multiple cell interactions are required for fate specification during male spicule development in *Caenorhabditis elegans*. Development 118: 297 – 324.

Chen, T. A., and G. Y. Wen. 1972. Ultrastructure of the feeding apparatus of *Pratylenchus penetrans*. Journal of Nematology 4: 155 – 161.

Chen, T. A., and G. Y. Wen. 1980. Electron microscopy of the stomatostylet and esophagus of *Criconemoides curvatum*. Journal of Nematology 12: 72 – 83.

Chitwood, B. G. 1930. Studies on some physiological functions and morphological characters of *Rhabditis* (Rhabditidae, Nematodes). Journal of Morphology and Physiology 49: 251 – 275.

Chitwood, B. G., and M. B. Chitwood. 1950. Introduction to Nematology. Baltimore, MD: University Park Press.

Chitwood, B. G., and E. E. Wehr. 1934. The value of cephalic structures as characters in nematode classification with special reference to the superfamily Spiuroidea. Zeitschrift fur Parasitenkunde 7: 273 – 335.

Clark, W. C. 1960. The oesophago-intestinal junction in the Mononchidae (Enoplida, Nematoda). Nematologica 5: 178 – 183.

Cliff, G., and J. G. Baldwin. 1985. Fine structure of body wall cuticle of females of eight genera of Heteroderidae. Journal of Nematology 17: 286 – 296.

Commonwealth Institute of Helminthology. 1974. CIH keys to the nematode parasites of vertebrates. Anderson, R. C., A. G. Chabaud, and S. Willmott, eds. Farnham Royal. Commonwealth Agricultural Bureaux.

Coomans, A. 1962. The spicular muscles in males of the subfamily Hoplolaiminae (Tylenchida, Nematoda). Biologisch Jaarboek Dodonaea 30: 313 – 315.

Coomans, A. 1963. Stoma structure in members of the Dorylaimina. Nematologica 9: 587 – 601.

Coomans, A. 1979. The anterior sensilla of nematodes. Revue de Nématologie 2: 259 – 283.

Coomans, A., and A. De Grisse. 1981. Sensory structures. Pp. 127 – 174. In B. M. Zuckerman and R. A. Rhode, eds. Plant Parasitic Nematodes vol. III. New York: Academic Press.

Coomans, A., and M. B. Lima. 1965. Description of *Anatonchus amiciae* n. sp. (Nematoda: Mononchidae) with observations on its juvenile stages and anatomy. Nematologica 11: 413 – 431.

Coomans, A., and P. A. A. Loof. 1986. Observations on the glands of the male reproductive system in dorylaims and its phylogenetic importance. Revue de Né- matologie 9: 261 – 265.

Coomans, A., and A. Van der Heiden. 1971. Structure and formation of the feeding apparatus in *Aporcelaimus* and *Aporcelaimellus* (Nematoda: Dorylaimidae). Zeitschrift fur Morphologie der Tiere 70: 103 – 118.

Cordero, D. A., and J. G. Baldwin. 1990. Effect of age on body wall cuticle morphology of *Heterodera schachtii* Schmidt females. Journal of Nematology 22: 356 – 361.

Costa, M., B. W. Draper, and J. R. Priess. 1997. The role of actin filaments in patterning the *Caenorhabditis elegans* cuticle. Development Biology 184: 373 – 384.

Cox, G. N., and D. Hirsh. 1985. Stage-specific patterns of collagen gene expression during development of *Caenorhabditis elegans*. Molecular and Cellular Biology 5: 363 – 372.

Cox, G. N., M. Kusch, and R. S. Edgar. 1981a. Cuticle of *Caenorhabditis elegans*: Its isolation and partial characterization. Journal of Cell Biology 90: 7 – 17.

Cox, G. N., S. Staprans, and R. Edgar. 1981b. The cuticle of *Caenorhabditis elegans* II. Stage-specific changes in ultrastructure and protein composition during postembryonic development. Developmental Biology 86: 456 – 470.

Croll, N. A., and A. R. Maggenti. 1968. A peripheral nervous system in Nematoda with a discussion of its functional and phylogenetic significance. Proceedings of the Helminthological Society of Washington 35: 108 – 115.

Croll, N. A., and J. M. Smith. 1974. Nematode setae as mechanoreceptors. Nematologica 20: 291 – 296.

Croll, N. A., L. Slater, and J. M. Smith. 1972. *Ancylostoma tubaeforme*: osmoregulatory ampulla of larvae. Experimental Parasitology 31: 356 – 360.

Croll, N. A., A. A. F. Evans, and J. M. Smith. 1975. Comparative nematode photoreceptors. Comparative Biochemical Physiology 51: 139 – 143.

Curtis, R. H. C., I. Segers, and K. Evans. 1996. Observations on the specificity of *Globodera pallida* and *G. rostochiensis* diagnostic proteins to the amphids of *Globodera* species. Nematologica 42: 466 – 480.

Davey, K. G. 1966. Neurosecretion and molting in some parasitic nematodes. American Zoologist 6: 243 – 249.

Davey, K. G. 1976. Hormones in nematodes. Pp. 273 – 291. In N. Croll, ed. The

Organization of Nematodes. New York: Academic Press.

Davey, K. G., and S. P. Kan. 1968. Molting in a parasitic nematode, *Phocanema decipiens*. IV. Ecdysis and its control. Canadian Journal of Zoology 46: 893 – 898.

Davies, K. G., B. R. Kerry, and C. A. Flynn. 1988. Observations on the pathogenicity of *Pasteuria penetrans*, a parasite of root-knot nematodes. Annals of Applied Biology 112: 491 – 501.

Davis, E. L., L. M. Aron, L. H. Pratt, and R. S. Hussey. 1992. Novel immunization procedures used to develop antibodies that bind to specific structures in *Meloidogyne* spp. Phytopathology 82: 1244 – 1250.

De Coninck, L. A. P. 1942. De symmetrie-verhoudingen aan het vooreinde der vrijlevende nematoden. Natuurwetenschappelijk tijdschrift 24: 29 – 68.

De Ley, P., A. De Grisse, M. C. Van de Velde, and A. Coomans. 1993. Ultrastructure of the stoma in *Tylopharynx*. Mededelingen van de Faculteit Landbouwwetenschappen, Rijksuniversiteit Gent 58: 763 – 777.

De Ley, P., M. C. Van de Velde, D. Mounport, P. Baujard, and A. Coomans. 1995. Ultrastructure of the stoma in Cephalobidae, Panagrolaimidae, and Rhabditidae, with a proposal for a revised stoma terminology in Rhabditida (Nematoda). Nematologica 11: 153 – 182.

Decraemer, W. 1995. The Family Trichodoridae: Stubby Root and Virus Vector Nematodes. London: Kluwer Academic.

Decraemer, W., and D. De Waele. 1981. Taxonomic value of the position of oesophageal gland nuclei and of oesophageal gland overlap in the Trichodoridae (Diphtherophorina). Nematologica 27: 82 – 94.

Delves, C. J., R. E. Howells, and R. J. Post. 1986. Gametogenesis and fertilization in *Dirofilaria immitis* (Nematoda: Filarioidea). Parasitology 92: 181 – 197.

Dick, T. A., and K. A. Wright. 1973. The ultrastructure of the cuticle of the nematode *Syphacia obvelata* (Rudoiphi, 1802). II. Modifications of the cuticle in the head end. Canadian Journal of Zoology 51: 172 – 202.

Dick, T. A., and K. A. Wright. 1974a. The ultrastructure of the cuticle of the nematode *Syphacia obvelata* (Rudolphi, 1802). III. Cuticle associated with the male reproductive structures. Canadian Journal of Zoology 52: 179 – 182.

Dick, T. A., and K. A. Wright. 1974b. The ultrastructure of the cuticle of the nematode *Syphacia obvelata* (Rudolphi, 1802). IV. The cuticle associated with the female excretory pore, vulva, and vagina vera. Canadian Journal of Zoology 52: 245 – 250.

Dolinski, C., G. Borgonie, R. Schnabel, and J. G. Baldwin. 1998. Buccal capsule development as a consideration for phylogenetic analysis of Rhabditida (Nemata). Development Genes and Evolution 208: 495 – 503.

Driscoll, M., and J. Kaplan. 1997. Mechanotransduction. Pp. 645 – 678. In D. L. Riddle, T. Blumenthal, B. J. Meyer, and J. R. Priess, eds. *C. elegans* II. New York: Cold Spring Harbor Laboratory Press.

Dusenbery D. B. 1974. Analysis of chemotaxis in the nematode *Caenorhabditis elegans* by countercurrent separation. Journal of Experimental Zoology 188: 41 – 48.

Eizinger, A., and R. J. Sommer. 1997. The homeotic gene *lin-39* and the evolution of nematode epidermal cell fates. Science 278: 452 – 455.

Emmons, S. W., and P. W. Sternberg. 1997. Male development and mating behaviour.

Pp. 295 – 334. In D. L. Riddle, T. Blumenthal, B. J. Meyer, and J. R. Priess, eds. *C. elegans* II. New York: Cold Spring Harbor Laboratory Press.

Endo, B. Y. 1978. Feeding plug formation in soybean roots infected with the soybean cyst nematode. Phytopathology 68: 1022 – 1031.

Endo, B. Y. 1984. Ultrastructure of the esophagus of larvae of the soybean cyst nematode, *Heterodera glycines*. Proceedings of the Helminthological Society of Washington 5: 1 – 24.

Endo, B. Y. 1985. Ultrastructure of the head region of molting second-stage juveniles of *Heterodera glycines* with emphasis on stylet formation. Journal of Nematology 17: 112 – 123.

Endo, B. Y., and W. R. Nickle. 1994. Ultrastructure of the buccal cavity region and oesophagus of the insect parasitic nematode, *Heterorhabditis bacteriophora*. Nematologica 40: 379 – 398.

Endo, B. Y., and W. R. Nickle. 1995. Ultrastructure of anterior and mid-regions of infective juveniles of *Steinernema feltiae*. Fundamental and Applied Nematology 18: 271 – 294.

Endo, B. Y., and W. P. Wergin. 1977. Ultrastructure of anterior sensory organs of the root-knot nematode, *Meloidogyne incognita*. Journal of Ultrastructure Research 59: 231 – 249.

Endo, B. Y., and W. P. Wergin. 1988. Ultrastructure of the second-stage juvenile of the root-knot nematode, *Meloidogyne incognita*. Proceedings of the Helminthological Society of Washington 55: 286 – 316.

Endo, B. Y., U. Zunke, and W. P. Wergin. 1997. Ultrastructure of the lesion nematode, *Pratylenchus penetrans* (Nemata: Pratylenchidae). Journal of the Helminthological Society of Washington 64: 59 – 95.

Endo, B. Y., U. Zunke, and W. P. Wergin. 1998. Ultrastructure of the male gonad and spermatogenesis in the lesion nematode, *Pratylenchus penetrans* (Nemata: Pratylenchidae). Journal of the Helminthological Society of Washington 65: 227 – 242.

Evans, A. A. F., and J. M. Fisher. 1970. The excretory systems of three *Ditylenchus* species. Journal of Nematology 2: 193 – 276.

Evans, A. A. F. 1998. Reproductive mechanisms. Pp. 133 – 154. In R. N. Perry, and D. J. Wright, eds. The Physiology and Biochemistry of Free-living and Plant-parasitic Nematodes. Wallingford, UK: CAB International.

Felix, M. A., and P. W. Sternberg. 1996. Symmetry breakage in the development of one-armed gonads in nematodes. Development 122: 2129 – 2142.

Fire, A., S. Q. Xu, M. K. Montgomery, S. A. Kostas, S. E. Driver, and C. G. Mello. 1998. Potent and specific genetic interference by double-stranded RNA in *Caenorhabditis elegans*. Nature 391: 806 – 811.

Fitch, D. H. A. 1997. Evolution of male tail development in rhabditid nematodes related to *Caenorhabditis elegans*. Systematic Biology 46: 145 – 179.

Fitch, D. H. A., and S. W. Emmons. 1995. Variable cell positions and cell contacts underlie morphological evolution of the rays in the male tails of nematodes related to *Caenorhabditis elegans*. Developmental Biology 170: 564 – 582.

Foor, W. E. 1970. Spermatozoan morphology and zygote formation in nematodes. Biology of Reproduction 2: 177 – 202.

Forrest, J. M. S. 1985. The occurrence of glycoproteins on juvenile heads of the potato cyst nematodes. Scottish Crop Research Institute, Annual Report for 1985: 107 – 108.

Forrest, J. M. S., and W. M. Robertson. 1986. Characterization and localization of saccharides on the head region of four populations of the potato cyst nematodes *Globodera rostochiensis* and *G. pallida*. Journal of Nematology 18: 23 – 26.

Francis, G. R., and R. H. Waterston. 1985. Muscle cell attachment in *Caenorhabditis elegans*. Journal of Cell Biology 114: 465 – 479.

Freckman, D. W., and J. G. Baldwin. 1990. Nematoda. Pp. 155 – 200. In D. L. Dindal, ed. Soil Biology Guide. New York: John Wiley & Sons.

Fujimoto, D., and S. Kanaya. 1973. Cuticlin, a noncollagen structural protein from *Ascaris* cuticle. Archives of Biochemistry and Biophysics 157: 1 – 6.

Geraert, E. 1965. The head structures of some tylenchs with special attention to the amphidial apertures. Nematologica 11: 131 – 136.

Geraert, E. 1983. The use of the female reproductive system in nematode systematics. Pp. 73 – 93. In A. R. Stone, H. M. Platt, and L. F. Khalil, eds. Concepts in Nematode Systematics. London: Academic Press.

Geraert, E., and A. De Grisse. 1981. The male copulatory system in tylenchid taxonomy (Nematoda). Nematologica 27: 432 – 442.

Goldschmidt, R. 1908. Das nervensystem von *Ascaris lumbricoides* und *megalocephala*. I. Zeitschrift fur wissenschaftliche Zoologie 90: 73 – 136.

Goldstein, B., L. M. Frisse, and W. K. Thomas. 1998. Embryonic axis specification in nematodes. Evolution of the first step in development. Current Biology 8: 157 – 160.

Goldstein, P., and A. C. Triantaphyllou. 1980. The ultrastructure of sperm development in the plant-parasitic nematode *Meloidogyne hapla*. Journal of Ultrastructure Research 71: 143 – 153.

Grootaert, P., and A. Coomans. 1980. The formation of the anterior feeding apparatus in dorylaims. Nematologica 26: 406 – 431.

Grootaert, P., and P. L. Lippens. 1974. Some ultrastructural changes in cuticle and hypodermis of *Aporcelaimellus* during the first moult (Nematoda: Dorylaimoidea). Zeitschrift fur Morphologie der Tiere 79: 269 – 282.

Grootaert, P., and U. Wyss. 1978. Ultrastructure and function of the anterior feeding apparatus in *Labronema* sp. Nematologica 24: 243 – 250.

Gu, G., G. A. Caldwell, and M. Chalfie. 1996. Genetic interactions affecting touch sensitivity in *Caenorhabditis elegans*. Proceedings of the National Academy of Sciences of the USA 93: 6577 – 6582.

Hall, D. H., and R. L. Russell. 1991. The posterior nervous system of the nematode *Caenorhabditis elegans*: Serial reconstruction of identified neurons and complete pattern of synaptic interactions. Journal of Neuroscience 11: 1 – 22.

Hess, R., and G. O. Poinar, Jr. 1989. Sperm development in the nematode *Neoaplectana intermedia* (Steinernematidae: Rhabditida). Journal of Submicroscopeic Cytology and Pathology 21: 543 – 555.

Hirschmann, H. 1960. Reproduction of nematodes. Pp. 40 – 167. In J. N. Sasser, and W. R. Jenkins, eds. Nematology, Fundamentals and Recent Advances with Emphasis on Plant Parasitic and Soil Forms. Chapel Hill, NC: The University of North Carolina Press.

Hirschmann, H. 1971. Comparative morphology and anatomy. Pp. 11 – 61. In B. M. Zuckerman, W. F. Mai, and R. A. Rohde, eds. Plant Parasitic Nematodes, Vol. I. New York: Academic Press.

Hirschmann, H., and A. C. Triantaphyllou. 1967. Mode of reproduction and development of the reproductive system of *Helicotylenchus dihystera*. Nematologica 13: 558 – 574.

Hirumi, H., T. A. Chen, K. J. Lee, and K. Maramorosch. 1968. Ultrastructure of the feeding apparatus of the nematode *Trichodorus christiei*. Journal of Ultrastructure Research 24: 434 – 453.

Hirumi, H., D. J. Raski, and N. O. Jones. 1971. Primitive muscle cells of nematodes: Morphological aspects of platymyarian and shallow coelomyarian muscles in two plant parasitic nematodes, *Trichodorus christiei* and *Longidorus elongatus*. Journal of Ultrastructure Research 34: 517 – 543.

Högger, C., and G. W. Bird. 1974. Secondary male sex characteristics of *Hoplolaimus galeatus*. Journal of Nematology 6: 12 – 16.

Hong, K., and M. Driscoll. 1994. A transmembrane domain of the putative channel subunit MEC-4 influences mechanotransduction and neurodegeneration in *C. elegans*. Nature 367: 470 – 473.

Hope, W. D. 1988. Ultrastructure of the feeding apparatus of *Rhabdodemania minima* Chitwood, 1936 (Enoplida: Rhabdodemaniidae). Journal of Nematology 20: 118 – 140.

Hopper, B. E., and S. P. Meyers. 1966. Aspects of the life cycle of marine nematodes. Helgolânder wiss Meeresunters 13: 444 – 449.

Huang, M., G. Gu, E. L. Ferguson, and M. Chalfie. 1995. A stomatin-like protein necessary for mechanosensation in *C. elegans*. Nature 378: 292 – 295.

Hussey, R. S. 1989. Disease-inducing secretions of plant-parasitic nematodes. Annual Review of Phytopathology 27: 123 – 141.

Hussey, R. S., and F. M. W. Grundler. 1998. Nematode parasitism of plants. Pp. 213 – 243. In R. N. Perry, and D. J. Wright, eds. The Physiology and Biochemistry of Free-living and Plant-parasitic Nematodes. Wallingford, UK: CAB International.

Hyman, L. H. 1951. The Invertebrates: Acanthocephala, Aschelminthes, and Entoprocta. The Pseudocoelomate Bilateria. Vol. III. New York: McGraw-Hill Book.

Inglis, W. G. 1964a. The comparative anatomy of the Ascaridoid cuticle (Nematoda). Bulletin de la Sociét Francaise 89: 317 – 338.

Inglis, W. G. 1964b. The structure of the nematode cuticle. Proceedings of the Zoological Society of London 1: 465 – 502.

Jacobs, P. J. F., J. Heyns, A. Coomans, and G. C. Loots. 1987. Studies on the ultrastructure of the pharyngeal region in Longidoridae (Nematoda). Nematologica 33: 199 – 212.

Jamuar, M. P. 1966. Studies of spermiogenesis in a nematode, *Nippostrongylus brasiliensis*. Journal of Cell Biology 31: 381 – 396.

Jansson, H. B., A. Jeyaprakash, R. A. Damon, Jr, and B. M. Zuckerman. 1984. *Caenorhabditis elegans* and *Panagrellus redivivus*: Enzyme-mediated modification of chemotaxis. Experimental Parasitology 58: 270 – 277.

Johnson, I. L., and J. D. Barry. 1996. Temporal reiteration of a precise gene expression pattern during nematode development. EMBO Journal 15: 3633 – 3639.

Johnson, P. W., S. D. Van Gundy, and W. W. Thomson. 1970. Cuticle formation in

Hemicycliophora arenaria, *Aphelenchus avenae* and *Hirschmanniella gracilis*. Journal of Nematology 2: 59 – 79.

Johnstone, I. L. 1994. The cuticle of the nematode *Caenorhabditis elegans*: A complex collagen structure. BioEssays 16: 171 – 178.

Jones, G. M. 1975. Ultrastructure of the amphids and amphidial glands of *Syngamus trachea* and *Cyathostoma lari* (Nematoda: Syngamidae). Ph. D. thesis, University of Wales, Aberystwyth, UK.

Jones, J. T., R. N. Perry, M. R. L. Johnston, and P. R. Burrows. 1992. Investigations of the secretions of nematode sense organs using molecular biological techniques. European Society of Nematologists, 21st International Nematology Symposium. Albufeira, Portugal.

Jones, J. T., R. N. Perry, and M. R. L. Johnston. 1993. Changes in the ultrastructure of the cuticle of the potato cyst nematode, *Globodera rostochiensis*, during development and infection. Fundamental and Applied Nematology 16: 433 – 445.

Jones, J. T., R. N. Perry, and M. R. L. Johnston. 1994. Changes in the ultrastructure of the amphids of the potato cyst nematode, *Globodera rostochiensis*, during development and infection. Fundamental and Applied Nematology 17: 369 – 382.

Jones, J. T., V. Blok, H. Popeijus, A. Prior, M. Phillips, and G. Smant. 2000. ESTs (expressed sequence tags) and parasitism-sequences and functions. Pp. 228 – 237. In C. T. Griffin, A. M. Burnell, M. J. Downes, and R. Muldar, eds. Developments in Entomopathogenic Nematode/Bacterial Research. Luxembourg: European Commission.

Kaplan, J. M., and H. R. Horvitz. 1993. A dual mechanosensory and chemosensory neuron in *Caenorhabditis elegans*. Proceedings of the National Academy of Sciences of the USA 90: 2227 – 2231.

Kennedy, E., H. E. Welch, and T. A. Dick. 1973. Cordon and spine ultrastructure of the waterfowl parasite *Echinuria uncinata* (Rudolphi 1819) Soloviev 1912 (Spirurida: Nematoda). Canadian Journal of Zoology 51: 1133 – 1137.

Kiefer, E., W. Rudin, and H. Hecker. 1989. Cytochemical demonstration of lectin binding-sites in the cuticle and tissues of *Acanthocheilonema viteae* (Filaroidea). Acta Tropica 46: 3 – 15.

Kimble, J., and W. J. Sharrock. 1983. Tissue-specific synthesis of yolk proteins in *Caenorhabditis elegans*. Developmental Biology 96: 189 – 196.

Klass, M. R., and D. Hirsh. 1981. Sperm isolation and biochemical analysis of major sperm protein from *Caenorhabditis elegans*. Developmental Biology 84: 299 – 312.

Kramer, J. M. 1997. Extracellular matrix. Pp. 471 – 500. In D. L. Riddle, T. Blumenthal, B. J. Meyer, and J. R. Priess, eds. *C. elegans* II. New York: Cold Spring Harbor Laboratory Press.

Kruger, J. C. de W. 1991. Ultrastructure of sperm development in the plant-parasitic nematode *Xiphinema theresiae*. Journal of Morphology 210: 163 – 174.

Kulikov, V. V., O. I. Dashchenko, T. V. Koloss, and V. V. Yuskin. 1998. A description of the free-living marine nematode *Euchromadora robusta* sp. n. (Nematoda: Chromadorida) with observations on the ultrastructure of the body cuticle. Russian Journal of Nematology 6: 103 – 110.

Laufer, J. S., P. Bazzicalupo, and W. B. Wood. 1980. Segregation of developmental potential in early embryos of *Caenorhabditis elegans*. Cell 19: 569 – 577.

Lee, D. L. 1965. The cuticle of adult *Nipplstrongylus brasiliensis*. Parasitology 55: 173 – 181.

Lee, D. L. 1970a. The ultrastructure of the cuticle of adult female *Mermis nigrescens* (Nematoda). Journal of Zoology, London 161: 513 – 518.

Lee, D. L. 1970b. The fine structure of the excretory system in adult *Nippostrongylus brasiliensis* (Nematoda) and a suggested function for the excretory glands. Tissue and Cell 2: 225 – 231.

Lee, D. L., and A. O. Anya. 1967. The structure and development of the spermatozoon of *Aspiculuris tetraptera* (Nematoda). Journal of Cell Science 2: 537 – 44.

Lewis, J. A., and J. A. Hodgkin. 1977. Specific neuroanatomical changes in chemosensory mutants of the nematode *Caenorhabditis elegans*. Journal of Comparative Neurology 172: 489 – 510.

Li, J., F. T. Ashton, H. R. Gamble, and G. A. Schad. 2000. Sensory neuroanatomy of a passively ingested nematode parasite, *Haemonchus contortus*: Amphidial neurons of the first stage larva. Journal of Comparative Neurology 417: 299 – 314.

Li, J., X. D. Zhu, F. T. Ashton, H. R. Gamble, and G. A. Schad. 2001. Sensory neuroanatomy of a passively ingested nematode parasite, *Haemonchus contortus*: Amphidial neurons of the third-stage larva. Journal of Parasitology 87: 65 – 72.

Lippens, P. L., A. Coomans, A. T. De Grisse, and A. Lagasse. 1974. Ultrastructure of the anterior body region in *Aporcelaimellus obtusicaudatus* and *A. obscurus*. Nematologica 20: 242 – 256.

Lopez de Mendoza, M. E., R. H. C. Curtis, and S. Gowen. 1999. Identification and characterisation of excreted-secreted products and surface coat antigens of animal and plant-parasitic nematodes. Parasitology 118: 397 – 405.

Luong, L. T., E. G. Platzer, P. De Ley, and W. K. Thomas. 1999. Morphological, molecular, and biological characterization of *Mehdinema alii* (Nematoda: Diplogasterida) from the decorated cricket (*Gryllodes sigillatus*). Journal of Parasitology 85: 1053 – 1064.

Maggenti, A. R. 1962. The production of the gelatinous matrix and its taxonomic significance in *Tylenchulus* (Nematoda: Tylenchulinae). Proceedings of the Helminthological Society of Washington 29: 139 – 144.

Maggenti, A. R. 1979. The role of cuticular strata nomenclature in the systematics of Nemata. Journal of Nematology 11: 94 – 98.

Maggenti, A. R. 1981. General Nematology. New York: Springer-Verlag.

Malakhov, V. V. 1994. Nematodes. Structure, Development, Classification and Phylogeny. Washington, DC: Smithsonian Press.

Mansir, A., and J. -L. Justine. 1995. Centrioles with ten singlets in spermatozoa of the parasitic nematode *Helignosomoides polygyrus*. Advances in Spermatozoal Phylogeny and Taxonomy 166: 119 – 128.

Mapes, C. J. 1965. Structure and function in the nematode pharynx. I. The structure of the pharynges of *Ascaris lumbricoides, Oxyuris equi, Aplectana brevicaudata,* and *Panagrellus silusiae*. Parasitology 55: 269 – 284.

McClure, M. A., and A. F. Bird. 1976. The tylenchid (Nematoda) egg shell: Formation of the egg shell in *Meloidogyne javanica*. Parasitology 72: 29 – 39.

McClure, M. A., and B. A. Stynes. 1988. Lectin binding sites on the amphidial exudates of *Meloidogyne*. Journal of Nematology 20: 321 – 326.

McLaren, D. J. 1973. The structure and development of the spermatozoon of *Dipetalonema viteae* (Nematoda: Filarioidea). Parasitology 66: 447 – 463.

McLaren, D. J. 1974. The anterior glands of adult *Necator americanus* (Nematoda: Strongyloidea). I. Ultrastructural studies. International Journal for Parasitology 4: 25 – 37.

McLaren, D. J. 1976a. Nematode sense organs. Pp. 194 – 265. In B. Dawes, ed. Advances in Parasitology. London: Academic Press.

McLaren, D. J. 1976b. Sense organs and their secretions. Pp. 139 – 161. In N. A. Croll, ed. The Organization of Nematodes. London: Academic Press.

McLaren, D. J., J. S. Burt, and B. M. Ogilvie. 1974. The anterior glands of adult *Necator americanus* (Nematoda: Strongyloidea). II. Cytochemical and functional studies. International Journal for Parasitology 4: 39 – 46.

Meissner, G. 1853. Beitrage zur anatomie und physiologie von *Mermis albicans*. Zeitschrift fur wissenschaftliche Zoologie 5: 207 – 284.

Mendel, J. E., H. C. Korswagen, K. S. Liu, Y. M. Hajdu-Cronin, M. I. Simon, R. H. A. Plasterk, and P. W. Sternberg. 1995. Participation of the protein Go in multiple aspects of behavior in *C. elegans*. Science 267: 1652 – 1655.

Moerman, D. G., and A. Fire. 1997. Muscle: Structure, function and development. Pp. 417 – 470. In D. L. Riddle, T. Blumenthal, B. J. Meyer, and J. R. Priess, eds. *C. elegans* II. New York: Cold Spring Harbor Press.

Mori, I., and Y. Ohshima. 1995. Neural regulation of thermotaxis in *Caenorhabditis elegans*. Nature 376: 344 – 348.

Mounport, D., P. Baujard, P. De Ley, M. C. Van de Velde, A. Coomans, and J. G. Baldwin. 2000. Interspecific ultrastructural variation of cuticular layers in Cephalobinae (Nematoda: Rhabditida). Journal of Nematology 32: 13 – 19.

Murrell, K. D., C. E. Graham, and M. McGreevy. 1983. *Strongyloides ratti* and *Trichinella spiralis*: Net charge of epicuticle. Experimental Parasitology 55: 331 – 339.

Musso, R. 1930. Die Genitalröhren von *Ascaris lumbricoides* und *megalocephala*. Zeitschrift fur wissenschaftliche Zoologie 137: 274 – 363.

Narang, H. K. 1970. The excretory system of nematodes: Structure and ultrastructure of the excretory system of *Enoplus brevis*. Nematologica 16: 517 – 522.

Narang, H. K. 1972. The excretory system of nematodes: structure and ultrastructure of the excretory system of *Panagrellus redivivus, Ditylenchus myceliophagus* with some observations on *D. dipsaci* and *Heterodera rostochiensis*. Parasitology 64: 253 – 268.

Nelson, F. K., and D. L. Riddle. 1984. Functional study of the *Caenorhabditis elegans* secretory-excretory system using laser microsurgery. Journal of Experimental Zoology 231: 45 – 56.

Nelson, F. K., P. S. Albert, and D. L. Riddle. 1983. Fine structure of the *Caenorhabditis elegans* secretory-excretory system. Journal of Ultrastructure Research 82: 156 – 171.

Nelson, G. A., and S. Ward. 1980. Vesicle fusion, pseudopod extension and amoeboid motility are induced in nematode spermatids by the ionophore monensin. Cell 19: 457 – 64.

Nelson, G. A., T. M. Roberts, and S. Ward. 1982. *Caenorhabditis elegans* spermatozoan locomotion: amoeboid movement with almost no actin. Journal of Cell Biology 92: 121 – 131.

Nicholas, A. C., and W. L. Stewart. 1997. Observations on the ultrastructure of *Gonionchus australis* (Xalidae, Nematoda). Journal of Nematology 29: 133 – 143.

Nourysrairi, N., N. Gourbault, and J. L. Justine. 1993. The development and evolution of actin-containing organelles during spermiogenesis of a primitive nematode. Biology of the Cell 79: 231 – 241.

O'Leary, S. A., A. M. Burnell, and J. R. Kusel. 1998. Biophysical properties of the surface of desiccation-tolerant mutants and parental strain of the entomopathogenic nematode *Heterorhabditis megidis* (strain UK211). Parasitology 117: 337 – 345.

Ogilvie, B. M., and V. E. Jones. 1971. *Nippostrongylus brasiliensis*: A review of immunity and the host/parasite relationship in the rat. Experimental Parasitology 29: 138 – 177.

Otto, A. 1816. Ueber das nervensystem der eingeweide-wurmer. Magazin fur die neuesten Entdeckungen in der gesammten Naturkunde 7: 223 – 233.

Perkins, G. A., C. W. Renken, J. Y. Song, and T. G. Frey. 1997. Electron tomography of large, multicomponent biological structures. Journal of Structural Biology 120: 219 – 227.

Perkins, L. A., E. M. Hedgecock, J. N. Thomson, and J. G. Culotti. 1986. Mutant sensory cilia in the nematode *Caenorhabditis elegans*. Developmental Biology 117: 456 – 487.

Perry, R. N. 1996. Chemoreception in plant-parasitic nematodes. Annual Review of Phytopathology 34: 181 – 199.

Perry, R. N. 1999. Desiccation survival of parasitic nematodes. Parasitology 119: S19 – S30.

Perry, R. N. 2002. Hatching. Pp. 147 – 169. In D. L. Lee, ed. The Biology of Nematodes. London: Taylor and Francis.

Perry, R. N., and J. Aumann. 1998. Behaviour and sensory responses. Pp. 75 – 102. In R. N. Perry, and D. J. Wright, eds. The Physiology and Biochemistry of Free-living and Plant-parasitic Nematodes. Wallingford, UK: CAB International.

Perry, R. N., D. Knox, and J. Beane. 1992. Enzymes released during hatching of *Globodera rostochiensis* and *Meloidogyne incognita*. Fundamental and Applied Nematology 15: 283 – 288.

Pertel, R., N. Paran, and C. F. T. Mattern. 1976. *Caenorhabditis elegans*: Localisation of cholinesterase associated with anterior nematode structures. Experimental Parasitology 39: 401 – 414.

Platt, H. M., and R. M. Warwick. 1983. Free-living marine nematodes Part I. British Enoplids. In D. M. Kermack, and R. S. K. Barnes, eds. Synopses of the Bristish Fauna (new series) No. 28. 307 Pp. Cambridge, UK: Cambridge University Press.

Platzer, A., and E. G. Platzer. 1988. Early cuticle formation in an Adenophorean nematode. International Journal for Parasitology 18: 793 – 801.

Premachandran, D., N. von Mende, R. S. Hussey, and M. A. McClure. 1988. A method for staining nematode secretions and structures. Journal of Nematology 20: 70 – 78.

Pritchard, D. I. 1993. Why do some parasitic nematodes secrete acetylcholinesterase

(AChE)? International Journal for Parasitology 23: 549 – 550.

Race, J. R., J. H. Martin, J. E. Larsh, Jr, H. W. Huizinga, and J. A. Lynn. 1968. An electron microscope study of the digestive tract of the third-stage larva of *Contracaecum multipapillatum*. The Journal of the Mitchell Society summer 84: 293 – 302.

Rankin, C. H. 1991. Interactions between two antagonistic reflexes in the nematode *C. elegans*. Journal of Comparative Physiology 169: 59 – 67.

Raski, D. J., and N. O. Jones. 1973. Ultrastructure of the cuticle of *Bunonema* spp. (Nematoda: Bunonematidae). Proceedings of the Helminthological Society of Washington 40: 216 – 227.

Reger, J. R. 1966. The fine structure of fibrillar components and plasma membrane contacts in esophageal myoepithelium of *Ascaris lumbrcoides* (var. suum). Journal of Ultrastructure Research 14: 602 – 617.

Reuter, G., J. Aumann, U. Wyss, H.-B. Jansson, and R. Schauer. 1991. *Panagrellus redivivus*: Failure to find evidence for the occurrence and biosynthesis of sialic acids. Experimental Parasitology 73: 389 – 395.

Rhoads, M. L., and R. D. Romanowski. 1974. The secretory nature of the excretory gland cells of *Strephanurus dentatus*. 3. Proteinase inhibitors. Experimental Parasitology 35: 363 – 368.

Riding, I. L. 1970. Microvilli on the outside of a nematode. Nature 226: 179 – 180.

Riemann, F. 1986. *Nicascolaimus punctatus* gen. et sp. n. (Nematoda, Axonolaimoidea, with note on sperm dimorphism in free-living marine nematodes. Zoologica Scripta 15: 119 – 124.

Riga, E., R. N. Perry, J. Barrett, and M. R. L. Johnston. 1995. Investigation of the chemosensory function of amphids of *Syngamus trachea* using electrophysiological techniques. Parasitology 111: 347 – 351.

Riga, E., R. N. Perry, J. Barrett, and M. R. L. Johnston. 1996. Electrophysiological responses of the potato cyst nematodes, *Globodera rostochiensis* and *G. pallida*, to their sex pheromones. Parasitology 112: 239 – 246.

Riga, E., D. R. Holdsworth, R. N. Perry, J. Barrett, and M. R. L. Johnston. 1997. Electrophysiological analysis of the responses of males of the potato cyst nematode, *Globodera rostochiensis*, to fractions of their homospecific sex pheromones. Parasitology 115: 311 – 316.

Robinson, A. F., G. L. Baker, and C. M. Heald. 1990. Transverse phototaxis by infective juveniles of *Agamermis* sp. and *Hexamermis* sp. Journal of Parasitology 76: 147 – 152.

Rogers, R., D. A. Denham, and G. S. Nelson. 1974. Studies with *Brugia pahangi*. 5. Structure of the cuticle. Journal of Helminthology 48: 113 – 117.

Roggen, D. R., D. J. Raski, and N. O. Jones. 1966. Cilia in nematode sensory organs. Science 152: 515 – 516.

Rohde, E. 1885. Beitrage zur kenntniss der anatomie der nematoden. Zoologische Beitrage 1: 11 – 32.

Rohde, R. A. 1960. Acetylcholinesterase in plant parasitic nematodes and an acetylcholinesterase from asparagus. Proceedings of the Helminthological Society of Washington 27: 121 – 123.

Roman, J., and H. Hirschmann. 1969. Embryogenesis and postembryogenesis in species of *Pratylenchus* (Nematoda: Tylenchidae). Proceedings of the Helminthological Society of Washington 36: 164 – 174.

Romanowski, R. D., M. L. Rhoads, and G. M. Malakatis. 1973. The secretory nature of the excretory gland cells of *Stephanurus dentatus*. II. Presence of hydrolytic enzymes. Experimental Parasitology 34: 40 – 43.

Rudin, W. 1990. Comparison of the cuticular structure of parasitic nematodes recognised by immunocytochemical and lectin binding studies. Acta Tropica 47: 255 – 268.

Samoiloff, M. R. 1980. Action of chemical and physical agents on free-living nematodes. Pp. 81 – 98. In B. M. Zuckerman, ed. Nematodes as Biological Models, Vol. 2. Aging and Other Model Systems. New York: Academic Press.

Sasser, J. N., and C. C. Carter. 1985. An Advanced Treatise on *Meloidogyne*. Raleigh, NC: North Carolina State University Graphics.

Scott, A. L. 1996. Nematode sperm. Parasitology Today 12: 425 – 430.

Searcy, D. G., M. J. Kisiel, and B. M. Zuckerman. 1976. Age-related increase of cuticle permeability in the nematode *Caenorhabditis briggsae*. Experimental Aging Research 2: 293 – 301.

Ségalat, L., D. A. Elkes, and J. M. Kaplan. 1995. Modulation of serotonin-controlled behaviors by Go in *Caenorhabditis elegans*. Science 267: 1648 – 1651.

Sengupta, P., J. H. Chou, and C. I. Bargmann. 1996. *odr-10* encodes a seven transmembrane domain olfactory receptor required for the responses to the odorant diacetyl. Cell 84: 899 – 909.

Seymour, M. K. 1983. Some implications of helical fibres in worm cuticles. Journal of the Zoological Society of London 199: 287 – 295.

Seymour, M. K., and A. M. Shepherd. 1974. Cell junctions acting as intestinal valves in nematodes. Journal of Zoology (London, England) 173: 517 – 523.

Shepherd, A. M., and S. A. Clark. 1976a. Structure of the anterior alimentary tract of the passively feeding nematode *Hexatylus viviparus* (Neotylenchidae: Tylenchida). Nematologica 22: 332 – 342.

Shepherd, A. M., and S. A. Clark. 1976b. Spermatogenesis and the ultrastructure of sperm and of the male reproductive tract of *Aphelenchoides blastophthorus* (Nematoda: Tylenchida, Aphelenchina). Nematologica 22: 1 – 9.

Shepherd, A. M., and S. A. Clark. 1983a. A re-examination of oesophageal ultrastructure in *Ditylenchus dipsaci* (Nematoda, Tylenchida) with some observations on intestinal structure. Nematologica 29: 151 – 170.

Shepherd, A. M., and S. A. Clark. 1983b. Spermatogenesis and sperm structure in some *Meloidogyne* species (Heteroderoidea, Meloidogynidae) and comparison with those in some cyst nematodes (Heteroderoidea, Heteroderidae). Revue de Nématologie 6: 17 – 32.

Shepherd, A. M., S. A. Clark, and P. J. Dart. 1972. Cuticle structure in the genus *Heterodera*. Nematologica 18: 1 – 17.

Shepherd, A. M., S. A. Clark, and D. J. Hooper. 1980. Structure of the anterior alimentary tract of *Aphelenchoides blastophthorus* (Nematoda: Tylenchida, Aphelenchina). Nematologica 26: 313 – 357.

Siddiqi, M. R. 1986. Tylenchida Parasites of Plants and Insects. London: Commonwealth

Agricultural Bureaux.

Siddiqui, I. A., and D. R. Viglierchio. 1970. Ultrastructure of photoreceptors in the marine nematode *Deontostoma californicum*. Journal of Ultrastructure Research 32: 558 – 571.

Siddiqui, I. A., and D. R. Viglierchio. 1971. Ultrastructure of the lamellated cytoplasmic inclusions in schwann cells of the marine nematode, *Deontostoma californicum*. Journal of Nematology 3: 264 – 275.

Siddiqui, I. A., and D. R. Viglierchio. 1977. Ultrastructure of the anterior body region of marine nematode *Deontostoma californicum*. Journal of Nematology 9: 56 – 81.

Singh, R. N., and J. E. Sulston. 1978. Some observations on moulting in *Caenorhabditis elegans*. Nematologica 24: 63 – 71.

Smith, J. M. 1974. Ultrastructure of the hemizonid. Journal of Nematology 6: 53 – 55.

Smith, L., and A. M. Stephenson. 1970. The hypodermis of *Enoplus brevis* and other related nematodes; no structural evidence for a peripheral nerve net. Nematologica 16: 572 – 576.

Smith, V. P., M. Selkirk, and K. Gounaris. 1996. *Brugia malayi*: Ultrastructural morphology of the cuticular surface membranes of adult parasites. Experimental Parasitology 83: 304 – 313.

Sommer, R. J., and P. W. Sternberg. 1994. Changes of induction and competence during the evolution of vulva development in nematodes. Science 265: 114 – 118.

Sommer, R. J., A. Eizinger, K.-Z. Lee, B. Jungblut, A. Bubeck, and I. Schlak. 1998. The *Pristionchus* HOX gene *Ppa-lin*-39 inhibits programmed cell death to specify the vulva equivalence group and is not required during vulval induction. Development 125: 3865 – 3873.

Sommer, R. J., I. Carmi, A. Eizinger, K. Grandien, B. Jungbult, K.-A. Lee, H. Nguyen, A. Pires-Da Silva, I. Schlak, C. B. Sigrist, and J. Srinivasan. 2000. *Pristionchus pacificus*: A satellite organism in evolutionary developmental biology. Nematology 2: 81 – 88.

Souza, R. M., and J. G. Baldwin. 1999. Changes in esophageal gland activity during the life cycle of *Nacobbus aberrans* (Nemata: Pratylenchidae). Journal of Nematology 30: 275 – 290.

Spiegel, Y., and M. A. McClure. 1991. Stage-specific differences in lectin binding to the surface of *Anguina tritici* and *Meloidogyne incognita*. Journal of Nematology 23: 259 – 263.

Sprent, J. F. A., J. Lamina, and A. McKeown. 1983. The development of *Baylisascaris tasmaniensis*. Parasitology 67: 67 – 83.

Stanssens, P., P. W. Bergom, Y. Gansemans, L. Jespers, Y. Laroche, S. Huang, S. Maki, J. Messens, M. Lauwereys, M. Cappello, P. J. Hotez, I. Lasters, and G. P. Vlasuk. 1996. Anticoagulant repertoire of the hookworm *Ancylostoma caninum*. Proceedings of the National Academy of Sciences of the USA 93: 2149 – 2154.

Steiner, G. 1933. The nematode *Cylindrogaster longistoma* (Stefanski) Goodey, and its relationship. Journal of Parasitology 20: 66 – 68.

Stengl, M., H. Hatt, and H. Breer. 1992. Peripheral processes in insect olfaction. Annual Review of Physiology 54: 665 – 681.

Sternberg, P. W., and H. R. Horvitz. 1981. Gonadal cell lineages of the nematode

Panagrellus redivivus and implications for evolution by the modification of cell lineage. Developmental Biology 88: 147 – 166.

Stewart, A. C., and W. L. Nicholas. 1994. Fine structure of the head and cervical region of *Ceramonema carinatus* (Chromadoridae: Ceramonematidae). Journal of Nematology 26: 188 – 211.

Stewart, G. R., R. N. Perry, J. Alexander, and D. J. Wright. 1993a. A glycoprotein specific to the amphids of *Meloidogyne* species. Parasitology 106: 405 – 412.

Stewart, G. R., R. N. Perry, and D. J. Wright. 1993b. Studies on the amphid specific glycoprotein gp32 in different life-cycle stages of *Meloidogyne* species. Parasitology 107: 573 – 578.

Subbotin, S. A., V. N. Chizov, and N. N. Zakharenkova. 1994. Ultrastructure of the integument of parasitic females in entomogenous tylenchids. 1. Two species of the genus *Wachekitylenchus*, *Allantonema mirabile* and *Bradynema rigidum*. Russian Journal of Nematology 2: 105 – 112.

Sulston, J. E., and H. R. Horvitz. 1977. Post-embryonic cell lineages of the nematode *Caenorhabditis elegans*. Developmental Biology 56: 110 – 156.

Sulston, J. E., M. Dew, and S. Brenner. 1975. Dopaminergic neurons in the nematode *Caenorhabditis elegans*. Journal of Comparative Neurology 163: 215 – 226.

Sulston, J. E., D. G. Albertson, and J. N. Thomson. 1980. The *Caenorhabditis elegans* male: postembryonic development of nongonadal structures. Developmental Biology 78: 542 – 76.

Talluri, M. V., L. Paggi, P. Orecchia, and R. Dallai. 1986. Fine structure of buccal cavity and esophagus in *Toxocara canis* (Nematoda, Ascarididae) infective larvae. Journal of Ultrastructure and Molecular Structure Research 97: 144 – 157.

Taylor, C. E., and W. M. Robertson. 1971. Ultrastructure of the guide ring and guiding sheath in *Xiphinema* and *Longidorus*. Nematologica 17: 303 – 307.

Taylor, C. E., P. R. Thomas, W. M. Robertson, and I. M. Roberts. 1970. An electron microscope study of the oesophageal region. Nematologica 16: 6 – 12.

Timm, R. W. 1960. The widespread occurence of the hemizonid. Nematologica 5: 150.

Trechert, G. 1977. The ultrastructure of the marine gastrotrich *Turbanella cornuta* Remane (Macrodasyoidea) and its functional and phylogenetical importance. Zoomorphologie 88: 189 – 246.

Trett, M. W., and R. N. Perry. 1985a. Functional and evolutionary implications of the anterior sensory anatomy of species of root-lesion nematodes (genus *Pratylenchus*). Revue de Nématologie 8: 341 – 55.

Trett, M. W., and R. N. Perry. 1985b. Effects of the carbamoyloxime, aldicarb, on the ultrastructure of the root lesion nematode, *Pratylenchus penetrans* (Nematoda, Pratylenchidae). Nematologica 31: 321 – 334.

Triantaphyllou, A. C. 1960. Sex determination in *Meloidogyne incognita* Chitwood, 1949 and intersexuality in *M. javanica* (Treub, 1885) Chitwood, 1949. Annales de L'Institut Phytopathologique Benaki 3: 12 – 31.

Triantaphyllou, A. C. 1983. Cytogenetic aspects of nematode evolution. Pp. 55 – 71. In A. R. Stone, H. M. Platt, and L. F. Khalil, eds. Concepts in Nematode Systematics. London: Academic Press.

Troemel, E. R., J. H. Chou, N. D. Dwyer, H. A. Colbert, and C. I. Bargmann.

1995. Divergent seven transmembrane receptors are candidate chemosensory receptors in *C. elegans*. Cell 83: 207 – 218.

Troemel, E. R., B. E. Kimmel, and C. I. Bargmann. 1997. Reprogramming chemotaxis responses: Sensory neurones define olfactory preferences in *C. elegans*. Cell 91: 161 – 169.

Turpeenniemi, T. A. 1998. Ultrastructure of spermatozoa in the nematode *Halalaimus dimorphus* (Nemata: Oxystominidae). Journal of Nematology 30: 391 – 403.

Turpeenniemi, T. A., and H. Hyvarinen. 1996. Structure and role of the renette cell and caudal glands in the nematode *Sphaerolaimus gracilis* (Monhysterida). Journal of Nematology 28: 318 – 327.

Tyler, S. 1979. Distinctive features of cilia in metazoans and their significance for systematics. Tissue and Cell 11: 385 – 400.

Van de Velde, M. C., and A. Coomans. 1987. Ultrastructure of the excretory system of the marine nematode *Monhystera disjuncta*. Tissue and Cell 19: 713 – 725.

Van de Velde, M. C., and A. Coomans. 1988. Ultrastructure of the photoreceptor of *Diplolaimella* sp. (Nematoda). Tissue and Cell 20: 421 – 430.

Van de Velde, M. C., and A. Coomans. 1991. The ultrastructure of the buccal cavity of the monhysterid nematodes *Geomonhystera disjuncta* and *Diplolaimella dievengatensis*. Revue de Nématologie 14: 133 – 146.

Van de Velde, M. C., A. Coomans, J. Heyns, and M. Claeys. 1990. Ultrastructure of the female reproductive system of *Xiphinema merideianum* (Nematoda). Revue de Nématologie 13: 211 – 223.

Van de Velde, M. C., P. De Ley, D. Mounport, P. Baujard, and A. Coomans. 1994. Ultrastructure of the buccal cavity and the cuticle of three Cephalobidae (Nematoda: Rhabditida). Nematologica 40: 541 – 563.

Vincent, A. L., L. R. Ash, and S. P. Frommes. 1975. The ultrastructure of adult *Brugia malayi* (Brug, 1927) (Nematodea: Filariodidea). Journal of Parasitology 61: 499 – 512.

Vogt, R. G., and L. M. Riddiford. 1981. Pheromone binding and inactivation by moth antennae. Nature 293: 161 – 163.

Vogt, R. G., G. D. Prestwich, and M. R. Lerner. 1991a. Odorant binding subfamilies associated with distinct classes of olfactory receptor neurones in insects. Journal of Neurobiology 22: 74 – 84.

Vogt, R. G., R. Rybczynski, and M. R. Lerner. 1991b. The biochemistry of odorant reception and transduction. Pp. 33 – 96. In D. Schild, ed. Chemosensory Information Processing. Berlin: Springer-Verlag.

Waddell, A. H. 1968. The excretory system of the kidney worm *Stephanurus dentatus* (Nematoda). Parasitology 58: 907 – 919.

Waetzig, G. H., M. Sobczak, and F. M. W. Grundler. 1999. Localization of hydrogen peroxide during the defence response of *Arabidopsis thaliana* against the plant parasitic nematode *Heterodera glycines*. Nematology 1: 681 – 686.

Wang, K. C., and T. A. Chen. 1985. Ultrastructure of the phasmids of *Scutellonema brachyurum*. Journal of Nematology 17: 175 – 186.

Ward, S. 1973. Chemotaxis by the nematode *Caenorhabditis elegans*: Identification of attractants and analysis of the response by use of mutants. Proceedings of the National Academy of Sciences of the United States of America 70: 817 – 821.

Ward, S., and J. S. Carrel. 1979. Fertilization and sperm competition in the nematode *Caenorhabditis elegans*. Developmental Biology 73: 304 – 321.

Ward, S., and M. Klass. 1982. The location of the major protein in *Caenorhabditis elegans* sperm and spermatocytes. Developmental Biology 92: 203 – 208.

Ward, S., N. Thomson, J. G. White, and S. Brenner. 1975. Electron microscopical reconstruction of the anterior sensory anatomy of the nematode *Caenorhabditis elegans*. Journal of Comparative Neurology 160: 313 – 337.

Ward, S. R., T. M. Roberts, G. A. Nelson, and Y. Argon. 1982. The development and motility of *Caenorhabditis elegans* spermatozoa. Journal of Nematology 14: 259 – 266.

Ward, S., E. Hogan, and G. A. Nelson. 1983. The initiation of spermiogenesis in the nematode *Caenorhabditis elegans*. Developmental Biology 98: 70 – 79.

Ware, R. W., D. Clark, K. Crossland, and R. L. Russell. 1975. The nerve ring of the nematode *Caenorhabditis elegans*: Sensory input and motor output. Journal of Comparative Neurology 160: 71 – 110.

Weise, R. W. 1977. A light and electron microscopic study of the dorsal buccal lancet of *Haemonchus contortus*. Journal of Parasitology 63: 854 – 857.

Wergin, W. P., and B. Y. Endo. 1976. Ultrastructure of a neurosensory organ in a root-knot nematode. Journal of Ultrastructure Research 56: 258 – 276.

Wharton, D. A. 1979. *Ascaris lumbricoides*: Water loss during desiccation of embryonating eggs. Experimental Parasitology 48: 398 – 406.

Wharton, D. A., and T. Jenkins. 1978. Structure and chemistry of the egg-shell of a nematode (*Trichuris suis*). Tissue and Cell 10: 427 – 440.

White, J. 1988. The anatomy. Pp. 81 – 122. In W. B. Wood, ed. The Nematode *Caenorhabditis elegans*. New York: Cold Spring Harbor Laboratory Press.

White, J. G., E. Southgate, J. N. Thomson, and S. Brenner. 1976. The structure of the ventral nerve cord of *Caenorhabditis elegans*. Philosophical Transactions of the Royal Society (London) B, Biological Sciences 275: 327 – 348.

Wilfred, M., and D. L. Lee. 1981. Observations on the buccal capsule and associated glands of adult *Bunostomum trigonocephalum* (Nematoda). International Journal of Parasitology 11: 485 – 492.

Winnepenninckx, B., T. Backeljau, L. Y. Mackey, J. M. Brooks, R. De Wachter, S. Kumar, and J. R. Garey. 1995. 18S rRNA data indicate that the aschelminthes are polyphyletic and consist of at least three distinct clades. Molecular Biology and Evolution 12: 1132 – 1137.

Wolf, N., D. Hirsh, and J. R. McIntosh. 1978. Spermatogenesis in males of the free-living nematode, *Caenorhabditis elegans*. Journal of Ultrastructure Research 63: 155 – 169.

Wright, D. J. 1998. Respiratory physiology, nitrogen excretion and osmotic and ionic regulation. Pp. 103 – 131. In R. N. Perry, and D. J. Wright, eds. The Physiology and Biochemistry of Free-living and Plant-parasitic Nematodes. Wallingford, UK: CAB International.

Wright, D. J., and D. R. Newall. 1976. Nitrogen excretion, osmotic, and ionic regulation in nematodes. Pp. 163 – 210. In N. A. Croll, ed. The Organization of Nematodes. London: Academic Press.

Wright, D. J., and R. N. Perry. 1998. Musculature and neurobiology. Pp. 49 – 73. In R. N. Perry, and D. J. Wright, eds. The Physiology and Biochemistry of Free-living and Plant-parasitic Nematodes. Wallingford, UK: CAB International.

Wright, E. J., and R. I. Sommerville. 1985. Spermatogenesis in a nematode *Nippostrongylus brasiliensis*. International Journal for Parasitology 15: 283 – 299.

Wright, K. A. 1972. The fine structure of the esophagus of some trichuroid nematodes I. The stichosome of *Capillaria hepatica, Trichuris myocasteri,* and *Trichuris vulpis*. Canadian Journal of Zoology 50: 319 – 324.

Wright, K. A. 1974. The fine structure of the esophagus of some trichuroid nematodes. II. The buccal capsule and anterior esophagus of *Capillaria hepatica* (Bancroft, 1893). Canadian Journal of Zoology 52: 47 – 58.

Wright, K. A. 1975. Cephalic sense organs of the rat hookworm, *Nippostrongylus brasiliensis* – form and function. Canadian Journal of Zoology 53: 1131 – 1146.

Wright, K. A. 1976. Functional organization of the nematode's head. Pp. 71 – 105. In N. A. Croll, ed. The Organization of Nematodes. New York: Academic Press.

Wright, K. A. 1980. Nematode sense organs. Pp. 237 – 295. In B. M. Zuckerman, ed. Nematodes as Biological Models, vol. 2. Aging and Other Model Systems. New York: Academic Press.

Wright, K. A., and R. Carter. 1980. Cephalic sense organs and body pores of *Xiphinema americanum* (Nematoda: Dorylaimoidea). Canadian Journal of Zoology 58: 1439 – 1451.

Wright, K. A., and W. D. Hope. 1968. Elaborations of the cuticle of *Acanthonchus duplicatus* Wieser, 1959 (Nematodea: Cyatholaimidae) as revealed by light and electron microscopy. Canadian Journal of Zoology 46: 1005 – 1011.

Wright, K. A., and J. Chan. 1974. A sense organ in the ventral anterior body wall of the nematode *Capillaria hepatica*. Canadian Journal of Zoology 52: 21 – 22.

Wright, K. A., and J. N. Thomson. 1981. The buccal capsule of *Caenorhabditis elegans* (Nematoda: Rhabditoidea): An ultrastructural study. Canadian Journal of Zoology 59: 1952 – 1961.

Wright, K. A., R. F. Carter, and W. M. Robertson. 1983. The musculature of the anterior feeding apparatus of *Xiphinema* species (Nematoda: Dorylaimoidea). Nematologica 29: 49 – 64.

Wright, K. A., D. L. Lee, and R. R. Shivers. 1985. A freeze-fracture study of the digestive tract of the parasitic nematode *Trichinella*. Tissue and Cell 17: 189 – 198.

Yeates, G. W. 1972. The oesophago-intestinal junction in the Dorylaimoidea (Nematoda). Journal of Natural History 6: 343 – 355.

Yushin, V. V., and A. Coomans. 2000. Ultrastructure of sperm development in the free-living marine nematodes of the family Chromadoridae (Chromadorida: Chromadorina). Nematology 2: 285 – 296.

Yushin, V. V., and V. V. Malakhov. 1994a. Ultrastructure of the body cuticle of free-living marine nematodes. Russian Journal of Nematology 2: 83 – 98.

Yushin, V. V., and V. V. Malakhov. 1994b. Ultrastructure of sperm cells in the female gonoduct of free-living marine nematodes from genus *Enoplus* (Nematoda: Enoplida). Fundamental and Applied Nematology 17: 513 – 519.

Zacharuk, R. Y. 1980. Ultrastructure and function of insect chemosensilla. Annual Review of Entomology 25: 27 – 47.

Zhang, Y. C., and J. G. Baldwin. 1999. Ultrastructure of the esophagus of *Diplenteron* sp. (Diplogasterida) to test hypotheses of homology with Rhabditida and Tylenchida.

Journal of Nematology 31: 1 – 19.

Zhang, Y. C., and J. G. Baldwin. 2000. Phylogenetic implications of ultrastructure of the postcorpus of *Zeldia punctata* (Cephalobina) with comparisons to *Caenorhabditis elegans* (Rhabditina) and *Diplenteron* sp. (Diplogasterina). Philosophical Transactions of the Royal Society (London) B 267: 129 – 1238.

Zhang, Y. C., and J. G. Baldwin. 2001. Ultrastructure of the postcorpus of the esophagus of *Teratocephalus lirellus* (Teratocephalida) for interpreting character evolution in Secernentea (Nematoda). Canadian Journal of Zoology 79: 16 – 25.

Zhang, Y. C., J. H. Chou, J. Bradley, C. I. Bargmann, and K. Zinn. 1998. The *Caenorhabditis elegans* seven transmembrane protein ODR-10 functions as an odorant receptor in mammalian cells. Proceedings of the National Academy of Sciences of the USA 94: 12162 – 12167.

Zuckerman, B. M., and J. Esnard. 1994. Biological control of plant nematodes – Current status and hypotheses. Japanese Journal of Nematology 24: 1 – 13.

Zuckerman, B. M., and I. Kahane. 1983. *Caenorhabditis elegans*: Stage specific differences in cuticle surface carbohydrates. Journal of Nematology 15: 535 – 538.

Zunke, U. 1986. Zur bildung der subkriustallinen Schicht bei *Heterodera schachtii* unter aseptischen bedingungen. Nematologica 31: 117 – 120.

5 Nematode Esophageal Glands and Plant Parasitism

Richard S. Hussey and Eric L. Davis

5.1 Introduction

Plant-parasitic nematodes are biotrophic parasites and can only obtain nutrients for development and reproduction from the cytoplasm of living plant cells. These nematodes have evolved diverse parasitic strategies and feeding relationships with their host plants to obtain the nutrients that are necessary for their survival. Depending on species, they feed from the cytoplasm of unmodified living plant cells or have evolved to modify root cells into elaborate discrete feeding cells (Fig. 5. 1). Plant-parasitic nematodes use a hollow, protrusible stylet to penetrate the wall of a plant cell, inject gland secretions into the cell, and withdraw nutrients from the cytoplasm. Migratory-feeding nematodes remove cytoplasm from the parasitized cell, frequently causing cell death, and then move to another cell to repeat the feeding process. Other plant parasites become sedentary and feed from a single cell or a group of cells for prolonged periods of time. For this sustained feeding, the sedentary parasites dramatically modify the root cells of susceptible hosts into elaborate feeding cells, including modulating complex changes in cell morphology, function, and gene expression. These feeding cells become the sole source of nutrients for sedentary endoparasites such as *Meloidogyne*, *Globodera*, and *Heterodera* species. Sedentary ectoparasites, such as *Criconemella xenoplax*, use a single feeding cell as a nutrient source for several days before the nematode moves on to establish another feeding cell (Hussey and Grundler, 1998).

Evolutionary adaptations of nematodes for plant parasitism led to the development of the protrusible stylet as well as marked morphological and physiological modifications of the esophagus (Bird, 1971; Maggenti, 1987). Secretory gland cells in the esophagus are the principal sources of secretions involved in plant parasitism, and these gland cells enlarged considerably as nematodes evolved from free-living nematodes to obtain nutrients from plants. Likewise the function of the secretions produced by the esophageal gland cells

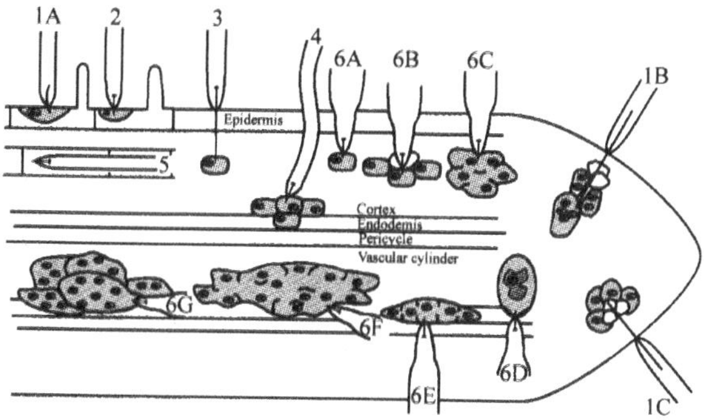

Figure 5.1 Schematic representation of feeding sites of selected root parasitic nematodes. (1) Dorylaimid migratory ectoparasites: 1A *Trichodorus* spp.; 1B *Xiphinema index*; 1C *Longidorus elongatus*. (2) – (6) Tylenchid nematodes. (2) Migratory ectoparasite: *Tylenchorhynchus dubius*. (3) Sedentary ectoparasites: *Criconemella xenoplax*. (4) Migratory ecto-endoparasites: *Helicotylenchus* spp. (5) Migratory endoparasites: *Pratylenchus* spp. (6) Sedentary endoparasites: 6A *Trophotylenchulus obscurus*; 6B *Tylenchulus semipenetrans*; 6C *Verutus volvingentis*; 6D *Cryphodera utahensis*; 6E *Rotylenchulus reniformis*; 6F *Heterodera* spp.; 6G *Meloidogyne* spp. (from Wyss, 1997).

also must have evolved to enable the nematodes to feed on plant cells or modify them into complex feeding cells.

5.2 Esophageal Gland Cell Morphology

Tylenchid nematodes are well adapted for plant parasitism. In addition to the stylet, parasitic tylenchid nematodes have a well developed esophagus for feeding on plants (Hussey, 1989a) (Fig. 5.2). The esophagus has a muscular metacorpus containing a triradiate pump chamber and three large and complex specialized secretory gland cells (Endo, 1984; Hussey and Mims, 1990). During feeding, rapid maximum dilation of the pump chamber creates the suction necessary for the nematode to ingest nutrients from the feeding cell through the lumen of the stylet and the esophagus and force their passage into the intestine.

The three large transcriptionally active esophageal gland cells, one dorsal and two subventral, produce secretory proteins involved in plant parasitism. The complexity and secretory function of these specialized glands are revealed in ultrastructural studies of the esophagi of pre-parasitic and parasitic

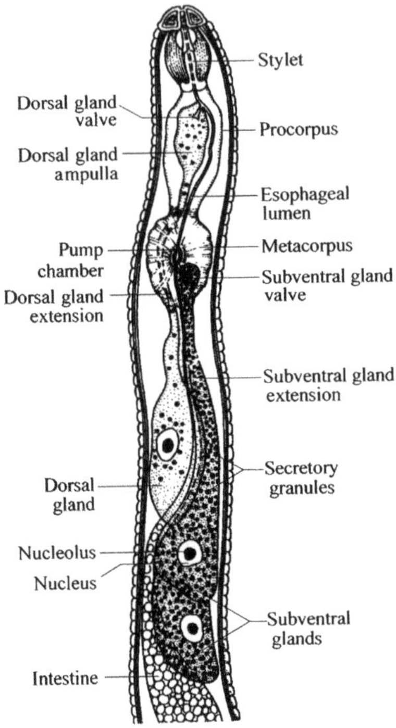

Figure 5.2 The anterior end of a second-stage juvenile of a plant-parasitic nematode (from Hussey, 1989a).

second-stage juveniles and adult females of *Meloidogyne* (Endo and Wergin, 1988; Hussey and Mims, 1990). Each esophageal gland cell overlaps the anterior end of the intestine and contains a large lobed nucleus with a prominent nucleolus, abundant Golgi bodies, rough endoplasmic reticulum, secretory granules, and other organelles typical of secretory cells. The dorsal gland cell has a long cytoplasmic extension that runs anteriorly through the metacorpus to terminate in an ampulla, a collecting reservoir for secretory granules, in the esophagus near the stylet knobs (Fig. 5.2). In contrast, the subventral gland cells have short cytoplasmic extensions that terminate in ampullae at the base of the pump chamber in the metacorpus. A sclerotized duct and an elaborate valve with a membrane-delineated end-sac connects each ampulla to the esophageal lumen (Anderson and Byers, 1975; Endo and Wergin, 1988; Hussey and Mims, 1990) (Figs 5.3(A), (B)). Secretory proteins are synthesized in the nuclear region of the gland cells and sequestered in spherical Golgi-derived membrane-bounded granules, which are transported along microtubules in the cytoplasmic extension to accumulate near the valves in the ampulla prior to their contents being secreted (Endo, 1987; Hussey and Mims, 1990) (Fig. 5.3). The association of neural processes and neurosecretory cells with the gland cytoplasmic extension and ampulla indicates that the regulated secretion of glandular proteins is controlled by the nervous system (Anderson

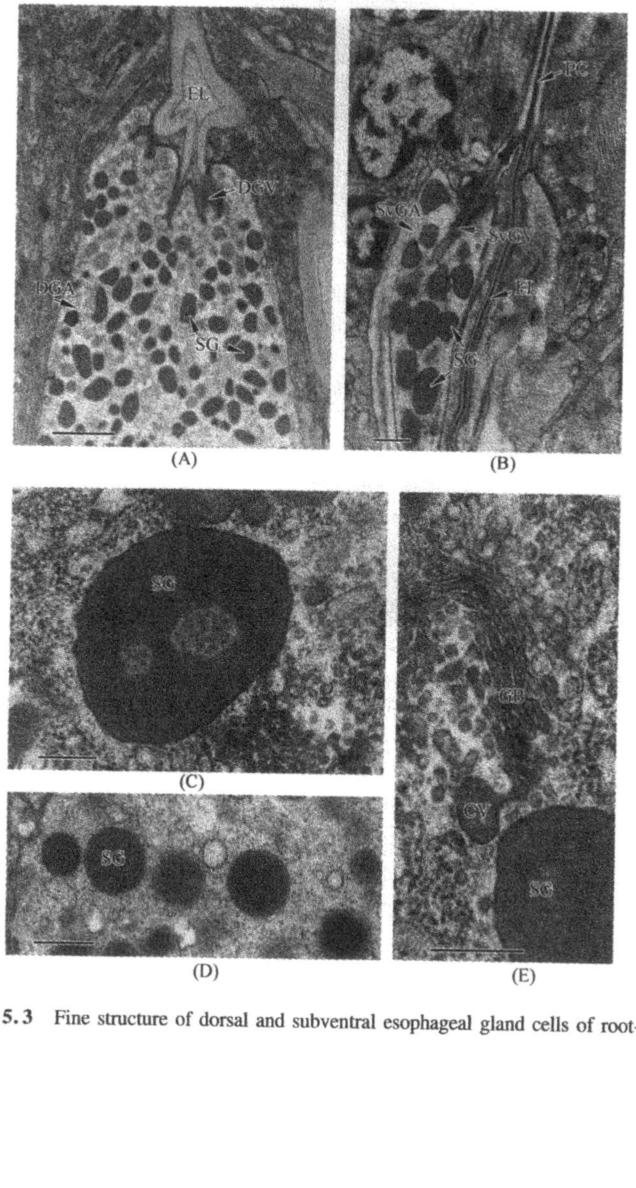

Figure 5.3 Fine structure of dorsal and subventral esophageal gland cells of root-knot

nematode. (A) Longitudinal section of dorsal gland ampulla (DGA) and valve (DGV) of an adult female. The branches of the dorsal gland valve duct converge to a common channel, which connects to the esophageal lumen (EL). Secretory granules (SG) accumulate in the ampulla prior to discharging their contents into the membrane-delineated end-sac, which encloses the DGV. Bar: 1.0 μm. (B) Longitudinal section of subventral gland ampulla (SvGA) and a subventral gland valve (SvGV) connecting the ampulla to the esophageal lumen (EL) immediately behind the pump chamber (PC) in the metacorpus of a second-stage juvenile. Numerous secretory granules (SG) are present in the SvGA. Bar: 0.5 μm. (C) Membrane-bounded secretory granule (SG) formed in a subventral gland cell of a second-stage juvenile. Bar: 0.25 μm. (D) Secretory granules (SG) formed in the dorsal gland cell of an adult female. Bar: 0.25 μm. (E) Condensing vesicle (CV) budding from a hypertrophied cisterna of a Golgi body (GB) adjacent to a mature secretory granule (SG) in a subventral gland cell. Bar: 0.25 μm. (from Hussey and Mims, 1990).

and Byers, 1975; Endo, 1984). During secretion, the gland cell is triggered to rapidly release the secretory proteins in the granules by exocytosis into the membranous end-sac of the valve. From this sac, the proteins pass through a duct to enter the lumen of the esophagus to be injected through the stylet into host tissue.

The spherical secretory granules vary in size, composition, and morphology among nematode genera and species. The secretory granules also differ between life stages of the same species and between the dorsal and subventral gland cells. Even different types of granules can be found within a gland cell in a specific life stage (Endo, 1993; Hussey and Mims, 1990). In *Meloidogyne* species, secretory granules produced in the subventral gland cells are usually less electron dense and larger (ca. 800 nm diam.) than those formed in the dorsal gland cell (ca. 500 nm diam.) (Bird, 1968a, 1969; Hussey and Mims, 1990). In addition, the composition of the matrix of the two types of granules differs (Figs 5.3(C), (D)). The matrix of subventral gland granules has an electron-transparent core containing spherical vesicles surrounded by a finely granular, electron-dense matrix (Hussey and Mims, 1990; McClure and von Mende, 1987). Dorsal gland granules, on the other hand, contain a homogeneous electron-dense matrix. Secretory granules produced in a regulated secretory pathway in other organisms contain at least two types of molecules: passenger proteins (secretory proteins) destined for export out of the cell and specific proteins involved in granule transport and exocytosis (Burgess and Kelly, 1987; Moore, et al., 1988).

During parasitic development of sedentary endoparasitic species (e. g.,

Meloidogyne), the esophageal gland cells and secretory granules undergo distinct morphological changes. As second-stage juveniles of *Meloidogyne* spp. establish a feeding relationship with host tissue and increase in body width, the dorsal gland increases and the subventral glands decrease in size (Bird, 1969, 1983). Secretory granules in the subventral glands in parasitic second-stage juveniles decrease in number (Wyss, et al., 1992), appear to be in various stages of degeneration and, in adult females, are smaller and greatly reduced in number (Bird, 1968a; Hussey and Mims, 1990). On the other hand, the dorsal gland cell in pre-parasitic second-stage juveniles contains few secretory granules (Bird, 1967; Hussey and Mims, 1990). However, soon after second-stage juveniles penetrate roots, secretory granules form in the dorsal gland cell and accumulate in its ampulla (Bird and Saurer, 1967; Wyss, et al., 1992). In adult females, the dorsal gland cell predominates and becomes the functional secreting gland cell whereas the subventral gland cells are greatly reduced in size (Bird, 1969) (Fig. 5.4). The changes in esophageal gland cell morphology, secretory granule morphology, and secretory antigens during parasitism by *Meloidogyne*, *Heterodera*, and *Globodera* species (Bird, 1967, 1983; Davis, et al., 1992; Endo, 1993; Hussey, 1989a, 1989b; Hussey and Mims, 1990; Smant, et al, 1997; Wyss, et al., 1992) indicate a changing role for the esophageal gland cells and their secretions at different stages of the nematode's life cycle.

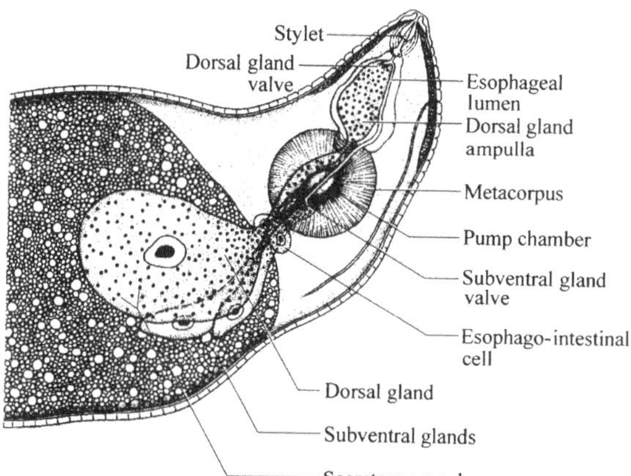

Figure 5.4 The esophagus of an adult female root-knot nematode. (from Hussey, et al., 1994).

5.3 Nature of Stylet Secretions

The bioactive molecules synthesized in the esophageal gland cells and secreted through the nematode's protrusible stylet regulate plant-nematode interactions. These stylet secretions may function in nematode penetration and migration through root tissue, modification and maintenance of root cells as feeding cells, formation of feeding tubes, and/or digestion of host cell cytoplasm to facilitate nutrient acquisition by the nematode (Hussey, 1989a). Secretory molecules from sedentary endoparasites are particularly intriguing because of the complex changes in plant cell phenotype, function, and gene expression that they modulate.

During parasitism of a plant cell, the nematode's stylet penetrates the cell wall but does not pierce the plasma membrane, which becomes invaginated around the stylet tip (Fig. 5.5). Esophageal gland cell secretions injected through the stylet by sedentary parasites transform root cells in susceptible plants into metabolically active feeding cells. These unidentified gland secretions modify, directly or indirectly, gene expression to induce profound morphological, physiological, and molecular changes in the recipient cells to enable them to function as a continuous source of nutrients for the nematode parasitic stages. Although the mechanism(s) by which these nematodes alter plant gene expression is unknown, studies suggest that factor(s) in the nematode stylet secretions induce novel gene regulatory cascades causing the parasitized root cells to differentiate into the unique feeding cells (Bird, 1996). The gland secretions may be deposited outside the plasma membrane or injected directly into the cytoplasm of the recipient cell through a perforation in the plasma membrane at the stylet orifice (Fig. 5.5). In either case, specific compounds in the secretions could bind to plant cell receptors to elicit a signal transduction cascade to modulate gene expression in the cell. Alternatively, the secretions may function as transcription factors and enter the nucleus and directly modify gene expression in the recipient plant cell.

The first evidence that nematodes inject secretions directly into host cells was published by Linford (1937), who suggested that stylet secretions from root-knot nematodes were responsible for development of the giant cells. In a detailed early study of the nature of the contents of esophageal gland cells, secretion proteins but no nucleic acids were detected in the subventral gland cell extensions in pre-parasitic *Meloidogyne javanica* second-stage juveniles (Bird and Saurer, 1967). Chemical changes detected in subventral gland granules in second-stage juveniles following invasion of root tissue led Bird (1968a) to

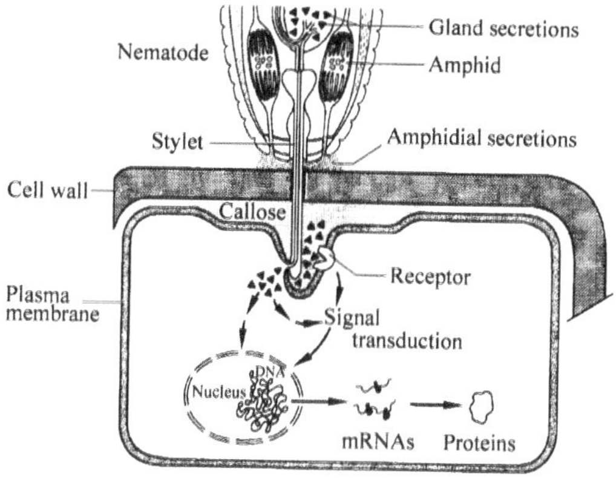

Figure 5.5 Schematic model of interactions of a sedentary parasitic nematode with its feeding cell. The anterior end of the nematode is shown with the stylet protruded through the plant cell wall but not piercing the plasma membrane which becomes invaginated around the stylet tip. Gland secretions ▼ originating from the esophageal gland cells of the nematode may be deposited outside the plasma membrane or injected directly into the cytoplasm of the recipient cell through a perforation in the plasma membrane at the stylet orifice. In either case, specific compounds in the secretions could bind to plant cell receptors to elicit a signal transduction cascade to modulate gene expression in the cell. Alternatively, the secretions may enter the nucleus and directly modify gene expression in the recipient plant cell (from Williamson and Hussey, 1996).

suggest that secretions synthesized in these gland cells following the onset of parasitism might be responsible for inducing a feeding site in a susceptible host. Ultrastructural cytochemical analyses performed on granules in subventral gland cells of pre-parasitic second-stage juveniles of *Meloidogyne incognita* were positive for acid phosphatase activity and negative for peroxidase, cellulase, DNase, RNase, and nucleic acids (Sundermann and Hussey, 1988). Cellulase (Bird, et al., 1975) and a cytokinin-like compound (Bird and Loveys, 1980) were detected in solutions in which second-stage juveniles of *M. javanica* were incubated.

Adult females of *Meloidogyne* species dissected from galls and placed in an aqueous solution produce stylet secretions that precipitate at the lip region (Bird, 1968b; Hussey, et al., 1990; Veech, et al., 1987). The fine structure of stylet secretions from adult females consists of electron-dense strands that usually surround a finely granular core (Bird, 1969; Davis, et al., 1994). These secretions stain positive for basic protein but stain negative for several enzymes and nucleic acids (Bird and Saurer, 1967). Dorsal gland cell secretory granules stained positive for nucleic acid and peroxidase but not for acid phosphatase, glucuronidase, DNase, catalase, polyphenoloxidase, or cellulase activity (Sundermann and Hussey, 1988). Although secretory granules were positive for nucleic acids using three reagents, the presence of nucleic acids in stylet secretions of adult females has not been confirmed with other techniques. Stylet secretions collected from adult females of *M. arenaria* and *M. incognita* separated into nine proteins on SDS-PAGE, three of which were identified as glycoproteins (Veech, et al., 1987).

Monoclonal antibody technology has been used to identify secretory components produced by plant-parasitic nematodes (Atkinson, et al., 1988; Davis, et al., 1992, 1994; de Boer, et al., 1996a, 1996b; Goverse, et al., 1994; Hussey, 1989b; Hussey, et al., 1990; Smant, et al., 1997). The specificities of these monoclonal antibodies reveal similarities and differences in the antigens sequestered in secretory granules formed in the subventral and dorsal gland cells (Davis, et al., 1992; de Boer, et al., 1996a). Several antibodies also have been produced that bind different antigens in stylet secretions of second-stage juveniles and adult females of root-knot and cyst nematodes (Davis, et al., 1994; Goverse, et al., 1994).

Critical unresolved questions in the study of nematode esophageal gland cell secretions are the nature and number of different secretory proteins packaged in the secretory granules and the temporal changes in the kinds of proteins secreted during the parasitic cycle by both types of gland cells. The core of a secretory granule is typically a large volume of highly concentrated protein and the number of different secretory proteins in the matrix can vary with gland cell type (Burgess and Kelly, 1987). Indeed, specific compartmentalization of a secretory protein within the matrix of granules formed in the subventral gland cells of *M. incognita* second-stage juveniles was determined by the distribution of gold labeling over the granules (Hussey, et al., 1990).

The relative importance of secretions from the dorsal gland cell versus the subventral gland cells in host-nematode interactions has been previously debated (Hussey, 1989a; Wyss, et al., 1992). The location of the dorsal gland valve near the stylet knobs readily allows dorsal gland cell secretions to be released through the stylet. Labeling of stylet secretions produced by adult females of *M. incognita* with a monoclonal antibody specific for dorsal gland granules (Hussey, et al., 1990) and video-enhanced observations of nematode secretory activity *in vivo* (Wyss and Zunke, 1986) provide evidence that dorsal gland cell secretions can be released through stylets of plant-parasitic nematodes. In contrast, the release of secretory proteins from the subventral gland cells into the esophageal lumen directly behind the triradiate pump chamber in the metacorpus and the rigid and circular lumen of the esophagus anterior to the pump chamber was considered to restrict anterior flow of subventral gland secretions (Doncaster, 1971). Therefore, secretions of the subventral gland cells were originally thought to pass only posteriorly in the esophageal lumen to the intestine to function in intracorporeal digestion (Hussey, 1989a; Wyss, et al., 1992). But studies using monoclonal antibodies specific for subventral gland secretions demonstrated that antigens in nematode stylet secretions originate in the subventral gland cells (Davis, et al., 1994; Goverse, et al., 1994; Smant, et al., 1997). In addition, the recent discovery that β-1, 4-endoglucanases (cellulase) are synthesized in subventral gland cells of cyst nematodes and secreted through the nematode's stylet *in planta* unequivocally establishes a role for subventral gland secretions in plant parasitism (Smant, et al., 1998; Wang, et al., 1999). The predicted morphological resistance to anterior flow of subventral gland secretions in the esophageal lumen during maximal pumping of the metacorpus (Doncaster, 1971) must be minimal during the secretion phase of a feeding cycle. Furthermore, production of stylet secretions *in vitro* by pre-parasitic juveniles and adult females involves very little movement of the metacorpal pump chamber (Hussey, unpublished observations; Smant, et al., 1997). Also, when the triradiate pump chamber is closed there is a slight gap between the sclerotized walls of the pump chamber that might permit anterior flow of subventral gland cell secretions during the secretion phase of the feeding cycle (Endo and Wergin, 1988).

5.4 Feeding Tubes

Removal of nutrients from feeding cells by sedentary nematodes may involve direct withdrawal of cytoplasmic components or withdrawal facilitated by a feeding tube produced by the parasite. In direct feeding, as documented with the sedentary ectoparasite *Criconemella xenoplax*, the nematode inserts its stylet into the food cell without piercing the plasma membrane, which becomes invaginated around the stylet tip (Fig. 5.6). Nutrients are withdrawn directly from the cytosol through a hole formed in the plasma membrane, which becomes tightly sealed against the wall of the stylet orifice in a manner resembling a patch-clamp electrode (Fig. 5.6(B)) (Hussey, et al., 1992). On the other hand, sedentary endoparasitic nematode species (*Globodera*, *Meloidogyne*, *Heterodera*, *Rotylenchulus* species) inject secretions that form unique tube-like structures called feeding tubes within the cytoplasm of the feeding cell. These structures were first observed by Nemec (1911) who later, without any biochemical evidence, described them as proteinaceous threads (Nemec, 1932). These feeding tubes apparently are required for efficient withdrawal of nutrients by the nematode from the feeding cell, although the mechanism for nutrient uptake is currently unknown. The morphology of feeding tubes produced by different sedentary nematodes differs greatly (Endo, 1991; Hussey and Mims, 1991; Razak and Evans, 1976; Rebois, 1980; Rumpenhorst, 1984; Wyss, et al., 1984). The electron dense feeding tube walls may be thin, e. g., tubes formed by *Rotylenchulus reniformis* (Fig. 5.8) (Rebois, 1980) and *Heterodera* species (Endo, 1991), or thick and crystalline in nature, e. g., tubes formed by *Meloidogyne* species (Fig. 5.7) (Hussey and Mims, 1991). Feeding tubes formed in giant cells by *M. incognita* adult females are approximately 1 μm wide and up to 110 μm long with a uniform lumen of 350 nm in diameter (Fig. 5.7) (Hussey and Mims,

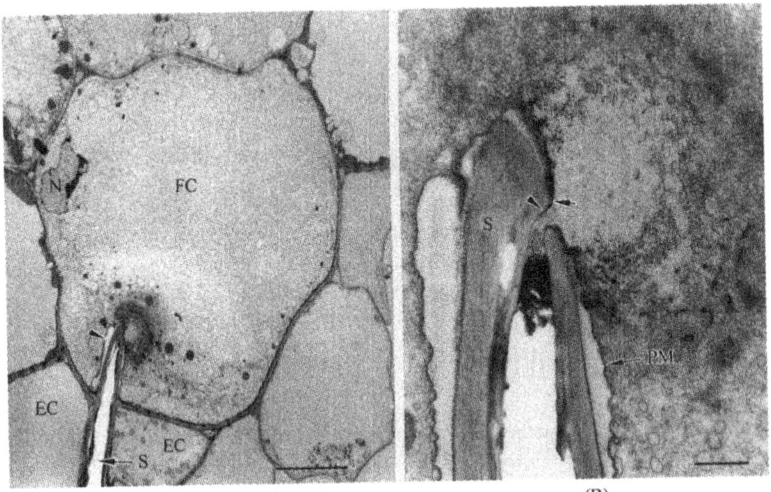

Figure 5.6 Fine structure of *Criconemella xenoplax* feeding cell relationship. (A) Cross-section of a feeding cell (FC) in the cortex of a tomato root. The nematode stylet (S) penetrated between two epidermal cells (EC) and was inserted into the feeding cell without penetrating the plasma membrane (arrowhead). A zone of the cytoplasm of the feeding cell is modified around the stylet tip. A profile of the cell's nucleus (N) is visible. Bar: 5 μm. (B) Detailed view of the opening created in the plasma membrane (PM), which is invaginated around the stylet (S) tip when the membrane (arrow) became tightly appressed (arrowhead) to the wall of the stylet orifice. Bar: 1 μm. (From Hussey, et. al., 1992.)

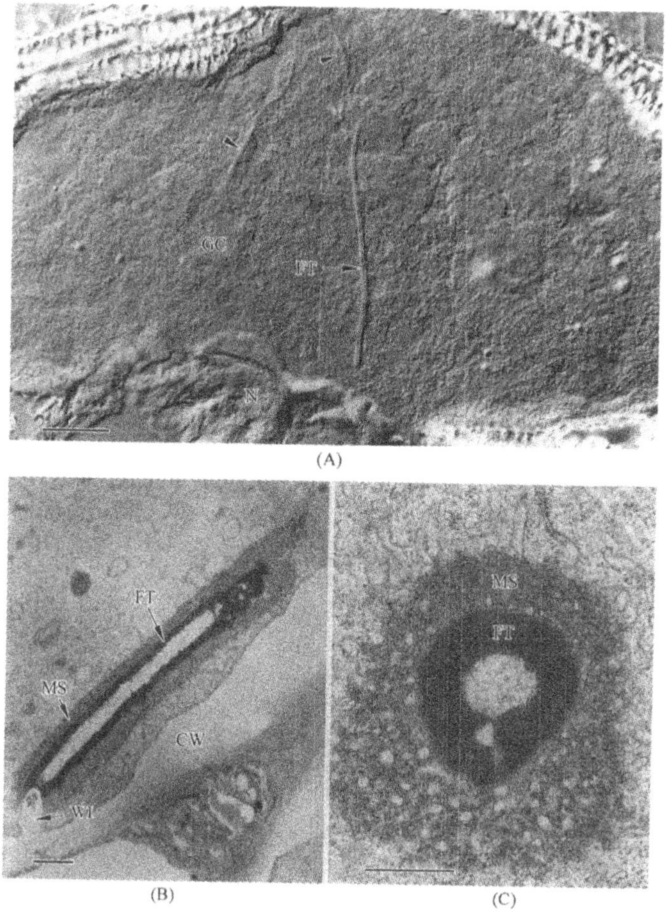

Figure 5.7 Feeding tubes formed by *Meloidogyne incognita*. (A) Nomarski light micrograph of a cryosection of a giant-cell (GC) from a root of a tomato plant infected by root-knot nematode. One long feeding tube (FT) and sections (arrowheads) of other feeding tubes are visible in the cytoplasm. A section of the head of the adult female nematode (N) that was feeding from the giant-cell is visible. Bar: 20 μm. (B) Longitudinal section through a feeding tube (FT) with its proximal end attached to a wall ingrowth (WI) where the nematode's stylet penetrated the cell wall (CW) of a giant-cell in a tomato root. The feeding tube is enveloped by a compact membrane system (MS). Bar: 1 μm. (C) Cross-section of electron-dense crystalline feeding tube (FT) wall surrounded by a dense membrane system (MS). Bar: 0.5 μm. (From Hussey and Mims, 1991.)

1991). The distal end of the feeding tube is sealed with wall material. A compact membrane system, formed from rearrangement of the giant-cell endomembrane system, forms around functioning feeding tubes and extends 450 – 500 nm outward from the tube into the cytoplasm of the giant cell (Fig. 5.7). The stimulus for reorganization of the giant-cell endoplasmic reticulum to form the compact membrane system appears to be associated with the stylet secretions that form the feeding tube (Hussey and Mims, 1991). The membrane system contiguous to the wall of the feeding tube is similar in appearance to smooth endoplasmic reticulum while ribosomes are bound to tubular elements at the periphery of the system. The enveloping membrane system might function in synthesizing and/or transporting nutrients in the metabolically active giant cells to the feeding tube for ingestion by the female (Hussey and Mims, 1991). Feeding by root-knot nematodes occurs in cycles and a new tube is formed each time the nematode removes its stylet and reinserts it into a giant cell to initiate a new feeding cycle. Crystalline walls of feeding tubes no longer being used by feeding nematodes persist in the cytoplasm and result in numerous feeding tubes being visible in a giant cell (Hussey and Mims, 1991; Rumpenhorst, 1984) (Fig. 5.7).

The manner by which a stylet removes nutrients through the feeding tube is unclear. Sedentary parasitic nematodes establish prolonged biotrophic relationships with the feeding cell, feeding in cycles and for an extended period of time from the same modified cell (Wyss, 1992). When the nematode feeds from the modified cell, the stylet is inserted through the cell wall without piercing the plasma membrane, which becomes invaginated around the stylet tip. The nematode then withdraws nutrients from the cytosol of the parasitized cell through a minute hole created in the plasma membrane at the stylet orifice (Fig. 5.8). To function in feeding, the lumen of a feeding tube on one side of the plasma membrane would have to be aligned and tightly appressed with the stylet orifice on the other side of the plasma membrane through the minute pore in the membrane. Feeding tubes produced by root-knot nematodes frequently project from short wall ingrowths through which the nematode's stylet is inserted, which may facilitate alignment of the stylet orifice with the feeding tube lumen (Fig. 5.7). Since the distal ends of feeding tubes apparently are sealed and tube walls do not have noticeable pores, the walls need to be permeable to soluble assimilates. Indeed, recent studies show that feeding tubes probably function as a molecular sieve during ingestion of nutrients from the feeding cell by the nematode (Böckenhoff and Grundler, 1994). These microinjection studies with fluorescent probes of different molecular weights showed that dextrans of a molecular weight between 20 and 40 kDa were not taken up by feeding *H. schachtii*, but dextrans less than

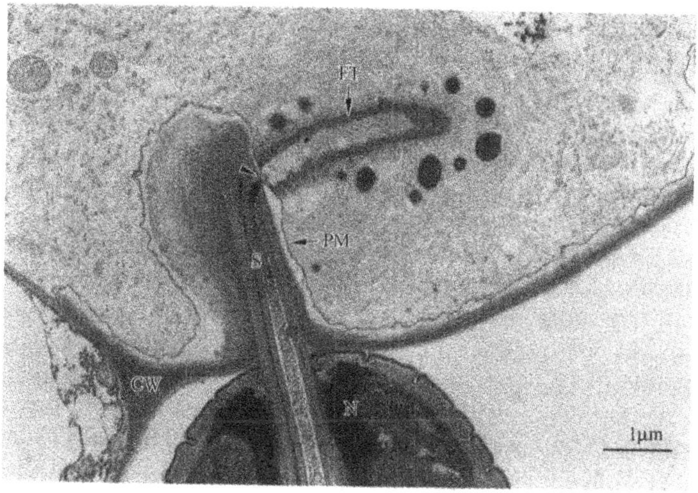

Figure 5.8 Longitudinal section of stylet (S) of *Rotylenchulus reniformis* penetrating the wall (CW) of a cortical cell in a cotton root. The plasma membrane (PM) of the cortical cell is invaginated around the inserted stylet tip. An opening (arrowhead) in the plasma membrane occurs between the stylet orifice and the lumen of a feeding tube (FT) formed from nematode secretions injected into the cytoplasm of the parasitized cell. A section of the head of the nematode (N) is visible (from Rebois, 1980).

20 kDa were ingested. The authors concluded that the feeding tube formed by *H. schachtii* had a size exclusion limit between 20 to 40 kDa. In another study, green-fluorescent protein (M_r 28 kDa) expressed in feeding cells in *Arabidopsis thaliana* was ingested by *M. incognita* but not *H. schachtii*, suggesting that the exclusion limit for globular proteins differs for feeding tubes produced by these two nematode species (Urwin, et al., 1997). The potato cyst nematode, on the other hand, ingested 32 kDa 35S-mGFP4 from syncytia of transgenic potato suggesting a large exclusion limit for feeding tubes produced by this nematode (Goverse, et al., 1998). Regardless of the manner in which feeding tubes function, more than likely they are essential for efficient ingestion of nutrients by nematodes from large feeding cells. For example, in *M. incognita* parasitism, where the stylet is inserted only 2 – 3 μm through the wall of the parasitized cell without piercing the plasma membrane, a long feeding tube would provide the nematode access to a greater area of the giant-cell cytoplasm than could be accessible with only the stylet (Fig. 5.7). However, conclusive evidence of an essential role for feeding tubes in nutrient

acquisition by sedentary endoparasites awaits experiments demonstrating that perturbation of feeding tube formation has a detrimental effect on nematode development.

The nature of nematode secretions that form feeding tubes remains unknown, although it is likely that the secretions are produced in the dorsal gland cell. A crystalline secretory component with fine structure similar to that of a feeding tube wall was observed in an open dorsal gland valve in an adult female of *M. incognita* (Hussey and Mims, 1990). Clearly, compounds forming these large tubes must be the principal component of stylet secretions of the life stages that produce feeding tubes. Evidence that feeding tubes form directly from nematode stylet secretions injected into the cytoplasm of parasitized cells includes: (1) the identical fine structure of feeding tubes formed by *Meloidogyne* species in giant cells formed in different plant species (Hussey and Mims, 1991); (2) the absence of any subcellular structures of morphology similar to feeding tubes being found in healthy cells of any plant species or plant cells parasitized by other organisms; (3) the rapid formation of feeding tubes in parasitized cells following the insertion of a stylet and initiation of the secretion phase of a feeding cycle (Wyss and Zunke, 1986); and (4) the fine structure of feeding tubes formed by different nematode genera parasitizing the same plant species (e. g., *M. incognita* and *Heterodera glycines* on soybean, Hussey, unpublished observations). Although secretions injected into the cytoplasm of giant cells by adult female, root-knot nematodes polymerize into feeding tubes, when these stylet secretions are produced *in vitro* they form a very viscous, insoluble precipitate that adheres to the nematode's lip region (Davis, et al., 1994). Apparently interaction of the stylet secretion with some component of the giant-cell cytoplasm, or perhaps the pH of the cytoplasm, induces the rapid polymerization of the secretion into the unique tube.

5.5 Secretions of Animal-parasitic Nematodes

Gland secretions of the animal-parasitic nematode *Trichinella spiralis*, appear to have a regulatory function similar to that of *M. incognita* esophageal gland cell secretions. Secretions formed in the stichosome, a large multicellular esophageal gland in *T. spiralis*, transform parasitized muscle cells into nurse cells (Despommier, 1990; Jasmer, 1995). Some of the changes induced in the parasitized muscle cells resemble the changes induced in plant cells by root-knot nematodes. The nurse cells characteristically have several hypertrophied nuclei with enlarged nucleoli and a proliferation of subcellular organelles. Like the esophageal gland cell secretory proteins of root-knot nematodes, secretions of *T. spiralis* are thought to be involved in the transformation of muscle cells, since they have been immunolocalized in the nucleoplasm of the infected cells (Despommier, 1990). This notion is supported by experiments showing that injection of secretions collected *in vitro* into rat muscles mimics cellular changes that occur *in vivo* (Ko, et al., 1992).

5.6 From Protein to Gene

Development of monoclonal antibodies that bind to secretory antigens within the esophageal gland cells of cyst and root-knot nematodes is critical in the identification of secreted proteins from cyst and root-knot nematodes (Davis, et al., 2000; Hussey and Grundler, 1998). Not only do the monoclonal antibodies bind to secretions from the nematode stylet, but the binding of the monoclonal antibodies is specific for different secretory proteins. The monoclonal antibodies have been used to monitor the expression of different esophageal antigens at various stages of nematode development (Atkinson and Harris, 1989; Davis, et al., 1994; Goverse, et al., 1994; Smant, et al., 1997). It has been demonstrated that some esophageal gland antigens are expressed in pre-parasitic second-stage juveniles of cyst nematodes, but these same antigens are absent from adult females. Two monoclonal antibodies (9C2 and 6A3) bound to a subventral gland antigen in pre-parasitic second-stage juveniles of *H. glycines* and bound to the antigen in both dorsal and subventral glands of adult females (Goverse, et al., 1994). In a more detailed study (Davis, et al., 1994), monoclonal antibodies were used to monitor the developmental expression of esophageal antigens in *M. incognita* (Table 5.1). One monoclonal antibody (6D4) bound to an antigen in both the dorsal and subventral glands of *M. incognita* through the pre-parasitic and early parasitic second-stage juvenile stages, and then was detected only within the dorsal gland of later developmental stages. The tissue specificity of antibody binding also changed for some monoclonal antibodies during nematode development. For example, a monoclonal antibody (7A9) that bound to muscle cells in *M. incognita* juveniles bound exclusively within the subventral esophageal gland cells of adult females.

Electrophoretic separations of nematode proteins have also been used to demonstrate differences in the specificity of monoclonal antibodies that bind to esophageal gland antigens. In fact, protein fractions separated electrophoretically were used to generate the monoclonal antibodies that bound to different esophageal antigens in *G. rostochiensis*, and these differential antigens were confirmed on Western blots (de Boer, et al., 1996a, 1996b). A

Table 5.1 Binding of monoclonal antibodies[a] to parasitic stages and stylet secretions of *Meloidogyne incognita*

Monoclonal Antibody	Pre-parasitic[b] J2	Early J2	Late J2	Early Female	Mature Female	Stylet[c] Secretions
3F4	subventral glands[d], esophageal lumen	subventral glands, esophageal lumen	subventral glands, esophageal lumen	subventral glands, esophageal lumen	subventral glands, esophageal lumen	J2, female
6D4	dorsal gland, subventral gland	dorsal gland, subventral gland	dorsal gland	dorsal gland	dorsal gland	J2, female
3H 11	subventral glands	ND	ND	ND	ND	J2
4B6	ND	ND	ND	dorsal gland	dorsal gland	ND
12H7	hypodermal cords	hypodermal cords	hypodermal cords	dorsal gland	dorsal gland	Female
7A9	muscle	muscle	subventral glands	subventral glands	subventral glands	ND

[a] Monoclonal antibody binding was confirmed by immunofluorescence microscopy using fluorescein isothiocyanate (FITC) conjugated anti-mouse IgG and IgM as second antibody.

[b] Pre-parasitic second-stage juveniles (J2) and mature females were tested previously (Davis, et al, 1992; Hussey, 1989a). Early J2 = sexually undifferentiated J2 removed from tomato roots 6 days after inoculation; Late J2 = sexually differentiated J2 removed from tomato roots 12 days after inoculation; Early female = adult female without eggs, removed from tomato roots 20 days after inoculation.

[c] Binding of monoclonal antibodies to stylet secretions isolated *in vitro* from J2 and females of *M. incognita*.

[d] Antibody binding localized to dorsal or subventral esophageal glands, other structure, or not detected (ND) in *M. incognita*.

high molecular weight (> 200 kDa) antigen was detected on Western blots of *M. incognita* protein probed with a gland-specific (6D4) monoclonal antibody (Hussey, 1989b). Monoclonal antibodies 9C2 and 6A3 recognized the same subventral gland antigen, an apparently heavily glycosylated protein of greater than 100 kDa, on Western blots of protein from *H. glycines* (Fig. 5.9). A dorsal gland antigen of slightly larger than 100 kDa was recognized by monoclonal antibody 5B9 on the same Western blot. A monoclonal antibody (MGR 48) raised to antigens in *G. rostochiensis*, bound to a single 49 kDa protein in *H. glycines*. MGR 48 has been demonstrated to bind specifically within the subventral gland cells of several species of cyst nematodes, but not to any other nematode species tested (de Boer, et al., 1996b). Western blots of protein from developmental stages of *G. rostochiensis* demonstrated that the subventral gland antigen recognized by MGR 48 was present in pre-parasitic and parasitic second-stage juveniles, but absent in any other sedentary developmental stages (Smant, et al., 1997). Interestingly, however, the MGR 48 antigen was also detected on Western blots of protein extracted from adult males of *G. rostochiensis*.

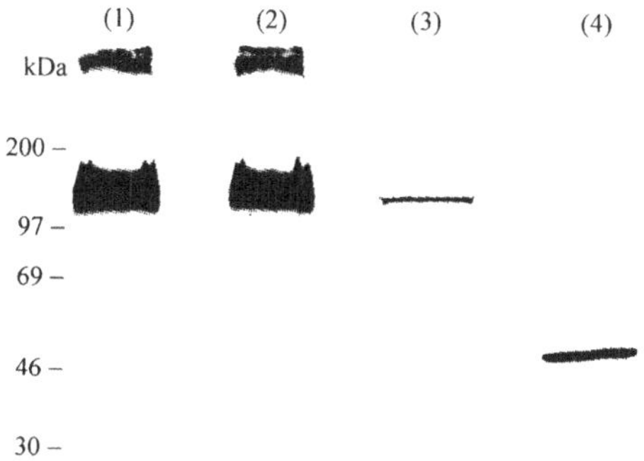

Figure 5.9 Western blot of a SDS-PAGE separation of total proteins extracted from second-stage juveniles of *Heterodera glycines* and probed with monoclonal antibodies that bind to antigens within the *H. glycines* esophageal gland cells (see Goverse, et al., 1994). Lane (1) Probed with monoclonal antibody 9C2 specific for *H. glycines* subventral esophageal gland cells (SvG); (2) Probed with monoclonal antibody 6A3 specific for *H. glycines* SvG; (3) Probed with monoclonal antibody 5B9 specific for the *H. glycines* dorsal esophageal gland cell; and (4) Probed with monoclonal antibody MGR 48 specific for SvG in several species of cyst nematodes (see de Boer, et al., 1996b). Molecular markers (kilodaltons) appear on the left.

To identify the esophageal gland proteins, the monoclonal antibodies have been used to affinity-isolate specific esophageal gland antigens directly, and to isolate cDNA clones of expressed nematode genes that encode peptides that are recognized by the antibodies. This latter technique was successful to isolate a cDNA clone (*sec-1*) from *M. incognita* using a monoclonal antibody (7A9) that was specific for a subventral esophageal gland antigen in adult females (Ray, et al., 1994). The sequence of *sec-1* predicted a protein of 61 kDa that had its strongest homologies in database searches to the rod portion of myosin heavy chains. This agrees with the finding that monoclonal antibody 7A9 appeared to bind to muscle filaments in second-stage juveniles (Davis, et al., 1994), and it is postulated that SEC-1 may be involved in the trafficking of secretory granules within *M. incognita* esophageal gland cells during the secretion process. An interesting feature of the *sec-1* mRNA was the presence of 5' spliced leader sequence (SL1) that is found in a majority of transcripts in *Caenorhabditis elegans* (Blumenthal and Steward, 1997). A single base substitution was discovered in the *sec-1* spliced leader, which was designated as SL1M (Ray, et al., 1994).

Monoclonal antibodies specific to secretory proteins within the esophageal glands of *H. glycines* have also been used to screen cDNA libraries constructed from mRNA isolated from second-stage juveniles of *H. glycines*. A subventral gland-specific monoclonal antibody (9C2) isolated a cDNA clone (*svg-1*) from *H. glycines* that encoded a protein with a predicted sequence of 412 amino acids to produce a full-length protein of 41 kDa (X. Wang and E. L. Davis, unpublished). The predicted SVG-1 protein has homology to mucins, a family of secreted proteins that have a relatively small protein core that is very heavily O-glycosylated. The pattern of the 9C2 antigen on Western blots (Fig. 5.9) is similar to that of a glycosylated protein, which may have accounted for the apparent retardation in electrophoretic mobility of the SVG-1 protein extracted from nematodes compared to the predicted 41 kDa peptide. Mucins secreted by the animal-parasitic nematode *Toxocara canis* are believed to change the nematode's surface coat and be involved in evasion of host defense (Gems and Maizels, 1996). It is possible that mucins secreted from the stylet of *H. glycines* during migration within plant roots are deposited on the nematode surface as it migrates forward through the secreted material. A second cDNA clone (*drs-1*) was isolated from an *H. glycines* expression library (Yan and Davis, 1997) using a dorsal gland-specific monoclonal antibody (5B9). The *drs-1* cDNA was 1276 bp with a predicted open reading frame (ORF) of about 300 amino acids, and database searches of this sequence had no significant homology to any reported genes. The *drs-1* probe bound to a transcript of approximately 2.7 kb on Northern blots of RNA extracted from *H. glycines* second-stage juveniles, and the size of the protein predicted from this transcript matched the size of the antigen recognized by monoclonal antibody 5B9 on Western blots (Fig. 5.9). Isolation of the full-length *drs-1* cDNA may reveal whether this protein has homology to other reported sequences.

Primers designed from amino acid sequences of *H. glycines* and *G. rostochiensis* subventral gland antigens that were immunoaffinity-isolated with monoclonal antibody MGR 48, were used to obtain two cDNA clones from each nematode that had very strong homology to β-1, 4-endoglucanases (cellulases) from bacteria (Smant, et al., 1998). The larger of the two cellulase cDNAs in each cyst species (*hg-eng-1* and *gr-eng-1*) was about 1.5 kb and encoded a predicted secretion signal peptide, cellulase catalytic domain, small peptide linker, and a cellulase binding domain (CBD). The smaller cellulase cDNA in *G. rostochiensis* (*gr-eng-2*) was missing only the CBD, and the *hg-eng-2* cellulase from *H. glycines* was missing both the CBD and peptide linker. mRNA *in situ* hybridization and anti-ENG polyclonal sera were used to confirm that *eng-1* and *eng-2* were expressed exclusively within the subventral esophageal gland cells of *H. glycines* and *G. rostochiensis* (de Boer, et al., 1998; Smant, et al., 1998). Genomic clones of *eng-1* and *eng-2* had intron/exon organization (Fig. 5.10) and putative promoter elements that were typical of eukaryotic genes (Yan, et al., 1998). The combined evidence demonstrated that the β-1, 4-endoglucanases were endogenous in cyst nematodes, the first report of cellulase genes within any animal. The strong

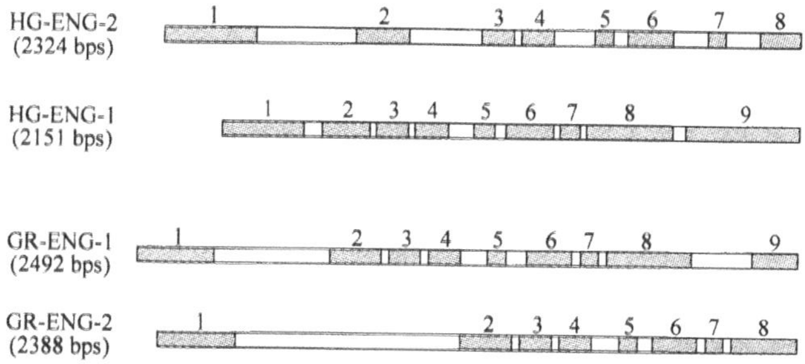

Figure 5.10 Exon/intron organization of four cyst nematode endoglucanase genes. *Heterodera glycines* genes *HG-eng-1* and *HG-eng-2*, and *Globodera rostochiensis* genes *GR-eng-1* and *GR-eng-2*. Exons (dark boxes) and introns (open boxes) are drawn proportionally to indicate the relative position and size of the introns (from Yan, et al., 1998).

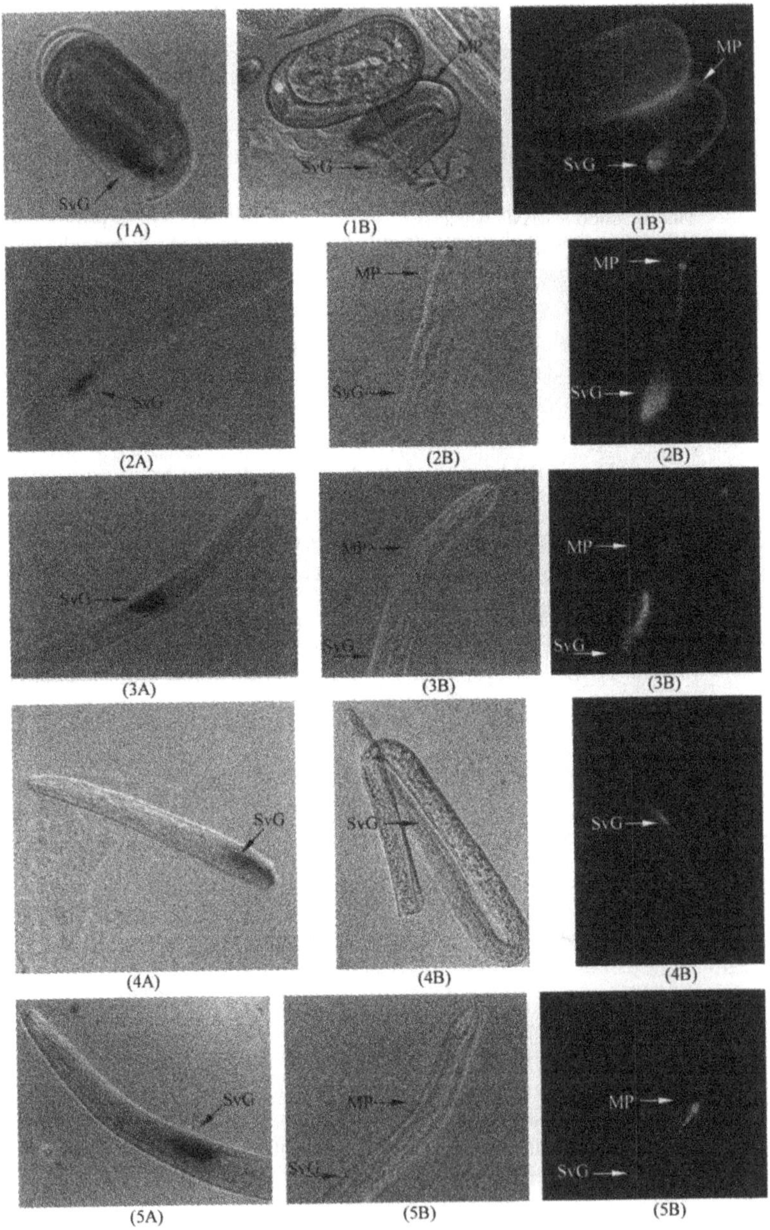

Figure 5. 11 Localization of transcripts of a *Globodera tabacum* endoglucanase gene

(*GT-eng-1*) and the endoglucanase protein (GT-ENG-1) within the nematode throughout development. All panels labeled "A" are mRNA *in situ* hybridizations of a digoxygenin-labeled antisense riboprobe of *GT-eng*-1 to transcripts within the nematode subventral esophageal gland cells (SvG). Panels labeled "B" represent a differential interference contrast (DIC) light micrograph (middle) and immunofluorescence (FITC second antibody) micrograph of GT-ENG-1 antisera binding within the SvG of the same specimen (right). (1) Second-stage juvenile within the eggshell – arrows indicate position of SvG and metacorpal pump (MP). (2) Hatched, pre-parasitic second-stage juvenile. (3) Parasitic second-stage juvenile. (4) Fourth-stage male. (5) Adult male. (From Goellner, et al., 2000.)

homology of the cyst cellulases to those from bacteria, and the weak homology of the cyst cellulases to other eukaryotic genes, suggests that the cellulase genes may have been acquired via horizontal gene transfer to an ancestor of cyst nematodes (Davis, et al., 2000; Keen and Roberts, 1998; Smant, et al., 1998; Yan, et al., 2001). Similar cellulase genes in nematodes have more recently been isolated from *Heterodera schachtii* (de Meutter, et al., 1998), *Globodera tabacum* (Goellner, et al., 2000, 2001), *Meloidogyne incognita* (Rosso, et al., 1999; Bera-Maillet, et al., 2000), and *Pratylenchue penetrans* (Uehara, et al., 2001) putatively from *Pratylenchus agilis*, *Ditylencus dipsaci*, *Bursaphelenchus xylophilus*, *Rotylenchulus reniformis*, and *Paratrichodorus minor* (Y. Yan and E. L. Davis, unpublished). The expression of cyst nematode cellulases (Fig. 5.11) has been detected in juveniles within eggs, in hatched juveniles, in parasitic second-stage juveniles, but not within third-stage juvenile males or females, nor in any subsequent female cyst nematode life stage (de Boer, et al., 1999; Goellner, et al., 2000; Smant, et al., 1997). These data are consistent with recent evidence that endogluconases of plant origin, rather than of nematode origin, are associated with cell-wall modifications in developing syncytia (Goellner, et al., 2001). Interestingly, cellulase expression was reinitiated in late-stage cyst nematode males, suggesting a primary role of cellulases secreted from the stylet of cyst nematodes in penetration, migration within, and emergence from plant root tissue. Antisera to HG-ENG-2 has been used to demonstrate the secretion of cellulase by second-stage juveniles of *H. glycines* into plant tissue as they penetrate and migrate through plant roots (Wang, et al., 1999) – the first identification of a nematode secretion *in planta*.

5.7 Expressed Esophageal Gland Genes

Analysis of differential nematode gene expression has been used as an alternative method to isolate genes expressed specifically within the esophageal gland cells of plant-parasitic nematodes. Esophageal gland regions from second-stage juveniles of *M. javanica* were excised and cDNA was prepared from this tissue by reverse-transcription-polymerase chain reaction (RT-PCR) (Lambert, et al., 1999). The cDNA pool was differentially screened against cDNA from the tail region to isolate genes that were upregulated or expressed specifically in the nematode esophageal gland region. A full-length cDNA clone was obtained from this procedure that had homology to a bacterial chorismate mutase (CM). Expression of Mj-cm-1 was localized within the esophageal gland cells of parasitic *M. javanica* by mRNA *in situ* hybridization and with antisera generated to the product of Mj-cm-1. Chorismate mutase, an enzyme associated with the shikimate pathway leading to the synthesis of phenylalanine and tyrosine, had not been previously detected in animals (as was the case with cellulase). Interestingly, the cloning and expression of *Mj-cm-1* in a chorismate mutase-deficient bacterial mutant was able to restore function. The secretion of Mj-cm-1 into the cytosol of a plant cell could potentially alter the spectrum of chorismate-dependent compounds, which, among other functions, are involved in cell wall formation, hormone biosynthesis, and synthesis of defense compounds (Doyle and Lambert, 2003).

RNA fingerprinting has been used to analyze differential gene expression between pre-parasitic and parasitic stages of *M. incognita*. A cDNA coding for a cellulose-binding protein (*Mi-cbp-1*) was isolated from *M. incognita* by this method (Ding, et al., 1998). The protein was specifically expressed in the subventral glands of *M. incognita*. The N-terminal region of the predicted peptide had no similarity to known proteins, but the C-terminus had strong homology to a CBD. *In vitro* analyses confirmed the secretion of Mi-CBP-1 through the nematode stylet. Although *in planta* secretion seems likely, the potential role of CBP in plant-nematode interactions is uncertain. A recombinant CBD derived from the cellulolytic bacterium *Clostridium cellulovorans* was found to modulate the elongation of different plant cells *in vitro* (Shpigel, et al., 1998). This finding may suggest a possible role of Mi-CBP-1 in plant cell modifications associated with root-knot nematode parasitism. A cDNA (*mi-msp-1*) encoding a putative secretory venom allergin

AG-5 like protein with strong similarity to *Ancylostoma*-secreted protein 2 (Hawdon, et al., 1999) was also obtained from *M. incognita* using this protocol (Ding, et al., 2000).

The sequencing of large numbers of cDNA clones to generate expressed sequence tags (ESTs) from parasitic nematodes represents a powerful and comprehensive genomic approach to isolate genes involved in nematode parasitism of plants. The most straightforward approach would be to construct cDNA libraries from whole nematodes of a chosen parasitic stage and to sequence as many random cDNA clones as possible (Bird, et al., 1999; Dautova, et al., 2001; Davis, et al., 2000; McCarter, et al., 2000; Popeijus, et al., 2000a). The ESTs could serve as anchors on physical maps of the nematode genome, and in addition, database searches may quickly reveal expressed genes that have an apparent role in parasitism. This was accomplished by analysis of ESTs from a pre-parasitic second-stage juvenile cDNA library of *G. rostochiensis* (Popeijus, et al., 2000b). A full-length cDNA that encoded a predicted peptide of 261 amino acids, with a signal peptide at its amino terminus was obtained that had strong homology to Class III pectate lyases of bacteria and fungi. A putative pectate lyase cDNA has also been cloned from *H. glycines* and *M. javanica* (de Boer, et al., 2002; Doyle and Lambert, 2002). Localization of transcripts to both pectate lyases to the esophageal gland cells in the nematodes indicates the potential for secretion of a pectate lyase from the nematode stylet during plant parasitism.

Sequences derived from nematode ESTs that have homology to genes with obvious (putative) functions in plant parasitism (i.e., cellulases, pectinases) can be relatively easily identified from ESTs generated from preparations of whole nematodes if abundant transcripts are present at the parasitic stage chosen for analysis. Rarely-expressed messages or genes with obscure functions in plant parasitism will be more difficult to isolate since essentially no idea of their identity exists at present. Limiting the ESTs generated to specific nematode tissues (i.e., esophageal gland cells) that are likely to be involved in the host-nematode interaction can narrow this focus. Several molecular approaches have recently been used to clone secretion genes expressed in the esophageal gland cells of plant-parasitic nematodes. Differential screening to identify genes expressed in the anterior gland region versus posterior of a nematode or between pre-parasitic and parasitic stages has resulted in cloning of two putative parasitism genes expressed in root-knot nematode gland cells (Ding, et al., 1998; Lambert, et al., 1999). A cDNA-AFLP analysis of different life stages of *G. rostochiensis* yielded three putative parasitism genes that are expressed in the dorsal gland cell (Qin, et al., 2000). However, the ability to generate cDNAs from individual cells by RT-PCR (Karrer, et al., 1995) has been coupled with a microaspiration technique used to obtain the cytoplasm of individual esophageal gland cells (Gao, et al., 2001a; Shields, et al., 1998; Wang, et al., 2001; Huang, et al., 2003). mRNA isolated from transcriptionally-active gland cells of a range of parasitic stages of *M. incognita* and *H. glycines* have been pooled and used to generate esophageal gland-specific cDNA libraries that provide a comprehensive profile of nematode esophageal gland gene expression during plant parasitism. ESTs may be sequenced directly from random clones within each library, or alternatively, genes encoding secreted peptides can be selected from these libraries using a secretion-specific vector expressed in yeast (Klein, et al., 1996). Cloning esophageal gland cell cDNAs into a signal peptide selection vector in yeast identified nine unique cDNA clones that encoded putative extracellular proteins, and four of these hybridized to transcripts within the dorsal gland cell of *H. glycines* parasitic stages (Wang, et al., 2001). Predicted proteins of two of the dorsal gland-specific clones had no similarity with known proteins in the databases; however, one protein was rerently shown to be similar to the peptide signal CLAVATA3(Olsen and Skriver, 2003) The other two predicted extracellular proteins had similarity with guanylyl cyclase from *H. glycines* (Yan and Davis, 2002) and ERp99 from *Mus musculus*. In another study, the direct method of microaspiration of cytoplasm of the esophageal gland cells to provide expressed mRNA for RT-PCR was combined with suppression subtractive hybridization (SSH) using cDNAs generated from contents microaspirated from the intestinal region of *H. glycines* to isolate genes expressed preferentially in the esophageal gland cells of parasitic stages (Gao, et al., 2001a). cDNAs encoding secretory proteins were identified by macroarraying the SSH products onto nylon membranes and high-throughput *in situ* screening of selected clones in *H. glycines* specimens. Twenty-three unique cDNA sequences from the SSH cDNA library were identified and all hybridized to genomic DNA of *H. glycines* in Southern blots. Full-length cDNAs of 21 clones were obtained by screening a gland cell long distance (LD) PCR cDNA library generated from the same esophageal gland cell cytoplasm. Deduced proteins of ten cDNA clones were preceded by a signal peptide for secretion (Nielsen, et al., 1997), and PSORT II (Nakai and Horton, 1999) computer analysis predicted eight of these proteins as extracellular, one as nuclear, and one as plasmalemma-localized. The presence of a signal peptide identifies a putative secretory protein that may have a role in SCN pathogenesis. *In situ* hybridization with digoxigenin-labeled antisense cDNA probes showed four of the predicted extracellular clones were specifically expressed in the dorsal gland cell, one in the subventral gland cells, and three (proteinases) in the intestine in *H. glycines* (Gao, et al., 2001a). Two of the predicted extracellular proteins of the clones expressed in the dorsal gland cell had homology with known proteins based on BLASTp analysis. One clone had strong similarity to salivary proline-rich glycoproteins from rat and human, whose functions remain unknown. Another cDNA clone was previously cloned from the dorsal gland cell of *H. glycines* parasitic stages but its function and any related proteins remain unknown (Wang, et al., 2001). The other dorsal gland cell specific clone, predicted by PSORT II to be extracellular, had no significant matches with any proteins in the databases. The predicted extracellular clone, expressed in the subventral gland cells, had strong similarity to the AG-5 like proteins, which are the major excretory/secretory proteins released by infective third-stage juveniles in the animal-parasitic nematode, *Ancylostoma caninum*. Interestingly, another clone, expressed in the subventral gland cells, was predicted by PSORT II analysis to be nuclear localized, but since the predicted encoded protein is preceded by a signal peptide for secretion, the product of this clone is a candidate for a *H. glycines* secretory protein that might enter the plant nucleus if secreted into a parasitized root cell. This possibility makes this clone a prime candidate for further functional analyses.

Additional ESTs in the gland cell-specific LD PCR cDNA library were analyzed by large-scale sequencing of 3711 cDNA clones macroarrayed on nylon membranes (Gao, et al., 2002C). Of these, deduced protein sequences of 261 cDNAs were preceded by a putative signal peptide for secretion based on Signal P analysis (Nielsen, et al., 1997). However, the presence of a signal peptide alone is not conclusive proof that the encoded protein functions as an extracellular protein, since membrane-associated proteins also possess signal peptides. Therefore, clones that were predicted to encode signal peptides were also analyzed by PSORT II to predict subcellular localizations (Nakai and Horton, 1999). Clones then predicted to be extracellular by PSORT II analysis are considered possible candidates for secretory proteins that could have roles in nematode parasitism. These clones are further characterized by high-throughput *in situ* hybridization with digoxigenin-labeled antisense cDNA probes to determine their developmental expression pattern in the life cycle of *H. glycines* (Gao, et al., 2001a). Of 261 cDNA clones encoding putative signal peptides that have been tested by *in situ* hybridizations with probes, 55 unique clones specifically hybridized with transcripts within the subventral (12 clones or 22%) or dorsal (43 clones or 78%) esophageal gland cells of *H. glycines*. These clones are considered prime candidates for further functional analyses. PSORT II predicted 15 of the deduced proteins to be extracellular and nuclear localized. The predicted nuclear localized proteins are additional candidates for

H. glycines secretory proteins that might target the plant nucleus if secreted into a parasitized root cell. In BLASTp analysis 77% of the predicted proteins had no similarity with known proteins in the databases. Some of the predicted proteins that had similarities with known proteins include a venom allergen-like protein (Gao, et al., 2001b), β-1,4-endoglucanases (Gao, et al., 2002), a pectate lyase, Chorismate mutase, annexin and a chitinase (Gao et al., 2002b). Noteworthy, only two of the 55 gland *in situ* positive clones had homologues in *C. elegans*. Also the developmental expression for many of the putative parasitism genes differed among the *H. glycines* parasitism stages, indicating the regulation of some parasitism genes requires stimulation by factors within soybean root tissue.

It is envisioned that improved bioinformatics systems such as PFAM peptide analysis (Eddy, 1998), and the development of efficient and definitive functional assays for putative plant-parasitic nematode parasitism genes, will advance rapidly within the next few years to confirm the identity of "pioneer" genes (cDNAs) isolated from the esophageal gland cells of plant-parasitic nematodes.

5.8 Conclusions and Future Research

The use of biotechnology to manage pests and pathogens in agriculture is coming to fruition, but information to specifically target nematode parasitism is lacking. A central problem in nematode molecular biology is identification of molecular triggers and elucidation of the molecular mechanisms that mediate specific host-plant responses. The key to understanding the molecular processes of plant parasitism by sedentary nematodes is the characterization of the parasitism genes that encode secretory proteins that are synthesized in the esophageal gland cells and released through the stylet into host tissue. By understanding these nematode signals that trigger essential complex molecular and developmental changes in host plant tissues necessary for nematode survival, we will have the basic knowledge needed to apply genetic engineering to the development of new resistant crop plants that interfere with the function of the molecular signals and thereby provide an unprecedented opportunity for limiting nematode damage to multiple crop species.

Although there has been significant progress made in the study of nematode secretions, our knowledge is still rather fragmentary. The nature and number of different components in stylet secretions must still be resolved. Most studies have focused on large macromolecules, but small molecules, e.g., oligopeptides, also could be important components of stylet secretions

(Goverse, et al., 1999; Huang, et al., 2003). Although it is possible that some nematode secretions might bind to receptors on the plasma membrane to modulate host cell changes via signal transduction pathways, more than likely most secretions are delivered directly into the cytoplasm via the stylet through a minute perforation in the plasma membrane at the stylet orifice (Hussey, et al., 1992) (Fig. 5.5).

Insights into function of parasitism genes may be inferred through homologies of the protein sequences predicted from the cDNAs with known proteins in the databases. Nevertheless, functional homologues in other organisms won't necessarily tell us what is the function of the gene product in the parasitic nematode. In addition, determining the role of proteins without functional homologues will be very challenging.

The biological activity of nematode secretory proteins needs to be determined before the role of any secretion in plant parasitism can be considered. Several experimental approaches may be utilized to elucidate the biological activity of nematode stylet secretions. When parasitism genes encoding stylet secretions have been cloned, overexpression of these genes will simplify purification of secretory proteins for use in assays of biological activity. Once sufficient quantities of a secretory protein are available, the protein can be microinjected into root protoplasts from susceptible and resistant hosts. An alternative approach is to determine the effect of expression of cloned secretion genes on the phenotype of transgenic plants. The availability of plant transformation techniques will enable nematode secretion genes, under the control of various promoters, e.g., a root specific promoter (Yamamoto, et al., 1991), to be easily transferred into plants. The effects of expression of cloned nematode secretory protein genes on cell phenotype in transgenic plants could provide important information on the ability of the gene product to modify plant tissue. However, these approaches would have limitations if transformation of plant cells requires more than a single nematode secretory protein.

Overexpression of secretory protein genes also will allow the development of polyclonal antibodies for immunolocalization studies *in planta* (Wang, et al., 1999). The polyclonal antisera for the products of cDNA clones can be used to immunolocalize nematode proteins secreted into root cells during pathogenesis and identify the subcellular targets that potentially interact with the secretions. Results of these immunocytochemical experiments on the spatial and temporal expression and localization of stylet secretions should provide evidence of their potential roles in pathogenesis.

Several aspects of plant-parasitic nematode biology make classical genetic studies difficult with these organisms. For those species, e.g., *M. incognita*, that reproduce by obligatory mitotic parthenogenesis, the opportunity to perform genetic crosses is not available. Therefore, proof-of-concept of gene function may be analyzed by gene knockout studies determining the loss-of-function phenotype. Recently it was discovered that microinjection of dsRNA can been used to manipulate gene expression (RNA-mediated interference of gene expression) in *C. elegans* (Fire, et al., 1998). Specific genes in the progeny are turned off by microinjecting dsRNA complementary to the coding region of the genes into parent nematodes. This dsRNA injection method provides for the first time a molecular genetic tool that should allow us to study gene function in plant-parasitic nematodes(Vrwin, et al., 2002). For cDNA clones with significant homology with genes with bacterial genes, the function of the nematode-derived putative genes can be assessed by complementation of bacterial strains, which are deficient in the gene (Lambert, et al., 1999).

Once disease-inducing stylet secretions are characterized, this knowledge will allow several approaches to be followed for developing transgenic resistant plants. Combinatorial libraries can be screened to identify synthetic bioactive peptides that bind to and interfere with the function of specific nematode stylet secretions *in vitro*, with the aim of expressing the peptide in transgenic plants as a novel form of nematode resistance. Synthetic peptide combinatorial libraries allow for rapid identification of peptide ligands for target molecules by an *in vitro* selection process (Reed, et al., 1997). Gene transfer techniques also have been used to achieve expression and assembly of mouse immunoglobulins (Hiatt, et al., 1989) and a synthetic gene encoding an antigen-binding single-chain F_v protein (Owen, et al., 1992) in transgenic plants. Engineering a plant with the ability to produce antibody fragments to specifically bind and inactivate a biologically active nematode stylet secretion has promise to open new avenues for novel plant resistance to these pathogens. Neutralization of a nematode stylet secretion following its injection into the cytoplasm of an antibody-producing root cell could inhibit, depending on the function of the antigen, feeding cell formation and/or nematode feeding, and subsequently suppress nematode development on the transgenic plant (Baum, et al., 1996; Rosso, et al., 1996).

Characterization of nematode stylet secretions synthesized in the esophageal gland cells is essential for defining what makes a nematode a plant parasite and for enhancing our understanding of the molecular genetics of nematode-plant interactions. Interspecific and intraspecific comparison of the structure of parasitism genes encoding stylet secretions that induce feeding cell formation will provide the knowledge that should lead to establishing a genetic basis for host range specificity among nematode species or races. Understanding this genetic variability will have an important positive effect on the development and deployment of sustainable nematode management strategies.

References

Anderson, R. V., and J. R. Byers. 1975. Ultrastructure of the esophageal procorpus in the plant parasitic nematode, *Tylenchorhynchus dubius*, and functional aspects in relation to feeding. Canadian Journal of Zoology 53: 1581 – 1595.

Atkinson, H. J., and P. D. Harris. 1989. Changes in nematode antigens recognized by monoclonal antibodies during early infections of soya beans with the cyst nematode *Heterodera glycines*. Parasitology 98: 479 – 487.

Atkinson, H. J., P. D. Harris, E. J. Halk, C. Novitski, J. Leighton-Sands, P. Nolan, and P. C. Fox. 1988. Monoclonal antibodies to the soya bean cyst nematode, *Heterodera glycines*. Annals of Applied Biology 112: 459 – 469.

Baum, T. J., A. Hiatt, W. A. Parrott, L. H. Pratt, and R. S. Hussey. 1996. Expression in tobacco of a functional monoclonal antibody specific to stylet secretions of the root-knot nematode. Molecular Plant-Microbe Interactions 9: 382 – 387.

Bera-Maillet, C., L. Arthaud, P. Abad, and M.-N. Rosso. 2000. Biochemical characterization of MI-ENG1, a family 5 endoglucanase secreted by the root-knot nematode *Meloidogyne incognita*. European Journal Biochemistry 267: 3255 – 3263.

Bird, A. F. 1967. Changes associated with parasitism in nematodes. I. Morphology and physiology of preparasitic and parasitic larvae of *Meloidogyne javanica*. Journal of Parasitology 53: 768 – 776.

Bird, A. F. 1968a. Changes associated with parasitism in nematodes. III. Ultrastructure of the egg shell, larval cuticle, and contents of the subventral esophageal glands in *Meloidogyne javanica*, with some observations on hatching. Journal of Parasitology 54: 475 – 489.

Bird, A. F. 1968b. Changes associated with parasitism in nematodes. IV. Cytochemical studies on the ampulla of the dorsal esophageal gland of *Meloidogyne javanica* and on exudations from the buccal stylet. Journal of Parasitology 54: 879 – 890.

Bird, A. F. 1969. Changes associated with parasitism in nematodes. V. Ultrastructure of the stylet exudation and dorsal esophageal gland contents of female *Meloidogyne javanica*. Journal of Parasitology 55: 337 – 345.

Bird, A. F. 1971. Specialized adaptation of nematodes to parasitism. Pp. 35 – 49. In B. M. Zuckerman, W. F. Mai, and R. A. Rohde, eds. Plant Parasitic Nematodes Vol. II New York: Academic Press.

Bird, A. F. 1983. Changes in the dimensions of the esophageal glands in root-knot nematodes during the onset of parasitism. International Journal of Parasitology 13: 343 – 348.

Bird, A. F., and B. R. Loveys. 1980. The involvement of cytokinins in a host-parasite relationship between the tomato (*Lycopersicon esculentum*) and a nematode (*Meloidogyne javanica*). Parasitology 7: 112 – 113.

Bird, A. F., and W. Saurer. 1967. Changes associated with parasitism in nematodes. II. Histochemical and microspectrophotometric analysis of preparasitic and parasitic larvae of *Meloidogyne javanica*. Journal of Parasitology 53: 1262 – 1269.

Bird, A. F., W. J. S. Downton, and J. S. Hawker. 1975. Cellulase secretion by second stage larvae of the root-knot nematode (*Meloidogyne javanica*). Marcellia 38: 165 – 169.

Bird, D. McK. 1996. Manipulation of host gene expression by root-knot nematodes. Journal of Parasitology 82: 881 – 888.

Bird, D. McK., C. H. Opperman, S. J. M. Jones, and D. L. Baillie. 1999. The *Caenorhabditis elegans* genome: A guide in the post genomics age. Annual Review of Phytopathology 37: 247 – 265.

Blumenthal, T., and K. Steward. 1997. RNA processing and gene structure, Pp. 117 – 145. In D. L. Riddle, T. Blumenthal, B. J. Meyer, and J. R. Priess, eds. *C. elegans* II. New York: Cold Spring Harbor Laboratory Press.

Böckenhoff, A., and F. M. W. Grundler. 1994. Studies on the nutrient uptake by the beet cyst nematode *Heterodera schachtii* by *in situ* microinjection of fluorescent probes into the feeding structures in *Arabidopsis thaliana*. Parasitology 109: 249 – 254.

Burgess, T. L., and R. B. Kelly. 1987. Constitutive and regulated secretion of proteins. Annual Review of Cell Biology 3: 243 – 293.

Dautova, M., M-N. Rosso, P. Abad, F. J. Gommers, J. Bakker, and G. Smant. 2001. Single pass cDNA sequencing-a powerful tool to analyze gene expression in preparasitic juveniles of the southern root-knot nematode *Meloidogyne incognita*. Nematology 3: 129 – 139.

Davis, E. L., L. M. Aron, L. H. Pratt, and R. S. Hussey. 1992. Novel immunization procedures used to develop monoclonal antibodies that bind to specific structures in *Meloidogyne* spp. Phytopathology 82: 1244 – 1250.

Davis, E. L., R. Allen, and R. S. Hussey. 1994. Developmental expression of esophageal gland antigens and their detection in stylet secretions of *Meloidogyne incognita*. Fundamentals and Applied Nematology. 17: 255 – 262.

Davis, E. L., R. S. Hussey, T. J. Baum, J. Bakker, A. Schots, M – N. Rosso, and P. Abad. 2000. Nematode parasitism genes. Annual Review of Phytopathology 38: 341 – 372.

de Boer, J. M., H. A. Overmars, G. Smant, H. Pomp, E. L. Davis, J. F. Zilverentant, A. Goverse, R. S. Hussey, J. Bakker, F. Gommers, and A. Schots. 1996a. Production and characterization of monoclonal antibodies to antigens from the second stage juveniles of the potato cyst nematode, *Globodera rostochiensis*. Fundamentals and Applied Nematology 19: 545 – 554.

de Boer, J. M., G. Smant, A. Goverse, E. L. Davis, H. Pomp, A. Overmars, M. Van Gent-Pelzer, J. F. Zilverentant, J. Stokkermans, R. S. Hussey, F. Gommers, J. Bakker, and A. Schots. 1996b. Secretory granule proteins from the subventral glands of the potato cyst nematode identified by monoclonal antibodies to a protein fraction from second-stage juveniles. Molecular Plant-Microbe Interactions 9: 39 – 46.

de Boer, J. M., Y. Yan, J. Bakker, E. L. Davis, and T. J. Baum. 1998. *In situ* hybridization to messenger RNA of *Heterodera glycines*. Journal of Nematology 30: 309 – 312.

de Boer, J. M., Y. Yan, X. Wang, G. Smant, R. S. Hussey, E. L. Davis, and T. J. Baum. 1999. Developmental expression of secretory β-1, 4-endoglucanases in the subventral esophageal glands of *Heterodera glycines*. Molecular Plant-Microbe Interactions 12: 663 – 669.

de Boer, J. M., J. P. McDermott, E. L. Davis, R. S. Hussey, G. Smant, and T. J.

Baum. 2002. Cloning of a putative pectate lyase gene expressed in the subventral esophageal glands of *Heterodera glycines*. Journal of Nematology 34: 9 – 11.

de Meutter J., T. Tytgat, E. van der Schueren, G. Smant, A. Schots, A. Coomans, M. van Montagu, and G. Gheysen. 1998. Analysis of *Heterodera schachtii* cellulase. Proceedings of the 24th International Nematology Symposium, Dundee, Scotland, 1998, p. 24 (Abstr.).

Despommier, D. D. 1990. *Trichinella spiralis:* The worm that would be a virus. Parasitology Today 6: 193 – 196.

Ding, X., J. P. Shields, R. I. Allen, and R. S. Hussey. 1998. A secretory cellulase-binding protein cDNA cloned from the root-knot nematode (*Meloidogyne incognita*). Molecular Plant-Microbe Interactions 11: 952 – 959.

Ding, X., J. P. Shields, R. Allen, and R. S. Hussey. 2000. Molecular cloning and characterization of a venom allergen AG5-like cDNA from *Meloidogyne incognita*. International Journal for Parasitology 30: 77 – 81.

Doncaster, C. C. 1971. Feeding in plant parasitic nematodes: Mechanisms and behavior. Pp. 137 – 157. In B. M. Zuckerman, W. F. Mai, and R. A. Rohde, eds. Plant Parasitic Nematodes Vol II. New York: Academic Press.

Doyle E. A., and K. N. Lambert. 2002. Cloning and characterization of an esophageal-gland-specific pectate lyase from the root-knot nematode *Meloidogyne javanica*. Molecular Plant-Microbe Interactions 15: 549 – 556.

Doyle, E. A. and K. N. Lambert. 2003. *Meloidogyne javanica* chorismate mutase 1 alters plant cell development. Molecular Plant-Microbe Interactions 16: 123 – 131.

Eddy, S. R. 1998. Multiple-alignment and sequence searches. Trends in Genetics Supplement: Trends Guide to Bioinformaticie 14: 15 – 18.

Endo, B. Y. 1984. Ultrastructure of the esophagus of larvae of the soybean cyst nematode, *Heterodera glycines*. Proceedings of the Helminthological Society of Washington 51: 1 – 24.

Endo, B. Y. 1987. Ultrastructure of esophageal gland secretory granules in juveniles of *Heterodera glycines*. Journal of Nematology 19: 469 – 483.

Endo, B. Y. 1991. Ultrastructure of initial responses of susceptible and resistant soybean roots to infection by *Heterodera glycines*. Revue de Nematologie 14: 73 – 94.

Endo, B. Y. 1993. Ultrastructure of subventral gland secretory granules in parasitic juveniles of the soybean cyst nematode, *Heterodera glycines*. Proceedings of the Helminthological Society of Washington 60: 22 – 34.

Endo, B. Y., and W. P. Wergin. 1988. Ultrastructure of the second-stage juvenile of the root-knot nematode, *Meloidogyne incognita*. Proceedings of the Helminthological Society of Washington 55: 286 – 316.

Fire, A., S. Xu, M. K. Montgomery, S. A. Kostas, S. E. Driver, and C. C. Mello. 1998. Potent and specific genetic interference by double-stranded RNA in *Caenorhabditis elegans*. Nature 391: 806 – 811.

Gao, B., R. Allen, T. Maier, E. L. Davis, T. J. Baum, and R. S. Hussey. 2001a. Identification of putative parasitism genes expressed in the esophageal gland cells of the soybean cyst nematode, *Heterodera glycines*. Molecular Plant-Microbe Interactions 14: 1247 – 1254.

Gao, B., R. Allen, T. Maier, E. L. Davis, T. J. Baum, and R. S. Hussey. 2001b.

Molecular characterization and expression of two venom allergen-like secretory proteins from the soybean cyst nematode, *Heterodera glycines*. International Journal for Parasitology 31: 1617 – 1625.

Gao, B., R. Allen, T. Maier, E. L. Davis, T. J. Baum, and R. S. Hussey. 2002a. Identification of a new β-1, 4-endoglucanase gene expressed in the subventral gland cells of *Heterodera glycines*. Journal of Nematology 33: 12 – 15

Gao, B., R. Allen, T. Maier, E. L. Davis, T. J. Baum, and R. S. Hussey. 2002b. Characterization and developmental expression of a chitinase gene in *Heterodera glycines*. International Journal for Parasitology 32: 1239 – 1300.

Gao, B., R. Allen, T. Maier, E. L. Davis, T. J. Baum, and R. S. Hussey. 2003. The parasitome of the phytonematode molecular Plant-microbe. Interactions *Heterodera glycines*. 16: 220 – 726 (Abstr.).

Gems D., and R. M. Maizels. 1996. An abundantly-expressed mucin-like protein from *Toxocara canis* infective larvae: The precursor of the larval surface coat protein. Proceedings of the National Academy of Sciences USA 93: 1665 – 1670.

Goellner, M., G. Smant, J. M. de Boer, T. J. Baum, and E. L. Davis. 2000. Isolation of beta-1, 4-endoglucanase genes of *Globodera tabacum* and their expression during parasitism. Journal of Nematology 32: 154 – 165.

Goellner, M., X. Wang, and E. L. Davis. 2001. Endo-beta-1, 4-glucanase expression in compatible plant-nematode interactions. Plant Cell 13: 2241 – 2255.

Goverse, A., E. L. Davis, and R. S. Hussey. 1994. Monoclonal antibodies to the esophageal glands and stylet secretions of *Heterodera glycines*. Journal of Nematology 26: 251 – 259.

Goverse, A., J. Biesheuvel, G.-J. Wijers, F. J. Gommers, J. Bakker, A. Schots, and J. Helder. 1998. *In planta* monitoring of the activity of two "constitutive" promoters, CaMv 35S and TR2', in developing feeding cells induced by *Globodera rostochiensis* using GFP in combination with confocal laser scanning microscopy. Physiological and Molecular Plant Pathology 52: 275 – 284.

Goverse, A., J. R. van der Voort, C. R. van der Voort, A. Kavelaars, G. Smant, A. Schots, J. Bakker, and J. Helder. 1999. Naturally-induced secretions of the potato cyst nematode costimulate the proliferation of both tobacco leaf protoplasts and human peripheral blood mononuclear cells. Molecular Plant-Microbe Interactions 12: 872 – 881.

Hawdon, J. M., S. Narasimhan, and P. J. Hotez. 1999. *Ancylostoma* secreted protein 2: Cloning and characterization of a second member of a family of nematode secreted proteins from *Ancylostoma caninum*. Molecular and Biochemical Parasitology 99: 149 – 165.

Hiatt, A., R. C. Cafferkey, and K. Bowdish. 1989. Production of antibodies in transgenic plants. Nature 342: 76 – 78.

Huang, G. B. Gao, R. Allen, T. Maier, E. L. Davis, T. J. Baum, and R. S. Hussey. 2003. A profile of parasitism genes expressed in the esophageal gland cells of the root-knot nematode *Meloidogyne incognita*. Molecular Plant-Microbe Interactions 16: 376 – 381

Hussey, R. S. 1989a. Disease-inducing secretions of plant parasitic nematodes. Annual Review of Phytopathology 27: 123 – 141.

Hussey, R. S. 1989b. Monoclonal antibodies to secretory granules in esophageal glands of

Meloidogyne species. Journal of Nematology 21: 392 – 398.

Hussey, R. S. and F. M. W. Grundler. 1998. Nematode parasitism of plants. Pp. 213 – 243. In R. N. Perry, and D. J. Wright, eds. Physiology and Biochemistry of Free-living and Plant-parasitic Nematodes. Wallingford, UK: CAB International.

Hussey, R. S., and C. W. Mims. 1990. Ultrastructure of esophageal glands and their secretory granules in the root-knot nematode *Meloidogyne incognita*. Protoplasma 156: 9 – 18.

Hussey, R. S., and C. W. Mims. 1991. Ultrastructure of feeding tubes formed in giant-cells induced in plants by the root-knot nematode *Meloidogyne incognita*. Protoplasma 162: 99 – 107.

Hussey, R. S., O. R. Paguio, and F. Seabury. 1990. Localization and purification of a secretory protein from the esophageal glands of *Meloidogyne incognita* with monoclonal antibodies. Phytopathology 80: 709 – 714.

Hussey, R. S., C. W. Mims, and S. W. Westcott. 1992. Ultrastructure of root cortical cells parasitized by the ring nematode *Criconemella xenoplax*. Protoplasma 167: 55 – 65.

Hussey, R. S., E. L. Davis, and C. Ray. 1994. *Meloidogyne* stylet secretions. Pp. 233 – 249. In F. Lamberti, C. De Giorgi, D. Mck Bird, eds. Advances in Molecular Plant Nematology. New York: Plenum Press.

Jasmer, D. P. 1995. *Trichinella spiralis*: Subversion of differentiated mammalian skeletal muscle cells. Parasitology Today 11: 185 – 188.

Karrer, E. E., J. E. Lincoln, S. Hogenhout, A. B. Bennett, R. M. Bostock, D. B. Martineau, W. J. Lucas, D. Gilchrist, and D. Alexander. 1995. *In situ* isolation of mRNA from individual plant cells: Creation of cell-specific cDNA libraries. Proceedings of the National Academy of Sciences USA 92: 3814 – 3818.

Keen, N. T., and P. A. Roberts. 1998. Plant parasitic nematodes: Digesting a page from the microbe book. Proceedings of the National Academy of Sciences USA 95: 4789 – 4790.

Klein, R. D., X. Gu, A. Goddard, and A. Rosenthal. 1996. Selection for genes encoding secreted proteins and receptors. Proceedings of the National Academy of Sciences USA 93: 7108 – 7113.

Ko, R. C., L. Fan, and D. L. Lee. 1992. Experimental reorganization of host muscle cells by excretory/secretory products of infective *Trichinella spiralis* larvae. Transactions of the Royal Society of Tropical Medicine and Hygiene 86: 77 – 78.

Lambert, K. N., K. D. Allen, and I. M. Sussex. 1999. Cloning and characterization of an esophageal-gland-specific chorismate mutase from the phytoparasitic nematode *Meloidogyne javanica*. Molecular Plant-Microbe Interactions 12: 328 – 336.

Linford, M. B. 1937. The feeding of the root-knot nematode in root tissue and nutrient solution. Phytopathology 27: 824 – 835.

McCarter, J., P. Abad, J. T. Jones, and D. Bird. 2000. Rapid gene discovery in plant-parasitic nematodes via expressed sequence tags. Nematology 2: 719 – 731.

McClure, M. A., and N. von Mende. 1987. Induced salivation in plant-parasitic nematodes. Phytopathology 77: 1463 – 1469.

Maggenti, A. R. 1987. Adaptive biology of nematode parasites. Pp. 188 – 195. In J. A. Veech, and D. W. Dickson, eds. Vistas on Nematology. Hyattsville, MD: Society of Nematologists.

Moore, H. H., L. Orci, and G. F. Oster. 1988. The biogenesis of secretory vesicles. Pp. 521 – 561. In R. C. Das, and P. M. Robbins, eds. Protein Transfer and Organelle Biogenesis. New York: Academic Press.

Nakai, K., and P. Horton. 1999. PSORT: A program for detecting sorting signals in proteins and predicting their subcellular localization. Trends in Biochemical Science 24: 34 – 35.

Nemec, B. 1911. Über die Nematodenkrankheiten der Zuckerr übe. Zeitschift für Pflanzenkrankheiten. 21: 1 – 10.

Nemec, B. 1932. Über die Gallen von *Heterodera schachtii* auf der Zuckerrübe. Studies from the Plant Physiological Laboratory of Charles University, Prague. Vol. IV 2: 1 – 14.

Nielsen, H., J. Engelbrecht, S. Brunak, and G. von Heijne. 1997. Identification of prokaryotic and eukaryotic signal peptides and prediction of their cleavage sites. Protein Engineering 10: 1 – 6.

Olsen, A. N., and K. Skriver. 2003. Ligand mimicry? Plant-parasitic nematode polypeptide with similarity to CLAVATA3. Trends in Plant Science 8: 55 – 57.

Owen, M., A. Gandecha, B. Cockburn, and G. Whitelam. 1992. Synthesis of a functional antiphytochrome single-chain Fv protein in transgenic tobacco. Biotechnology 10: 790 – 794.

Popeijus H., V. Blok, L. Cardle, E. Bakker, M. S. Phillips, J. Helder, G. Smant, and J. Jones. 2000a. Analysis of genes expressed in second stage juveniles of the potato cyst nematodes *Globodera rostochiensis* and *G. pallida* using the expressed sequence tag approach. Nematology 2: 567 – 574.

Popeijus H. E., H. Overmars, J. Jones, V. Blok, A. Goverse, J. Helder, A. Schots, J. Bakker, and G. Smant. 2000b. Degradation of plant cell walls by nematodes. Nature 406: 36 – 37.

Qin, L., H. Overmars, J. Helder, H. Popeijus, J. R. van der Voort, W. van Groenink, P. Koert, A. Schots, J. Bakker, and G. Smant. 2000. An efficient cDNA-AFLP-based strategy for the identification of putative pathogenicity factors from the potato cyst nematode *Globodera rostochiensis*. Molecular Plant-Microbe Interactions 13: 830 – 836.

Ray, C., A. G. Abbott, and R. S. Hussey. 1994. Trans-splicing of a *Meloidogyne incognita* mRNA encoding a putative esophageal gland protein. Molecular and Biochemical Parasitology 68: 93 – 101.

Razak, A. R., and A. A. F. Evans. 1976. An intracellular tube associated with feeding by *Rotylenchulus reniformis* on cowpea root. Nematologica 22: 182 – 189.

Rebois, R. V. 1980. Ultrastructure of a feeding peg and tube associated with *Rotylenchulus reniformis* in cotton. Nematologica 26: 396 – 405.

Reed, J. D., D. L. Edwards, and C. F. Gonzalez. 1997. Synthetic peptide combinatorial libraries: A method for the identification of bioactive peptides against phytopathogenic fungi. Molecule Plant-Microbe Interactions 10: 537 – 549.

Rosso, M-N, A. Schouten, J. Roosien, T. Borst-Vrenssen, R. S. Hussey, F. J. Gommers, J. Bakker, A. Schots, and P. Abad. 1996. Expression and functional characterization of a single-chain FV antibody directed against secretions involved in plant nematode infection process. Biochemistry and Biophyics Research Communications 220: 255 – 263.

Rosso, M-N., C. Piotte, B. Favery, L. Arthaud, R. S. Hussey, J. M. de Boer, T. J. Baum, and P. Abad. 1999. Isolation of a cDNA encoding a β-1, 4-endoglucanase in the root-knot nematode *Meloidogyne incognita* during plant parasitism. Molecular Plant-Microbe Interactions 12: 585 – 591.

Rumpenhorst, H. J. 1984. Intracellular feeding tubes associated with sedentary plant parasitic nematodes. Nematologica 30: 77 – 85.

Shields, J. P., X. Ding, and R. S. Hussey. 1998. Microaspiration of esophageal gland contents from plant-parasitic nematodes. Journal of Nematology 30: 515 (Abstr.).

Shpigel, E., L. Roiz, R. Goren, and O. Shoseyov. 1998. Bacterial cellulose-binding domain modulates *in vitro* elongation of different plant cells. Plant Physiology 117: 1185 – 1194.

Smant, G., A. Goverse, J. W. P. G. Stokkermans, J. M. de Boer, H. Pomp, J. F. Zilverentant, H. A. Overmars, J. Helder, A. Schots, and J. Bakker. 1997. Potato root difusate-induced secretion of soluble, basic proteins originating from the subventral esophageal glands of potato cyst nematodes. Phytopathology 87: 839 – 845.

Smant, G., Y. W. P. G. Stokkermans, J. Yan, J. M. de Boer, J. S. Baum, X. Wang, R. S. Hussey, F. J. Gommers, B. Henrise, E. L. Davis, J. Helder, A. Schots, and J. Bakker. 1998. Endogenous cellulases in animals: Isolation of β-1, 4-endoglucanase genes from two species of plant-parasitic cyst nematodes. Proceedings of the National Academy of Sciences USA, 95: 4906 – 4911.

Sundermann, C. A., and R. S. Hussey. 1988. Ultrastructural cytochemistry of secretory granules of esophageal glands of *Meloidogyne incognita.* Journal of Nematology 20: 141 – 149.

Uehara, T., A. Kushida, and Y. Momota. 2001. PCR-based cloning of two β-1, 4-endoglucanases from the root-lesion nematode *Pratylenchus penetrans.* Nematology 3: 335 – 341.

Urwin, P. E., S. G. Moller, C. J. Lilley, M. J. McPherson, and H. J. Atkinson. 1997. Continual green-fluorescent protein monitoring of cauliflower mosaic virus 35S promoter activity in nematode-induced feeding cells in *Arabidopsis thaliana.* Molecular Plant-Microbe Interactions 10: 394 – 400.

Urwin P. E., C. J. Lilley, and H. J. Atkinson. 2002. Ingestion of double-stranded RNA by preparasitic juvenile cyst nematodes leads to RNA interference. Molecular Plant-Microbe Interactions 15: 747 – 752.

Veech, J. A., J. L. Starr, and R. M. Nordgren. 1987. Production and partial characterization of stylet exudate from adult females of *Meloidogyne incognita.* Journal of Nematology 19: 463 – 468.

Wang, X., D. Meyers, Y. Yan, T. Baum, G. Smant, R. S. Hussey, and E. L. Davis. 1999. *In planta* localization of a β-1, 4-endoglucanase secreted by *Heterodera glycines.* Molecular Plant-Microbe Interactions 12: 64 – 67.

Wang, X., R. Allen, X. Ding, M. Goellner, T. Maier, J. M. de Boer, T. J. Baum, R. S. Hussey, and E. L. Davis, 2001. Signal peptide-selection of cDNA cloned directly from the esophageal gland cells of the soybean cyst nematode, *Heterodera glycines.* Molecular Plant-Microbe Interactions 14: 536 – 544.

Williamson, V. M., and R. S. Hussey. 1996. Nematode pathogenesis and resistance in plants. Plant Cell 8: 1735 – 1745.

Wyss, U. 1992. Observations on the feeding behavior of *Heterodera schachtii* throughout development, including events during molting. Fundamental and Applied Nematology 15: 75 – 89.

Wyss, U. 1997. Root parasitic nematodes: An overview. Pp 5 – 22. In C. Fenoll, F. M. W. Grundler, and S. A. Ohl, eds. Cellular and Molecular Aspects of Plant-nematode Interactions. Dordrecht: Kluwer Academic Publishers.

Wyss, U., and U. Zunke. 1986. Observations on the behavior of second stage juveniles of *Heterodera schachtii* inside host roots. Review de Nematologie 9: 153 – 165.

Wyss, U., C. Stender, and H. Lehmann. 1984. Ultrastructure of feeding sites of the cyst nematode *Heterodera schachtii* Schmidt in roots of susceptible and resistant *Raphanus sativus* L. var. *oleiformis* cultivars. Physiological Plant Pathology 25: 21 – 37.

Wyss, U., F. M. W. Grundler, and A. Munch. 1992. The parasitic behavior of second-stage juveniles of *Meloidogyne incognita* in roots of *Arabidopsis thaliana*. Nematologica 38: 98 – 111.

Yamamoto, Y. T., C. G. Taylor, G. N. Acedo, C. Cheng, and M. A. Conkling. 1991. Characterization of *cis*-acting sequences regulating root-specific gene expression in tobacco. Plant Cell 3: 371 – 382.

Yan, Y., and E. L. Davis. 1997. Characterization of secretory protein genes from the soybean cyst nematode. Phytopathology 87(6S): S106 (Abstr.).

Yan, Y., and E. L. Davis. 2002. Characterisation of guanylyl cyclase genes in the soybean cyst nematode, *Heterodera glycines*. International Journal for Parasitology 32: 65 – 72.

Yan, Y., G. Smant, J. Stokkermans, L. Qin, J. Helder, T. Baum, A. Schots, and E. Davis. 1998. Genomic organization of four β-1, 4-endoglucanase genes in plant-parasitic cyst nematodes and its evolutionary implications. Gene 220: 61 – 70.

Yan, Y., G. Smant, and E. L. Davis. 2001. Functional screening yields a new β-1, 4-endoglucanase gene from *Heterodera glycines* that may be the product of a recent gene duplication. Molecular Plant-Microbe Interactions 14: 63 – 71.

6 Surface Adhesion to Nematodes and its Consequences

Alan F. Bird

6.1 Introduction

It is a fact of scientific endeavor, and human behavior in general, that published observations and discoveries may, in time, become forgotten or even ignored and hence be unavailable in modern computer literature searches. One of the values of a historical review is that the reviewer can acknowledge the significant contributions of earlier workers and take into account the information and equipment available to them at the time of their research. These are often ignored by later workers whose references are based on the relatively limited resources of computer searches.

In the two volumes of this book, advances into research on nematodes start appropriately at the beginning of the 20th century. At that time, the stage had been set for the great discoveries in microbiology that were to result from refinements to the light microscope by Ernst Abbe and August Köhler at the turn of the century. These refinements resulted in the production of commercial instruments that had the capacity to resolve objects close to the theoretical limits. Further optical refinements led to increases in contrast so that the details of living material became resolvable. These occurred with the development of phase contrast microscopy by Zernike (1934) and differential interference contrast microscopy by Nomarski (1955). The fine structure of microorganisms became resolvable with the development of the electron microscope.

The transmission electron microscope (TEM) was developed by Ruska (1933) and the first commercial model was produced in 1939 by Siemans. The history of this research, development and production is described by Ernst Ruska in a 140-page booklet that was translated into English by Mulvey in 1980. Using Ruska's instrument and a negative staining technique, Krause (1937, cited in Mulvey's 1980 translation of Ruska's booklet) was able to photograph bacteria for the first time with better resolution than could be obtained with the light microscope. The first scanning electron microscope (SEM) was constructed by von Ardenne (1938a, 1938b) and it was first produced commercially by Cambridge Scientific Instruments in 1965. The history of the development of the first SEM is also given in Mulvey's translation of Ruska's booklet together with drawings and a photograph of von Ardenne's first model. These instruments paved the way for discoveries on the ultrastructure of bacteria and viruses and were necessary tools to establish the morphological nature of adhesion of microorganisms to nematodes. Various refinements to microscopy such as fluorescent, confocal and the environmental SEM further extend the usefulness of these instruments.

Interactions involving adhesion of organisms are many and varied in nature (Berkeley et al., 1980) and doubtless more will be discovered as studies continue into the 21st century. In this chapter I propose to examine a number of well documented examples of relationships between nematodes and adhering organisms and to highlight what I consider to be the most relevant fundamental research that paved the way to a better understanding of the nature and consequences of their interactions. They are all examples of symbiosis, that is, the living together of two or more phylogenetically unrelated species. The consequences of adhesion are various forms of symbiosis, namely, commensalism, in which no organism benefits at the expense of the other although one may benefit from the relationship, mutualism in which both organisms benefit, or parasitism in which one benefits at the expense of the other. It must be emphasized that these are just terms and that more than one degree of symbiosis may occur in some of the relationships described below. Before examining some of these interactions, I shall consider the surface of the nematode cuticle.

6.2 The Surface Coat of Nematodes

The surface coat of nematodes has been shown to function in a variety of ways and its chemical composition, source and method of secretion are less than clear and have been the subject of much research.

Some clues on the chemical composition of exudations from nematodes came in the first half of the 20th century from the work of Sarles (1938), who observed that when infective larvae of *Nippostrongylus muris* were placed in immune rat serum precipitates that were absent when the nematodes were placed in normal serum were observed at the nematodes' secretory-excretory pores. Later work by Jackson (1961) and others verified these results in different species of nematodes. Several years later, as a result of serological studies using fluorescent-antibody techniques (Bird, 1964), it was shown that antigenic material exuded from the secretory-excretory pore, the buccal stylet and rectal glands, and occurred to a lesser extent on the surface of the cuticle of different stages of the plant-parasitic nematode *Meloidogyne javanica*. The output of this material diminished with aging. Further evidence for the secretion of a material over the whole body of a nematode was provided by Green and Greet (1972), who demonstrated that a pheromone which attracted males was secreted by female cyst nematodes of the genera *Heterodera* and *Globodera* and covered the surfaces of their entire bodies. Green and Greet (1972) concluded that these pheromones were unlikely to be secreted through glands opening to the surface of the nematodes but were probably produced in the epidermis (hypodermis) and then moved through the cuticle to the surface.

The phenomenon of swarming has been observed in many nematodes. In *Tylenchorhynchus martini* this was thought by Hollis (1962) to be associated with stickiness on the surface of the cuticle that was inhibited by various proteolytic enzymes. Many years later a polypeptide aggregation pheromone was extracted from female *Nippostrongylus brasiliensis* and partially purified (Bone, et al., 1980). Three years later Nelson et al. (1983) suggested that pheromones could be secreted through the secretory-excretory duct and in 1984, during research on molting and growth in *Rotylenchulus reniformis*, in which several molts take place without the nematodes feeding, it was noticed that the secretory-excretory (excretory) system appeared to continue to function throughout the molts although the majority of structures in the pharyngeal (esophageal) region broke down during these processes (Bird, 1984). It was thought that this would enable *R. reniformis* to provide enzymes and lubricants that assist both in cuticle breakdown and in maintenance of the flexibility of the shed cuticles during molting. As a result of their research on the secretory-excretory system in *Caenorhabditis elegans*, Nelson et al. (1983) had suggested that this structure might "synthesize and release material through the excretory pore that coats the cuticle surface". Some evidence for these ideas was already available, namely the serological experiments on *M. javanica* (Bird, 1964). As a result of histochemical tests on the dauer second-stage larva (DL_2) of *Anguina funesta* using the vital dye Brilliant Blue G and a range of fluorescein-isothiocyanate labeled lectins coupled with inhibition by pretreatment of the lectin with the appropriate hapten sugar (Bird, et al., 1988), it was concluded that the "excretory" system in this nematode was responsible for the secretion of glycoprotein macromolecules that covered its surface and were known as the surface coat or glycocalyx. Shortly after these experiments and using the same material and fluorescence labeling techniques, it was shown that microscopic prints of the surface coat could be obtained

(Bird, 1988). These results were thought to be a reflection of the stickiness of the surface coat which was presumed to be "continuously secreted and lost during movement through the soil by the nematode". Experimental evidence for these ideas was provided a year later (Bird and Zuckerman, 1989) when it was shown that the adhesion of the bacterium *Rathayibacter toxicus* to the glycoprotein surface coat of the DL_2 of *Anguina funesta* was inhibited after treatment of the nematodes with the proteolytic enzymes pepsin and trypsin. However, the nematodes were able to regain most of their adhesive capacity within 18 hours of removal from the enzyme solutions, indicating that they had the capacity to renew their surface coats. In some nematodes the shed surface coat is renewed within minutes to hours (Blaxter and Robertson, 1998).

Maizels and Page (1990) and Page et al. (1992) have found that the surface coat of the infective larva of *Toxocara canis* consists of a number of proteoglycans and mucin-like glycoproteins that have the same carbohydrate determinants. Thus, the surface coat or glycocalyx, which is secreted by various glands (Bird and Bird, 1998), is largely made up of glycoproteins that resemble mammalian mucins in structure (Maizels and Page, 1990; Page, et al., 1992). The surface coat appears to be a common and multifunctional structure involved in adhesion, lubrication, evasion of host immunity by the shedding of surface-bound antibodies (Blaxter, et al., 1992), and the evasion of parasites by similarly shedding its disposable surface coat. If the latter is too slow, adhesion of a symbiont may result.

Research on the surface coats of nematodes has been extensively reviewed by Blaxter and Bird (1997) in their chapter in the book "*C. elegans* II", and the reader is referred to this work for further information.

6.3 The Surface of Bacteria

Both Gram-positive (*Rathayibacter toxicus*) and Gram-negative (*Xenorhabdus nematophilus*) bacteria adhere to nematodes and are used as examples in this chapter. *Rathayibacter toxicus* adheres to the cuticular surface of members of the genus *Anguina*, and *Xenorhabdus nematophilus* to members of the genus *Steinernema* within vesicles in the anterior portion of the intestine. Thus, a brief description of these bacterial surfaces in relation to adhesion is necessary.

The cell walls of Gram-negative bacteria consist, from the outer surface inwards, of an outer membrane, a thin peptidoglycan layer, a gel-like matrix (the periplasm), and the cell membrane. The cell walls of Gram-positive bacteria consist of a peptidoglycan layer covering the cell membrane, although they are less uniform in structure than those of Gram-negative bacteria. Many bacteria have an outer polysaccharide-containing secreted coat or glycocalyx, which may take the form of a slime layer or of a capsule such as is found in *Rathayibacter toxicus* where it is involved in adhesion.

The bacterial surface in relation to adhesion has been reviewed by Ward and Berkeley (1980) and the structure of Gram-negative cell walls by Beveridge (1999).

6.4 Adhesion

Adhesion of microorganisms to the nematode surface coat was first observed in the middle of the 20th century when commercial electron microscopes were first being used in biological research. In 1957, Bird and Deutsch observed and photographed *Pseudomonas aeruginosa* adhering to the grooves between the surface annuli of the cuticle of *Ascaris suum*. The nature of this adhesion was not discussed although the bacteria were photographed in the grooves both whole on the surface and as longitudinal sections cut through the grooves on the surface of the cuticle. The relationship between bacteria and nematodes did not appear to be pathological. These results and many others involving interactions between microorganisms and the surface of the nematode cuticle were reviewed by Bird (1980). Many of the interactions between microorganisms and nematodes are pathological and cause cuticular lesions and sometimes death of the nematode, but many others are not. Although in 1980 the nature of the forces that maintained this attraction was unclear, it was known that they could be localized, as in the case of bacteria in the transverse grooves of the surface cuticle of *Ascaris suum* (Bird and Deutsch, 1957) or distributed over the entire cuticular surface as in *Rathayibacter* on *Anguina* (Bird, 1980). The topic of microbial adhesion to surfaces became of wide general interest and in 1980 a multi-author book of this title was published for the Society of Chemical Industry (London). This included numerous chapters on topics that ranged from microbial adhesion to soil surfaces to the adhesion of bacteria to the surfaces of the human mouth (Berkeley, et al., 1980). The more specific topic of adhesion to nematodes was reviewed briefly in a chapter on receptors and recognition by Jansson (1987). In the last decade of the 20th century, this topic was reviewed by Bird and Bird (1991) and Blaxter and Bird (1997). Two processes appear to be involved in the adhesion of microorganisms to the surface coat of nematodes. The first is site recognition and the second is physical attachment. Site recognition, which may or may not result in disease, consists of a molecular interaction between a carbohydrate on the surface coat of the nematode and a protein, located on the surface of the adhering microorganism, that binds to this carbohydrate. An example of adhesion with no resulting pathological effect is that of the conidia of the fungus *Dilophospora alopecuri* to the surface coat of the DL_2 of the nematode *Anguina funesta*, which will be discussed in more detail later in this chapter. Examples of physical attachment that lead to parasitism of the nematode include *Rathayibacter toxicus* adhering to the surface of *Anguina funesta*, *Pasteuria penetrans* adhering to the surface of the second-stage larvae (L_2) of *Meloidogyne javanica* and *Arthrobotrys oligospora* adhering to the surface of *Panagrellus redivivus*. The history of the research into the two bacterial host-parasite relationships will be considered later in this chapter.

6.5 Commensalism as a Consequence of Adhesion

6.5.1 Plant Viruses and Their Nematode Vectors

An example of commensalism involving adhesion to nematodes is that of the plant viruses and their adenophorean nematode vectors. Details of these nematodes and the viruses that they transmit have been reviewed recently (Taylor and Brown, 1997) and the reader is referred to this work for further information.

Evidence that plant-parasitic nematodes act as vectors for plant viruses was first reported by Hewitt et al. in 1958 when it was shown that *Xiphinema index* was the vector for grape fanleaf nepovirus (GFLV). Since that time there has been a remarkable increase in the number of plant viruses known to be transmitted by nematodes and in the number of species of nematode vectors. Taylor and Brown (1997) estimated that by the end of the 20th century these nematode vectors would consist of about 120 species of *Longidorus*, 240 species of *Xiphinema*, 45 species of *Trichodorus*, and 34 species of *Paratrichodorus*. The viruses associated with these plant-parasitic nematodes do not multiply within the nematodes but use them as a mechanical means of invasion of the plant, although in the case of *Xiphinema*, where virus retention is relatively long, it was originally thought that there could be a more biological association between virus and nematode. The earliest studies with the electron microscope did not lead to the discovery of a site in the vector for these viruses. It was not until it was shown that GFLV did not persist in *X. index* after molting (Taylor and Raski, 1964) that attention was focused on internal cuticular parts that were shed during molting, which are the shed feeding components, as sites for viral retention. However, Taylor and Raski (1964) thought that there was evidence to suggest that there were basic differences in the mechanisms of virus transmission between *Xiphinema* and

Longidorus. The sites of virus retention in the nematode vectors of these virus diseases were established by Taylor and Robertson (1969, 1970a, 1970b), who examined thin sections of nematodes fed virus-infected plants, under the electron microscope. These workers found that, when *Longidorus* was the vector, the viruses were selectively adhering to the inner surface of the odontostyle or to the cuticular part of the guiding sheath whereas in *Xiphinema* species, viruses adhered as a monolayer to the cuticular lining of the pharynx; this was also the case for *Trichodorus* and *Paratrichodorus* although here most of the virus particles adhered in the glandular region of the pharynx. It has been shown (Taylor and Brown, 1997) that the surfaces to which virus particles adhere contain carbohydrate. It is suggested that virus retention in these nematodes involves a specific interaction between this carbohydrate and lectin-like molecules in the protein coat of the adhering viruses. This carbohydrate-containing surface coat could originate from the nematode's pharyngeal glands, as has been suggested above.

The subventral pharyngeal (esophageal) glands in secernentean nematodes such as *Meloidogyne javanica* L_2 have been shown to contain carbohydrate in the parasitic stage when feeding starts but not in the preparasitic stage (Bird and Saurer, 1967). Recently, M. A. McClure (pers. comm., 1999) using monoclonal antibodies to the L_2 of *M. incognita*, has detected glycoprotein on the surface coat of the cuticle, in the contents of the rectal dilation and in the subventral pharyngeal glands and in their exudations into the lumen of the pharynx, adding weight to the idea that the surface coat is produced by glands and secreted through various orifices. It seems likely, then, that glycoprotein coats cover not only the external surfaces of nematodes but also the internal invaginations of these surfaces.

Virus particles do not appear to adhere to the cuticular surface coats of the pharynx in secernentean nematodes as they do in adenophorean nematodes and so perhaps other factors, such as differences in composition of the pharyngeal gland secretions, or surface electrical charges (Taylor and Brown, 1997) may have roles in adhesion, although the evidence for the latter is fragmentary.

The consequence of the adhesion of viruses to the pharyngeal regions of the adenophorean nematodes listed above is the transmission of a range of viral diseases of plants. The viruses are classified into two groups based on their ultrastructure. The first group is known as nepoviruses, because it consists of nematode transmitted viruses with isometric, polyhedral particles, about 28 nm in diameter. The second group, known as tobraviruses, is tubular with a diameter of about 23 nm and with lengths up to 210 nm. Plant viruses cause a range of diseases, which are named after the visual symptoms they cause. In 1997 Taylor and Brown reported that 37 different types of nepoviruses but only three types of tobraviruses had been identified; all of the latter were transmitted by trichodorid nematodes. The topic of virus vectors will be considered in greater detail in Volume II of this book in the chapter by Brown et al.

6.5.2 *Dilophospora* and Its Vector Nematode

A complex example of a plant parasite that begins as a commensal of a nematode is the relationship between the fungus *Dilophospora* and its vector nematode, *Anguina funesta*, which also is the vector for the causal agent of the disease of sheep and cattle called "annual ryegrass toxicity" (ARGT). The relationship between the fungus and the nematode was first observed and described in detail by Atanasoff (1925), although the disease that *Dilophospora* causes in cereals had been observed almost a hundred years previously. *Dilophospora alopecuri* is a pathogen of cereals and grasses that relies, in the field, completely on a nematode species of the genus *Anguina* to carry its spores to its host plant. Invasion of the host plant by the fungus is not normally achieved without the nematode although it can be induced artificially under laboratory conditions. Eventually, *D. alopecuri* has an inhibitory effect on nematode growth and development in the plant. These observations, made by Atanasoff in Wageningen (The Netherlands) in 1925 on *D. alopecuri* and *Anguina tritici*, were verified 74 years later in Western Australia by Riley (1994), who, working with *D. alopecuri* and *A. funesta*, found that the fungus not only reduced the nematode populations but also was a naturally occurring antagonist of the bacterium *Rathayibacter toxicus*, the causative agent of ARGT (Riley, 1996). The potential of *D. alopecuri* as a biological control agent to reduce nematode numbers by preventing their development in galls was reported by McKay et al. (1981). Six years later, as a result of their studies on the adhesion of conidia of *Dilophospora* to the cuticle of the DL_2 of *Anguina*, Bird and McKay (1987) stated that their observations drew attention to the potential of *D. alopecuri* as an agent for the biological control of annual ryegrass toxicity. Details of this disease complex will be considered later in this chapter. In his field studies on *D. alopecuri* and decline in ARGT in Western Australia, Riley (1994) stated that his findings suggested that there was a potential to use *D. alopecuri* to control ARGT. During the process of adhesion, the fungal conidium appears to secrete an adhesive material that covers the surface of the nematode and fills the surface annulations of its cuticle (Bird and McKay, 1987). The attachment of the fungal spores, which have a characteristic shape (Fig. 6.1), does not give rise to any morphological changes to the cuticle of the nematode. Adhesive material is produced from the convoluted surfaces of the setulose appendages of the fungus. The nematode does not appear to be harmed although its movements through the soil may be

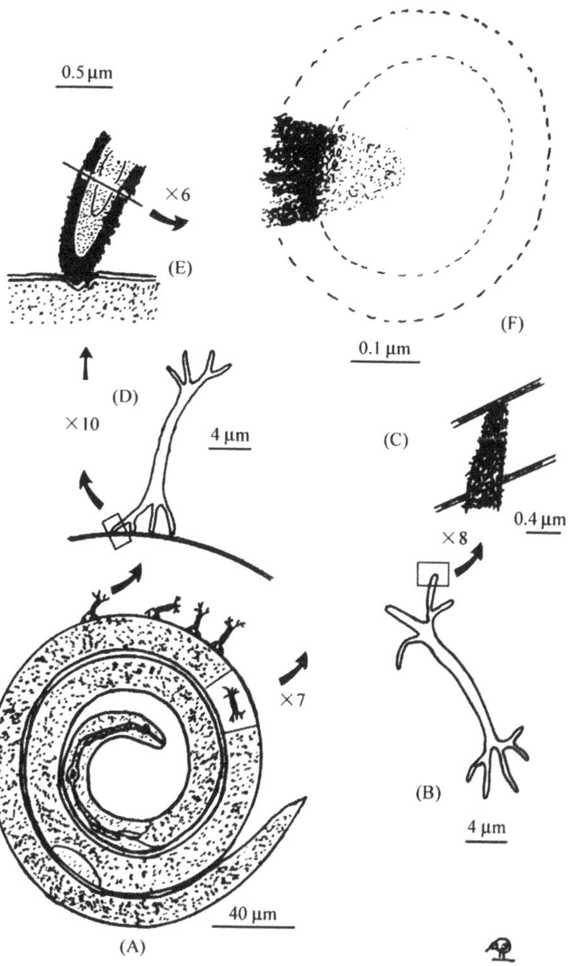

Figure 6.1 Scale diagram constructed from photo- and electron-micrographs showing adhesion of the conidia of the fungus *Dilophospora alopecuri* to the surface of the DL_2 of the nematode *Anguina funesta*. (A) The coiled DL_2 of *A. funesta* with several adhering conidia. (B), (D) Branched setulose appendages at higher magnification. (C) Surface view of an appendage viewed between two transverse annulations, showing its convoluted surface from which a mucilaginous fibrillar material is exuded and deposited in the annulations. (E) Longitudinal section of a setulose appendage at high resolution, showing no sign of damage to the nematode's epicuticle. (F) Cross-section of a branch of a setulose appendage further magnified, showing a cytoplasmic core surrounded by an electron-dense coat of fibrillar material.

slowed and, as mentioned above, the fungus inhibits nematode growth in the plant. This type of adhesion involves site recognition only and is much more tenuous than those types of adhesion which involve a morphological change at the host surface. The consequence of the adhesion of *D. alopecuri* to the surface of the cuticles of *Anguina* DL_2 is that the fungus, commonly called twist fungus because it causes twisting of the seedhead, is carried into ryegrass and colonizes the seedhead where it competes with both the nematode and the bacterium responsible for ARGT. From the economic point of view, damage to the seedheads by the fungus is offset by its potential as an agent for biocontrol of ARGT. Furthermore, despite its widespread natural distribution in Australia, there have been no reports of the pathogenicity of *D. alopecuri* on plant species of economic importance (Riley, 1996). Also, this fungus is not toxic to animals (I. T. Riley, pers. comm., 1999).

6.6 Mutualism as a Consequence of Adhesion

An example of adhesion leading to mutualism associated with a complex symbiotic relationship is that of certain species of the gram-negative bacterium *Xenorhabdus* and various species of the entomopathogenic nematode *Steinernema* (Akhurst, 1993; Boemare, et al., 1997, Forst, et al., 1997; Kaya, et al., 1993; Wouts, 1991).

In 1993, 14 different species of *Steinernema* were known, together with five species of *Xenorhabdus* (Kaya, et al., 1993), although the numbers of species of these two mutualistic genera will doubtless increase as new members are identified. However, each nematode species has a specific mutualistic association with only one species of bacterium, although a single bacterial species may have such an association with more than one nematode species.

The history of our understanding of this mutualistic behavior is interesting. It started in 1923 when Steiner dissected a nematode from a paper wasp. He called the nematode *Aplectana kraussei*; it is now known as *Steinernema kraussei*. The potential for this genus as a biological control agent for insect pests was not realized for a further 9 years until Glaser (1932) published the first of a series of papers, which clearly demonstrated the potential of *Neoaplectana* (*Steinernema*) *glaseri* as a biocontrol agent for the Japanese beetle, *Popillia japonica*. However, at that time the mutualistic relationship between the nematode and the bacterium that is the key to biocontrol, was not realized. Five years later in a Ph. D. thesis, Dutky (cited by Wouts, 1991) was able to show that bacteria had a mutualistic relationship with the nematode and that the bacteria were responsible for a disease of insects by causing septicemia (Dutky, 1959; Dutky and Hough, 1955). The symbiotic bacterium was given various names, notably *Achromobacter nematophilus* (Poinar and Thomas, 1965), which was changed to *Xenorhabdus* sp. (Thomas and Poinar, 1979). Subsequently, a number of species of this genus have been recorded (Akhurst and Boemare, 1988, 1990), as mentioned above.

Akhurst (1980) observed that when *Xenorhabdus* was grown on agar media, two distinct morphological types were produced, a phenomenon that he described as dimorphism. However, there are significant differences in the metabolism of these morphs and for these, and other reasons, the term phase variation was adopted (Boemare and Akhurst, 1988). Phase I bacteria are larger than Phase II bacteria and contain inclusions and have flagella and fimbriae and a thicker layer of capsular material, both of which are thought to be involved in adhesion as well as in avoidance of host immune defenses or bacteriophage infection (Akhurst, 1993; Brehelin, et al., 1993; Givaudan, et al., 1995; Moureaux, et al., 1995). Phase I bacteria provide better conditions for nematode reproduction and can always be isolated in the fields whereas Phase II bacteria are found in *in vitro* cultures and not from field isolates of *Xenorhabdus*.

By means of antibody/gold labeling (Binnington and Brooks, 1993), it has been shown that *Xenorhabdus* cells in the intestine of *S. carpocapsae* are fimbriated. Electron micrographs of sections cut through the intestinal vesicles of various species of *Steinernema* (Bird and Akhurst, 1983; Fig. 5) show that, in this figure of *Steinernema bibionis* (= *feltiae*), the bacterium is surrounded by a tangle of attached tubular structures that have the same morphological structure as fimbriae. Twelve years later in the study of the ultrastructure of *Steinernema feltiae*, Endo and Nickle (1995) confirmed the presence of bacteria in vesicles that appeared to open into the most anterior region of the intestinal lumen. Thus, bacteria in the vesicle appear to have continuity with bacteria located within or at the junction of the pharyngo-intestinal valve and intestinal lumen.

The consequences of adhesion of Phase I *Xenorhabdus* to the walls of the intestinal vesicles of *Steinernema* (Fig. 6.2) are that the nematode functions as a vector for the bacteria as it enters the digestive tracts of the larval stages of various insects. From here the nematode penetrates into the hemocoel and the *Xenorhabdus* are released into the hemolymph resulting in the death of the insect, usually within two days. The interaction between the bacterium and nematode is a good example of mutualism; the former provides the nutrients for the nematode and inhibits the growth of other bacteria while the latter secretes an inhibitor of antibacterial proteins put out by the insect (Gotz, et al., 1981). The presence of phages and bacteriocins in species of *Xenorhabdus*

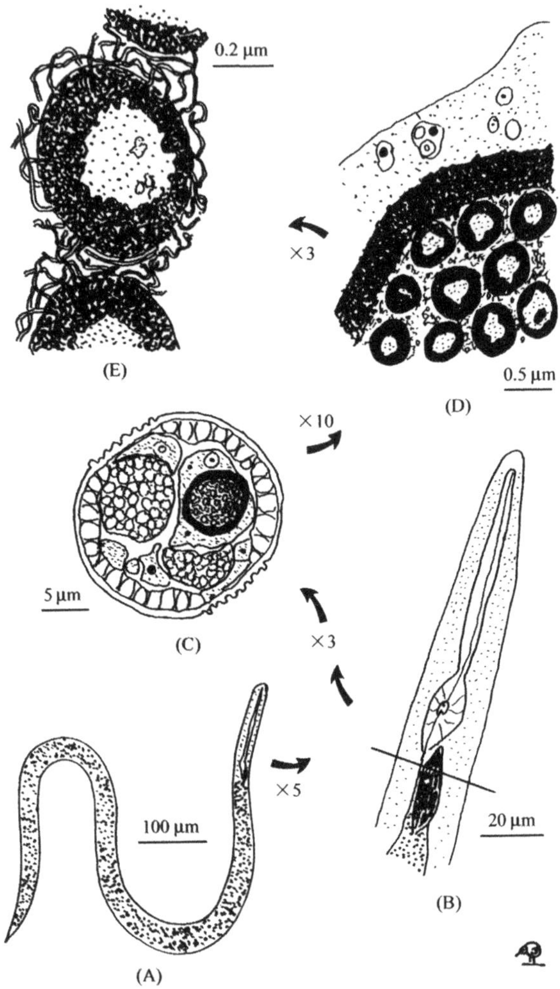

Figure 6. 2 Scale diagram constructed from photo- and electron-micrographs showing the intestinal vesicle of *Steinernema feltiae* with its contained adhering *Xenorhabdus nematophilus* at various magnifications. (A) The whole living nematode (×200). (B) The anterior part of this nematode (×1000) showing the site of the intestinal vesicle. (C) Transverse section cut through the vesicle (×3000) showing the position of the vesicle and some shrinkage of the nematode due to fixation and embedding. (D) A portion of a vesicle (×30,000) showing bacteria adhering to each other and to the capsular wall of the vesicle. (E) A single *X. nematophilus* (×90,000) showing its relationship with adjacent adhering bacteria surrounded by tangled fimbriae.

(Boemare, et al., 1993) has been shown to occur in the two phases, both of which are equally pathogenic. Since the presence of a phage has been shown to be linked to the toxicity of *Rathayibacter toxicus*, which also involves a nematode in a symbiotic relationship (Bird, et al., 1980; Ophel, et al., 1993), it seems reasonable to speculate that the phages in *Xenorhabdus* also may confer pathogenicity to the bacterium. However, there is no evidence to support this idea, and the role of phages and bacteriocins in this complex symbiosis may be to protect the nematode-bacterium association from contaminating bacteria (Boemare, et al., 1997).

6.7 Parasitism as a Consequence of Adhesion

I shall consider two examples of parasitism resulting from adhesion. The first is the association between *Anguina* and *Rathayibacter*. It consists of complex symbioses involving a nematode, a bacterium, a virus, and the inflorescence and seeds of a plant. The second only involves a nematode, a bacterium, and the roots of various plants and is the association between *Meloidogyne* and *Pasteuria*. This must, however, be one of the most sophisticated forms of parasitism that have been described.

6.7.1 *Anguina* and *Rathayibacter*

The seed gall nematode, *Anguina tritici*, was the first plant-parasitic nematode to be discovered. It was found in 1743 by the English theologian J. T. Needham. During the examination of seed galls of wheat, Needham observed a soft white substance within the galls which, when examined at higher magnification, was seen to consist of motionless fibers knotted together (Needham, 1743). When a drop of water was added to the fibers, to Needham's surprise, they began to move in an undulating fashion. Needham called these aquatic animals "worms, eels or serpents". He repeated these experiments several times and "gratified others with the sight of it". These observations were important since they revived interest in the concept of latent life, which Leeuwenhoek had demonstrated "in animalcules found in the sediment in gutters of the roofs of houses" 41 years earlier (Leeuwenhoek, 1702) and which had been forgotten. Leeuwenhoek, however, did not use terms such as "suspended animation" or "latent life" or discuss such concepts as life, resuscitation or death.

Needham's "worms" had the advantage over Leeuwenhoek's "animalcules" in that they were a single species and readily obtainable for experimental purposes. Needham may have regretted publishing his observations because they were criticized by Spallanzani, then Professor of Natural History at the University of Pavia, in northern Italy and one of the most highly thought of scientists of his time. He thought that Needham's worms were just dried fibers that moved only because water was penetrating them. This caused Needham to have doubts about what he had seen and to abandon his original ideas. However, Spallanzani was wrong and Needham had been right in this matter. This was clearly demonstrated by Roffredi and Fontana who independently established that Needham's "worms" were nematodes and attacked Needham for abandoning his original views (Keilin, 1959).

Although such matters are beyond the scope of this chapter, I mention them because they emphasize the important historical role that early studies on *Anguina* had on both science and philosophy. It was thought by some that Leeuwenhoek's animalcules (nematodes, rotifers, and tardigrades) had died and then were resurrected. Keilin (1959) has reviewed the early research in this field including the extraordinary debates that took place in Paris in the middle of the 19th century between resurrectionists and anti-resurrectionists. In the midst of these debates the publication of Darwin's work "On the Origin of Species by Means of Natural Selection" in 1859 was hardly noticed (Keilin, 1959).

Of course, animals in an anhydrobiotic state are not dead but have entered a state in which metabolic activity has been reduced to below a measurable level. In *Anguina* this leads to a number of distinct morphological changes that have been resolved at high resolution under the TEM after sectioning and freeze fracture (Bird and Buttrose, 1974). Briefly, the DL_2 assumes a characteristic coiled shape, loses 90% of its water content and undergoes a number of morphological changes such as shrinkage and distortion of lipid bodies, an orderly packing of muscle fibers and changes to the structure of the epicuticle (external cortical layer). This change involves a thickening of the osmiophilic outermost layer of the epicuticle which, together with the results of freeze-fracture studies, suggested that a surface lipid is involved (Bird and Buttrose, 1974). Almost a quarter of a century later, Womersley et al. (1998) provided SEM photographs of the surface of anhydrobiotic *Anguina* larvae that show "a thin lipophilic coating on the exterior and at the points at which the nematodes come into contact with each other", which masks the structure of the annuli beneath.

I have mentioned these changes in the anhydrobiotic larvae because they illustrate the extraordinary versatility of *Anguina* to change its cuticular surface components. Thus, in the anhydrobiotic state the outermost layer of the epicuticle of the DL_2 is covered by a lipophilic layer. However, as the larva becomes hydrated and starts to move through the soil, a glycocalyx, which is secreted from glands, covers the outside of the cuticle. This surface coat is presumably lost and secreted continuously as the nematode moves through the soil (Bird, 1988). A better example of the functional polymorphism of the outermost layer of the nematode cuticle has yet to be found. Surprisingly little information is available about the manner in which the lipid surface layer is laid down. It seems likely to me that, like the glycocalyx, it is secreted from glands. Is there a relationship between the changes that take place in the morphology of the lipid bodies as the DL_2 becomes anhydrobiotic and the production of an outermost lipid layer? If so, how does this come about? The answers to these questions depend on a thorough study of the surface changes in the DL_2 of these gall-forming *Anguina* spp., together with studies on the physiology of the glands that may be responsible for these different secretions.

Just as the anhydrobiotic capabilities of the genus *Anguina* have led to controversy, so has the early work on their taxonomy and that of their host plants. They were first described by a botanist, J. G. Steinbuch (1799) who collected specimens of a grass in a wood in the vicinity of Erlangen in Germany. Steinbuch collected a grass that he thought was *Agrostis silvatica* until he dissected it and found that, in fact, the enlarged structure of the inflorescence was not natural but had been caused by parasitic nematodes and that the grass was actually *Agrostis capillaris*. He found and described female nematodes and embryonated eggs and larvae in the seed galls and called the nematode *Vibrio agrostis*. It is now known as *Anguina agrostis*. The confusion to botanical taxonomists caused by parasitism of inflorescences by nematodes such as *Anguina agrostis* has been well documented. A good example that has been illustrated by T. Goodey (1930) and later by B. Goodey (1959) is that of parasitized *Agrostis stolonifera* being named *Agrostis silvatica*. The taxonomy of the Australian ryegrass nematode has not been without some controversy. It was first described, in a Ph. D. thesis by Price (1973), as *Anguina lolii*, its specific name being derived from that of its host, *Lolium rigidum* (annual ryegrass). When the work was published (Price, et al., 1979), the name was changed to *Anguina funesta*. However, in the meantime, the bacterial disease of sheep and cattle known as annual ryegrass toxicity (ARGT) had become a problem in Western Australia and the bacterium's nematode vector had been referred to as *Anguina lolii* by Berry and Wise (1975) and Bird and Stynes (1977). Because the Western Australian nematode vector associated with ARGT had not been measured or identified, it was described and compared with *A. funesta* from South Australia and with populations of *A. agrostis* from England, America, and New Zealand (Stynes and Bird, 1980). The Australian populations were found to be morphologically indistinguishable from *A. agrostis* populations from the other countries mentioned above and, accordingly, *A. funesta* was deemed to be *A. agrostis*. However, subsequent allozyme electrophoresis studies (Riley, et al., 1988) have convinced these authors that "*A. funesta* is the valid name for the nematode associated with annual ryegrass toxicity" and that this nematode is distinct from *A. agrostis*. Further morphological studies (Southey, et al., 1990) supported the identity of *A. funesta* as a true species and, recently, Subbotin (S. Subbotin, pers. comm., 1999) as a result of sequence analyses of the ITS region of ribosomal DNA, considers *A. funesta* to be a distinct species different from *A. agrostis* and others. Accordingly, *A. funesta* is the most appropriate designation for the ryegrass nematode.

Rathayibacter toxicus adhering to *Anguina funesta* can grow rapidly at the expense of the nematodes, in nematode-induced seed galls on annual ryegrass. This leads to atrophy of the nematodes, and galls which are packed with the bacteria (Bird, 1981). However, a more important consequence of the adhesion of *Rathayibacter toxicus* to *Anguina funesta* than the effect of the bacteria on the nematodes, is the disease of livestock mentioned above and known as ARGT. This disease results in a characteristic staggering gait, which can be followed by collapse and death. It was first recorded in Oregon in the USA (Haag, 1945; Shaw and Muth, 1949) and was later described in more detail from that area (Galloway, 1961). At that time it was realized that a combination of the grass, Chewings fescue, and galls containing *Anguina agrostis* was responsible for the toxicity. It was suggested that the condition be called nematode poisoning, but the exact nature of the toxic substance remained unknown.

In Australia, the first outbreak of ARGT was reported in 1967 from the mid-north of South Australia (McIntosh, et al., 1967). These workers noted the involvement of a yellow bacterium and in 1976 the site of toxicity was located in the bacterial seed gall of *Lolium rigidum* (Lanigan, et al., 1976). The early history of the diseases in which nematodes of the genus *Anguina* and bacteria, originally identified as belonging to the genus *Corynebacterium*, are involved has been reviewed (Bird, 1981; McKay and Ophel, 1993).

Stock losses caused by a disease known as flood plain staggers (FPS) also caused by *R. toxicus* but growing in galls and seedheads of *Agrostis avenacea* and *Polypogon monspeliensis* are now known to be transmitted by a nematode that is not *A. funesta* (Bourke, et al., 1992; McKay and Ophel, 1993).

The morphology of the bacterium that is involved in ARGT and FPS was first examined in the hydrated and anhydrobiotic states under the electron microscope in 1977 (Bird and Stynes, 1977) and the bacterium was tentatively called *Corynebacterium* sp. Reference was made to its similarities to *C. rathayi*. The toxin became known (as it still is) as corynetoxin and the bacterium as *Corynebacterium rathayi*. Subsequently, it was recommended that this bacterium be placed in the genus *Clavibacter* (Riley, 1987) and it was identified as a new species, *Clavibacter toxicus* (Riley and Ophel, 1992). A year later a new generic name was proposed for some Gram-positive, aerobic, coryneform bacteria previously placed in the genus *Clavibacter*. This was *Rathayibacter* (Zgurskaya, et al., 1993), and it was pointed out that "in contrast to *Clavibacter* spp., most *Rathayibacter* strains are associated with nematodes belonging to the genus *Anguina*". Several years later and as a result of 2, 4-diaminobutyric acid isomer determination and menaquinone composition, *C. toxicus* was transferred to the genus *Rathayibacter* as *Rathayibacter toxicus* (Sasaki, et al., 1998). This bacterium can be transmitted to a wide range of grasses and several species of *Anguina* can be vectors. Thus, *A. tritici* has been shown in pot experiments to be a vector of *R. toxicus* that developed in wheat seed galls (Riley, 1992).

Corynetoxin poisoning or tunicaminyluracil toxicosis (Bourke, et al., 1992) is an often fatal neurological disease induced in various domestic animals by the ingestion of galls and seedheads colonized by *R. toxicus* infected with a bacteriophage. This bacteriophage was first observed under the TEM by Bird et al. (1980) in freeze-fractured preparations and in sections cut through bacteria in bacterial galls (Fig. 6.3). At that time, we stated "Perhaps the most interesting of our observations is the discovery of the particles located in and around the bacteria and particularly in the gall wall, a site which is known to contain the greatest amount of toxin. We have detected these structures in freeze-etch preparations of cultured corynebacteria which introduces the possibility that they might be corynephages. It is tempting to compare *C. rathayi* with *C. diphtheriae* in which a toxin is only produced by strains that are converted to toxin-producing forms by corynephages, which carry the tox structural gene." These speculations were further supported by the observations of Stynes and Bird (1983), who examined hundreds of sections under the TEM that had been cut through bacterial galls from the field and harvested at weekly intervals. Stynes and Bird (1983) showed that toxin production appeared to be related to the phage-like particles and to the age of the plant. Thirteen years after their initial detection, a positive correlation was found between bacteriophage presence in *R. toxicus* and corynetoxin production (Ophel, et al., 1993). It was speculated that toxin production by these infected bacteria might give these slow-growing bacteria a competitive advantage in ryegrass galls. Thus, the phages here may function in a similar manner to those in *Xenorhabdus* in helping to keep competing bacteria at bay.

Corynetoxins belong to the tunicamycin group of antibiotics and consist of N-acetylglucosaminyl-tunicaminyl-uracil in amide linkage with fatty acids

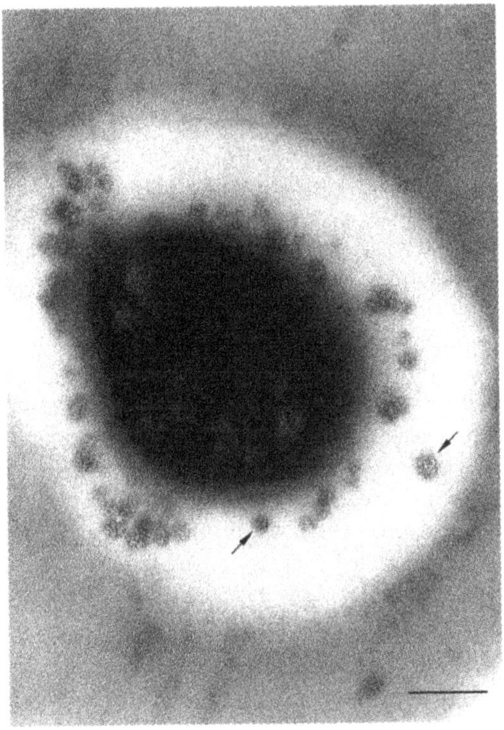

Figure 6. 3 High resolution TEM photomicrograph of a cross-section cut through *Rathayibacter toxicus* (× 180, 000) showing phage-head particles (~ 40 nm in diameter) embedded in the capsule (arrows). Scale bar = 0. 1 μm.

(Edgar, et al., 1982; Frahn, et al., 1984). Differences between individual corynetoxins and between corynetoxins and tunicamycins are due to differences in their fatty acid component.

The neurological symptoms of ARGT and FPS such as staggering and convulsions are brought about by oxygen deprivation to the brain caused by inhibition of protein glycosylation (Jago, et al., 1983). Corynetoxins do not appear to be toxic to nematodes nor do they appear to influence adhesion (Bird, et al., 1985). This complex disease is a direct consequence of the ability of *Rathayibacter toxicus* to adhere to a nematode of the genus *Anguina*.

Evidence for the role of *Anguina* in the bacterial seed gall disease syndrome was obtained early in the 20th century particularly with the disease of wheat known as tundu or yellow ear rot and involving *A. tritici* and *Rathayibacter tritici* (Bird, 1981).

Work on the life cycle of *A. funesta* including embryogenesis, post-embryonic growth, and development in the host plant (Bird and Stynes, 1981a, 1981b, 1981c) has shown that this nematode molts once in the egg and takes 9 - 10 days to complete embryogenesis and hatch at an optimum temperature of between 18℃ and 20℃. The larvae grow from freshly hatched L_2 about 550 μm in length to DL_2 that are close to 850 μm in length in about 9 days at 20℃ when grown on callus tissue culture. The DL_2 are larger, more active, contain numerous lipid granules, are much more resistant to detergents and anesthetics (Riddle and Bird, 1985), and have the capacity to undergo anhydrobiosis. The transition, after invasion of the host plant, of the DL_2 to the parasitic second-stage larva (PL_2) involves the loss of the lipid storage granules, the development of an intestine, and the loss of its anhydrobiotic capabilities. The PL_2 develops into an adult male or female after feeding and molting three times. In the field the nematode only completes one generation a year (Price, et al., 1979). The critical stage for bacterial adhesion is, of course, the DL_2 stage. The viability of 1-year- old DL_2 in galls is high and mortality was found to range from 5.1% to 8.3% (Bird, 1985). The average number of *A. funesta* DL_2 per gall was 1528. Exposure of these larvae to *R. toxicus*, also from South Australia, showed that the mean percentage per gall of DL_2 with bacteria adhering was 42%. However, there was much variability and in some galls no nematodes succumbed to bacterial adhesion (Bird, 1985). Subsequent examination of hundreds of populations from both South and Western Australia exposed to a single strain of *R. toxicus* (Riley and McKay, 1991) gave similar results for the South Australian population giving a mean percentage adhesion of 40% whereas that for the Western Australian population was much higher at 80%. These workers concluded that adhesion to the nematode in the bacterial strain used might be determined by a single recessive gene and that their results suggested that, in time, bacterial colonization of the galls might significantly reduce the adhesiveness of the nematode populations. These comments support earlier speculation (Bird, 1985) that natural selection of non-vector surviving DL_2 could be a factor in reducing the risks of ARGT. Wisely, in view of the limited amount of research yet conducted on the genetic basis for bacterial adhesion, Riley and McKay (1991) caution against relying on reducing the risk of ARGT by means of this biological control by *R. toxicus*. Nevertheless, it has been clearly shown that *R. toxicus* can inhibit movement and cause serious damage to the cuticle of its host (Bird, 1985; Bird and Riddle, 1984). The propagation and survival of this bacterium is dependent on its ability to adhere to the DL_2 of *Anguina* without causing undue damage to the nematode or inhibiting its movements, so that the bacterium cannot only be transported to its host plant but also can still enable the nematode to initiate a suitable seed gall in which it multiplies. In the field the bacterium and nematode are often much more widely dispersed than in the laboratory, and the chances of DL_2 reaching their host plants relatively lightly encumbered with adhering bacteria are, accordingly, much greater. Once contact has been made the forces of adhesion are strong enough to resist abrasion due to nematode movement through the soil and the surface coat of the nematode's cuticle, where adhesion occurs, is readily renewed by the nematode (Bird and Zuckerman, 1989).

The morphological nature of adhesion of *Rathayibacter toxicus* to the surface coat and cuticle of *Anguina* has been observed under the electron microscope (Bird, 1985; Bird and Riddle, 1984; McClure and Spiegel, 1991). The degree of adhesion of bacteria to nematodes varies depending on whether the nematodes are exposed to bacteria from galls or from laboratory cultures. Normal adhesion from field material leads to fusion of the bacterium capsular material with the surface coat of the nematode (Fig. 6.4). Slight swelling of the nematode's epicuticle occurs at the point of adhesion and cuticular damage, which is reflected in loss of the nematode's mobility, may also occur.

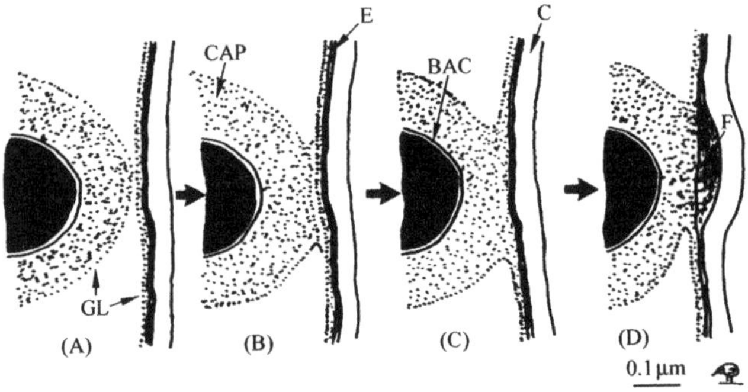

Figure 6.4 Diagram showing stages of adhesion of *Rathayibacter toxicus* to glycocalyx or surface coat of the DL_2 of *Anguina funesta*. Adapted from electronmicrographs of sections cut through the bacterium at its point of contact with the nematode. (A) The bacterium, enclosed in a capsule and covered with a glycocalyx, about to attach to the glycocalyx of the nematode's cuticle. (B) Adhesion between the respective glycocalyces has occurred but the nematode's glycocalyx remains discrete. (C) Fusion of the glycocalyces has occurred. (D) Fine fibers traversing this fused area have penetrated the epicuticle, which has become swollen. BAC = bacterium; CAP = capsule; C = cuticle; E = epicuticle; F = fibers; GL = glycocalyx.

6.7.2 *Meloidogyne* and *Pasteuria*

The relationship between nematodes of the genus *Meloidogyne* and bacteria of the genus *Pasteuria* is an example of the degree of refinement of parasitism that has been achieved in nature. Here, the nematode, itself a highly specialized plant parasite, is completely exploited by the bacterium, which channels all the energy required by the nematode for reproduction into the production of its own spores (Fig. 6. 5). The early history of the discovery of the nematode and the giant cells that it induces in its host plants has been reviewed (Bird, 1961, 1978).

One of the first root-knot nematodes to be identified was found in roots of sugarcane growing at Cheribon and Bogor (then Buitenzorg) in Java, Indonesia. It was named *Heterodera javanica* (Treub, 1886), later to become *Meloidogyne javanica* (Chitwood, 1949). Throughout the 20th century there have been numerous records published of root-knot nematodes and their hosts. They are now known to parasitize thousands of different species of plants throughout the world, including both monocotyledons and dicotyledons. The genus *Meloidogyne* includes more than 70 species, of which at least four are considered to be major pests, which are widely distributed in tropical, subtropical and temperate climates (Eisenback and Triantaphyllou, 1991). The feeding strategies and host responses to these and other plant-parasitic nematodes have been reviewed recently (Hussey and Grundler, 1998).

Meloidogyne is susceptible to parasites that can adhere to it as it migrates through the soil. The viability of the hatched L_2 is dependent on its energy resources in the form of lipid granules. It has been found that the infectivity of *M. javanica* declines sharply to less than 30% within 8 days in the soil at 27℃ (Van Gundy, et al., 1967). During its time in the soil, as it searches for the safety of its host, the L_2 is exposed to predators, parasites, competition at the root surface, and environmental fluctuations. The mycelial, endospore-forming bacterium, *Pasteuria penetrans*, is one of the hazards that the nematode faces.

The historical background to this most specific obligate parasite of nematodes yet discovered has been reviewed by Sayre and Starr (1985) in their description of this parasite, by Bird and Bird (1991) in the chapter "Nematode Pathology" and most recently by Chen and Dickson (1998). The first observations of *P. penetrans* were made early in the 20th century and were thought to be spores of a parasitic protozoan (Cobb, 1906; Micoletzky, 1925, both cited by Chen and Dickson, 1998 and Sayre and Starr, 1985). A similar parasite found in *Pratylenchus pratensis* was called *Duboscqia penetrans* (Thorne, 1940). The description and nomenclature persisted for a further 25 years until this parasite was re-examined under the electron microscope (Imbriani and Mankau, 1977; Mankau, 1975) and its true bacterial nature recognized. It was then renamed

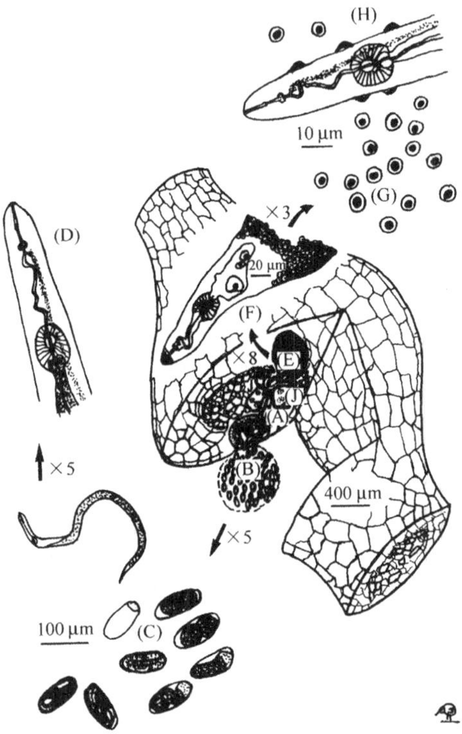

Figure 6. 5 Scale diagram showing the host-parasite relationship of the root-knot nematode (*Meloidogyne*) on the left- hand side of the drawing (A) – (D) and the actinomycete (*Pasteuria penetrans*) on the right- hand side of the drawing (E) – (H). Each nematode is feeding from giant cells (J) that they have initiated and maintained in the vascular tissue of the host plant's root. The uninfected female *Meloidogyne* (A) has laid eggs to form an egg mass (B) on the surface of the root. An enlarged view of these eggs (C) shows various stages of development up to the second-stage larva which hatches from the egg. An enlarged view of the head region of this stage (D) shows the buccal stylet that is used to penetrate a root and complete the life cycle. The infected female *Meloidogyne* (E) is packed with spores of *Pasteuria penetrans* that have grown at the expense of the nematode's reproductive system. These spores are liberated into the soil as the nematode atrophies (F) – (G) and adhere to passing larvae (H) and are taken into the roots of plants by their nematode host where the cycle is repeated.

Bacillus penetrans. However, 2 years later Sayre and Wergin (1977) published a paper on this parasite in which they pointed out that it resembled bacteria in the genus *Pasteuria* more closely than those in the genus *Bacillus*. Continuing their studies, Sayre et al. (1979) examined a bacterial parasite of a water flea that had a similar morphology and which had been named *Pasteuria ramosa* by Metchnikoff (1888, cited by Sayre and Starr, 1985). It was clear to Sayre and Starr (1985) and others that *Bacillus penetrans* should be called *Pasteuria penetrans* and in their paper they describe *Pasteuria penetrans* and present an emended description of the genus *Pasteuria* (Sayre and Starr, 1985). These workers compared the similarities and differences between *P. ramosa* and *P. penetrans* and provided a list of the different nematode hosts for *P. penetrans* and its geographic distribution, which is world wide (Chen and Dickson, 1998).

More recently, another species of *Pasteuria* parasitizing the root-lesion nematode *Pratylenchus brachyurus*, has been described (Sayre, et al., 1988). This bacterium differs from *P. penetrans* in its restricted host range, life stages, and morphology and has been named *Pasteuria thornei* (Starr and Sayre, 1988). Another mycelial, endospore-forming bacterium that clearly belonged to the genus *Pasteuria* was found by Nishizawa in 1986 and was subsequently named *Pasteuria nishizawae* (Sayre, et al., 1991). This species of *Pasteuria* can parasitize nematodes such as *Heterodera glycines* and *Globodera rostochiensis* but not species of *Meloidogyne* nor *Pratylenchus* (Sayre, et al., 1991).

Recently, a *Pasteuria* species has been described from the pea cyst nematode, *Heterodera goettingiana*, which differs from other members of this genus described so far, in that all the parasitic stages of the bacterium can be found in the L_2 of this nematode which thus can complete its life cycle in a preparasitic stage of the nematode. Here the endospores move right up to the anterior end of the L_2 and appear to obliterate the pharyngeal glands (Sturhan, et al., 1994). It is clear that some species of *Pasteuria* are much more host specific than others. For example, Kaplan (1994) has partially characterized a *Pasteuria* sp. that appears to be a specific parasite of the citrus nematode, *Tylenchulus semipenetrans*. Doubtless many more species of this bacterial parasite of nematodes will be discovered in the course of time. The list of nematode hosts is already known to include both sedentary and migratory endoparasites as well as ectoparasites, predators, and free-living nematodes (Chen and Dickson, 1998; Sayre and Starr, 1985). In their review of *P. penetrans* Chen and Dickson (1998) list 323 species of nematodes from 116 genera world wide.

As few as seven attached spores of *P. penetrans* per L_2 of *M. incognita* can significantly reduce larval mobility and subsequent development of the female nematode in the root (Davies, et al., 1991). It has been known for some time that temperature has a marked effect on adhesion and development of the bacterial parasite (Stirling, 1981). The optima for these processes appear to vary depending on the species. Approximately twice as many spores of *P. penetrans will attach to* L_2 of *M. javanica* at 27 ℃ than at 18 ℃ (Stirling, et al., 1990). Furthermore, gall formation is markedly reduced if these larvae have to migrate more than 20 mm in soil containing 10^5 bacterial spores g^{-1} soil.

The adhesion of *Pasteuria penetrans* spores to *Meloidogyne javanica* L_2 results, as mentioned above, in a unique form of parasitism in which the development of the bacterium is synchronized with that of its root-knot nematode host (Mankau, 1981).

Spores attached to the nematode's cuticle germinate about a week after the nematode has entered a root and started to feed on its plant host. A germ tube from the attached spore penetrates the cuticle and gives rise to a branched microcolony that fragments into daughter colonies that proliferate throughout the body cavity of the female nematode. These microcolonies give rise to sporangia in which the endospores develop. Examination and comparison of giant cell development in *M. javanica* in tomato roots with and without *Pasteuria penetrans* (Bird, 1986) has shown that the bacterium does not have a marked influence on the rate of growth of the female nematode in its early stages. This is because it does not invade or inactivate the pharyngeal region which is thought to provide stimuli for giant cell formation and growth; instead it develops at the expense of the reproductive system so that no egg or gelatinous matrix production takes place. Thus, in both instances the stimulus to the giant cells is maintained and normal giant cell development takes place. However, in the case of the female without *Pasteuria*, the energy from the giant cells is translated into eggs in a gelatinous matrix, and in the case of females with *Pasteuria*, this energy is translated into endospores retained in the female's body. As many as 2 million endospores can lie packed in the body of a *Meloidogyne* female (Fig. 6.5). These are released into the soil when the nematode dies and decomposes and there await adhesion to the surface coat of a migrating L_2.

Different populations of *P. penetrans* can differ from each other in size, adhesive capacity, electrophoretic mobility and phosphorus, sulfur and calcium content (Bird, et al., 1990). Bird and his co-workers found that, of three populations tested, two from the USA and one from Australia, all had similar responses to chemicals and stains and an isoelectric point close to pH4.0. How endospore electrophoretic mobility and elemental composition affect adhesion to the nematode's surface coat is unclear, although Afolabi et al. (1995) consider that adhesion of the spores to particular parts of the nematode cuticle is the result of interaction between attractive hydrophobic forces and repulsive electrostatic forces on the spore surface with the nematode cuticle. Using both physiochemical and immunological techniques, Davies et al. (1996) concluded that spore attachment of *P. penetrans* to *M. incognita* involves hydrophobic forces together with the presence of fibronectin, or similar extracellular matrix compounds that they have shown to be present in the nematode's cuticle. It is interesting to note that where fusion occurs between the bacterium and the nematode in both *Pasteuria* (Stirling, et al., 1986, Fig. 3F) and *Rathayibacter* (Bird, 1985, Figs 9 and 11), high resolution electron microscopy reveals fiber-like structures traversing the surface coats of both bacterium and nematode at the site of adhesion. More recently, Esnard et al. (1997) have suggested that the early stages of adhesion, which might be non-specific and reversible, could involve changes in membrane potentials at the interface of the surfaces of nematode and bacterium, which are followed by a stage of irreversible attachment. As a result of their experiments on the effects of monoclonal antibodies, cationized ferritin and other organic chemicals on adhesion of *P. penetrans* to *M. incognita*, Esnard et al. (1997) have concluded that this attachment "may involve a combination of at least two physico-chemical factors". One of these is surface charge and the other nematode movement.

Early studies on adhesion of *P. penetrans* spores to the surface of root-knot nematodes showed that these spores adhere both to the infective L_2 and the parasitic L_2 but not to males or females, which indicated that recognition sites for spore adhesion were changed during molting (Stirling, et al., 1986). Subsequently there have been reports of adhesion to males of *Meloidogyne hapla* (Carneiro, et al., 2000) and *M. arenaria* (Hatz and Dickson, 1992), but this is not common.

In field conditions, *Pasteuria penetrans* has been shown to occur in soils that significantly reduce egg mass production of root-knot nematodes added to those soils (Bird and Brisbane, 1988). *Pasteuria penetrans* can also increase yields of various crops when added to field plots infested with these nematodes (Brown, et al., 1985; Chen, et al., 1996).

6.8 Conclusions

The consequences of adhesion are many and varied, conferring advantages to the attaching organism but not to the nematode, so that their respective functions are markedly different.

Morphologically, however, there are similarities. The surface coats or glycocalyces of *Anguina* and *Rathayibacter*, for instance, are structurally similar and appear to fuse imperceptibly when viewed under the electron microscope (Fig. 6.4). The bacterial capsule and glycocalyx, in this instance, are much more extensive than is the glycocalyx of the nematode, so that many bacteria adhere to each other even when attached to the surface of the nematode. Furthermore, it seems unlikely that the bacterial and nematode surface coats are formed in the same way since the nematode surface coat is thought to arise in glands and to be spread over its surface through orifices such as the mouth, secretory-excretory pore, and anus. However, transcuticular secretion of the surface coat can also take place in some nematodes (Blaxter and Bird, 1997). The nematode surface coat varies in thickness, is synthesized almost continuously, and is replaced when shed. Active shedding of this coat by nematodes parasitic in animals is thought to act as a defense mechanism against immune attack from their host (Blaxter, et al., 1992). In plant-parasitic nematodes, it has been suggested that the surface coat may mimic host self-identity (Bird and Wilson, 1994), since antibodies raised to the surface of *M. incognita* cross-react specifically with host phloem cells.

Future studies aimed at understanding the molecular biology of bacterial adhesion to nematode surface coats will doubtless involve the use of *C. elegans* to test for the function of cuticle protein genes from plant-parasitic nematodes. The importance of the use of the *C. elegans* genome as a genetic tool in future research in studies on plant-parasitic nematodes has been clearly outlined in a recent review by Bird et al. (1999). With the completion of the genomic map for this nematode, the sequences of all the genes involved in the synthesis and maintenance of the cuticle of one nematode are known (Blaxter and Robertson, 1998) and make this nematode an important research tool to help analyze other nematode genomes. These genetic studies should enable us to understand the nature of the genes that control chemical and structural changes that take place, for instance, as members of the genus *Anguina* move into or out of the anhydrobiotic state or how a surface coat ceases to respond to the adhesion of a bacterium. An increase in our knowledge of the genomes of both parasitic bacteria and nematodes will eventually lead to more precise methods of controlling these organisms.

Genetic engineering and molecular biology will be the keys to our future understanding of the nature of diseases of animals (including humans) and plants. One area of research in the 21st century will be an attempt to understand the nature of adhesion in many different forms of viruses and prokaryotes to various parts of eukaryotes. An even more difficult part will be to extrapolate from the laboratory to the field. It is essential that ecological studies of the future involve closer collaboration with biotechnologists. The final joust is always with mother nature in the field.

References

Afolabi, P., K. G. Davies, and P. S. O'Shea. 1995. The electrostatic nature of the spore of *Pasteuria penetrans*, the bacterial parasite of root-knot nematodes. Journal of Applied Bacteriology 79: 244 – 249.

Akhurst, R. J. 1980. Morphological and functional dimorphism in *Xenorhabdus* spp., bacteria symbiotically associated with the insect pathogenic nematodes *Neoaplectana* and *Heterorhabditis*. Journal of General Microbiology 121: 303 – 309.

Akhurst, R. 1993. Bacterial symbionts of entomopathogenic nematodes – the power behind the throne. Pp. 127 – 135. In R. A. Bedding, R. Akhurst, and H. Kaya, eds. Nematodes and the Biological Control of Insects. Melbourne: CSIRO Publications.

Akhurst, R. J., and N. E. Boemare. 1988. A numerical taxonomic study of the genus *Xenorhabdus* (Enterobacteriaceae) and proposed elevation of the subspecies of *X. nematophilus* to species. Journal of General Microbiology 134: 1835 – 1845.

Akhurst, R. J., and N. E. Boemare. 1990. Biology and taxonomy of *Xenorhabdus*. Pp. 75 – 90. In R. Gaugler, and H. K. Kaya, eds. Nematodes in Biological Control. Boca Raton, FL: CRC Press.

Atanasoff, D. 1925. The *Dilophospora* disease of cereals. Phytopathology 15: 11 – 40.

Berkeley, R. C. W., J. M. Lynch, J. Melling, P. R. Rutter, and B. Vincent, eds. 1980. Microbial Adhesions to Surfaces. Chichester, UK: Ellis Horwood Ltd.

Berry, P. H., and J. L. Wise. 1975. Wimmera rye grass toxicity in Western Australia. Australian Veterinary Journal 51: 525 – 530.

Beveridge, T. J. 1999. Structures of gram-negative cell walls and their derived membrane vesicles. Journal of Bacteriology 181: 4725 – 4733.

Binnington, K. C., and L. Brooks. 1993. Fimbrial attachment of *Xenorhabdus nematophilus* to the intestine of *Steinernema carpocapsae*. Pp. 147 – 155. In R. A. Bedding, R. Akhurst, and H. Kaya, eds. Nematodes and the Biological Control of Insects. Melbourne: CSIRO Publications.

Bird, A. F. 1961. The ultrastructure and histochemistry of a nematode-induced giant cell. The Journal of Biophysical and Biochemical Cytology 11: 701 – 715.

Bird, A. F. 1964. Serological studies on the plant parasitic nematode, *Meloidogyne javanica*. Experimental Parasitology 15: 350 – 360.

Bird, A. F. 1978. Root-knot nematodes in Australia. Division of Horticultural Research Technical Paper No. 2. Australia: CSIRO.

Bird, A. F. 1980. The nematode cuticle and its surface. Pp. 213 – 236. In B. M. Zuckerman, ed. Nematodes as Biological Models. Vol. 2. New York: Academic Press.

Bird, A. F. 1981. The *Anguina-Corynebacteriun* association. Pp. 303 – 323. In B. M. Zuckerman, and R. A. Rohde, eds. Plant Parasitic Nematodes. Vol. III. New York: Academic Press.

Bird, A. F. 1984. Growth and moulting in nematodes: Moulting and development of the hatched larva of *Rotylenchulus reniformis*. Parasitology 89: 107 – 119.

Bird, A. F. 1985. The nature of the adhesion of *Corynebacterium rathayi* to the cuticle of the infective larva of *Anguina agrostis*. International Journal for Parasitology 15: 301 – 308.

Bird, A. F. 1986. The influence of the actinomycete, *Pasteuria penetrans*, on the host-parasite relationship of the plant-parasitic nematode *Meloidogyne javanica*. Parasitology 93: 571 – 580.

Bird, A. F. 1988. Cuticle printing of nematodes. International Journal for Parasitology 18: 869 – 871.

Bird, A. F., and R. K. Akhurst. 1983. The nature of the intestinal vesicle in nematodes of the family Steinernematidae. International Journal for Parasitology 13: 599 – 606.

Bird, A. F., and J. Bird. 1991. The Structure of Nematodes. 2nd ed. San Diego, CA: Academic Press.

Bird. A. F., and J. Bird. 1998. Introduction to functional organization. Pp. 1 – 24. In R. N. Perry, and D. J. Wright, eds. The Physiology and Biochemistry of Free-living and Plant- parasitic Nematodes. Wallingford, UK: CAB International.

Bird, A. F., and P. G. Brisbane. 1988. The influence of *Pasteuria penetrans* in field soils on the reproduction of root-knot nematodes. Revue de Nématologie 11: 75 – 81.

Bird, A. F., and M. S. Buttrose. 1974. Ultrastructural changes in the nematode *Anguina tritici* associated with anhydrobiosis. Journal of Ultrastructural Research 48: 177 – 189.

Bird, A. F., and K. Deutsch. 1957. The structure of the cuticle of *Ascaris lumbricoides* var. *suis*. Parasitology 47: 319 – 328.

Bird, A. F., and A. C. McKay. 1987. Adhesion of conidia of the fungus *Dilophospora alopecuri* to the cuticle of the nematode *Anguina agrostis*, the vector in annual ryegrass toxicity. International Journal for Parasitology 17: 1239 – 1247.

Bird, A. F., and D. L. Riddle. 1984. Effect of attachment of *Corynebacterium rathayi* on movement of *Anguina agrostis* larvae. International Journal for Parasitology 14: 503 – 511.

Bird, A. F., and W. Saurer. 1967. Changes associated with parasitism in nematodes. II. Histochemical and microspectrophotometric analyses of preparasitic and parasitic larvae of *Meloidogyne javanica*. The Journal of Parasitology 53: 1262 – 1269.

Bird, A. F., and B. A. Stynes. 1977. The morphology of a *Corynebacterium* sp. parasitic on annual ryegrass. Phytopathology 67: 828 – 830.

Bird, A. F., and B. A. Stynes. 1981a. The life cycle of *Anguina agrostis*: Embryogenesis. International Journal for Parasitology 11: 23 – 33.

Bird, A. F., and B. A. Stynes. 1981b. The life cycle of *Anguina agrostis*: Post-embryonic growth of the second-stage larva. International Journal for Parasitology 11: 243 – 250.

Bird, A. F., and B. A. Stynes. 1981c. The life cycle of *Anguina agrostis*: Development in host plant. International Journal for Parasitology 11: 431 – 440.

Bird, A. F., and B. M. Zuckerman. 1989. Studies on the surface coat (glycocalyx) of the dauer larva of *Anguina agrostis*. International Journal for Parasitology 19: 235 – 240.

Bird, A. F., B. A. Stynes, and W. W. Thomson. 1980. A comparison of nematode and bacteria-colonized galls induced by *Anguina agrostis* in *Lolium rigidum*. Phytopathology 70: 1104 – 1109.

Bird, A. F., M. V. Jago, and P. A. Cockrum. 1985. Corynetoxins and nematodes. Parasitology 91: 169 – 176.

Bird, A. F., I. Bonig, and A. Bacic. 1988. A role for the "excretory" system in secernentean nematodes. Journal of Nematology 20: 493 – 496.

Bird, A. F., P. G. Brisbane, S. G. McClure, and R. W. L. Kimber. 1990. Studies on the properties of the spores of some populations of *Pasteuria penetrans*. Journal of Invertebrate Pathology 55: 169 – 178.

Bird, D. M., and M. A. Wilson. 1994. Plant molecular and cellular responses to nematode infection. Pp. 181 – 195. In F. Lamberti, C. De Giorgi, and D. McK. Bird, eds. Advances in Molecular Plant Nematology. New York: NATO ASI. Plenum Press.

Bird, D. McK., C. H. Opperman, S. J. M. Jones, and D. L. Baillie. 1999. The *Caenorhabditis elegans* genome: A guide to the post genomics age. Annual Review of Phytopathology 37: 247 – 265.

Blaxter, M., and D. Bird. 1997. Parasitic nematodes. Pp. 851 – 878. In D. L. Riddle, T. Blumenthal, B. J. Meyer, and J. R. Priess, eds. *C. elegans* II. New York: Cold Spring Harbor Laboratory Press.

Blaxter, M. L., and W. M. Robertson. 1998. The cuticle. Pp. 25 – 48. In R. N. Perry, and D. J. Wright, eds. The Physiology and Biochemistry of Free-living and Plant-parasitic Nematodes. Wallingford, UK: CAB International.

Blaxter, M. L., A. P. Page, W. Rudin, and R. M. Maizels. 1992. Nematode surface coats: Actively evading immunity. Parasitology Today 8: 243 – 247.

Boemare, N. E., and R. J. Akhurst. 1988. Biochemical and physiological characterization of colony form variants in *Xenorhabdus* spp. (Enterobacteriaceae). Journal of General Microbiology 134: 1835 – 1845.

Boemare, N. E., M. H. Boyer-Giglio, J. O. Thaler, and R. J. Akhurst. 1993. The phages and bacteriocins of *Xenorhabdus* spp. and *Heterorhabditis* spp. Pp. 137 – 145. In R. A. Bedding, R. Akhurst, and H. Kaya, eds. Nematodes and the Biological Control of Insects. Melbourne: CSIRO Publications.

Boemare, N., A. Givaudan, M. Brehelin, and C. Laumond. 1997. Symbiosis and pathogenicity of nematode-bacterium complexes. Symbiosis 22: 21 – 45.

Bone, L. W., B. D. Hammock, L. E. Graston, S. K. Reed, and H. H. Shorey. 1980. Partial purification of the aggregation pheromone nippolure from female *Nippostrongylus brasiliensis* (Nematoda). Journal of Chemical Ecology 6: 297 – 308.

Bourke, C. A., M. A. J. Carrigan, and S. C. J. Love. 1992. Flood plain staggers, a tunicaminyl-uracil toxicosis of cattle in northern New South Wales. Australian Veterinary Association 69: 228 – 229.

Brehelin, M., A. Cherqui, L. Drif, L. Luciani, R. Akhurst, and N. Boemare. 1993. Ultrastructural study of surface components of *Xenorhabdus* sp. in different cell phases and culture conditions. Journal of Invertebrate Pathology 61: 188 – 191.

Brown, S. M., J. L. Kepner, and G. C. Smart. 1985. Increased crop yields following application of *Bacillus penetrans* to field plots infested with *Meloidogyne incognita*. Soil Biology and Biochemistry 17: 483 – 486.

Carneiro, R. M. D. G., O. Randig, L. G. Freitas, and D. W. Dickson. 2000. Attachment of endospores of *Pasteuria* to males and juveniles of *Meloidogyne* spp. Nematology 1: 267 – 271.

Chen, Z. X., and D. W. Dickson. 1998. Review of *Pasteuria penetrans*: Biology, ecology and biological control potential. Journal of Nematology 30: 313 – 340.

Chen, Z. X., D. W. Dickson, R. McSorley, D. J. Mitchell, and T. E. Hewlett. 1996. Suppression of *Meloidogyne arenaria* race 1 by soil application of endospores of *Pasteuria penetrans*. Journal of Nematology 28: 159 – 168.

Chitwood, B. G. 1949. Root-knot nematodes. I. A revision of the genus *Meloidogyne* Goeldi, 1887. Proceedings of the Helminthological Society of Washington 16: 90 – 104.

Davies, K. G., V. Laird, and B. R. Kerry. 1991. The motility, development and infection of *Meloidogyne incognita* encumbered with spores of the obligate hyperparasite *Pasteuria penetrans*. Revue de Nématologie 14: 611 – 618.

Davies, K. G., P. Afolabi, and P. S. O'Shea. 1996. Adhesion of *Pasteuria penetrans* to the cuticle of root-knot nematodes (*Meloidogyne* spp.) inhibited by fibronectin: A study of electrostatic and hydrophobic interactions. Parasitology 112: 553 – 559.

Dutky, S. R. 1959. Insect microbiology. Advances in Applied Microbiology 1: 175 – 200.

Dutky, S. R., and W. S. Hough. 1955. Notes on a parasitic nematode from codling moth larvae, *Carpocapsa pomonella*. Proceedings of the Entomological Society of Washington 57: 244.

Edgar, J. A., J. L. Frahn, P. A. Cockrum, N. Anderton, and M. V. Jago. 1982. Corynetoxins, causative agents of annual ryegrass toxicity; their identification as tunicamycin group antibiotics. Journal of the Chemical Societies Chemical Communications 4: 222 – 224.

Eisenback, J. D., and H. H. Triantaphyllou. 1991. Root-knot nematodes: *Meloidogyne* species and races. Pp. 191 – 274. In W. R. Nickle, ed. Manual of Agricultural Nematology. New York: Marcel Dekker.

Endo, B. Y., and W. R. Nickle. 1995. Ultrastructure of anterior and mid-regions of infective juveniles of *Steinernema feltiae*. Fundamental and Applied Nematology 18: 271 – 294.

Esnard, J., M. A. McClure, D. W. Dickson, T. E. Hewlett, and B. M. Zuckerman. 1997. Effects of monoclonal antibodies, cationized ferritin, and other organic molecules on adhesion of *Pasteuria penetrans* endospores to *Meloidogyne incognita*. Journal of Nematology 29: 556 – 564.

Forst, S., B. Dowds, N. Boemare, and E. Stackebrandt. 1997. *Xenorhabdus* and *Photorhabdus* spp.: Bugs that kill bugs. Annual Review of Microbiology 51: 47 – 72.

Frahn, J. L., J. A. Edgar, A. J. Jones, P. A. Cockrum, N. Anderton, and C. C. J. Culvenor. 1984. Structure of the corynetoxins, metabolites of *Corynebacterium rathayi* responsible for toxicity of annual ryegrass (*Lolium rigidum*) pastures. Australian Journal of Chemistry 37: 165 – 182.

Galloway, J. H. 1961. Grass seed nematode poisoning in livestock. Journal of the American Veterinary Medical Association 139: 1212 – 1214.

Givaudan, A., S. Baghdiguian, A. Lanois, and N. Boemare. 1995. Swarming and swimming changes concomitant with phase variation in *Xenorhabdus nematophilus*. Applied and Environmental Microbiology 61: 1408 – 1413.

Glaser, R. W. 1932. Studies on *Neoaplectana glaseri*, a nematode parasite of the Japanese beetle (*Popillia japonica*). New Jersey Department of Agriculture Circular No. 211.

Goodey, B. 1959. Gall-forming nematodes of grasses in Britain. Journal of the Sports Turf Research Institute 10: 1 – 7.

Goodey, T. 1930. *Tylenchus agrostis* (Steinbuch 1799). Journal of Helminthology 8:

197 – 210.

Gotz, P., A. Boman, and H. G. Boman. 1981. Interactions between insect immunity and an insect-pathogenic nematode with symbiotic bacteria. Proceedings of the Royal Society (London) B 212: 333 – 350.

Green, C. D., and D. N. Greet. 1972. The location of the secretions that attract male *Heterodera schachtii* and *H. rostochiensis* to their females. Nematologica 18: 347 – 352.

Haag, J. R. 1945. Toxicity of nematode infested Chewings fescue seed. Science 102: 406 – 407.

Hatz, B., and D. W. Dickson. 1992. Effect of temperature on attachment, development, and interactions of *Pasteuria penetrans* on *Meloidogyne arenaria*. Journal of Nematology 24: 512 – 521.

Hewitt, W. B., D. J. Raski, and A. C. Goheen. 1958. Nematode vector of soil-borne fanleaf virus of grapevines. Phytopathology 48: 586 – 595.

Hollis, J. P. 1962. Nature of swarming in nematodes. Nature 193: 798 – 799.

Hussey, R. S., and F. M. W. Grundler. 1998. Nematode parasitism of plants. Pp. 213 – 243. In R. N. Perry, and D. J. Wright, eds. The Physiology and Biochemistry of Free-living and Plant- parasitic Nematodes. Wallingford, UK: CAB International.

Imbriani, J. L., and R. Mankau. 1977. Ultrastructure of the nematode pathogen, *Bacillus penetrans*. Journal of Invertebrate Pathology 30: 337 – 347.

Jackson, G. J. 1961. The parasitic nematode, *Neoaplectana glaseri*, in axenic culture. I. Effects of antibodies and anthelminthics. Experimental Parasitology 11: 241 – 247.

Jago, M. V., A. L. Payne, J. E. Peterson, and T. J. Bagust. 1983. Inhibition of glycosylation by corynetoxin, the causative agent of annual ryegrass toxicity: A comparison with tunicamycin. Chemical and Biological Interactions 45: 223 – 234.

Jansson, H. -B. 1987. Receptors and recognition in nematodes. Pp. 153 – 158. In J. A. Veech, and D. W. Dickson, eds. Vistas on Nematology. Hyattsville, MD: Society of Nematologists.

Kaplan, D. T. 1994. Partial characterization of a *Pasteuria* sp. attacking the citrus nematode, *Tylenchulus semipenetrans*, in Florida. Fundamental and Applied Nematology 17: 509 – 512.

Kaya, H. K., R. A. Bedding, and R. J. Akhurst. 1993. An overview of insect-parasitic and entomopathogenic nematodes. Pp. 1 – 10. In R. A. Bedding, R. Akhurst, and H. Kaya, eds. Nematodes and the Biological Control of Insect Pests. Melbourne: CSIRO Publications.

Keilin, D. 1959. The problem of anabiosis or latent life: History and current concept. Proceedings of the Royal Society(London) B 150: 149 – 191.

Lanigan, G. W., A. L. Payne, and J. L. Frahn. 1976. Origin of toxicity in parasitized annual ryegrass (*Lolium rigidum*). Australian Veterinary Journal 52: 244 – 246.

Leeuwenhoek, A. van. 1702. On certain animalcules found in the sediment in gutters in the roofs of houses. Letter 144 In "The selected works of Antony van Leeuwenhoek" translated by Samuel Hoole. 1798. London 2: 207 – 213.

Maizels, R. M., and A. P. Page. 1990. Surface associated glycoproteins from *Toxocara canis* larval parasites. Acta Tropica 47: 355 – 364.

Mankau, R. 1975. *Bacillus penetrans* n. comb. causing a virulent disease of plant-parasitic nematodes. Journal of Invertebrate Pathology 26: 333 – 339.

Mankau, R. 1981. Microbial control of nematodes. Pp. 475 – 494. In B. M. Zuckerman, and R. A. Rhode, eds. Plant Parasitic Nematodes Vol. III. New York: Academic Press.

McClure, M. A., and Y. Spiegel. 1991. Role of the nematode surface coat in the adhesion of *Clavibacter* to *Anguina funesta* and *Anguina tritici*. Parasitology 103: 421 – 427.

McIntosh, G. H., R. Rac, and M. R. Thomas. 1967. Toxicity of parasitized Wimmera ryegrass. Australian Veterinary Journal 43: 349 – 353.

McKay, A. C., and K. M. Ophel. 1993. Toxigenic *Clavibacter/Anguina* associations infecting grass seedheads. Annual Review of Phytopathology 31: 151 – 167.

McKay, A. C., J. M. Fisher, and A. J. Dube. 1981. Ecological field studies on *Anguina funesta*, the vector in annual ryegrass toxicity. Australian Journal of Agricultural Research 32: 917 – 926.

Moureaux, N., T. Karjalainen, A. Givaudan, P. Bourlioux, and N. Boemare. 1995. Biochemical characterization and agglutinating properties of *Xenorhabdus nematophilus* Fl fimbriae. Applied and Experimental Microbiology 61: 2707 – 2712.

Mulvey, T. 1980. The Early Development of Electron Lenses and Electron Microscopy. Stuttgart: Hirzel. (English Translation of Ruska, E. 1933. Die electronen-microskopische abbildung elektronenbestrahlter oberflächen. Zeitschrift für Physik 83: 492 – 497.)

Needham, J. T. 1743. Concerning certain chalky concretions called malm; with some microscopical observations on the farina of Red Lilly, and worms discovered in Smuthy corn. Philosophical Transactions of the Royal Society (London) 42: 634.

Nelson, F. K., P. S. Albert, and D. L. Riddle. 1983. Fine structure of the *Caenorhabditis elegans* secretory-excretory system. Journal of Ultrastructural Research 82: 156 – 171.

Nomarski, G. 1955. Microinterféromètre différentielle à ondes polarisées. Journal de Physique et le Radium 16: 9S – 13S.

Ophel, K. M., A. F. Bird, and A. Kerr. 1993. Association of bacteriophage particles with toxin production by *Clavibacter toxicus*, the causal agent of annual ryegrass toxicity. Phytopathology 83: 676 – 681.

Page, A. P., W. Rudin, E. Fluri, M. L. Blaxter, and R. M. Maizels. 1992. *Toxocara canis*: A labile antigenic surface coat overlaying the epicuticle of infective larvae. Experimental Parasitology 75: 72 – 86.

Poinar, G. O. Jr. and G. M. Thomas. 1965. A new bacterium, *Achromobacter nematophilus* sp. nov. (Achromobacteriaceae: Eubacteriales) associated with a nematode. International Bulletin of Bacteriological Nomenclature and Taxonomy 15: 249 – 252.

Price, P. C. 1973. Investigation of a nematode-bacterium disease complex affecting Wimmera ryegrass. Ph. D. Thesis, University of Adelaide, Adelaide, Australia.

Price, P. C., J. M. Fisher, and A. Kerr. 1979. On *Anguina funesta* n. sp. and its association with *Corynebacterium* sp., in infecting *Lolium rigidum*. Nematologica 25: 76 – 85.

Riddle, D. L., and A. F. Bird. 1985. Responses of *Anguina agrostis* to detergent and anesthetic treatment. Journal of Nematology 17: 165 – 168.

Riley, I. T. 1987. Serological relationships between strains of coryneform bacteria responsible for annual ryegrass toxicity and other plant pathogenic bacteria. International Journal for Systematic Bacteriology 35: 153 – 159.

Riley, I. T. 1992. *Anguina tritici* is a potential vector of *Clavibacter toxicus*. Australian Plant Pathology 21: 147 – 149.

Riley, I. T. 1994. *Dilophospora alopecuri* and decline in annual ryegrass toxicity in Western Australia. Australian Journal of Agricultural Research 45: 841 – 850.

Riley, I. T. 1996. *Dilophospora alopecuri* on *Lolium rigidum* and *Holcus lanatus* in south-eastern Australia. Australian Plant Pathology 25: 255 – 259.

Riley, I. T., and A. C. McKay. 1991. Susceptibility of *Anguina funesta* (Nematoda: Anguinidae) populations to adhesion by the toxigenic *Clavibacter* sp. responsible for annual ryegrass toxicity. Nematologica 37: 439 – 446.

Riley, I. T., and K. M. Ophel. 1992. *Clavibacter toxicus* sp. nov., the bacterium responsible for annual ryegrass toxicity in Australia. International Journal of Systematic Bacteriology 42: 64 – 68.

Riley, I. T., T. B. Reardon, and A. C. McKay. 1988. Electrophoretic resolution of species boundaries in seed-gall nematodes. *Anguina* sp. (Nematoda: Anguinidae), from some graminaceous hosts in Australia and New Zealand. Nematologica 34: 401 – 411.

Ruska, E. 1933. Die electronen-microskopische abbildung elektronenbestrahlter oberflächen. Zeitschrift für Physik 83: 492 – 497. [English translation by T. Mulvey. 1980. The Early Development of Electron Lenses and Electron Microscopy. Hirzel: Stuttgart.]

Sarles, M. P. 1938. The *in vitro* action of immune rat serum on the nematode *Nippostrongylus muris*. Journal of Infectious Diseases 62: 337 – 348.

Sasaki, J. S., M. Chijimatsu, and K. Suzuki. 1998. Taxonomic significance of 2, 4-diaminobutyric acid isomers in the cell wall peptidoglycan of actinomycetes and reclassification of *Clavibacter toxicus* as *Rathayibacter toxicus* comb. nov. International Journal of Systematic Bacteriology 48: 403 – 410.

Sayre, R. M., and M. P. Starr. 1985. *Pasteuria penetrans* (ex Thorne, 1940) nom. rev., comb. n., sp. n., a mycelial and endospore-forming bacterium parasitic in plant-parasitic nematodes. Proceedings of the Helminthological Society of Washington 52: 149 – 165.

Sayre, R. M., and W. P. Wergin. 1977. Bacterial parasite of a plant nematode: Morphology and ultrastructure. Journal of Bacteriology 129: 1091 – 1101.

Sayre, R. M., J. R. Adams, and W. P. Wergin. 1979. Bacterial parasite of a cladoceran: Morphology, development *in vivo*, and taxonomic relationships with *Pasteuria ramosa* Metchnikoff 1888. International Journal of Systematic Bacteriology 29: 252 – 262.

Sayre, R. M., M. P. Starr, A. M. Golden, W. P. Wergin, and B. Y. Endo. 1988. Comparison of *Pasteuria penetrans* from *Meloidogyne incognita* with a related mycelial and endospore-forming bacterial parasite from *Pratylenchus brachyurus*. Proceedings of the Helminthological Society of Washington 55: 28 – 49.

Sayre, R. M., W. P. Wergin, T. Nishizawa, and M. P. Starr. 1991. Light and electron microscopical study of a bacterial parasite from the cyst nematode, *Heterodera glycines*. Journal of the Helminthological Society of Washington 58: 69 – 81.

Shaw, J. N., and O. H. Muth. 1949. Some type of forage poisoning in Oregon cattle and

sheep. Journal of the American Veterinary Medical Association 114: 315 – 317.
Southey, J. F., P. B. Topham, and D. J. F. Brown. 1990. Taxonomy of some species of *Anguina* Scopoli, 1777 (sensu Brzeski, 1981) forming galls on Gramineae: Value of diagnostic characters and present status of nominal species. Revue de Nématologie 13: 143 – 154.
Starr, M. P., and R. M. Sayre. 1988. *Pasteuria thornei* sp. nov. and *Pasteuria penetrans sensu stricto* emend., mycelial and endospore-forming bacteria parasitic, respectively, on plant-parasitic nematodes of the genera *Pratylenchus* and *Meloidogyne*. Annales de l'Institut Pasteur/Microbiologie 139: 11 – 31.
Steinbuch, J. G. 1799. Das Grasälchen, *Vibrio agrostis*. Der Naturforscher 28: 233 – 259.
Steiner, G. 1923. *Aplectana kraussei*, n. sp., eine in der Blattwespe, *Lyda* sp. parasiterende Nematodenform, nebst Bemerkungen ueber das Seitenorgan der parasitschen Nematoden. Zentralblatt fuer Balterologie, Parasitenkunde und Infectionskrankheiten II 59: 14 – 18.
Stirling, G. R. 1981. Effect of temperature on infection of *Meloidogyne javanica* by *Bacillus penetrans*. Nematologica 27: 458 – 462.
Stirling, G. R., A. F. Bird, and A. B. Cakurs. 1986. Attachment of *Pasteuria penetrans* spores to the cuticles of root-knot nematodes. Revue de Nématologie 9: 251 – 260.
Stirling, G. R., R. D. Sharma, and J. Perry. 1990. Attachment of *Pasteuria penetrans* spores to the root-knot nematode *Meloidogyne javanica* in soil and its effects on infectivity. Nematologica 36: 246 – 252.
Sturhan, D. R., R. Winkelheide, R. M. Sayre, and W. P. Wergin. 1994. Light and electron microscopical studies of the life cycle and developmental stages of a *Pasteuria* isolate parasitizing the pea cyst nematode, *Heterodera goettingiana*. Fundamental and Applied Nematology 17: 29 – 42.
Stynes, B. A., and A. F. Bird. 1980. *Anguina agrostis*, the vector of annual rye grass toxicity in Australia. Nematologica 26: 475 – 490.
Stynes, B. A., and A. F. Bird. 1983. Development of annual ryegrass toxicity. Australian Journal of Agricultural Research 34: 653 – 660.
Taylor, C. E., and D. J. F. Brown. 1997. Nematode Vectors of Plant Viruses. Wallingford, UK: CAB International.
Taylor, C. E., and D. J. Raski. 1964. On the transmission of grape fanleaf by *Xiphinema index*. Nematologica 10: 489 – 495.
Taylor, C. E., and W. M. Robertson, 1969. The location of raspberry ringspot and tomato black ring viruses in the nematode vector, *Longidorus elongatus*. Annals of Applied Biology 64: 233 – 237.
Taylor, C. E., and W. M. Robertson. 1970a. Sites of virus retention in the alimentary tract of the nematode vectors, *Xiphinema diversicaudatum* (Micol.) and *X. index* (Thorne and Allen). Annals of Applied Biology 66: 375 – 380.
Taylor, C. E., and W. M. Robertson. 1970b. Location of tobacco rattle virus in the nematode vector *Trichodorus pachydermis* Seinhorst. Journal of General Virology 6: 179 – 182.
Thomas, G. M., and G. O. Poinar, Jr. 1979. *Xenorhabdus* gen. nov., a genus of entomopathogenic, nematophilic bacteria of the family Enterobacteriaceae. International Journal of Systematic Bacteriology 29: 352 – 360.

Thorne, G. 1940. *Duboscqia penetrans* n. sp. (Sporozoa: Microsporidia, Nosematidae) a parasite of the nematode *Pratylenchus pratensis* (de Man) Filipjev. Proceedings of the Helminthological Society of Washington 7: 51 – 53.

Treub, M. 1886. Quelques mots sur les effets dur parasitisme de l'*Heterodera javanica* dans les racines de la canne a sucre. Annales du Jardin Botanique de Buitenzorg 6: 93.

Van Gundy, S. D., A. F. Bird, and H. R. Wallace. 1967. Aging and starvation in larvae of *Meloidogyne javanica* and *Tylenchulus semipenetrans*. Phytopathology 57: 559 – 571.

von Ardenne, M. 1938a. Das Electronen-Rastermikroskop. Theoretische Grundlagen. Zeitschrift für Physik 109: 553 – 572.

von Ardenne, M. 1938b. Das Electronen-Rastermikroskop. Praktische Ausführung. Zeitschrift für Technisch Physik 19: 407 – 416.

Ward, J. B., and R. C. W. Berkeley. 1980. Microbial cell surface and adhesion. Pp. 47 – 66. In R. C. W, Berkeley, J. M. Lynch, J. Melling, P. R. Rutter, and B. Vincent, eds. Microbial Adhesions to Surfaces. Chichester, UK: Ellis Horwood Ltd.

Womersley, C. Z., D. A. Wharton, and L. M. Higa. 1998. Survival biology. Pp. 271 – 302. In R. N. Perry, and D. J. Wright, eds. The Physiology and Biochemistry of Free-living and Plant-parasitic Nematodes. Wallingford, UK: CAB International.

Wouts, W. M. 1991. *Steinernema* (*Neoaplectana*) and *Heterorhabditis* species. Pp. 855 – 897. In W. R. Nickle, ed. Manual of Agricultural Nematology. New York: Marcel Dekker.

Zernike, F. 1934. Das phasenkontrastver fahren bei der mikroskopischen beobachtung. Zeitschrift für Physik 36: 848.

Zgurskaya, H. I., L. I. Evtushenko, V. N. Akimov, and L. V. Kalakoutskii. 1993. *Rathayibacter* gen. nov., including the species *Rathayibacter rathayi* comb. nov., *Rathayibacter tritici* comb. nov., *Rathayibacter iranicus* comb. nov., and six strains from annual grasses. International Journal of Systematic Bacteriology 43: 143 – 149.

Acknowledgements I thank Ian Riley for reading this chapter and for his helpful comments and Jean Bird for her editorial skills and constructive criticism.

7 Nematode Behavior and Migrations through Soil and Host Tissue

A. Forest Robinson

7.1 Introduction

Although a relatively unexplored aspect of biology, nematode behavior has been examined in at least 500 publications, including about 50 review articles and book chapters. The first report was Needham's (1743) elated descriptions of *Anguina tritici* escaping wetted wheat cockles. Other anecdotal accounts of nematode movements are found in early monographs. Examples include: Khalil's (1922) experiments on thermotaxis in hookworm larvae to understand skin penetration; Peters' (1928) detailed studies on the vinegar eelworm; Spindler's (1934) tests for responses to light by larvae of the swine kidney worm; Christie's (1936) observations of foliar migrations by infective juveniles of the grasshopper parasite, *Agamermis decaudada*; and Baunacke's (1922) experiments on host preferences of the sugar beet cyst nematode, *Heterodera schachtii*.

During the last 40 years, major topics of study in nematode behavior have included: (1) responses of ecologically diverse nematodes to non-specific stimuli such as light, temperature, carbon dioxide (at least 50 papers); (2) root finding by plant-parasitic species (*ca.* 130 papers since 1958); (3) foraging behavior by dauer larvae of *Steinernema* and *Heterorhabditis* species (about 40 papers reviewed by Ishibashi and Kondo in 1990, with dozens of papers since); (4) the migrations of nematodes through plant and animal tissues (a long-studied yet poorly understood topic); (5) pheromone responses in *ca.* 20 genera of plant, insect, and free-living nematodes (*ca.* 80 papers beginning in 1964); and (6) several dozen studies related to the attraction or repulsion of nematodes by nematophagous fungi. Recently there has also been intense research in many labs on the behavioral genetics and neurophysiology of *Caenorhabditis elegans* (Avery and Thomas, 1997; Bargmann and Mori, 1997; Driscoll and Kaplan, 1997; Emmons and Sternberg, 1997; Jorgensen and Rankin, 1997; Ruvkun, 1997).

This chapter will provide background information needed to interpret nematode behavior experiments and briefly review central questions and advances in major areas of research on nematode behavior during the last 40 years. Emphasis is placed on the migrations of free-living stages of plant and insect parasites through soil and host tissues, but other kinds of behavior, in particular foraging, host seeking, and mate finding behavior by mycophagous, bacteriophagous, marine, and animal-parasitic nematodes, is briefly discussed where useful for comparisons. Unfortunately, much literature must be omitted and some topics, such as predation, copulation, defecation, and hatching, are ignored. Previous extensive reviews, however, are cited with each specific area of research as appropriate. Additional reviews of a more general nature include those of Dropkin (1965), Van Gundy (1965), Wallace (1968a), Croll (1970, 1972a, 1975a, 1976a), Croll and Sukhdeo (1981), Dusenbery (1980a, 1983, 1992), Aumann (1993), Perry (1994), Perry and Aumann (1998), and Ashton et al. (1999).

7.2 Basis of Behavior

7.2.1 How Nematodes Move

7.2.1.1 Types of Movement

Since basic nematode movements are constrained anatomically, it is significant that the overall morphology and neuroanatomy of nematodes (reviewed in Chapter 4, this volume by Baldwin and Perry) are highly conserved. Of particular importance to behavior is the retention of a vermiform body shape and one or more microscopic juvenile stages inhabiting soil or water. Saccate adult females of genera in the order Tylenchida (*Cactodera*, *Globodera*, *Heterodera*, *Meloidodera*, *Meloidogyne*, *Nacobbus*, *Rotylenchulus*, *Sphaeronema*, *Trophotylenchulus*, *Tylenchulus*, and *Verutus*) are notable exceptions to the vermiform shape, but these genera include only a small portion of all plant-parasitic nematode species (Siddiqi, 1986).

Croll and Sukhdeo (1981) listed 22 kinds of nematode behavior (Table 7.1). Movements range from single motions to coordinated repetitive movements (of the body wall, esophagus, and stoma) and complex sequences of actions. Most activities clearly have evolved to achieve locomotion, feeding, copulation, oviposition, defecation, or perforation of barriers during hatching, skin penetration, or tissue migrations. Other activities with less obvious adaptive value seem to play roles in predator avoidance, desiccation survival, host finding, and host infection. Behaviors also differ in the extent to which they are endogenously or exogenously controlled. The most commonly observed motions were considered by Croll and Sukhdeo (1981) to show five activity phases, which they referred to as inactive, refractory, head-waving, feeding, and active locomotion.

Table 7.1 Nematode activities involving or affecting behavior (from Croll and Sukhdeo, 1981).

1.	Movement in the egg
2.	Hatching
3.	Feeding
4.	Defecation
5.	Molting
6.	Exsheathment
7.	Migration and dispersal
8.	Swarming
9.	Penetration
10.	Nictating
11.	Site location in hosts
12.	Osmoregulation
13.	Capturing prey
14.	Leaping
15.	Escape from predators
16.	Air breathing
17.	Sampling in orientation
18.	Attachment by caudal glands
19.	Bursal flapping
20.	Copulation
21.	Clumping
22.	Oviposition

7.2.1.2 Undulatory Propulsion

Most nematodes move forward primarily by undulatory propulsion (Gray, 1953; Wallace, 1968b, 1969) consisting of sinusoidal dorso-ventral waves of the entire body that are rhythmically propagated backward from the anterior end (Fig. 7.1(A)). In three-dimensional matrices, lateral steering by the anterior end may also occur and could be the primary response contributing to directed movement in three-dimensional stimulus fields, but is impossible when nematodes lie on their sides on an agar plate and so far has been investigated only in *Mermis nigrescens*. When phototaxing on horizontal surfaces, the head of *M. nigrescens* is held elevated above the surface and the dorsal-ventral body waves lie above the surface except at the ventral contact points. On moist felt, the head periodically pushes through the surface fibers to lace the body under fiber loops, against which body waves press in order to propel the body. When climbing through grass, the body waves press against blades or stems (Gans and Burr, 1994).

On agar surfaces, nematodes typically lie on their sides, which probably is unnatural, and leave a sinusoidal track (Figs7.1(B), 7.2(B) - (E), 7.3(B) - (C)). Body undulations also occur in water but more quickly than in agar, and if nematodes in a series of agar concentrations are compared, the speed of backward waves along the body and the wavelength both decrease as agar concentration, and presumed resistance of the substrate, increase (Fig. 7.1(B)) (Ahmad and Jairajpuri, 1982; Bilgrami, 1996; Khan, et al., 1985; Wallace and Doncaster, 1964).

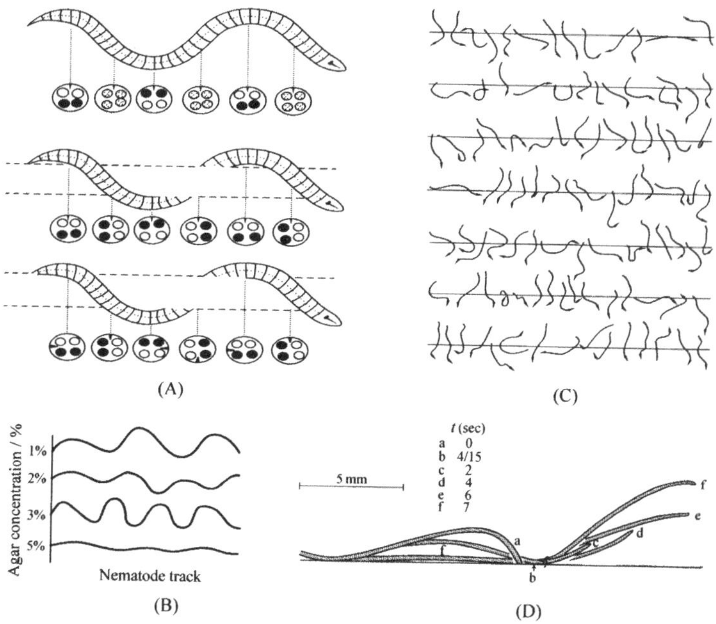

Figure 7.1 Some fundamental aspects of nematode movement. (A) Sequences of longitudinal muscle contraction facilitating undulatory propulsion by two- and three-dimensional body waves (from Wallace, 1968a). (B) Tracks showing different body wavelengths of *Tylenchorhynchus brassicae* at different agar concentrations (from Khan, et al., 1985). (C) Vertical orientation of 100 vermiform *Rotylenchulus reniformis* relative to a horizontal reference line as they settle in water (from Robinson and Heald, 1993). Arrowheads indicate anterior ends. (D) Progression and vertical elevation of *Mermis nigrescens* on a damp felt substrate (from Gans and Burr, 1994).

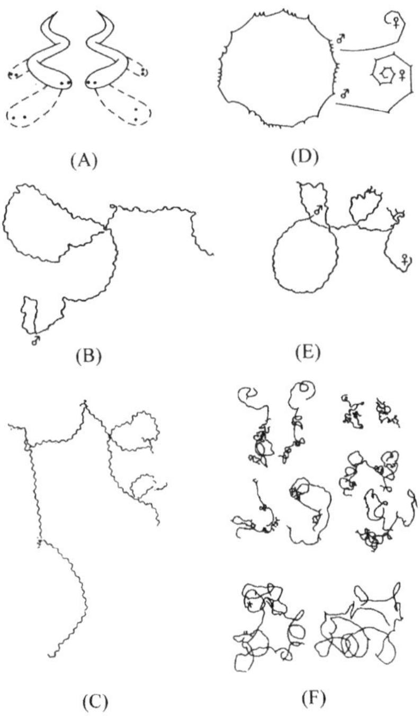

Figure 7. 2 Some basic features of nematode movement on an agar surface. (A) Comparative orientation of bilateral sensors in snakes (left) and nematodes (right) when moving on a horizontal flat surface (from Croll, 1972b). (B) Tracks of a male of *Globodera rostochiensis* showing slew and intermittent reversals (from Croll, 1970). (C) Tracks of an adult female of *Caenorhabditis elegans* showing slew and reversals (from Croll, 1975a). (D) Tracks generated by computer simulation of a male nematode displaying slew (left) and a male approaching a female when slew curvature or path length are allowed to change with increased chemical stimulus from a female (from Green, 1977). (E) Tracks left by a male of *Heterodera schachtii* while approaching a female (from Croll, 1970). (F) Idiosyncrasies of tracks of *Trichonema* spp. infective larvae observed when each of several nematodes was sequentially released twice on an agar surface (from Croll, 1975a).

At less frequent intervals, determined by the species, substrate, and other factors, forward waves can be propagated from the posterior end. Forward waves may result when the anterior end encounters an obstacle (Croll, 1976b; Wicks and Rankin, 1997) or can occur spontaneously at more or less regular intervals. The typical pattern seen on agar is a quick withdrawal for one or

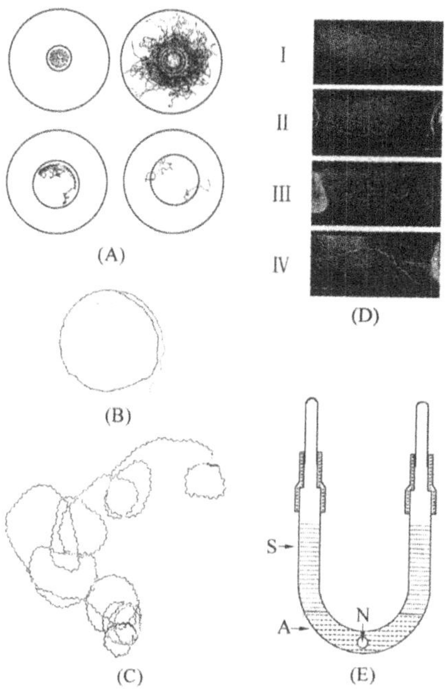

Figure 7. 3 Additional agar tracks showing responses or absences of responses to stimuli. (A) Presence (upper left) and absence (upper right) of osmotic avoidance, contrasted with presence (lower left) and absence (lower right) of Na^+ avoidance in *Caenorhabditis elegans* (from Culloti and Russell, 1978). (B) Thermotaxing by *C. elegans* in a radial gradient field (from Rutherford and Croll, 1979). (C) Persistent unilateral bias in track left by *Aneylosotma tubaeforme* larva (from Croll, 1975a). (D) (I–IV) Three-dimensional nematode tracks left by *Meloidogyne* second-stage juveniles (J2) moving through agar in response to salt gradients (from Prot, 1978c) (E) tubular apparatus used by Prot (1978a, 1978c) to allow *Meloidogyne* J2 to move 3-dimensionally through agar "A" in response to salt gradients achieved by submersing the two ends of the agar with salt solution "S" or distilled water. Nematodes "N" were introduced at bottom of tube.

more body waves, hesitation, and resumed forward movement in a new direction. This behavior pattern in different organisms has been called a *withdrawal response* (Croll and Sukhdeo, 1981), *reversal bout* (Croll, 1976b; Dusenbery, 1980b; Goode and Dusenbery, 1985), *pirouettes* (Pierce-Shimomura, et al., 1999), *avoidance response*, *shock response*, *phobic response*, *trial and error response*, *tumbling*, and *twiddles* (Burr,

1984a). Burr considered most of these terms to have misleading implications and so coined a new term, *ecclitic response* to refer to this response sequence. The Greek meaning of ecclisis (εκκλισιζ), to turn out of one's course, emphasizes the significant result of the sequence: a change in direction of movement.

A kink, considered a refractory state by Croll and Sukhdeo (1981), can be formed when backward and forward waves meet. A distinct wave form, called an *omega wave*, was observed periodically in *C. elegans* (Croll, 1975b) and *Meloidogyne javanica* (Wallace, 1969). Similar waveforms have been observed by the author thousands of times while watching the infective juveniles of the foliar nematode *Ditylenchus phyllobius* in water. While in a more or less linear posture, the anterior end and tail arch sharply, pointing ventrally, then sweep together, nearly touching as the middle bends concave ventrally. The tail and anterior end then arch and point dorsally (in an "Ω" shape) as the middle abruptly straightens.

Croll (1975b) recorded 6674 backward waves, forward waves, head lifting bouts, omega waves, and pauses in *C. elegans* under various conditions. He found that when *C. elegans* nematodes were transferred from agar to water, the frequency of spontaneous, anteriorly initiated backward waves increased six-fold whereas the frequency of spontaneous, posteriorly initiated coils and kinks remained constant, as if forward and backward waves were controlled by separate clocks. These and additional observations *in vitro* and from the literature were presented to support Croll's *two-generator model* of nematode behavior, which hypothesized that separate ganglia, or collections of ganglia, in the anterior and posterior of the body initiated backward and forward waves, with the anterior end exerting an interruptible inhibitory influence over the posterior, so that the nematode was constantly primed for rapid withdrawal. Video analysis of 100 glass bead encounters by the anterior end of *C. elegans* showed no habituation in the withdrawal response, which was interpreted to support Croll's hypothesis of inhibitory control of reversals (Croll and Sukhdeo, 1981).

The tap withdrawal response of *C. elegans*, that results upon tapping the side of the culture plate, unlike bead encounters, does show habituation, but this affects neither the magnitude nor the frequency of spontaneous reversal activity, nor the reversals evoked by thermal stimuli (Wicks and Rankin, 1997). Thus Croll's hypothesis seems to explain *C. elegans'* movements partly but not completely. Control of backward and forward locomotion in *C. elegans*, which now is understood at the level of neural circuitry, is discussed further by Driscoll and Kaplan (1997). Jorgensen and Rankin (1997) review current knowledge regarding neural plasticity and habituation in *C. elegans* within a molecular and anatomical context.

7.2.1.3 Other Movements Associated with Propulsion

A complex mode of undulatory propulsion in which different portions along the body respond independently to pressures from the substrate, reminiscent of the locomotion of arboreal snakes, is exhibited by the 10 cm – long adult females of *M. nigrescens* on foliar surfaces (Gans and Burr, 1994). When *D. phyllobius* swims or burrows through semisolid dilute agar preparations (0.25% –0.75% water agar), an element of three-dimensional spiraling or corkscrew motion is sometimes superimposed onto the dorso-ventral waves exhibited on impenetrable surfaces (Robinson, pers. obs.). Three-dimensional undulatory propulsion (Fig. 7.1(A)) also has been observed for *Hemicycliophora* sp. and *M. javanica* during root penetration, and for larvae of *Nippostrongylus* sp. climbing hairs of the animal host (Wallace, 1968a, 1968b).

Other types of locomotion include jumping by dauer larvae of *Steinernema* spp. (Campbell and Kaya, 1999), forward propulsion by forwardly propagated peristaltic contractions in criconematids (Croll, 1970; Streu, 1961), and undulatory swimming by the diverse taxa including vinegar eelworm *Turbatrix aceti* (Peters, 1928), aquatic mermithids (Petersen, 1985), and marine nematodes (Chitwood and Murphy, 1964). A most unusual form of locomotion is exhibited by *Sphaerolaimus gracilis*, which utilizes adhesive secretions from a caudal gland system and an anteriorly positioned renette cell to release and refasten posterior and anterior ends alternately to the substrate at sequential points in a controlled fashion within water currents (Turpeenniemi and Hyvärinen, 1996).

7.2.2 Morphological and Physiological Basis of Behavior in Nematodes

Baldwin and Perry (Chapter 4, this volume) review the comparative structure and function of nematode sensilla, central and peripheral nervous systems, and major muscle groups. Additional reviews are given by Aumann (1993), Bird and Bird (1991), Driscoll and Kaplan (1997), Perry (1994, 1996), and Ruvkun (1997).

There is much indirect evidence for specific sensory roles in the numerous sensilla found in nematodes. Based on data from laser ablation and behavioral mutant experiments with *C. elegans*, various touch sensors have been identified conclusively and specific nerve endings in the amphids can be assigned roles in perception of volatiles, aqueous solutes, and temperature (Bargmann and Mori, 1997). It can be summarized that sensory structures consistent with tropotaxis, klinotaxis, orthokinesis, and klinokinesis (Table 7.2) in response to chemical, touch, and thermal stimuli are known in *C. elegans* and highly probable in other nematode groups. Ocelli, needed for phototaxis, also have been described in marine species (see 7.4.1.1), and behavioral evidence for an ocellus that provides directional information in mermithids is irrefutable (Burr and Babinszki, 1990; Robinson, et al., 1990). Negative phototaxis occurs in marine nematodes (Burr, 1979), positive phototaxis in *Mermis nigrescens* (Burr, et al., 1990), and transverse phototaxis in *Agamermis* and *Hexamermis* spp. (Robinson, et al., 1990). *C. elegans* is light sensitive though lacking an ocellus (Burr, 1985). Statocysts are absent.

Table 7.2 Behavior terminology given by Burr (1984a) and adopted by Dusenbery (1992).

Term	Definition	
Kinesis	Behavior comprised of undirected responses that are dependent on the intensity or temporal change in intensity of a stimulus. Undirected responses have no bias with respect to the orientation of the stimulus or stimulus field.	
	Orthokinesis	A kinesis in which translational motion is affected.
	Klinokinesis	A kinesis in which rotational motion is affected. Rotation refers to a change of the direction of locomotion.
Taxis	Migration oriented with respect to the stimulus direction or gradient which is established and maintained by directed turns. Directed turns are ones that are biased in some way with respect to the orientation of the stimulus field.	
	Klinotaxis	A taxis which results from directed responses to sequential samples of stimulus intensity or direction.
	Tropotaxis	A taxis which results from directed responses to two (or in three dimensions, three) simultaneous samples of stimulus spatial distribution or directional distribution.
	Teleotaxis	A taxis due to directed responses to information gathered by a raster of many receptors. Directed turning occurs until the fixation area of the raster is exposed to the directional stimulus. An extreme form of tropotaxis. (Added by Dusenbery, 1992.)

Very little is known about how quiescent nematodes are stimulated into activity or how active nematodes utilize mechanosensory inputs to coordinate movements into complex activity patterns, such as undulatory propulsion, hatching, copulation, and host infection. Evidence for learning in *H. schachtii* is presented by Clemens et al. (1994). Neural plasticity in *C. elegans* in relation to development, modulatory peptides, and experience (sensory adaptation, habituation, sensitization, and associative learning) is discussed by Jorgensen and Rankin (1997).

7.2.3 Theory and Terminology of Directed Locomotion

The world often turns out different from what we expect. One result is controversy in behavior terminology (Burr, 1984a; Croll, 1970; Diehn et al., 1977; Dusenbery, 1992; Fraenkel and Gunn, 1961; Gunn, 1975; Gunn, et al., 1937; Perry and Aumann, 1998). The definitions adopted here (Table 7.2) are those proposed by Burr (1984a) and adopted by Dusenbery (1992), and differ only in minor details from those of Perry and Aumann (1998). The most important concepts are *migration*, *taxis*, and *kinesis*. All three are exhibited by nematodes. Migration is the trend of movement of an individual or population in a direction oriented with respect to a stimulus gradient or direction. This can be accomplished by a taxis or kinesis. Taxis is migration that is established and maintained by directed turns, whereas kinesis is behavior comprised of undirected responses (Table 7.2). Taxis and kinesis are root words to which prefixes are often added to denote specific attributes of the response, such as path curvature, method of information acquisition, and operative stimulus. A taxis may be up gradient (positive), down gradient (negative), or across gradient (transverse). Prefixes that have been used for operative stimuli affecting nematodes include chemo- (chemical), galvano- (electric field), geo- (gravity), magneto- (magnetic field), photo- (light), rheo- (fluid flow), thermo- (temperature), and thigmo- (touch). Other environmental variables shown or proposed to influence directional movement by nematodes include electric current, osmotic pressure, pH, redox potential, soil moisture, surface tension, and soil texture.

Some concepts worth emphasizing are that orientation (Table 7.2) may result from comparing stimulus intensity simultaneously at two points in space (tropotaxis) or at two points in time (klinotaxis). In either case the perceived intensity may differ from the absolute intensity (Dusenbery, 1980a). Moreover, signals among a group of animals (pheromones, calls, visual cues) can influence net movement on a stimulus gradient so that the path of the group differs entirely from the typical path of an individual.

Computer models of nematode migration and host infection that consider some of the above factors are explored by Anderson et al. (1997a, 1997b), Bohan and Hominick (1995), Dusenbery (1988a, 1989a, 1989b), Green (1977), Hay and Fenlon (1995), Pierce-Shimomura, et al. (1999), and

Westerman and Van der Werf (1998).

7.3 Environmental Limits

7.3.1 Factors Affecting Activity

Many factors affect the rate at which nematodes move. Infective stages of plant-parasitic nematodes typically exhibit uninterrupted spontaneous movements, which are several times more rapid in foliar than in root parasites, whereas in other groups, notably bacteriophagous nematodes and free-living stages of animal parasites, activities are largely controlled by exogenous factors (Croll and Sukhdeo, 1981). Often, certain developmental stages in a species are behaviorally quiescent until stimulated to move by light (Robinson, et al., 1990), heat (Croll, 1971), vibration (Wicks and Rankin, 1997), or carbon dioxide (Gaugler, et al., 1980). Many small nematodes require thin moisture films for movement, lack the strength to dislodge soil particles, and easily become trapped in excessively thick films of water (Wallace, 1959c). If all other factors are equal, large nematodes migrate faster than small nematodes (Fig. 7.4). Effects of soil moisture and water films on foliage are discussed in 7.3.3.1.

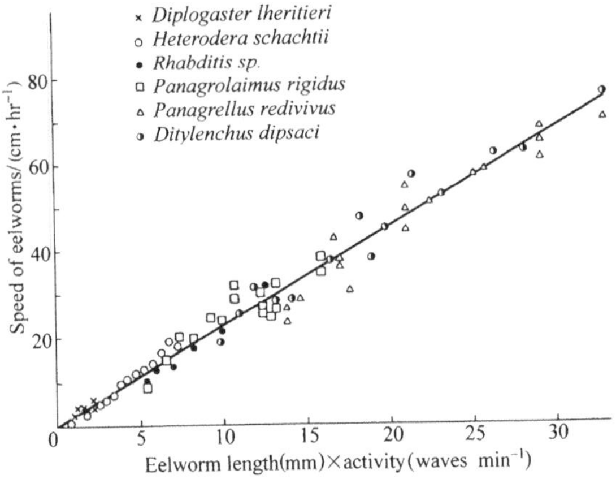

Figure 7.4 Speed of locomotion by morphologically and ecologically diverse nematodes on a moist glass surface (from Wallace, 1958c).

7.3.1.1 Temperature

Where data are available, the thermal optimum for motility of a nematode appears generally similar to the temperature of its natural habitat when it is active. Croll (1975a) compared data for nematodes from diverse trophic groups and showed that thermal optima for the juvenile stages of a stem parasite, a root parasite, and two animal parasites that ascend foliage were all around 20℃ - 25℃ whereas optima for active penetrators of warm-blooded hosts were near 40℃ (Fig. 7. 5A). Robinson (1989) found that when nematodes from different niches in the same climatic region (southern Texas) were compared, infective juveniles of two foliar parasites, *Ditylenchus dipsaci* and *D. phyllobius*, exhibited optima for motility 10℃ cooler than the root parasites, *Rotylenchulus reniformis* and *Tylenchulus semipenetrans* (Fig. 7.5 (B)). This observation is consistent with the ecological need for the foliar parasites to be maximally active on foliar surfaces during the cool rainy periods in the summer when the foliar moisture films required for infection of foliar buds are present. Optimal temperature ranges for host infection, establishment within the host, and reproduction were separately measured for 12 species and strains of entomopathogenic nematodes from different latitudes around the world (Grewal, et al., 1994b) and found to be broadly similar to the climatic conditions of their original locality. Trudgill (1995a, 1995b) discusses base

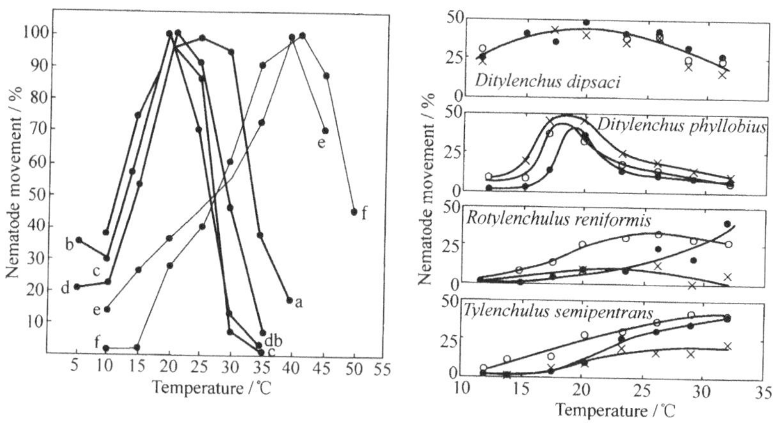

Figure 7.5 Thermal optima for activity. (A) Optima for skin-penetrating stages of warm-blooded host parasites (thin lines) near 40℃ contrasted with optima for plant parasites (thick lines) at 20℃ - 25℃ (from Croll, 1975a). (B) Optima for dispersal of the foliar parasites *Ditylenchus dipsaci* and *D. phyllobius* below 20℃, compared with higher optima (22℃ - 32℃) observed for the root parasites *Rotylenchulus reniformis* and *Tylenchulus semipenetrans* from the same latitude (from Robinson, 1989).

temperatures and thermal constants for embryogenesis and development in relation to niche temperature and reproductive strategy for more than 20 plant-parasitic, animal-parasitic, and free-living nematodes.

Temperature, of course, also affects the rate of release and diffusion of behaviorally active chemicals. For example, temperature markedly altered the release of attractants guiding predatory nematodes to plant-parasitic nematode prey (Pervez and Bilgrami, 2000).

7.3.1.2 Osmotic Pressure

Being small, nematodes have a large surface to volume ratio as well as a body covering permeable to water. They occupy a wide range of fundamentally aquatic habitats. Nematodes in soil and on foliar surfaces must be enclosed within a film of water to move effectively, and parasitic nematodes are alternately exposed to hyper and hyposmotic environments during development. Since nematodes require body turgor for normal movement, osmoregulation is essential to behavior in all species. Nematodes within soil are, in addition, subjected to the powerful attraction of soil water for charged clay micelle surfaces.

Some definitions are given here to facilitate comparisons in the literature. Osmotic pressure can be expressed in units of *osmolarity* (osmoles per liter), *osmolality* (osmoles per kilogram), or *osmosity* (the molar concentration of a NaCl solution with the same freezing point) (Schmidt-Nielsen, 1990; Weast, et al., 1984). An *osmole* (abbreviated Osm) is the osmotic pressure resulting from 1 mole of an ideal solute.

The direction and rate of water movement through any matrix is the cumulative result of differences in osmotic pressure, hydrostatic pressure, gravity, vapor pressure, and the affinity of water for colloidal surfaces. The rate at which water moves is determined by these differences and the diffusive or hydraulic resistance of the regions through which it moves. To understand how such factors influence the water status of plants, plant physiologists (Kramer, 1983; Milburn, 1979) have simplified water movement through the soil-plant-air continuum (Campbell, 1985) by separating the total Gibbs, free energy of water (its chemical potential, usually abbreviated Ψ) into components of *osmotic potential* (Ψ_o, equivalent to osmotic pressure), *matric potential* (Ψ_m, resulting from affinity of water for colloidal soil micelle surfaces), *gravitational potential* (Ψ_g), and *turgor potential* (Ψ_p, hydrostatic pressure inside plant cells). These various components can be expressed in the same units to predict the direction in which water will move. In soil, air, and biological tissues, the *water potential* is usually a negative value given in bars or Pascals (Pa); 1 bar = 0.1 MPa = 10^6 dynes cm^{-2} and approximates the pressure exerted by Earth's atmosphere at sea level, equivalent to a 10 m water column (Weast, et al., 1984). An osmotic potential of -1 bar equals an osmotic pressure of 44 mOsm kg^{-1}. The osmotic potential of sea water is approximately -22.7 bar, which is equivalent to an osmotic pressure of 1000 mOsm kg^{-1} and about three times the osmotic potential of mammalian blood. The vapor pressure of water with this osmotic potential is equal to that of air at 98.3% relative humidity and 20℃ (Papendick and Campbell, 1981). Relative humidity and vapor pressure also can be expressed in bars or Pascals.

The relative abilities of 20 nematode species to regulate their body water contents in solutions with different osmotic pressures were compared by Wright (1998) and Wright and Newall (1976). Four additional species were examined by Forster (1998). Upper intertidal species regulate water contents in both hyperosmotic and hyposmotic conditions, whereas sublittoral marine species generally lack the ability to osmoregulate in hyposmotic conditions. Among freshwater and soil nematodes, the internal osmotic pressures that have been measured for some species (Wright, 1998; Wright and Newall, 1976) are 50 – 100 mM NaCl (equivalent to -2.2 bar to -4.2 bar), and thus are several times greater than typical ionic concentrations in soil water (0 bar to -0.5 bar). These species are generally able to regulate their body water contents well in hyposmotic environments. This is consistent with the low osmotic potential of soil water, which except in salinized agricultural soils usually varies over a narrow range between 0 bar and -0.5 bar (22 mOsm kg^{-1}). Osmotic potential has little effect on the movement of water through soil in comparison with gravity and matric potential (Kramer, 1983).

In glucose, mannitol, and polyethylene glycol solutions, vermiform stages of various plant-parasitic nematodes remain motile and suffer no obvious volume loss at 110 mOsm kg^{-1} (-2.5 bar) (Castro and Thomason, 1973; Robinson, et al., 1984a, 1984b; Robinson and Carter, 1986; Viglierchio, et al., 1969; Wyss, 1970). Therefore, if soil matric potential and osmotic potential are similar in their effects on nematode water balance, as has been shown in plants (Kramer, 1983) and microflora (Papendick and Campbell, 1981), migration by nematodes in drying soil probably is constrained by the physical forces of water films before physiological water loss occurs, as proposed by Blake (1961). Since external osmotic potentials exceeding the classical permanent wilting point for plants (660 mOsm kg^{-1} = -15 bar) are required to extract most water from plant-parasitic nematodes, plant death and the loss by plant-parasitic nematodes of body water regulation appear to occur under approximately the same conditions. On the other hand, nematodes inhabiting gastrointestinal tracts, feces, plant residue, decaying meat, living animal tissues, and marine environments are often exposed to osmotic potentials exceeding those normally encountered by plant-parasitic species and they likely have greater tolerance to hyperosmotic conditions.

7.3.1.3 Specific Salts

The specific ion composition of the aqueous solution bathing a nematode can critically alter water balance, motility, and survival independently of osmotic pressure. Although nematodes may seem unaffected over a wide range of external ion concentrations, the body wall is highly permeable to water and can be highly permeable to ions too if ion pumps fail (Castro and Thomason, 1973; Marks, et al., 1968; Robinson and Carter, 1986). Nematodes differ in ability to regulate specific internal ion concentrations within physiologically tolerable ranges (Wright, 1998).

Wright (1998) cautioned against the use of unnatural single salt solutions, especially NaCl solutions lacking K^+ and Ca^{2+}, to measure osmoregulation without recognizing the abnormal side effects that single salt solutions may have on membrane-bound ion pumps, nerves, and muscles (Schmidt-Nielsen, 1990). Robinson et al. (1984b) showed that monocationic solutions of Mg^{2+} or Na^+ with Cl^- or NO^{3-} present as the anion, stopped all movement by the normally highly spontaneously motile J4 of the foliar parasite *D. phyllobius* in water at 100 meq l^{-1}. At equivalent molality, sucrose, mannitol, and a synthetic soil solution had no measurable toxic or osmotic effects. No volume loss was observed at toxic concentrations of Mg^{2+} or Na^+ and nematodes fully recovered from Na^+ but not Mg^{2+} exposure when returned to a synthetic soil solution.

7.3.2 Movement on Agar

7.3.2.1 Constraints of Agar

Most published studies on nematode behavior have examined responses to test stimuli partly or exclusively on agar surfaces (Robinson, 2000). When directly comparing the behaviors of soil-borne stages of various species differing substantially in size on a range of concentrations of agar, I have observed the robust 650 and 1100 μm long infective juveniles of *D. dipsaci* and *Steinernema glaseri* to burrow more easily through 0.75% agar and become trapped in water films less frequently than the smaller (350 – 400 μm long) infective juveniles of *Meloidogyne incognita*, *R. reniformis*, and *T. semipenetrans*. In the same agar preparation some nematode species may be constrained to the surface, lying flat on their sides, while others occasionally or always burrow through the agar in random angles of rotation about the body axis. Some species (*S. glaseri* and *D. phyllobius*, for example) readily and others never ascend the walls of plastic Petri dishes. Much seems to depend on nematode strength, agar rigidity, and water film thickness on the agar surface, but other, unknown factors are also involved. Most studies of nematode behavior on agar have been done with agar preparations too tough for nematodes to penetrate.

7.3.2.2 Slew and Klinokinesis on Agar

Analyses of nematode tracks on the surface of agar or agarose show many species to produce curved tracks, exhibiting slew caused by dorso-ventral asymmetry of the laterally (and unnaturally) positioned body (Figs7.2(B) – (F), 7.3(C)); slew is particularly marked for males with ventrally curved tails (Croll and Sukhdeo, 1981; Green, 1977). Simulation models showed that changes in slew curvature in response to changes in stimulus intensity within a stimulus field could allow males to find females (Fig. 7.2(D)) (Green, 1977). This mechanism can be considered a type of klinokinesis (Table 7.2). Slew does not seem to account for the circular tracks made by thermotaxing wild-type *C. elegans* on radial temperature gradients on agar (Fig. 7.3(B)) (Bargmann and Mori, 1997; Hedgecock and Russell, 1975; Mori and Ohshima, 1995). Slew is difficult to discern or absent in three-dimensional tracks left by root-knot nematode juveniles burrowing through agar (Figs7.3(D) –(E)) (Prot, 1978c). The significance of klinokinetic slew in soil is unknown.

I wish to make a difficult point here about morphological and environmental requirements for tropotaxis (Table 7.2). Bilaterally symmetrical organisms typically exhibit tropotaxis and nematodes are fundamentally bilaterally symmetrical, with the primary chemo- and thermosensors, the amphids, situated on either side near the anterior end. Their precise position varies among species. The distance between the two amphids is always adequate, though, to detect edaphically realistic gradients (Campbell, 1985; Nye, 1981) by comparing input from left and right amphids, given the theoretical limits of thermo- and chemosensory function (Dusenbery, 1988c), and the measured sensitivity of nematodes to abrupt changes in temperature (10^{-4}℃) (Dusenbery, 1988b) and solutes (10^{-10} M) (Balan, 1985; Terrill and Dusenbery, 1996). Investigators of nematode behavior have long been confronted, however, with the dilemma that the only easy way to see most nematodes move is on the surface of a transparent gel where surface tension and dorso-ventral body waves compel nematodes to remain on their sides within the plane of a two-dimensional gradient field, i.e., in an orientation where it is simultaneously impossible to fulfill either of two absolute requirements for using the morphological faculties of nematodes to achieve tropotaxis (Fig. 7.2 (A)). First, they cannot achieve lateral steering without leaving the agar surface, and second, they cannot simultaneously compare stimulus intensities at two points along the gradient because their primary sensors are aligned perpendicular to and consequently positioned isocratically (or isothermally) within the stimulus field. Thus nematodes on agar are handicapped (they are both blind and hobbled) such that the only behaviors we are likely to see are kineses and klinotaxes, unless the nematodes under study are strong enough to rotate on the agar surface, or burrow through it.

Klinotaxis by nematodes on agar is interesting because it may also occur in natural habitats. Klinotaxis requires temporally sequential comparisons whereas tropotaxis requires simultaneous spatial comparisons. The ability to make the sequential comparisons required for klinotaxis in response to chemical stimuli was demonstrated with tethered *C. elegans* (Dusenbery, 1980b). Klinotaxis in response to temperature was confirmed in *C. elegans* indirectly by observing thermotaxis after killing the neurons in one amphid (Bargmann and Mori, 1997), eliminating tropotaxis as a mechanism. These experiments, however, do not test for the occurrence of tropotaxis, which is a more direct way than klinotaxis is to orient and one for which nematodes appear well suited. We must not fall into the trap of thinking we understand how nematodes move when we have not observed them under conditions that allow a natural range of movement.

7.3.2.3 Comparisons between Agar and Other Substrates

Some studies have examined horizontal movement in shallow containers of sephadex beads, sand, or agar coated with a thin or partial layer of sand to provide heterogeneity if not a third dimension (Robinson, 2000). A sand layer on agar stimulates nictation by *Steinernema* spp. (Campbell and Kaya, 1999) and alters the pattern of movement by *C. elegans* (Anderson, et al., 1997a, 1997b). In other cases, agar just dilute enough for the nematodes under study to burrow through, has been poured into molds (Figs 7.3(D) – (E)) (Prot, 1978a, 1978c, 1979a), sometimes with the same dimensions as containers used in other experiments to examine movement through columns of sand or soil (Robinson, 2000; Robinson and Heald, 1993). As will be discussed, substrate geometry can have a negligible or profound influence on the behavior observed.

7.3.3 Constraints of Natural Substrates

7.3.3.1 Factors Limiting Movement on Foliage

For numerous plant, insect, and vertebrate parasites, the ascension of foliar surfaces is a normal link in the life cycle. Investigations of factors influencing foliar migrations typically have been limited to macroscopic variables such as ambient temperature, relative humidity, illumination, and rainfall (Buckley, 1940; Castellanos-Coutino and Alfredo, 1980; Choo, et al., 1995; Christie, 1936, 1937; Rees, 1950; Robinson, et al., 1978, 1979, 1990; Rogers, 1940; Solomon, et al., 1997; Spindler, 1934; Stromberg, 1997). Such studies have shown clearly that moisture films from dew or rain are usually essential for foliar migration, and the linear geometry of grass blades appears to increase the efficiency of vertical movement. Foliar surfaces vary greatly in pubescence, waxiness, and exposure to sun and wind, all of which influence the development, retention, and collapse of moisture films.

The contemporary consensus is that most nematodes ascending foliage do not sense gravity. However, the movements of the 10-cm-long gravid females of *M. nigrescens* (Fig. 7.1(D)), which nictate and cantelever up to one-third of the anterior body while ascending grass stems (Gans and Burr, 1994), suggest that proprioceptors could allow exceptionally large nematodes to do so. Light appears to be the key stimulus guiding *M. nigrescens* and two other mermithids on foliage (Robinson, et al., 1990). Passive vertical orientation in water as a result of differences between specific gravities of anterior and posterior halves of the nematode body have been observed for three species that do not crawl on foliage, viz. *T. aceti* (Peters, 1952), *Globodera rostochiensis* (Wallace, 1960), and *R. reniformis* (Fig. 7.1(C)) (Robinson and Heald, 1993).

Voss (1930) suggested that *Aphelenchoides ritzemabosi* (the chrysanthemum foliar nematode) moves up stems by orienting rheotactically to a downward current of moisture on the stem surface. Wallace (1959b) could not confirm this effect with the same nematode or with *D. dipsaci*, nor could Barraclough and French (1965). Wetted wick crawling experiments by Adamo et al. (1976) showed *Aphelenchoides besseyi* (rice white tip nematode) to move consistently to the wetter end of wicks independently of gravity, which generally but not always was against a water current.

7.3.3.2 Factors Limiting Movement Through Soil

Most of what we know regarding fundamental principles governing undirected nematode dispersal in soil came from the extensive and elegant experiments of Wallace (1958a, 1958b, 1958c, 1959a, 1959b, 1959c, 1960, 1968a). As introduced in Section 7.3.1.2, major advances during the 1950s and 1960s in the understanding of plant and soil water relationships emerged from the realization that water movement through plants could be explained in terms of the Gibbs' free energy of water at the leaf-air interface, within leaf cells, in roots, and in the soil (Kramer, 1983; Milburn, 1979). The water status of plants is directly affected not by the quantity of water in soil, but rather by the

Gibbs' free energy of water (the water potential), which in soil is determined almost entirely by the attraction of soil particles, primarily colloidal clay micelles, for water molecules. The osmotic component of soil water potential resulting from dissolved salts was found to be too small to be of physiological significance to plants in most cases. Concurrent with these developments was the incorporation into plant physiology of concepts and notation from electrical engineering to explain and predict the direction and rate of movement of water in soil and plants. As discussed in 7.3.1.2, this notation partitioned the total Gibbs'free energy or water potential, into matric, osmotic, gravitational, and turgor potential. Wallace's visionary studies considered nematode movement in the context of these then novel though now well-established concepts.

The results of Wallace's nematode movement experiments relate primarily to the complex geometry and Gibbs'free energy of water within soil interstices. When he compared nematode species and stages of 15 different mean body lengths ranging from 186 to 2000μm in four soil fractions ranging from 75 to 1000μm, he found not surprisingly that large nematodes require soil with larger particles than small nematodes do, and the optimum particle size is linearly related to nematode length (Wallace, 1958c). When moving on a moistened glass surface, the speed along the surface, also not surprising, was found to bear a perfectly linear relation to the product of body length and wave frequency (Fig. 7.4). Not obvious, however, was Wallace's important finding that although particle size limits movement in sand, when complex soil containing significant amounts of colloidal clay and silt is dried and sieved into crumb size classes and then rewetted, it is crumb size not soil particle size that is critical, i.e., in most natural soils where fine particles can stick together to form much larger crumbs, soil structure is the primary determinant (Figs 7.6 (A) –(B)).

Wallace's second fundamental finding was that nematode movement was fastest not in water-saturated soil, but rather when soil water content was near field capacity (approximately −0.05 bars = 50 cm water), i.e., when soil pores had drained just to the extent achievable by gravity (Figs 7.6(A) – (B)). More precisely, movement was fastest in soil with a moisture content corresponding to the inflection point of the sigmoid curve obtained when soil is equilibrated at various pressures within a pressure plate extractor and moisture content is plotted against the log of extraction pressure (referred to by Wallace as "inflection point of the moisture characteristic"). This effect was shown with infective juveniles of both *H. schachtii* and the much larger *D. dipsaci* (Wallace, 1956, 1958a, 1958b). The important lesson is that the optimum moisture for nematode movement is directly related to the matric component of soil water potential and corresponds to different water contents in different

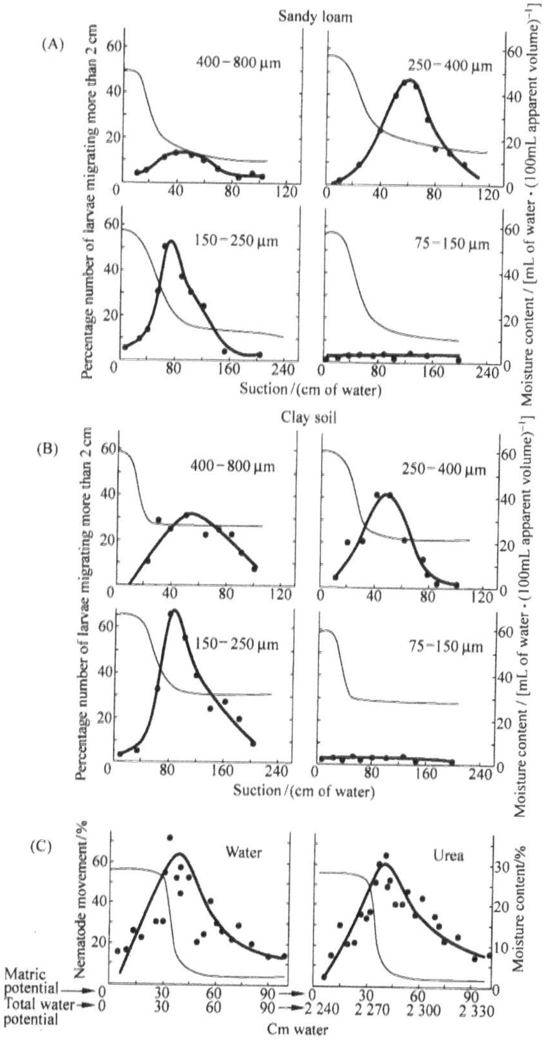

Figure 7.6 Key factors influencing the rate of movement of nematodes through soil. (A) – (B) Similarity of effects on *Globodera rostochiensis* movement exerted by soil structure (crumb diameter in μm) and water potential (cm of water) for two soils of markedly different textural classes (from Wallace, 1966). (C) The marked influence of soil matric potential on movement by *Ditylenchus dipsaci* (left) is unaltered when 0.1 M urea is added to elevate greatly the osmotic and thereby the total water potential of the soil (from Blake, 1961).

soils. It also was observed that movement usually occurred over a wider range of water potential in finely textured soils than in sand. Finally, in any soil at soil matric potentials dryer than *ca.* −0.5 bar, nematode movement virtually stopped even though most nematodes remain turgid and fully motile in water at equivalent osmotic pressures (Wright, 1998; see 7.3.1.2 for additional re-ferences). This last important interpretation was proven in *D. dipsaci* by Blake (1961) (Fig. 7.6(C)).

Wallace's work on gradients of soil moisture is reviewed in 7.4.2.1.

7.4 Research Topics on Nematode Behavior

7.4.1 Isolated Stimuli Affecting Migrations of Diverse Taxa

7.4.1.1 Light

There is evidence for a generally stimulatory effect of light or dark on various nematode species (Bloom, 1964; Burr, 1985; Croll, 1966, 1970; Rees, 1950; Rogers, 1940). Such responses likely are triggered by specific peripheral receptors rather than by a diffuse dermal sensitivity (Steven, 1963). Ocelli (reviewed by Baldwin and Perry, Chapter 4, this volume) that can provide nematodes with directional information about light appear to be limited to marine nematodes and mermithid parasites of insects.

Information about the direction of a light source can be enhanced in ocelli by a lens or a shading structure that restricts receptor illumination to a specific angle (Burr, 1984a). Proven or putative photoreceptors associated with shading pigment and without a lens, and sometimes called *eyespots* or *chromatropes*, have been described in *Araeolaimus elegans* and *Chromadorina* sp. (Croll, 1967b; Croll, et al., 1972, 1975), *Diplolaimella* sp. (Van de Velde and Coomans, 1988), *Deontostoma californicum* (Siddiqui and Viglierchio, 1970a, 1970b), *Enoplus anisopiculus* (Burr, 1984b), *Leptosomatides* spp. (Kito and Hope, 1999), and *Oncholaimus vesicarus* (Burr and Burr, 1975). Compelling behavioral evidence without ultrastructural confirmation exists for photoreceptors in the positively phototaxic gravid females of *M. nigrescens* (Burr and Babinszki, 1990; Burr, et al., 1990) and the transversely phototaxic infective juveniles of *Agamermis catedecaudata* and *Agamermis* spp. *aff. sinuosa* (Baker and Poinar, 1995; Robinson, et al., 1990). Directional resolution for positive phototaxis in *M. nigrescens* is provided by an open-ended, hollow reddish cylinder (Figs 7.7(A) − (B)), which surrounds an undescribed photoreceptor near the anterior end of the body (Burr and Harosi, 1985).

No photoreceptor pigment has been identified yet in nematodes. Cobb

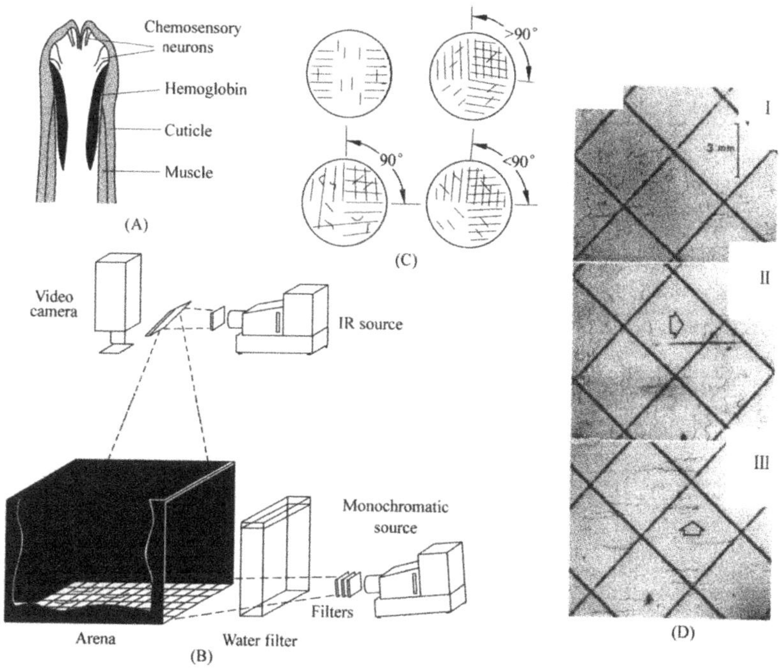

Figure 7.7 Research on nematode phototaxis. (A) Hollow hemoglobin cylinder (so-called chromatrope) in the anterior region of *Mermis nigrescens* (from Burr, et al., 1989). (B) Apparatus used to study phototaxis in *M. nigrescens* (from Burr, et al., 1989). (C) Orientation of *Agamermis catedecaudata* to dual light sources aligned as indicated by straight lines outside circles (from Robinson, et al., 1990). (D) Orientation of *A. catedecaudata* in the dark (I), to light directed as indicated by the fat arrow (II), and several seconds after rotating light source 90° (III) (from Robinson, et al., 1990).

(1929) hypothesized that the outer tissues of the neck "of *M. nigrescens* and other free-living" nematodes served as a lens to focus light on "the pigmented tissues" (Cobb's words in quotes). Croll et al. (1972) also proposed this mechanism in *Chromadorina bioculata*. Histochemical investigations and *in situ* microspectrophotometry of the pigments of various species, however, showed eyespots to contain hemoglobin or melanin, not likely photosensitive. Comparisons between the absorbance spectrum obtained by *in situ* microspectrophotometry and the action spectrum of the response for *M. nigrescens* further indicated that the hemoglobin of the hollow pigmented cylinder was not photosensitive (Burr, et al., 1975, 1989). The hemoglobin has an amino acid sequence typical of nematode hemoglobins (Burr, et al., 2000a) and is over-expressed, accumulating in the hypodermal cells of the eye region during the 1 – 2 year maturation of the female in the soil (Burr, et al., 2000b). Experiments with strobed unidirectional light synchronized to nematode undulations provided evidence that a sensor with photosensitive pigment lies entirely within and shaded by the hemoglobin cylinder (Burr and Babinszki, 1990). A rhodopsin photoreceptor pigment would be consistent with available data, and need not be present in visually observable quantities (Burr, et al., 1990).

Early in the morning, the positively phototaxic females of *M. nigrescens* crawl and lay eggs on moist foliage. Eggs are eaten by prospective insect hosts. Phototaxis may mediate these egg-laying migrations because behaviorally effective light intensities are achieved in the sky 30 minutes before sunrise, when nematodes are active (Burr, et al., 1989). On moistened black felt in the laboratory, females raise the anterior end above the felt, pointing into the light (like a cobra) and move forward, by dorsoventral undulations but with the ventral surface oriented downward, so that the dorsal surface is elevated as the wave progresses along the body (Figs 7.1(D), 7.7(B)). This is quite unlike the movements typically observed for other nematodes lying on their sides on agar.

Head waving in both horizontal and vertical dimensions causes the forward directed ocellus to scan the directional distribution of light in the environment (Burr, et al., 1990). This, together with ocellar morphology, with the yet to be identified photoreceptor inside the pigment cylinder, provides a unique mechanism for positive phototaxis (Burr, et al., 1990). The cyclic illumination and shadowing of the sensor and the cyclic bending of the anterior 2 mm, the "head", go out of phase when the "neck" deviates from the direction towards the light. This is used as a signal during phototaxis to maintain orientation towards the light. This scanning motion was confirmed by shuttering the light stimulus so as to artificially shift the phase and thus provide a nematode optical illusion. This biased phototaxis in the predicted way (Burr and Babinszki, 1990). The experiment also confirms that a photoreceptor must reside inside the cylinder.

New hatched infective juveniles of *Agamermis* spp. *aff. sinuosa* and *A. catedecaudata*, which parasitize orthopterans in the tablelands of New South Wales, exhibit a remarkable orientation precisely perpendicular to light (Figs 7.7(C) – (D)), called *transverse phototaxis* (Robinson, et al., 1990). The response was observed on numerous occasions under various circumstances by projecting light from a fiber optic or from the sun *via* mirrors horizontally onto shallow 2 ×2 cm square dishes containing 0.75% water agar and nematodes. The response does not fatigue readily. One day, 30 nematodes moved parallel to each other without interruption for 8 hours back and forth across the arena, perpendicular to the direction of light. A similar but transient transverse phototaxis has also been seen in newly molted, unpigmented virgin females of *M. nigrescens* (Gans and Burr, 1994). However, older unpigmented females are not significantly phototaxic and female larvae within 2 days of emergence from the host are slightly negatively phototaxic, on average (Burr, et al., 2000a).

The precise location of the photoreceptor in *Agamermis* spp. *aff. sinuosa* and *A. catedecaudata* is unknown and there is no observable pigment in the infective juveniles of these nematodes to provide a clue as to its location. However, decaudate juveniles responded to light and shading the anterior 1/3 of the body blocked the response, indicating the receptor is near the anterior end of the body. Dual light sources yielded precise orientation parallel to the bisector of the angle between them if it was greater than 90° but perpendicular to the bisector of the angle if less than 90° (Fig. 7.7(C)), consistent with light being focused on a receptor near the body axis, possibly through a cylindrical lens effect of the body wall, as proposed by Cobb (1929). This interpretation was further supported by the observation that when nematodes were put in a thin layer of dilute agar through which they could burrow freely, orientation of the body axis perpendicular to light within 10° was maintained during forward movement regardless of rotation about the body axis, and regardless of gravity.

The females of *Agamermis* spp. *aff. sinuosa* and *A. catedecaudata*, unlike those of *Mermis nigrescens*, lay eggs in the soil. Upon hatching, it is the infective juveniles rather than the gravid females that must ascend foliage. The threshold for a half-maximal response by *Agamermis* spp. *aff. sinuosa* and *A. catedecaudata* when orienting perpendicular to light at maximally effective wavelengths was 0.03 mW cm^{-2} (equivalent to a fluence of *ca.* 1014 photons sec^{-1} cm^{-2}), and thus a full response required direct sunlight, which was confirmed outdoors by transient loss of the response when clouds or neutral density filters blocked the sun. Action spectra obtained for *A. catedecaudata* with a series of twenty 10 nm bandpass filters showed maximally effective light was blue (440 nm) with almost no response above 525 nm. By tilting agar plates every possible angle indoors and outdoors at sunrise, it was found that horizontal light from fiber optic illuminators and from the sun drove nematodes to the highest point on the surface by the most direct route, regardless of plate orientation. Thus, the response must disperse infective nematodes vertically on moist grass blades, facilitating encounters with the grasshopper nymphs they infect (Baker and Poinar, 1995).

Christie (1936) did not report transverse phototaxis in his careful descriptions of all developmental stages of the North American *A. decaudata*. I found no record of phototaxis in *Agamermis unka*, which was investigated by Choo et al. (1995) as a biological control agent for the brown rice hopper in Korea. Phototaxis also appears absent in aquatic mermithids (Petersen, 1985).

7.4.1.2 Temperature

It is believed that directional behavioral responses of nematodes to temperature are common. Of the 20,000 described species of nematodes, I was able to list 23 that were reported to migrate on a temperature gradient (Table 7.3). Additional species respond when hot objects are placed near them. Although major taxonomic groups are unrepresented, the general diversity of taxa and ecological groups within which oriented responses to temperature are found suggests a widespread occurrence. As indicated in 7.2.3, strong morphological evidence supports this interpretation. Laser ablation studies on *C. elegans* (Mori and Ohshima, 1995) identified two interneurons, denoted AIY and AIZ, respectively, which synapse with the thermoreceptor neuron AFD, and are separately essential for cryophilic and thermophilic responses. Neurological similarities in animal-parasitic nematodes are reviewed by Ashton et al. (1999). Considering the conservation of neuroanatomy among nematodes, homologous morphology can be expected in other nematode groups.

Table 7.3 Nematodes migrating on temperature gradients.

Order	Species	Reference
Rhabditida		
	Caenorhabditis elegans	Hedgecock and Russell, 1975
	Heterorhabditis bacteriophagous	Robinson, 1994
	Steinernema carpocapsae	Burman and Pye, 1980
	Steinernema glaseri	Robinson, 1994
Spirurida		
	Dirofilaria immitis	Mok, et al., 1986
Strongylida		
	Ancylostoma ceylanicum	Khalil, 1922
	Ancylostoma duodenale	Khalil, 1922
	Ancylostoma tubaeformae	Croll and Smith, 1972
	Galoncus perniciosus	Khalil, 1922
	Necator americanus	Khalil, 1922

Continued

Order	Species	Reference
Strongylida		
	Nippostrongylus brasiliensis	Parker and Haley, 1960
	Strongyloides stercoralis	Khalil, 1922
	Trichonema spp.	Croll, 1971
	Trichostrongylus douglassi	Khalil, 1922
Trichocephalida		
	Terranova decipiens	Ronald, 1960
Tylenchida		
	Ditylenchulus dipsaci	Wallace, 1961
	Ditylenchulus phyllobius	Robinson, 1989
	Globodera rostochiensis	Rode, 1969a
	Meloidogyne incognita	Pline, et al., 1988
	Pratylenchus penetrans	El-Sherif and Mai, 1969
	Rotylenchulus reniformis	Robinson, 1989
	Tylenchorhynchus claytoni	El-Sherif and Mai, 1969
	Tylenchulus semipenetrans	Robinson, 1989

Classification proposed by Maggenti (1981).

The first responses to temperature noticed in nematodes were the attractions of certain hookworm and filariform nematodes to heat, which appeared to assist penetration of vertebrate hosts (Al-Hadithi and Habash, 1979; Croll and Smith, 1972; Gupta, 1963; Khalil, 1922; McCue and Thorson, 1964; Mok, et al., 1986; Parker and Haley, 1960; Ronald, 1960). Subsequently, at least seven root parasites, two foliar parasites, two insect parasites, and two bacteriophagous nematodes from soil were found to move to one end or the other of temperature gradients (Table 7.3). Responses were interpreted to indicate a true taxis in larval hookworms (Croll and Smith, 1972) and a klinokinesis (see Section 7.2.2) in infective juveniles of potato cyst (Rode, 1970) and root-knot (Dusenbery, 1988a) nematodes.

Animals commonly exhibit a preferred temperature, or preferendum (Figs 7.8(A) – (C)), and on a gradient move toward cold if placed above the preferendum but toward heat if below it. Although some early experiments showed some animal-parasitic nematodes are suicidally attracted to heat (McCue and Thorson, 1964), the seal and fish parasite *T. decipiens* (Ronald, 1960) exhibits a preferendum. Preferenda also have been reported for the four root parasites *Globodera rostochiensis* (Rode, 1969a, 1969b), *M. incognita* (Diez and Dusenbery, 1989a), *R. reniformis* (Robinson, 1989; Robinson and Heald, 1989, 1993), and *T. semipenetrans* (Robinson, 1989), the two foliar parasites *D. dipsaci* (Croll, 1967a; Robinson, 1989; Wallace, 1961) and *D. phyllobius* (Robinson, 1989), the entomopathogenic *Steinernema carpocapsae* (Burman and Pye, 1980), and the bacteriophagous *C. elegans* (Hedgecock and Russell, 1975).

Some definitions are needed here. Schmidt-Nielsen (1990) recommended the term *acclimation* for physiological adjustments to artificial changes in ambient conditions in the laboratory and the term *acclimatization* for adjustments to changes in climate under natural conditions. The term *thermotactic adaptation*, which also has been used to describe nematode acclimation (Robinson, 1989), may risk confusion with the unrelated phenomenon of evolutionary adaptation.

In most cases tested, the thermal preferendum of nematodes was shifted within 2 hours to 30 days by acclimating to a new ambient temperature (Fig. 7.8(B)). Thermal preferenda and optima may also be altered by mutations (Dusenbery and Barr, 1980). Interpretations of thermotaxic acclimation probably have been oversimplified. Although the preferendum may coincide with the culture or storage temperature (Croll, 1967a; Hedgecock and Russell, 1975), the preferendum can also occur above or below it (Diez and Dusenbery, 1989a; Robinson, 1989) or may only partially shift during acclimation (Robinson, 1989; Rode, 1969b).

Data for the four plant parasites examined by Robinson (1989) suggest, furthermore, that the thermal optimum for motility and the thermotaxic preferendum are not necessarily equivalent; and it seems reasonable that the kinetics of enzymes controlling a complex behavioral response may differ from those involved in spontaneous muscular contractions. The biggest differences in the Robinson (1989) study, however, were associated with niche as discussed in 7.3.1.1. Although all four species were from the same latitude, preferenda and thermal optima for motility for the foliar parasites on average were 10℃ cooler than for the root parasites, consistent with the need for infective juveniles of the foliar parasites to be maximally active and attracted to moisture films chilled by evaporation on foliar surfaces during summer rains (Fig. 7.5(B)).

Soil inhabiting bacteriophagous, mycophagous, and predatory nematodes, and the soil phases of foliar, insect, and vertebrate parasites are constantly subjected to dynamically fluctuating, vertical temperature gradients caused by surface heating and cooling during the day (Fig. 7.9(B)). Most nematodes on agar are unresponsive to the direct effects of gravity (Croll, 1970) and so it is significant that these gradients in soil might be utilized by nematodes to achieve vertical orientation. A less obvious possible utilization by nematodes of fluctuating vertical gradients to detect roots horizontally is discussed in 7.4.2.1.

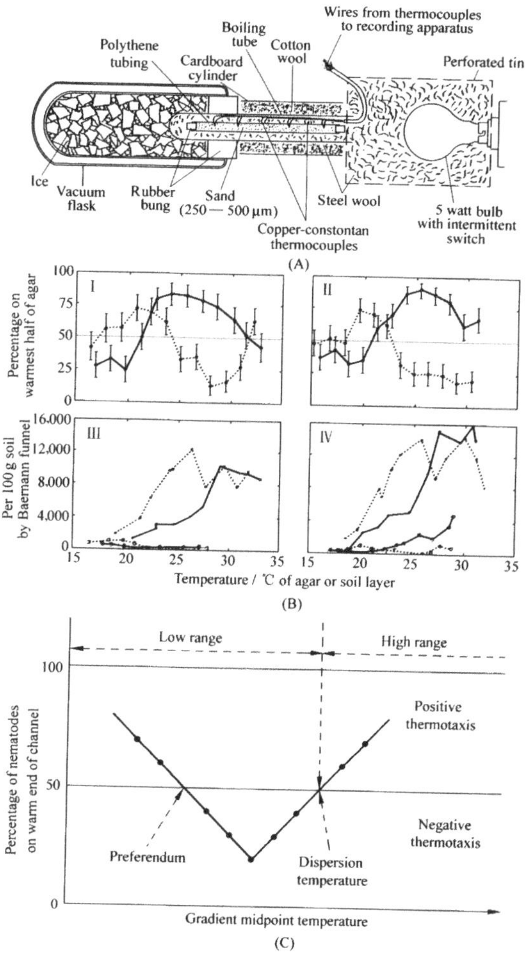

Figure 7.8 Research on nematode responses to temperature. (A) Primitive but effective apparatus used by Wallace to discover the thermal preferendum in *Ditylenchus dipsaci* (from Wallace, 1961). (B) (I) Thermal preference of *Rotylenchulus reniformis* on agar when stored previously at 30℃ (solid line) and 20℃ (dashed line) (from Robinson and Heald, 1989). (II) Optimal ambient temperatures for Baermann funnel extraction of *R. reniformis* from soil stored previously at 30℃ (upper solid line) and 20℃ (upper dashed line) (Robinson and Heald, 1989). (II – IV). Experiments B(I – III) repeated. (C) Diagrammatic representation of a thermal preferendum response when nematodes are subjected to a temperature gradient within a linear channel of agar (from Robinson, 1989).

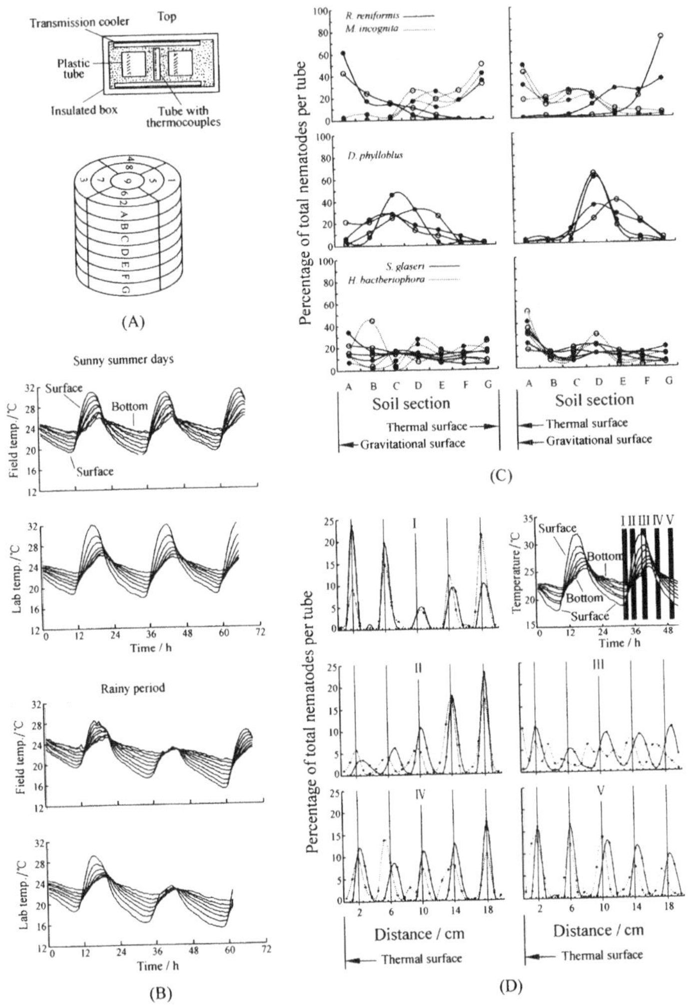

Figure 7.9 Responses to dynamically fluctuating vertical temperature gradients in soil (from Robinson, 1994). (A) Apparatus used to reproduce natural temperature fluctuations. (B) Temperature fluctuations measured in a cotton field and simulated in the laboratory. (C) Responses of five species simultaneously injected into the center of a soil cylinder subjected to the sunny day simulation shown in B after thermally equilibrating nematodes for 24 hours within aerated water buried at the injection point. (D) Differential vertical movement of *Rotylenchulus reniformis* (solid line) and *Meloidogyne incognita* (dashed line) 30 minutes after simultaneous injection at five depths, five times during a 24 hour period.

The measured magnitudes of soil vertical gradients during the afternoon on a typical sunny summer day in southern Texas were 0.3℃ cm^{-1} and 0.15℃ cm^{-1} at depths of 10 cm and 25 cm, respectively (Heald and Robinson, 1987; Robinson, 1994). The progressive pattern of vertical temperatures in soil during the course of the day, most days of the year (Fig. 7.9(B)), shows a heat wave starting with maximum amplitude at the surface during the daytime, dampening as it moves downward 2 – 3 cm/hour (Dusenbery, 1988a, 1989c; Robinson, 1994).

Dusenbery (1988a) developed a computer simulation model to explore the theoretical effects of various rates of nematode movement and adaptation and found that by imposing hypothetical combinations of conditions, nematodes could be induced to move away from the surface, toward the surface, or to a preferred depth. For illustration, it can be asked where an organism would go if it sought the thermal condition to which it was last exposed. When in a warming environment (i. e., in the soil during the daytime), it would go toward cool, which most of the time would be downward. On the other hand, in a cooling environment (at night, after the heat wave had passed) it would go toward heat, again downward. Thus it would always go down. By inverse logic, an organism avoiding the last condition experienced would go up, or to the surface if in a tree trunk (Dusenbery, 1988a).

Dusenbery's (1988a, 1989c) predictions were tested experimentally by Robinson (1994) by subjecting two entomopathogenic nematodes, two root parasites, and a foliar parasite in the laboratory to gradient fluctuations in soil that precisely mimicked those measured at hourly intervals in 2.5 cm increments in a cotton field in Texas during rainy and sunny summer days (Figs 7.9(A) – (D)). Nematodes in the laboratory were thermally equilibrated for 24 hours under fluctuating temperatures identical to those at the depth in soil at which nematodes were released, and at intervals following release nematode distributions in soil were monitored. The most pronounced oriented responses were obtained for the two slowest species, the root parasites, and their responses were opposite one other. *Meloidogyne incognita* moved toward and *R. reniformis* moved away from the thermal surface throughout the 24-hour cycle regardless of the depth where released (down to 20 cm) (Fig. 7.9(D)) and regardless of the orientation of gravity relative to the direction of heat wave propagation. The ecological importance of these response patterns can only be speculated without further studies. However, the direction of movement is consistent with the occurrence of *R. reniformis* but not *M. incognita* at highest population densities 50 – 120 cm deep in Texas cotton fields (Robinson, unpubl. data).

An additional, perhaps unlikely but unexplored possibility that I will add here, is that nematodes could utilize the horizontal temperature gradient components generated when moisture patterns and therefore heat capacity gradients around roots horizontally warp the daily downward heat wave.

7.4.1.3 Carbon Dioxide and Ammonia

Carbon dioxide and ammonia are discussed together because both gases are biogenic, are potentially toxic, and react with water to form highly soluble chemical species. One or both are released abundantly by living or decaying plant and animal tissues, providing a pair of volatile cues that jointly could assist plant-parasitic as well as saprophagous nematodes to find food. One or both are known to attract certain insects. As gases, they can diffuse greater distances through interstices between soil particles than can most substances in soil.

Nematode attraction to known sources of CO_2 *in vitro* was observed (or refuted) in several laboratories about 40 years ago (Bird, 1959, 1960; Johnson and Viglierchio, 1961; Klingler, 1959; Rohde, 1960) and has been examined in several studies subsequently (Figs 7.10(A) – (H)). Today, however, the central ecological question remains unanswered regarding how, and if, the same respiratory gas can effectively guide bacterial feeders (Balan and Gerber, 1972; Dusenbery, 1985; Viglierchio, 1990), insect parasites (Gaugler, et al., 1980), root parasites (Bird, 1960; Johnson and Viglierchio, 1961; Pline and Dusenbery, 1987; Robinson, 1995; Robinson and Heald, 1991), tree trunk parasites (Miyazaki, et al., 1978a, 1978b), foliar parasites (Klingler, 1959, 1961, 1970, 1972), free-living marine nematodes (Riemann and Schragge, 1988), and vertebrate parasites (Granzer and Haas, 1991) to find food (Table 7.4), although theoretical considerations suggest it is possible (Dusenbery, 1987).

Table 7.4 Nematodes attracted to carbon dioxide.

Order	Species	Reference
Aphelenchida		
	Aphelenchoides fragariae	Klingler, 1970
	Bursaphelenchus lignicolus	Miyazaki, et al., 1978a
Enoplida		
	Adoncholaimus thalassophygas	Riemann and Schragge, 1988
Rhabditida		
	Caenorhabditis elegans	Dusenbery, 1985
	Panagrellus redivivus	Balan and Gerber, 1972
	Panagrellus silusiae	Viglierchio, 1990

Continued

Order	Species	Reference
Rhabditida		
	Steinernema carpocapsae	Gaugler, et al., 1980
	Steinernema glaseri	Robinson, 1995
Tylenchida		
	Ditylenchus dipsaci	Klingler, 1961
	Heterodera schachtii	Bird, 1960
	Meloidogyne hapla	Johnson and Viglierchio, 1961
	Meloidogyne incognita	Pline and Dusenbery, 1987
	Meloidogyne javanica	Bird, 1960
	Pratylenchus minyus	Bird, 1960
	Pratylenchus penetrans	Klingler, 1972, Edmunds and Mai, 1967
	Rotylenchulus reniformis	Robinson and Heald, 1991
Strongylida		
	Ancylostoma caninum	Granzer and Haas, 1991

Classification proposed by Maggenti (1981).

As was noted for temperature (Section 7.4.1.2), the ecological diversity of responsive nematodes suggests that CO_2 is a universal nematode attractant. It must be noted, however, that as for temperature the number of species known to be attracted (16) is small and most nematode species have not been tested. With the exception of the marine nematode *Adoncholaimus thalassophygas* and possibly *Ancylostoma caninum* (Granzer and Haas, 1991), all responsive nematodes reported so far belong to the orders Tylenchida and Rhabitida, and so most nematode orders are unrepresented. There are few comparative studies (Klingler, 1972; Robinson, 1995). Comparison of the root parasites *M. incognita* and *R. reniformis* (Figs 7.10(G) –(H)) showed that the kinetics of attraction to CO_2 in soil may differ sufficiently between two ecologically similar species for attraction to be detected in one but not the other even though both are responsive (Robinson, 1995). Most nematodes are partly or completely anaesthetized in water equilibrated against air containing CO_2 concentrations exceeding 5% (v/v) (Fig. 7.10(D)), and in several published studies nematodes were subjected to flow rates of pure CO_2 that probably produced anaesthetic concentrations in water agar (Robinson, 2000).

Insects known to be attracted to CO_2 include *Sinella coeca* (order Collembola) (Klingler, 1959), termites (Isoptera) (Bjostad, 1999), *Helicoverpa armigera* (Lepidoptera) (Rasch and Rembold, 1994), various Dipterans including fruit flies (Robacker, et al., 1996), mosquitoes (Gillies, 1980). and tabanid flies (Hayakawa, 1994), and at least four beetles and

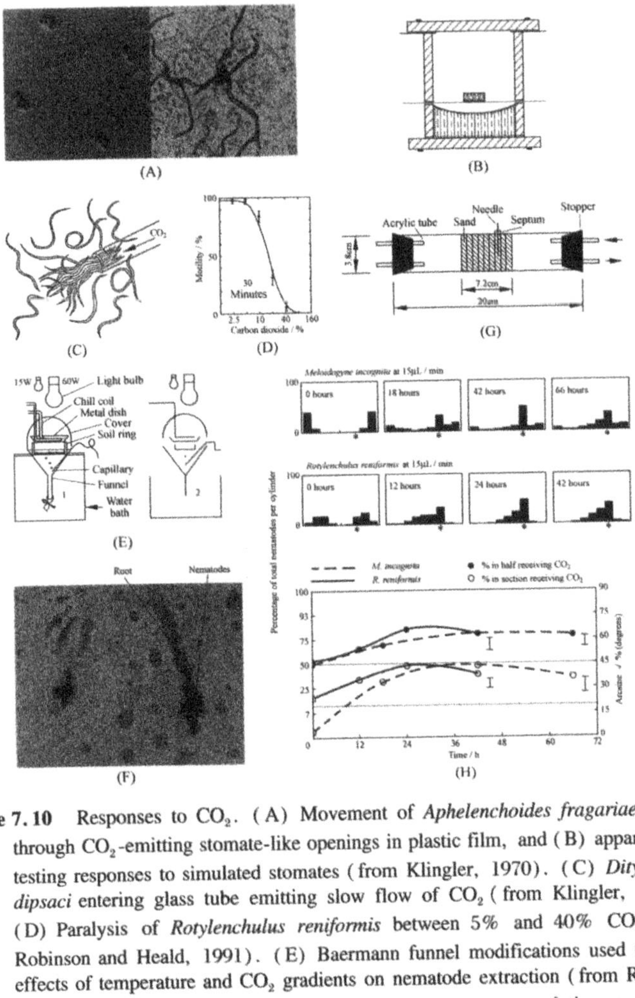

Figure 7.10 Responses to CO_2. (A) Movement of *Aphelenchoides fragariae* to and through CO_2-emitting stomate-like openings in plastic film, and (B) apparatus for testing responses to simulated stomates (from Klingler, 1970). (C) *Ditylenchus dipsaci* entering glass tube emitting slow flow of CO_2 (from Klingler, 1959). (D) Paralysis of *Rotylenchulus reniformis* between 5% and 40% CO_2 (from Robinson and Heald, 1991). (E) Baermann funnel modifications used to study effects of temperature and CO_2 gradients on nematode extraction (from Robinson and Heald, 1991). (F) Juveniles of *Meloidogyne* spp. accumulating around root tips of *Portulaca oleracea* in quartz sand, probably in response to metabolically generated CO_2 (from Linford, 1939). (G) Type of sand cylinder used to investigate nematode responses to CO_2 in moist sand either by continuously purging the head space in both ends of the cylinder with hydrated air at different CO_2 concentrations or by continuously injecting CO_2 into the center of the sand through a needle and purging both ends with hydrated air (from Robinson, 1995). (H) Movement of *M. incognita* and *R. reniformis* toward point source of CO_2 within sand cylinder.

weevils (Coleoptera) including wireworms (Doane, et al., 1976; Klingler, 1965), the western corn rootworm (Bernklau and Bjostad, 1998a, 1998b), *Cryptolestes ferrugineus* (Cucujidae) (White, et al., 1993), and *Otiorrhynchus sulcatus* (Klingler, 1958). As noted for nematodes, attracted insect species are ecologically diverse but are a very small percentage of all insects and many insect orders are unrepresented.

Attraction of at least some insects to CO_2 appears modulated by additional signals. Mosquitoes, for example, are attracted to lactic acid (Acree, et al., 1968), an abundant blood constituent of homeotherms that volatilizes from sweat in behaviorally effective concentrations, and additional modifiers in sweat have been demonstrated. Tabanid flies are attracted to both CO_2 and ammonia, a combination that may help them find optimal oviposition sites (Hribar, et al., 1992). Fruit flies are strongly attracted to ammonia and acetic acid, both produced abundantly by rotting fruit (Robacker, et al., 1996).

There is some evidence for modulatory effects in nematodes similar to those in insects. Several studies have shown ammonia may play a key role. Grewal et al. (1993b) found that entomopathogenic *Heterorhabditis bacteriophora*, *S. carpocapsae*, and *S. glaseri*, but not *Steinernema scapterisci*, were specifically repelled by feces from German cockroaches, though not by host insect species (Fig. 7.11(A)). Cockroach feces have high concentrations of ammonia and removal of ammonia by mixing boric acid with feces eliminated repellency. As another example, Grewal and Wright (1992) observed bacteria species-specific differences in the attraction of *C. elegans* to volatiles released by 10 bacteria and postulated that multiple concentrations or kinds of compounds modulated the known attraction of this nematode to carbon dioxide (Dusenbery, 1985). *Arthrobotrys dactyloides* (and other nematophagous fungi) produce CO_2 and several additional but unknown attractants for *Panagrellus redivivus* (and other nematodes, see 7.4.3.2), and release ammonia as a nematicidal venom in response to nematode contact (Balan and Gerber, 1972). Ammonia, however, was not thought to be one of the attractants. The root-knot nematode *M. incognita* is strongly repelled by ammonia and several nitrogenous salts (Castro, et al., 1989, 1990, 1991), which may help it discern plant roots as sources of CO_2 from bacterial and fungal sources. The reductions in root invasion by root-knot and other plant-parasitic nematodes that have been observed in pots and in field studies following inundative release (see 7.4.2.2) of Steinernematid or Heterorhabditid nematodes (Bird and Bird, 1986; Gouge, et al., 1994; Grewal, et al., 1997b; Ishibashi and Kondo, 1986, 1987; Perry, et al., 1998; Smitley, et al., 1992), in fact, seem to result, based on experiments conducted by Grewal et al. (1999), not from physical jamming of the rhizosphere by Steinernematids as suspected in early studies

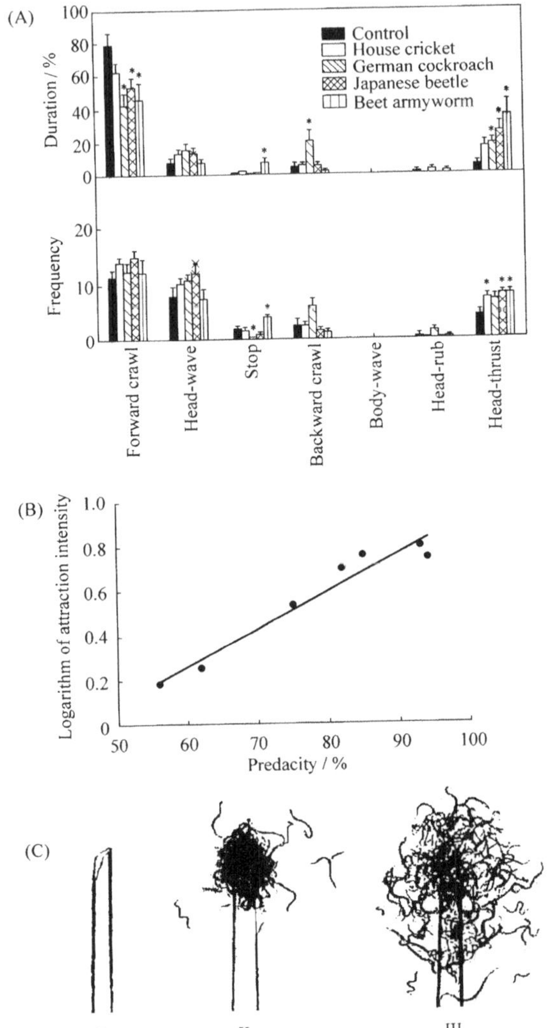

Figure 7.11 Interactions with other organisms. (A) Avoidance movements exhibited by *Heterorhabditis bacteriophora* in response to German cockroach feces (Grewal, et al., 1993b). (B) Relation between predacity of nematophagous fungi against *Panagrellus redivivus* and the degree to which fungi attracted nematodes (from Jansson, 1982b). (C) Aggregation of *Neotylenchus linfordi* around a drawn capillary containing pigments extracted from a culture of the fungus *Gliocladium roseum* (from Klink, et al., 1970).

(Bird and Bird, 1986), but rather from repulsion of root-knot nematodes by ammonia released by the *Xenorhabdus* bacteria symbiotically associated with the Steinernematids.

Chemosensation of CO_2 is not understood at the neurological level and early investigators of the possible role of CO_2 in root-finding (Bird, 1959; Johnson and Viglierchio, 1961; Klingler, 1961, 1965) considered the complex effects that CO_2 released by roots may have on gradients within soil of redox potential, pH, carbonic acid, bicarbonate, and carbonate. In unbuffered, enzyme-free solutions at standard atmospheric pressure, dissolved CO_2 reacts slowly with water to form carbonic acid, which in turn dissociates (pK = 6.1) to yield bicarbonate and a proton, tending to lower the pH. However, buffers, enzymes, pressure, and temperature can greatly alter the ratios of CO_2, H_2CO_3, HCO_3^{-1}, $CO_3^{-2}$and H^{+1} present (Schmidt-Nielsen, 1990). The current consensus seems to be that either dissolved CO_2 or carbonic acid is the nematode attractive component.

Difficulties similar to those encountered in trying to explain oriented responses to CO_2 in soil, arose in attempts to explain the physiological effects of CO_2 on ingested animal parasites. The addition of carbon dioxide in the gas phase to salt solutions bathing resistant stages of various trichostrongylid, ascarid, trichurid, and oxyurid nematodes *in vitro* triggers hatching or exsheathment (Sommerville and Rogers, 1987). Interpretations of the significance of these effects *in vivo* are confounded not only by the possible forms of CO_2 present in the host gut but also by known stimulatory effects of HCl, NH_4^+, bile, pepsin, and carbonic anhydrase. The latter enzyme plays an important role in respiratory gas exchange in many organisms by catalytically accelerating the interconversion of $CO_2$and HCO_3^{-1} (Lindskog, et al., 1971). Carbonic anhydrase has been added to nematode incubation media to study hatching and exsheathment and can be inactivated by oxidation or restored with reducing agents. After comparing rate constants and pKs with experimental data on the effects on nematodes of pH, specific acids, and carbonic anhydrase, Sommerville and Rogers (1987) concluded that "readily available" undissociated carbonic acid is the most likely biologically active form. They also noted, however, that carbonic acid occurs at very low concentrations in vertebrate guts and tissues; in fact, over a wide range of pH values the ratio of dissolved CO_2 to H_2CO_3 at equilibrium is 400:1. However, in soil, where carbonic anhydrase is absent, equlibration is slow and a different ratio could exist under changing conditions.

Limited data are available relating effective stimulus intensities to gradients encountered in natural habitats. Klingler (1963) obtained attraction of *D. dipsaci* on agar in response to CO_2 from a capillary delivery tube positioned within a thin layer of air sandwiched between the agar and a glass cover (Fig. 7.10(C)). The minimum effective gradient in the air, as determined by gas chromatography, was approximately 1 mmol CO_2 mol^{-1} cm^{-1}.

In cylinders of moist sand along which linear gradients were achieved by enclosing the cylinder in a plastic tube and equilibrating the two ends against air masses containing controlled CO_2 concentrations (Fig. 7.10(G)), *M. incognita*, *R. reniformis*, and *Steinernema glaseri* migrated up-gradient in response to a change of 0.2% cm^{-1} (2 mmol mol^{-1} cm^{-1}) at a mean CO_2 concentration of 1.2% (12 mmol mol^{-1}), which is a relative change of about 16% cm^{-1} or 1% per nematode body length (Robinson, 1995; Robinson and Jaffee, 1996). For comparison, the first instar larvae of the western corn rootworm can distinguish 1.12 mmol mol^{-1} from 0.99 mmol mol^{-1}, i.e., a 12% relative difference in concentration (Bernklau and Bjostad, 1998a, 1998b).

Effective release rates for attracting *M. incognita* and *R. reniformis* to a point source in sand ranged from 6 to 35 μl $minute^{-1}$, with an optimal flow of 15 μl $minute^{-1}$. Intervals of 40 and 29 hours, respectively, were required to achieve maximal attraction of the two species from a distance of 52 mm (Fig. 7.10(H)) (Robinson, 1995). Enough gas to achieve the same level of attraction was theoretically achievable with a germinating sunflower seed or 1 mg of baker's yeast. It was proposed that co-pelletization of a nematicide and a CO_2 releasing agent might have practical potential for management of plant-parasitic species in crop fields. Several ways to accomplish this were discussed by Bernklau and Bjostad (1998a, 1998b).

Pline and Dusenbery (1987) employed computer tracking to analyze responses of *M. incognita* on agar exposed to horizontal bi-laminar flow of air from two parallel gas jets emitting different concentrations of CO_2 so as to achieve a different atmosphere over each half of the agar plate. They found that the threshold gradient varied with the ambient concentration, i.e., the nematodes became more sensitive as the ambient concentration dropped and thus could detect about the same relative change at any ambient concentration. This effect is very important ecologically because it enables nematodes to detect gradients at far greater distances from the source than would be possible with a fixed concentration differential threshold.

The threshold for *M. incognita* (Pline and Dusenbery, 1987) corresponded to a relative change of about 3% cm^{-1}, which subsequently was calculated (Dusenbery, 1987) to allow detection of roots from a distance of 5 to 500 cm, with a best estimate of 1 – 2 meters. Vertical gradients measured in apple orchards in New York state in a previous study (Boynton, 1941) were sufficient, according to Dusenbery's (1987) calculations, to expect downward movement to a depth of 1 meter. As emphasized by Dusenbery, these predictions sharply contrast with the previous estimate by Prot (1980) that CO_2 is unlikely to help nematodes find roots from more than 1 or 2 cm away. A reexamination of Klingler's (1965, see his page 6) review, however, suggests that Prot's (1980) estimate of 1 – 2 cm probably came from a benchmark experiment (Klingler, 1963) in which germinating seeds on agar could be observed to attract *Ditylenchus dipsaci* from 1 – 2 cm only if a glass cover slip was placed 1 mm above the surface, to prevent air currents and provide an opportunity for a CO_2 gradient to establish in the air. These conditions cannot be compared directly with the baffled but continuous air interstices within soil and it was Klingler's clearly stated view (Klingler, 1965, his page 15) that the evidence available in 1965 pointed to CO_2 as the most important long-distance nematode attractant released by roots.

Pline and Dusenbery (1987) also introduced the concept that CO_2 might serve as a *collimating stimulus*, providing a directional reference for other responses.

7.4.2 Migrations

7.4.2.1 Migrations through Soil to Plants

Host finding by plant-parasitic nematodes is examined in over 100 published studies, too many to discuss here. Among the most extensive classical reviews are those of Steiner (1925), Jones (1960), Klingler (1965), and Prot (1980). Kaplan and Keen (1980) examine nematode behavior in relation to host plant resistance. Wynn (1981) gives a mycologist's perspective. A short review is given by Robertson and Forrest (1989). Excellent contemporary treatments are presented by Perry (1996, 1997) and by Perry and Aumann (1998).

All root-parasitic and most foliar-parasitic nematodes have one or more infective stages that occur in the soil. These stages typically must find roots or find stems emerging from the soil to ascend foliage, in order to complete the life cycle. Exceptions are parasites in Aphelenchida that may feed on fungi in soil or wood.

Perry (1997) suggested that signals affecting nematodes in the soil are likely to be water soluble, facilitating the establishment of concentration gradients in soil water. I would add that they are also likely to be volatile, and many substances are both. As noted by Campbell (1985), the fact that diffusivities in the liquid phase are four orders of magnitude smaller than those in the gas phase indicates that respiratory gas exchange in a soil profile without a continuous air phase is, for all practical purposes, zero, and gas transport in unsaturated soil profiles can be assumed to occur only through the gas phase. Gradients of volatile substances dissolved in the soil water surrounding nematodes, like those of respiratory gases, can also be expected to result primarily not from diffusion through the soil water, but rather through equilibrium between the air and aqueous phase within the soil interstices, with movement of volatile molecules significant distances through the soil occurring primarily through the air phase.

Since roots respire, extract water, differentially take up salts, and release various organic compounds, they can significantly modify local soil chemistry. As roots are approached, soil moisture decreases, Na^+ increases, K^+ and NO_3^- decrease, O_2 decreases, CO_2 increases, and amino acids, sugars, secondary metabolites, and psiderophores increase. Very near roots, temperature should increase slightly. Other soil gases (ammonia, ethylene, methane), pH, redox potential, and electrical potential may increase or decrease.

Most research on effects of soil variables on nematode movement has been fragmentary, and few studies test hypotheses in soil based on known effects in simpler matrices. The cost of studying soil systems undoubtedly has been an important obstacle. Apparently no study has measured nematode movement in relation to the entire complex of physical and chemical gradients in a natural soil profile.

What we know now is probably best conveyed in an historical context. Although Steiner (1925) postulated that nematodes are attracted to roots, the well-known nematode accumulations near roots on agar plates (Fig. 7.10(F)) (Linford, 1939; Lownsbery and Viglierchio, 1960; Widdowson, et al., 1958; Wieser, 1955, 1956) were still thought by some researchers in the early 1960s to result only from chance encounters and physical trapping (Kühn, 1959; Sandstedt and Schuster, 1962; Sandstedt, et al., 1961). Numerous contemporary and subsequent studies, however, demonstrated directed movement toward roots in soil, toward living but not dead roots on agar plates, and toward soil without roots but in which roots had been grown (Prot, 1980). Thus it became accepted that plant-parasitic nematodes were attracted to roots and actively had to locate and penetrate roots in response to stimuli in order to feed and complete the life cycle. Nematodes, however, were not thought to travel more than a few centimeters in their lives.

By the mid 1960s, contradictory results were obtained regarding nematode responses under controlled conditions to most substances known to occur as gradients around roots (Klingler, 1965; Prot, 1980). Although some studies showed secondary metabolites to correlate with host specificity, most evidence pointed to general attractants. The only consistent directions of movement were toward CO_2 and the wet end of a soil moisture gradient (Wallace, 1960).

Movement toward moisture vertically would help nematodes avoid desiccation but horizontally would take them away from roots. Therefore, CO_2 seemed the prime candidate as a root signal.

Subsequent investigations of responses to salts (usually *unbalanced* salts, see 7.3.1.3) in or on gels, sand, and soil (Abou-Setta and Duncan, 1998; Bilgrami and Jairajpuri, 1984; Castro, et al., 1990, 1991; Prot, 1978a, 1978c, 1979a, 1979b; Riddle and Bird, 1985) made salt gradients appear increasingly important (Figs 7.3 (A), (D)-(E), 7.12(A)-(D)). Experiments on long-distance movement by many species (Prot, 1980; Rode, 1962; Thomas, 1981), including the notoriously sluggish juveniles of root-knot nematodes (Dickson and Hewlett, 1986; Johnson and McKeen, 1973; Pinkerton, et al., 1987; Prot, 1975, 1976, 1978b; Prot and Netscher, 1978; Prot and Van Gundy, 1980), showed that nematodes frequently moved more than 15 cm and sometimes 1 meter in less than 1 month. Movement by the root-knot nematode *M. javanica* toward moisture in tubes of sand was reversed when the concentration of Hoagland's solution and other salts was highest at the wet end (Prot, 1979b). Salts affected different species somewhat differently, however, and while two root-knot nematodes were repelled by a wide range of salts, *Heterodera oryzae* was repelled only by sodium, and *Scutellonema cavenessi* was unaffected. Movement away from salts vertically usually would lead nematodes to deeper regions of higher water content. The role of salts as a root-finding cue, however, was not clear because sodium, which occurs in highest concentrations near roots, was particularly repellent.

During the late 1980s and 1990s, temperature and CO_2 gradients were reexamined under more stringent conditions than in the 1960s. The extreme sensitivity of *M. incognita* (Dusenbery, 1988b; Pline, et al., 1988) to small temperature changes (0.001℃) rekindled the hypothesis that metabolic heat from roots could attract nematodes (El-Sherif and Mai, 1969) while studies under thermally natural conditions (Robinson, 1994) confirmed that diurnally fluctuating vertical gradients could shift the vertical distributions of nematode taxa differently within the soil profile. Nematodes were confirmed to be attracted to minute sources of CO_2 in the soil and behaviorally effective release rates of the gas from a point source were shown to be achievable by roots and other biological sources. In pot studies, ammonium was shown to repel nematodes from tomato plants (Castro, et al., 1991).

The consensus today is that most responses of nematodes to plants are general and lead to similar rates of root penetration in good and poor hosts. Thus, general stimuli are probably commonly used because the host cannot easily avoid generating them. Gradients of salts and CO_2 probably play the most important roles but many other factors affect movement. Holistic comparative studies in natural soil profiles are still badly needed to sort out the relative importance of various stimuli to different nematode species.

Some plants repel or fail to attract nematodes, but nematode responses often do not correlate well with host specificity. Roots of cucumber plants carrying the *Bi* (bitter) locus for triterpenoid curcurbitacins, for example, are reported to attract root-knot nematodes less than roots of other cucumber plants (Kaplan and Keen, 1980), but plants with the *Bi* gene, nonetheless, can become infected and gall heavily in response to nematode feeding. The resistant grass *Aegilops variabilis* was less attractive to juveniles of *Meloidogyne naasi* than susceptible barley (*Hordeum vulgare*) and wheat (*Triticum aestivum*) (Balhadere and Evans, 1994), but susceptible and resistant barley cultivars were similarly attractive. Griffin (1969) observed a nematode-susceptible alfalfa genotype to attract two very different nematodes, *D. dipsaci* and *M. incognita*, less than a resistant genotype did, but apparently was unable to show this effect for additional resistant and susceptible cultivars. Roots of two very different plants, cabbage and carrot, perhaps not surprisingly, attracted *Hirschmanniella oryzae* more than roots of a third very different plant, onion (Bilgrami, et al., 1985). Repellents may cause some of these differences. In a search for both volatile and nonvolatile attractants of *M. incognita* to host roots, the only attractive chemical found was CO_2, although the presence of complex unidentified repellent chemicals was demonstrated (Diez and Dusenbery, 1989b; McCallum and Dusenbery, 1992).

In some cases, a high degree of host specificity seems irrefutable. This situation is perhaps not surprising since it parallels that in plant-feeding insects. Potato root exudate increases the motility of the obligate potato parasite, *G. rostochiensis*, in sand (Clarke and Hennessy, 1984) and triggers electrophysiological activity in the anterior end of the nematode (Rolfe, et al., 2000). Lee and Evans (1973) observed a strong correlation between nematode reproduction and the attractiveness of 16 genotypes of rice for *A. besseyi*. The foliar nematode *D. phyllobius*, with which the author has worked extensively, has a restricted host range limited to foliage of certain *Solanum* spp. (Robinson, et al., 1978) and the nematode is strongly attracted to some unknown compound that is apparently unique to *Solanum* spp. (Robinson, et al., 1979). The latter case has been studied in some detail. The type host, *S. elaeagnifolium*, accumulates attractant in all foliar tissues throughout development, and the attractant can be collected by rinsing foliage with water (Robinson, 1992). The infective fourth-stage juveniles in soil can utilize gradients of the compound around the bases of stems during rainy weather to locate stems, which they then ascend, invading foliar buds up to 60 cm above the soil surface. Attractant activity is retained during freeze drying and is freely soluble in water (Robinson and Saldaña, 1989), and so seems well suited for this purpose.

As a final example of specific chemical responses that appear ecologically relevant, pinewood nematodes (*Bursaphelenchus* spp.) exhibit attraction to terpenes (in particular β-myrcene) predominant in the diet of their monochamid beetle vectors as well as within tissues of the pine tree trunks they inhabit (Futai, 1979, 1980; Shuto and Watanabe, 1987; Tominaga, et al., 1983, 1984). Responses appear differentially tuned to changes in the transport and diet needs of the nematode during development (Stamps and Linit, 1998). β-myrcene is an oil with a pleasant odor, practically insoluble in water (Windholz, et al., 1976), that occurs in hundreds of plants. It is widely used as a scenting agent in cosmetics, soaps, detergents, and foods (Delgado, et al., 1993; Lorente, et al., 1989), and is a precursor or inducer of male pheromone in the bark beetle *Ips paraconfusus* (Byers, et al., 1979).

Only a few studies have explored interference with plant-parasitic nematode behavior as a crop protection tool. These studies have examined co-formulation of the sex attractant, vanillic acid, with biological control agents in pots and in the field (Meyer and Huettel, 1996; Meyer, et al., 1997) to manage *Heterodera glycines*, release of CO_2 to increase the efficacy of pelletized mycophagous fungi (Robinson and Jaffee, 1996), manipulation of CO_2 and temperature gradients to increase Baermann funnel extraction efficiency (Robinson and Heald, 1989, 1991), application of lectins to block chemosensory function (Marbán-Mendoza, et al., 1987), shielding of plant roots with ammonium nitrate and other repellent fertilizer salts (Castro, et al., 1991), and application of tannic acid attractants to the soil as confusants (Hewlett, et al., 1997). All of these studies have shown at least some promise but much more work is needed.

7.4.2.2 Dispersal and Host Seeking Behavior of Entomopathogenic Nematodes

The term entomopathogenic nematode is used in the literature to refer almost exclusively to species of *Steinernema* and *Heterorhabditis*. The dispersal and behavior of these nematodes were reviewed by Ishibashi and Kondo (1990), and more recently by Downes and Griffin (1996) with emphasis on population biology. Reviews of a more general nature are given by Petersen (1985), Poinar (1990), Popiel (1992), Georgis and Hom (1992), Tanada and Kaya (1993), and in the three chapters in this book by Nguyen and Smart (Vol. II), Giblin-Davis et al. (Vol. II), and Kaya. I have tried to avoid unneeded repetition.

Research on entomopathogenic nematodes is recent and has been driven by strong economic incentives. Commercialization of entomopathogenic nematodes in the 1980s and 1990s stimulated searches for isolates optimally suited to many new target pests and environments. Populations of *Steinernema* and *Heterorhabditis* species collected around the world often were found to differ in unforeseen ways from those first discovered, in their thermal optima, reproductive behavior, host finding behavior, and survival strategies. As cultures of wild-type isolates accumulated, nematode breeding and transformation projects were also initiated to develop nematode genotypes with novel trait combinations that would increase the efficacy of inundative releases yet decrease ecological competence of subsequent generations to improve ecological safety (Gaugler and Campbell, 1991; Gaugler, et al., 1991).

Various dispersal and infectivity assays were devised to evaluate effects of soil temperature (Grewal, et al., 1994b; Steiner, 1996; Westerman, 1997), soil texture (Georgis and Poinar, 1983; Hsiao and All, 1996; Kamionek, et al., 1997), soil moisture (Wang, et al., 1995), soil pH (Fischer and Führer, 1990), and pesticides (Barbarossa, et al., 1996; Ishibashi and Kondo, 1987; Patel and Wright, 1996). Global exchange of experimental isolates created a need for a standardized battery of assays to characterize newly discovered isolates, in addition to the routine tests already in place for monitoring batch quality during commercial production and storage (Koppenhöfer, et al., 1998; Ricci, et al., 1996). Simulation models also are being developed to understand and predict isolate performance (Bohan and Hominick, 1995; Westerman, 1998; Westerman and Van der Werf, 1998).

Carbon dioxide released by insects activates, attracts, and in some cases induces nictating in entomopathogenic nematodes (Gaugler, et al., 1980; Schmidt and All, 1978, 1979). The nematodes also respond to various, mostly volatile, known and unknown substances in insect blood (Khlibsuwan, et al., 1992), excreta (Grewal, et al., 1993b; Schmidt and All, 1978), and aqueous external washings. Several species accumulate around plant roots, presumably in response to CO_2, as discussed in Section 7. 4. 1. 3. Host colonization is modulated by responses of infective juveniles in the soil to volatiles from insects already infected with the same or a different species or sex. *Steinernema* males tend to infect hosts first and are thought to make insects more accessible to infective females (Grewal, et al., 1993c), although infected insects in general are less attractive than uninfected insects (Fairbairn, et al., 1999; Grewal, et al., 1997b; Wang and Ishibashi, 1998). Reduced attractiveness of infected insects is thought to play an important role in interspecific competition (Grewal, et al., 1997a; Koppenhöfer and Kaya, 1996) since concomitant infections by two species can lead to death of all nematodes in the insect due to incompatibilities with foreign *Xenorhabdus* symbionts (Downes and Griffin, 1996).

Heterorhabditis bacteriophora and *S. carpocapsae* are the most extensively investigated species of the two genera and have been suggested to lie at opposite extremes of a continuum of foraging strategies among entomopathogenic species (Campbell and Gaugler, 1993, 1997). The terms *cruizer* and *ambusher* are used to denote the contrasting host-finding strategies of *H. bacteriophora* and *S. carpocapsae*, respectively (Grewal, et al., 1994a; Lewis and Gaugler, 1994). *Heterorhabditis bacteriophora* differs from *S. carpocapsae* by greater horizontal and vertical dispersal in soil (Ferguson, et al., 1995; Koppenhöfer, et al., 1996; Schroeder and Beaver, 1987), a tooth used for perforating insect cuticles, which extends the host range to include insects whose spiracles are too small to enter (Bedding and Molyneux, 1982), and a hermaphroditic rather than amphimictic infective juvenile, which allows a single nematode to establish a reproducing population within a host (Downes and Griffin, 1996). *Steinernema carpocapsae* is distinguished by a lower metabolic rate (Lewis, et al., 1995b), the ability to lie quiescent on the soil surface until stimulated by insect cues, and the ability to nictate (Ishibashi, et al., 1995; Kondo and Ishibashi, 1985, 1986) and jump (Campbell and Kaya, 1999), facilitating attachment to insects moving across the soil surface. There is evidence that the level of complexity of sequential insect cues required to trigger insect penetration behavior (Grewal, et al., 1993a) also shows a gradation across nematode species (Lewis, et al., 1995a, 1996).

Several recent studies have examined temporal changes in infectivity within entomopathogenic nematode populations from individual insects, sometimes referred to as *phased infectivity* (Bohan and Hominick, 1996; Hay and Fenlon, 1995). Direct application of infected insect cadavers to soil is problematic on a large scale but can yield higher infection rates than equivalent numbers of nematodes in aqueous suspension (Shapiro and Glazer, 1996). Slow emergence over a period of days probably contributes to greater infection. It also has been observed, however, that the early emergers tend to have ambusher traits and are heat tolerant, whereas the late emergers have cruizer traits, have shorter bodies, and are desiccation tolerant (O'Leary, et al., 1998; Wang, et al., 1997). Thus, a species may increase its chances for survival by producing individuals from the same host that exploit different host-seeking strategies. Another, related question is whether the infective juveniles, which have always been considered developmentally arrested, typically change their host-seeking behavior as they age or starve, and it seems this is still being investigated. It is known at this point that a sizable proportion of the population in an aqueous suspension of Steinernematids may be non-infective at the time of application, and this proportion may or may not change its behavior in time, depending on the species and conditions (Campbell, et al., 1999; Ishibashi, et al., 1995; Lewis, et al., 1995b, 1997).

7.4.2.3 Host Tissue Migrations and Their Evolution in Plant and Animal Parasites

The tissue migrations, feeding sites, and reproduction of animal and plant parasites have been studied in some cases for more than 100 years. The last two decades have seen increased interest in the evolution of parasitic behaviors, particularly in relation to host finding, site finding, and feeding site establishment. Molecular evidence has stimulated new theories and insights. A recent conceptual advance has been to explain differential migrations in terms of a small number of fixed action patterns (Tinbergen, 1951) elicited by different stimuli in different species or at different times during development (Sukhdeo, 1997). Examples include the characteristic sequences of activities exhibited by plant-parasitic nematodes during hatching and root invasion, the resumption of spontaneous activity by *S. carpocapsae* and *A. caninum* when vibrated or by *Trichonema* sp. and *A. catedecaudata* when illuminated, nictating by *S. carpocapsae* and *A. caninum* in response to CO_2, conversion from tortuous to straight locomotion by males of *P. redivivus* on agar when exposed to sex attractant (Samoiloff, et al., 1973), and accelerated movement by animal parasites in response to bile (Sukhdeo, 1997).

Von Mende (1997) reviewed movement of juvenile stages of sedentary root parasites through roots, emphasizing work with the model plant, *Arabidopsis thaliana* (Sijmons, et al., 1991). Wyss (1997) examined literature for feeding site establishment in roots across a continuum of parasitic specialization among 14 illustrated types of feeding patterns or nematode-induced feeding sites. Scheres et al. (1997) discussed nematode-elicited cellular and tissue modifications in plants in relation to contemporary genetic analysis of cell determination in roots.

Based on interpretations of tissue migrations by animal parasites it seems plausible that a small group of fixed action patterns evoked in plant-parasitic nematodes at key points during plant tissue development or invasion may be sufficient to guide nematodes through tissue and regulate their involvement in feeding site establishment. Doncaster and Seymour (1973) analyzed behavior leading to feeding in 11 species of Tylenchida. Contact with roots led to a predictable sequence of actions referred to as local exploration (lip rubbing and stylet probing), followed by a cell-wall cutting cycle (body immobilization and rhythmic stylet thrusting to cut a slit in the cell against which the lips are appressed), and then cell penetration. When a cell to be fed on was encountered, a different sequence occurred. Feeding by *Aphelenchus avenae* (Fisher, 1975) in liquid culture showed that membrane penetration is not necessary for feeding, but several activities are: cessation of body movement, followed by bending of the head, stylet thrusting, and protracted stylet exsertion during pumping of the esophageal metacorpus.

A given taxon will typically exhibit the same migration and feeding pattern in a wide range of plant hosts, and a small set of stimuli and responses likely guide these activities. Carbon dioxide has been suggested to mediate the acropetal migration of *Meloidogyne* spp. intercellularly through the root cortex following root entry in cotton (McClure and Robertson, 1973). After reaching the meristematic tissue just behind the growing tip, these nematodes turn basipetally to rest and feed on one or several cells which are induced to develop into enlarged hypermetabolic nurse cells usually called giant cells. The same migration pattern is seen in many hosts (Von Mende, 1997; Wyss, et al., 1992). As one of many possible contrasting examples, root entry by the infective, vermiform females of the reniform nematode (*R. reniformis*) is largely intracellular rather than intercellular through the cortex, and perpendicular rather than parallel to the root axis, in a wide range of hosts, with the infective female usually coming to rest and feed permanently on an endodermal cell, eliciting the formation of a nurse syncytium comprised of several contiguous pericycle cells, without the nematode ever perforating the pericycle. This same feeding site establishment behavior, like that of root-knot nematodes, is exhibited in dozens of unrelated plant hosts (Robinson, et al., 1997).

The tissue migrations of animal parasites have received more study than those of plant parasites, but they are still inadequately understood. Wilson (1982) presented a colorful argument that theories of tissue migrations by skin penetrators presented in textbooks at the time were in most cases supported by inconclusive or misleading evidence and that migration routes for additional species often were assumed with almost no evidence at all. Wilson gave as examples *Strongyloides* spp. in dogs and *Nippostrongylus brasiliensis* in rats. The classical migration route concluded from experiments near the turn of the century (Fülleborn, 1914; Looss, 1905) was from the skin penetration site through the blood or lymph to the heart and from there in the blood stream to the capillary beds of the lung alveoli, where the nematodes broke through into the airways, to be coughed up, swallowed, and carried with ingesta through the stomach to the intestine, where the nematodes rested and developed to maturity.

Wilson (1982) argued this route had been concluded by inference and without proper experimental controls, by identifying tissues along the presumed route with "high impedance", i. e. tissues where nematodes accumulated due to obstructed movement. Wilson contended that loci where nematodes accumulated were inadequate proof of the route taken by successful parasites unless all individuals were accounted. He gave a recent example for *N. brasiliensis* where the normal route taken in mice was found to be through the lungs eventually but by way of an initial subdermal migration to the head rather than to the heart; migration to the intestine could be blocked by cutting and suturing a ring of skin around the neck to the underlying muscle (Tada, et al., 1979). As a second example, he noted that the normal migration route of 95% of the homogonic strain of *Strongyloides ratii* could be reproducibly diverted subdermally to the mammaries, in response to some unknown signal produced by suckling, with almost none of the nematodes ever reaching the intestine of the mother (Wilson, 1982).

The ecological value of tissue migrations of nematodes in animal hosts is often perplexing, and there may be some lessons here for students of plant-parasitic nematode behavior. In *Ascaris* and *Strongylus*, larvae may undergo extensive migrations that begin and end in the same place. The orthodox explanation is that these nematodes are developmentally locked into such behavior following evolutionary loss of skin penetration or intermediate hosts. Read and Skorping (1995) recently presented a provocative alternative hypothesis by comparing published data for 13 pair of sister taxa, selected on a phylogenetic basis such that one sister of each pair did and one did not exhibit a larval tissue phase outside the gastrointestinal tissues or lumen. Based on comparisons between nematode body size, growth rate, and fecundity, Read and Skorping proposed that the migratory phase is, in fact, selectively advantageous because the tissue environment favors faster development of ultimately larger and more fecund adults than possible for sister taxa developing continuously in the intestinal environment.

Another fresh perspective of animal tissue migrations is given by Sukhdeo (1997), who suggests we tend to misperceive the primary dimensions of variability within the vertebrate host environment, by failing to recognize its extreme topological and chemical predictability when compared with terrestrial and aquatic habitats. Sukhdeo (1997) begins with the observation that animal parasites show the remarkable tissue specificity in their final sites of development within the host, regardless of the particular host they are in or the routes taken to get there. He follows by suggesting that given this predictability, a relatively small number of fixed action patterns should be sufficient to guide each parasite to its final destination. Examples include 200 gastrointestinal parasite species that change their behavior in response to bile, and the topological certainty that liver flukes moving along the egg-shaped abdominal wall will encounter the liver.

Another important step, perhaps, in ultimately understanding the genetic control and evolution of fixed action patterns in plant and animal parasites comes from studies of the enzymes and genes regulating *clumping* behavior in *C. elegans* (Thomas, 1998). In this case, a single nucleotide change appears sufficient to change the dispersal foraging behavior pattern found in some wild-type populations into an aggregation behavior pattern, referred to as clumping, that is found in other populations.

7.4.3 Behavioral Interactions

7.4.3.1 Behavioral Interaction with Congeneric Nematodes – Pheromones

This topic has been reviewed by Bone and Shorey (1978), Green (1980), Bone (1982a, 1982b, 1982c, 1987), Huettel (1986), Haseeb and Fried (1988), Greet and Perry (1992), and Emmons and Sternberg (1997). Other related reviews include Karlson and Butenandt (1959), Yamamoto and Ogawa (1989), Schneider (1992), Huettel and Meyer (1992), and Perry and Riga (1995).

Experimental evidence of pheromone activity in nematodes was first seen in *Panagrolaimus rigidis* (Greet, 1964). Research on nematode pheromones appeared to increase and then decrease between 1970 and 1990, perhaps due to limiting technology. Recently developed neurophysiological techniques (Jones, et al., 1991; Perry and Riga, 1995; Rolfe, et al., 2000) for quickly measuring the responses of individual nematodes to minute quantities of test material have led to renewed and rapid progress (Fig. 7. 12(E)) (Riga, et al., 1996a, 1996b, 1997a, 1997b).

Based on knowledge from insects and other groups, nematodes might be expected to produce pheromones that trigger or regulate mate finding, copulation, alarm responses, aggregation and dispersal behavior, or developmental events. Haseeb and Fried (1988) listed 32 species for which there was experimental evidence for sex attractants, and several additional species have been reported (Huang and Becker, 1999; Huettel, et al., 1982; Neves, et al., 1998; Riga and MacKinnon, 1987; Tahseen and Bilgrami, 1993). An unidentified developmental pheromone regulates dauer larva formation in *C. elegans* (Golden and Riddle, 1982, 1984) and similar pheromones are suspected in entomopathogenic species.

Most sex pheromones are produced by females and attract males; male-produced female attractants appear more common among animal parasites (Haseeb and Fried, 1988). A female homosexual attractant is apparent in *Pelliodítis pellio* (Eveland and Fried, 1990). Studies with *Heterodera* spp. have indicated partial interspecific activity (Green and Plumb, 1970). The only pheromone whose structure is known, vanillic acid, attracts males of the soybean cyst nematode (*H. glycines*) (Jaffe, et al., 1989; Meyer and

Huettel, 1996) and males of the bacteriophagous *P. pellio* (at 10^{-5} M) (Eveland and Haseeb, 1993), but apparently not other cyst nematode species (Aumann and Hashem, 1993). There is limited evidence for interspecific activity among animal-parasitic nematodes (Roberts and Thorson, 1977).

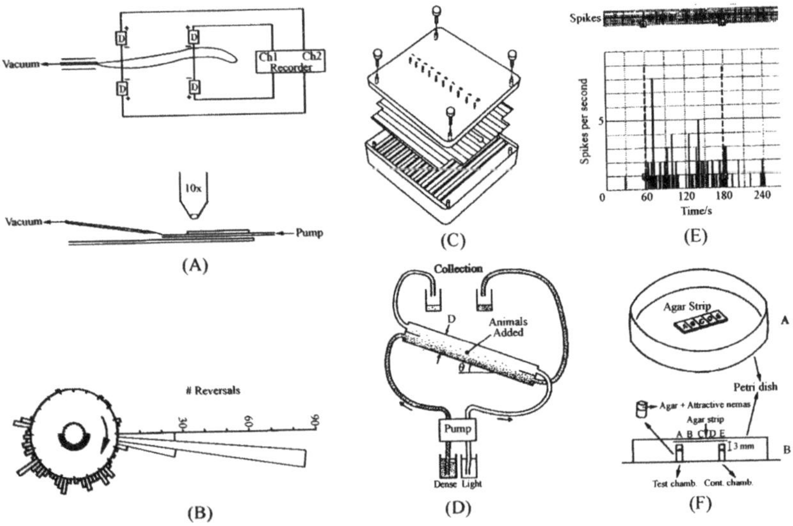

Figure 7.12 Research on chemical stimuli. (A) Tethered nematode apparatus used to investigate responses of *Caenorhabditis elegans* and *Meloidogyne incognita* to various substances (from Dusenbery, 1980b). (B) Cumulative data obtained with the apparatus shown in (A), for repeated exposures of *M. incognita* to NaCl (from Goode and Dusenbery, 1985). Black semicircle shows duration of NaCl exposure. (C) Agar channel used by Castro et al. (1989) to measure repulsion of *M. incognita* by inorganic salts. (D) Countercurrent separator used by Dusenbery (1973) to study chemotaxis in *Caenorhabditis elegans*. (E) Extracellular recordings from the anterior of an individual juvenile of *Globodera rostochiensis* when exposed to potato root diffusate, starting at "a" when time = 60 seconds and after being washed off with tap water at "b" when time = 180 seconds (from Rolfe, et al., 2000). (F) Apparatus used by Kiyohara (1982) to study responses of males to volatile sex pheromone components released by females of *Bursaphelenchus xylophilus*.

Nematode pheromones are largely uncharacterized and may turn out to differ entirely from the well-studied pheromones of insects, especially when it is considered that many nematodes occupy fundamentally aquatic habitats where the types of compounds most effective in chemical communication are apt to differ from those that work best in air. In this light it is interesting to note that the first sex pheromone to be identified for a marine polychaete (*Platynereis dumerilii*) has a basic structure that shows a close relationship to insect pheromones (Zeeck, et al., 1988).

Insect pheromones typically have multiple components that must be present in appropriate relative concentrations for maximal effectiveness (Schneider, 1992). Preliminary evidence for nematodes suggested two components for *H. glycines* (Bone, 1986), *P. redivivus* (Balakanich and Samoiloff, 1974), and *N. brasiliensis* (Bone, 1986).

Most insect pheromones are volatile, allowing dispersal through air by diffusion or convection (Dusenbery, 1992; Schneider, 1992). Volatile components could be expected to work well for nematodes within the soil with a continuous air phase, where there is essentially no stirring, especially when it is considered that substances can diffuse 10^4 times as fast through the air phase in soil as they can through the aqueous phase (see 7.4.2.1). In animal tissue or intestinal habitats, however, water solubility would seem a much more important characteristic than volatility. Current evidence points to both volatile and nonvolatile components, although few studies have employed techniques that facilitated detection of volatile attractants (Fig. 7.12(F)).

Greet et al. (1968) gave evidence for volatiles in the sex pheromones from *G. rostochiensis* and *H. schachtii*. The volatile, and heat stable, nature of female sex pheromone from *H. schachtii* permitted purification by direct gas chromatography by injecting methanol and pentane extracts of dried, vulva-secreted gelatinous matrices collected from females the previous day (Aumann, et al., 1998). The *H. glycines* pheromone, vanillic acid, though water soluble, is also slightly volatile (sublimes undecomposed, soluble in 860 parts water) as is the related compound, vanillin, which has limited pheromone activity in *H. glycines* (soluble in 100 parts water, pleasant aromatic vanilla odor) (Windholz, et al., 1976). Sex pheromone from *G. rostochiensis* appears to be composed of several weakly basic polar compounds (Riga, et al., 1997a). Pheromones of *Bursaphelenchus xylophilus* were both volatile and water soluble (Kiyohara, 1982).

7.4.3.2 Behavioral Interactions with Nematophagous Fungi

Reviews on this topic are given by Zuckerman (1983), Zuckerman and Jansson (1984), Nordbring-Hertz (1988), Jansson and Nordbring-Hertz (1988), Stirling (1991), and Zuckerman and Esnard (1994). Recent developments are presented in the numerous articles by Jaffee and co-workers (Jaffee, 1998, 1999, 2000; Jaffee and Muldoon, 1997; Jaffee, et al., 1997, 1998).

During the 1970s and 1980s, research projects seeking biological control agents against plant-parasitic nematodes discovered that nematophagous fungi attracted nematodes far more frequently that did non-nematophagous fungi (Field and Webster, 1977; Jansson and Nordbring-Hertz, 1979, 1980; Klink, et al., 1970; Townshend, 1964). The predacity of eight species of nematophagous fungi against the nematode *P. redivivus* was highly correlated ($r = 0.98$) with the ability of the fungi to attract nematodes *in vitro* (Fig. 7.11(B)), suggesting that nematophagous fungi had evolved a means to lure their prey with cues normally useful to the nematodes for finding food (Jansson, 1982b). Initially, only bacteriophagous and mycophagous nematodes (Fig. 7.11(C)) were studied but, in time, experiments were extended to include Tylenchid parasites of plants as well (Jansson and Nordbring-Hertz, 1980). Some obligate plant parasites were attracted and some were not.

The fungal attraction phenomenon was confounded partly by nematode species-specific effects and partly by a need of the fungi to be stimulated by the presence of nematodes before undergoing transformations required to attack nematodes. These changes included the formation of the trapping structures, or development of sticky residues on infective spores (Jansson, 1982a, 1982b), and localized release of ammonia upon nematode contact, which intoxicated nematodes rendering them defenseless against fungal adhesion and invasion (Balan, et al., 1974, 1976). At the time, the strong repellent activity of ammonia (Grewal, et al., 1999) and ammonium ion (Castro, et al., 1990) against root-knot nematodes was not known.

The sticky residues were revealed to include lectins, and sticky fungal structures bound to specific regions of the glycoproteinaceous cuticular coat secreted by the amphids (Jansson and Nordbring-Hertz, 1984). When applied to nematodes directly, lectins obstructed nematode chemosensory function, as did sticky fungal conidia (Spiegel and McClure, 1995; Zuckerman, 1983; Zuckerman and Jansson, 1984). Further investigations revealed species-specific interactions between lectins produced by nematode antagonists and nematode surface chemistry. Thus, several intriguing avenues of inquiry that were of both fundamental and applied interest were opened up within a short time.

The possible utilization of lectins without nematophagous fungi as agricultural nematicides was explored with variable success (Marbán-Mendoza, et al., 1987). Efforts to isolate the attractants indicated CO_2 was and ammonia was not one of several attractive components from the fungal culture medium (Balan and Gerber, 1972). Some other components were partially purified but not identified (Balan, et al., 1976; Monoson, et al., 1973).

Some years later, nematode behavior experiments in sand revealed a marked repulsion rather than attraction of *M. incognita* by three kinds of calcium alginate pellets containing nematophagous fungi (Lackey, et al., 1993; Robinson and Jaffee, 1996). This observation underlined the long-standing unanswered question regarding the potential repellency of ammonia released during trap formation. Timed release of minute quantities of CO_2 into the center of the regions in sand where pellets were placed eliminated the repellent effect and attracted nematodes, leading to high rates of infection and nematode mortality (Robinson and Jaffee, 1996). Repellent activity was water-extractable and recoverable after freeze drying, but the repellent was not identified. Unfortunately, the identification and species-specificity of naturally occurring attractants and repellents released by nematophagous fungi still remain largely unresolved and the practical potential of these compounds as a management tool remains untapped.

7.5 Conclusions

Most investigations of nematode behavior have examined root finding by plant parasites, host foraging by entomopathogenic nematodes, migrations through host tissues by plant and animal parasites, and responses *in vitro* to non-specific stimuli, pheromones, and nematophagous fungi. We have only begun to explore behaviorally-based opportunities for managing plant nematodes and increasing the efficacy of entomopathogenic nematodes as biological control agents. Future applications include baited traps, confusants, superior entomopathogenic nematode strains, and modified soil management practices to suppress nematode populations. The greatest barrier to these goals is inadequate funding.

Funding for research on nematodes in general is driven by economic impact and guided by the view that nematodes are propagules or pathogens, rather than individual animals responding in specific ways to environmental cues to complete the life cycle. Funding for nematode behavior research has always been incommensurate with the ecological diversity of nematodes and complexity of niches occupied and as a result most experiments have been on agar, utilizing inexpensive materials, and responses seldom have been confirmed within soil and host tissues.

Improvements are needed in the sophistication with which natural gradients of volatiles, solutes, and temperature are measured and mimicked experimentally. When migrations are investigated, nematodes *in vitro* need freedom to move in three dimensions if that is the way they move in nature. More electrophysiological and laser ablation studies are needed to reveal precisely where and how nematodes sense and respond to specific stimuli. Significant future advances will come from studies designed to integrate molecular, anatomical, physiological, and behavioral data, explain migrations of populations in terms of individual responses, and measure within natural matrices the relative importance of responses originally detected separately *in vitro*.

References

Abou-Setta, M. M., and L. W. Duncan. 1998. Attraction of *Tylenchulus semipenetrans* and *Meloidogyne javanica* to salts *in vitro*. Nematropica 28: 49 – 59.

Acree, F., R. B. Turner, H. K. Gouck, M. Beroza, and N. Smith. 1968. l-Lactic acid: A mosquito attractant isolated from humans. Science 161: 1346 – 1347.

Adamo, J. A., C. P. Madamba, and T. A. Chen. 1976. Vertical migration of the rice white-tip nematode, *Aphelenchoides besseyi*. Journal of Nematology 8: 146 – 152.

Ahmad, I., and M. S. Jairajpuri. 1982. Locomotory characteristics of *Chiloplacus symmetricus* juveniles. Indian Journal of Nematology 12: 175 – 177.

Al-Hadithi, I. A. W., and A. H. Habash. 1979. The influence of some physical factors on infective larvae of *Ancylostoma duodenale*. Bulletin of the Natural History Research Centre 7: 23 – 32.

Anderson, A. R. A., B. D. Sleeman, I. M. Young, and B. S. Griffiths. 1997a. Nematode movement along a chemical gradient in a structurally heterogeneous environment. 2. Theory. Fundamental and Applied Nematology 20: 165 – 172.

Anderson, A. R. A., I. M. Young, B. D. Sleeman, B. S. Griffiths, and W. M. Robertson. 1997b. Nematode movement along a chemical gradient in a structurally heterogeneous environment. 1. Experiment. Fundamental and Applied Nematology 20: 157 – 163.

Ashton, F. T., J. Li, and G. A. Schad. 1999. Chemo- and thermosensory neurons: Structure and function in animal parasitic nematodes. Veterinary Parasitology 84: 297 – 316.

Aumann, J. 1993. Chemosensory physiology of nematodes. Fundamental and Applied Nematology 16: 193 – 198.

Aumann, J., and M. Hashem. 1993. Studies on substances with sex pheromone activity produced by *Heterodera schachtii* females. Fundamental and Applied Nematology 16: 43 – 46.

Aumann, J., H. Ladehoff, and S. Rutengrantz. 1998. Gas chromatographic characterization of the female sex pheromone of *Heterodera schachtii* (Nematoda: Heteroderidae). Fundamental and Applied Nematology 21: 119 – 122.

Avery, L., and J. H. Thomas. 1997. Feeding and defecation. Pp. 679 – 716. In D. L. Riddle, T. Blumenthal, B. J. Meyer, and J. R. Priess, eds. *C. elegans* II. New York: Cold Spring Harbor Laboratory Press.

Baker, G. L., and G. O. Poinar, Jr. 1995. *Agamermis catadecaudata* n. sp. (Nematode: Mermithidae), a parasitoid of Orthoptera in south-eastern Australia. Fundamental and

Applied Nematology 18: 139 – 148.
Balakanich, S., and M. R. Samoiloff. 1974. Development of nematode behavior: Sex attraction among different strains of the free-living *Panagrellus redivivus*. Canadian Journal of Zoology 52: 835 – 845.
Balan, J. 1985. Measuring minimal concentrations of attractants detected by the nematode *Panagrellus redivivus*. Journal of Chemical Ecology 11: 105 – 111.
Balan, J., and N. N. Gerber. 1972. Attraction and killing of the nematode *Panagrellus redivivus* by the predaceous fungus *Arthrobotrys dactyloides*. Nematologica 18: 163 – 173.
Balan, J., L. Krizkova, P. Nemec, and V. Vollek. 1974. Production of nematode-attracting and nematicidal substances by predacious fungi. Folia Microbiol. 19: 512 – 519.
Balan, J., L. Krizkova, P. Nemec, and A. Kolozsvary. 1976. A qualitative method for detection of nematode attracting substances and proof of production of three different attractants by the fungus *Monacrosporium rutgeriensis*. Nematologica 22: 306 – 311.
Balhadere, P., and A. A. F. Evans. 1994. Characterization of attractiveness of excised root tips of resistant and susceptible plants for *Meloidogyne naasi*. Fundamental and Applied Nematology 17: 527 – 536.
Barbarossa, B., R. Frulla, G. Grazzi, and L. Rovesti. 1996. Compatibility of *Steinernema carpocapsae* with pesticides. Bulletin OILB/SROP. 19: 136 – 139.
Bargmann, C. I., and I. Mori. 1997. Chemotaxis and thermotaxis. Pp. 717 – 737. In D. L. Riddle, T. Blumenthal, B. J. Meyer, and J. R. Priess, eds. *C. elegans* II. New York: Cold Spring Harbor Laboratory Press.
Barraclough, R. M., and N. French. 1965. Observations on the orientation of *Aphelenchoides ritzemabosi* (Schwartz). Nematologica 11: 199 – 206.
Baunacke, W. 1922. Untersuchungen zur Biologie und Bekämpfung des Rübennematoden *Heterodera schachtii* Schmidt. Arbeiten der Biologie der Reichanstalt für Landwirtschaft und Forstwirtschaft 11: 185 – 288.
Bedding, R. A., and A. S. Molyneux. 1982. Penetration of insect cuticle by infective juveniles of *Heterorhabditis* spp., (Heterorhabditidae: nematoda). Nematologica 28: 354 – 359.
Bernklau, E. J., and L. B. Bjostad. 1998a. Behavioral responses of first-instar western corn rootworm (Coleoptera: Chrysomelidae) to carbon dioxide in a glass bead bioassay. Ecology and Behavior 91: 444 – 456.
Bernklau, E. J., and L. B. Bjostad. 1998b. Reinvestigation of host location by western corn rootworm larvae (Coleoptera: Chrysomelidae): CO_2 is the only volatile attractant. Ecology and Behavior 91: 1331 – 1340.
Bilgrami, A. L. 1996. Effect of agar concentration, thickness, coloured lights, temperature and distance on the attraction of *Tylenchorhynchus mashhoodi* towards cabbage roots. Annals of Plant Protection Sciences 4: 29 – 34.
Bilgrami, A. L., and M. S. Jairajpuri. 1984. The responses of *Mononchus aquaticus* to chemicals and pH. Indian Journal of Nematology 14: 171 – 174.
Bilgrami, A. L., I. Ahmad, and M. S. Jairajpuri. 1985. Responses of adult *Hirschmanniella oryzae* towards different plant roots. Revue de Nematologie 8: 265 – 272.

Bird, A. F. 1959. The attractiveness of roots to the plant parasitic nematodes *Meloidogyne javanica and M. hapla*. Nematologica 4: 322 – 335.

Bird, A. F. 1960. Additional notes on the attractiveness of roots to plant parasitic nematodes. Nematologica 5: 217.

Bird, A. F., and J. Bird. 1986. Observations on the use of insect parasitic nematodes as a means of biological control of root-knot nematodes. International Journal for Parasitology 16: 511.

Bird, A. F., and J. Bird. 1991. The Structure of Nematodes. New York: Academic Press.

Bjostad, B. J. 1999. Insect pest management with infochemicals. Colorado AES Project COL00622-1999 – 2000. Ft. Collins, CO: Colorado State University.

Blake, C. D. 1961. Importance of osmotic potential as a component of the soil water on the movement of nematodes. Nature 191: 144 – 145.

Bloom, J. R. 1964. Photonegative reaction of the chrysanthemum foliar nematode. Phytopathology 54: 118 – 119.

Bohan, D. A., and W. M. Hominick. 1995. Examination of the *Steinernema feltiae* (site 76 strain) infection interaction with the *Galleria mellonella* host, using an infection model. Parasitology 111: 617 – 625.

Bohan, D. A., and W. M. Hominick. 1996. Investigations on the presence of an infectious proportion amongst populations of *Steinernema feltiae* (site 76 strain) infective stages. Parasitology 112: 113 – 118.

Bone, L. W. 1982a. Chemotaxis of parasitic nematodes. Pp. 52 – 66. In W. S. Bailey, ed. Cues that Influence Behavior of Internal Parasites. New Orleans, LA: United States Department of Agriculture, Agricultural Research Service, Southern Region.

Bone, L. W. 1982b. Reproductive chemical communications of helminths. I. Platyhelminthes. International Journal of Invertebrate Reproduction 5: 261 – 268.

Bone, L. W. 1982c. Reproductive chemical communication of helminths. II. Aschelminthes. International Journal of Invertebrate Reproduction 5: 311 – 321.

Bone, L. W. 1986. Fractionation of the female's pheromone of the soybean cyst nematode, *Heterodera glycines*. Proceedings of the Helminthological Society of Washington 53: 132 – 134.

Bone, L. W. 1987. Pheromone communication in nematodes. Pp. 147 – 152. In J. A. Veech, and D. W. Dickson, eds. Vistas on Nematology. Hyattsville, MD: Society of Nematologists.

Bone, L. W., and H. H. Shorey. 1978. Nematode sex pheromones. Journal of Chemical Ecology 4: 595 – 612.

Boynton, D. 1941. Soils in relation to fruit-growing in New York. Part IV. Seasonal and soil influences on oxygen and carbon-dioxide levels of New York orchard soils. Bulletin 763. Ithaca, NY: Cornell University Agricultural Experiment Station.

Buckley, J. J. C. 1940. Observations on the vertical migrations of infective larvae of certain Bursate nematodes. Journal of Helminthology 18: 173 – 182.

Burman, M., and A. E. Pye. 1980. *Neoaplectana carpocapsae*: Movements of nematode populations on a thermal gradient. Experimental Parasitology 49: 258 – 265.

Burr, A. H. 1979. Analysis of phototaxis in nematodes using directional statistics. Journal of Comparative Physiology 134: 85 – 93.

Burr, A. H. 1984a. Photomovement behavior in simple invertebrates. Pp. 179 – 215. In M. A. Ali, ed. Photoreception and Vision in Invertebrates. New York: Plenum Press.

Burr, A. H. 1984b. Evolution of eyes and photoreceptor organelles in the lower phylla. Pp. 131 – 178. In M. A. Ali, ed. Photoreception and Vision in Invertebrates. New York: Plenum Press.

Burr, A. H. 1985. The photomovement of *Caenorhabditis elegans*, a nematode which lacks ocelli. Proof that the response is to light not radiant heating. Photochemistry and Photobiology 41: 577 – 582.

Burr, A. H. J., and C. P. F. Babinszki. 1990. Scanning motion, ocellar morphology and orientation mechanisms in the phototaxis of the nematode *Mermis nigrescens*. Journal of Comparative Physiology A 167: 257 – 268.

Burr, A. H., and C. Burr. 1975. The amphid of the nematode *Oncholaimus vesicarius*: Ultrastructural evidence for a dual function as chemoreceptor and photoreceptor. Journal of Ultrastructure Research 51: 1 – 15.

Burr, A. H., and F. I. Harosi. 1985. Naturally crystalline hemoglobin of the nematode *Mermis nigrescens*. Biophysical Journal 47: 527 – 536.

Burr, A. H., R. Schiefke, and G. Bollerup. 1975. Properties of a hemoglobin from the chromatrope of the nematode *Mermis nigrescens*. Biochimica et Biophysica Acta 405: 404 – 411.

Burr, A. H., D. K. Eggleton, R. Patterson, and J. T. Leutscher-Hazelhoff. 1989. The role of hemoglobin in the phototaxis of the nematode *Mermis nigrescens*. Photochemistry and Photobiology 49: 89 – 95.

Burr, A. H. J., C. P. F. Babinszki, and A. J. Ward. 1990. Components of phototaxis of the nematode *Mermis nigrescens*. Journal of Comparative Physiology A 167: 245 – 255.

Burr, A. H. J., P. Hunt, D. R. Wagar, S. Dewilde, M. L. Blaxter, J. R. Vanfleteren, and L. Moens. 2000a. A hemoglobin with an optical function. Journal of Biological Chemistry 275: 4810 – 4815.

Burr, A. H. J., D. Wagar, and P. Sidhu. 2000b. Ocellar pigmentation and phototaxis in the nematode *Mermis nigrescens*: Changes during development. Journal of Experimental Biology 203: 1341 – 1350.

Byers, J. A., D. L. Wood, L. E. Browne, R. H. Fish, B. Piatek, and L. B. Hendry. 1979. Relationship between a host plant compound, myrcene, and pheromone production in the bark beetle *Ips paraconfusus*. Journal of Insect Physiology 25: 477 – 482.

Campbell, G. S. 1985. Soil Physics with Basic Transport Models for Soil-Plant Systems. New York: Elsevier.

Campbell, J. F., and R. Gaugler. 1993. Nictation behaviour and its ecological implications in the host search strategies of entomopathogenic nematodes (Heterorhabditidae and Steinernematidae). Behaviour 126: 3 – 4.

Campbell, J., and R. Gaugler. 1997. Inter-specific variation in entomopathogenic nematode foraging strategy: Dichotomy or variation along a continuum. Fundamental and Applied Nematology 20: 393 – 398.

Campbell, J. F., and H. K. Kaya. 1999. How and why a parasitic nematode jumps. Nature 397: 485 – 486.

Campbell, J. F., A. M. Koppenhöfer, H. K. Kaya, and B. Chinnasri. 1999. Are there temporarily non-infectious dauer stages in entomopathogenic nematode populations: A test of the phased infectivity hypothesis. Parasitology 188: 499 – 508.

Castellanos-Coutiño, C., and J. Alfredo. 1980. Migración vertical de larvas de nematodos gastroentéricos de bovinos en pasto del trópico. Veterinaria Mexico 11: 110.

Castro, C. E., and I. J. Thomason. 1973. Permeation dynamics and osmo-regulation in *Aphelenchus avenae*. Nematologica 19: 100 – 108.

Castro, C. E., N. O. Belser, H. E. McKinney, and I. J. Thomason. 1989. Quantitative bioassay for chemotaxis with plant parasitic nematodes. Journal of Chemical Ecology 15: 1297 – 1309.

Castro, C. E., N. O. Belser, H. E. McKinney, and I. J. Thomason. 1990. Strong repellency of the root knot nematode, *Meloidogyne incognita* by specific inorganic ions. Journal of Chemical Ecology 16: 1199 – 1205.

Castro, C. E., H. E. McKinney, and S. Lux. 1991. Plant protection with inorganic ions. Journal of Nematology 23: 409 – 413.

Chitwood, B. G., and D. G. Murphy. 1964. Observations on two marine monhysterids-their classification, cultivation, and behavior. Transactions of the American Microscopical Society 83: 311 – 329.

Choo, H. Y., H. K. Kaya, and H. H. Kim. 1995. Biological studies on *Agamermis unka* (Nematoda: Mermithidae), a parasite of the brown planthopper *Nilaparvata lugens*. Biocontrol 5: 209 – 223.

Christie, J. R. 1936. Life history of *Agamermis decaudata*, a nematode parasite of grasshoppers and other insects. Journal of Agricultural Research 52: 161 – 198.

Christie, J. R. 1937. *Mermis subnigrescens*, a nematode parasite of grasshoppers. Journal of Agricultural Research 55: 353 – 364.

Clarke, A. J., and J. Hennessy. 1984. Movement of *Globodera rostochiensis* (Wollenweber) juveniles stimulated by potato-root exudate. Nematologica 30: 206 – 212.

Clemens, C. D., J. Aumann, Y. Spiegel, and U. Wyss. 1994. Attractant-mediated behaviour of mobile stages of *Heterodera schachtii*. Fundamental and Applied Nematology 17: 569 – 574.

Cobb, N. A. 1929. The chromatropism of *Mermis subnigrescens*, a nemic parasite of grasshoppers. Journal of the Washington Academy of Science 19: 159 – 166.

Croll, N. A. 1966. Activity of the orthokinetic response of *Trichonema* larvae in light. Parasitology 56: 307 – 312.

Croll, N. A. 1967a. Acclimatization in the eccritic thermal response of *Ditylenchus dipsaci*. Nematologica 13: 385 – 389.

Croll, N. A. 1967b. The mechanism of orientation in nematodes. Nematologica 13: 17 – 22.

Croll, N. A. 1970. The Behavior of Nematodes: Their activity, senses, and responses. New York: St. Martin's Press.

Croll, N. A. 1971. Movement patterns and photosensitivity of *Trichonema* spp. infective larvae in non-directional light. Parasitology 62: 467 – 478.

Croll, N. A. 1972a. Behavioral activities of nematodes. Helminthological Abstracts A41: 359.

Croll, N. A. 1972b. Behavior of larval nematodes. Pp. 31 –52. In E. U. Canning, and C. A. Wright, eds. Behavioural Aspects of Parasite Transmission. London: Academic Press.

Croll, N. A. 1975a. Behavioral analysis of nematode movement. Pp. 71 – 122. In B. Dawes, ed. Advances in Parasitology. New York: Academic Press.

Croll, N. A. 1975b. Components and Patterns in the behaviour of the nematode *Caenorhabditis elegans*. Journal of Zoology London 176: 159 –176.

Croll, N. A. 1976a. Behavioral coordination of nematodes. Pp. 343 – 364. In N. A. Croll, ed. The Organization of Nematodes. London: Academic Press.

Croll, N. A. 1976b. When *Caenorhabditis elegans* (Nematoda: Rhabditidae) bumps into a bead. Canadian Journal of Zoology 54: 566 –570.

Croll, N. A., and J. M. Smith. 1972. Mechanism of thermopositive behavior in larval hookworms. Journal of Parasitology 58: 891 –896.

Croll, N. A., and M. V. K. Sukhdeo. 1981. Hierarchies in nematode behavior. Pp. 227 –251. In B. M. Zuckerman, and R. A. Rohde, eds. Plant Parasitic Nematodes, Volume III. New York: Academic Press.

Croll, N. A., I. L. Riding, and J. M. Smith. 1972. A nematode photoreceptor. Comparative Biochemistry and Physiology 42a: 999 – 1009.

Croll, N. A., A. A. F. Evans, and J. M. Smith. 1975. Comparative nematode photoreceptors. Comparative Biochemistry and Physiology 51A: 139 – 143.

Culloti, J. G., and R. L. Russell. 1978. Osmotic avoidance defective mutants of the nematode *Caenorhabditis elegans*. Genetics 90: 243 –256.

Delgado, I. F., A. C. Nogueira, C. A. Souza, M. A. Costa, L. H. Figuierdo, I. Chahoud, and F. J. Paumgartten. 1993. Peri- and post-natal development toxicity of β-myrcene in the rat. Food Chemistry and Toxicology 31: 623 –628.

Dickson, D. W., and T. E. Hewlett. 1986. Vertical migration of *Meloidogyne arenaria* in soil colums in the field. Revue de Nematologie 9: 295 (abstr.).

Diehn, B., M. Feinleib, W. Haupt, E. Hildebrand, F. Lenci, and W. Nultsch. 1977. Terminology of behavioral responses of motile microorganisms. Photochemistry and Photobiology 26: 559 –560.

Diez, J. A., and D. B. Dusenbery. 1989a. Preferred temperature of *Meloidogyne incognita*. Journal of Nematology 21: 99 – 104.

Diez, J. A., and D. B. Dusenbery. 1989b. Repellent of root-knot nematodes from exudate of host roots. Journal of Chemical Ecology 15: 2445 –2455.

Doane, J. F., Y. W. Lee, and N. D. Westcot. 1976. Carbon dioxide-a potent attractant for wireworms (*Ctenicera destructor*, field crop pest). Canadian Agriculture 21: 12 – 14.

Doncaster, C. C., and M. K. Seymour. 1973. Exploration and selection of penetration site by Tylenchida. Nematologica 19: 137 – 145.

Downes, M. J., and C. T. Griffin. 1996. Dispersal behaviour and transmission strategies of the entomopathogenic nematodes *Heterorhabditis* and *Steinernema*. Biocontrol 6: 347 –356.

Driscoll, M., and J. Kaplan. 1997. Mechanotransduction. Pp. 645 – 678. In D. L. Riddle, T. Blumenthal, B. J. Meyer, and J. R. Priess, eds. *C. elegans* II. New York: Cold Spring Harbor Laboratory Press.

Dropkin, V. H. 1965. Behavior of nematodes in relation to that of other invertebrates. Nematologica 11: 1 – 3.

Dusenbery, D. B. 1973. Countercurrent separation: A new method for studying behavior of small aquatic organisms. Proceedings of the National Academy of Sciences USA 70: 1349 – 1352.

Dusenbery, D. B. 1980a. Behavior of free-living nematodes. Pp. 127 – 158. In B. M. Zuckerman, ed. Nematodes as Biological Models. New York: Academic Press.

Dusenbery, D. B. 1980b. Responses of the nematode *Caenorhabditis elegans* to controlled chemical stimulation. Journal of Comparative Physiology 136: 327 – 331.

Dusenbery, D. B. 1983. Chemotactic behavior of nematodes. Journal of Nematology 15: 168 – 173.

Dusenbery, D. B. 1985. Video camera-computer tracking of nematode *Caenorhabditis elegans* to record behavioral responses. Journal of Chemical Ecology 11: 1239 – 1247.

Dusenbery, D. B. 1987. Theoretical range over which bacteria and nematodes locate plant roots using carbon dioxide. Journal of Chemical Ecology 13: 1617 – 1625.

Dusenbery, D. B. 1988a. Avoided temperature leads to the surface: Computer modeling of slime mold and nematode thermotaxis. Behavioral Ecology and Sociobiology 22: 219 – 223.

Dusenbery, D. B. 1988b. Behavioral responses of *Meloidogyne incognita* to small temperature changes. Journal of Nematology 20: 351 – 355.

Dusenbery, D. B. 1988c. Limits of thermal sensation. Journal of Theoretical Biology 131: 263 – 271.

Dusenbery, D. B. 1989a. Efficiency and the role of adaptation in klinokinesis. Journal of Theoretical Biology 136: 281 – 293.

Dusenbery, D. B. 1989b. The value of asymmetric signal processing in klinokinesis. Biological Cybernetics 61: 401 – 404.

Dusenbery, D. B. 1989c. A simple animal can use a complex stimulus pattern to find a location: Nematode thermotaxis in soil. Biological Cybernetics 61: 431 – 437.

Dusenbery, D. B. 1992. Sensory Ecology, How Organisms Acquire and Respond to Information. New York: W. H. Freemann and Company.

Dusenbery, D. B., and J. Barr. 1980. Thermal limits and chemotaxis in mutants of the nematode *Caenorhabditis elegans* defective in thermotaxis. Journal of Comparative Physiology Series A 137: 353 – 356.

Edmunds, J. E., and W. F. Mai. 1967. Effect of *Fusarium oxysporum* on movement of *Pratylenchus penetrans* toward alfalfa roots. Phytopathology 57: 468 – 471.

El-Sherif, M., and W. F. Mai. 1969. Thermotactic response of some plant parasitic nematodes. Journal of Nematology 1: 43 – 48.

Emmons, S. W., and P. W. Sternberg. 1997. Male development and mating behavior. Pp. 295 – 334. In D. L. Riddle, T. Blumenthal, B. J. Meyer, and J. R. Priess, eds. *C. elegans* II. New York: Cold Spring Harbor Laboratory Press.

Eveland, L. K., and B. Fried. 1990. Chemoattraction in *Pellioditis pellio* (Nematoda: Secernentia) in the presence and absence of barriers. Journal of Chemical Ecology 16: 2859 – 2863.

Eveland, L. K., and M. A. Haseeb. 1993. *Pellioditis pellio*: Substituted phenolic compounds as chemoattractants. Journal of Chemical Ecology 19: 2543 – 2546.

Fairbairn, J. P., A. Fenton, R. A. Norman, and P. J. Hudson. 1999. The invasion efficiency of the entomopathogenic *Steinernema feltiae* (Rhabditida: Steinernematidae). Aspects of Applied Biology 53: 83 – 88.

Ferguson, C. S., P. C. Schroeder, and E. J. Shields. 1995. Vertical distribution, persistence, and activity of entomopathogenic nematodes (Nematoda: Heterorhabditidae and Steinernematidae) in alfalfa snout beetle (Coleoptra: Curculionidae)-infested fields. Environmental Entomology 24: 146 – 158.

Field, J. I., and J. Webster. 1977. Traps of predacious fungi attract nematodes. Transactions of the British Mycological Society 68: 467 – 469.

Fischer, P., and E. Führer. 1990. Effect of soil acidity on the entomopathogenic nematode *Steinernema kraussei* Steiner. Biology and Fertility of Soils 9: 174 – 177.

Fisher, J. M. 1975. Chemical stimuli for feeding by *Aphelenchus avenae*. Nematologica 21: 358 – 364.

Forster, S. J. 1998. Osmotic stress tolerance and osmoregulation of intertidal and subtidal nematodes. Journal of Experimental Marine Biology and Ecology 224: 109 – 125.

Fraenkel, G. S., and D. L. Gunn. 1961. The Orientation of Animals. London: Constable and Company Limited.

Fülleborn, F. 1914. Untersuchungenüb er de Infectionsweg bei *Strongyloides* und *Ancylostomen* und die Biologie dieser Parasiten. Archiv für Schiffs- und Tropenhygiene 18: 182 – 236.

Futai, K. 1979. Responses of two species of *Bursaphelenchus* to the extracts from pine segments and to the segments immersed in different solvents. Japanese Journal of Nematology 9: 54 – 59.

Futai, K. 1980. Host preference of *Bursaphelenchus lignicolus* (Nematode: Aphelenchoididae) and *Bursaphelenchus mucronatus* shown by their aggregation to pine saps. Applied Entomology and Zoology 15: 193 – 197.

Gans, C., and A. H. J. Burr. 1994. Unique locomotory mechanism of *Mermis nigrescens*, a large nematode that crawls over soil and climbs through vegetation. Journal of Morphology 222: 133 – 148.

Gaugler, R., and J. F. Campbell. 1991. Selection for enhanced host-finding of scarab larvae (Coleoptera: Scarabaeidae) in an entomopathogenic nematode. Environmental Entomology 20: 700 – 706.

Gaugler, R., L. LeBeck, B. Nakagaki, and G. M. Boush. 1980. Orientation of the entomogenous nematode *Neoaplectana carpocapsae* to carbon dioxide. Environmental Entomology 9: 649 – 652.

Gaugler, R., J. F. Campbell, and P. Gupta. 1991. Characterization and basis of enhanced host-finding in a genetically improved strain of *Steinernema carpocapsae*. Journal of Invertebrate Pathology 57: 234 – 241.

Georgis, R., and A. Hom. 1992. Introduction of entomopathogenic nematode products into Latin America and the Caribbean. Nematropica 22: 81 – 98.

Georgis, R., and G. O. Poinar, Jr. 1983. Effect of soil texture on the distribution and infectivity of *Neoaplectana carpocapsae* (Nematoda: Steinernematidae). Journal of Nematology 15: 308 – 311.

Gillies, M. T. 1980. The role of carbon dioxide in host-finding by mosquitoes (Diptera: Culicidae). Bulletin of Entomological Research 70: 525 – 532.

Golden, J. W., and D. L. Riddle. 1982. A pheromone influences larval development in the nematode *Caenorhabditis elegans*. Science 218: 578 – 580.

Golden, J. W., and D. L. Riddle. 1984. The *Caenorhabditis elegans* Dauer larva: Developmental effects of pheromone, food, and temperature. Developmental Biology 102: 368 – 378.

Goode, M., and D. B. Dusenbery. 1985. Behavior of tethered *Meloidogyne incognita*. Journal of Nematology 17: 460 – 464.

Gouge, D. H., A. A. Otto, A. Schirocki, and N. G. M. Hague. 1994. Effects of Steinernematids on the root-knot nematode *Meloidogyne javanica*. Tests of Agrochemicals and Cultivars (Annals of Applied Biology Supplement) 124: 134 – 135.

Granzer, M., and W. Haas. 1991. Host-finding and host recognition of infective *Ancylostoma caninum* larvae. International Journal for Parasitology 21: 429 – 440.

Gray, J. 1953. Undulatory propulsion. Quarterly Journal of Microscopical Science 94: 551 – 578.

Green, C. D. 1977. Simulation of nematode attraction to a point in a flat field. Behaviour 61: 130 – 146.

Green, C. D. 1980. Nematode sex attractants. Helminthological Abstracts, Series A 49: 81 – 93.

Green, C. D., and C. Plumb. 1970. The interrelationships of some *Heterodera* species indicated by the specificity of the male attractants emitted by their females. Nematologica 16: 39 – 46.

Greet, D. N. 1964. Observations on sexual attraction and copulation in the nematode *Panagrolaimus rigidus* (Schneider). Nature 204: 96 – 97.

Greet, D. N., and R. N. Perry. 1992. Nematoda and Nematomorpha. Pp. 147 – 173. In K. G. Adiyodi, and R. G. Adiyodi, eds. Reproduction Biology of Invertebrates. New Delhi: Oxford IBH Publishers.

Greet, D. N., C. D. Green, and M. E. Poulton. 1968. Extraction, standardization, and assessment of the volatility of the sex attractants of *Heterodera rostochiensis* Woll. and *H. schachtii*. Annals of Applied Biology 61: 511 – 519.

Grewal, P. S., and D. J. Wright. 1992. Migration of *Caenorhabditis elegans* (Nematoda: Rhabditidae) larvae towards bacteria and the nature of the bacterial stimulus. Fundamental and Applied Nematology 15: 159 – 166.

Grewal, P. S., R. Gaugler, and E. E. Lewis. 1993a. Host recognition behavior by entomopathogenic nematodes during contact with insect gut contents. Journal of Parasitology 79: 495 – 503.

Grewal, P. S., R. Gaugler, and S. Selvan. 1993b. Host recognition by entomopathogenic nematodes: Behavioral response to contact with host feces. Journal of Chemical Ecology 19: 1219 – 1231.

Grewal, P. S., S. Selvan, E. E. Lewis, and R. Gaugler. 1993c. Male insect-parasitic nematodes: A colonizing sex. Experientia 49: 605 – 608.

Grewal, P. S., E. E. Lewis, R. Gaugler, and J. F. Campbell. 1994a. Host finding behavior as a predictor of foraging strategy in entomopathogenic nematodes. Parasitology 108: 207 – 215.

Grewal, P. S., S. Selvan, and R. Gaugler. 1994b. Thermal adaptation of entomopathogenic nematodes: niche breadth for infection, establishment, and reproduction. Journal of Thermal Biology 19: 245 – 253.

Grewal, P. S., E. E. Lewis, and R. Gaugler. 1997a. Response of infective stage parasites (Nematoda: Steinernematidae) to volatile cues from infected hosts. Journal of Chemical Ecology 23: 503 – 515.

Grewal, P. S., W. R. Martin, R. W. Miller, and E. E. Lewis. 1997b. Suppression of plant-parasitic nematode populations in turfgrass by application of entomopathogenic nematodes. Biocontrol Science and Technology 7: 393 – 399.

Grewal, P. S., E. E. Lewis, and S. Venkatachari. 1999. Allelopathy: A possible mechanism of suppression of plant-parasitic nematodes by entomopathogenic nematodes. Nematology 1: 735 – 743.

Griffin, G. D. 1969. Attractiveness of resistant and susceptible alfalfa to stem and root-knot nematodes. Journal of Nematology 1: 9 (abstr.).

Gunn, D. L. 1975. The meaning of the term "Klinokinesis". Animal Behavior 23: 409 – 412.

Gunn, D. L., J. S. Kennedy, and D. P. Pielou. 1937. Classification of taxes and kineses. Nature 140: 1064.

Gupta, S. P. 1963. Mode of infection and biology of infective larvae of *Molineus barbatus* Chandler, 1942. Experimental Parasitology 13: 252 – 255.

Haseeb, M. A., and B. Fried. 1988. Chemical communication in helminths. Advances in Parasitology 27: 169 – 207.

Hay, D. B., and J. S. Fenlon. 1995. A modified binomial model that describes the infection dynamics of the entomopathogenic nematode *Steinernema feltiae* (Steinernematidae: Nematoda). Parasitology 111: 627 – 633.

Hayakawa, H. 1994. Relationship between the volume of carbon dioxide and the effectiveness in collecting tabanid flies. Japanese Journal of Sanitary Zoology 39: 313 – 315.

Heald, C. M., and A. F. Robinson. 1987. Effects of soil solarization on *Rotylenchulus reniformis* in the Lower Rio Grande Valley of Texas. Journal of Nematology 19: 93 – 103.

Hedgecock, E. M., and R. L. Russell. 1975. Normal and mutant thermotaxis in the nematode *Caenorhabditis elegans*. Proceedings of the National Academy of Sciences of the United States of America 72: 4061 – 4065.

Hewlett, T. E., E. M. Hewlett, and D. W. Dickson. 1997. Response of *Meloidgyne* spp., *Heterodera glycines*, and *Radopholus similis* to tannic acid. Supplement to Journal of Nematology 29: 737 – 741.

Hribar, L. J., D. J. Leprince, and L. D. Foil. 1992. Ammonia as an attractant for adult *Hybomitra lasiophthalma* (Diptera: Tabanidae). Journal of Medical Entomology 29: 346 – 348.

Hsiao, W.-F., and J. N. All. 1996. Effects of temperature and placement site on the dispersal of the entomopathogenic nematode, *Steinernema carpocapsae* in four soils. Chinese Journal of Entomology 16: 95 – 106.

Huang, X., and J. O. Becker. 1999. Life cycle and mating behavior of *Belonolaimus longicaudatus* in gnotobiotic culture. Journal of Nematology 31: 70 – 74.

Huettel, R. N. 1986. Chemical communicators in nematodes. Journal of Nematology 18: 3 – 8.

Huettel, R. N., and S. L. F. Meyer. 1992. Novel management strategies for plant parasitic nematodes. Pp. 273 – 276. In E. S. Tjamos, ed. Biological Control for Plant Diseases. New York: Plenum Press.

Huettel, R. N., D. W. Dickson, and D. T. Kaplan. 1982. Sex attractants and behavior in the two races of *Radopholus similis*. Nematologica 28: 360 – 369.

Ishibashi, N., and E. Kondo. 1986. *Steinernema feltiae* (DD-136) and *S. glaseri*: Persistence in soil and bark compost and their influence on native nematodes. Journal of Nematology 18: 310 – 316.

Ishibashi, N., and E. Kondo. 1987. Dynamics of the entomogenous nematode *Steinernema feltiae* applied to soil with and without nematicide treatment. Journal of Nematology 19: 404 – 412.

Ishibashi, N., and E. Kondo. 1990. Behavior of infective juveniles. Pp. 139 – 150. In R. Gaugler, and H. K. Kaya, eds. Entomopathogenic Nematodes in Biological Control. Boca Raton, FL: CRC Press.

Ishibashi, N., X. D. Wang, and E. Kondo. 1995. *Stienernema carpocapsae*: Post-storage infectivity and sex ratio of invading infective juveniles. Japanese Journal of Nematology 24: 60 – 68.

Jaffe, H., R. N. Huettel, A. B. DeMilo, D. K. Hayes, and R. V. Rebois. 1989. Isolation and identification of a compound from soybean cyst nematode, *Heterodera glycines*, with sex pheromone activity. Journal of Chemical Ecology 15: 2031 – 2043.

Jaffee, B. A. 1998. Susceptibility of a cyst and a root-knot nematode to three nematode-trapping fungi. Fundamental and Applied Nematology 21: 695 – 703.

Jaffee, B. A. 1999. Enchytraeids and nematophagous fungi in tomato fields and vineyards. Phytopathology 89: 398 – 406.

Jaffee, B. A. 2000. Augmentation of soil with the nematophagous fungi *Hirsutella rhossiliensis* and *Arthrobotrys haptotyla*. Phytopathology 90: 498 – 504.

Jaffee, B. A., and A. E. Muldoon. 1997. Suppression of the root-knot nematode *Meloidogyne javanica* by alginate pellets containing the nematophagous fungi *Hirsutella rhossiliensis*, *Monacrosporium cionopagum*, and *M. ellipsosporum*. Biocontrol Science and Technology 7: 203 – 217.

Jaffee, B. A., P. F. Santos, and A. E. Muldoon. 1997. Suppression of nematophagous fungi by enchytraeid worms: A field exclosure experiment. Oecologia 112: 412 – 423.

Jaffee, B. A., H. Ferris, and K. M. Scow. 1998. Nematode-trapping fungi in organic and conventional cropping systems. Phytopathology 88: 344 – 350.

Jansson, H. -B. 1982a. Attraction of nematodes to endoparasitic nematophagous fungi. Transactions of the British Mycological Society 79: 25 – 29.

Jansson, H. -B. 1982b. Predacity by nematophagous fungi and its relation to the attraction of nematodes. Microbial Ecology 8: 233 – 240.

Jansson, H., and B. Nordbring-Hertz. 1979. Attraction of nematodes to living mycelium of nematophagous fungi. Journal of General Microbiology 112: 89 – 93.

Jansson, H. -B., and B. Nordbring-Hertz. 1980. Interactions between nematophagous fungi and plant-parasitic nematodes: Attraction, induction of trap formation and capture. Nematologica 26: 383 – 389.

Jansson, H. -B., and B. Nordbring-Hertz. 1984. Involvement of sialic acid in nematode chemotaxis and infection by an endoparasitic nematophagous fungus. Journal of General

Microbiology 130: 39 – 43.

Jansson, H. -B., and B. Nordbring-Hertz. 1988. Infection events in the fungus-nematode system. Pp. 59 – 72. In G. O. Poinar, and H. -B. Jansson, eds. Diseases of Nematodes, Volume II. Boca Raton, FL: CRC Press.

Johnson, P. W., and C. D. McKeen. 1973. Vertical movement and distribution of *Meloidogyne incognita* (Nematoda) under tomato in a sandy loam greenhouse soil. Canadian Journal of Plant Science 53: 837 – 841.

Johnson, R. N., and D. R. Viglierchio. 1961. The accumulation of plant parasitic nematode larvae around carbon dioxide and oxygen. Proceedings of the Helminthological Society of Washington 28: 171 – 174.

Jones, J. G. W. 1960. Some observations and reflections on host finding by plant nematodes. Mededelingen LandbHoogeschool Gent 25: 1009 – 1024.

Jones, J. T., R. N. Perry, and M. R. L. Johnston. 1991. Electrophysiological recordings of electrical activity and responses to stimulants from *Globodera rostochiensis* and *Syngamus trachea*. Revue de Nematologie 14: 467 – 473.

Jorgensen, E. M., and C. Rankin. 1997. Neural plasticity. Pp. 769 – 790. In D. L. Riddle, T. Blumenthal, B. J. Meyer, and J. R. Priess, eds. *C. elegans* II. New York: Cold Spring Harbor Laboratory Press.

Kamionek, M., E. Pezowicz, and G. Weclewska. 1997. The effect of ecological conditions on the behaviour of entomopathogenic nematodes (Steinernematidae) in the soil environment. Annals of Warsaw Agricultural University 33: 3 – 13.

Kaplan, D. T., and N. T. Keen. 1980. Mechanisms conferring plant incompatibility to nematodes. Revue de Nematologie 3: 123 – 134.

Karlson, P., and A. Butenandt. 1959. Pheromones (ectohormones) in insects. Annual Review of Entomology 4: 39 – 58.

Khalil, M. 1922. Thermotropism in ankylostome larvae. Proceedings of the Royal Society of Medicine 15: 16 – 18.

Khan, M. A. A., A. Hasan, and A. M. Khan. 1985. Movement and locomotion of *Tylenchorhynchus brassicae* Siddiqi, 1961 on water agar medium. Indian Journal of Nematology 15: 30 – 32.

Khlibsuwan, W., N. Ishibashi, and E. Kondo. 1992. Response of *Steinernema carpocapsae* infective juveniles to the plasma of three insect species. Journal of Nematology 24: 156 – 159.

Kito, K., and W. D. Hope. 1999. *Leptosomatides brevicaudatus* n. sp. and a redescription of *Leptosomatides marinae* Platonova, 1967 (Enoplida: Leptosomatidae). Journal of Nematology 31: 460 – 474.

Kiyohara, T. 1982. Sexual attraction in *Bursaphelenchus xylophilus*. Japanese Journal of Nematology 11: 7 – 12.

Klingler, J. 1958. Die Bedeutung der Kohlendioxyd-Ausscheidung der Wurzeln für die Orientierung der Larven von *Otiorrhynchus sulcatus* F. und anderer bodenbewohnender phytophager Insektenarten. Bulletin de la Societe Entomologique Suisse 31: 205 – 269.

Klingler, J. 1959. Anziehung von Collembolen und Nematoden durch Kohlendioxyd-quellen. Mitteilungen der Schweizerischen Entomologischen Gesellschaft Bulletin de la Societe Entomologique Swisse 2: 311 – 316.

Klingler, J. 1961. Anziehungsversuche mit *Ditylenchus dipsaci* unter Berucksichtigung der

Wirkung des Kohlendioxyds, des Redoxpotentials und anderer Faktoren. Nematologica 6:69 – 84.

Klingler, J. 1963. Die Orientierung von *Ditylenchus dipsaci* in gemessenen kunstilichen und biologischen CO2-Gradienten. Nematologica 9:185 – 199.

Klingler, J. 1965. On the orientation of plant nematodes and of some other soil animals. Nematologica 11:4 – 18.

Klingler, J. 1970. The reaction of *Aphelenchoides fragariae* to slit-like micro-openings and to stomatal diffusion gases. Nematologica 16:417 – 422.

Klingler, J. 1972. The effect of single and combined heat and CO_2 stimuli at different ambient temperatures on the behavior of two plant-parasitic nematodes. Journal of Nematology 4:95 – 100.

Klink, J. W., V. H. Dropkin, and J. E. Mitchell. 1970. Studies on the host finding mechanisms of *Neotylenchus linfordi*. Journal of Nematology 2:106 – 117.

Kondo, E., and N. Ishibashi. 1985. Effects of soil moisture on the survival and infectivity of the entomogenous nematode, *Steinernema feltiae* (DD-136). Proceedings of the Kyushu Plant Protection Association 31:186 – 190.

Kondo, E., and N. Ishibashi. 1986. Nictating behavior and infectivity of the entomogenous nematode, *Steinernema* spp., to the larvae of common cutworm, *Spodoptera litura* (Lepidoptera: Noctuidae), on the soil surface. Applied Entomology and Zoology 21: 553.

Koppenhöfer, A. M., and H. K. Kaya. 1996. Coexisting of entomopathogenic nematode species (Steinernematidae and Heterorhabditidae) with different foraging behavior. Fundamental and Applied Nematology 19:175 – 183.

Koppenhöfer, A. M., M. E. Baur, and H. K. Kaya. 1996. Competition between two Steinernematid nematode species for an insect host at different soil depths. Journal of Parasitology 82:34 – 40.

Koppenhöfer, A. M., J. F. Campbell, H. K. Kaya, and R. Gaugler. 1998. Estimation of entomopathogenic nematode population density in soil by correlation between bait insect mortality and nematode penetration. Fundamental and Applied Nematology 21: 95 – 102.

Kramer, P. J. 1983. Water Relations of Plants. Orlando, FL: Academic Press.

Kühn, H. 1959. Zum Problem der Wirtsfindung phytopathogener Nematoden. Nematologica 4:165 – 171.

Lackey, B. A., A. E. Muldoon, and B. A. Jaffee. 1993. Alginate pellet formulation of *Hirsutella rhossiliensis* for biological control of plant-parasitic nematodes. Biological Control 3:155 – 160.

Lee, Y.-B., and A. A. F. Evans. 1973. Correlation between attractions and susceptibilities of rice varieties to *Aphelenchoides besseyi* Christie, 1942. Korean Journal of Plant Protection 12:147 – 151.

Lewis, E. E., and R. Gaugler. 1994. Entomopathogenic nematode (Rhabditida: Steinernematidae) sex ratio relates to foraging strategy. Journal of Invertebrate Pathology 64:238 – 242.

Lewis, E. E., P. S. Grewal, and R. Gaugler. 1995a. Heirarchical order of host cues in parasite foraging strategies. Parasitology 110:207 – 213.

Lewis, E. E., S. Selvan, J. F. Campbell, and R. Gaugler. 1995b. Changes in foraging behaviour during the infective stage of entomopathogenic nematodes. Parasitology 110: 583 – 590.

Lewis, E. E., M. Ricci, and R. Gaugler. 1996. Host recognition behaviour predicts host suitability in the entomopathogenic nematode *Steinernema carpocapsae* (Rhabditida: Steinernematidae). Parasitology 113: 573 – 579.

Lewis, E. E., J. F. Campbell, and R. Gaugler. 1997. The effects of aging on the foraging behaviour of *Steinernema carpocapsae* (Rhabditida: Steinernematidae). Nematologica 43: 355 – 362.

Lindskog, S., L. E. Henderson, K. K. Kannan, A. Liljas, P. O. Nyman, and B. Strandberg. 1971. Carbonic anhydrase. Pp. 587 – 665. In P. D. Boyer, ed. The Enzymes, Volume 5. New York: Academic Press.

Linford, M. B. 1939. Attractiveness of roots and excised shoot tissues to certain nematodes. Proceedings of the Helminthological Society of Washington 6: 11 – 18.

Looss, A. 1905. Wanderung der *Ancylostomum-* und *Strongyloides* Larven von der Haut nach dem Darm. Berne, 1904: Comptes Rendus de 63 Congrés International de Zoologie, Pp. 225 – 233.

Lorente, I., M. A. Ocete, A. Zarzuelo, M. M. Cabo, and J. Jimenez. 1989. Bioactivity of the essential oil of *Bupleurum fructicosum*. Journal of Natural Products 52: 267 – 272.

Lownsbery, B. F., and D. R. Viglierchio. 1960. Mechanism of accumulation of *Meloidogyne incognita acrita* around tomato seedlings. Phytopathology 50: 178 – 179.

Maggenti, A. 1981. General Nematology. New York: Springer-Verlag.

Marbán-Mendoza, N., A. Jeyaprakash, H.-B. Jansson, H. A. Damon, and B. M. Zuckerman. 1987. Control of root-knot nematode on tomato by lectins. Journal of Nematology 19: 331 – 335.

Marks, C. F., I. J. Thomason, and C. E. Castro. 1968. Dymanics of the permeation of nematodes by water, nematocides and other substances. Experimental Parasitology 22: 321 – 337.

McCallum, M. E., and D. B. Dusenbery. 1992. Computer tracking as a behavioral GC detector: Nematode responses to vapor of host roots. Journal of Chemical Ecology 18: 585 – 592.

McClure, M. A., and J. Robertson. 1973. Infection of cotton seedlings by *Meloidogyne incognita* and a method of producing uniformly infected root segments. Nematologica 19: 428 – 434.

McCue, J. F., and R. E. Thorson. 1964. Behavior of parasitic stages of helminths in a thermal gradient. Journal of Parasitology 50: 67 – 71.

Meyer, S. L. F., and R. N. Huettel. 1996. Application of a sex pheromone, pheromone analogs, and *Verticillium lecanii* for management of *Heterodera glycines*. Journal of Nematology 28: 36 – 42.

Meyer, S. L. F., G. Johnson, M. Dimock, J. W. Fahey, and R. N. Huettel. 1997. Field efficacy of *Verticillium lecanii*, sex pheromone, and pheromone analogs as potential management agents for soybean cyst nematode. Journal of Nematology 29: 282 – 288.

Milburn, J. A. 1979. Water Flow in Plants. London: Longman Group.

Miyazaki, M., A. Yamaguchi, and K. Oda. 1978a. Behaviour of *Bursaphelenchus lignicolus* in response to carbon dioxide. Journal of the Japanese Forestry Society 60: 203 – 208.

Miyazaki, M., A. Yamaguchi, and K. Oda. 1978b. Behaviour of *Bursaphelenchus lignicolus* in response to carbon dioxide released by respiration of *Monochamus alternatus* pupa. Journal of the Japanese Forestry Society 60: 249 – 254.

Mok, M., D. Abraham, R. B. Grieve, and C. B. Thomas. 1986. Thermotaxis in third and fourth-stage *Dirofilaria immitis* larvae. Journal of Helminthology 60: 61 – 64.

Monoson, H. L., A. G. Galsky, J. A. Griffin, and E. J. McGrath. 1973. Evidence for and partial characterization of a nematode attraction substance. Mycologia 65: 78 – 86.

Mori, I., and Y. Ohshima. 1995. Neural regulation of thermotaxis in *Caenorhabditis elegans*. Nature 376: 344 – 348.

Needham, T. 1743. A letter from Mister Turbevil Needham to the president; concerning certain chalky tubulous concretions, called malm: with some observations on the farina of the red lily, and of worms discovered in smutty corn. Philosophical Transactions of the Royal Society (London) B 42: 634 – 641.

Neves, J. M., N. Simoes, and M. Mota. 1998. Evidence for a sex pheromone in *Steinernema carpocapsae*. Nematologica 44: 95 – 98.

Nordbring-Hertz, B. 1988. Ecology and recognition in the nematode-nematophagous fungus system. Advances in Microbial Ecology 10: 81 – 114.

Nye, P. H. 1981. Changes of pH across the rhizosphere induced by plants. Plant and Soil 61: 7 – 26.

O'Leary, S. A., C. M. Stack, M. A. Chubb, and A. M. Burnell. 1998. The effect of day of emergence from the insect cadaver on the behavior and environmental tolerances of infective juveniles of the entomopathogenic nematode *Heterorhabditis megidis* (strain UK211). Journal of Parasitology 84: 665 – 672.

Papendick, R. I., and G. S. Campbell. 1981. Theory and measurement of water potential. Pp. 1 – 22. In D. M. Kral, and M. K. Cousin, eds. Water Potential Relations in Soil Microbiology. Madison, WI: Soil Science Society of America.

Parker, J. C., and A. J. Haley. 1960. Phototactic and thermotactic responses to the filariform larvae of the rat nematode *Nippostrongylus muris*. Experimental Parasitology 9: 92 – 97.

Patel, M. N., and D. J. Wright. 1996. The influence of neuroactive pesticides on the behaviour of entomopathogenic nematodes. Journal of Helminthology 70: 53 – 61.

Perry, R. N. 1994. Studies on nematode sensory perception as a basis for novel control strategies. Fundamental and Applied Nematology 17: 199 – 202.

Perry, R. N. 1996. Chemoreception in plant parasitic nematodes. Annual Review of Phytopathology 34: 181 – 199.

Perry, R. N. 1997. Plant signals in nematode hatching and attraction. Pp. 38 – 50. In C. Fenoll, F. M. W. Grundler, and S. A. Ohl, eds. Cellular and Molecular Aspects of Plant-nematode Interactions. Dordrecht: Kluwer Academic Publishers.

Perry, R. N., and J. Aumann. 1998. Behaviour and sensory responses. Pp. 75 – 102. In R. N. Perry, and D. J. Wright, eds. The Physiology and Biochemistry of Free-living and Plant-parasitic Nematodes. Wallingford, UK: CAB International.

Perry, R. N., and E. Riga. 1995. Electrophysiological analysis of sensory responses of parasitic nematodes. Japanese Journal of Nematology 25: 61 – 69.

Perry, R. N., W. M. Hominick, J. Beane, and B. Briscoe. 1998. Effects of the entomopathogenic nematodes, *Steinernema feltiae* and *S. carpocapsae*, on the potato cyst nematode, *Globodera rostochiensis*, in pot trials. Biocontrol Science and Technology 8: 175 – 180.

Pervez, R., and A. L. Bilgrami. 2000. Some factors influencing chemoattraction behaviour of dorylaim predators, *Laimydorus baldus* and *Discolaimus major*, towards prey kairomones. International Journal of Nematology 10: 41 – 48.

Peters, B. G. 1928. On the bionomics of the vinegar eelworm. Journal of Helminthology 4: 1 – 38.

Peters, B. G. 1952. Toxicity tests with vinegar eelworms. I. Counting and culturing. Journal of Helminthology 26: 97 – 110.

Petersen, J. J. 1985. Nematodes as biological control agents: Part I. Mermithidae. Advances in Parasitology 24: 307 – 344.

Pierce-Shimomura, J. T., T. M. Morse, and S. R. Lockery. 1999. The fundamental role of pirouettes in *Caenorhabditis elegans* chemotaxis. Journal of Neuroscience 19: 9557 – 9569.

Pinkerton, J. N., H. Mojtahedi, G. S. Santo, and J. H. O'Bannon. 1987. Vertical migration of *Meloidogyne chitwoodi* and *Meloidogyne hapla* under controlled temperature. Journal of Nematology 19: 152 – 157.

Pline, M., and D. B. Dusenbery. 1987. Responses of plant-parasitic nematode *Meloidogyne incognita* to carbon dioxide determined by video camera- computer tracking. Journal of Chemical Ecology 13: 873 – 888.

Pline, M., J. A. Diez, and D. B. Dusenbery. 1988. Extremely sensitive thermotaxis of the nematode *Meloidogyne incognita*. Journal of Nematology 20: 605 – 608.

Poinar, G. O., Jr. 1990. Biology and taxonomy of Steinernematidae and Heterorhabditidae. Pp. 23 – 61. In R. Gaugler, and H. K. Kaya, eds. Nematodes in Biological Control. Boca Raton, FL: CRC Press.

Popiel, I. 1992. Nematodes as biological control agents: Part II. Advances in Parasitology 31: 381 – 433.

Prot, J. 1975. Recherches concernant le deplacement des juveniles de *Meloidogyne* species vers les racines. Cahiers ORSTOM serie Biologie 3: 251 – 262.

Prot, J. -C. 1976. Amplitude et cinetique des migrations du nematode *Meloidogyne javanica* sous l'influence d'un plant de tomate. Cahiers ORSTOM serie Biologie 11: 157 – 166.

Prot, J. -C. 1978a. Influence of concentration gradients of salts on the movement of second stage juveniles of *Meloidogyne javanica*. Revue de Nematologie 1: 21 – 26.

Prot, J. -C. 1978b. Vertical migration of four natural populations of *Meloidogyne*. Revue de Nematologie 1: 109 – 112.

Prot, J. -C. 1978c. Behavior of juveniles of *Meloidogyne javanica* in salt gradients. Revue de Nematologie 1: 135 – 142.

Prot, J. -C. 1979a. Influence of concentration gradients of salts on the behaviour of four plant parasitic nematodes. Revue de Nematologie 2: 11 – 16.

Prot, J. -C. 1979b. Horizontal migrations of second-stage juveniles of *Meloidogyne javanica* in sand in concentration gradients of salts and in a moisture gradient. Revue de Nematologie 2: 17 – 21.

Prot, J. 1980. Migration of plant-parasitic nematodes towards plant roots. Revue de

Nematologie 3: 305 – 318.
Prot, J.-C., and C. Netscher. 1978. Improved detection of low population densities of *Meloidogyne*. Nematologica 24: 129 – 132.
Prot, J.-C., and S. D. Van Gundy. 1980. Effect of soil texture and the clay component on migration of *Meloidogyne incognita* second-stage juveniles. Journal of Nematology 13: 213 – 217.
Rasch, C., and H. Rembold. 1994. Carbon dioxide – highly attractive signal for larvae of *Helicoverpa armigera*. Naturwissenschaften 81: 228 – 229.
Read, A. F., and A. Skorping. 1995. The evolution of tissue migration by parasitic nematode larvae. Parasitology 111: 359 – 371.
Rees, G. 1950. Observations on the vertical migrations of the third-stage larva of *Haemonchus contorus* (Rud.) on experimental plots of *Lollum perenne* S24, in relation to meteorological and micrometeorlogical factors. Parasitology 40: 127 – 142.
Ricci, M., I. Glazer, J. F. Campbell, and R. Gaugler. 1996. Comparison of bioassays to measure virulence of different entomopathogenic nematodes. Biocontrol 6: 235 – 245.
Riddle, D. L., and A. F. Bird. 1985. Responses of the plant parasitic nematodes *Rotylenchulus reniformis*, *Anguina agrostis* and *Meloidogyne javanica* to chemical attractants. Parasitology 91: 185 – 195.
Riemann, F., and M. Schragge. 1988. Carbon dioxide as *n* attractant for the free living marine nematode *Aconcholaimus thalassophygas*. Marine Biology 98: 81 – 85.
Riga, E., and B. M. MacKinnon. 1987. Sex and aggregation attractants of *Heligmosomoides polygrus* (Nematoda: Trichostrongylidae). Canadian Journal of Zoology 65: 1842 – 1846.
Riga, E., R. N. Perry, and J. Barrett. 1996a. Electrophysiological analysis of the response of males of *Globodera rostochiensis* and *G. pallida* to their female sex pheromones and to potato root diffusate. Nematologica 42: 493 – 498.
Riga, E., R. N. Perry, J. Barrett, and M. R. L. Johnston. 1996b. Electrophysiological responses of males of the potato cyst nematodes, *Globodera rostochiensis* and *G. pallida* to their sex pheromones. Parasitology 112: 239 – 246.
Riga, E., D. R. Holdsworth, R. N. Perry, J. Barrett, and M. R. L. Johnston. 1997a. Electrophysiological analysis of the response of males of the potato cyst nematode, *Globodera rostochiensis*, to fractions of their homospecific sex pheromone. Parasitology 115: 311 – 316.
Riga, E., R. N. Perry, J. Barrett, and M. R. L. Johnston. 1997b. Electrophysiological responses of male potato cyst nematodes, *Globodera rostochiensis* and *G. pallida*, to some chemicals. Journal of Chemical Ecology 23: 417 – 428.
Robacker, D. C., D. S. Moreno, and A. B. DeMilo. 1996. Attractiveness to Mexican fruit flies of combinations of acetic acid with ammonium/amino attractants with emphasis on effects of hunger. Journal of Chemical Ecology 22: 499 – 511.
Roberts, T. M., and R. E. Thorson. 1977. Pairing between adults of *Nippostrongylus brasiliensis* and other species of nematodes *in vitro*. Journal of Parasitology 63: 764 – 766.
Robertson, W. M., and J. M. S. Forrest. 1989. Factors involved in host recognition by plant-parasitic nematodes. Aspects of Applied Biology 22: 129 – 133.
Robinson, A. F. 1989. Thermotactic adaptation in two foliar and two root-parasitic

nematodes. Revue de Nematologie 12: 125 – 131.

Robinson, A. F. 1992. Cambios en la atracción de *Ditylenchus phyllobius* hacia tejidos de *Solanum elaeagnifolium* durante el desarrollo de la planta. Nematropica 22: 37 – 45.

Robinson, A. F. 1994. Movement of five nematode species through sand subjected to natural temperature gradient fluctuations. Journal of Nematology 26: 46 – 58.

Robinson, A. F. 1995. Optimal release rates for attracting *Meloidogyne incognita, Rotylenchulus reniformis*, and other nematodes to carbon dioxide in sand. Journal of Nematology 27: 42 – 50.

Robinson, A. F. 2000. Techniques for studying nematode movement and behavior on physical and chemical gradients. Pp. 1 – 20. In Society of Nematologists Ecology Committee, ed. Methods of Nematode Ecology. Hyattsville, MD: Society of Nematologists.

Robinson, A. F., and W. W. Carter. 1986. Effects of cyanide ion and hypoxia on the volumes of second-stage juveniles of *Meloidogyne incognita* in polyethylene glycol solutions. Journal of Nematology 18: 563 – 570.

Robinson, A. F., and C. M. Heald. 1989. Accelerated movement of nematodes from soil in Baermann funnels with temperature gradients. Journal of Nematology 21: 370 – 378.

Robinson, A. F., and C. M. Heald. 1991. Carbon dioxide and temperature gradients in Baermann funnel extraction of *Rotylenchulus reniformis*. Journal of Nematology 23: 28 – 38.

Robinson, A. F., and C. M. Heald. 1993. Movement of *Rotylenchulus reniformis* through sand and agar in response to temperature, and some observations on vertical descent. Nematologica 39: 92 – 103.

Robinson, A. F., and B. A. Jaffee. 1996. Repulsion of *Meloidogyne incognita* by alginate pellets containing hyphae of *Monacrosporium cionopagum, M. ellipsosporum*, or *Hirsutella rhossiliensis*. Journal of Nematology 28: 133 – 147.

Robinson, A. F., and G. Saldaña. 1989. Characterization and partial purification of attractants for nematode *Orrina phyllobia* from foliage of *Solanum elaeagnifolium*. Journal of Chemical Ecology 15: 481 – 495.

Robinson, A. F., C. C. Orr, and J. R. Abernathy. 1978. Distribution of *Nothanguina phyllobia* and its potential as a biological control agent for silver-leaf nightshade. Journal of Nematology 10: 362 – 366.

Robinson, A. F., C. C. Orr, and J. R. Abernathy. 1979. Behavioral responses of *Nothanguina phyllobia* to selected plant species. Journal of Nematology 11: 73 – 77.

Robinson, A. F., C. C. Orr, and C. E. Heintz. 1984a. Activity and survival of *Orrina phyllobia*: preliminary investigations on the effects of solutes. Journal of Nematology 16: 26 – 30.

Robinson, A. F., C. C. Orr, and C. E. Heintz. 1984b. Effects of the ionic composition and water potential of aqueous solution on the activity and survival of *Orrina phyllobia*. Journal of Nematology 16: 30 – 37.

Robinson, A. F., G. L. Baker, and C. M. Heald. 1990. Transverse phototaxis by infective juveniles of *Agamermis* sp. and *Hexamermis* sp. Journal of Parasitology 76: 147 – 152.

Robinson, A. F., R. N. Inserra, E. P. Caswell-Chen, N. Vovlas, and A. Troccoli. 1997. *Rotylenchulus* species: Identification, distribution, host ranges, and crop plant resistance. Nematropica 27: 127 – 180.

Rode, H. 1962. Untersuchungen über das Wandervermögen von Larven des Kartoffelnematoden (*Heterodera rostochiensis* Woll.) in Modellversuchen mit verschiedenen Bodenarten. Nematologica 7: 74 – 82.

Rode, H. 1969a. Über das Thermopräferendum von Larven des Kartoffelnematoden (*Heterodera rostochiensis* Woll.) unter Berüksichtigung ihrer Wanderaktivitätim Reizgefälle. Pedobiologia 9: 405 – 425.

Rode, H. 1969b. Über Verhalten und Reaktionsempfindlichkeit von Larven des Kartoffelnematoden gegenüber thermischen Reizgefällenim überoptimalen Temperaturbereich. Nematologica 15: 510 – 524.

Rode, H. 1970. Zur Orientierungsweise von Larven des Kartoffelnematoden in Temperaturgeflen. Nematologica 16: 258 – 266.

Rogers, W. P. 1940. The effects of environmental conditions on the accessibility of third stage trichostrongyle larvae to grazing animals. Parasitology 32: 208 – 226.

Rohde, R. A. 1960. The influence of carbon dioxide on respiration of certain plant-parasitic nematodes. Proceedings of the Helminthological Society of Washington 27: 160 – 164.

Rolfe, R. N., J. Barrett, and R. N. Perry. 2000. Analysis of chemosensory responses of second stage juveniles of *Globodera rostochiensis* using electrophysiological techniques. Nematology 2: 523 – 533.

Ronald, K. 1960. The effects of physical stimuli on the larval stage of *Terranova decipiens* (Krabbe, 1878) (Nematoda: Anisakidae). I. Temperature. Canadian Journal of Zoology 38: 623 – 642.

Rutherford, T. A., and N. A. Croll. 1979. Wave forms of *Caenorhabditis elegans* in a chemical attractant and repellent and in thermal gradients. Journal of Nematology 11: 232 – 240.

Ruvkun, G. 1997. Patterning the nervous system. Pp. 543 – 581. In D. L. Riddle, T. Blumenthal, B. J. Meyer, and J. R. Priess, eds. *C. elegans* II. New York: Cold Spring Harbor Laboratory Press.

Samoiloff, M. R., S. Balakanich, and M. Pertovich. 1973. Evidence for the two-state model of nematode behavior. Nature 247: 73 – 74.

Sandstedt, R., and M. L. Schuster. 1962. Liquid trapping of *Meloidogyne incognita* about roots in agar medium. Phytopathology 52: 174 – 175.

Sandstedt, R., T. Sullivan, and M. L. Schuster. 1961. Nematode tracks in the study of movement of *Meloidogyne incognita*. Nematologica 6: 261 – 265.

Scheres, B., P. C. Sijmons, C. Van den Berg, G. DeVrieze, V. Willemsen, and H. Wolkenfelt. 1997. Root anatomy and development, the basis for nematode parasitism. Pp. 25 – 37. In C. Fenoll, F. M. W. Grundler, and S. A. Ohl, eds. Cellular and Molecular Aspects of Plant-nematode Interactions. Dordrecht: Kluwer Academic Publishers.

Schmidt, J., and J. N. All. 1978. Chemical attraction of *Neoaplectana carpocapsae* (Nematoda: Steinernematidae) to insect larvae. Environmental Entomology 7: 605 – 607.

Schmidt, J., and J. N. All. 1979. Attraction of *Neoaplectana carpocapsae* (Nematoda: Steinernematidae) to common excretory products of insects. Environmental Entomology 8: 55 – 61.

Schmidt-Nielsen, K. 1990. Animal Physiology: Adaptation and Environment. Cambridge: Cambridge University Press.

Schneider, D. 1992. One-hundred years of pheromone research. Naturwissenschaften 79: 241 – 250.

Schroeder, W. J., and J. B. Beaver. 1987. Movement of the entomogenous nematodes of the families Heterorhabditidae and Steinernematidae in soil. Journal of Nematology 19: 257 – 259.

Shapiro, D. I., and I. Glazer. 1996. Comparison of entomopathogenic nematode dispersal from infected hosts versus aqueous suspension. Environmental Entomology 25: 1455 – 1461.

Shuto, Y., and H. Watanabe. 1987. Attractants from a vector, *Monochamus alternatus*, for the pine wood nematode. Agricultural and Biological Chemistry 51: 1457 – 1458.

Siddiqi, M. R. 1986. Tylenchida, Parasites of Plants and Insects. Farnham Royal, UK: Commonwealth Agricultural Bureaux.

Siddiqui, I. A., and D. R. Viglierchio. 1970a. Fine structure of photoreceptors in *Deontostoma californicum*. Journal of Nematology 2: 274 – 276.

Siddiqui, I. A., and D. R. Viglierchio. 1970b. Ultrastructure of photoreceptors in the marine nematode *Deontostoma californicum*. Journal of Ultrastructure Research 32: 558 – 571.

Sijmons, P. C., F. M. W. Grundler, N. von Mende, P. R. Burrows, and U. Wyss. 1991. *Arabidopsis thaliana* as a new model for plant-parasitic nematodes. Plant Journal 1: 245 – 254.

Smitley, D. R., F. W. Warner, and G. W. Bird. 1992. Influence of irrigation and *Heterorhabditis bacteriophora* on plant-parasitic nematodes in turf. Journal of Nematology 24: 637 – 641.

Solomon, A., I. Paperna, I. Glazer, and P. U. Alkon. 1997. Migratory behaviour and desiccation tolerance of protostrongylid nematode first-stage larva. International Journal for Parasitology 27: 1517 – 1522.

Sommerville, R. I., and W. P. Rogers. 1987. The nature and action of host signals. Advances in Parasitology 26: 239 – 293.

Spiegel, Y., and M. A. McClure. 1995. The surface coat of plant-parasitic nematodas: Chemical composition, origin, and biological role – A review. Journal of Nematology 27: 127 – 134.

Spindler, L. A. 1934. Field and laboratory studies on the behavior of the larvae of the swine kidney worm, *Stephanurus dentatus*. USDA Technical Bulletin 405: 1 – 17.

Stamps, W. T., and M. J. Linit. 1998. Chemotactic response of propagative and dispersal forms of the pinewood nematode *Bursaphelenchus xylophilus* to beetle and pine derived compounds. Fundamental and Applied Nematology 21: 243 – 250.

Steiner, G. 1925. The problem of host selection and host specialization of certain plant-infesting nemas and its application in the study of nemic pests. Phytopathology 15: 499 – 534.

Steiner, W. A. 1996. Dispersal and host-finding ability of entomopathogenic nematodes at low temperatures. Nematologica 42: 243 – 261.

Steven, D. M. 1963. The dermal light sense. Biological Reviews 38: 204 – 240.

Stirling, G. R. 1991. Biological Control of Plant Parasitic Nematodes. Wallingford, UK:

CAB International.

Streu, H. T. 1961. Nematodes Associated with Carnations, Bulletin 800. Rutgers, NJ: New Jersey Agricultural Experiment Station.

Stromberg, B. E. 1997. Environmental factors influencing transmission. Veterinary Parasitology 72: 247 – 264.

Sukhdeo, M. V. K. 1997. Earth's third environment: The worm's eye view. Bioscience 47: 141 – 149.

Tada, I., T. Mimori, and M. Nakai. 1979. Migration route of *Strongyloides ratti* in albino rats. Japanese Journal of Parasitology 28: 219 – 227.

Tahseen, Q., and A. L. Bilgrami. 1993. Chemo-oriented sex attraction in a plant parasitic nematode, *Hirschmanniella oryzae*. Indian Journal of Nematology 23: 87 – 91.

Tanada, Y., and H. K. Kaya. 1993. Insect Pathology. New York: Academic Press.

Terrill, W. F., and D. B. Dusenbery. 1996. Threshold chemosensitivity and hypothetical chemoreceptor function of the nematode *Caenorhabditis elegans*. Journal of Chemical Ecology 22: 1463 – 1475.

Thomas, J. H. 1998. Social life and the single nucleotide: Foraging behavior in *C. elegans*. Cell 94: 549 – 550.

Thomas, P. K. 1981. Migration of *Longidorus elongatus*, *Xiphinema diversicaudatum* and *Ditylenchus dipsaci* in soil. Nematologia Mediterranea 9: 75 – 81.

Tinbergen, N. 1951. The Study of Instinct. New York: Oxford University Press.

Tominaga, Y., A. Nagase, Y. Kuwahara, and R. Sugawara. 1983. Behavioral responses of *Bursaphalenchus lignicolus* (Nematoda: Aphelenchoididae) to bitter and pungent substances. Applied Entomolgy and Zoology 18: 106 – 110.

Tominaga, Y., M. Yamamoto, Y. Kuwahara, and R. Sugawara. 1984. Behavioral responses of the pine wood nematode to terpenes. Agricultural and Biological Chemistry 48: 519 – 520.

Townshend, J. L. 1964. Fungus hosts of *Aphelenchus avenae* Bastian, 1965 and *Bursaphelenchus fungivorus* Franklin & Hooper, 1962 and their attractiveness to these nematode species. Canadian Journal of Microbiology 10: 727 – 737.

Trudgill, D. L. 1995a. An assessment of the relevance of thermal time relationships to nematology. Fundamental and Applied Nematology 18: 407 – 417.

Trudgill, D. L. 1995b. Why do tropical poikilothermic organisms tend to have higher threshold temperatures for development than temperate ones? Functional Ecology 9: 136 – 137.

Turpeenniemi, T. A., and H. Hyvärinen. 1996. Structure and role of the renette cell and caudal glands in the nematode *Sphaerolaimus gracilis* (Monhysterida). Journal of Nematology 28: 318 – 327.

Van Gundy, S. D. 1965. Nematode behavior. Nematologica 11: 19 – 32.

Van de Velde, M. C., and A. Coomans. 1988. Ultrastructure of the photoreceptor of *Diplolaimella* sp. (Nematoda). Tissue and Cell 20: 421 – 429.

Viglierchio, D. R., N. A. Croll, and J. H. Gortz. 1969. The physiological response of nematodes to osmotic stress and an osmotic treatment for separating nematodes. Nematologica 15: 15 – 21.

Viglierchio, D. R. 1990. Carbon dioxide sensing by *Panagrellus silusiae* and *Ditylenchus dipsaci*. Revue de Nematologie 13: 425 – 432.

Von Mende, N. 1997. Invasion and migration behavior of sedentary nematodes. Pp. 51 - 64. In C. Fenoll, F. M. W. Grundler, and S. A. Ohl, eds. Cellular and Molecular Aspects of Plant-nematode Interactions. Dordrecht: Kluwer Academic Publishers.

Voss, W. 1930. Beiträge aur Kenntnis der Aelchenkrankheit der Chrysanthemen. Zeitschrift für Parasitenkeit 2: 310 - 356.

Wallace, H. R. 1956. Migration of nematodes. Nature 177: 287 - 288.

Wallace, H. R. 1958a. Movement of eelworms I. The influence of pore size and moisture content of the soil on the migration of larvae of the beet eelworm, *Heterodera schachtii* Schmidt. Annals of Applied Biology 46: 74 - 85.

Wallace, H. R. 1958b. Movement of eelworms II. A comparative study of the movement in soil of *Heterodera schachtii* Schmidt and of *Ditylenchus dipsaci* (Kuhn) Filipjev. Annals of Applied Biology 46: 86 - 94.

Wallace, H. R. 1958c. Movement of eelworms III. The relationship between eelworm length, activity and mobility. Annals of Applied Biology 46: 662 - 668.

Wallace, H. R. 1959a. Movement of eelworms. IV. The influence of water percolation. Annals of Applied Biology 47: 131 - 139.

Wallace, H. R. 1959b. Movement of eelworms. V. Observations of *Aphelenchoides ritzemabosi* (Schwartz, 1912) Steiner, 1932 on florists/chrysanthemums. Annals of Applied Biology 47: 350 - 360.

Wallace, H. R. 1959c. The movement of eelworms in water films. Annals of Applied Biology 47: 366 - 370.

Wallace, H. R. 1960. Movement of eelworms. VI. The influences of soil type, moisture gradients and host plant roots on the migration of the potato-root eelworm *Heterodera rostochiensis* Wollenweber. Annals of Applied Biology 48: 107 - 120.

Wallace, H. R. 1961. The orientation of *Ditylenchus dipsaci* to physical stimuli. Nematologica 6: 222 - 236.

Wallace, H. R. 1966. The influence of moisture stress on the development, hatch and survival of eggs of *Meloidogyne javanica*. Nematologica 12: 57 - 69.

Wallace, H. R. 1968a. The dynamics of nematode movement. Annual Review of Phytopathology 6: 91 - 114.

Wallace, H. R. 1968b. Undulatory locomotion of the plant parasitic nematode, *Meloidogyne javanica*. Parasitology 58: 377 - 391.

Wallace, H. R. 1969. Wave formation by infective larvae of the plant parasitic nematode *Meloidogyne javanica*. Nematologica 15: 65 - 75.

Wallace, H. R., and C. C. Doncaster. 1964. A comparative study of the movement of some microphagous, plant parasitic and animal parasitic nematodes. Parasitology 54: 313.

Wang, Q. L., J. Y. Cui, Z. X. Li, and C. Q. Liu. 1995. Effect of soil moisture on the motion of *Steinernema glaseri* (Nematoda: Steinernematidae). Chinese Journal of Biological Control 11: 150 - 152.

Wang, X. D., and N. Ishibashi. 1998. Effects of precedent infection of entomopathogenic nematodes (Steinernematidae) on the subsequent invasion of infective juveniles. Japanese Journal of Nematology 28: 8 - 15.

Wang, X. D., E. Kondo, and N. Ishibashi. 1997. Infectivity of entomopathogenic nematode, *Steinernema carpocapsae*, as affected by the emergence time from the host cadaver. Japanese Journal of Nematology 27: 1 – 6.

Weast, R. C., M. J. Astle, and W. H. Beyer. 1984. CRC Handbook of Chemistry and Physics. Boca Raton, FL: CRS Press.

Westerman, P. 1997. Comparative vertical migration of insect parasitic nematodes *Heterorhabditis* spp. and *Steinernema* spp. in sand at 9℃. Fundamental and Applied Nematology 20: 405 – 408.

Westerman, P. R. 1998. Penetration of the entomopathogenic nematode *Heterorhabditis* spp. into host insects at 9 and 20℃. Journal of Invertebrate Pathology 72: 197 – 205.

Westerman, P. R., and W. Van der Werf. 1998. Modelling nematode behaviour to identify the most promising traits and strategies to improve biocontrol by *Heterorhabditis* spp. Bulletin OILB/SROP. 21: 193 – 196.

White, N. D. G., R. N. Sinha, D. S. Jayas, and W. E. Muir. 1993. Movement of *Cryptolestes ferrugineus* (Coleoptera: Cucujidae) through carbon dioxide gradients in stored wheat. Journal of Economic Entomology 86: 1846 – 1851.

Wicks, S. R., and C. H. Rankin. 1997. Effects of tap withdrawal response habituation on other withdrawal behaviors: The localization of habituation in the nematode *Caenorhabditis elegans*. Behavioral Neuroscience 111: 342 – 353.

Widdowson, E., C. C. Doncaster, and D. W. Fenwick. 1958. Observations on the development of *Heterodera rostochiensis* in sterile root cultures. Nematologica 3: 308 – 314.

Wieser, W. 1955. The attractiveness of plants to larvae of root-knot nematodes. I. The effect of tomato seedlings and excised roots on *Meloidogyne hapla* Chitwood. Proceedings of the Helminthological Society 22: 106 – 112.

Wieser, W. 1956. The attractiveness of plants to larvae of root-knot nematodes. II. The effect of excised bean, eggplant, and soybean roots on *Meloidogyne hapla* Chitwood. Proceedings of the Helminthological Society 23: 59 – 64.

Wilson, P. A. G. 1982. Roundworm juvenile migration in mammals: The pathways of skin-penetrators reconsidered. Pp. 459 – 485. In E. Meerovitch, ed. Aspects of Parasitology. Montreal: Institute of Parasitology, McGill University.

Windholz, M., S. Budavari, L. Y. Stroumtsos, and M. N. Fertig. 1976. The Merck Index. Rahway, NJ: Merck.

Wright, D. J. 1998. Respiratory physiology, nitrogen excretion and osmotic and ionic regulation. Pp. 103 – 131. In R. N. Perry, and D. J. Wright, eds. The Physiology and Biochemistry of Free-living and Plant-parasitic Nematodes. Wallingford, UK: CAB International.

Wright, D. J., and D. R. Newall. 1976. Nitrogen excretion, osmotic and ionic regulation in nematodes. Pp. 163 – 210. In N. A. Croll, ed. The Organization of Nematodes. London: Academic Press.

Wynn, W. K. 1981. Tropic and taxic responses of pathogens to plants. Annual Review of Phytopathology 19: 237 – 255.

Wyss, U. 1970. Zur Toleranz wandernder Wukrzelnematoden gegenüber zunehmender Austrocknung des Bodens und hohen osmotischen Drücken. Nematologica 16: 63 – 73.

Wyss, U., F. M. W. Grundler, and A. Munch. 1992. The parasitic behaviour of second-stage juveniles of *Meloidogyne incognita* in roots of *Arabidopsis thaliana*. Nematologica 38: 98 – 111.

Wyss, U. 1997. Root parasitic nematodes: An overview. Pp. 5 – 22. In C. Fenoll, F. M. W. Grundler, and S. A. Ohl, eds. Cellular and Molecular Aspects of Plant-nematode Interactions. Dordrecht: Kluwer Academic Publishers.

Yamamoto, A., and K. Ogawa. 1989. Chemistry and commercial production of pheromones and other behaviour-modifying chemicals. Pp. 123 – 148. In A. R. Jutsum, and R. F. S. Gordon, eds. Insect Pheromones in Plant Protection. New York: John Wiley & Sons.

Zeeck, E., J. Hardege, H. Bartels-Hardege, and G. Wesselmann. 1988. Sex pheromone in a marine polychaete: Determination of the chemical structure. Journal of Experimental Zoology 246: 285 – 292.

Zuckerman, B. M. 1983. Hypotheses and possibilities of intervention in nematode chemoresponses. Journal of Nematology 15: 173 – 182.

Zuckerman, B. M., and J. Esnard. 1994. Biocontrol of plant nematodes-current status and hypotheses. Japanese Journal of Nematology 24: 1 – 13.

Zuckerman, B. M., and H.-B. Jansson. 1984. Nematode chemotaxis and possible mechanisms of host/prey recognition. Annual Review of Phytopathology 22: 95 – 113.

Acknowledgements The author wishes to thank A. H. J. Burr, D. B. Dusenbery, P. S. Grewal, and R. N. Perry for critically reviewing an earlier draft of the manuscript, but the author assumes full responsibility for any errors in interpretation and presentation of the literature. The author thanks Steven-Koether for compiling the figures.

8 Background for Nematode Ecology in the 21st Century

Gregor W. Yeates and Brian Boag

8.1 Introduction

The diversity of nematodes in plants, soils, and aquatic habitats has meant that many workers have sought to understand how those species, and their populations, are distributed in space and time. The habitat lists of Micoletzky (1925) represent an early inventory approach, and Winslow's (1960) chapter a compilation of population information. However, it was not until overviews that clearly related distributions to soil physical characteristics and populations to plant growth (Wallace, 1963) and, subsequently, related crop loss to the interaction of physical, nutrient, and nematode stresses on plants (Wallace, 1973), that the significance of integration was apparent. Nielsen (1949) made the first major attempt to relate plant and soil nematodes to ecological processes, and recent developments in this area have been summarized in Coleman and Hendrix (2000). Cadisch and Giller (1997) have reviewed terrestrial decomposition processes as a whole. Recent publications in the general area of nematode ecology include Barker et al. (1998), Bongers and Bongers (1998), Cook and Yeates (1993), Ettema (1998), Yeates (1998, 1999), and Yeates and Bongers (1999). A few researchers (e. g., Bongers and Ferris, 1999; Freckman and Ettema, 1993; Warwick and Clarke, 1998) cover the topic of application of various indices of the nematode fauna as indicators of environmental change.

At the beginning of the 21st century it is appropriate that this overview of nematode ecology should treat nematodes as one group of animals, and relate them to general ecological principles. Only through an integrated understanding can the role of nematodes and their importance in global biological processes, on which our future depends, be related to current and future topics such as biodiversity, climate change, environmental impact assessment, biological production systems, sustainability, and crop production.

This review deals with the resources nematodes use, and the environmental characteristics and variations that affect them and their populations. Nematodes are heterotrophic organisms ultimately dependent on food resources provided by autotrophs. Nematode populations depend on physical access to these resources and may further be limited by temperature or moisture conditions. An integrated approach to nematode ecology means that relevant information is drawn from nematodes of plants, invertebrates, and vertebrates, and from those in soil, freshwater, and marine substrates.

8.2 Nematode Substrates and Food Resources

It is necessary to distinguish between the substrate (medium or habitat) in which nematodes live and the food resource from which they obtain nutrition. The feeding types of plant and soil nematodes as reviewed and classified by Yeates et al. (1993a) focused on mechanisms rather than on the food resources used. Historically, the term "free-living" has been applied to bacterial feeding, fungal feeding, and predacious nematodes in soil (and freshwater and marine substrates). Quiescent and migratory stages of parasites of plants, invertebrates, and vertebrates also have, implicitly or explicitly, been included in this term. Further, many nematode species have two distinct reproductive cycles (e. g., *Deladenus* as a fungal feeder or insect parasite; *Strongyloides* as a bacterial feeder or parasite in the gut mucosa of vertebrates; various Diplogasterida as bacterial feeders or insect parasites) (Anderson, 1992; Bedding, 1973; Poinar, 1975). As current knowledge does not clearly define the resource used by nematode parasites of invertebrates and vertebrates, such nematodes are commonly regarded as using the host as a whole as a resource, rather than specifying the actual tissue, organ or site from which nutrition is obtained.

The degree of association between an animal and its food varies (the mutual benefits of symbiosis, commensalism, or poorly defined associations are not discussed). A loose association is commonly regarded as feeding or grazing (e. g., *Rhabditis* on a bacterial population; sheep on a pasture sward; a caterpillar on leaves of plants); a permanent association between two individuals is regarded as parasitism (e. g., *Heterodera* in a root; *Ascaris* in a vertebrate gut). Grazing may stimulate activity or growth of the microbial or plant resource (Ingham, et al., 1985; Langer, 1990). Leakage from the feeding site of parasites may stimulate other biota (Yeates, et al., 1999b). In other cases the host plants or animals fed on, or parasitized, show reduced growth, ill-thrift, or other reaction; this is regarded as pathogenicity. In many cases the host can support organisms without showing any obvious damage.

This may reflect a highly developed host-parasite relationship, infection by a parasite with low virulence, tolerance (Trudgill, 1986), or a host not subject to other significant stresses.

8.2.1 Substrates Occupied by Nematodes

While nematodes inhabit niches in a range of substrates, free water, appropriate temperature, and other conditions are necessary for them to be active. During their life cycle, individuals of a given species may inhabit different substrates and niches, and the following list indicates some of this complexity:

1. Non-living substrates (mineral soil; marine sediments; freshwater sediments; organic layers; plant residues; animal cadavers)
   – bacterial-feeding nematodes; fungal-feeding nematodes; algal-feeding nematodes; predators; migratory stages of sedentary and endoparasitic plant nematodes; ectoparasitic nematodes; plant-associated nematodes; quiescent, dormant, inactive, inactive phoretic, and migratory stages of parasites of invertebrates and vertebrates.
2. Living substrates
   (1) Plant tissues
   – migratory and sedentary stages of plant-feeding nematodes;
   – microbial feeders (especially bacterial feeders) often occurring in roots and seeds.
   (2) Invertebrates (tissues and body cavities)
   – parasites of invertebrates; may be intermediate hosts of parasites of vertebrates;
   – microbial feeders may be "associates" of invertebrates.
   (3) Vertebrates (tissues and body cavities)
   – parasites of vertebrates; may also be intermediate hosts of parasites of other vertebrates.

8.2.2 Food Resources (Sources of Nutrition) and Nematode Feeding Types

The feeding types used generally follow those of Yeates et al. (1993a, 1993b) and Bongers and Bongers (1998). The following list includes the food source of various groups of nematodes:

1. Microbes (in terrestrial (soils), marine, and freshwater sediments; on plant roots, stems, and leaves; and detrital substrates)
   (1) Bacterial-feeding nematodes } (1 + 2) microbivores
   (2) Fungal (hyphal)-feeding nematodes }

(3) Algal (unicellular eukaryotes)-feeding nematodes

2. Plants

(4) Plant-feeding nematodes (utilize roots or shoots; sometimes pathogenic)

(4a) Plant-parasites

(4aa) Sedentary endoparasites

(4ab) Migratory endoparasites

(4ac) Semi-endoparasites

(4ad) Ectoparasites

(4b) Plant-associated nematodes (typically in rhizosphere; no significant growth reduction)

(4ba) Epidermal and root-hair feeders

3. Invertebrates

(5) Predators

(5a) Ingesters

(5b) Piercers

(6) Parasites (sometimes pathogenic)

4. Vertebrates

(7) Vertebrate parasites (many pathogenic)

5. Non-feeding

(8) Quiescent, dormant, inactive, cryptobiotic, inactive phoretic stages

(9) Migratory stages

8.3 Temporal Distribution

Investigation of the temporal distribution of nematode populations often reveals spatial variability within an area. Historically, the most studies used a sampling pattern to produce a mean nematode population density (Yeates and Bongers, 1999), whereas a few studies of economically important species have analyzed spatial distributions using geostatistical techniques (Schomaker and Been, 1992; Webster and Boag, 1992). However, recent ecological investigations have begun examining micro-distribution of both nematodes and environmental factors (Ettema, et al., 2000; Savin, et al., 2001).

Temporal changes are well documented for certain species, especially cyst nematodes that show one or two distinct reproductive periods, normally when their host plants are actively growing. Examples include the spring and autumn reproductive pulses of *Heterodera trifolii* on white clover in perennial pastures (Yeates, 1973a) and the single reproductive period of *Globodera rostochiensis* on annual potato crops (Brodie, et al., 1993). Microbial feeding and predacious nematodes show a diverse range of temporal patterns (Ettema, et al., 1998; Yeates, 1973b), and in studies of discrete patches of organic matter have demonstrated the marked effects of the spatial distribution of the resource on nematode populations (Sohlenius and Bostrom, 1984; Wasilewska, et al., 1981). Long-term studies may show regular population pulses, reflecting natural or management (e.g., irrigation) conditions. At some times or seasons of the year, certain populations are likely to be less active - these times will commonly be those of lower plant growth, lower moisture, or lower temperature. Sampling and sample analysis should recognize this; unless repeated samples are sufficiently frequent (e.g., less than monthly) to detect reproductive pulses, a single sampling at an appropriate time is probably preferable.

Long-term studies of grazed pastures have demonstrated marked year-to-year variability in nematode populations (McSorley, 1996; Yeates, 1984). Much of the year-to-year variability in populations appears to be related to annual variability in rainfall. The integration of rainfall and temperature to measure nematode activity (Jones, 1975) was a valuable advance. The principle should be applied through the relations of soil moisture and food resources to a wide range of nematodes and habitats.

Although crop rotations have been studied extensively, a factor that has not received significant attention is how long the nematode assemblage in a management system takes to reach some degree of equilibrium. Large pathogenic populations of nematodes have been studied but the survival of small, trace populations that have significance for both parasite epidemiology and maintenance of biodiversity are much less studied. A 7-year study of nematodes under annual (maize) and perennial (asparagus) crops in 0 - 10 cm of a sandy loam (Yeates, et al., 1999c) showed both significant year-to-year variability in many components of the nematode fauna under five management regimes as well as the need to sample nematode populations under management regimes that have been in place for many growing seasons. The range of treatments suggested that the quality of organic matter inputs could confer stability on soil biological properties, including nematode populations (Wardle, et al., 1999b).

Seasonal changes occur in soil microbial communities in fields (Bardgett, et al., 1999a). Some of the variations reported may reflect the impact of nematode grazing; it will also reflect changes in the quantity and quality of the microbial resource available to microbial-feeding nematodes and their predators. Regular estimates of environmental factors, microbial populations, and nematode populations can give information on how populations of organisms are regulated. Experimental data showed a complex interplay of top-down and bottom-up forces controlling populations in decomposer food webs (Wardle et al., 1999a, 1999b). The chronology of nematodes and associated populations needs to be closely followed to understand temporal effects at the finest scale. These can be both at the species and functional group levels (Gunapala, et al., 1998; Savin, et al., 2001; Yeates, et al., 1999a, 1999c). Information on specific developmental rates (Ferris, et al., 1996; Venette and Ferris, 1997), feeding activity, and food preferences (Ruess, et al., 2000; Venette and Ferris, 1998) are keys to such understanding.

8.4 Geographical and Spatial Distribution

The global distribution of nematode species reflects their origin, natural dispersal (whether by plate tectonics, self-dispersal, in air or water, in or on hosts), and human mediated dispersion. The potential distribution of each species is limited by its biological characteristics. Only by embracing the ecological extremes can the true ranges of species be determined. A survey of 14 European countries, for example, showed some longidorids to have a limited geographic range (e. g., *Longidorus macrosoma* to southern England and Spain) while others had a wider distribution (e. g., *L. elongatus* from north of the Arctic circle to the south of Spain) (Alphey and Taylor, 1986).

On a regional or national scale, assessment of the distribution of nematode species should reflect not only the range of land use and vegetation but also climate, topography, and soils. Neher et al. (1998) compared nematode communities of cultivated agricultural soils of three land resource regions in North Carolina (126, 180 km^2) and Nebraska (199, 120 km^2). They found greater differences in the maturity index (MI) and trophic structure between states than among the land resource regions within a state, and recommended that when calibrating maturity and diversity indices for geographic areas, areas no smaller than states be used. However, the variable scale of soil patterns (USDA, 1981), the selection of soils with particular characteristics for agriculture, and the potentially different types of nematode communities found in association with natural and agricultural soils should be considered in extending such work.

On a field and woodlot scale it is essential to sample nematodes adequately from the range of soil texture, drainage, higher plant species diversity, location on each species of higher plant (roots, stems, leaves, inflorescences, epiphytic algae, etc.), distance from plants (e. g., row vs. inter-row), and crop vigor. Whether such subsets are analyzed separately or pooled will depend on the objective of the study; geostatistical techniques may be appropriate for both design and analysis of such data sets (Ettema, et al., 2000; Webster and Boag, 1992). In a field, the "patches" sampled for nematodes may well be on the scale applicable to other groups of organisms and such "common" samples are necessary for the study of interdependence of the various populations (Ettema, et al., 1998; Heneghan and Bolger, 1998; Wardle, et al., 1999a, 1999b). A microscale is probably appropriate for detailed studies of populations and interactions under particular conditions, such as longidorids in cracks (Jones, et al., 1969), and rhabditids in pockets of decomposing organic matter (Sohlenius and Bostrom, 1984; Wasilewska, et al., 1981). Information from a microscale study can be scaled up to field, regional, and national bases; however, the opposite is not appropriate.

8.5 Food Resources vs. Temperature Regimes in Determining Terrestrial Nematode Distribution

Moisture, temperature, and soil type, as well as presence of suitable host plants in the case of plant-parasitic nematodes, are known to be of paramount importance among the factors affecting the distribution of terrestrial nematodes. Of these factors, temperature alters in the most predictable way over the surface of the earth. This means that knowing the temperature requirements of a nematode species should allow its potential world distribution to be predicted (Boag, et al., 1997). Perceptions of the geographic distribution of terrestrial nematodes are largely based on the distribution of plant-pathogenic nematodes and that of their host plants. This reflects the desire to understand their potential distribution, their economic impact, and to have effective quarantine controls.

Conventional knowledge of relationships between nematodes and temperature is embraced in studies such as the degree-days (Jones, 1975), thermal time (Trudgill, 1995), and matching distributions to isotherms (Boag, et al., 1991). Such work must use some measurements of a basal temperature, below which a species, or stage, will not be active or develop. Decreased activity (Boag, 1980), paralysis, and death (Fig. 27 in Wallace, 1963) may follow after exposure to higher (lethal) temperatures; such temperature-induced mortality may be used as a nematode control measure (Wallace, 1963). As there are an increasing number of exceptions to typical, isotherm-related species distribution (even after altitude has been taken into account), this section examines some potential alternative factors controlling nematode distribution on a global scale. Understanding such alternative factors is of particular importance if attempts are made to model potential changes in the distribution of particular plant and soil nematodes with postulated global climate change.

Distribution of organisms is dynamic. Although the historic, natural patterns for nematode taxa may have been modified by human-mediated dispersal, natural geological and climatological events still underpin basic geographic distribution patterns. Klingler et al. (1983) noted a very close correlation between the present day distribution of *Xiphinema diversicaudatum* on the Swiss Plateau and the boundary of glaciation in the last ice age that occurred approximately 10,000 years ago. Conventional wisdom is that *Heterodera* spp. are generally temperate in distribution, *Meloidogyne* spp. more tropical, and *Pratylenchus* spp. more cosmopolitan. For some genera, the presumed center of speciation is apparent (e. g., *Radopholus* in Australia and New Zealand) (Loof, 1991), while in some species and genera morphological groups correspond to vegetation types (e. g., forms of *Criconema* under grasslands vs. forests and of *Nothocriconema* under deciduous trees vs. other vegetation types) (Boag and Orton Williams, 1976; Yeates, et al., 1997b).

Undoubtedly there is also ongoing natural change in the ranges of nematode taxa as existing species modify their natural range in response to past and present changes in climate and vegetation (or other food resources), and as new species evolve. Further, with the ability to biochemically discriminate sympatric sibling species, previously continuous distribution patterns are being fragmented; these may reflect emerging natural species or selection imposed by agricultural plant varieties and their management (Hominick, 1982).

Nematode species illustrating classic human-mediated distribution of plant-pathogenic nematodes include *Globodera rostochiensis* and *G.* pallida, which have been transferred with *Solanum tuberosum* (Evans, et al., 1975), *Heterodera schachtii* with *Beta vulgaris* (Cooke, 1993), and the virus-vector *Xiphinema index* whose recorded distribution closely shadows that of its major host, the grapevine (*Vitis* spp.) (Hunt, 1993; Taylor, 1962). The less host-specific *Longidorus elongatus*, also a vector of plant viruses, has been found (often associated with grapevines) in a number of countries outside Europe, e. g., New Zealand (Boag, et al., 1997). Another nematode that apparently moved with grapes is the predacious mononchid *Anatonchus killicki*. It was described from New Zealand but its natural range appears to be in Spain (Pena Santiago, et al., 1992; Yeates, et al., 1994).

In The Netherlands there are many cases, not generally published, where nematodes have escaped from glasshouses and survived in local pockets (T. Bongers, pers. comm.). Many *Pratylenchus* species occur in temperate climates and mountainous regions in the tropics, while others are widespread in the tropics (Loof, 1991). However, in Norway a population of *Pratylenchus vulnus* has been established on a southern, sun-facing slope (Bongers, 1988). The genetic constitution of the colonising propagules, whether dispersed naturally or human-mediated, leads to founder effects. For economic reasons, knowledge is greatest about plant-pathogenic forms such as the races of *Globodera* spp. Further, once a species starts reproducing in a location there may be rapid selection favoring those nematodes that are most successful in the new location; the development of an early maturing race of *G. rostochiensis* on potatoes in Ayrshire, Scotland (Hominick, 1982) is a good example. Genetically, such selection does not differ from the increasing dominance or emergence of resistance-breaking races of plant-pathogenic nematodes (e. g., selection of *G. rostochiensis* race Ro2, which overcomes the H_1 gene for resistance to *G. rostochiensis,* in New York state; increasing dominance of *G. pallida* over *G. rostochiensis* in the United Kingdom when cultivars with the H_1 gene are grown) (Brodie, 1999; Lane and Trudgill, 1999). Among zoo-parasitic nematodes, it was demonstrated that when sheep were moved to warmer areas of the world, the temperature requirements of microbial feeding and the infective stages of the sheep parasite *Ostertagia circumcincta* increased (Crofton and Whitlock, 1965).

Among plant-feeding nematodes, the most narrow, best-defined distributions appear to occur in host-specific forms; this seems to be the case at both generic and species-specific levels. Among bacterial-feeding nematodes, many genera appear globally distributed, with both *Plectus* and *Panagrolaimus* occurring at many sites in continental Antarctica as well as on all other land masses – their bacterial food resource being cosmopolitan (Freckman and Virginia, 1995; Wharton and Brown, 1989). The distinctive aphelenchid *Aprutides guidettii* was originally described from Italy but has subsequently been reported from Turkey, The Netherlands, and South Australia (Bird and Yeates, 1994; Bongers, 1988). Insect-associated nematodes can be host-specific, with parasites having global distributions similar to that of their range of hosts (e. g., *Bursaphelenchus xylophilus*); potential specificity of a vector arthropod is also a factor. Just as *G. rostochiensis* adapted to local climatic conditions in Scotland, it has been suggested that polyploidy may provide a mechanism for single species of bacterial-feeding, predacious, or plant-parasitic nematode to adapt and exploit otherwise unused food resources in a locality (Yeates, 1986). Certainly various levels of polyploidy occur in *Meloidogyne* (e. g., *M. microcephala*) (Triantophyllou and Hirschmann, 1997). The traditional management tool for control of plant-pathogenic nematodes is crop rotation – removal of the food resource for the problem species.

Thus, while climate-based studies are of clear importance in assessing the potential economic effects of those pest nematode species with human-mediated distribution and climate change, the availability of a suitable food resource is paramount in determining the potential distribution of all nematode species. Given an adequate supply of bacteria, *Plectus* and *Panagrolaimus* species can have a generation spanning the climatically favourable windows in successive Antarctic summers; *G. rostochiensis* can rapidly adapt to early maturing potatoes in Scotland; and *Aprutides guidettii* can maintain populations from arid South Australia to cool, northern Europe. Providing nematodes can survive the most adverse stresses in a locality (e. g., low temperature, extreme aridity, oxygen deprivation – the one year in ten event may prove critical) it will establish a viable population as long as a suitable food resource is available.

8.6 Impact of Substrate Dimensions

It is axiomatic that actively migrating nematodes, taken to be vermiform and of circular cross-section, can only pass through apertures (generally pores in the substrate) that have a diameter equal to or greater than nematode body width. Traditionally, various nematodes have been associated with particular soil textures (Table 8. 1), but this may more reflect the soils in which particular crops have been grown, and the economic imperative to understand the plant-parasitic nematodes involved. While the pioneering work of Wallace showed how this relationship interacted with soil moisture and soil type to affect "the bionomics of free-living stages of zoo-parasitic and phyto-parasitic nematodes" (Wallace, 1961), its importance has not been consistently followed. Basic information on the relationship between water-filled pores (i. e., those through which nematodes can migrate) in a range of substrate textures at various moisture contents is given in Table 8. 2. Freshwater and marine sediments are normally saturated with all pores being available to nematodes. Numerous studies carefully relate marine nematode assemblages to substrate texture (Tietjen, 1989), although texture, latitude, organic inputs, and disturbance interact in such a way that there is no clear overall picture (Boucher, 1990). Moisture content varies in terrestrial matrices (i. e., mineral soils, organic soil horizons, and synthetic rooting media) and with it the ability of nematodes to migrate and reach suitable food resources. For the eight species listed in Table 8. 3, the body width of migratory stages increases from an average of 19 μm in stage 1 to 54 μm in females. However, it should be noted that migratory females of *H. pseudorobustus* and *A. bicaudatus* are narrower than stage 1 of *L. elongatus*. When assessing the ability of nematodes to migrate through the

Table 8.1 Plant and soil nematodes associated with soils of various types in The Netherlands (from Bongers, 1988).

Dune sand	
	Criconema loofi
	Hemicycliophora epicharoides
	Alaimus elongatus
	Longidorus kuiperi
	Tylencholaimus formosus
Fertile ground generally	
	Coslenchus costatus
	Tylenchorhynchus claytoni
	Pratylenchus penetrans
	Criconemella macrodora
	Hemicycliophora triangulum
	Paratylenchus projectus
	Wilsonema otophorum
	Longidorus attenuatus
	Trichodorus similis
	Paratrichodorus pachydermus
Moist fertile ground	
	Macropsothonia xenoplax
Sandy loam	
	Pratylenchus fallax
	Longidorus proximus
Loam	
	Aprutides guidetti
	Pratylenchus neglectus
Sandy clay	
	Paratylenchus tateae
	Trichodorus primitivus
Clay	
	Tylenchorhynchus clarus
	Helicotylenchis varicaudatus
	Pratylenchus pratensis
	Criconema demani
	Paratylenchus goodeyi
	Longidorus macrosoma
	Trichodorus cylindicus
Swampy clay	
	Hirschmanniella gracilis
Brackish ground	
	Hemicycliophora thornei
	Doryllium uniforme
Peat soil	
	Chronogaster boettgeri

substrate and utilize resources, species and stage are highly significant. Comparison of nematode body widths (Table 8.3) with the diameter of water-filled substrate pores (Table 8.2) indicates that once soil moisture tension exceeds 33 kPa, few pores, and their contained resources, are accessible to nematodes.

Table 8.2 Ideal relationship between soil water content and diameter of water-filled pores in soils (derived from McLaren and Cameron, 1996).

Soil water potential (kPa)	Soil water suction		Approximate maximum pore diameter full of water (μm)
	(bars)	(cm H_2O)	
0	0	0	(all pores)
-1	0.01	10	300
-5	0.05	51	60
-10	0.1	102	30[a]
-33	0.33	337	10
-50	0.5	510	6
-100	1	1,020	3[b]
-300	3	3,060	1
-1,500	15	15,300	0.2[c]

[a] Field capacity in many soils

[b] Readily available water for plant growth is between 10 and 100 kPa (inclusive)

[c] Wilting point in many soils

Table 8.3 Typical body widths (μm) of the developmental stages and females of eight nematode species associated with sediments, soils or plants (data from various sources).

Stage	*Steineria piolosa*	*Panagrellus redivivus*	*Strongyloides thylacis* (bacterivorous)	*Diplenteron colobocercus*	*Meloidogyne incognita*	*Helicotylenchus pseudorobustus*	*Aphelenchoides bicaudatus*	*Longidorus elongatus*
	Monhystera	Rhabditida	Rhabditida	Diplogasterida	Tylenchida	Tylenchida	Aphelenchida	Dorylaimida
1	22	19	15	23	e[1]	e	12	23
2	29	24	17	22	15	12	19	32
3	44	33	19	34	79s[2]	16	22	40
4	43	50	23	46	103s	22	25	57
♀	52	65	65	72	82s	23	16	60

[1] e: within egg, non-migratory

[2] s: sedentary, non-migratory

It is also important to accept that nematodes are one group of animals using resources, and that they may interact with other soil organisms in a positive manner. Studies by Elliott et al. (1980) showed that when an amoeba was present it was able to pass through the narrower pore necks of a finer-textured soil, and graze on bacteria; on returning to larger pores, the amoeba was grazed on by *Mesodiplogaster*, whose populations showed greater increase in such soils (Table 8.4).

Table 8.4 Proportional increase in numbers and biomass of *Mesodiplogaster* in the presence of bacteria (*Pseudomonas*), as influenced by the presence of amoebae (*Acanthamoeba*) at two soil textures in microcosms incubated for 33 days. *Mesodiplogaster* can feed on both the bacteria and amoebae (after Elliott, et al., 1980).

Nematode measurement	Soil texture	Proportional increase when amoebae added
Juvenile population	Coarse	1.3
	Fine	2.3
Adult population	Coarse	0.6
	Fine	8.7
Total biomass	Coarse	1.1
	Fine	2.9

The non-living environments of nematodes have small-scale heterogeneity. Modern tools are being used to study the response of nematodes to chemical gradients and food "hot spots" at a scale reflecting the size of particles and interstices (Anderson, et al., 1997; Young, et al., 1996).

As Jones et al. (1969) discussed, many experiments have used repacked or synthetic soils that lack the aggregation of primary soil particles into structural units. Such aggregation has a significant effect on pore-size distribution and thus on pores accessible to nematodes. Although the use of bulk density has been suggested (Jones and Thomasson, 1976), its use also may not reflect structural characteristics. Soil structural characteristics change with land use. With increasing concern about maintenance of soil structure and quality (Doran and Parkin, 1994), it is important for us to know the ability of a range of plant-feeding, microbivorous, and predacious nematodes to utilize resources to which access may be physically limited. This needs to be done in natural soils with a range of mineralogies, textures, and structures.

The matrix affects not only nematode activity and resource utilization but also chemical and biological properties. In aquatic sediments, anaerobic conditions can be critical, with oxybiotic and thiobiotic nematode faunae having been found in such environment (Jensen, 1987). Also, Hogue and

Miller (1981) showed small-scale spatial variability of meiobenthic nematodes in response to sediment microtopography. Further, Lambshead et al. (2000) found nematodes to be very resistant to physical disturbance of marine sediments, compared to larger animals. The growth and nutrient content of microbes (i. e., potential food for, or pathogens of, nematodes) are affected by moisture stress (Malajczuk and Theodorou, 1979; van Veen and Paul, 1979). In terrestrial systems plant growth, the main source of energy for plant-, bacterial-, and fungal-feeding nematodes in soil, is affected by root exploration of the soil, which in turn is influenced by soil mineralogy, texture, structure, and nutrient and water availability. In marine environments, the spatial and temporal variation of the falling phytodetritus affects resources available to nematodes across a range of substrate texture (Rice and Lambshead, 1994). In rivers and streams, the current transfers food resources from further upstream to nematodes. Wetlands are increasingly being used to remove nutrients from effluent. Determining what proportion of their spatial variability is due to the matrix and textural mosaic may play an important part in understanding the complexity – both spatial and specific – of these and other nematode communities (Ettema, et al., 1998).

8.7 Diversity of Nematode Faunae in Terrestrial and Aquatic Sediments

The scale of sampling affects the range of habitats and microhabitats from which nematodes are recovered, and thus the number of nematode species found. While this section utilizes data from field-scale samples (of approximately 1 ha) with relatively uniform soil and vegetation, it must be noted that much greater diversity has been recorded in regional samples with more diverse soil and vegetation (Boag and Yeates, 1998). Regional samples are more likely to represent a faunal collection of species, while on a 1 – 10 cm scale the species are more likely to be part of an interacting community; marine collections are commonly described by the more neutral term "assemblage". The total number of taxa identified is regarded as the diversity, and this may be calculated in various ways. Due to varying levels of identification, selectivity of extraction procedures, varying inclusion of parasites of vertebrates and invertebrates, and weightings used in calculations, extreme care must be taken in comparing any indices which may be given for nematode faunae. Family or genus level identifications may give similar discrimination among indices as those at species level (Warwick, 1988). There is also evidence that the nematode faunae of different regions may vary in the contribution of different *c-p* classes as defined by Bongers (1990), with Dorylaimida apparently contributing a greater proportion of the nematode fauna in New Zealand than in Europe (Yeates, 1994). Such differences may reflect geological history (Ferris, et al., 1976).

A substrate that offers a greater range of physical, chemical, and biological resources (e. g., texture, nutrients, qualitative, and quantitative biological inputs) is likely to provide a greater range of niches by which the nematodes, and other organisms, can partition the total resource and co-exist. The more restricted any of the resource components, the less diverse the nematode fauna may be. Not only must these issues be considered in appropriate (seasonal) time frames, but bias of the nematode extraction method (cyst flotation, passive sieving, active migration through paper tissue, collapse during centrifugation) and the weighting put on the individuals recovered (1 heteroderid cyst vs. about 100 juveniles in it; active *Rhabditis* vs. inactive dauer larvae) are also significant. While the diversities listed in Table 8. 5 attempt to reflect the total nematode fauna present when the contribution of the nematode fauna to ecological processes is being considered, the contribution of the various feeding types, or some other analysis in terms of the resource used and possibly responding to the grazing, is also desirable. The various faunae in Table 8. 5 are grouped to illustrate trends in diversity with substrate texture, management regime, etc. In Welsh pasture diversity was greatest in sand, but within each texture greater under organic than conventional management. Grassland in New South Wales had greater nematode diversity under a native than introduced sward, but the age of a *Picea* stand in the Czech Republic or a *Pinus* forest in Poland had no effect. However, when the Polish forest had increased tree diversity, nematode diversity was increased. New Zealand pastures on loamy sand had greater nematode diversity than those on silt loam, but the latter was increased with summer irrigation. In the Puerto Rico Trench, diversity of marine nematodes was greatest at the site containing 77% clay (Table 8.5).

Many groups of organisms show hot spots of biodiversity. Although species-rich faunae have been described from tropical forest (Lawton, et al., 1998) and there are some ideas about centers of speciation (e. g., *Globodera* in the Andes; *Radopholus* in Australia and New Zealand), comparative knowledge of nematode biodiversity is poor. Stohlgren et al. (1999) have reported that sites with high plant diversity are susceptible to establishment of invading plant species – although this may be a function of resources available at the sites. What the situation is in nematode faunae is unknown. So far we are not aware of the impact on the nematode fauna of the establishment of invasive plants in otherwise undisturbed indigenous vegetation.

Table 8.5 Total nematode taxa recorded from a series of ecosystems. The sets of samples have been selected to show management or texture effects; as many underlying procedures differed, comparisons of actual values between sets are not possible.

Ecosystem, Region and Location	Management	Texture	Taxa[1] (*n*)	Diversity (H')	Reference
Grazed pasture, Wales (0—7.5 cm)	Organic	Sand	32.3	2.23	Yeates, et al., 1997a
	Conventional	Sand	30.3	1.77	
	Organic	Loam	28.8	2.03	
	Conventional	Loam	27.1	1.97	
	Organic	Silt	26.8	1.95	
	Conventional	Silt	22.6	1.89	
Grassland, New South Wales, Australia (0—10 cm)	Native	Fine sandy clay loam – sandy loam	44	2.36	Yeates and King, 1997
	Improved		31	2.07	
Picea forest, Bilý k⊆i, Czech Republic	5 years old		60	2.67	Háněl, 1996
	105 years old		59	2.55	
	134 years old		57	2.16	
Pinus on dunes, Kampinos, Poland	6 years old	Sand	34	—	Wasilewska, 1970
	10 years old	Sand	52	—	
	135 years old	Sand	45	—	
Pinus / Quercus / Carpinus on dunes, Kampinos, Poland	135 years old	Sand	67	—	
Mixed forest, Mt Etna, Italy	Uncut forest	Volcanic sand	41	2.87	Clausi and Vinciguerra, 1999
	Cut forest		38	1.96	
	Clearing		28	2.42	

Continued

Ecosystem, Region and Location	Management	Texture	Taxa[1] (*n*)	Diversity (H')	Reference
Chionochloa tussock grassland, Otago, New Zealand (0—20 cm)	Unburned grassland	Silt	36	2.33	Yeates and Saggar, 1998
	Burned grassland		32	2.20	
	Pinus plantation		28	1.68	
Grazed pasture, Tikitere, New Zealand (0—10 cm)	Dryland; 1600 mm rain	Loamy sand	35	2.74	Yeates, 1984
Grazed pasture, Kaitoke, NZ (0—10 cm) (3 year means)	Dryland; 1673 mm rain	Silt loam	30	2.24	
Grazed pasture, Massey Univ., NZ(0—10 cm)	Dryland; 1015 mm rain	Silt loam	30	2.48	
	Summer irrigated	Silt loam	33	2.67	
Grazed pasture, Otiake, NZ (0—10 cm)	Dryland; 550 mm rain	Silt loam	26	2.04	
Hatteras Abyssal Plain	Depth 5411 m	77% clay	116	4.10	Tietjen, 1989
Puerto Rico Trench	Depth 2217 m	70% sand	70	3.97	
	Depth 7460 m	83% clay	61	3.58	
	Depth 8189 m	81% clay	52	3.58	
	Depth 8380 m	79% clay	47	3.33	

[1] The level of discrimination varies between the various studies but is uniform within each.

8.8 Microbial and Detrital Substrates

Nematologists have long considered the rhizosphere as the site of most nematological interactions in soil. However, as we seek to understand not only nematode – plant interactions but also decomposition processes, biological population interactions, and the distribution of activity at various scales, there is a need to study a range of zones (e.g., rhizosphere, organic soil horizons, bulk mineral soil, and marine phytodetritus deposits), and within those zones particular sites of interactions. Many disciplines are making advances in these areas (Cadisch and Giller, 1997; Coleman and Hendrix, 2000).

Other soil microfauna are also involved in substrate processes, and in the field their populations interact. Bacterial populations are grazed by organisms of many phyla (e.g., Protozoa, Rotifera, and Nematoda), and just as there is typically a suite of interacting nematode species so other phyla often have competing species (Umorin, 1992). Protozoa, Tardigrada, Collembola, and mites are among the potential competitors that predacious nematodes face in searching for prey, but evidence suggests that in a particular horizon one of the groups may be dominant (Yeates and Foissner, 1995; Yeates and Wardle, 1996).

Nematode access to microbes is partly controlled by pore size – in litter environments a simple water film may be present. Pore size is also critical in processes controlling carbon dynamics, microbial populations, and enzymatic activity (Gunapala, et al., 1998; Kandeler, et al., 1999). Microbial populations also vary seasonally (Bardgett, et al., 1999b), the variation partly reflecting the impact of nematode grazing but also changes in the microbial resource available to microbivorous nematodes. Comparison of nematode populations under differing management regimes has shown that under integrated management of winter wheat fields in The Netherlands there was significantly more carbon in bacterial-feeding nematodes than under conventional management (Bloem, et al., 1994; Bouwman and Zwart, 1994). In a Californian silty loam soil, Gunapala et al. (1998) found there was increased abundance of bacterial-feeding nematodes under organic farming systems during crop growth. However, such differences were not achieved when soils were amended with organic matter (air-dried vetch), with both bacterial- and fungal-feeding numbers being significantly greater in amended soils at times during the 42 days incubation period. The nematode response to added vetch involved similar species (*Acrobeloides bodenheimeri*, *Bursilla labiata*, *Chiloplacus sp.*, *Cruznema tripartitum*, and *Panagrolaimus detritophagus*) to those in soils from organic and conventional field plots.

From the many microcosm studies, two have been selected to show the impact of nematode grazing on nutrient cycling. In what is now a classic paper, Ingham et al. (1985) showed that while adding the bacterial-feeding nematode *Acrobeloides* sp. to microcosms led to a significant increase in NH_4^+-N in the sandy loam soil (C/N ~ 10), addition of fungus and *Aphelenchus avenae* gave no significant response. Shoot biomass also increased significantly in the case of *Acrobeloides* sp. (Table 8.6). In more recent studies, Ferris and his co-workers (Table 8.7) have shown that grazing by three of four bacterial-feeding nematodes significantly increased N-mineralization at soil C: N ratios of 10 – 45, while N-mineralization was only increased by fungal-feeding nematodes at C: N≥15 (Table 8.7).

Table 8.6 Effect of increasing microfloral and microfaunal diversity (i.e., food web complexity) on soil N and growth of a grass (*Bouteloua gracilis*) in microcosms (after Ingham, et al., 1985).

Organisms Present	NH_4^+-N (μg / g soil)	Shoot Biomass (mg)
plant	33.0	2.1
plant + bacteria	32.2	2.6
plant + bacteria + bacterial-feeding nematode	45.5	6.6
plant + fungus	48.6	7.5
plant + fungus + fungal-feeding nematode	49.1	8.6
plant + bacteria + fungus + b/f nematode + f/f nematode	46.4	13.5
Tukey's honestly significant difference ($p < 0.05$)	10.2	3.0

Table 8.7 Substrate C: N ratios at which 6 nematodes significantly ($p < 0.05$) increased N-mineralization in experimental columns (*E. coli* and *Rhizoctonia solani* were food sources) (after Chen and Ferris, 1999; Ferris, et al., 1998).

Substrate C: N	Bacterial-feeding Nematodes				Fungal-feeding Nematodes	
	Acrobeloides buetschlii	*Bursilla labiata*	*Cephalobus persegnis*	*Cruznema tripartitum*	*Aphelenchus avenae*	*Aphelenchoides composticola*
11	+	+	+	+	n. s.[1]	n. s.
15	+	+	+	+	+	n. s.
20	+	+	+	n. s.	+	+
25	+	n. s.	+	n. s.	n. s.	+
30	+	+	+	n. s.	+	+
35	+	+	+	n. d.[2]	+	+
40	+	+	+	n. s.	+	+
45	+	+	+	n. s.	+	+

[1] not significant ($p > 0.05$)

[2] not determined

The significance of marine phytodetritus inputs to deep-sea nematode populations is now appreciated and is being added to parameters explaining the abundance and diversity of marine nematode assemblages (Rice and Lambshead, 1994). Long-term changes in intertidal diatom assemblages have been related not only to changes in organic waste discharges but also to grazing by the nematode *Eudiplogaster pararmatus* on the diatom *Navicula pygmaea* (Peletier, 1996). Using freshwater microcosms, Traunspurger et al. (1997) found that grazing by *Caenorhabditis elegans* stimulated bacterial activity, confirming work such as that of Ingham et al. (1985) in soils.

The relative importance of the bacterial and fungal pathways of decomposition varies among ecosystems, and in space and time within ecosystems, with dominance of the fungal pathway often reflecting lower rates of decomposition (Cadisch and Giller, 1997). Nematode populations and functional groups provide one way of assessing the balance between pathways in terrestrial ecosystems (Ruess, et al., 1999; Wardle, et al., 1999a, 1999b) – marine habitats are not known to support populations of fungal-feeding nematodes.

8.9 Impact of Environmental Pressures on Nematode Faunae

A range of indices has been used to assess the impact of various environmental changes on nematode faunae. Bongers and Bongers (1998) and Bongers and Ferris (1999) have reviewed the widely used maturity index (MI), and a broader range of indices was covered by Freckman and Ettema (1993), Warwick and Clarke (1998), and Yeates (1999). These papers should be consulted for the impact of environmental changes not discussed below.

8.9.1 Soil Tillage

Soil tillage is generally intended to alter soil structure to improve crop establishment. In addition to the direct effects on soil structure there may be direct effects on nematodes (Jones, et al., 1969) and other soil organisms. It has been advocated that no-till practices can improve soil quality and land use sustainability (Karlen, et al., 1994). Initially, tillage should increase plant yield and thus the resource for plant- and microbial-feeding nematodes. In the longer term, there may be degradation of soil structure that may impact on both plant growth and soil biota. In a fine loamy soil in Michigan, Freckman and Ettema (1993) found total nematode abundance and the plant parasite index (PPI) to be higher in soil under annual crops, while predacious nematodes and the maturity index (MI) were both lower. A study that tested the assumption that larger dorylaimid nematodes should be reduced in structurally degraded conditions found that, in two of four soils, the predacious mononchids actually increased (Yeates, et al., 1998b). Apparently the bacteria→bacterial-feeding nematode→predacious nematode pathway was altered to a bacteria→bacteria-feeding protozoan→predacious nematode pathway. This alteration occurred in soils in which earthworm populations became insignificant and which had lower soil aggregate stability than those soils in which earthworm populations persisted.

8.9.2 Plant Cultivars

The use of nematode-resistant plant cultivars is seen as an environmentally benign way to control pest nematode populations. Effectively removing a host plant by using a cultivar genetically resistant to the pest results in a vacant niche that may be exploited by other nematodes. In the United Kingdom, use of potato cultivars, such as Maris Piper, resistant to *Globodera rostochiensis* but not to *G. pallida*, led to an increase in the percentage of potato cyst nematode populations composed solely of *G. pallida* from 44% to 74% between 1993 and 1997 (Lane and Trudgill, 1999). Analysis of the seasonal pattern of white clover root infestation by *Heterodera trifolii*, *Meloidogyne* sp., and *Pratylenchus* sp. in New Zealand pastures suggested that a successful *H. trifolii*-resistant white clover would provide a root resource to be exploited by the other two nematodes (Yeates, et al., 1985). We are unaware of how such changes affect the wider soil biological community. However, given the concentration of microorganisms in the rhizosphere (Hofman and s'Jacob, 1989), and the general increase of leakage of plant photosynthate into the rhizosphere in the presence of root-feeding nematodes (Yeates, et al., 1999b), any effect is likely to be significant.

8.9.3 Global Climate Change

The dependence of biological processes on temperature and moisture means they will be affected by global climate change. There have been some studies mapping the expected change in distribution of plant-feeding nematodes with predicted temperature changes (Boag, et al., 1997; Yeates, et al., 1998a). Although nematodes may not disperse rapidly naturally, plant-feeding nematodes and their hosts probably will move in parallel. How much new distributions will be due to migration or increase of trace populations already present is uncertain.

Microbivorous nematodes have not been included in such distributional studies – probably because their present distributions are insufficiently well known. However, on two soils in subarctic Sweden, increasing temperature and/or NPK fertilization have been found to change the composition of the nematode fauna, generally in favor of fungal- and plant-feeding species (Ruess, et al., 1999). Providing their populations can survive the increased frequency of extreme events (particularly variation in precipitation), nematodes and other soil microfauna that play such important roles in decomposition processes and nutrient cycling probably have sufficient specific diversity among them to ensure that processes continue despite species distributions changing. While artificial modification of conditions (e.g., by irrigation and heating) can give indications of possible faunal responses, the greater time frame and larger areas involved mean that colonization processes will be more important than local adaptation.

8.9.4 Elevated Carbon Dioxide

Experimental studies in this area have centered on plants grown under elevated CO_2 and the results reflect the growth and decomposition of novel (i.e., first generation) plant material and the behavior of nematode populations under the simulated conditions. However, evidence suggests that organic material synthesized under historic elevated CO_2 may have differing decomposition characteristics from that grown under novel conditions (Gorissen, et al., 1995). There are natural sites where the flora and fauna have been exposed to elevated CO_2 for decades (e.g., near volcanic vents or CO_2-rich springs) (Raschi, et al., 1998; Yeates, et al., 1999a) but they commonly show a mosaic of soils. In grassland, Yeates et al. (1999a) found a general decrease in nematode abundance and diversity as long-term atmospheric CO_2 concentrations increased, but an increase in the dominance and abundance of bacterial-feeding nematodes. However, the responses of the bacterial-feeding nematodes differed between the organic and gley soils at the site, reflecting soil carbon and soil microbial carbon concentrations. They considered that any long-term response to CO_2 enrichment would be an integration of responses in the various soils contributing to the local soil pattern.

8.10 Nematode Population Genetics

Studies of nematode faunae as a whole have to rely on traditional morphological species rather than using taxa defined in terms of gene flow (biological species) or biochemistry. Among the plant-parasitic cyst nematodes (e.g., *Heterodera carotae*, *H. cruciferae*, and *H. goettingiana*) and in the bacterial-feeding nematodes (*Rhabditis mariannae* and *R. synpapillata*), it has been found that morphological species are indeed reproductively isolated (Rehfeld and Sudhaus, 1985; Yeates, 1970). Supplementing the morphological concept with a study of gene flow between *Haemonchus contortus,* which tends to parasitize sheep, and *H. placei*, which tends to parasitize cattle, confirmed that host specificity was not absolute, and mixed infections produced hybrids (Le Jambre, 1981). Le Jambre considered that rapid separation between these forms was apparently occurring. A more recent study, using ribosomal DNA, has indicated that three morphological forms of trichostrongylid parasites (*Teladorsagia circumcincta*, *T. davtiana*, and *T. trifurcata*) represent a single species (Stevenson, et al., 1996). In plant-feeding taxa, similar patterns are shown by races, pathotypes, and intraspecific variants among *Heterodera*, *Globodera*, *Meloidogyne*, *Pratylenchus*, and *Radopholus* (Brodie, 1999; France and Brodie, 1996; Windham and Pederson, 1989), while isolates of *Heterorhabditis* from Europe show a range of differentiation (Griffin, et al., 1999).

Although many changes in intraspecific variants of nematodes will have secondary effects on substrate biological activity, via altered nematode or host inputs, the impact of some characteristics (e.g., the effect of differing cuticle structure, or chemistry, on predators and parasites; altered nutrition and excretion) is unknown.

8.11 Overview

1. Nematodes occupy plant, invertebrate, and vertebrate tissues as well as non-living substrates, use a range of food resources and have a wide range of feeding types. Their feeding may stimulate host and other populations; in some cases there is pathogenicity.

2. Nematode populations change with time, reflecting species succession within patches of substrate, and seasonal and annual patterns.

3. Global spatial distribution of nematodes reflects natural and human-mediated distribution. In order to understand their relationship to environmental factors, and contribution to ecosystem processes, distribution at the microscale must be understood.

4. It appears that access to suitable food resources is the critical factor determining the survival of a nematode population at a site; temperature seems secondary.

5. Matrix dimensions, and the water occupying them, affect nematodes directly and through their food resources.

6. In a given non-living environment, nematode diversity tends to increase

as the substrate becomes sandier, but between patches the magnitude of the food resource can over-ride this trend.

7. Grazing by nematodes significantly affects both bacterial and fungal populations, although the effect varies with ecosystem, space, and time. Nematode populations and functional groups provide one way of assessing the balance between decomposition pathways in terrestrial ecosystems.

8. Environmental pressures are reflected in nematode faunae. Knowledge of their relationship to environmental factors and ecosystem processes is a key to understanding and predicting how human impacts, such as tillage, plant cultivars, and elevated carbon dioxide may impact on long-term productivity of ecosystems.

9. The genetics of nematode populations are changing. Use of molecular techniques can demonstrate the impact of natural speciation, of human-mediated redistribution of nematodes and hosts, and of development of nematode "resistant" crop plants and grazing animals.

References

Alphey, T. J. W., and C. E. Taylor. 1986. European Atlas of the Longidoridae and Trichodoridae. Invergowrie: Scottish Crop Research Institute.

Anderson, A. R. A., I. M. Young, B. D. Sleeman, B. S. Griffiths, and W. M. Roberston. 1997. Nematode movement along a chemical gradient in a structurally heterogeneous environment. 1. Experiment. Fundamental and Applied Nematology 20: 157 – 163.

Anderson, R. C. 1992. Nematode Parasites of Vertebrates: Their Development and Transmission. Wallingford, UK: CAB International.

Bardgett, R. D., R. Cook, G. W. Yeates, and C. S. Denton. 1999a. The influence of nematodes on below-ground processes in grassland ecosystems. Plant and Soil 212: 23 – 33.

Bardgett, R. D., R. D. Lovell, P. J. Hobbs, and S. C. Jarvis. 1999b. Seasonal changes in soil microbial communities along a fertility gradient of temperate grasslands. Soil Biology and Biochemistry 31: 1021 – 1030.

Barker, K. R., G. A. Pedersen, and G. L. Windham, eds. 1998. Plant and Nematode Interactions. Madison, WI: American Society of Agronomy.

Bedding, R. A. 1973. Biology of *Deladenus siridicola* (Neotylenchidae) an entomophagous-mycetophagous nematode parasitic in siricid woodwasps. Nematologica 18: 482 – 493.

Bird, A. F., and G. W. Yeates. 1994. Studies on *Aprutides guidettii* (Nematoda: Seinuridae) isolated from soil at Northfield, South Australia. Transactions of the Royal Society of South Australia 118: 261 – 266.

Bloem, J., G. Lebbink, K. B. Zwart, L. A. Bouwman, S. L. G. E. Burgers, J. A. de Vos, and P. C. de Ruiter. 1994. Dynamics of microorganisms, microbivores and nitrogen mineralisation in winter wheat fields under conventional and integrated management. Agriculture, Ecosystems and Environment 51: 129 – 143.

Boag, B. 1980. Effect of temperature on rate of feeding of the plant parasitic nematodes *Rotylenchus robustus*, *Xiphinema diversicaudatum* and *Hemicycliophora conida*. Journal of Nematology 12: 193 – 195.

Boag, B., and K. J. Orton Williams. 1976. The Criconematidae of the British Isles. Annals of Applied Biology 84: 361 – 369.

Boag, B., and G. W. Yeates. 1998. Soil nematode biodiversity in terrestrial ecosystems. Biodiversity and Conservation 7: 617 – 630.

Boag, B., J. W. Crawford, and R. Neilsen. 1991. The effect of potential climatic changes on the geographical distribution of the plant-parasitic nematodes *Xiphinema* and *Longidorus*. Nematologica 37: 312 – 323.

Boag, B., K. A. Evans, G. W. Yeates, D. J. F. Brown, and R. Neilsen. 1997. Global potential distribution of European longidorid virus-vector nematodes. Nematologica 43: 99 – 106.

Bongers, T. 1988. De Nematoden van Nederland. Utrecht: KNNV.

Bongers, T. 1990. The maturity index: an ecological measure of environmental disturbance based in nematode species composition. Oecologia 83: 14 – 19.

Bongers, T., and M. Bongers. 1998. Functional diversity of nematodes. Applied Soil Ecology 10: 239 – 251.

Bongers, T., and H. Ferris. 1999. Nematode community structure as a bioindicator in environmental monitoring. TREE 14: 224 – 228.

Boucher, G. 1990. Pattern of nematode species diversity in temperate and tropical subtidal sediments. Marine Ecology 11: 133 – 146.

Bouwman, L. A., and K. B. Zwart. 1994. The ecology of bacterivorous protozoans and nematodes in arable soil. Agriculture, Ecosystems and Environment 51: 145 – 160.

Brodie, B. B. 1999. Classical and molecular approaches for managing nematodes affecting potato. Canadian Journal of Plant Pathology 21: 222 – 230.

Brodie, B. B., K. Evans, and J. Franco. 1993. Nematode parasites of potatoes. Pp. 87 – 132. In K. Evans, D. L. Trudgill, and J. M. Webster, eds. Plant Parasitic Nematodes in Temperate Agriculture. Wallingford, UK: CAB International.

Cadisch, G., and K. E. Giller, eds. 1997. Driven by Nature – Plant Litter Quality and Decomposition. Wallingford, UK: CAB International.

Chen, J., and H. Ferris. 1999. The effects of nematode grazing on nitrogen mineralization during fungal decomposition of organic matter. Soil Biology and Biochemistry 31: 1265 – 1279.

Clausi, M., and M. T. Vinciguerra. 1999. Changes in nematode communities of forest soil in relation to clear-cutting. Nematologia Mediterranea 27: 315 – 322.

Coleman, D. C., and P. F. Hendrix, eds. 2000. Invertebrates as Webmasters in Ecosystems. Wallingford, UK: CAB International.

Cook, R., and G. W. Yeates. 1993. Nematode pests of grasslands and forage crops. Pp. 305 – 350. In K. Evans, D. L. Trudgill, and J. M. Webster, eds. Plant Parasitic Nematodes in Temperate Agriculture. Wallingford, UK: CAB International.

Cooke, D. 1993. Nematode parasites of sugarbeet. Pp. 133 – 169. In K. Evans, D. L. Trudgill, and J. M. Webster. Plant Parasitic Nematodes in Temperate Agriculture.

Wallingford, UK: CAB International.

Crofton, H. D., and J. H. Whitlock. 1965. Ecology and biological plasticity of sheep nematodes. III. Studies on *Ostertagia circumcincta* (Stadelman, 1894). Cornell Veterinarian 55: 259 – 262.

Doran, J. W., and T. B. Parkin. 1994. Defining and assessing soil quality. Pp. 3 – 21. In J. W. Doran, D. C. Coleman, D. F. Bezdicek, and B. A. Stewart, eds. Defining Soil Quality for a Sustainable Environment. Madison, WI: Soil Science Society of America.

Elliott, E. T., R. V. Anderson, and D. C. Coleman. 1980. Habitable pore space and microbial trophic interactions. Oikos 35: 327 – 335.

Ettema, C. H. 1998. Soil nematode diversity: Species coexistence and ecosystem function. Journal of Nematology 30: 321 – 328.

Ettema, C. H., D. C. Coleman, G. Vellidis, R. Lowrance, and S. L. Rathbun. 1998. Spatiotemporal distributions of bacterivorous nematodes and soil processes in a restored riparian wetland. Ecology 79: 2721 – 2734.

Ettema, C. H., S. L. Rathbun, and D. C. Coleman. 2000. On spatiotemporal patchiness and the coexistence of five species of *Chronogaster* (Nematoda: Chrongasteridae) in a riparian wetland. Oecologia 125: 444 – 452.

Evans, K., J. Franco, and M. M. De Scurrah. 1975. Distribution of species of potato cyst nematodes in South America. Nematologica 21: 365 – 369.

Ferris, H., M. Eyre, R. C. Venette, and S. S. Lau. 1996. Population energetics of bacterial-feeding nematodes: Stage-specific development and fecundity. Soil Biology and Biochemistry 28: 271 – 280.

Ferris, H., R. C. Venette, H. R. van der Meulen, and S. S. Lau. 1998. Nitrogen mineralization by bacterial-feeding nematodes: Verification and measurement. Plant and Soil 203: 159 – 171.

Ferris, V. R., C. G. Goseco, and J. M. Ferris. 1976. Biogeography of free-living soil nematodes from the perspective of plate tectonics. Science 193: 508 – 510.

France, R. A., and B. B. Brodie. 1996. Characterization of *Pratylenchus penetrans* from ten geographically isolated populations based on their reaction on potato. Journal of Nematology 28: 520 – 526.

Freckman, D. W., and C. H. Ettema. 1993. Assessing nematode communities is agroecosystems of varying human intervention. Agriculture, Ecosystems and Environment 45: 239 – 261.

Freckman, D. W., and R. A. Virginia. 1995. Extraction of nematodes from Dry Valley Antarctic soils. Polar Biology 13: 483 – 487.

Gorissen, A., J. H. van Ginkel, J. J. B. Keurentjes, and J. A. van Veen. 1995. Grass root decomposition is retarded when grass has been grown under elevated CO_2. Soil Biology and Biochemistry 27: 117 – 120.

Griffin, C. T., I. Dix, S. A. Joyce, A. M. Burnell, and M. J. Downes. 1999. Isolation and characterization of *Heterorhabditis* spp. (Nematoda: Heterorhabditidae) from Hungary, Estonia and Denmark. Nematology 1: 321 – 332.

Gunapala, N., R. C. Venette, H. Ferris, and K. M. Scow. 1998. Effects of soil management history on the rate of organic matter decomposition. Soil Biology and Biochemistry 30: 1917 – 1927.

Háněl, L. 1996. Soil nematodes in five spruce forests of the Beskydy Mountains, Czech Republic. Fundamental and Applied Nematology 19: 15 – 24.

Heneghan, L., and T. Bolger. 1998. Soil microarthropod contribution to forest ecosystem processes: the importance of observational scale. Plant and Soil 205: 113 – 124.

Hofman, T. W. and J. J. s'Jacob. 1989. Distribution and dynamics of mycophagous and microbivorous nematodes in potato fields and their relationship to some food sources. Annals of Applied Biology 115: 291 – 298.

Hogue, E. W., and C. B. Miller. 1981. Effects of sediment microtopography on small-scale spatial distributions of meiobenthic nematodes. Journal of Experimental Marine Biology and Ecology 53: 181 – 191.

Hominick, W. M. 1982. Selection of a rapidly maturing population of *Globodera rostochiensis* by continuous cultivation of early potatoes in Ayrshire, Scotland. Annals of Applied Biology 100: 345 – 351.

Hunt, D. J. 1993. Aphelenchida, Longidoridae and Trichodoridae: Their Systematics and Bionomics. Wallingford, UK: CAB International.

Ingham, R. E., J. A. Trofymow, E. R. Ingham, and D. C. Coleman. 1985. Interactions of bacteria, fungi, and their nematode grazers: effects on nutrient cycling and plant growth. Ecological Monographs 55: 119 – 140.

Jensen, P. 1987. Differences in microhabitat, abundance, biomass and body size between oxybiotic and thiobiotic free-living marine nematodes. Oecologia 71: 564 – 567.

Jones, F. G. W. 1975. Accumulated temperature and rainfall as measures of nematode development and activity. Nematologica 21: 62 – 70.

Jones, F. G. W., and A. J. Thomasson. 1976. Bulk density as an indicator of pore space in soils usable by nematodes. Nematologica 22: 133 – 137.

Jones, F. G. W., D. W. Larbey, and D. M. Parrott. 1969. The influence of soil structure and moisture on nematodes, especially *Xiphinema*, *Longidorus*, *Trichodorus* and *Heterodera* spp. Soil Biology and Biochemistry 1: 153 – 165.

Kandeler, E., S. Palli, M. Stemmer, and M. H. Gerzabek. 1999. Tillage changes microbial biomass and enzyme activities in particle-size fractions of a Haplic Chernozem. Soil Biology and Biochemistry 31: 1253 – 1264.

Karlen, D. L., N. C. Wollenhaupt, D. C. Erbach, E. C. Berry, J. B. Swan, N. S. Eash, and J. L. Jordahl. 1994. Long-term tillage effects on soil quality. Soil and Tillage Research 32: 313 – 327.

Klingler, J., O. Güntzel, and P. Kunz. 1983. *Xiphinema*- und *Longidorus*-Arten (Nematoda) im schweizerischen Mittelland. Vierteljahrsschrift der Naturforschenden Gesellschaft in Zürich 128: 89 – 114.

Lambshead, P. J. D., J. Tietjen, T. Ferrero, and P. Jensen. 2000. Latitudinal diversity gradients in the deep-sea with special reference to North Atlantic nematodes. Marine Ecology Progress Series 194: 159 – 167.

Lane, S., and D. Trudgill. 1999. Potato Cyst Nematode: A Management Guide. London: MAFF Publications.

Langer, R. H. M. 1990. Pastures: Their Ecology and Management. Auckland, NZ: Oxford University Press.

Lawton, J. H., D. E. Bignell, B. Bolton, G. F. Bloemers, P. Eggleton, P. M. Hammond, M. Hodda, R. D. Holt, T. B. Larsen, N. A. Mawdsley, and N. E.

Stork. 1998. Bioindicator inventories, indicator taxa and effects of habitat modification in tropical forest. Nature, London 391: 72 – 76.

Le Jambre, L. F. 1981. Hybridization of Australian *Haemonchus placei* (Place, 1893), *Haemonchus contortus cayugensis* (Das & Whitlock, 1960) and *Haemonchus contortus* (Rudolphi, 1803) from Louisiana. International Journal for Parasitology 11: 323 – 330.

Loof, P. A. A. 1991. The family Pratylenchidae Thorne, 1949. Pp. 363 – 421. In W. R. Nickle, ed. Manual of Agricultural Nematology. New York: Marcel Dekker.

Malajczuk, N., and C. Theodorou. 1979. Influence of water potential on growth and cultural characteristics of *Phytophthora cinnamomi*. Transactions of the British Mycological Society 72: 15 – 18.

McLaren, R. G., and K. C. Cameron. 1996. Soil Science: Sustainable Production and Environmental Protection. Auckland, NZ: Oxford University Press.

McSorley, R. 1996. Correlation of nematode densities and rainfall in a Florida pasture. Soil and Crop Sciences Society of Florida, Proceedings 56: 55 – 57.

Micoletzky, H. 1925. Die freilebenden Süsswasser- und Moornematoden Dänemarks nebst anhangüber amöbosporidien und andere parasiten bei freilebenden Nematoden. Memoires de l'Academic Royale des Sciences et des Lettres de Danemark, Copenhaguen, Section des Sciences, serie 8, tome 10 (2): 55 – 310, 13 plates.

Neher, D. A., K. N. Easterling, D. Fiscus, and L. Campbell. 1998. Comparison of nematode communities in agricultural soils of North Carolina and Nebraska. Ecological Applications 8: 213 – 223.

Nielsen, C. O. 1949. Studies on the soil microfauna. II. The soil inhabiting nematodes. Natura Jutlandica 2: 1 – 132.

Peletier, H. 1996. Long-term changes in intertidal estuarine diatom assemblages related to reduced input of organic waste. Marine Ecology Progress Series 137: 265 – 271.

Pena Santiago, R., D. Jimenez Guirado, R. Quijano, M. Peralta, and G. Liebanas. 1992. Observations on the postembryonic development in species of the genus *Anatonchus* (Cobb, 1969) de Coninck, 1939 (Nematoda: Anatonchidae). Nematologica 38: 428.

Poinar, G. O. 1975. Entomogenous Nematodes: a Manual and Host List of Insect-Nematode Associations. Leiden: E. J. Brill.

Raschi, A., F. Miglietta, R. Tognetti, and P. van Gardingen, eds. 1998. Plant Responses to Elevated CO_2 – Evidence from Natural Springs. Cambridge, UK: Cambridge University Press.

Rehfeld, K., and W. Sudhaus. 1985. Vergleichende Untersuchung des Sexualverhaltens eines Zwillingsarten-Paares von *Rhabditis* (Nematoda). Zoologischer Jahrbucher Systematik 112: 435 – 454.

Rice, A. L., and P. J. D. Lambshead. 1994. Patch dynamics in deep-sea benthos: The role of heterogeneous supply of organic matter. Pp. 469 – 499. In P. S. Giller, A. G. Hildrew, and D. G. Raffaelii, eds. Aquatic Ecology: Scale, Pattern and Process. Oxford, UK: Blackwell Scientific Publications.

Ruess, L., A. Michelsen, and S. Jonasson. 1999. Simulated climate change in subarctic soils: Responses in nematode species composition and dominance structure. Nematology 1: 513 – 526.

Ruess, L., E. J. Garcia Zapata, and J. Dighton. 2000. Food preferences of a fungal-feeding *Aphelenchoides* species. Nematology 2:223 – 230.

Savin, M. C., J. H. Görres, D. A. Neher, and J. A. Amador. 2001. Biogeophysical factors influencing soil respiration and mineral nitrogen content in an old field soil. Soil Biology and Biochemistry 33:429 – 438.

Schomaker, C. H., and T. H. Been. 1992. Sampling strategies for the detection of potato cyst nematodes; developing and evaluating a model. Pp. 182 – 194. In F. J. Gommers, and P. W. Th. Maas, eds. Nematology from Molecule to Ecosystem. Dundee, UK: European Society of Nematologists.

Sohlenius, B., and S. Bostrom. 1984. Colonization, population development and metabolic activity of nematodes in buried barley straw. Pedobiologia 27:67 – 78.

Stevenson, L. A., R. B. Gasser, and N. B. Chilton. 1996. The ITS-2DNA of *Teladorsagia circumcincta, T. trifurcata* and *T. davtiana* (Nematoda, Trichostrongylidae) indicates that these taxa are one species. International Journal for Parasitology 26:1123 – 1126.

Stohlgren, T. J., D. Binkley, G. W. Chong, M. A. Kalkhan, L. D. Schell, K. B. Bull, Y. Otsuki, G. Newman, M. Bashkin, and Y. Son. 1999. Exotic plant species invade hot spots of native plant diversity. Ecological Monographs 69:25 – 46.

Taylor, C. E. 1962. Transmission of raspberry ringspot virus in *Longidorus elongatus* (de Man) (Nematoda: Dorylaimoidea). Virology 17:493 – 494.

Tietjen, J. H. 1989. Ecology of deep-sea nematodes from the Puerto Rico Trench area and Hatteras Abyssal Plain. Deep-Sea Research 36:1579 – 1594.

Traunspurger, W., M. Bergtold, and W. Goedkoop. 1997. The effects of nematodes on bacterial activity and abundance in a freshwater sediment. Oecologia 112:118 – 122.

Triantophyllou, A. C., and H. Hirschmann. 1997. Evidence of direct polyploidization in the mitotic parthenogenetic *Meloidogyne microcephala*, through doubling of its somatic chromosome number. Fundamental and Applied Nematology 20:385 – 391.

Trudgill, D. L. 1986. Concepts of resistance, tolerance and susceptibility in relation to cyst nematodes. Pp. 179 – 209. In F. Lamberti, and C. E. Taylor, eds. Cyst Nematodes. London: Plenum Press.

Trudgill, D. L. 1995. An assessment of the relevance of thermal time relationships to nematology. Fundamental and Applied Nematology 18:407 – 417.

Umorin, P. P. 1992. Competition between bacterivorous flagellates and bacterivorous ciliates for food resources. Oikos 63:175 – 179.

United States Department of Agriculture. 1981. Land resource regions and major resource areas of the United States. USDA-SCS Agriculture Handbook 296.

van Veen, J., and E. A. Paul. 1979. Conversion of biovolume measurements of soil organisms, grown under various moisture tensions, to biomass and their nutrient content. Applied and Environmental Microbiology 37:686 – 692.

Venette, R. C., and H. Ferris. 1997. Thermal constraints to population growth of bacterial-feeding nematodes. Soil Biology and Biochemistry 29:63 – 74.

Venette, R. C., and H. Ferris. 1998. Influence of bacterial type and density on population growth of bacterial-feeding nematodes. Soil Biology and Biochemistry 30:949 – 960.

Wallace, H. R. 1961. The bionomics of the free-living stages of zoo-parasitic and phyto-parasitic nematodes – a critical survey. Helminthological Abstracts 30:1 – 22.

Wallace, H. R. 1963. The Biology of Plant Parasitic Nematodes. London: Edward Arnold.

Wallace, H. R. 1973. Nematode Ecology and Plant Disease. London: Edward Arnold.

Wardle, D. A., K. I. Bonner, G. M. Barker, G. W. Yeates, K. S. Nicholson, R. D. Bardgett, R. N. Watson, and A. Ghani. 1999a. Plant removals in perennial grassland: Vegetation dynamics, decomposers, soil biodiversity and ecosystem properties. Ecological Monographs 69: 535 – 568.

Wardle, D. A., G. W. Yeates, K. S. Nicholson, K. I. Bonner, and R. N. Watson. 1999b. Response of soil microbial biomass dynamics, activity and plant litter decomposition to agricultural intensification over a seven-year period. Soil Biology and Biochemistry 31: 1707 – 1720.

Warwick, R. M. 1988. The level of taxonomic discrimination required to detect pollution effects on marine benthic communities. Marine Pollution Bulletin 19: 259 – 268.

Warwick, R. M., and K. R. Clarke. 1998. Taxonomic distinctiveness and environmental assessment. Journal of Applied Ecology 35: 532 – 543.

Wasilewska, L. 1970. Nematodes of the dune sands in the Kampinos Forest. I. Species structure. Ekologia Polska 18: 429 – 443.

Wasilewska, L., E. Paplinska, and J. Zienlinski. 1981. The role of nematodes in decomposition of plant material in a rye field. Pedobiologia 21: 182 – 191.

Webster, R., and B. Boag. 1992. Geostatistical analysis of cyst nematodes in soil. Journal of Soil Science 43: 583 – 595.

Wharton, D. A., and I. M. Brown. 1989. A survey of terrestrial nematodes from the McMurdo Sound region, Antarctica. New Zealand Journal of Zoology 16: 467 – 470.

Windham, G. L., and G. A. Pederson. 1989. Aggressiveness of *Meloidogyne incognita* host races on white clover. Nematropica 19: 177 – 183.

Winslow, R. D. 1960. Some aspects of the ecology of free-living and plant-parasitic nematodes. Pp. 341 – 415. In J. N. Sasser, and W. R. Jenkins, eds. Nematology: Fundamentals and Recent Advances with Emphasis on Plant Parasitic and Soil Forms. Chapel Hill, NC: University of North Carolina Press.

Yeates, G. W. 1970. Failure of *Heterodera carotae*, *H. cruciferae* and *H. goettingiana* to interbreed *in vitro*. Nematologica 16: 153 – 154.

Yeates, G. W. 1973a. Annual cycle of root nematodes on white clover in pasture. I. *Heterodera trifolii* in a yellow-grey earth. New Zealand Journal of Agricultural Research 16: 569 – 574.

Yeates, G. W. 1973b. Nematoda of a Danish beech forest. II. Production estimates. Oikos 24: 179 – 185.

Yeates, G. W. 1984. Variation in pasture nematode populations over thirty-six months in a summer moist silt loam. Pedobiologia 27: 207 – 219.

Yeates, G. W. 1986. Stylet and body lengths as nice dimensions in plant-parasitic nematodes. Zoologischer Anzeiger 216: 327 – 337.

Yeates, G. W. 1994. Modification and qualification of the nematode maturity index. Pedobiologia 38: 7 – 101.

Yeates, G. W. 1998. Feeding in free-living soil nematodes: A functional approach. Pp. 245 – 269. In R. N. Perry, and D. J. Wright, eds. Physiology and Biochemistry of

Free-living and Plant-parasitic Nematodes. Wallingford, UK: CAB International.
Yeates, G. W. 1999. Effects of plants on nematode community structure. Annual Review of Phytopathology 37: 127 – 149.
Yeates, G. W., and T. Bongers. 1999. Nematode diversity in agroecosystems. Agriculture, Ecosystems and Environment 74: 113 – 135.
Yeates, G. W., and W. Foissner. 1995. Testate amoebae as predators of nematodes. Biology and Fertility of Soils 20: 1 – 7.
Yeates, G. W., and K. L. King. 1997. Soil nematodes as indicators of the effect of management of grasslands in the New England Tablelands (NSW): Comparison of native and improved grasslands. Pedobiologia 41: 526 – 536.
Yeates, G. W., and S. Saggar. 1998. Comparison of soil microbial properties and fauna under tussock-grassland and pine plantation. Journal of the Royal Society of New Zealand 28: 523 – 535.
Yeates, G. W., and D. A. Wardle. 1996. Nematodes as predators and prey: Relationships to biological control and soil processes. Pedobiologia 40: 43 – 50.
Yeates, G. W., R. N. Watson, and K. W. Steele. 1985. Complementary distribution of *Meloidogyne, Heterodera* and *Pratylenchus* (Nematoda : Tylenchida) in roots of white clover. Proceedings of the Australasian Conference on Grassland Invertebrate Ecology 4: 71 – 79.
Yeates, G. W., T. Bongers, R. G. M. de Goede, D. W. Freckman, and S. S. Georgieva. 1993a. Feeding habits in soil nematode families and genera – an outline for soil ecologists. Journal of Nematology 25: 315 – 331.
Yeates, G. W., D. A. Wardle, and R. N. Watson. 1993b. Relationships between nematodes, soil microbial biomass and weed management strategies in maize and asparagus cropping systems. Soil Biology and Biochemistry 25: 869 – 876.
Yeates, G. W., B. Boag, and R. W. Small. 1994. Species diversity and biogeography of Mononchoidea (Nematoda). Russian Journal of Nematology 2: 45 – 54.
Yeates, G. W., R. D. Bardgett, R. Cook, P. J. Hobbs, P. J. Bowling, and J. F. Potter. 1997a. Faunal and microbial diversity in three Welsh grassland soils under conventional and organic management regimes. Journal of Applied Ecology 34: 453 – 470.
Yeates, G. W., P. A. A. Loof, and W. M. Wouts. 1997b. Criconematidae (Nematoda: Tylenchida) from the New Zealand region: Analysis of and key to *Criconema (Nothocriconemella)* species. New Zealand Journal of Zoology 24: 153 – 162.
Yeates, G. W., B. Boag, K. A. Evans, and R. Neilsen. 1998a. Impact of climatic changes on the distribution of *Paratrichodorus minor* (Nematoda: Trichodoridae) as estimated using "CLIMEX". Nematologica 44: 293 – 301.
Yeates, G. W., T. G. Shepherd, and G. S. Francis. 1998b. Contrasting response to cropping of populations of earthworms and predacious nematodes in four soils. Soil and Tillage Research 48: 255 – 264.
Yeates, G. W., P. C. D. Newton, and D. J. Ross. 1999a. Response of soil nematode fauna to naturally elevated CO_2 levels influenced by soil pattern. Nematology 1: 285 – 293.
Yeates, G. W., S. Saggar, C. B. Hedley, and C. F. Mercer. 1999b. Increase in ^{14}C –

carbon translocation to the soil microbial biomass when five plant-parasitic nematode infect roots of white clover. Nematology 1: 295 – 300.

Yeates, G. W., D. A. Wardle, and R. N. Watson. 1999c. Responses of soil nematode populations, community structure and temporal variability to agricultural intensification over a seven-year period. Soil Biology and Biochemistry 31: 1721 – 1733.

Young, I. M., B. G. Griffiths, and W. M. Robertson. 1996. Continuous foraging by bacterial-feeding nematodes. Nematologica 42: 378 – 382.

Acknowledgements This work was partly funded by the New Zealand Foundation for Research, Science and Technology. It was developed with the aid of a British Council "Higher Education Link" and discussions with DrTom Bongers (Wageningen, The Netherlands). The Scottish Crop Research Institute is funded by the Scottish Executive Rural Affairs Department.

9 Marine Nematode Biodiversity

P. John D. Lambshead

9.1 Introduction

9.1.1 Overview

When I first joined The Natural History Museum in 1976, the free-living nematode collection was located at the back of the Parasitic Worms storeroom with some other small marine metazoa in a cupboard labeled *Minor Phyla*. No zoologist ever doubted the importance of plant- and animal-parasitic nematode worms, their impact on human bodies and human affairs was all too evident, but free-living nematodes had slipped to the status of minor country cousins. The zoological texts from which I was taught suggested that "Nematodes are so remarkably alike that a description of *Ascaris* roughly fits almost any other roundworm" (Buchsbaum, 1968).

In recent years, research has demonstrated that nematodes are far from alike (see Baldwin and Perry, Chapter 4, this volume) and *Caenorhabditis elegans* has replaced *Ascaris* as the best known example of the phylum. *Caenorhabditis elegans* has become an experimental animal quite separated from its relatives in the collective mind of science to the point where a researcher recently felt moved to publish a paper entitled "*Caenorhabditis elegans* is a nematode" (Blaxter, 1998). Many, possibly most, nematodes are primarily free-living and the sediments of the oceans are one of the greatest reservoirs of these animals. Indeed, nematodes are arguably the most successful free-living metazoans; they are abundant, ubiquitous, and both functionally and taxonomically diverse.

This chapter will assess the current state of our understanding of the biodiversity of marine free-living nematodes and where possible, marine nematode biodiversity will be compared with that of soil and aquatic nematodes. The chapter focuses on marine members of the phylum because rather more is known about their biodiversity than the otherwise better known terrestrial nematodes. A comparison of the two domains of terrestrial soils and marine sediments has the potential to yield exciting insights (Freckman, et al.,

1997). I will try to demonstrate how knowledge of nematode biodiversity can advance our understanding of the biodiversity processes that affect all metazoans.

9.1.2 What Is Biodiversity?

If this chapter were concerned with nematode ecology, it would be considered eccentric to devote a section to defining what might be meant by "ecology". It is a reflection of the confusion surrounding the word "biodiversity" that it is first essential to discuss what the word means. Biodiversity is a contraction of "biological diversity" and the first use of this phrase in a modern context appears to have been by Norse and McManus (1980). The contraction is considered to have been invented in 1985 by Rosen (Wilson and Peters, 1988), but was subsequently popularized by a number of authors, most notably E. O. Wilson (see Ghilarov, 1996). Biodiversity has always been a political issue because the aims of biodiversity research are to provide information for the conservation and long-term social and economic utilization of the diversity of life. In this chapter, I propose to restrict discussion solely to scientific issues.

One of the problems of defining the word biodiversity is that the meaning has evolved considerably over two decades. Originally, the term was restricted to a marriage of the study of ecological diversity and the new discipline of genetic diversity. Most biodiversity researchers would now find this far too restrictive. Unfortunately, authors incorporating the powerful new "buzzword" into a variety of disciplines have muddied the waters further.

Article 2, Use of Terms, in the Convention on Biological Diversity, defines biodiversity as "The variability among living organisms from all sources, including inter alia, terrestrial, marine and other aquatic ecosystems and the ecological complexes of which they are part; this includes diversity within species, between species and of ecosystems." This definition represents three aspects of diversity, namely ecosystems, species, and genes. So the study of biodiversity draws primarily upon the disciplines of taxonomy and classification, biogeography, ecology, genetics, population biology, and evolutionary biology.

The importance of scale is implicit to the definition. Biologists rightly focus on the species as the most useful point on the taxonomic scale because in the absence of phylogenetic resolution higher taxonomic classifications are often artificial. However, recent work has shown the utility of looking at higher taxonomic levels when analyzing the biodiversity of marine nematodes (Warwick and Clarke, 1998). The main problem of using higher taxonomic levels of marine nematodes is that it is not clear how closely the current classification reflects phylogeny, in other words it remains to be established whether the current classification represents real clades, and we have little idea of branch lengths.

There has been little work on the biodiversity of marine nematodes at taxonomic levels below that of a species, i. e., metapopulations. Increasingly, molecular biology will be important in addressing these questions.

Spatial scale is extremely important, even simple statements about biodiversity must be qualified by spatial scale. For example, a square meter of European chalk grassland is more botanically diverse than a square meter of lowland Amazonian rainforest; the reverse is true for a square kilometer (Groombridge, 1992). Spatial scale must be related to the size of the animal and temporal scale to its life span. Spatial and temporal scales tend to be interrelated; ecological processes will tend to dominate over short time scales and local distances while evolutionary processes are more important over large time scales and regional/global distances.

9.1.3 Measuring Biodiversity

It cannot be overstated that biodiversity is not a "real" property of the environment in the sense that temperature, pressure, water content, or concentration of sodium ions are real. Biodiversity is a concept in a similar sense to "quality of life" or "elegance of form". Such conceptual measures are notoriously difficult to tie down to numbers. Unfortunately as scientists, we are obliged to boil down complex biodiversity information into a numerical format for analysis and interpretation. I do not propose to list all the various formulae that are currently employed for this purpose, there are a number of excellent books that address this (e. g., Magurran, 1988). Instead, I will try to discuss the principles involved and show how marine and terrestrial scientists have tended to employ different approaches for practical reasons connected with the type of data they can gather.

There are essentially two ways to measure diversity, species richness or ecological diversity. Species richness, a count of the total number of recorded species, is the simplest and "purest" of the measures. But such a species count is only meaningful if it is related to some external measure of sampling size such as number of individuals sampled or area of habitat sampled. Clearly, if one samples a larger area, or collects more animals, one would expect to find more species. In practice, the two ways of measuring species richness can give surprisingly similar results (e. g., Gray, 1994) because the two parameters tend to correlate for the same data. Species richness works best when analyzing medium-scale, well-defined, "closed" areas that have a limited number of species. Samples are taken until the rate of discovery of new species drops sharply or no new species are encountered, the classic museum collector's curve (Fig. 9.1A). An example might be the species richness of a lake's nematodes.

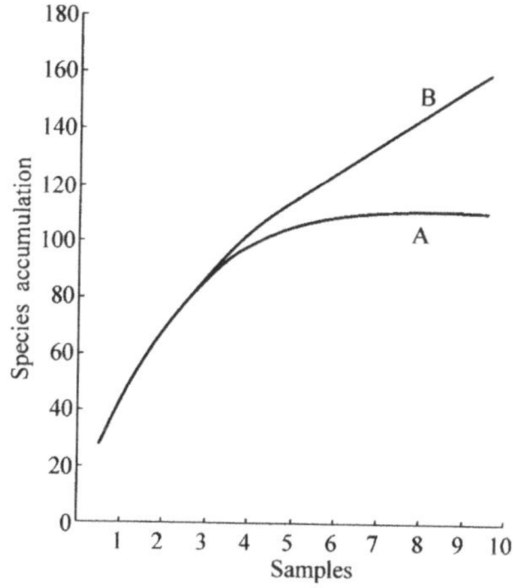

Figure 9.1 Collector's curves (i.e., the accumulation of new species with additional samples) for two regions. Region A is well defined with a limited number of species so the curve reaches an asymptote and the maximum number of species in the region can be censused. Region B is an open ended or highly species rich region and the curve never reaches an asymptote. Deep-sea and rainforest regions tend to give results similar to curve B.

Species richness runs into difficulties as a measure where the region to be assessed is too large to be properly sampled, is open (i.e., has no biogeographic boundaries), or is highly species rich. Classically, in this situation the collectors curve never comes to an asymptote but, after a settling down period, each new sample adds consistently to the estimated species richness (Fig. 9.1B). No upper limit to the number of species can be determined. Species richness is not a good measure for analyzing the effect of localized ecological factors on biodiversity because it does not incorporate quantitative information about the relative abundance of each species. Species richness counts also require a reasonably good knowledge of the taxonomy of the group in question, as it is essential to determine conspecificity in samples taken over a whole region or area.

Ecological diversity has traditionally been interested in two properties of samples, the number of species in a sample (species richness) and the proportional abundance of each species (equitability). These two inter-correlated parameters are amalgamated into a single number as an ecological diversity index; it is this amalgamation that is problematic. There are a number of ways of calculating a diversity index and they are not guaranteed to place samples in the same rank order of diversity.

Two popular diversity indices are the Shannon-Weiner index and Simpson's index. The Shannon-Weiner index (H') is calculated by the formula:

$$H' = -\sum p_i \ln p_i$$

Simpson's index (D) is calculated by the formula:

$$D = \sum [n_i(n_i - 1)]/[N(N-1)]$$

where p_i is the proportional abundance of the ith species (n_i/N), n_i is the number of individuals in the ith species, and N is the total number of individuals.

Deep-sea nematologists commonly use Sanders' (1968; as modified by Hurlbert, 1971) rarefaction graphs or the rarefaction index derived from them to measure diversity. This index gives an estimation of number of species per abundance so it is robust to sample size, which is an extremely valuable property when working along abundance gradients. When used as an index it is normally expressed as $ES(n)$, i. e., the expected number of species per n number of individuals, where n is given by the size of the smallest sample.

$$ES(n) = \sum \left\{1 - \left[\binom{N - N_i}{n} \Big/ \binom{N}{n}\right]\right\}$$

where ES = expected number of species, n = standardized sample size, N = total number of individuals recorded, and N_i = number of individuals in the ith species.

Table 9.1 shows four littoral nematode samples from Northern Ireland. The Shannon-Weiner and Simpson's indices are both popular statistics of a type known as intrinsic diversity indices (Lambshead, et al., 1983). The two indices do not rank the four nematode samples in the same order because the Shannon index is weighted towards species richness while Simpson's index is weighted towards equitability. It is meaningless to ask which index is "right", that depends on whether species richness or equitability is perceived as the more important parameter, which depends on the scientific question being tackled and the nature of the available data.

An analogy may help demonstrate the point. The tourist industry commonly quotes both temperature (hot is good) and rainfall (dry is good) when describing the weather at a resort. If tour operators were unwise enough to

Table 9.1 The diversity of four intertidal nematode samples from Northern Ireland showing that Simpson's diversity index and the Shannon-Wiener diversity index ($\log_e$) can rank samples in different orders (Lambshead, et al., 1983).

Sample	Number of individuals	Number of species	Shannon-weiner index	Sample rank	Simpson's index	Sample rank
July	158	17	1.86	4	0.28	3
January	398	30	1.95	3	0.35	4
October	410	23	1.96	2	0.26	2
March	417	34	2.59	1	0.14	1

amalgamate these properties into a single index then Antarctica (Cold, Dry) might have the same rating on a "weather index" as a rainforest (Hot, Wet) and both might be indistinguishable from a British seaside town (Cool, Damp).

The first reaction of an ecological community to a small disturbance is likely to be a change in equitability between the species present. An equitability-weighted index would be a sensitive indicator of this. As the disturbance becomes stronger, the species present will change. In this situation, it is impossible to predict what might happen to equitability and a species richness-weighted index will probably give a more coherent picture (Lambshead, 2001). Similarly, equitability-weighted indices are likely to be useful for investigations at a small scale while species richness-weighted indices are likely to give a better result for large-scale studies (Boucher and Lambshead, 1995).

The unpredictable nature of equitability must be considered whenever a diversity index is employed. It is always misleading to physically add together non-contiguous samples and then express the diversity of the amalgamated sample using a diversity index. This is because the behavior of the equitability component of the index is wildly variable depending on the exact species present in each of the samples that are amalgamated. For example, if the area is dominated by a single overriding ecological factor then it is possible that each sample will have the same dominant species present (because the factor favors it) plus a random assortment drawn from the regional pool. In this case, the amalgamated sample will have a high dominance (i.e., low equitability) and hence a low diversity. Conversely, if each sample is dominated by a different species (as is common in deep-sea samples) then the amalgamated sample will have an artificially high equitability and hence a high diversity.

It is essential to consider the likely behavior of the equitability component of diversity before accepting the results indicated by an index at face value. For example, if a factor in the environment causes the population of a few species to grow, then a diversity index will signal a drop in diversity because equitability has fallen, even if the remaining species are quite unaffected.

Terrestrial biodiversity researchers have tended to use species richness as their primary method for analyzing diversity because terrestrial organisms tend to be reasonably well classified, so assessing conspecificity between samples is possible. In addition, terrestrial sampling methods tend to be semi-quantitative and hence unsuitable for assessing equitability. This is of little relevance for species richness calculations but prejudices the employment of diversity indices. Marine benthic researchers can obtain good quantitative samples from corers but the taxonomy of many marine invertebrates is problematic, making assessment of conspecificity difficult. For these reasons, marine benthic biodiversity research has tended to concentrate on the ecological diversity of samples using diversity indices for measurement.

In conclusion, diversity is not a "real" property of ecosystems and is only meaningful when employed correctly to address specific questions at a given scale.

9.2 Global Issues

9.2.1 Distribution and Abundance

The first astonishing feature of nematode biodiversity that strikes a new student is that free-living nematodes are ubiquitous; they are found everywhere. The author has been involved with studies of European soils, African tropical rainforest soils, Loch Ness, polluted estuaries, sandy beaches, the deep-sea abyssal plain, Antarctic sediments, and frozen coal dust from the Arctic circle – not once have we failed to find nematodes. One can do no better than quote that great student of nematology, Nathan Cobb. "[Nematodes] occur in arid deserts and at the bottom of lakes and rivers, in the waters of hot springs and in the polar seas where the temperature is constantly below the freezing point of fresh water. They were thawed out alive from Antarctic ice in the far south by members of the Shackleton Expedition. They occur at enormous depths in Alpine lakes and in the ocean."

Another of Cobb's quotes has become one of the most famous statements in zoological literature. "If all matter in the universe except the nematodes were swept away, our world would still be dimly recognisable, and if, as disembodied spirits, we could then investigate it, we should find its mountains, hills, vales, rivers, lakes and oceans represented by a film of nematodes" (Cobb, 1914).

Free-living nematodes are usually small; the average marine free-living, deep-sea nematode is 0.5 – 0.75 mm long, and sediment-soil dwelling. Their sinusoidal movement method of locomotion works best when pushing against a particulate substrate. A few aquatic nematodes can swim, for example monhysterids, and some terrestrial nematodes can move along plants through thin films of water, but most nematodes are infaunal. In soils, lack of water can limit nematode distributions, but in general, the limitations of an apparently simple morphology seem to be a minor impediment to the phylum.

Table 9.2 compares nematode abundance per area from a variety of biotopes. Obtaining data for such a comparison is made difficult by the incompatible methods for measuring abundance employed by marine and terrestrial nematologists. Marine nematologists commonly use abundance per area, which is the most accurate method given the vertical distribution of nematodes in marine sediments. Terrestrial colleagues commonly favor abundance per weight or volume of soil. A recurring theme in the literature is that nematodes are "predominantly terrestrial" (e.g., Groombridge, 1992). The abundance data show no clear preference of nematodes for either environment, indeed the highest abundance is found in coastal muds and muddy saltings where the sea and land meet. This thin smear of salty mud around the edge of the continents seems to be the most productive nematode habitat.

Table 9.2 Examples of nematode abundance, as millions per m^2, from a variety of biotopes ([1]Boucher and Lambshead, 1995; [2]Brown, 1998; [3]Platt, 1981; [4]Boucher, 1997; [5]David, 1998; [6]de Goede and Bongers, 1998; [7]Bloemers, et al., 1997).

Biotope	Abundance (10^6 m^{-2})
[1]North Atlantic abyssal plain	0.3
[2]Central Pacific abyssal plain	0.1
[1]West European bathyal slope	0.6
[1]West European sandy sublittoral	3.9
[1]West European muddy sublittoral	84.6
[3]West European coarse sandy beach	0.1
[3]West European fine sandy beach	3.7
[3]West European muddy estuary	12.4
[3]West European saltmarsh	24.7
[4]SW Pacific coral lagoon mud	0.7
[4]SW Pacific coral lagoon sand	1.4

Continued

Biotope	Abundance (10^6 m^{-2})
[5]North European recent lake, shallow mud	1.9
[5]North European recent lake, deep sediment	0.4
[6]West European grassland	3.1
[6]West European pasture soil	7.9
[6]South European heathland	0.1
[6]Central European mountain grassland	1.7
[6]Central European grassland	6.7
[6]Central European peat meadow	1.4
[6]North American pasture	2.0
[7]African primary rainforest	2.4
[7]African secondary rainforest	2.4
[7]African slash and burn "rainforest"	0.8
[7]African deforested "rainforest"	0.3

Possibly the most interesting conclusion of this limited analysis is that nematode abundance in the two major domains, marine and terrestrial, is surprisingly similar. Impoverished habitats, including the extensive deep-sea abyssal plains, have 10^5 nematodes per square metre while productive habitats contain 10^6 or, exceptionally, 10^7 possibly with an upper limit of 10^8 animals. Given that the surface of the earth is about 3.62×10^8 km^2, then there are at least 10^{19} free-living nematode individuals in the world (based on a conservative estimate of 10^5 nematodes m^{-2}). The large estimate is to some degree a direct consequence of the small size of these animals. Nevertheless, compared to other organisms, nematodes are astonishingly common. It has been estimated that 80% to 90% of all metazoa are nematode worms (Bongers, 1988; Jairajpuri and Ahmad, 1992).

The distribution of marine nematodes in space is similar to other benthic organisms in that it is resource based. Abundance is highest in coastal waters, especially organically rich muds, and gradually declines proportionally with distance from the continents and with increased depth (the two factors are correlated; e.g., Lambshead, et al., 1994; Vincx, et al., 1994). This is because production in marine systems primarily depends on the photic zone or on terrigenous (terrestrial) input. The photic zone only extends for 200 m down into the water column; below that region life is almost entirely dependent on the organic flux, which weakens as it passes through deeper layers of water containing hungry stomachs. Marine nematode abundance, in particular, is highly correlated with food in the sediments. Figure 9.2 shows the log of nematode abundance regressed against the hydrogen index, a measure of food quality and quantity, for four stations located at depths from 400 m to 3400 m in the Arabian Sea ($R^2 = 0.996$, $P = 0.002$, Cook, et al., 1999). This close association is all the more surprising because the depth transect runs through an oxygen minima zone, demonstrating another property of marine nematodes – their ability to overcome environmental factors that would act as major disturbances to other organisms.

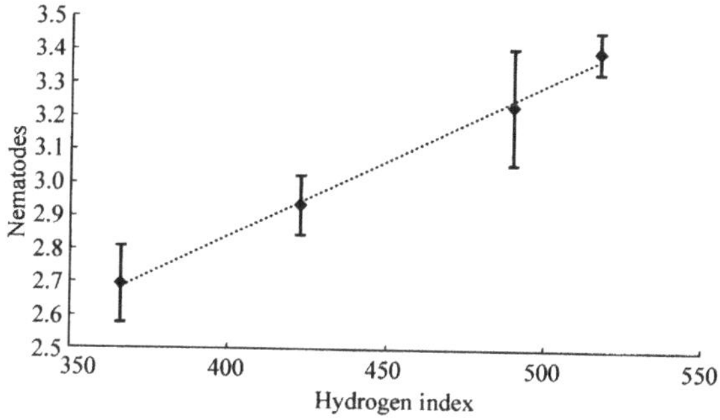

Figure 9.2 Linear regression of $\log_{10}$ mean nematode abundance plotted against mean hydrogen index (mg hydrocarbon g^{-1} total organic carbon, i.e., food) and the 95% confidence of the means for four deep stations from the Arabian Sea. Note the close association between nematode abundance and food (Cook, et al., 1999).

9.2.2 Global Species Richness

The natural history of marine nematodes suggests that this group might have a high global species richness (Lambshead, 1993). They are small infaunal organisms with limited mobility, a conservative reproductive method (i.e. internal fertilization with limited egg production), and no obvious dispersal phase in their life history. These are characteristics normally associated with animals that have restricted biogeographic ranges and high speciation rates (Castillo-Fernandez and Lambshead, 1990).

There are three ways of estimating global species richness for groups for which taxonomists have not yet produced a comprehensive global taxonomic census (i.e., most taxa). The first might be described as the "Trust me, I'm a museum taxonomist" approach, the second might be called "Let's extrapolate", and the third, "Playing the ratio game".

The one clear truth about marine nematode global diversity is that taxonomists have singularly failed to produce a comprehensive global census. Only 4000 or so species of marine nematodes have been described (Warwick, et al., 1998). Many types are lost and many descriptions are inadequate to the point where it is difficult to reconcile them with newly collected specimens. The global coverage is very uneven; around 10% of species records come from the coasts of Great Britain and Ireland. In general, northwest European coasts are reasonably known with scattered information from other European coastlines and North America. Most of the rest of the world is unknown territory. For example, Jensen (1988) found that only 4% of the species in his Norwegian Sea samples could be given species names and Tietjen (1984) reported that only 1% of species were previously described from some Venezuela Basin deep-sea samples. The situation for soil nematodes is possibly not quite as depressing. Ecologists have often loosely applied European names to non-European species, so the marine nematode global biogeographic record is unreliable.

The taxonomist approach to estimating species richness rests on the concept that a museum taxonomist knows the species richness of a number of regions and has seen a sufficient sample from less well censused regions to estimate how many species remain to be discovered. This method was used to estimate the global nematode species richness at 10^6 (May, 1988). It is unreliable for nematodes because not enough regions have been adequately censused to give any baseline for reasonable estimation, especially since the regions with supposedly the highest species richness are the least explored.

The second method for estimating global species richness is to use an extrapolation from a single set of data from a single region to the world. The idea is to collect a series of samples along a transect until the collectors curve settles down to a steady rise (see Fig. 9.1). This is then used to calculate the rate of collection of new species with distance. This figure can then be applied to the total size of the environment to get a maximum species richness. No marine nematode data set exists which could be used for such a calculation. However, Grassle and Maciolek (1992) calculated the rate of increase of the collectors curve for marine macroinvertebrates along a bathyal transect off the east coast of North America at one new species per kilometre. Taking this figure as one new species per square kilometre they calculated the global diversity of marine macrofauna as 10^8. They reduced this figure to 10^7 because there is reason to believe that samples from bathyal depths are unusually diverse (see regional diversity section). Lambshead (1993) pointed out that in any one location, nematode abundance was greater than macrofauna abundance by about 10^2, and that this was commonly reflected in the species richness by about 10^1 more species. This would give marine nematodes a global species diversity of 10^8 using Grassle and Maciolek's (1992) calculation as a guide. This approach is problematic in that it makes the assumption that the conditions in the censused region are repeated throughout the environment, which is why Grassle and Maciolek (1992) conservatively modified their final conclusion. Another potential problem with extended extrapolation is that tiny changes in the initial assumptions can cause large variations in the final estimate.

A third way to look at this problem is to divide the total global abundance, a figure we can be moderately confident about, by the estimated number of species and see whether the resulting mean number of individuals per species is reasonable. The common krill (*Euphausia superba*) is considered to be the most abundant metazoan species on earth with around 10^{14} individuals (Price, pers. comm.). If we take this figure as the maximum mean abundance per nematode species we would estimate a global species richness of about 10^5. If *E. superba* really is the most abundant metazoan then this estimate is too low. Human beings have a global abundance of about 10^{10}. Taking this figure as the mean nematode abundance per species gives an estimated global species richness for free-living nematodes at around 10^9. This is probably too high, human beings are a highly successful cosmopolitan terrestrial species, which would inflate their abundance, but on the other hand people are considerably larger than any nematode so can pack fewer individuals into a given area.

So we are left with an estimate of between 10^6 and 10^8, which at least has the merit of consistency with the other estimation methods. My instincts incline on current evidence more towards the 10^6 end of the range, assuming our grasp of what is a nematode species is secure.

It is by no means clear that we do have an adequate classificatory grip on the marine nematode species concept, currently species are defined purely on morphological criteria. The morphological species concept is a purely operational "museum" concept that has limited validity as a true delimiter of species (Mayden, 1999). Morphological criteria are best suited to large organisms with hard skeletal parts; nematodes are small and mostly soft bodied. There have been no molecular studies of the marine nematode species concept but molecular studies of some other marine invertebrate species have shown that what were considered to be distinct species have been found to be clusters of siblings (Butman and Carlton, 1993). Some of these species were thought to be extremely well known, e.g., the polychaetes *Capitella capitata* and *Streblospio benedecti* (Lambshead, 1993). We may be suspicious that some of our apparently well-described nematode species are sibling groups.

There is even some morphological evidence that questions our current understanding of the nematode species concept (Warwick and Robinson, 2000). The sibling species problem does not seem to be such a problem with terrestrial invertebrates. It is not clear to me whether this is because of some fundamental biodiversity distinction between the fauna of marine and terrestrial domains or whether it is simply that the morphology of terrestrial invertebrates has been more thoroughly investigated. The latter proposition is certainly true for soil and marine nematodes.

Sibling species in marine nematodes and other fauna are an example of how a single operational species concept will fail to accurately census all biodiversity compared to a more fundamental evolutionary species concept (Mayden, 1999). The disinterest marine nematologists have historically shown for moving beyond the morphological species concept and incorporating some measure of history is a serious limitation of biodiversity studies because nematode biodiversity is, and should be, a taxon-based science as opposed to the process-based biodiversity research appropriate to some other groups (Ferris, 1999). The work of Warwick and colleagues on taxonomic distinctness (e.g., Warwick and Clarke, 1998) is conceptually a step in the right direction.

The question arises, why are there so many nematode species in the world? The phylum appears to have a similar order of magnitude of species richness as the Arthropoda. An answer might be that nematodes as a taxon show great physiological tolerance of different environmental conditions. But this is a better explanation of their ubiquity and abundance rather than their species richness. One might put it another way and ask what arthropods and nematodes share in common since the two phyla appear to share few characteristics, apart from being molting animals and hence members of the Ecdysozoa. One taxon is a soft-bodied worm with little apparent scope for specialization while the other has a hard exoskeleton that is preadapted for the development of specialized appendages.

Traditional equilibrial ecological theory stipulated that each species that was not physically separated biogeographically must have some unique specialization to exploit a different niche from all surrounding organisms. A taxonomic examination of marine data demonstrates that a number of co-existing species are generic siblings. For example, 13% of genera in the Clyde Estuary data consisted of three or more species. This is even more characteristic of the deep sea, for example, at a single station in the Porcupine Abyssal Plain data, nine species of *Halalaimus* and nine species of *Acantholaimus* were identified (Rice and Lambshead, 1994). Jensen (1988) reported that four species of *Acantholiamus* from the Arctic deep sea had different buccal cavity morphologies, which he considered evidence of food resource partitioning. However, such neat partitioning is the exception rather than the rule. It is not unknown for co-existing species to be only distinguishable on male reproductive characters (Castillo-Fernandez and Lambshead, 1990). For example, the Clyde data included four species of *Neochromadora*, of which three were morphologically similar. This problem of morphological similarity between co-existing species caused Tietjen and Lee (1977) to suggest that marine nematodes used physiological resource partitioning, but there is little supporting evidence.

Modern ecological theory suggests that co-existing sibling species over ecological time scales are possible, assuming a mosaic of non-equilibrium patches (see local diversity section). This still raises the issue of how so many species came into being over evolutionary time scales. Castillo-Fernandez and Lambshead (1990) pointed out that the complicated male reproductive structures, which seem to be unique to each species, coupled with an internal reproductive system may be indicative of sexual selection in marine nematodes through a "key-lock" mechanism. Sexual selection would act to increase speciation through localization and vicariance events and maintain reproductive barriers between similar species. It may be significant that the reproductive system of arthropods is not dissimilar to nematodes.

Seven years on, I find little reason to change the conclusion expressed in Lambshead (1993): "modern research in tropical rainforests and the marine benthos, particularly the deep sea, leads us to speculate that the number of... species in the world is far higher than had previously been considered, although we may regard the highest figures with some considerable degree of suspicion."

The essential missing data that prevent an accurate estimate of global species richness is that we still have no idea of the biogeographic range of marine nematode species.

9.2.3 Species Richness in Major Domains

Free-living nematodes are unique in that they are found in abundance in the three domains: marine, freshwater, and terrestrial. Unfortunately, terrestrial and marine scientists traditionally use incompatible collecting methods. However, recently two studies were carried out in a rainforest and a deep European lake using marine techniques. For the first time this allows a comparison between domains.

Table 9. 3 lists the number of different species (based on morphological criteria) that have been reported from a variety of locations in the three domains plotted against number of specimens used in each analysis for each location (Fig. 9.3). This is the methodology adopted by Lambshead et al. (2000) for measuring nematode species richness. The marine locations separate into two groups, a low diversity intertidal group and a high diversity offshore group. This division is unexpected because the deep sea is normally considered to be more diverse than the shallow water environment although Gray (1994) has produced macrofaunal species richness data (as opposed to diversity index data) to challenge this view. This nematode data supports Gray's controversial conclusion. It is also noteworthy that it does not seem to matter which ocean the data are derived from, Fig. 9.3 uses offshore data from the Pacific and Atlantic Oceans and the Mediterranean Sea.

Table 9.3 The species richness (number of species) reported from different depths from the world's oceans along with the number of specimens used to generate the data ([1]Lambshead, 1986; [2]Warwick and Price, 1979; [3]Warwick and Gee, 1984; [4]Boucher and Lambshead, 1995; [5]Lambshead, et al., 1994; [6]Tietjen, 1984; [7]Thistle, et al., 1995; [8]Rice and Lambshead, 1994; [9]Brown, 1998; [10]Soetaert, et al., 1991; [11]David, 1998; [12]Bloemers, et al., 1997).

Location	Depth (m)	Species richness	No. of specimens
Beach sand, Clyde Estuary[1]	0	113	9755
Lynher Estuary[2]	0	39	4575
Tamar Estuary, upper[3]	0	30	527
Tamar Estuary, mid[3]	0	32	977
Tamar Estuary, lower[3]	0	46	1015
Irish Sea[4]	39 – 56	178	2253
Rockall Trough[5]	545	81	304
Rockall Trough[5]	835	83	292
Rockall Trough[5]	1474	93	334
Venezuela Basin[6]	3517 – 5045	136	1004
HEBBLE[7]	4626	205	2467
Madeira Abyssal Plain[8]	4856	72	578
Porcupine Abyssal Plain[8]	4850	125	1401
San Diego Trough[5]	1050	116	1335
Equatorial Central Pacific[9]	4301 – 4328	128	477
Central Pacific 2°N[9]	4409 – 4414	106	386

Continued

Location	Depth (m)	Species richness	No. of specimens
Central Pacific 5°N[9]	4418 – 4447	60	498
Central Pacific 9°N[9]	4986 – 4994	90	291
Central Pacific 23°N[9]	4871 – 4884	81	407
Western Mediterranean[10]	160	176	720
Western Mediterranean[10]	280	212	818
Western Mediterranean[10]	530	155	608
Western Mediterranean[10]	820	160	741
Western Mediterranean	990	189	808
Western Mediterranean[10]	1220	145	535
Loch Ness[11]	0 – 211	95	6184
Cameroon rainforest[12]	–	431	5000

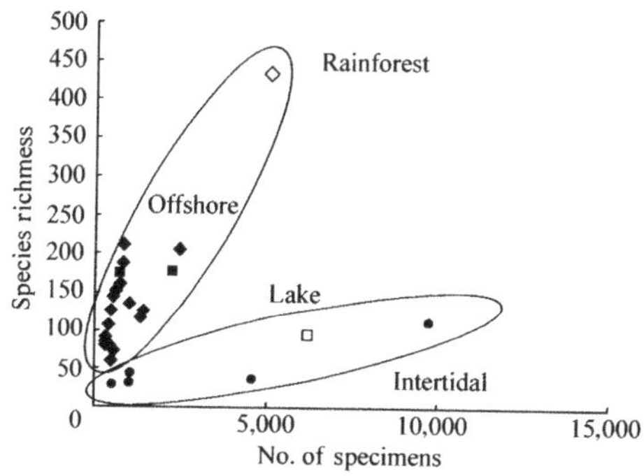

Figure 9. 3 Nematode species richness plotted against the number of specimens identified for a number of deep-water offshore marine locations (filled diamond), shallow-water marine locations (filled square), marine littoral locations (filled circle), deep lake (open square) and a tropical rainforest location (open diamond). See Table 9.3 for description of data.

The lower species richness found in the intertidal data compared with offshore locations suggests that it is perhaps unfortunate that, for understandable logistical reasons, a disproportionate amount of marine research centers on this environment. The single rainforest location lies within the variation shown by the offshore marine group implying similar species richness, while the single lake location shows a biodiversity similar to the littoral marine group.

9.3 Regional Diversity

9.3.1 Productivity Gradients

Rosenzweig (1995) concluded that "The connection between productivity and diversity – long thought to be intimate – remains murky". The problem is that the relationship between productivity and diversity does not appear to be linear, the highest species richness comes at some intermediate level of productivity. Clearly there is little mystery why extremely low levels of productivity are associated with low diversity, because an impoverished resource supply supports few species. Increasing productivity raises diversity in this situation because greater numbers of species can be supported. However, it is a common pattern that past a certain point, further increases in productivity decrease diversity (Huston, 1994; Rosenzweig, 1995). There have been many explanations put forward to explain this commonplace observation (Rosenzweig, 1995) but the more convincing is based on interspecies competition. As productivity increases, population sizes grow and interspecific competition reduces diversity through competitive displacement (Huston, 1994). One problem with this explanation is that there is little evidence for interspecific competition among organisms, and this is certainly true of nematodes. Partly, this may be because it is difficult to establish competition in field experimentation (and laboratory experiments are always suspect) and partly because in most situations, interspecific competition may be often reduced in importance by disturbance (see later).

The best known potentially productivity-mediated diversity gradients are latitudinal gradients, arguably the oldest known diversity pattern in the world (Lambshead, et al., 2000). The deep sea is a particularly useful environment to look for large-scale productivity-mediated diversity gradients because (1) there is little variation in ecological factors such as temperature, (2) all production is secondary and derived from the organic flux to the seabed, and (3) seasonality is minimal. The only ocean for which suitable data exist is the North Atlantic, which is unfortunate as this ocean is divided into basins that, by deep-sea standards, have variable ecological conditions (Lambshead, et al., 2000, 2001). The consequence of this is that if the deep North Atlantic mean nematode diversity, as measured by a diversity index such as the species richness-weighted modified Sanders rarefaction index $ES(X)$, is plotted against latitude no significant productivity gradient can be discerned (Fig. 9.4).

If the data are replotted as species richness (a count of the number of species) against the number of specimens used in the analysis then it may be concluded that the Norwegian Sea is noticeably less diverse than that of the rest of the North Atlantic (Fig. 9.5). This may be attributed to small size, geographic isolation, and the Quaternary glaciation (Lambshead, et al., 2000). Plotted against latitude, the species richness data show a significant rise

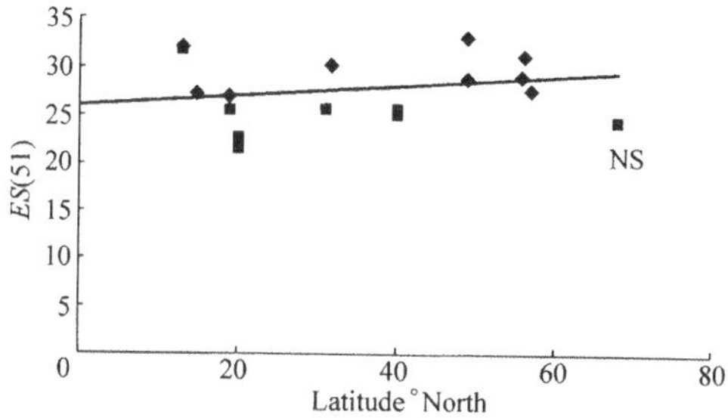

Figure 9.4 The rarefaction ($ES(51)$) means derived from replicate samples of marine nematodes from a number of sites in the North Atlantic plotted against latitude (Lambshead, et al., 2000). Squares indicate stations subject to physical disturbance and NS is the Norwegian Sea site. Excluding the Norwegian Sea the regression equation is: $ES(51) = 26.1 + 0.0471$ Latitude. $R^2 = 0.053$, $P = 0.393$.

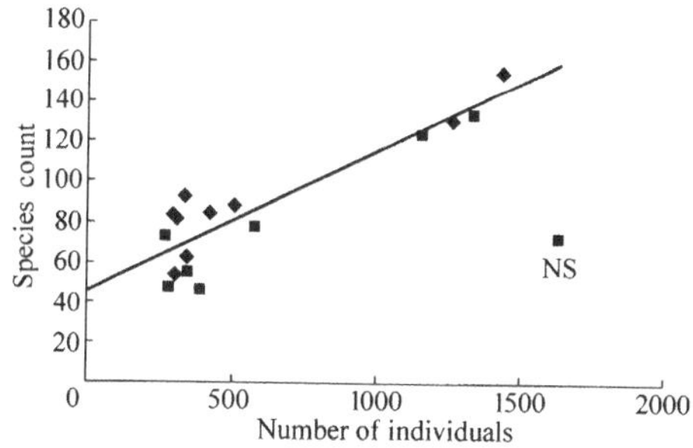

Figure 9.5 Count of marine nematode species (y) per North Atlantic site plotted against the number of individuals (x) in the samples at that site (Lambshead, et al., 2000). Squares indicate stations subject to physical disturbance and NS is the Norwegian Sea site. Excluding the Norwegian Sea the regression equation is: $y = 44.8 + 0.0705x$. $R^2 = 0.818$, $P < 0.001$.

in diversity from the equator to the edge of the Norwegian Sea (Fig. 9.6). Organic input into the North Atlantic is complicated (Lambshead, et al., 2000), but in general productivity also increases from the equator northwards. The regression line is still significant when sample size variation, the different number of specimens used at each location to make a species count, is accounted for (Table 9.4).

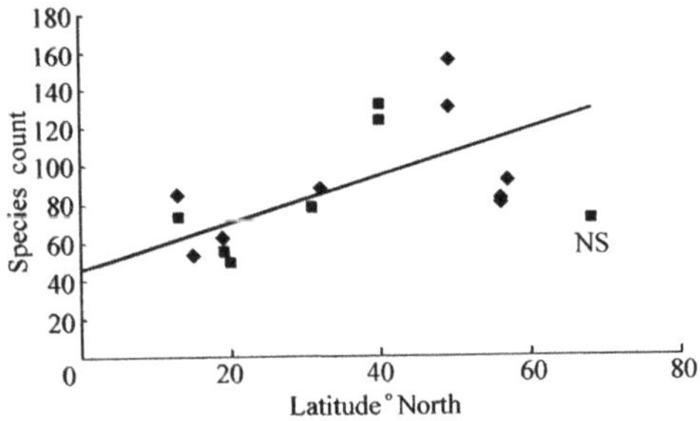

Figure 9.6 Count of marine nematode species (y) plotted against latitude (x) for a number of sites in the North Atlantic. Squares indicate stations subject to physical disturbance and NS is the Norwegian Sea site (Lambshead, et al., 2001). Excluding the Norwegian Sea the regression equation is: $y = 46.0 + 1.24x$. $R^2 = 0.379$, $P = 0.011$.

Table 9.4 Regression of Species Count (SC) on Number of Individuals (NI) and Latitude (LT)

SC = 30.2 + 0.0611 NI + 0.611(s.d =.195) LT
$R^2 = 0.896$, $P = 0.008$

9.3.2 Disturbance Impacts

In modern ecology, disturbance like productivity is considered to have a non-linear relationship with diversity. High levels of disturbance will cause low diversity because species populations are devastated or species even eliminated. So diversity should increase as disturbance falls. But at some critical point, low levels of disturbance allow species populations to grow to the point where interspecific competition acts to reduce diversity. In other words, diversity peaks at some intermediate level of disturbance similar to the way diversity peaks at an intermediate level of productivity (Huston, 1994; Rosenzweig, 1995). So the actual diversity measured for a given taxon in a given location will be the product of a complex non-linear interplay between diversity and productivity.

It is worth noting that a disturbance effect must be clearly distinguished from the disturber, i.e. the disturbance agent. Disturbance can only be measured by its impact, which is to reduce the abundance of organisms by killing them, inhibiting their reproduction or driving them away. The disturber can be any physical, chemical, or biological process, and a disturber can only be detected by its impact on organisms. The same agent may cause a disturbance in one situation or at one frequency but not at others, and to one group of organisms but not others.

In the previous section, I demonstrated how it was impossible to detect a large-scale productivity gradient using a diversity index and suggested the index was sensitive to local ecological effects. Some of the locations in this data set have been subject to physical disturbance (Lambshead, et al., 2000, 2001). For example, the HEBBLE site in the deep North Atlantic off Newfoundland is subject to severe benthic storms. The diversity of this site is significantly lower than undisturbed reference sites in the deep North Atlantic (Table 9.5). Diversity is measured as the modified Sanders rarefaction index, which is robust to sample size, in this case the expected number of species per 51 individuals, *ES*(51), is used.

Table 9.5 Comparison of the mean *ES*(51) diversity at the disturbed HEBBLE site compared to reference sites (Lambshead, et al., 2001).

Sites	No. of samples	Mean *ES*(51)	SD/%
HEBBLE	17	25.11	2.21
Reference sites	18	30.49	2.61
Comparison	T	DF	P
	-6.60	32	0.00

Another example is the Madeira Abyssal Plain site that was subject to a turbidite (a mass of unstable sediment sliding down a submarine slope) approximately 1000 years ago (Table 9.6) (Lambshead, et al., 2001). The Madeira turbidite covers a substantial area(Fig. 9.7).

Both physical disturbances cause a 17% reduction in diversity. The fall in

Table 9.6 Comparison of the mean *ES*(51) diversity at the madeira abyssal plain (MAP) turbidite site compared to reference sites (Lambshead, et al., 2001).

Sites	No. of Samples	Mean *ES*(51)	SD/%
MAP	6	25.39	1.21
Reference Sites	18	30.49	2.61
Comparison	T	DF	P
	-6.47	18	0.00

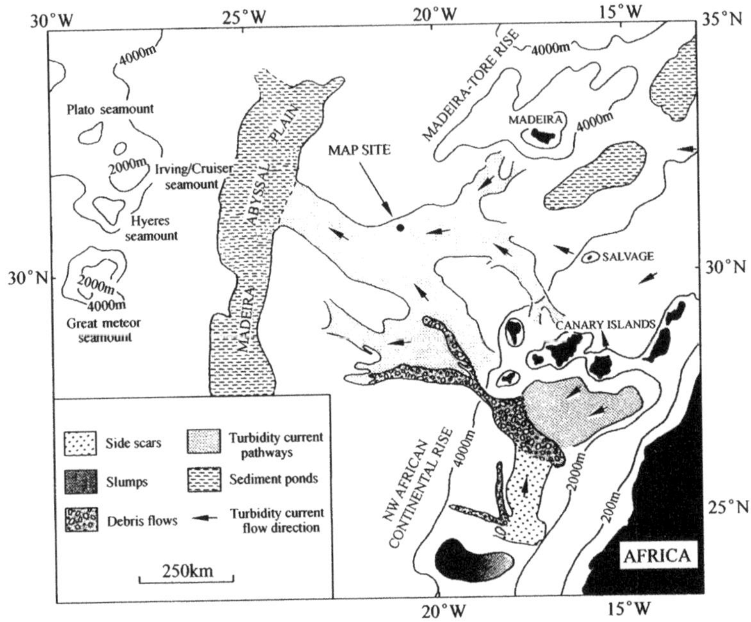

Figure 9.7 Map of the Madeira Abyssal Plain (MAP) showing the location of the collecting site and the extent of the turbidite (Lambshead, et al., 2001).

diversity is modest in both cases but nematodes are known to be robust to physical disturbance (e.g., Austen, et al., 1998; Warwick, et al., 1990). Of particular interest is the long duration of the impact of the turbidite. The diversity index successfully detects a significant drop in diversity caused by a local physical disturbance in both these examples. The studies demonstrate how effective diversity indices can be when employed to address a suitable question at the ecological, or sample diversity, scale.

9.3.3 Diversity in Biotopes

Boucher and Lambshead (1995) investigated the ecological or sample diversity of some major marine biotopes for which data were available, temperate estuarine, tropical sublittoral, temperate sublittoral, bathyal depths, abyssal depths, and hadal depths (Fig. 9.8). The index is the modified Sanders rarefaction index set at expected number of species per 91 individuals.

The results at this scale show a somewhat different pattern than for species richness (Fig. 9.3) on substantially the same data. The bathyal and abyssal

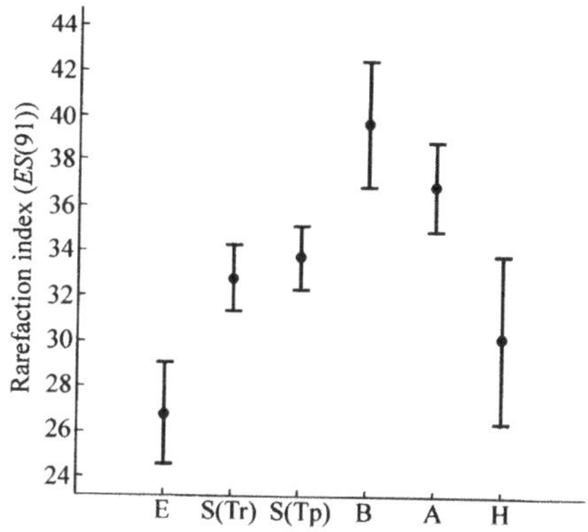

Figure 9.8 The mean, and standard error of the mean, sample diversity of six biotopes as measured by the modified Sanders rarefaction index, *ES*(91); E (temperate estuarine), S (Tr) (sublittoral tropical), S (Tp) (sublittoral temperate), B (bathyal), A (abyssal), and H (hadal) (Boucher and Lambshead, 1995).

deep-sea samples are significantly more diverse than the sublittoral coastal samples. The estuarine samples have a significantly lower diversity than the offshore samples in this analysis showing a similar pattern to their low species richness in Fig. 9.3. Added to this analysis are hadal (deep trench) samples with a mean diversity not significantly different from the estuarine biotope. It is noteworthy that the tropical coastal samples are not more diverse than the temperate coastal samples.

The data show a classic intermediate diversity curve with diversity peaking at bathyal depths. There is certainly a productivity gradient, the organic flux to the seabed declines with depth, but there are also different levels of disturbance. To nematodes, estuaries are physically challenging environments and in general physical disturbance declines with depth. However, trenches can also be disturbed, as the sides tend to fall in. Any explanation of these results has to consider the interaction between productivity and disturbance.

Huston (1994) has proposed a theoretical 3-dimensional graph to summarize the effect of productivity and disturbance on diversity. The two ecological factors make up the horizontal axis of the graph and diversity is measured on a third vertical axis by contour lines. The result is the "Huston Hill" with diversity peaking at some intermediate point in the centre of the graph. I have attempted to place the nematode diversity from biotopes of different depths onto a Huston curve (Fig 9.9). Estuarine environments are highly productive and disturbed leading to low sample diversity. With increasing depth, both diversity and disturbance decrease allowing an increase in sample diversity to some optimum point, probably at a bathyal depth. Further decreases in disturbance and productivity cause a concomitant decrease in diversity due to increasing competitive exclusion. In deep, low productivity trenches (not all trenches are low productivity), diversity falls further through a combination of usually slow growth rates and high disturbance – the trench walls tend to collapse.

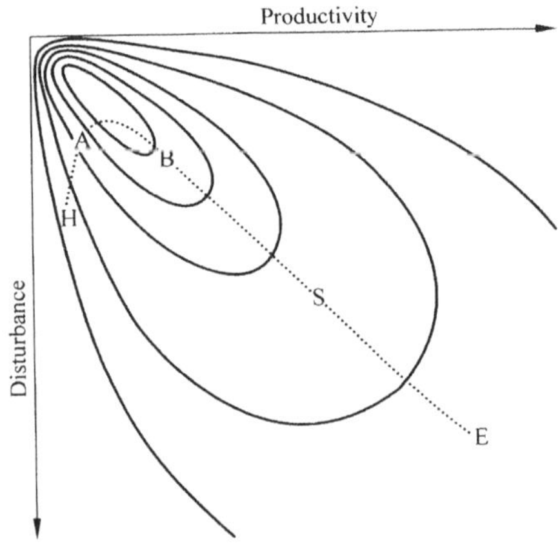

Figure 9.9 Change of nematode ecological diversity with depth from E (estuarine) to S (sublittoral), B (bathyal), A (abyssal), and H (hadal) biotopes (dotted line) plotted on the Huston diversity graph (Huston, 1994) to show how ecological diversity changes with depth. This is a 3-dimensional graph, diversity is measured on the third vertical axis as indicated by the contour lines.

9.4 Local Diversity

9.4.1 Patch Dynamic Theory

So far, we have discussed the processes controlling biodiversity in a way more or less independent of scale. Clearly large-scale processes are mediated through local scale impacts acting at the level of the individual organism. At the local scale, it is a commonplace observation that the environment is patchy and that there is a large body of literature discussing the importance of patch dynamics in the maintenance of diversity over ecological time scales. Grassle (1989) convincingly summarized the role of non-equilibrium patch effects on biodiversity, as an explanation of how such patches maintain high species richness in apparently featureless deep-sea sediment. Grassle (1989) noted three key features of the deep sea: (1) it is a large habitat with few barriers to dispersal, (2) it has low productivity, and (3) large-scale environmental disturbance is rare (i. e., it is not physically controlled). Grassle's spatial temporal mosaic theory postulates that in such a system, small-scale discrete disturbance and fluctuations in food supply maintain high diversity by initiating patches, each of which runs through unique non-equilibrium cycles over different time scales depending on exact start-up conditions. Species richness is high because for every species, somewhere within dispersion distance is a patch offering acceptable conditions. And when it becomes unsuitable, another acceptable patch will have become available within a reasonable distance.

This theory is illustrated in Fig. 9. 10. Patch A is disturbed by a physical event at time T_1, wiping out all organisms by T_2. The patch is recolonized and supports a diverse assemblage through T_3 and T_4. Patch B is not impacted and its diversity gradually falls from T_1 to T_4 because of competitive exclusion. Patch C is biologically disturbed at T_2 and is recolonized with a diverse assemblage by

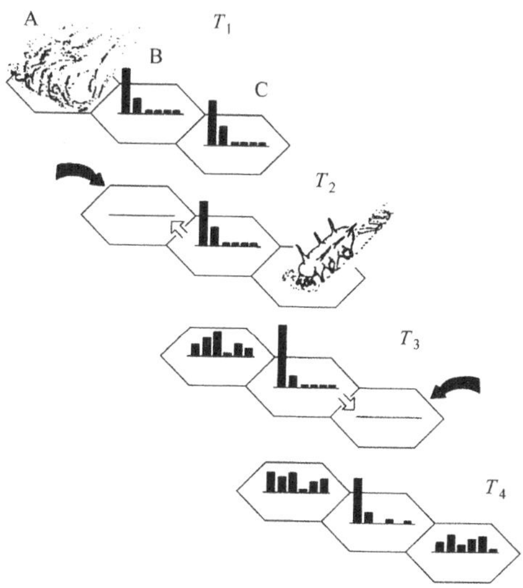

Figure 9. 10 The spatial temporal mosaic theory (Grassle, 1989). Three deep-sea patches, A, B, and C are subject to different events from time T_1 to T_4. Patch A is physically disturbed, patch B undisturbed, and patch C biologically disturbed. The bars within the patches indicate the number of species present in the patch and the height of each bar indicates the abundance of each species (after Paterson, 1993). Arrows indicate sources of colonizing species. See text for full explanation.

T_4. However, Patch C has had a different history compared to Patch A, so the two patches are unlikely to have identical assemblages.

Grassle (1989) suggested that the spatial temporal mosaic theory might be equally applicable to rainforests, so it is perhaps significant that rainforests and the deep sea show the same nematode species richness (Fig. 9.3). If this theory is the explanation for a high level of nematode species richness then Fig. 9.3 suggests that patch dynamic effects are important for all offshore benthic nematode communities. It is noteworthy that marine intertidal and lake nematode communities, which seem to form a separate low species richness group, fail to meet a number of Grassle's (1989) criteria because they are in environments that are fragmented with barriers that impede dispersal and that have high levels of physical disturbance. Grassle's criterion of low productivity does not appear to be important. The species rich group is not completely derived from low abundance fauna.

9.4.2 Food Controlled Patches

Grassle (1989) suggested that discrete falls of food, such as phytodetritus, would be one of the agents causing increased patchiness and hence increased diversity. This offered a test of the spatial temporal mosaic theory. Rice and Lambshead (1994) and Lambshead and Hodda (1994) tested the theory by comparing the aggregation of nematodes at the phytodetritus-impacted Porcupine Abyssal Plain site (PAP) and the Madeira Abyssal Plain site (MAP) which is not subject to a seasonal input of phytodetritus. A χ^2 test showed that around 72% of the species at PAP were significantly aggregated but only 28% at MAP. It is also noteworthy that the species in the patches caused by the phytodetritus at PAP were discordantly distributed, i. e., they tended to be aggregated in different patches, as predicted by the spatial temporal mosaic hypothesis.

Brown (1998) compared mean-sample diversity, using the Shannon-Weiner index $H'(\log_2)$, for phytodetritus enriched stations in the Atlantic (PAP1 = 5.086; PAP2 = 5.643) and the Central Pacific (0° = 5.174; 2°N = 5.085; 5°N = 5.022) against stations not enriched by phytodetritus in the Atlantic (MAP = 4.606) and the Central Pacific (9°N = 4.992; 23°N = 4.798). The phytodetritus stations were significantly more diverse ($P = 0.0369$, using a Mann-Whitney U-test) indicating that the phytodetritus caused both increased patchiness and increased sample diversity as predicted by the spatial-temporal mosaic hypothesis.

9.4.3 Disturbance Controlled Patches

Schratzberger and Warwick (1998) investigated the response of intertidal nematodes to different frequencies of small-scale physical disturbance, by laboratory experiments with microcosms. They reported different results between sand and mud nematode assemblages. The sand assemblages were more robust to physical disturbance than the mud community. This is probably because nematodes in sandy sediments would normally experience, and be adapted to, higher levels of physical disturbance through wave and water currents than nematodes adapted to live in sheltered mudflats.

The sand assemblages disturbed in the experiment showed a reduction in diversity but the mud assemblages responded in a more complicated way. Diversity first increased, peaking at an intermediate frequency of disturbance, and then declined at higher frequencies of disturbance. These changes in diversity can be interpreted as classic non-equilibrium reactions to frequency of disturbance (Huston, 1994; see Fig. 9.11). The sand assemblage is already disturbed to the point where physical impact has reduced diversity from the theoretical optimum so further disturbance correspondingly reduces diversity. The sheltered mud community on the other hand has a diversity below optimum because of competitive interactions between the species. An increase in disturbance actually raises diversity by preventing such interactions. However, as the frequency of disturbance continues to rise, the impact of disturbance

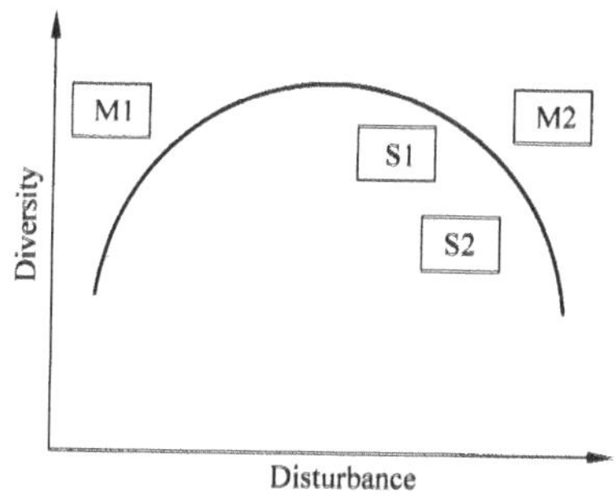

Figure 9. 11 Mud and sand intertidal nematode communities exposed to increasing frequencies of disturbance in a microcosm experiment. The graph shows the diversity of the sheltered mud community (M) and naturally disturbed sand community (S) at the start of the experiment (1) and after disturbance (2). Further disturbance causes a reduction in diversity of the sand community but the sheltered mud community follows a more complicated pattern with diversity peaking at an intermediate level of disturbance where frequency of disturbance is sufficient to prevent competitive exclusion. At higher frequencies of disturbance, diversity drops as in the sand experiment. Note that the mud and sand diversities do not share the same scales on the axis; they are overlaid here for convenience (after Schratzberger and Warwick, 1998).

becomes sufficient to reduce diversity as in the sand experiment.

9.4.4 Biologically Controlled Patches

Austen et al. (1998) carried out a series of mesocosm experiments on the impact of the bivalves *Nuculoma tenuis* and *Abra abra*, and the sea urchin *Brissopsis lyrifera* on subtidal shallow nematode communities. The results are shown in Fig. 9.12; the actions of the larger animals always increased diversity but in a manner consistent with the intermediate disturbance hypothesis. Note that the impact of *Abra* is more intense than *Nuculoma* in that *Abra*-low density disturbance results in the highest diversity compared to *Nuculoma*-medium density disturbance. *Abra* is a noticeably larger animal than *Nuculoma* and so could be expected to have a higher impact. Examination of the actual species present in the experimental samples by multivariate analysis showed that the type of disturbance (i. e., *what* caused it) was more important than the intensity of disturbance. This is an elegant experimental demonstration of how apparently similar biological disturbances can cause variability in the resulting patches leading to a spatial temporal mosaic.

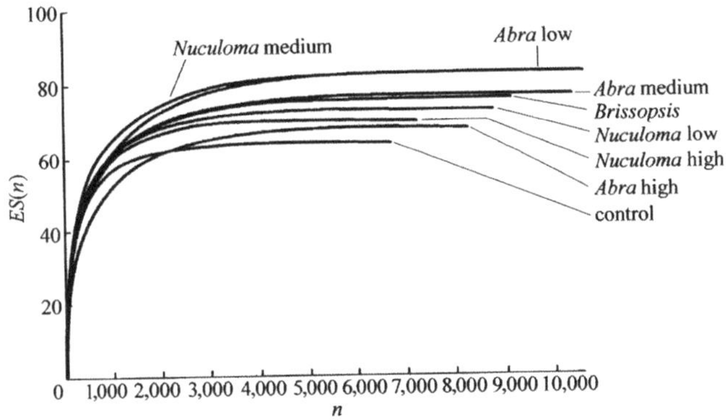

Figure 9.12 A mesocosm experiment demonstrating the impact of larger organisms on nematode diversity, as measured by modified Sanders rarefaction graphs. *ES*(*n*) is the Expected number of nematode Species (*ES*) per number (*n*) of nematodes. "High", "medium", and "low" indicate the abundance of the larger organisms (*Abra*, *Brissopsis*, and *Nuculoma*) acting as a disturbance in the mesocosms. The controls have no large organisms. Note that the activities of the larger organisms always increase nematode diversity but there is evidence of a non-equilibrium intermediate disturbance curve (after Austen, et al., 1998).

9.5 Conclusions and Overview

Free-living nematodes are ubiquitous, abundant, and species rich. Most habitats have between 10^5 and 10^6 nematodes per m^2 and total global abundance is approximately 10^{19}. Global and regional species richness is difficult to estimate because of our ignorance of biogeographic ranges and lack of confidence in the morphological criteria used to define species, but global species richness is probably somewhere between 10^6 and 10^8. Molecular analysis of free-living nematodes is undoubtedly necessary to clarify these issues. The conservative reproductive strategy of free-living nematodes should encourage speciation through sexual selection over evolutionary time scales.

In large, unfragmented, non-physically controlled environments such as deep sea and offshore coastal marine and rainforests, local to regional species richness is high, mediated by non-equilibrium local patch dynamic interactions. Patch-forming agents include irregular distribution of food or other resources, physical disturbance, or the activities of other organisms. In fragmented, physically-controlled environments such as marine intertidal and lakes, species richness is demonstrably lower because of a reduction in patch dynamics. Productivity gradients do influence species richness over large scales but productivity does not seem to be an overwhelmingly important factor for determining nematode species richness.

Sample diversity in marine biotopes is linked to the depth gradient, diversity peaking at bathyal depths. I would postulate that the differences between sample diversity and species richness are the result of the impact of local "within-patch" disturbance and productivity which is measured by sample diversity analysis but not by species richness per location analysis. This is probably why mean sample diversity analysis tends to produce intermediate disturbance curves, reflecting the non-equilibrium dynamics within a patch.

The properties of abundance, richness, and ubiquity, that make nematodes unique among the Metazoa, also make them ideal for testing biodiversity theory at a variety of scales and in a variety of habitats. Once the problems involving species identification criteria and biogeographic ranges are solved by the application of molecular techniques, I envisage nematode experimentation releasing a flood of information about the processes that produce and maintain biodiversity. Nematodes are in no danger of extinction but they can help provide the information to save less robust species.

References

Austen, M. C., S. Widdicombe, and N. Villano-Pitacco. 1998. Effects of biological disturbance on diversity and structure of meiobenthic nematodes communities. Marine Ecology Progress Series 174: 233 – 246.

Blaxter, M. 1998. *Caenorhabditis elegans* is a nematode. Science 282: 2041 – 2046.

Bloemers, G. F., M. Hodda, P. J. D. Lambshead, J. H. Lawton, and F. R. Wanless. 1997. The effects on diversity of tropical soil nematodes. Oecologia 111: 575 – 582.

Bongers, T. 1988. De Nematoden van Nederland. Utrecht: Koninklijke Nederlandse Natuurhistorische Vereniging.

Boucher, G. 1997. Structure and biodiversity of nematode assemblages in the SW lagoon of New Caledonia. Coral Reefs 16: 177 – 186.

Boucher, G., and P. J. D. Lambshead. 1995. Marine nematode ecological biodiversity in samples from temperate, tropical and deep-sea regions. Conservation Biology 9: 1594 – 1604.

Brown, C. J. 1998. Effects of a phytodetrital input on nematode communities of the abyssal, equatorial Pacific. Ph. D. Thesis, Southampton University, Southampton, UK.

Buchsbaum, R. 1968. Animals Without Backbones: 2. Harmondsworth, UK: Penguin Books.

Butman, C. A., and J. T. Carlton. 1993. Biological Diversity in Marine Systems (BIOMAR). Washington, DC: The National Science Foundation.

Castillo-Fernandez, D., and P. J. D. Lambshead. 1990. Revision of the genus *Elzalia* Gerlach, 1957 (Nematoda: Xyalidae) including three new species from an oil producing zone in the Gulf of Mexico, with a discussion of the sibling species problem. Bulletin of the British Museum of Natural History 56: 63 – 71.

Cobb, N. A. 1914. Nematodes and their Relationships. US Department of Agriculture Yearbook.

Cook, A. A., P. J. D. Lambshead, L. E. Hawkins, N. Mitchell, and L. A. Levin. 1999. Nematode abundance at the oxygen minimum zone in the Arabian Sea. Deep Sea Research 47: 75 – 85.

David, R. 1998. An investigation of biodiversity patterns and processes in nematode populations of Loch Ness. Ph. D. Thesis, Southampton University, Southampton, UK.

de Goede, R. G. M., and T. Bongers. 1998. Nematode Communities of Northern Temperate Grassland Ecosystems. Giessen, UK: Focus Verlag.

Ferris, V. R. 1999. Species concepts do matter in nematology. Journal of Nematology 31: 93 – 94.

Freckman, D. W., T. H. Blackburn, L. Brussaard, P. Hutchings, M. A. Palmer, and V. R. Snelgrove. 1997. Linking biodiversity and ecosystem functioning of soils and sediments. Ambio 26: 556 – 562.

Ghilarov, A. 1996. What does biodiversity mean – scientific problem or convenient myth. Trends in Ecology and Evolution 11: 304 – 306.

Grassle, J. F. 1989. Species diversity in deep-sea communities. Trends in Ecology and Evolution 4: 12 – 15.

Grassle, J. F., and N. J. Maciolek. 1992. Deep-sea species richness: Regional and local diversity estimates from quantitative bottom samples. American Naturalist 139: 313 – 341.

Gray, J. S. 1994. Is deep-sea biodiversity really so high? Species diversity of the Norwegian continental shelf. Marine Ecology Progress Series 112: 205 – 209.

Groombridge, B. 1992. Global Biodiversity. London: Chapman and Hall.

Hurlbert, S. H. 1971. The non-concept of species diversity: a critique and alternative parameters. Ecology 52: 577 – 586

Huston, M. A. 1994. Biological Diversity. Cambridge: Cambridge University Press.

Jairajpuri, M. S., and W. Ahmad. 1992. Dorylaimida: Free-living, Predaceous and Plant-parasitic nematodes. New Delhi: Oxford University Press and IBH.

Jensen, P. 1988. Nematode assemblages in the deep-sea benthos of the Norwegian Sea. Deep-Sea Research 35: 1173 – 1184.

Lambshead, P. J. D. 1986. Sub-catastrophic sewage and industrial waste contamination as revealed by marine nematode faunal analysis. Marine Ecology Progress Series 29: 247 – 260.

Lambshead, P. J. D. 1993. Recent developments in marine benthic biodiversity research. Oceanis 19: 5 – 24.

Lambshead, P. J. D. 2001. Numerical biodiversity analysis with the Natural History Museum and Scottish Association for Marine Sciences BioDiversity Professional PC package. Oceanis 24: 375 – 391.

Lambshead, P. J. D., and M. Hodda. 1994. The impact of disturbance on measurements of variability in marine nematode populations. Vie et Milieu 44: 21 – 27.

Lambshead, P. J. D., H. M. Platt, and K. M. Shaw. 1983. Detection of differences among assemblages of marine benthic species based on an assessment of dominance and diversity. Journal of Natural History 17: 859 – 874.

Lambshead, P. J. D., B. J. Elce, D. Thistle, J. E. Eckman, and P. R. O. Barnett. 1994. A comparison of the biodiversity of deep-sea marine nematodes from three stations in the Rockall Trough, Northeast Atlantic, and one station in the San Diego Trough, Northeast Pacific. Biodiversity Letters 2: 95 – 107.

Lambshead, P. J. D., J. Tietjen, T. Ferrero, and P. Jensen. 2000. Latitudinal diversity gradients in the deep-sea with special reference to North Atlantic nematodes. Marine Ecology Progress Series 194: 159 – 167.

Lambshead, P. J. D., J. Tietjen, A. Glover, D. Thistle, and A. Gooday. 2001. The impact of large-scale natural physical disturbance on the diversity of deep-sea North Atlantic nematodes. Marine Ecology Progress Series 214: 121 – 126.

Magurran, A. E. 1988. Ecological Diversity and its Measurement. London: Chapman and Hall.

May, R. M. 1988. How many species are there on earth? Science 241: 1441 – 1449.

Mayden, R. L. 1999. Consilience and a hierarchy of species concepts: Advances toward closure on the species puzzle. Journal of Nematology 31: 95 – 116.

Norse, E. A., and R. E. McManus. 1980. Ecology and living resources biological diversity. Pp. 31 – 80. In The Eleventh Annual Report of the Council On Environmental Quality. Washington, DC: Council On Environmental Quality.

Paterson, G. L. J. 1993. Patterns of polychaete assemblage structure from bathymetric transects in the Rockall Trough, N. E. Atlantic Ocean. Ph. D. Thesis, University of Wales, UK.

Platt, H. M. 1981. Meiofaunal dynamics and the origin of the metazoa. Pp. 207 – 216. In P. L. Forey, ed. The Evolving Biosphere, Chance, Change, and Challenge.

Cambridge: Press Syndicate of the University of Cambridge.

Rice, A. L., and P. J. D. Lambshead. 1994. Patch dynamics in the deep-sea benthos: the role of a heterogeneous supply of organic matter. Pp. 469 – 499. In P. S. Giller, A. G. Hildrew, and D. G. Raffaelli, eds. Aquatic Ecology: Scale, Pattern, and Process. 34th Symposium of the British Ecological Society. Oxford: Blackwell Scientific Publications.

Rosenzweig, M. L. 1995. Species Diversity in Time and Space. Cambridge: Cambridge University Press.

Sanders, H. L. 1968. Marine benthic diversity: a comparative study. American Naturalist 102: 243 – 282

Schratzberger, M., and R. M. Warwick. 1998. Effects of physical disturbance on nematode communities in sand and mud: a microcosm experiment. Marine Biology 130: 643 – 650.

Soetaert, K., C. Heip, and M. Vincx. 1991. Diversity of nematode assemblages along a Mediterranean deep sea transect. Marine Ecology Progress Series 75: 275 – 282.

Thistle, D., P. J. D. Lambshead, and K. M. Sherman. 1995. Changes in proportional abundance of nematode buccal cavity and tail-shape groups in three different regions of the North Atlantic deep sea. Vie et Milieu 45: 107 – 115.

Tietjen, J. H., and J. J. Lee. 1977. Feeding behaviour of marine nematodes. Pp. 21 – 35. In B. C. Coull, ed. Ecology of Marine Benthos. Columbia, SC: South Carolina Press.

Tietjen, J. H. 1984. Distribution and species diversity of deep-sea nematodes in the Venezuela Basin. Deep-Sea Research 31: 119 – 132.

Vincx, M., B. J. Bett, A. Dinet, T. Ferrero, A. J. Gooday, P. J. D. Lambshead, O. Pfannkuche, T. Soltwedel, and A. Vanreusel. 1994. Meiobenthos of the deep north-east Atlantic. Advances in Marine Biology 30: 1 – 88.

Warwick, R. M., and K. R. Clarke. 1998. Taxonomic distinctness and environmental assessment. Journal of Applied Ecology 35: 532 – 543.

Warwick, R. M., and J. M. Gee. 1984. Community structure of estuarine meiobenthos. Marine Ecology Progress Series 18: 97 – 111.

Warwick, R. M., and R. Price. 1979. Ecological and metabolic studies on freeliving nematodes from an estuarine mudflat. Estuarine and Coastal Marine Science 9: 257 – 271.

Warwick, R. M, and J. Robinson. 2000. Sibling species in the marine pollution indicator genus *Pontonema* Leidy (Nematoda: Oncholaimidae) with a description of *P. mediterranea sp. nov*. Journal of Natural History 34: 641 – 662

Warwick, R. M., H. M. Platt, K. R. Clarke, J. Agard, and G. Gobin. 1990. Analysis of macrobenthic and meiobenthic community structure in relation to pollution and disturbance in Hamilton harbour, Bermuda. Journal of Experimental Biology and Ecology 138: 119 – 142.

Warwick, R. M., H. M. Platt, and P. J. Somerfield. 1998. Free-living Marine Nematodes: Part III Monhysterids. Shrewsbury, UK: Field Studies Council.

Wilson, E. O., and F. M. Peters. 1988. Biodiversity. Washington, DC: National Academy Press.

10 Population Dynamics

Robert McSorley and Larry W. Duncan

10.1 Population Dynamics

Population dynamics is the study of changes in nematode population size and structure over time. Knowledge of population dynamics is essential to predict nematode population growth and to anticipate nematode damage. Nematodes cannot migrate readily from site to site during a growing season. Therefore, it is possible to predict the final nematode population size (*Pf*) from the initial population size (*Pi*) present within a site, without complications from migration. *Pf* can be measured at harvest of a crop, or after some specified time interval. The subjects of nematode population dynamics and corresponding models for predicting population changes have been reviewed by various authors (Duncan and McSorley, 1987; Ferris and Noling, 1987; McSorley, 1998a; McSorley and Phillips, 1993; Nusbaum and Barker, 1971; Seinhorst, 1970).

Nematologists have used various types of models to predict population size (Ferris, 1981). Many of these are based on classical models that are well known from population ecology and modified for application to nematology. Simulation models, some of which are based on classical population models, are used to update population estimates through a series of time intervals (Duncan and McSorley, 1987; Ferris, 1976, 1981; McSorley and Phillips, 1993). Simulated time increments may range from seconds to years, depending on the situation.

10.2 Classical Models

10.2.1 Theoretical Models

Relatively simple classical models have been used to describe population growth for a variety of organisms (Andrewartha and Birch, 1954; Poole, 1974), and these are applicable to nematode populations as well.

10.2.1.1 Exponential Growth

Exponential growth, which is independent of population size, is a special case that applies only when environmental conditions are optimal and the food supply is unlimited. Under these conditions

$$Pf = Pi\ e^{rt} \tag{10.1}$$

where Pf = final population size after a specified time interval, Pi = initial population size, t = the time interval between Pi and Pf, r = the intrinsic rate of increase of the population (a constant), and e = the natural logarithm base. According to this model, population size increases exponentially over time (Fig. 10.1(A)). An equivalent form of Eq. 10.1 is:

$$\ln Pf = \ln Pi + rt \tag{10.2}$$

Note that if ln Pf is plotted against ln Pi, the graph is linear (Fig. 10.1(B)) for a constant time interval. While the assumption of an unlimited food source is unrealistic, a large root system is an almost limitless resource to a single microscopic nematode. Therefore, low nematode numbers in highly favorable biotic and abiotic environments may increase exponentially, as observed on several occasions (Basson, et al., 1991; McSorley and Dickson, 1989; Rodríguez-Kábana, et al., 1986).

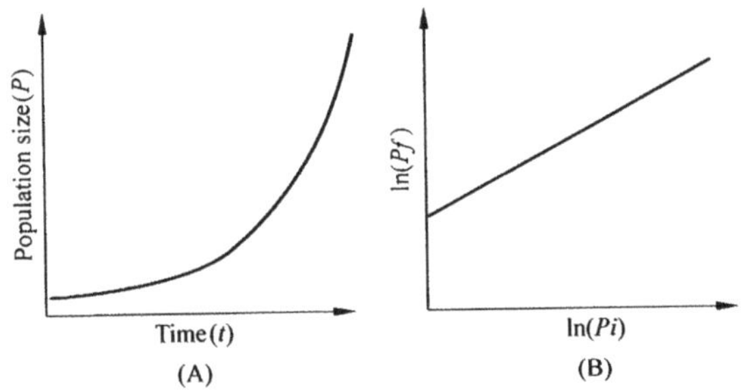

Figure 10.1 Exponential growth. (A) Increase in population size over time. (B) Relationship between Pf and Pi.

The rate of increase, r, for a particular set of conditions can be estimated from Eq. 10.2 when ln Pf is plotted against ln Pi. Life table studies can be especially useful in estimating r (Southwood, 1978), and life table data have been used to calculate r for *Rhabditis marina* (Vranken and Heip, 1983). However, life tables can be most easily constructed and used if the population can be divided into cohorts, or groups of individuals of similar age. Therefore this approach may be more useful with nematodes that deposit large numbers of eggs at once and have synchronized generations, rather than with nematodes that have overlapping generations and deposit a few eggs at a time over several weeks.

10.2.1.2 Logistic Growth

In reality, food sources become limited as populations grow, and so the logistic growth model, used to describe *density-dependent* population growth, is often applied to nematode population growth (McSorley, 1998a; Seinhorst, 1966). According to the logistic model:

$$\frac{\mathrm{d}P}{\mathrm{d}t} = rp\frac{E-P}{E} \tag{10.3}$$

where P = population size, $\mathrm{d}P/\mathrm{d}t$ = change in population density with time (i.e., the growth rate of the population), r = intrinsic rate of increase, and E = equilibrium density. A low population size may grow initially at an exponential rate, but the growth rate slows as population size increases, until population size levels off over time to the equilibrium density (Fig. 10.2(A)). This upper limit to population size for a specific set of conditions, termed equilibrium density by nematologists (Seinhorst, 1966, 1970), is called carrying capacity and given the symbol K by most other disciplines (Andrewartha and Birch, 1954).

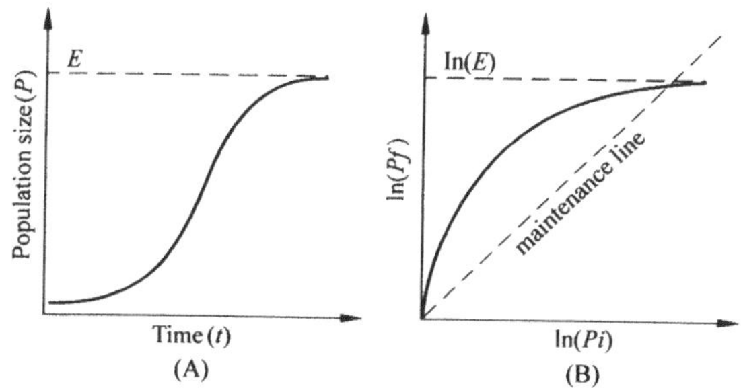

Figure 10.2 Logistic growth. (A) Increase in population size over time. E = equilibrium density. (B) Relationship between Pf and Pi, shape varies with host. Pf = Pi on the maintenance line.

The logistic model is based on a number of assumptions (Poole, 1974) which often are not met, yet this model has been successfully applied to nematode population growth on many occasions (Seinhorst, 1970). In practice, probably the most critical assumption is that of a constant environment, since a change in environmental conditions will affect r and will result in a change in the equilibrium density to a value of E appropriate for the new set of conditions.

Solution of Eq. 10.3 after a specified time interval has been expressed by Seinhorst (1966) and by Jones et al. (1967) in the following form:

$$Pf = \frac{aEPi}{(a-1)Pi + E} \qquad (10.4)$$

where $a = e^{rt}$ and the other terms are as defined previously. A typical shape of the relationship between ln Pf and ln Pi is illustrated (Fig. 10.2(B)), but varies with the nematode and crop involved as well as environmental conditions. Populations increase as long as the curve remains above the *maintenance line*, where $Pf = Pi$.

10.2.1.3 Exponential Decline

Although the previous models are descriptive of population growth, nematode populations can be expected to decline on nonhost crops, during clean fallow, or during winter. In these instances, an exponential decline equation may be appropriate

$$Pf = a\,e^{bt} \qquad (10.5)$$

where Pf = population size, t = time, and a and b are constant, for a particular set of conditions, with $b < 0$. Population size decreases over time (Fig. 10.3), so that $Pf < Pi$. This decline model or related models have been applied to a variety of nematological situations (Atkinson, et al., 2001; Ferris, 1985; Jeger and Starr, 1985; MacGuidwin and Forge, 1991; McSorley, 1998a; Rodríguez-Kábana, et al., 1986; Windham and Barker, 1988). It is possible that winter survival is density dependent (Duncan and McSorley, 1987; Ferris, 1985; MacGuidwin and Forge, 1991), and in that case alternative models incorporating Pi could be sought. However, it is evident that the constant a in Eq. 10.5 is equivalent to Pi, since it defines the initial height of the decline curve (Fig. 10.3).

10.2.2 Application of Classical Models

10.2.2.1 Critical Point Models Relating *Pf* and *Pi*

Critical point models are those in which Pf is forecast from an initial estimate of Pi (Ferris, 1981). Relationships such as that shown in Fig. 10.2(B) are fundamental in estimating Pf from Pi for specific nematodes on specific crops. At high values of Pi, populations are likely to reach E (Fig. 10.2(B)), and at

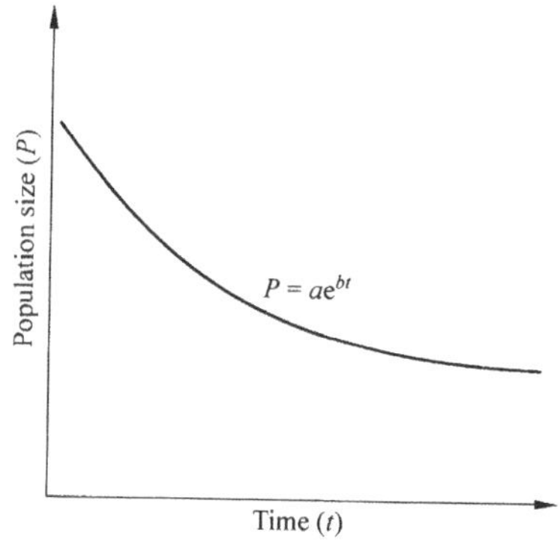

Figure 10.3 Exponential decline of population size over time, with $b < 0$.

extremely high nematode densities, root damage may be so severe that E cannot be maintained. In the latter case, Fig. 10.2(B) may take on a quadratic appearance (McSorley and Gallaher, 1993; Seinhorst, 1970). In most situations, however, the relationship between *Pf* and *Pi* at low to moderate *Pi* is particularly useful in evaluating plant host status (Seinhorst, 1970) and the potential for population increase under field conditions.

Data relating *Pf* to *Pi* can be obtained from greenhouse, field, or microplot experiments (McSorley, 1998a; McSorley and Phillips, 1993). In greenhouse or microplot experiments, inoculation can be used to establish a range of *Pi* levels from which corresponding *Pf* values can be measured. In the field, a range in *Pi* can be obtained by pooling plot data from a variety of locations (Ferris, 1985; McSorley and Gallaher, 1993) or by establishing small plots within a large field in which nematode distribution is uneven (Noe, 1986). Spatial heterogeneity is common in field populations of plant-parasitic nematodes (McSorley, 1987). Nevertheless, it can sometimes be difficult to encounter a range in *Pi* in the field that is sufficiently broad to establish the complete shape of the relationship between *Pf* and *Pi* (Fig. 10.2(B)). If only low and moderate levels of *Pi* are present, E may not be measured, and the shape of the relationship may approach linearity (McSorley and Dickson, 1989).

Once a graph relating *Pf* and *Pi* is available, approximations of *Pf* can be estimated from various *Pi* values by inspection. Derivation of an equation relating *Pf* and *Pi* is relatively straightforward if the relationship is linear. Derivation of such equations for more typical curved relationships (Fig. 10.2(B), Eq. 10.4) is more complex since approximations of E and a are required. Methods for deriving expressions similar to Eq. 10.4 are described in more detail elsewhere (Ferris, 1985; Ferris and Noling, 1987; McSorley, 1998a; McSorley and Gallaher, 1993).

Of course, any derived relationship between *Pf* and *Pi* is specific to a particular set of conditions. The shape of the curve (Fig. 10.2(B)) varies with crop cultivar, and therefore relationships between *Pf* and *Pi* provide useful quantitative descriptions of host status (Seinhorst, 1970). It is likely that the nematode isolate present will affect curve shape, since differential reproduction of various nematode isolates on the same host cultivar(s) is the basis for characterization of nematode races and pathotypes (Dropkin, 1988; Kort, et al., 1977; Riggs and Schmitt, 1988). A wide variety of environmental factors, both physical and biological, affect the rate of nematode reproduction and the relationship between *Pf* and *Pi* (McSorley, 1998a). Those factors include temperature (Duncan and McSorley, 1987; Ferris, 1976; Noe, 1991), moisture (Jones, 1975; Sultan and Ferris, 1991; Wallace, 1973), numerous edaphic factors (Ferris, 1976; McSorley, 1998a; Norton, 1978, 1979; Wallace, 1971, 1973; Yeates, 1981), other nematodes (Duncan and Ferris, 1982; Eisenback, 1985; Norton, 1978), nematode antagonists (Kerry, 1987; Stirling, 1991), and a variety of agricultural practices (Duncan and Noling, 1998; McSorley, 1998a; Noe, 1998).

For these reasons, an individual equation or graph relating *Pf* to *Pi* may be of limited value for predictive purposes, since it depends on a specific set of conditions. More robust relationships, indicative of the existing variability, may be obtained if data are pooled across several seasons, locations, or environments (McSorley and Gallaher, 1993). Furthermore, it should be emphasized that any relationship depicted by Eqs 10.1, 10.2, and 10.4 (Figs 10.1(B), 10.2(B)) represents *only one point in time*. The precise time when *Pf* is measured defines this time interval. Theoretically, *Pf* could be measured at any point in time during a season, making it possible to obtain an infinite array of *Pf/Pi* relationships similar to Fig. 10.2(B). If Fig. 10.2(B) represented a *Pf* measured at 100 days, then some of these other possibilities may give very different impressions of the *Pf/Pi* relationship. For example, if *Pf* was measured prematurely, for instance at 20 days when little reproduction had occurred, populations may still be so low that growth was exponential, resulting in a linear relationship between *Pf* and *Pi* (similar to Fig. 10.1(B)). Such an extensive misinterpretation of the shape of the *Pf/Pi* relationship might be unlikely in practice, but more subtle errors are possible. For instance, a *Pf* measured at "harvest" might be different if harvest occurred at 100 days or at 120 days. Thus, while *Pf* is an arbitrary term, it is always a function of sampling time.

Some workers use the term "reproductive factor", or "Rf value", to represent the ratio *Pf/Pi*. This provides a convenient measure for comparing *Pf/Pi* across a range of treatments (often different crop cultivars) in an inoculation test with fixed *Pi*, although a comparison of *Pf* values alone should accomplish the same thing. It can be risky to compare *Pf/Pi* ratios from one experiment with those obtained in another. Experimental conditions will likely differ, affecting *Pf/Pi*. In addition, *Pf/Pi* varies with *Pi* so that different responses might be expected at different *Pi* levels. *Pf/Pi* tends to be greatest at low *Pi* values, where $Pf/Pi > 1.0$ on a good host. However, on the same host, at *Pi* values greater than *E*, a *Pf* approaching *E* is forecast (Seinhorst, 1970), so that $Pf/Pi < 1.0$. Therefore $Pf/Pi > 1.0$ and $Pf/Pi < 1.0$ are both possible on the same host, as a result of a difference in the initial inoculum levels.

10.2.2.2 Simulation Models

In contrast to critical point models that use initial conditions to predict conditions at a designated "endpoint", simulation models use iterative algorithms to predict conditions at a sequence of points in time. Sequence intervals may be any length, from an instant to years. Thus, critical point models that are used in algorithms to predict nematode population change during a multi-year crop rotation cycle (described below) are components of a simulation model. However, a distinct advantage of modeling shorter time intervals than occur with successive crops is that various factors known to affect populations can be incorporated in the models (Ferris, 1976; Schmidt, et al., 1993; Schneider and Ferris, 1987). Theoretically, a model that simulates the influence of temperature, soil moisture, soil oxygen, root growth, etc. on nematode population development can provide more reliable predictions for specific cases than can critical point models whose parameters are derived from a unique set of conditions. Nevertheless, few remediation tactics are available to manage nematode damage to crops on a daily basis. Because intervention generally must occur prior to planting, critical point models usually provide the most cost-effective information for management decisions (Been and Schomaker, 1998; Haydock and Evans, 1998a, 1998b; Jones and Kempton, 1978; Kinloch, 1986; Noe, 1998).

Simulation models can be used to understand nematode biology and to identify aspects amenable to management intervention. They have been used to validate hypothetical relationships and to identify areas needing additional research. Examination of forecasts from a temperature driven model of *Heterodera schachtii* population development reveal uncertainty about the effects of some physical factors on root and nematode development, about distribution of nematodes and roots in the soil profile, about hatching and survival rates of eggs in relation to their position in the cyst, and about movement of nematodes between soil strata (Caswell, et al., 1986). Simulation models may use distributed delay functions to simulate nematode development through each phase of the life cycle (Duncan and McSorley, 1987; Manetsch, 1976). If the basic dynamics of a model conform reasonably to published data, it can be used to evaluate further refinements based on future research. A model of root-knot nematode (*Meloidogyne* spp.) development on grape (*Vitis vinifera*) was a close fit to field data during eight months, although end-of-season discrepancies revealed a deficiency in understanding nematode dynamics during cool months when plant physiology is slowed (Ferris, 1976). Alternatively, models that simulate crop growth can incorporate functions relating nematode population development and damage and be used to evaluate cropping systems (Sperow and Lybecker, 1998).

10.2.2.3 Design of Cropping Systems

If models for forecasting *Pf* from *Pi* are available for a nematode on a particular crop, then note that *Pf* at harvest of that crop will become *Pi* for a crop planted immediately after harvest. In this way, nematode population levels in a sequence of crops can be linked and modeled over time (Noe, 1986). This approach can be used to design cropping systems and crop sequences for optimal nematode management. For example, it has provided a convenient means to evaluate the effects of various three-year sequences of maize (*Zea mays*), resistant soybean (*Glycine max*), and susceptible soybean on *Meloidogyne incognita* populations (Kinloch, 1986), or of cotton (*Gossypium hirsutum*) and soybean on *Hoplolaimus columbus* (Noe, et al., 1991).

Rather specific data are required to make forecasts of *Pf* from *Pi* or to design cropping systems for a specific location. Review articles that include crop rotation information can be a useful starting point for locating this information (Duncan, 1991; Duncan and Noling, 1998; Good, 1968; Johnson, 1982; McSorley, 1996, 1998a, 1998b, 2001; McSorley and Duncan, 1995; Noe, 1998; Nusbaum and Ferris, 1973; Trivedi and Barker, 1986). Some recent studies have examined peanut (*Arachis hypogaea*) – cotton rotations (Johnson, et al., 1998), potato (*Solanum tuberosum*) – cotton rotations (Crow et al., 2000a, 2000b), or the potato system (Chen, et al., 1995) in detail. Data on bermudagrass (*Cynodon dactylon*) or on sorghum (*Sorghum bicolor*) and related crops as rotation crops have been provided by recent work as well (Johnson, et al., 1997; McSorley, et al., 1994; Trevathan and Robbins, 1995; Weaver, et al., 1995). Because rotation crops can shift the proportions of plant-parasitic nematodes present in mixed populations (Fortnum and Currin, 1993; Johnson, et al., 1996), rotation plans and crop sequences should be designed to focus on the key pest present in a particular situation (McSorley, 1996; McSorley and Gallaher, 1992). Recent studies have suggested that crop rotation sequences can be optimized to favor parasites of key nematode pests (Chen and Reese, 1999; Ciancio and Queneherve, 2000).

Since nematode population fluctuations continue even when major crops are not present, it is important to have data on nematode dynamics during winter or on off-season cover crops or weeds when designing crop sequences. Population declines during the winter have been examined in a variety of situations (Ferris, 1985; Jeger and Starr, 1985; Windham and Barker, 1988), and may fit the exponential decline model (Eq. 10.5). Crops not commonly grown in a particular location may be grown as cover crops to suppress nematodes during seasons when main cash crops cannot be grown (Caswell, et al., 1991; McSorley, 1996, 1998b, 2001; McSorley and Dickson, 1995; McSorley and Gallaher, 1992; McSorley and Parrado, 1983; McSorley, et al., 1994; Reddy, et al., 1986; Rhoades and Forbes, 1988; Rodríguez-Kábana, 1992; Rodríguez-Kábana, et al., 1988, 1989; Trivedi and Barker, 1986; Vargas-Ayala and Rodríguez-Kábana, 2001). Recent work provides some examples of winter cover crops (McSorley, 1994; McSorley and Gallaher, 1992), summer cover crops in warm temperate locations (McSorley and Dickson, 1995; McSorley, et al., 1994; Weaver, et al., 1998), and off-season crops in tropical locations (Ko and Schmitt, 1996; Quénéhervé, et al., 1998; Sipes and Arakaki, 1997) that have shown some efficacy in suppressing key nematode pests such as *Meloidogyne* spp. or *Rotylenchulus reniformis*.

10.3 Issues in Population Dynamics

Population forecasts may be derived from simulation models, classical models, critical point *Pf/Pi* models, or simply based on data from previous experiments. Regardless of the source of nematode population estimates, a variety of important issues affects their quality and use.

10.3.1 Sample Variability and Forecasting

Population dynamics models and nematode forecasts are almost always based on an initial estimate of *Pi*, obtained by sampling in the field. Unfortunately, highly irregular horizontal dispersion is typical of nematode populations in the field (Barker and Campbell, 1981; Goodell and Ferris, 1980; McSorley, 1987). This causes difficulty in sampling and results in a high degree of error (lack of precision) in estimates of *Pi* (Barker and Campbell, 1981; Goodell and Ferris, 1981; McSorley, 1987).

While it is convenient to obtain a mean estimate of *Pf* from a mean of *Pi*, in reality the broad confidence interval associated with *Pi* will result in a broad confidence interval around *Pf*, i. e., a vague forecast (McSorley, 1992, 1998a). This problem can be addressed by including stochastic or probabilistic statements in population models (Ferris, 1984b; McSorley, 1998a; McSorley and Phillips, 1993), although this is rarely done in practice. Repeated runs of simulation models can give an impression of the range in variability for a final estimate (Goodell and Ferris, 1981; McSorley, 1992). It is also possible to group population data into several density classes (Ferris, 1984a), and then, given a confidence interval associated with *Pi*, to determine the probability with which the final estimate of *Pf* will fall within the various density classes (McSorley, 1992, 1998a). The effects of spatial heterogeneity in *Pi* on forecast results for potato cyst nematode (*Globodera* spp.) populations have been examined in detail by Perry (1983). The high level of variability associated with estimates of *Pi* is a reminder that population models cannot provide precise forecasts, but rather a somewhat general estimate or impression of future population trends.

10.3.2 Population Dynamics and Pest Management Strategies

The logistic model of population growth over time (Fig. 10.2(A)) provides a basic frame of reference in which the effects of pest management strategies and tactics over time can be examined (Powers and McSorley, 2000). For example, if a growing nematode population is treated with a nematicide at one point in time, a rapid decrease in population size might be expected (Fig. 10.4). Ecologically, the effect is to shift the population to a lower point on the growth curve, but exponential and later logistic growth can still initiate from this lower point. Eventually *E* will be reached, but at a later time than that for a scenario without nematicide application. Hopefully, the depression in the population growth curve (Fig. 10.4) would result in improved crop performance and yield, but if the crop were maintained for a long enough time following treatment, populations would still be expected to recover to *E*.

An alternative management strategy might be to attempt to change *E*. This parameter depends on a constant set of environmental conditions, but agricultural management practices often alter the crop environment. Therefore this violation of the "constant environment" assumption could be expected to result in a new value of *E* appropriate to the new set of environmental conditions. A common example would be the replacement of a susceptible crop cultivar with one that is more resistant and has a lower *E* for a particular nematode. A wide variety of environmental factors may affect nematode dynamics (McSorley, 1998a). In one example, changes in irrigation practices

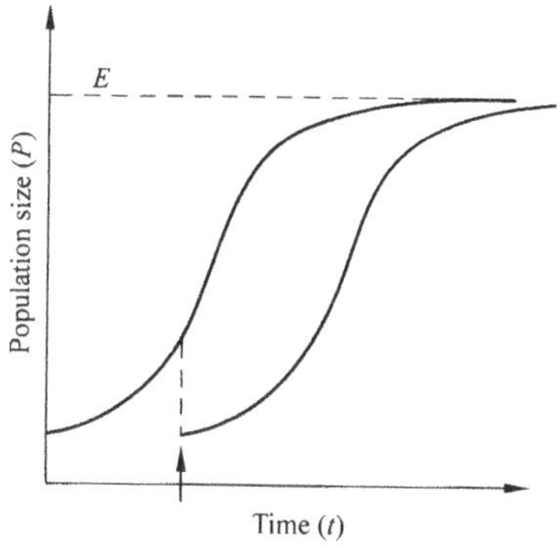

Figure 10.4 Effect of population reduction on logistic growth. A nematicide applied at the point indicated by the arrow reduces population size to a lower level and shifts the logistic growth curve in time.

caused *Tylenchulus semipenetrans* populations to achieve different population densities over time (Duncan and El-Morshedy, 1996). In another case, increased levels of *Pasteuria penetrans* inoculated into microplots resulted in lower levels of *Meloidogyne arenaria* over time (Chen, et al., 1997). Even in the simple nematicide example presented in the previous paragraph, environmental conditions may be altered so that E is changed over the long term. Evidence for this exists in the well-known resurgence of trichodorid nematodes to high levels following soil fumigation (Weingartner, et al., 1983).

Although the previous examples have focused on reduction of E through management practices, r may also be affected by management, or both E and r may be affected (Duncan and Noling, 1998). If nematode population growth on two crop cultivars were illustrated (Fig. 10.5), then E for the more resistant cultivar is lower as shown (curve B in Fig. 10.5), but r may be lower as well, meaning that it will take the population more time to reach E on the resistant cultivar.

Modifications to exponential decline models (Fig. 10.3) are also possible. Some practices may result in changes to the parameter b of Eq. 10.5. For example, cultivation intensity during the cropping season was found to affect lipid reserves and consequently the rate of overwinter survival of *Scutellonema cavenessi* (Duncan, 1986). Trap crops are used in Europe to stimulate hatch of overwintering potato cyst nematodes, which are then prevented from reproducing by early crop destruction (Haydock and Evans, 1998b).

Understanding the population dynamics of a nematode in a specific situation

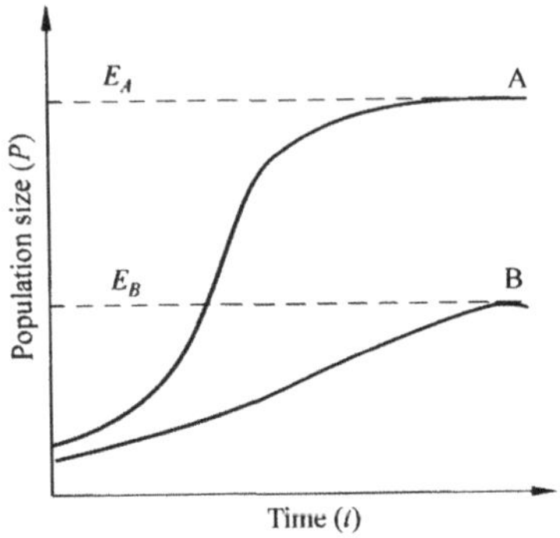

Figure 10.5 Effect of changes in r and E on logistic growth. Curve B has lower r and lower E than curve A. E_A, E_B = equilibrium densities corresponding to curves A and B respectively.

improves our ability to develop an effective management strategy. Knowledge of typical E values and of the factors affecting E is essential in designing systems that will maintain nematode populations at relatively low levels over time with minimal intervention.

10.3.3 Population Dynamics and Sustainable Agriculture

Sustainable agriculture is an evolving paradigm designed to maintain food production at necessary levels with minimal environmental degradation and resource depletion, by developing methods to reduce inappropriate or overuse of agricultural resources (Allen, et al., 1991; Duncan and Noling, 1998; Hamblin, 1995; McSorley and Porazinska, 2001; Powers and McSorley, 2000). Knowledge about biological processes and derivation of methods to predict the outcomes of agricultural activities will provide the means to develop agricultural systems that can be sustained for future generations. Application of these principles for pest control is called integrated pest management (IPM). Simply put, IPM consists of: (1) assessing pest population densities and determining whether densities need to be reduced below an economic threshold, (2) effecting the necessary changes with the most appropriate available tactic(s), and (3) continually developing new tactics compatible with economic and environmental requirements (Geier, 1966). Societies effect IPM by regulating (through laws and consumer-imposed restrictions) tactics available to growers, in response to the broad concerns of sustainable agriculture. At the grower level, IPM decisions are necessarily based on economic considerations (Duncan, 1999). For nematodes and many other pests and diseases, population models are used to estimate optimal solutions to the concerns of both growers and social planners.

To achieve maximum information from the use of population models, two broad generalizations should be recognized. The first is that stochastic models, those that estimate the likelihood of various possible outcomes, are preferable to deterministic models, which predict only the most likely response (Headley, 1975). A grower may choose not to use a management tactic if the probability of crop failure is 20%, even if the expected average response is very profitable. For reasons mentioned earlier, few nematode crop systems have been studied sufficiently to accurately predict the variation in expected responses. Second, the use of models to predict nematode population behavior and crop responses in a series of successive crops (cropping systems analysis) has a better potential to provide optimum solutions for nematode management than if models are used to predict population development during the current crop only (Noe, 1998). The reason for this is that all pest management practices affect not only the current crop, but also successive crops in terms of the pest population that remains at the end of the current growing season. In addition to models of population growth and survival rates, cropping system simulation requires an awareness of the relationship between nematode population density and crop yield (see below). Again, only a few systems (e.g., Schmidt, et al., 1993) have been studied extensively enough to provide data to adequately evaluate the reliability of various cropping systems models. One notable exception is the management of potato cyst nematodes (PCN) on potato in Europe. Population modeling has been central to characterizing and understanding many aspects of managing these nematodes, and growers increasingly rely on this information to manage PCN.

10.3.4 Toward Sustainable Management: Potato Cyst Nematodes

Nematode management in potato provides some clear examples of competing issues in sustainable agriculture and purposeful development of IPM. In the United Kingdom, traditional rotations of potato with non-host crops permitted potato to be grown for one year in eight and maintained population densities below damaging levels (Haydock and Evans, 1998a). Although crop rotation can be a highly effective management tactic, available agricultural land is insufficient to produce adequate quantities of potatoes while relying exclusively on rotation. Therefore, growers shorten the traditional rotations by integrating the use of additional management tactics that include use of nematicides, resistant potato varieties, and trap crops. The adoption of multiple tactics was the impetus for increased reliance on modeling to help design the most profitable cropping systems, and more recently to minimize environmental risk from overuse of nematicides.

Equation 10.4 was first proposed for PCN (Jones, et al., 1967) in order to study the effects of resistant potatoes to manage field infestations. The model has been modified to reflect variation caused by interspecific competition, density dependence of both food supply (Jones and Kempton, 1978) and sex ratio, as well as cultivar tolerance to damage (Phillips, et al., 1991). Models were also used to understand the contribution of volunteer potato plants on population carry-over (den Ouden, 1967), nematicide efficacy and its effect on resistance selection pressure (Jones and Perry, 1978; Perry and Clark, 1983), and density-dependent parasitism of nematodes by natural enemies (Perry, 1978).

The continuous refinement and validation of population models of PCN have resulted in their routine use, in combination with damage functions, for developing regional potato IPM programs, and increasingly for making management decisions at the farm level. As with population growth, yield reduction due to nematodes is also density dependent. A variety of damage functions have been used to describe the relationship between preplant nematode density (Pi) and crop yield, but the most widely used has been that of Seinhorst (1965), expressing relative yield (y = yield in the presence of nematodes / yield in their absence) as a function of the host tolerance limit (T = the highest initial population density below which damage does not occur), a minimum yield (m) which occurs even at high nematode density, and a rate determining parameter z (the proportion of the root system that is not damaged by a single nematode):

$$y = m + (1 - m)\, z^{Pi - T} \tag{10.6}$$

Seinhorst's model was derived from the competition curve of Nicholson (1933) and reflects that proportionately less root material remains to be damaged by successively invading nematodes. The curve is sigmoid when Pi is log-transformed (Fig. 10.6).

PCN management was the focus of the first cropping systems' models (Jones and Kempton, 1978) and recent analyses of cropping systems in the Netherlands incorporate stochastic elements by estimating the variances associated with the nematode population growth model (Been, et al., 1998). For example, the estimated yield reduction resulting from rotation of non-host crops with potato varieties having various levels of resistance in various sequences for 5 years is shown in Table 10.1. Although use of resistance is the management tactic in this analysis, other tactics such as use of nematicides, trap crops, or combinations of tactics can be considered with this approach.

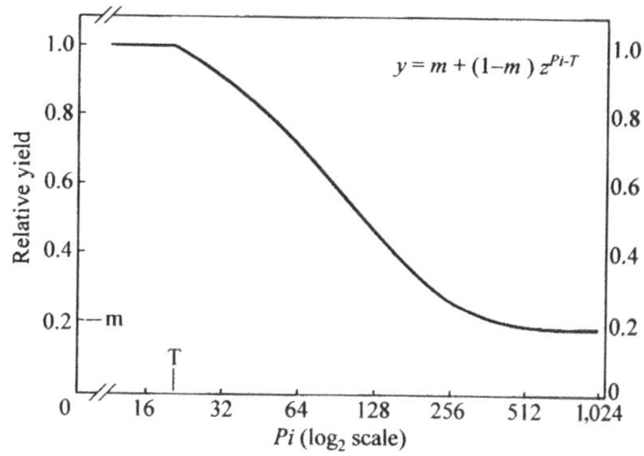

Figure 10.6 The general relationship between initial nematode population density (*Pi*) and relative crop yield. *T* and *m* refer to the tolerance limit and minimum yield, respectively.

Table 10.1 Predicted potato yield reduction due to potato cyst nematodes (*Pi* = 5 eggs/g soil in first year) during various rotation cycles using non-host crops and potato cultivars with various levels of relative susceptibility (RS = 0% – 100% of susceptible cultivars). Table reprinted from Been and Schomaker, 1998.

Cultivar RS and frequency of use in rotation cycle*					Average yield reduction	Probability (%) of yield reduction larger than					Yield reduction in 95% of farms smaller than
1:1	1:2	1:3	1:4	1:5	(%)	0%	5%	10%	12.5%	15%	(%)
1	2	3	5	7	0.0	0.0	0.0	0.0	0.0	0.0	0.0
2	4	6	9	14	0.0	0.1	0.0	0.0	0.0	0.0	0.0
3	6	9	14	20	0.1	3.7	0.0	0.0	0.0	0.0	0.0
4	8	12	18	27	0.4	18.0	1.2	0.0	0.0	0.0	2.6
5	10	15	23	34	1.2	37.8	7.4	0.0	0.0	0.0	5.8
6	12	18	27	41	2.5	60.0	20.0	1.8	0.0	0.0	9.6
7	14	21	33	47	4.2	70.9	35.4	8.9	2.1	0.0	11.1
8	16	24	36	54	6.1	86.4	51.1	21.9	9.2	2.3	13.7
9	18	27	41	61	8.1	90.5	63.8	37.0	21.4	7.9	16.1
10	20	30	45	67	10.1	93.3	74.0	51.0	35.4	19.5	18.2
11	22	33	50	74	12.2	95.7	81.4	62.7	49.7	32.9	20.2
12	24	36	54	81	14.1	96.8	86.6	72.3	61.6	47.0	22.2
13	26	39	59	88	16.1	100.0	91.0	79.3	70.8	59.5	24.2
14	28	42	63	94	17.9	100.0	93.8	84.4	77.8	69.3	25.7
15	30	45	68	100	19.7	100.0	95.6	88.4	83.5	76.6	27.5

* Frequency ranges from 1:1 (cultivar used every year) to 1:5 (cultivar used once in every five years).

The estimates of yield reduction were derived by assuming a starting population density of 5 PCN eggs/g soil (*Pi*), predicting the *Pf* after the first crop, estimating survival until the next spring, and so on, through the cropping sequence. *Pi* for each year in which potatoes are grown was used in the damage function to estimate annual yield loss, and *Pi* was used in the population growth model to estimate *Pf*. Damage and reproduction models were adjusted depending on the relative damage susceptibility and relative reproduction rate for the appropriate cultivar. The stochastic elements were achieved from data providing an estimate of the variance associated with *a* and *E*. Because evidence suggests that the relative tolerances of the various cultivars are independent of external conditions, population growth was considered to be the main source of variation in the simulation. The added value to growers of stochastic compared to deterministic predictions is evident. Growers with large financial resources can afford to use the average projected yield reductions to estimate long-term economic thresholds based on profitabilities of the various treatments (McSorley and Duncan, 1995). Growers who cannot afford to sustain major short-term losses may choose much more conservative management practices and forego the probability of higher long-term gains.

Improved methods of estimating population densities of PCN are being used to accurately predict the need for nematicide applications in fields based on predicted yield responses to the nematode densities in those fields (Been and Schomaker, 1998; Haydock and Evans, 1998a). Between 1984 and 1998, fumigant application for control of PCN in The Netherlands declined by 85% as a result of these programs (Been and Schomaker, 1998). Further efficiency in sample processing (Davies, et al., 1996) may make it profitable to use global positioning technology and geostatistical analyses to modulate nematicide delivery across individual fields based on the patterns of nematode densities in those fields (Evans, et al., 1998). To save on sampling costs, nematode spatial patterns could be adjusted for several years based on predicted population changes. Again, however, it is preferable to estimate the likely effects of various tactics on future crops by forecasting the effects on population growth. For example, major food retail outlets in the United Kingdom require potato growers to follow specific IPM guidelines for PCN due to public concern about pesticide use. The guidelines include the use of crop rotation, resistant varieties, and nematicides when *Pi* is high enough to damage the current potato crop. However, nematode reproductive factors are density dependent and nematicides may be more efficient in the long term if they are applied when nematode population densities are below the economic threshold for the current crop (Table 10.2). Increasing *Pf* by inappropriate timing of nematicides can result in more frequent use of fumigants (Haydock and Evans, 1998b), lower profits, and increased frequency of resistance-breaking genes in the nematode population (Perry and Clark, 1983).

Table 10.2 Interaction of *Pi* and nematicides on *Pf* of potato cyst nematodes. (Data from Haydock and Evans, 1998b.)

	Pi	
	1 egg/g of soil	80 eggs/g of soil
Pf/*Pi*	40	2
Pf	40	160
Pi following nematicide (80% kill)	0.2	16
Pf/*Pi* following nematicide	40	12
Pf following nematicide	8	192
$Pf_{\text{not treated}}/Pf_{\text{treated}}$	0.2	1.2

10.4 Outlook

Provided that unrealistic levels of precision are not demanded, the analysis of nematode population dynamics can provide useful approximations of forecast population trends. However, applied population dynamics requires a strong base of field data. Nematode population dynamics and the effect of management practices on nematode dynamics vary with local conditions (McSorley, 1998a, 1998b). In examining the literature of applied nematology, it becomes apparent that sizeable bases of field data have been developed around centers that have strong programs in applied nematology, whereas in many other areas, little or no information is available on the dynamics of local nematode populations. Thus, the ability to develop effective forecasts and to make informed management decisions based on nematode population levels varies greatly from place to place.

The consequences of such uneven distribution of applied nematology databases could become more serious in the future. Increased restrictions on the use of chemical nematicides (Noling and Becker, 1994) should mean an increasing reliance on nonchemical methods of nematode management. Many of these methods, particularly those that involve the design of effective cropping systems, depend on an extensive knowledge of nematode dynamics on the crops and cultivars grown in an area. Without an available database, nematode management in a site could become guesswork, in adapting systems effective in other regions but unproven locally. On the other hand, a strong local database can provide at least a starting point from which to design cropping and management systems that hopefully could maintain key nematode pests at low levels over time and with minimal adverse impact on crops.

References

Allen, P. A., D. Van Dusen, J. L. Lundy, and S. R. Gliessman. 1991. Integrating social, enviornmental, and economic issues in sustainable agriculture. American Journal of Alternative Agriculture 6: 34 – 39.

Andrewartha, H. G., and L. C. Birch. 1954. The Distribution and Abundance of Animals. Chicago, IL: University of Chicago Press.

Atkinson, H. J., R. A. Holz, E. Riga, G. Main, R. Oros, and J. Franco. 2001. An algorithm for optimizing rotational control of *Globodera rostochiensis* on potato crops in Bolivia. Journal of Nematology 33: 121 – 125.

Barker, K. R., and C. L. Campbell. 1981. Sampling nematode populations. Pp. 451 – 474. In B. M. Zuckerman, and R. A. Rohde, eds. Plant Parasitic Nematodes. Vol. 3. New York: Academic Press.

Basson, S., D. G. M. A. DeWaele, and A. J. Meyer. 1991. Population dynamics of *Ditylenchus destructor* on peanut. Journal of Nematology 23: 485 – 490.

Been, T. H., and C. H. Schomaker. 1998. Quantitative Studies on the Management of Potato cyst Nematodes (*Globodera* spp.) in the Netherlands. Wageningen, The Netherlands: DLO-Research Institute for Plant Protection (IPO-DLO).

Been, T. H., C. H. Schomaker, and J. W. Seinhorst. 1998. An advisory system for the management of potato cyst nematodes (*Globodera* spp.). Pp. 213 – 231. In T. H. Been, and C. H. Schomaker, eds. Quantitative Studies on the Management of Potato Cyst Nematodes (*Globodera* spp) in the Netherlands. Wageningen, The Netherlands: DLO-Research Institute for Plant Protection (IPO-DLO).

Caswell, E. P., A. E. MacGuidwin, K. Milne, C. E. Nelsen, I. J. Thomason, and G. W. Bird. 1986. A simulation model of *Heterodera schachtii* infecting *Beta vulgaris*. Journal of Nematology 18: 512 – 519.

Caswell, E. P., J. DeFrank, W. J. Apt, and C. S. Tang. 1991. Influence of nonhost plants on population decline of *Rotylenchulus reniformis*. Journal of Nematology 23: 91 – 98.

Chen, J., G. W. Bird, and R. L. Mather. 1995. Impact of multi-year cropping regimes on *Solanum tuberosum* tuber yields in the presence of *Pratylenchus penetrans* and *Verticillium dahliae*. Supplement to the Journal of Nematology 27: 654 – 660.

Chen, S., and C. D. Reese. 1999. Parasitism of the nematode *Heterodera glycines* by the fungus *Hirsutella rhossiliensis* as influenced by crop sequence. Journal of Nematology 31: 437 – 444.

Chen, Z. X., D. W. Dickson, D. J. Mitchell, R. McSorley, and T. E. Hewlett. 1997. Suppression mechanisms of *Meloidogyne arenaria* race 1 by *Pasteuria penetrans*. Journal of Nematology 29: 1 – 8.

Ciancio, A., and P. Queneherve. 2000. Population dynamics of *Meloidogyne incognita* and infestation levels by *Pasteuria penetrans* in a naturally infested field in Martinique.

Nematropica 30: 77 – 86.

Crow, W. T., D. P. Weingartner, and D. W. Dickson. 2000a. Effects of potato-cotton cropping systems and nematicides on plant-parasitic nematodes and crop yields. Journal of Nematology 32: 297 – 302.

Crow, W. T., D. P. Weingartner, R. McSorley, and D. W. Dickson. 2000b. Population dynamics of *Belonolaimus longicaudatus* in a cotton production system. Journal of Nematology 32: 210 – 214.

Davies, K. G., R. H. Curtis, and K. Evans. 1996. Serologically based diagnostic and quantification tests for nematodes. Pesticide Science 47: 81 – 87.

den Ouden, H. 1967. The influence of volunteer potato plants in oats on the population density of *Heterodera rostochiensis*. Nematologica 13: 325 – 335.

Dropkin, V. H. 1988. The concept of race in phytonematology. Annual Review of Phytopathology 26: 145 – 161.

Duncan, L. W. 1986. Effects of bare fallow on plant-parasitic nematodes in the Sahelian zone of Senegal. Revue de Nematologie 9: 75 – 81.

Duncan, L. W. 1991. Current options for nematode management. Annual Review of Phytopathology 29: 469 – 490.

Duncan, L. W. 1999. The basis of citrus health management. Pp. 3 – 13. In L. W. Timmer, and L. W. Duncan, eds. Citrus Health Management. St. Paul, MN: American Phytopathological Society.

Duncan, L. W., and M. M. El-Morshedy. 1996. Population changes of *Tylenchulus semipenetrans* under localized versus uniform drought in the citrus root zone. Journal of Nematology 28: 360 – 368.

Duncan, L., and H. Ferris. 1982. Interactions between phytophagous nematodes. Pp. 29 – 51. In D. W. Freckman, ed. Nematodes in Soil Ecosystems. Austin, TX: University of Texas Press.

Duncan, L. W., and R. McSorley. 1987. Modeling nematode populations. Pp. 377 – 389. In J. A. Veech, and D. W. Dickson, eds. Vistas on Nematology. Hyattsville, MD: Society of Nematologists.

Duncan, L. W., and J. W. Noling. 1998. Agricultural sustainability and nematode integrated pest management. Pp. 251 – 287. In K. R. Barker, G. A. Pederson, and G. L. Windham, eds. Plant and Nematode Interactions. Madison, WI: ASA, CSSA, and SSSA.

Eisenback, J. D. 1985. Interactions among concomitant populations of nematodes. Pp. 193 – 213. In J. N. Sasser, and C. C. Carter, eds. An Advanced Treatise on *Meloidogyne*. Vol. 1. Raleigh, NC: North Carolina State University Graphics.

Evans, K., J. Stafford, R. Webster, P. Halford, M. Russell, A. Barker, and S. Griffin. 1998. Mapping potato cyst nematode populations for modulated application of nematicide. Aspects of Applied Biology 52: 101 – 108.

Ferris, H. 1976. Development of a computer-simulation model for a plant-nematode system. Journal of Nematology 8: 255 – 263.

Ferris, H. 1981. Mathematical approaches to the assessment of crop damage. Pp. 405 – 420. In B. M. Zuckerman, and R. A. Rohde, eds. Plant Parasitic Nematodes. Vol. 3. New York: Academic Press.

Ferris, H. 1984a. Nematode damage functions: The problems of experimental and sampling

error. Journal of Nematology 16: 1 – 9.
Ferris, H. 1984b. Probability range in damage predictions as related to sampling decisions. Journal of Nematology 16: 246 – 251.
Ferris, H. 1985. Density-dependent nematode seasonal multiplication rates and overwinter survivorship: A critical point model. Journal of Nematology 17: 93 – 100.
Ferris, H., and J. W. Noling. 1987. Analysis and prediction as a basis for management decisions. Pp. 49 – 85. In R. H. Brown, and B. R. Kerry, eds. Principles and Practice of Nematode Control in Crops. Orlando, FL: Academic Press.
Fortnum, B. A., and R. E. Currin. 1993. Crop rotation and nematicide effects on the frequency of *Meloidogyne* spp. in a mixed population. Phytopathology 83: 350 – 355.
Geier, P. W. 1966. Management of insect pests. Annual Review of Entomology 11: 471 – 490.
Good, J. M. 1968. Relation of plant parasitic nematodes to soil management practices. Pp. 113 – 138. In G. C. Smart, Jr., and V. G. Perry, eds. Tropical Nematology. Gainesville, FL: University of Florida Press.
Goodell, P., and H. Ferris. 1980. Plant-parasitic nematode distributions in an alfalfa field. Journal of Nematology 12: 136 – 141.
Goodell, P. B., and H. Ferris. 1981. Sample optimization for five plant-parasitic nematodes in an alfalfa field. Journal of Nematology 13: 304 – 313.
Hamblin, A. 1995. The concept of agricultural sustainability. Pp. 1 – 19. In J. H. Andrews, and I. C. Tommerup, eds. Advances in Plant Pathology. Vol. 11. San Diego, CA: Academic Press.
Haydock, P. P. J., and K. Evans. 1998a. Integrated crop management (ICM) protocols and the management of potato cyst nematodes. Aspects of Applied Biology 52: 361 – 366.
Haydock, P. P. J., and K. Evans. 1998b. Management of potato cyst nematodes in the UK: An integrated approach? Outlook on Agriculture 27: 253 – 260.
Headley, J. C. 1975. The economics of pest management. Pp. 75 – 99. In R. L. Metcalf, and W. H. Luckmann, eds. Introduction to Insect Pest Management. New York: John Wiley & Sons.
Jeger, M. J., and J. L. Starr. 1985. A theoretical model of the winter survival dynamics of *Meloidogyne* spp. eggs and juveniles. Journal of Nematology 17: 257 – 260.
Johnson, A. W. 1982. Managing nematodes in crop production. Pp. 193 – 203. In R. D. Riggs, ed. Nematology in the Southern Region of the United States. Southern Cooperative Series Bulletin 276, Arkansas Agricultural Experiment Station, Fayetteville, AR.
Johnson, A. W., C. C. Dowler, N. C. Glaze, and Z. A. Handoo. 1996. Role of nematodes, nematicides, and crop rotation on the productivity and quality of potato, sweet potato, peanut, and grain sorghum. Journal of Nematology 28: 389 – 399.
Johnson, A. W., G. W. Burton, D. R. Sumner, and Z. Handoo. 1997. Coastal bermudagrass rotation and fallow for management of nematodes and soilborne fungi on vegetable crops. Supplement to Journal of Nematology 29: 710 – 716.
Johnson, A. W., N. A. Minton, T. B. Brenneman, J. W. Todd, G. A. Herzog, G. J. Gascho, S. H. Baker, and K. Bondari. 1998. Peanut-cotton-rye rotations and soil chemical treatment for managing nematodes and thrips. Journal of Nematology 30:

211 – 225.
Jones, F. G. W. 1975. Accumulated temperature and rainfall as measures of nematode development and activity. Nematologica 21: 62 – 70.
Jones, F. G. W., and R. A. Kempton. 1978. Population dynamics, population models and integrated control. Pp. 333 – 361. In J. F. Southey, ed. Plant Nematology Bulletin 201. London: Ministry of Agriculture, Fisheries, and Food.
Jones, F. G. W., and J. N. Perry. 1978. Modeling populations of cyst-nematodes. (Nematoda: Heteroderidae). Journal of Applied Ecology 15: 349 – 371.
Jones, F. G. W., D. M. Parrott, and G. J. S. Ross. 1967. The population genetics of the potato cyst-nematode *Heterodera rostochiensis*: Mathematical models to simulate the effects of growing eelworm-resistant potatoes bred from *Solanum tuberosum* ssp. *andigena*. Annals of Applied Biology 60: 151 – 171.
Kerry, B. R. 1987. Biological control. Pp. 233 – 263. In R. H. Brown, and B. R. Kerry, eds. Principles and Practice of Nematode Control in Crops. Orlando, FL: Academic Press.
Kinloch, R. A. 1986. Soybean and maize cropping models for the management of *Meloidogyne incognita* in the Coastal Plain. Journal of Nematology 18: 451 – 458.
Ko, M. P., and D. P. Schmitt. 1996. Changes in plant-parasitic nematode populations in pineapple fields following inter-cycle cover crops. Journal of Nematology 28: 546 – 556.
Kort, J., H. Ross, H. J. Rumpenhorst, and A. R. Stone. 1977. An international scheme for identifying and classifying pathotypes of potato cyst nematodes *Globodera rostochiensis* and *G. pallida*. Nematologica 23: 333 – 339.
MacGuidwin, A. E., and T. A. Forge. 1991. Winter survival of *Pratylenchus scribneri*. Journal of Nematology 23: 198 – 204.
Manetsch, T. J. 1976. Time-varying distributed delays and their use in aggrevative models of large systems. IEEE Transactions on Systems, Man and Cybernetics 6: 547 – 553.
McSorley, R. 1987. Extraction of nematodes and sampling methods. Pp. 13 – 47. In R. H. Brown, and B. R. Kerry, eds. Principles and Practice of Nematode Control in Crops. Orlando, FL: Academic Press.
McSorley, R. 1992. Applied population modeling: Fact or fiction? Pp. 170 – 181. In F. J. Gommers, and P. W. T. Maas, eds. Nematology from Molecule to Ecosystem. Invergowrie, UK: European Society of Nematologists.
McSorley, R. 1994. Changes in population densities of *Meloidogyne* spp. and *Paratrichodorus minor* on winter rye cover crops. Nematropica 24: 151 – 160.
McSorley, R. 1996. Impact of crop management practices on soil nematode populations. Soil and Crop Science Society of Florida Proceedings 55: 63 – 66.
McSorley, R. 1998a. Population dynamics. Pp. 109 – 133. In K. R. Barker, G. A. Pederson, and G. L. Windham, eds. Plant and Nematode Interactions. Madison, WI: ASA, CSSA, and SSSA.
McSorley, R. 1998b. Alternative practices for managing plant-parasitic nematodes. American Journal of Alternative Agriculture 13: 98 – 104.
McSorley, R. 2001. Multiple cropping systems for nematode management: A review. Soil and Crop Science Society of Florida Proceedings 60: 132 – 142.
McSorley, R., and D. W. Dickson. 1989. Nematode population density increase on cover crops of rye and vetch. Nematropica 19: 39 – 51.

McSorley, R., and D. W. Dickson. 1995. Effects of tropical rotation crops on *Meloidogyne incognita* and other plant-parasitic nematodes. Supplement to Journal of Nematology 27: 535 – 544.

McSorley, R., and L. W. Duncan. 1995. Economic thresholds and nematode management. Pp. 147 – 171. In J. H. Andrews, and I. C. Tommerup, eds. Advances in Plant Pathology. Vol. 11. San Diego, CA: Academic Press.

McSorley, R., and R. N. Gallaher. 1992. Managing plant-parasitic nematodes in crop sequences. Soil and Crop Science Society of Florida Proceedings 51: 42 – 45.

McSorley, R., and R. N. Gallaher. 1993. Population dynamics of plant parasitic nematodes on cover crops of corn and sorghum. Journal of Nematology 25: 446 – 453.

McSorley, R., and J. L. Parrado. 1983. Influence of summer management strategies on nematode populations in a subtropical agroecosystem. Nematropica 13: 1 – 8.

McSorley, R., and M. S. Phillips. 1993. Modeling population dynamics and yield losses and their use in nematode management. Pp. 61 – 85. In K. Evans, D. L. Trudgill, and J. M. Webster, eds. Plant Parasitic Nematodes in Temperate Agriculture. Wallingford, UK: CAB International.

McSorley, R., and D. L. Porazinska. 2001. Elements of sustainable agriculture. Nematropica 31: 1 – 9.

McSorley, R., D. W. Dickson, J. A. de Brito, and R. C. Hochmuth. 1994. Tropical rotation crops influence nematode densities and vegetable yields. Journal of Nematology 26: 308 – 314.

Nicholson, A. J. 1933. The balance of animal populations. Journal of Animal Ecology 2: 132 – 178.

Noe, J. P. 1986. Cropping Systems Analysis for Limiting Losses due to Plant-parasitic Nematodes: Guide to research methodology. Raleigh, NC: North Carolina State University and United States Agency for International Development.

Noe, J. P. 1991. Development of *Meloidogyne arenaria* on peanut and soybean under two temperature cycles. Journal of Nematology 23: 468 – 476.

Noe, J. P. 1998. Crop- and nematode-management systems. Pp. 159 – 171. In K. R. Barker, G. A. Pederson, and G. L. Windham, eds. Plant and Nematode Interactions. Madison, WI: ASA, CSSA, SSSA.

Noe, J. P., J. N. Sasser, and J. L. Imbriani. 1991. Maximizing the potential of cropping systems for nematode management. Journal of Nematology 23: 353 – 362.

Noling, J. W., and J. O. Becker. 1994. The challenge of research and extension to define and implement alternatives to methyl bromide. Supplement to Journal of Nematology 26: 573 – 586.

Norton, D. C. 1978. Ecology of Plant-parasitic Nematodes. New York: John Wiley & Sons.

Norton, D. C. 1979. Relationship of physical and chemical factors to populations of plant-parasitic nematodes. Annual Review of Phytopathology 17: 279 – 299.

Nusbaum, C. J., and K. R. Barker. 1971. Population dynamics. Pp. 303 – 323. In B. M. Zuckerman, W. F. Mai, and R. A. Rohde, eds. Plant Parasitic Nematodes. Vol. 1. New York: Academic Press.

Nusbaum, C. J., and H. Ferris. 1973. The role of cropping systems in nematode population management. Annual Review of Phytopathology 11: 423 – 440.

Perry, J. N. 1978. A population model for the effect of parasitic fungi on numbers of the cereal cyst-nematode, *Heterodera avenae*. Journal of Applied Ecology 15: 781 – 787.

Perry, J. N. 1983. Effects of spatial heterogeneity on Jones' model for cyst-nematode population dynamics and crop root damage. Journal of Applied Ecology 20: 549 – 586.

Perry, J. N., and S. J. Clark. 1983. Use of population models to estimate the proportion of cyst-nematode eggs killed by fumigation. Journal of Applied Ecology 20: 857 – 864.

Phillips, M. S., C. A. Hackett, and D. L. Trudgill. 1991. The relationship between the initial and final population densities of the potato cyst nematode *Globodera pallida* for partially resistant potatoes. Journal of Applied Ecology 28: 109 – 119.

Poole, R. W. 1974. An Introduction to Quantitative Ecology. New York: McGraw-Hill.

Powers, L. E., and R. McSorley. 2000. Ecological Principles of Agriculture. Albany, NY: Delmar Publishers.

Quénéhervé, P., P. Topart, and B. Martiny. 1998. *Mucuna pruriens* and other rotational crops for control of *Meloidogyne incognita* and *Rotylenchulus reniformis* in vegetables in polytunnels in Martinique. Nematropica 28: 19 – 30.

Reddy, K. C., A. R. Soffes, G. M. Prine, and R. A. Dunn. 1986. Tropical legumes for green manure. II. Nematode populations and their effects on succeeding crop yields. Agronomy Journal 78: 5 – 10.

Rhoades, H. L., and R. B. Forbes. 1986. Effects of fallow, cover crops, organic mulches, and fenamiphos on nematode populations, soil nutrients, and subsequent crop growth. Nematropica 16: 141 – 151.

Riggs, R. D., and D. P. Schmitt. 1988. Complete characterization of the race scheme for *Heterodera glycines*. Journal of Nematology 20: 392 – 395.

Rodríguez-Kábana, R. 1992. Cropping systems for the management of phytonematodes. Pp. 219 – 233. In F. J. Gommers, and P. W. T. Maas, eds. Nematology from Molecule to Ecosystem. Invergowrie, UK: European Society of Nematologists.

Rodríguez-Kábana, R., C. F. Weaver, D. G. Robertson, and E. L. Snoddy. 1986. Population dynamics of *Meloidogyne arenaria* juveniles in a field with Florrunner peanut. Nematropica 16: 185 – 196.

Rodríguez-Kábana, R., P. S. King, D. G. Robertson, and C. F. Weaver. 1988. Potential of crops uncommon to Alabama for management of root-knot and soybean cyst nematodes. Supplement to Journal of Nematology 2: 116 – 120.

Rodríguez-Kábana, R., D. G. Robertson, L. Wells, P. S. King, and C. F. Weaver. 1989. Crops uncommon to Alabama for the management of *Meloidogyne arenaria* in peanut. Supplement to Journal of Nematology 21: 712 – 716.

Schmidt, K., R. A. Sikora, and O. Richter. 1993. Modelling the population dynamics of the sugar beet cyst nematode *Heterodera schachtii*. Crop Protection 12: 490 – 496.

Schneider, S. M., and H. Ferris. 1987. Stage-specific population development and fecundity of *Paratrichodorus minor*. Journal of Nematology 19: 395 – 403.

Seinhorst, J. W. 1965. The relationship between nematode density and damage to plants. Nematologica 11: 137 – 154.

Seinhorst, J. W. 1966. The relationship between population increase and population density in plant parasitic nematodes. I. Introduction and migratory nematodes. Nematologica 12: 157 – 169.

Seinhorst, J. W. 1970. Dynamics of populations of plant parasitic nematodes. Annual

Review of Phytopathology 8: 131 – 156.

Sipes, B. S., and A. S. Arakaki. 1997. Root-knot nematode management in dryland taro with tropical cover crops. Supplement to the Journal of Nematology 29: 721 – 724.

Southwood, T. R. E. 1978. Ecological Methods, 2nd edn. London: Chapman and Hall.

Sperow, M., and D. W. Lybecker. 1998. Modeling the alfalfa stem nematode in irrigated alfalfa. Agricultural Systems 58: 555 – 570.

Stirling, G. R. 1991. Biological Control of Plant-parasitic Nematodes. Wallingford, UK: CAB International.

Sultan, S. A., and H. Ferris. 1991. The effect of soil moisture and soil particle size on the survival and population increase of *Xiphinema index*. Revue de Nematologie 13: 345 – 351.

Trevathan, L. E., and J. T. Robbins. 1995. Yield of sorghum and soybean, grown as monocrops and in rotation, as affected by insecticide and nematicide applications. Nematropica 25: 125 – 134.

Trivedi, P. C., and K. R. Barker. 1986. Management of nematodes by cultural practices. Nematropica 16: 213 – 236.

Vargas-Ayala, R., and R. Rodríguez-Kábana. 2001. Bioremediative management of soybean nematode population densities in crop rotations with velvetbean, cowpea, and winter crops. Nematropica 31: 37 – 46.

Vranken, G., and C. Heip. 1983. Calculation of the intrinsic rate of natural increase, r_m, with *Rhabditis marina* Bastian 1865 (Nematoda). Nematologica 29: 468 – 477.

Wallace, H. R. 1971. Abiotic influences in the soil environment. Pp. 257 – 280. In B. M. Zuckerman, W. F. Mai, and R. A. Rohde, eds. Plant Parasitic Nematodes. Vol. 1. New York: Academic Press.

Wallace, H. R. 1973. Nematode Ecology and Plant Disease. London: Edward Arnold.

Weaver, D. B., R. Rodríguez-Kábana, and E. L. Carden. 1995. Comparison of crop rotation and fallow for management of *Heterodera glycines* and *Meloidogyne* spp. in soybean. Supplement to the Journal of Nematology 27: 585 – 591.

Weaver, D. B., R. Rodríguez-Kábana, and E. L. Carden. 1998. Velvetbean and bahiagrass as rotation crops for management of *Meloidogyne* spp. and *Heterodera glycines* in soybean. Supplement to the Journal of Nematology 30: 563 – 568.

Weingartner, D. P., J. R. Shumaker, and G. C. Smart, Jr. 1983. Why soil fumigation fails to control potato corky ringspot disease in Florida. Plant Disease 67: 130 – 134.

Windham, G. L., and K. R. Barker. 1988. Winter survival of *Meloidogyne incognita* in six soil types. Journal of Nematology 20: 113 – 117.

Yeates, G. W. 1981. Nematode populations in relation to soil environmental factors: A review. Pedobiologia 22: 312 – 338.

11 Entomophilic Nematode Models for Studying Biodiversity and Cospeciation

Robin M. Giblin-Davis, Kerrie A. Davies, Gary S. Taylor, and W. Kelley Thomas

11.1 Introduction

Nematodes are among the most abundant, ubiquitous, and genetically diverse multicellular organisms in the world (Blaxter, et al., 1998; Lambshead, 1993; Platt, et al., 1984). A vastly incomplete taxonomic inventory and the lack of an efficient system for nematode identification have curtailed our understanding of their contribution to worldwide biological diversity. Estimates of the number of species vary widely depending upon the assumptions of the estimators and current systematic understanding includes only about 15,000 recognized species strongly biased towards species of agronomic importance.

Patterns and mechanisms of speciation in nematodes must be understood in order to form effective and testable hypotheses about nematode diversity. A major question regarding nematode diversity is the relative importance of associations with other organisms. Nematode associations with other organisms are common (Hunt, 1993; Poinar, 1975; Rühm, 1956; Siddiqi, 1986a). Increases in nematode biodiversity may result from genetic isolation or breaks in homeostasis caused by ecological associations with other animal and/or plant hosts as a consequence of association by descent (Mitter and Brooks, 1983). These associations have the potential to become isolating mechanisms for speciation raising the possibility that the species diversity of plants and animals can directly affect associated nematode species diversity. Nematode symbionts can have short generation times, larger populations, and isolated subpopulations that result in a high evolutionary potential allowing for the tracking of host evolutionary changes (Price, 1980). Coevolution is probably an important force in tightly aligned specific host-symbiont relationships common to the Nemata (Price, 1996; Thompson, 1989).

Thompson (1989) classifies reciprocal change in interacting species (coevolution) into two non-exclusive categories and five modes and discusses the relationship of parallel cladogenesis to coevolution. The first category involves three modes that yield anagenetic evolution (changes within an existing lineage that do not produce new species). These modes are: (1) gene-for-gene coevolution involving interacting species pairs where parasite or pathogen and hosts have complementary loci for virulence and resistance, respectively; (2) specific coevolution involving interacting species pairs where competing species or converging mutualists alternatively select for polygenic traits; and (3) guild coevolution (= diffuse coevolution) where the interacting partners involve groups rather than pairs of species (e. g., mutualistic nectar for pollination scenarios) (Fig. 11.1).

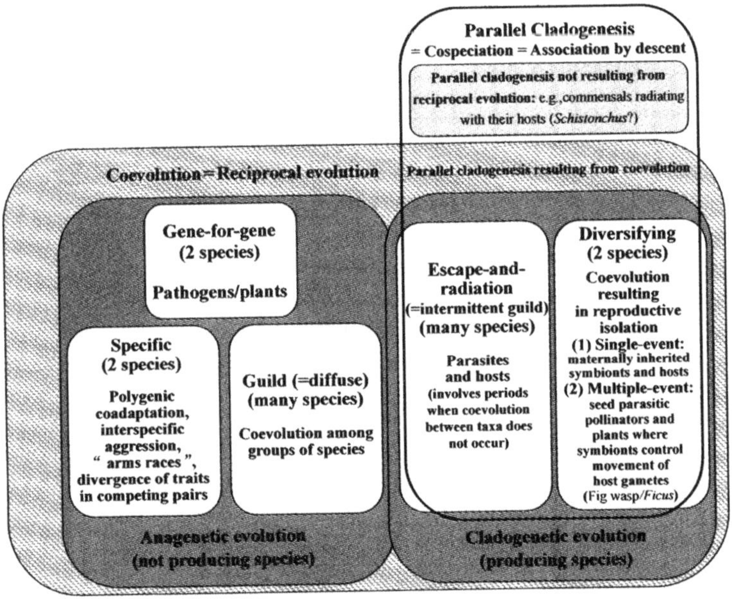

Figure 11.1 Schematic diagram illustrating the categories and modes of coevolution and types of parallel cladogenesis (Thompson, 1989). Thompson argues that cospeciation would be the best word for situations where diversifying coevolution yields reciprocal speciation. However, he recognizes that the word has become entrenched in systematics literature as synonymous for parallel cladogenesis regardless of whether coevolution is operating or not.

The second category of coevolution encompasses two modes that can yield anagenetic and/or cladogenetic (splitting of existing lineages or species producing) evolution (Fig. 11. 1). These are diversifying coevolution and escape-and-radiation coevolution. Diversifying coevolution occurs when reproductive isolation is caused by selective pressure from one of the two coevolutionary partners. Single-event diversifying coevolution is best illustrated where symbiont-induced speciation occurs in hosts with maternally-inherited intracellular symbionts, whereas multiple-event diversifying coevolution is best exemplified where multiple-events of reciprocal speciation have occurred. Examples of multiple-event diversifying coevolution are uncommon. They include relationships between plants and their pollinators where the pollinator larvae act as "host specific seed parasites on the same plants that the adults pollinate", i. e., the figs and fig wasps, yuccas and yucca moths, and globeflowers and *Chiastocheta* flies (Thompson, 1989). Escape-and-radiation coevolution is intermittent guild coevolution where one group that is overexploited by another group of associates evolves a novel defense response that allows for escape from the attackers, expansion into a new adaptive zone while decoupled from the attackers, and subsequent radiation. Reciprocal change in a population of associates that counteracts the novel defense allows for radiation of the successful attacking associate onto the radiated host group. This mode of coevolution is punctuated by long periods of little or no interaction during host radiation. Thus, unlike diversifying coevolution where speciation in the host is caused by the interaction, host speciation in escape-and-radiation coevolution hypothetically occurs when hosts are released from the attacking pressure of the symbionts (Thompson, 1989).

The first step in proving cladogenetic coevolution has often involved looking for parallel radiation of each of the interacting taxa. Cladistic methods can test whether or not plants and their herbivores or pollinators evolve in parallel (parallel cladogenesis = cospeciation = association by descent) (Mitter and Brooks, 1983) but such congruence in cladograms cannot be used as definitive proof for coevolution (Thompson, 1989). This is because vicariance events and species-specific host-tracking adaptations could produce similar results (Gould, 1988). Thus, some parallel radiation of symbionts and their hosts is not due to coevolution, because parallel cladogenesis can occur without reciprocal change in interacting species, e. g., commensals speciating with their hosts (Fig. 11.1).

There are over 3000 known entomophilic nematodes that have associations with their insect hosts ranging from loose phoresy to strict species-specific parasitism (Giblin, 1986; Poinar, 1975). We suspect that this is a gross under-representation of true species abundance of this grouping of nematodes. There are over a million species of insects in the world with projections of greater than five million species. Many of these insects spend most or part of their life cycles in environments conducive to nematode growth and reproduction. In fact, the insect itself represents a potential food and shelter resource for nematodes that can overcome defense mechanisms. Nematodes are aquatic invertebrates with limited powers of dispersion relative to their terrestrial-adapted insect hosts. Those that develop associations with insects stand to benefit from more predictable resource availability and an accompanying dispersal mechanism. We hypothesize that the tighter the association between nematode and insect host (high degree of host specificity) the greater the likelihood for cladogenic origins that are a consequence of association by descent with the host. Insect radiations involving isolated moist environments with large amounts of temporally available resources (fungi, plants, bacteria, or other invertebrates) may be the most suitable for biodiversity and coevolutionary study. Although dung and carrion offer such opportunities, these environments also offer many chances for host-switching which may increase the opportunities for gene flow thereby minimizing biodiversity.

The best-studied assemblage of insect-associated nematodes may be those associated with bark beetles (Scolytinae), their plant hosts, and fungal symbionts (Massey, 1974; Rühm, 1956). However, even this grouping requires much work. In examples where host specificity has been tested, bark beetle-associated nematodes (e. g., *Parasitorhabditis* spp.) appear to be host specific with evidence of association by descent (Korenchenko, 1987). The well-known entomopathogenic nematodes, i. e., *Steinernema* and *Heterorhabditis*, may not necessarily be good models for biodiversity because they are involved in mutualistic complexes with their associated bacteria that render most insects as good hosts (wide host range) for bacterial and nematode reproduction.

We propose two broad complexes involving entomophilic nematodes in tight (tritrophic) associations with insects and plants to help elucidate how symbiotic associations might affect nematode evolution and species diversity. The best example of cospeciation in the phylum Nemata may occur in conjunction with the dramatic radiation of the fig genus *Ficus* (*Moraceae*). The *Ficus* radiation is well known for its associated mutualism with pollinator wasps (Agaonidae). Less well known, but equally interesting, are the nematodes *Schistonchus* (Aphelenchoididae) and *Parasitodiplogaster* (Diplogasteridae) that are vectored by pollinator wasps in apparently species-specific associations (Giblin-Davis, et al., 1995). Another equally provocative tritrophic association involving nematodes, flies, and plants is the *Fergusobia/Fergusonina/Myrtaceae* complex (Currie, 1937). These tritrophic complexes are interesting in how differences between associations with other organisms may be involved in creating opportunities for nematode biodiversity. The *Ficus* complex entails a mutualism between a plant and an insect with the nematode symbionts being parasitic or phoretic on said hosts (Fig. 11.2(A)), whereas the *Fergusobia* system involves a unique mutualism between an insect and a nematode that parasitizes plants (Fig. 11.2(B)).

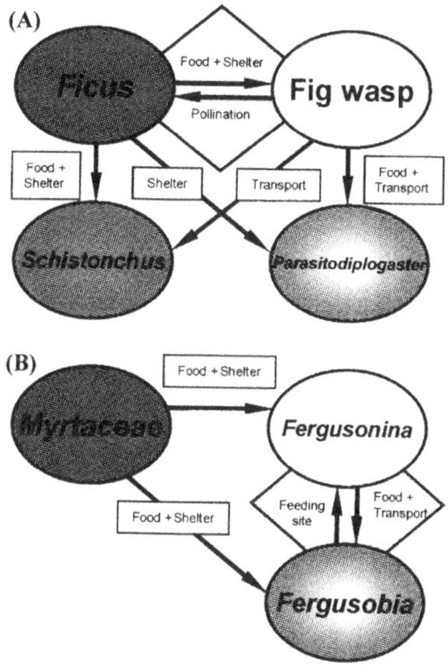

Figure 11.2 Schematic diagrams of the flow of benefits and the organisms sharing mutualisms (connected by a diamond) in two model nematode systems. (A) Tritrophic association between *Ficus*, fig wasps (Agaonidae), and nematodes (*Schistonchus* and/or *Parasitodiplogaster*). (B) Association between myrtaceous plant host, fly (*Fergusonina*), and nematode (*Fergusobia*).

In the *Ficus* system, gene flow for both the plant and nematode populations is controlled by the fig wasps (pollinators) resulting in possible breaks in species cohesion for the nematodes through genetic drift. Host specificity between the wasps and figs creates ecological allopatry for the nematode symbionts thereby creating opportunities for cospeciation. Interestingly, recent data suggests that *Schistonchus* can have a negative effect on the mutualism by attacking the anthers of the fig host thereby acting as a selective agent in the association (Center, et al., 1999; Giblin-Davis, et al., 1995). Furthermore, *Schistonchus* may provide selective pressure as an internal associate (phoretic or parasitic) of the wasps. Thus, specific or guild coevolution may play a role in the evolution of *Schistonchus*. *Parasitodiplogaster* virulence may be a selective factor that interacts with pollinator population structure and parasite transmission opportunities and may be affected by the plant host morphology (Herre, 1995) allowing for specific or guild coevolution.

In the *Fergusobia* system, the insect (fly) also controls gene flow for nematode populations. However, in this situation the nematode/fly complex functions as an herbivore attacking the reproductive structures (inflorescence or flower buds) and/or new shoots (leaf buds) of its plant host having less effect on gene flow of the plant host compared with the pollinator-based mutualism in *Ficus*. The *Fergusobia* complex might provide evolutionary impetus for the plant host through overexploitation and the plant might respond with improved defensive mechanisms in a coevolutionary spiral (specific or escape-and-radiation coevolution). However, Futuyma (1983) argues that proof for these sorts of scenarios in nature is lacking. The possible struggle between overexploitive parasitic nematodes and overzealous defenses by their fly hosts is the most plausible scenario for specific coevolution, although this is speculative and not easily studied.

11.2 Methodology

11.2.1 Investigating Nematode Species Diversity with Gene Sequences

Species-level systematic studies of nematodes are often confounded by a dearth of easily accessible morphological characters that vary between species. As a consequence, molecular approaches are often employed and have proven informative. The criteria for molecular approaches to nematode taxonomy were recently evaluated by the National Science Foundation (NSF) Workshop on Systematics and Inventory of Soil Nematodes (1997). The conclusion of this workshop was that regions of the nuclear ribosomal RNA (rRNA), specifically the D2/D3 expansion segments of 28S coding gene and the internal transcribed spacer (ITS) regions, have the greatest promise to aid species-level nematode taxonomy. The multicopy rRNA genes of many eukaryotic organisms appear to evolve in a concerted pattern whereby the vast majority of the genes within a genome and within a species are of a single sequence. By contrast, sequence differences are found between species (Brown, et al., 1972). Empirical evidence suggests that a concerted pattern is also found in nematode rRNA evolution (Baldwin, et al., 1997a, 1997b; Blaxter, et al., 1998; Fitch, et al., 1995). Highly conserved sequences found in the rRNA genes are useful for the design of "universal" primers (Arnheim, et al., 1990; Kocher, et al., 1989; White, et al., 1989). Primers designed to highly conserved regions will amplify the homologous fragment from most taxa with those priming sites and provide potentially homologous sequences for analysis.

We have chosen for our studies the most rapidly evolving coding region of the large subunit of the rRNA genes (the D2 and D3 expansion segments). This region will allow unambiguous communication and cost-effective resolution of the potentially large number of nematode species that may comprise the *Parasitodiplogaster* and *Schistonchus* assemblages associated with *Ficus* and the *Fergusobia* associated with the *Myrtaceae*. We have empirically tested the practical utility of the D2/D3 sequence to discriminate nematodes to the species level and created a preliminary sequence database. We amplified the D2/D3 expansion segment from diverse taxa including sequences from over 260 independent isolates representing 62 genera of 18 families from several orders and both nematode classes (Thomas, et al., unpubl. data). Two important criteria for the use of D2/D3 sequences as a species-level systematic tool are: (1) the sequences should uniquely identify different species; and (2) the sequences should be the same for individuals of a single species (minimal intra-specific variation). In our preliminary data set, there are 25 cases where multiple species have been analyzed from a single genus. In all cases, each species was found to have a unique sequence. Furthermore, in cases where multiple independent isolates of a single species were examined, nearly all individuals of a single species had the same sequence. In one of the largest surveys of conspecific nematodes to date, the D2/D3 region was sequenced from 169 *Scottnema lindsayae* from Antarctica and found only a few variable nucleotides, whereas numerous mitochondrial haplotypes were found among the same individuals (Courtright, et al., 2000). Recent work with *Pratylenchus coffeae* and closely related species confirm the utility of D2/D3 sequences for species resolution in nematodes (Duncan, et al., 1999).

11.2.2 An Operational Species Concept

In the past, nematologists have recognized several criteria for defining species including discontinuities in morphology (Davis and Heywood, 1963) and reproductive isolation as in the biological species concept (Mayr, 1969). While operationally a morphological or phenetic species concept is often applied as the most practical approach for species definition, these methods are limited in their application across Nemata (Andrássy, 1984; Appendix 2 of NSF Workshop on Systematics and Inventory of Soil Nematodes, 1997; Triantaphyllou, 1985). Most importantly, they fail to incorporate a historical context for species consistent with an understanding of their evolutionary/phylogenetic framework (Ferris, 1983; Nelson and Platnick, 1981; Wiley, 1981).

A critical aspect of investigating the association of each fig and fig wasp with a unique nematode species and each *Fergusonina* fly with a unique *Fergusobia* is the requirement that we have an operation to recognize nematode species. Recently, Adams (1998) proposed a set of operations to delimit nematode species. This approach is an amalgamation of the phylogenetic and evolutionary species concepts (Frost and Kluge, 1994) and defines species as the smallest discernable phylogenetic unit with an autapomorphy, a uniquely derived character. The practicality and unambiguous nature of this approach make it particularly useful in our investigations. Our choice of the D2/D3 region reflects in part the desire to generate data that are useful for phylogenetic definitions while providing the necessary resolution to distinguish nematodes at the species level. The product of these molecular investigations will be an "operational species". We define an operational species as a unit of study around which evidence may be developed to support its entity as an independent phylogenetic trajectory. The "operational species" is a testable hypothesis that will be further addressed in independent monographic studies within a phylogenetic context incorporating morphology, other biological information, and within the current taxonomy.

11.2.3 Investigating Parallel Cladogenesis with Gene Sequences

The first requirement for a meaningful investigation of cospeciation is to have sound and independent evolutionary frameworks for the putative evolutionary partners. In the *Ficus* system, we are using the established molecular frameworks for figs and fig wasps (Herre, 1995; Herre, et al., 1996; Machado, et al., 1996) (Fig. 11.3). Our early nematode molecular studies are focusing on a subset of sample lineages included in those studies. In the *Fergusobia* system, we are using established molecular and morphological frameworks of the *Myrtaceae* (Hill and Johnson, 1995, 1998; Ladiges and Humphries, 1983, 1986; Ladiges, et al., 1983, 1987, 1995; Pryor and Johnson, 1981; Sale, et al., 1993). We are generating our own *Fergusobia* and *Fergusonina* molecular trees for comparisons of congruence (Giblin-Davis, et al., unpubl. data; Scheffer, et al., unpubl. data). The ultimate generation of sound phylogenetic hypotheses involves many competing factors, meaningful characters, and analyses and interpretations. In our molecular analyses of nematodes, we plan to include the D2/D3 rRNA for all species collected, and the complete 18s rRNA loci, as well as a portion of the coding region from the gene for RNAP II for some of the nematodes for deep lineage comparisons. We have employed these loci in several previous studies (Baldwin, et al., 1997a, 1997b; Blaxter, et al., 1998).

The levels of divergence may vary widely and the molecular approach must

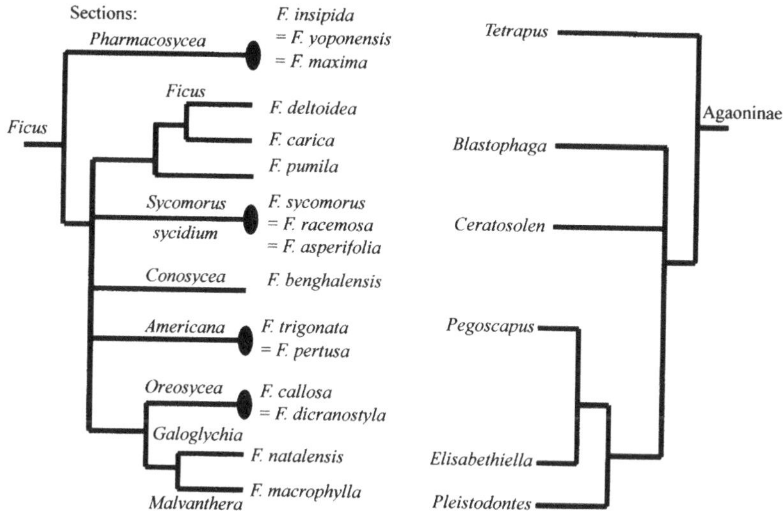

Figure 11.3 Molecular phylogeny of the *Ficus* sequenced by Herre et al. (1996). Branching order is inferred from rbcl and tRNA spacer sequences. *Ficus deltoides*, *F. carica*, and *F. pumila* are gynodioecious fig species; all others are monoecious.

accommodate this possibility. The D2/D3 sequence data that will be collected for all taxa has proven useful for closely related nematode taxa and will contribute to the overall phylogenetic analysis (Baldwin, et al., 1997b). The 18S rDNA sequences have proven valuable for generating a molecular phylogenetic hypothesis for the phylum Nemata (Baldwin, et al., 1997a, 1997b; Blaxter, et al., 1998; Fitch, 1997; Fitch and Emmons, 1995; Fitch and Thomas, 1997; Fitch, et al., 1995). Several laboratories continue to add 18S sequences to the current phylogeny (Kampfer, et al., 1998; Nadler and Hudspeth, 1998). In all likelihood, this locus will continue to function as a common metric for nematode phylogeny across broad taxonomic levels. Nevertheless, nematode 18S sequences vary dramatically in both evolutionary rate and base composition and may confound phylogenetic inference of some groups. Some taxa, such as *Pelodera*, have extremely different 18S sequences and are often eliminated from inclusion in 18S molecular phylogenies (Blaxter, et al., 1998). While other molecules will contribute to the resolution of nematode relationships (Thomas and Wilson, 1991; Vanfleteren, et al., 1994), based on preliminary analysis, we intend to focus our efforts on the collection of D2/D3, 18S rRNA and large subunit RNAP II sequence data.

11.2.4 Molecular Phylogenetic Analysis

Sequences collected for host isolates of *Schistonchus*, *Parasitodiplogaster*, and *Fergusobia* are aligned by eye using ESEE (Cabot and Beckenbach, 1989). Regions of the ambiguous alignment are removed from the analysis (Sidow and Thomas, 1994; Swofford, et al., 1996). Maximum Parsimony (MP), Neighbor-joining (NJ), and Maximum Likelihood (ML) methods (as implemented in PAUP* (Swofford, 1998), MEGA (Kumar, et al., 1993), and PHYLIP (Felsenstein, 1993)) are employed to infer phylogenies. In MP searches, we use 100 random input orders, nearest neighbor interchanges, and branch swapping to increase the probability of recovering the best tree. Weighting schemes consider the work of others (Hollar and Springer, 1996; Krajewski, et al., 1997; Lavergne, et al., 1996; Springer, et al., 1995; Stanhope, et al., 1996). For the NJ method, sequence divergence is estimated using a variety of models (e.g., Jukes-Cantor (Jukes and Cantor, 1969), Tamura-Nei (Tamura and Nei, 1993), ML (Swofford, et al., 1996), and logdet method (Lockhart, et al., 1994)). ML employs empirical base frequencies, and where possible, ML estimates of ts/tv ratio. We test phylogenetic hypotheses using winning sites (Prager and Wilson, 1988), Templeton (1983), and Kishino-Hasegawa (Kishino and Hasegawa, 1989) tests, as well as bootstrapping (Felsenstein, 1985; Hillis and Bull, 1993; Hillis, et al., 1996; Huelsenbeck, et al., 1996; Sanderson, 1989, 1995).

11.2.5 Combining Molecular Datasets

Both taxonomic congruence and total evidence strategies for analysis of multiple datasets have advantages and disadvantages (Bromham, et al., 1996; Huelsenbeck, et al., 1996; Kluge, 1989; Leignel, et al., 1997). We recognize that the separate analysis of multiple genes (i. e., 18S, and RNAP II) provides independent support for specific clades; however, in other cases resolving power is increased with combined datasets. Recently, Ballard et al. (1998) proposed a potentially useful approach to analysis of multiple datasets. Morphological data will initially be evaluated in the context of the independent molecular framework.

11.2.6 Expected Results

If the null hypothesis is that each tritrophic association originates just once and that cospeciation is strict, then we expect that the evolutionary histories will be congruent. Also, will the similarities observed between histories exceed that which would be expected by chance. We recognize that even in the event that we find widespread species-specific nematode associations with either the *Ficus* or *Myrtaceae* radiations, that congruent patterns may not result from long term cospeciation. High levels of nematode species diversity may simply result from the isolation afforded by recent vicariance events or unilateral host-tracking adaptations (Futuyma and Slatkin, 1983). Also, nematodes may shift associations at a frequency greater than the level necessary to maintain evidence of common evolutionary histories and this may or may not be a response to coevolution (Gould, 1988). Behaviors currently supporting the isolation of nematode lineages may not have existed in the ancestral lineages. Studies in other taxa such as the cospeciation of gophers and parasitic lice (Hafner, et al., 1994) or between fungus-growing ants and their fungi (Chapela, et al., 1994; Hinkle, et al., 1994) have shown that such comparisons are fruitful opportunities to study potential coevolution-derived patterns. In these examples cospeciation is clearly evident, but exceptions are also found which represent important biological phenomena such as the repeated acquisition of symbionts (Chapela, et al., 1994) and putative cases of host switching (Hafner, et al., 1994). Statistical tests of host-parasite cospeciation (Huelsenback, et al., 1997) between the nematodes and the *Ficus* evolutionary histories and between *Fergusobia* and *Fergusonina* and myrtaceous plants will provide valuable biological understanding of the factors affecting nematode speciation and the propensity for host-switching. In the case of the *Fergusobia* and *Fergusonina* associated with the introduced and invasive weed species *Melaleuca quinquenervia* in Florida, this information will have relevance in decisions about the safety of releasing them as an introductive biological control complex.

11.3 *Ficus/*Agaonid/*Schistonchus* and *Parasitodiplogaster* Associations

11.3.1 The *Ficus* Radiation

The fig genus *Ficus* represents an excellent opportunity to investigate the potential for the evolutionary radiation of a plant species to bring about the radiation of associated nematodes. There are about 750 species of *Ficus* worldwide (Berg, 1989). This genus represents an interesting and well-characterized case of potential strict cospeciation with its associated fig wasp pollinators (Brooks, 1988; Herre, et al., 1996; Janzen, 1979; Mitter and Brooks, 1983; Ramirez, 1970a, 1970b, 1974). Also associated with this radiation are two genera of nematodes, *Schistonchus* and *Parasitodiplogaster* (Giblin-Davis, et al., 1995). The biology of these nematode taxa and

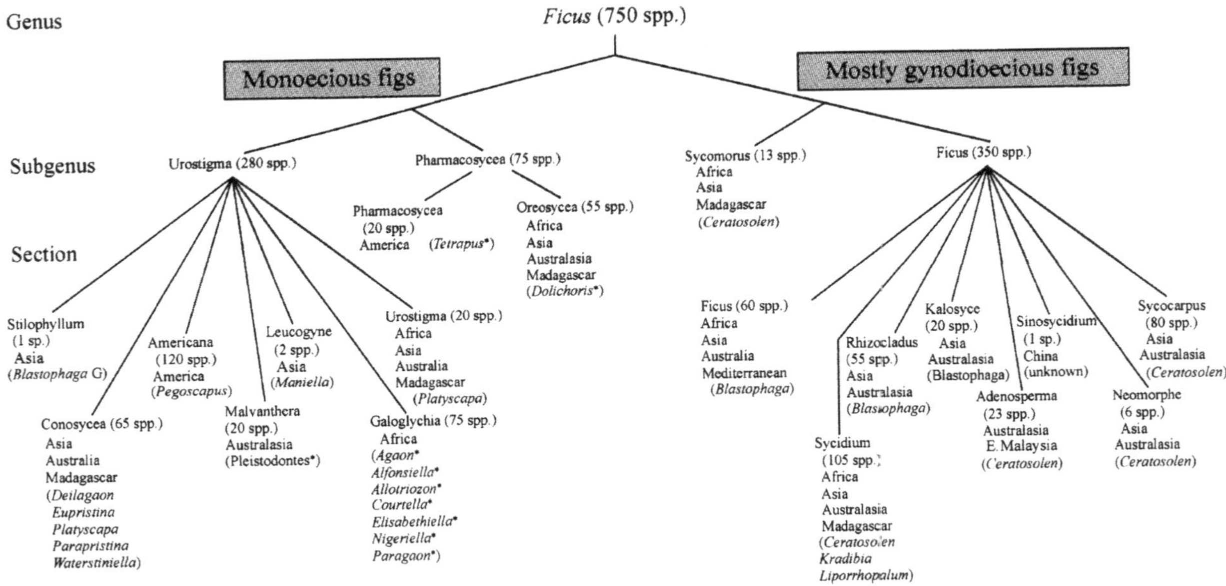

Figure 11.4 Classical taxonomy of *Ficus* based upon Corner (1965) and Berg (1989). Pollinator wasps with an * are of the subfamily Blastophaginae; all others are of the subfamily Agaoninae.

preliminary molecular data suggest that the "habitat" isolation afforded by the fig/fig wasp mutualism has the potential to drive speciation in these nematodes and the radiation of vast numbers of nematode species.

In *Ficus*, monoecy is hypothesized to be ancestral and gynodioecy derived (Fig. 11.4). In monoecious figs, each fruit (syconium) has both male and female florets functioning to produce pollen and viable seed. Some of the florets are parasitized to produce wasps that carry the pollen to other trees (Fig. 11.5). In gynodioecious figs, separate trees produce syconia creating either viable seed (functionally female trees) or pollen and wasps (functionally male trees). Obligatory outcrossing has been hypothesized to be the selective force driving the evolution of imperfect heterostyly, pronounced protogyny, and intermittent synchronous flowering in monoecious species, and perfect heterostyly, asynchronous flowering, or overlapping fig crops in gynodioecious species of *Ficus* (Bawa and Crisp, 1980; Berg, 1990). Seed production and rearing of pollinators are correlated in space and time occurring in the same fig on the same tree in monoecious species.

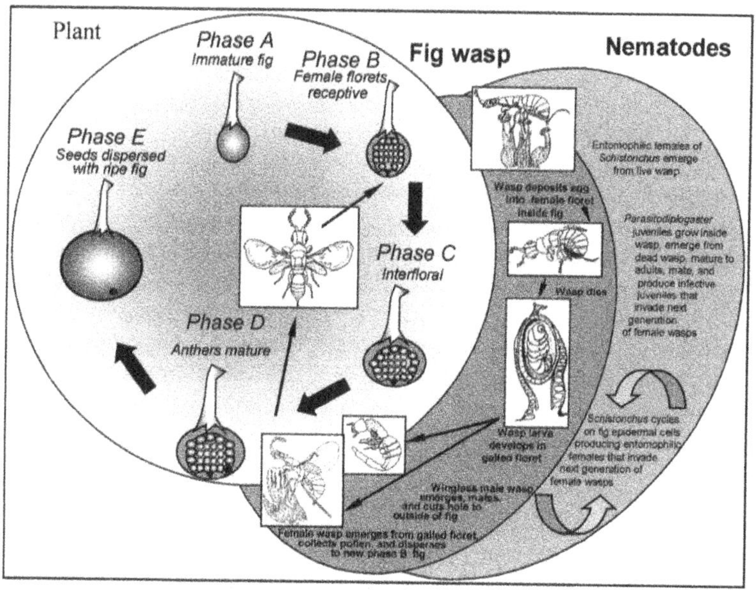

Figure 11.5 Diagram of the successive phases of development of a typical monoecious *Ficus* sycone (fig) and the synchronized association of its mutualistic fig wasp pollinator and *Schistonchus* and *Parasitodiplogaster* nematode associates.

Of the species of *Ficus*, about 500 occur in Asia, Australasia, and Malaysia (Fig. 11.4). Most of these are gynodioecious (Berg, 1989) suggesting that evolution of gynodioecy may have been centered there and then moved into Australia, Madagascar and Africa competing with and displacing established monoecious lines. There are no gynodioecious species in the Neotropics and only 10 species in Africa (Berg, 1989). According to recent molecular studies on a limited number of *Ficus*, it appears that the New World subgenus *Pharmacosycea* is genetically distinct from all other lineages (Fig. 11. 3) (Herre, et al., 1996). These studies also suggest that either dioecy evolved twice (in the subgenus *Sycomorus* and the subgenus *Ficus*, section Sycidium) or that it has been lost once in the Sycomorus group (Fig. 11. 3) (Berg, 1989; Herre, et al., 1996).

Like *Shorea* and *Eugenia* in Malaysia and *Eschweilera*, *Licaniua*, and *Protium* in tropical America, *Ficus* often occurs as numerous, very similar, sympatric species (Richards, 1996). The presence of these apparently sympatric species may reflect differences in floral biology, seed dispersal mechanisms, phenology, and micro-habitats (Ashton, 1969; Gentry, 1992; Rogstad, 1990). By contrast, *Ficus* speciation may be a consequence of allopatric isolation mechanisms. The diversity of fig species may at a primary level be a consequence of inherent instability of tropical habitats during climatic shifts. However, speciation stimulated by the allopatry of geoclimatically-caused refugia has been challenged (Gentry, 1992) and may not explain the extraordinary species radiation of *Ficus* and their current sympatry (Richards, 1996).

Fossil evidence (Collinson, 1989; Weier, et al., 1974) supports the existence of *Ficus* forms by Mid-Eocene, about 50 million years before present (mybp) (Poinar and Herre, 1991). The discovery of a fig-pollinating wasp with nematodes in Dominican amber (Herre, 1995) suggests that the nematode/fig wasp/fig associations are quite ancient. Taken together and consistent with molecular data (Herre, et al., 1996) these observations support a potentially ancient divergence of *Ficus* lineages.

11.3.2 The Role of Fig Wasps

The radiation of *Ficus* is thought to be intricately associated with fig wasp pollinators (Brooks, 1988). In fact, the modern fig fruit (syconium) is thought to have evolved in response to over-exploitative beetle or fly pollinators, thus, selecting the seed-predatory chalcid wasp ancestors of the present day fig wasps (Agaonidae) to occupy a unique pollination niche as the beetle or flies were screened out (Berg, 1990). If this is true, we suspect that *Schistonchus* and *Parasitodiplogaster* may be remnant associates of those original pollinators that developed a new association with fig wasps during the evolution of the syconium (Fig. 11.6). Fig-associated wasps in the family Agaonidae belong to the superfamily Chalcidoidea, which is the largest superfamily in the Hymenoptera with estimates of over 60,000 species (Gordh, 1979). By 1975, however, only two nematode associates had been reported for the entire superfamily and these were from agaonid fig pollinators (Poinar, 1975). Since then, an additional 15 species have been reported (Giblin-Davis, et al., 1995; Lloyd and Davies, 1997; Poinar, 1979; Poinar and Herre, 1991; Vovlas, et al., 1998) but they have all been limited to fig-associated pollinators and inquilines in the Agaonidae (Table 11.1). This suggests that the association between the Agaonidae and nematodes is the result of a unique evolutionary history within the Chalcidoidea or that nematologists and hymenopterists have not looked very carefully. Such a hypothesis would predict that divergence within either *Parasitodiplogaster* and/or *Schistonchus* traces back to the origins of syconia and the genus *Ficus* (Fig. 11.6).

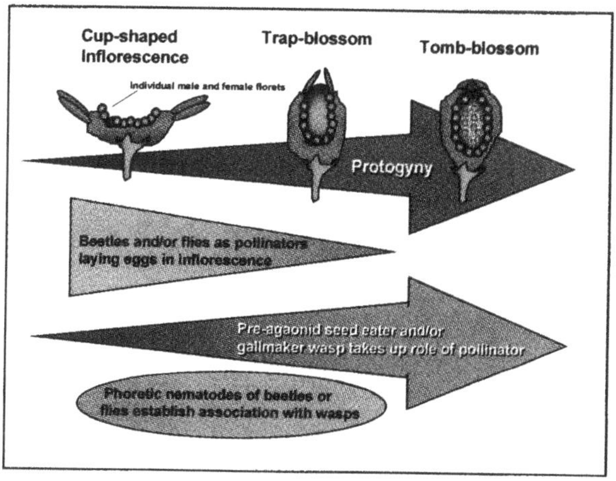

Figure 11.6 Possible scenario for the evolution of the *Ficus* syconium (fig) and the mutualistic association with fig wasps and their associated nematodes, *Schistonchus* and *Parasitodiplogaster*.

The fig wasps are potentially important for both primary isolation and the reinforcement of species boundaries for both the figs and the associated nematodes. First, the wasps are small (generally < 1 mm long) and short-lived, restricting active long distance gene flow for geographically isolated *Ficus* (Bronstein, 1992). Second, while habitat shifts and ecological barriers may explain initial genetic separation of *Ficus* populations, the wasp pollinators are assumed to be a key isolating mechanism in the fig/fig wasp radiation

Table 11.1 Nominal and known undescribed species of *Schistonchus* and *Parasitodiplogaster* nematodes.

Ficus Species	Agaonidae Species	Location	*Schistonchus* sp.	*Parasitodiplogaster* sp.	Reference
***FICUS*(subgenus) *Ficus*(section)**					
F. carica	*Blastophaga psenes*	Italy	*S. caprifici*		(1)
FICUS Sycidium					
F. copiosa v. *copiosa*	*Blastophaga* sp.	Australia	n. sp.		(3)
FICUS Sycocarpus					
F. hispida	*Ceratosolen solmsi*	India	*S. hispida*		(4)
F. hispida	*Ceratosolen solmsi*	Australia	n. sp.		(3)
SYCOMORUS Sycomorus					
F. racemosa	*Ceratosolen fusciceps*	India	*S. racemosa*		(8)
F. racemosa	*Ceratosolen fusciceps*	Australia	n. sp.		(3)
PHARMACOSYCEA Pharmacosycea					
F. insipida(= *glabrata*)	*Pegoscapus* sp.	Panama	n. sp.	n. sp.	(3)
F. maxima	*Pegoscapus* sp.	Panama	n. sp.	*P. maxinema*	(3)(7)
F. yoponensis	*Pegoscapus ecuatorianis*	Panama	n. sp.	*P. yoponema*	(3)(7)
UROSTIGMA Americana					
F. aurea	*Pegoscapus mexicanus*	Florida, U. S. A.	*S. aureus*		(2)
F. columbrinae	*P. orozcoi*	Panama		n. sp. "columbrinema"	(3)(7)
F. dugandii	*Pegoscapus* sp.	Panama	n. sp.	*P. duganema*	(3)(7)
F. hemsleyana(= *F. citrifolia*)	*P. tonduzi*	Costa Rica	n. sp.	*P. citrinema*	(3)(7)
F. jimenezi	*P. jimenezi*	Costa Rica	n. sp.		(3)
F. laevigata(= *F. citrifolia*)	*Pegoscapus* sp.	Florida, U. S. A.	*S. laevigatus*	n. sp.	(2)(3)
F. morazaniana	*Pegoscapus* sp.	Costa Rica		n. sp.	(3)
F. nymphaeifolia	*P. piceipes*	Panama		*P. nymphanema*	(7)
F. obtusifolia	*P. hoffmeyerii*	Panama		*P. obtusinema*	(7)
F. paraensis	*P. herrei*	Panama		*P. paranema*	(7)
F. perferata	*P. insularis*	Panama		n. sp. "perfonema"	(3)(7)

Continued

Ficus Species	Agaonidae Species	Location	*Schistonchus* sp.	*Parasitodiplogaster* sp.	*Reference*
F. pertusa	*P. silvestrii*	Panama	n. sp.	*P. pertanema*	(3)(7)
F. poponoei	*P. gemellus*	Panama	n. sp	*P. popenema*	(7)
F. trigonata	*Pegoscapus* sp.	Panama	n. sp.	*P. trigonema*	(3)(7)
F. near *trigonata*	*P. lopesi*	Panama		n. sp. "trinema"	(3)(7)
F. turbinata(=*F. citrifolia*)	*Pegoscapus* sp.	Panama	n. sp.	n. sp	(3)
UROSTIGMA Conosycea					
F. benjamini v. *benjamini*	*Eupristina adempta*	Australia	n. sp.		(3)
UROSTIGMA Galoglychia					
Chlamydodorae(subsection)					
F. thonningii	*Elisabethiella stuckenbergi*	So. Africa	*S. africanus*	*P. sycophilon*	(9)(6)
F. iliciana	*E. enriquesi*	So. Africa		n. sp.	(3)
UROSTIGMA Malvanthera					
F. macrophylla	*Pleistodontes forggatti*	Australia	*S. altermacrophylla*		(5)
F. macrophylla	*Pleistodontes froggatti*	Australia	*S. macrophylla*		(5)
F. baileyana	*Pleistodontes* sp.	Australia	n. sp.		(3)(5)
F. brachypoda	*Pleistodontes* sp.	Australia	n. sp.		(3)
F. obliqua	*Pleistodontes greenwood*	Australia	n. sp.		(3)
F. platypoda	*Pleistodontes cuneatus*	Australia	n. sp.		(3)(5)
F. rubiginosa	*Pleistodontes imperialis*	Australia	n. sp.		(3)(5)
F. pleurocarpa	*Pleistodontes regalis*	Australia	n. sp.		(3)
F. watkinsiana	*Pleistodontes* sp.	Australia	n. sp.		(3)
UROSTIGMA Urostigma					
F. virens v. *sublanceolata*	*Blastophaga coronata*	Australia	n. sp.		(3)
TOTAL			30	21	

References: (1) Gasparinni(1864); (2) DeCrappeo and Giblin-Davis(2001); (3) Giblin-Davis et al. (Unpub. Obs.); (4) Kumari and Reddy (1984); (5) Lloyd and Davies (1997); (6) Poinar (1979); (7) Poinar and Herre (1991); (8) Reddy and Rao (1986); and (9) Vovas et al. (1998).

(Berg, 1989; Ramirez, 1970a, 1970b, 1974; Wiebes, 1979). Once figs have been sufficiently isolated to reduce compatibility, selection will strongly favor wasp pollinators that can recognize conspecific *Ficus*. If two species of figs exist in geographic proximity, mistakes by the wasp can result in aborted syconia and terminate the mistaken wasp's lineage as well as that of the associated nematodes. If wasps that make mistakes die, then mechanisms of exact species recognition are favored. Recent evidence suggests that species-specific semiochemicals produced by phase B syconia are important in exact species-specific pollination in *Ficus* (Bronstein, 1992). Contrary to this generalized scenario, a few contemporary patterns of fig/fig wasp associations suggest that *Ficus* species have been colonized by wasp pollinators that displaced resident wasps establishing new fig/fig wasp lineages. If such events are rare in the evolutionary history of *Ficus*, our investigations of cospeciation should detect such host-switching phenomena (Berg, 1989; Compton, 1990; Herre, et al., 1996; Ramirez, 1970a, 1970b, 1974; Wiebes, 1979, 1982).

11.3.3 Nematode Associates

There are two genera of nematodes (*Schistonchus* and *Parasitodiplogaster*) that are found associated with phase B through D figs of *Ficus* species and their fig wasp pollinators (Giblin-Davis, et al., 1995) (Figs 11.2, 11.5; Table 11.1). Both genera are only known from figs and fig wasps. *Schistonchus* is apparently distributed worldwide with only six described species (Table 11.1). It is a stylet-bearing nematode in the order Aphelenchida, family Aphelenchoididae. It shares subfamily ranking with the notorious fungal and plant-parasitic genus, *Aphelenchoides* and the mycophagous and beetle-associated genera, *Laimaphelenchus* and *Sheraphelenchus* (Hunt, 1993). In the New World, entomophilic females of *Schistonchus* are carried in the hemocoel of the fig wasp; the association with the wasp appears to be one of internal phoresy (Giblin-Davis, et al., 1995). *Schistonchus caprifici* from Europe can be carried as eggs, juveniles and adults in the host suggesting parasitism (Vovlas, et al., 1992). The nematode leaves the hemocoel of the wasp upon entering the lumen of a phase B syconium (female phase fig; female florets are receptive to pollination) and parasitizes epidermal cells of the pedicels or the syconial wall (*Schistonchus aureus* from *F. aurea*) (Center, et al., 1999) or the anthers of the fig (*Schistonchus laevigatus* from *F. laevigata*) (Center, et al., 1999; DeCrappeo and Giblin-Davis, 2001; Giblin-Davis, et al., 1995) (Fig. 11.5). At least one generation of *Schistonchus* develops at the expense of the fig during phase C (interfloral phase fig; where developing fig embryos and wasp larvae and pupae occur). After about 30 days, infested phase D syconia (male phase fig; male florets with mature pollen) contain adult male and female *Schistonchus*, adult male and female wasps, and maturing fig seeds. Presumably after mating, entomophilic female nematodes enter the hemocoels of emerging female wasps of the next generation. Wasps collect pollen, exit the syconium through an emergence hole created by the wingless male wasps, and seek out a new phase B syconium for oviposition. In the process, she vertically transmits nematodes to the next generation of figs and fig wasps (Fig. 11.5) (Giblin-Davis, et al., 1995). Amphimictic reproduction is presumed.

Parasitodiplogaster is a genus with 11 nominal species that exclusively parasitizes fig wasps in an otherwise mostly freeliving or phoretic group of omnivorous (predatory, bacterial and fungal feeding) nematodes in the order Rhabditida, subfamily Diplogasterinae. *Parasitodiplogaster* has been reported only in African and New World *Ficus* (Fig. 11.4, Table 11.1). However, unidentified diplogasterids have been recently discovered associated with *F. racemosa*, *F. copiosa*, and *F. virens* and their pollinators in Australia (Giblin-Davis, unpubl. data). The third stage (assuming one molt in the egg) infective juvenile (dauer larva) invades the abdominal hemocoel of the emerging female fig wasp. The nematode appears to develop at the expense of the host wasp and becomes greatly enlarged during its fourth stage. Nematode development appears to be timed after the fig wasps' difficult migration through the ostiole of the fig and after pollination and oviposition have occurred (Giblin-Davis, et al., 1995). Herre (1993, 1995) argues that fig wasp population structure and the resultant opportunities for nematode parasite transmission influence the evolution of species interactions such that host wasp populations with the greatest opportunities for horizontal transmission (larger numbers of female foundresses) select for *Parasitodiplogaster* (and possibly *Schistonchus*) species that are more virulent parasites. Nematodes emerge from the dead or dying wasp as adults or fourth stage juveniles that molt to adults that then mate and produce eggs (Fig. 11.5). Eggs give rise to second and then infective third stage juveniles that wait for the next generation of female wasps to complete metamorphosis, infecting them as they emerge from their galls inside the fig. These life history scenarios for *Parasitodiplogaster* and *Schistonchus* suggest the potential for strict isolation of nematodes from different *Ficus* species (Fig. 11.5).

11.3.4 Morphological Variation in Associated Nematode Taxa

Very little has been published that elucidates evolutionary patterns in morphological characters of either of these two nematode genera. However, the few morphological differences that have been identified are consistent with major *Ficus* lineages and the potential for long-term evolutionary associations.

Specifically, there appears to be a trend of morphological divergence between *Schistonchus associated with Ficus* of Gondwanan origin (c. f. *S. altermacrophylla*, *S. africanus*, *S. laevigatus*, and *S. aureus*) and of Laurasian origin (*S. caprifici*, *S. macrophylla*, *S. hispida*, and *S. racemosa*). We recently finished scanning electron microscopic (SEM) comparisons of *S. caprifici*, *S. laevigatus*, and *S. aureus* from Florida, and a new, undescribed species of *Parasitodiplogaster* from Florida (DeCrappeo and Giblin-Davis, 2001; Giblin-Davis and DeCrappeo, unpubl. data). The transmission electron microscope (TEM) was used to study the buccal capsule of the *Parasitodiplogaster* sp. from Florida (Giblin-Davis and Williams, unpubl. data). Major morphological issues that should be examined from a wider sampling of *Schistonchus* include: (1) the trend for the presence of a labial disc on Laurasian species versus no labial disc on Gondwanan isolates; (2) the trend for the excretory pore occurring at or near the level of the metacorpus in Laurasian species versus an extremely anterior opening in Gondwanan isolates; and (3) the trend for the presence of flagellate sperm in Laurasian species versus amoeboid sperm in Gondwanan isolates. As more species are observed these trends may be challenged.

Major morphological issues for the evolutionary history of *Parasitodiplogaster* include the trend for a more pleisomorphic spicule and gubernaculum shape in *Parasitodiplogaster* from the presumed less derived Neotropical *Ficus* subgenus and section *Pharmacosycea* (strikingly similar to generalist diplogasterids such as *Diplogaster*). This contrasts with *Parasitodiplogaster* from the Neotropical and African *Ficus* subgenus *Urostigma* in the respective sections of Americana and Galoglychia that possess spicules with a wide lamina and an enlarged capitulum. In addition, Poinar and Herre (1991) suggest that morphology of the stoma, pharynx and intestine of *Parasitodiplogaster* reflect the type of nourishment and the nature of the physical environment encountered by these nematodes. Herre (1995) argues that *Ficus* species with pollinator population structures that result in more opportunities for parasite transmission select for more virulent species of *Parasitodiplogaster*. Movement from the ancestral phoretic condition of most other diplogasterids to the increasingly parasitic niche of some species of *Parsitodiplogaster* should allow correlations between the average number of wasp foundresses per fig and the morphological manifestation of the occupied niche for the corresponding species of *Parasitodiplogaster*. Using this logic, we would hypothesize that the assumed-derived parasitic characters of a small indistinct stoma, glandular pharynx, and large thin-walled intestinal cells with indistinct borders occurs in *Parasitodiplogaster* from fig species with proportions of single foundress broods closer to 0.3. Thus, we would expect for *Parasitodiplogaster* from members of the Pharmacosycea, figs with large numbers of foundresses because of the rather open ostiole, to trend towards the latter while we would expect considerable variation in morphology of *Parasitodiplogaster* from the *Urostigma* which have variable numbers of foundresses. The extreme morphological conservation of *Schistonchus* and *Parasitodiplogaster* is not uncommon in Nemata. We intend to investigate the morphological characters described above of selected taxa to evaluate patterns of morphological evolution when mapped on the molecular phylogeny.

11.3.5 Preliminary Molecular Data

As a preliminary test of the hypothesis that the dramatic radiation of *Ficus* species and associated fig wasps has affected nematode speciation, we collected nematodes from a representative subset of *Ficus* taxa, including up to three different isolates from each species. To test for the existence of unique nematode species associated with each *Ficus* and the potential for strict cospeciation, we examined the D2/D3 expansion segments of the 28S rRNA gene from each nematode. For the *Parasitodiplogaster*, nematode taxa from six different species of *Ficus* were examined (Fig. 11.7(A)). This preliminary data set clearly demonstrates the existence of genetic differences between nematodes associated with different *Ficus* species. By contrast, the isolates from the same species were identical in sequence. The application of a phylogenetic species concept, specifically the operational definition described by Adams (1998), can be applied to this data set and suggests the existence of six unique *Parasitodiplogaster* species (Fig. 11.7(A)). Only three of these are nominal species. A similar result is obtained from a smaller set of sequences from four *Schistonchus* (Fig. 11.7(B)). As with *Parasitodiplogaster*, all *Schistonchus* species from different *Ficus* species are genetically unique based on D3 sequences. However, with the reduced number of nucleotides employed (D3 region only) one putative meta-species (a new, undescribed species of *Schistonchus* ex *F. poponoei*) is not strictly defined. These preliminary data show that in 10 out of 10 cases examined, different *Ficus* species are associated with genetically different nematodes. Also of importance is the fact that the D2/D3 sequences appear to evolve rapidly enough to accumulate differences between nematodes associated with closely related *Ficus* taxa within a species group. This observation is consistent with other studies for nematodes (Duncan, et al., 1999).

The *Parasitodiplogaster* samples were collected from taxa representing a definable *Ficus*/pollinator relationship presenting an opportunity to test for long-term cospeciation by comparing *Ficus* and *Parasitodiplogaster* phylogenies. The *Ficus* represent two phylogenetically distinct subgenera, the

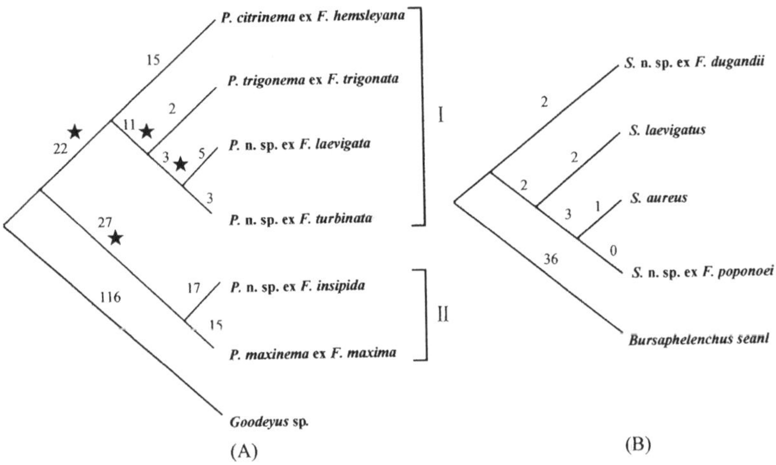

Figure 11. 7 Nematode relationships inferred from large subunit rRNA coding sequences. (A) Cladogram based on the unambiguous alignment of 546 nucleotides from species of *Parasitodiplogaster* (= *P.*). Tree is rooted using *Goodeyus* as an outgroup. The Neotropical *Ficus* (= *F.*) represent two phylogenetically distinct subgenera; the *Urostigma*, section *Americana* (Clade I), and the *Pharmacosycea*, section *Pharmacosycea* (Clade II). (B) Cladogram based on the unambiguous alignment of 289 nucleotides from species of *Schistonchus* (= *S.*). All sequenced representatives are from the Neotropical *Ficus* subgenus *Urostigma*, section *Americana*. Tree is rooted using *Bursaphelenchus* as an outgroup. In both cases the branching order presented is the most parsimonious and is also the same branching order inferred from a Neighbor-joining analysis. The star symbol (★) indicates branches that appear in 95% or more of the bootstrap analyses. Each number represents the number of apomorphies.

Urostigma, section *Americana* (Fig. 11. 7 (A), Clade I), and the *Phamacosycea*, section *Pharmacosycea* (Fig. 11. 7 (A), Clade II). If the evolutionary history of the *Parasitodiplogaster* strictly follows that of the *Ficus*, we expect to see two distinct clades among the nematode taxa. The phylogeny inferred for the nematode taxa based on the D2/D3 sequences also fall into two well-supported clades (Fig. 11. 7 (A)). This phylogenetic congruence is consistent with the potential existence of cospeciation.

11.4 *Myrtaceae/Fergusonina/Fergusobia* Associations

11.4.1 The *Myrtaceae* Radiation

The association between certain taxa of the family *Myrtaceae* and the gall-forming *Fergusonina/Fergusobia* complex represents a second model system to examine the relationship between evolutionary radiation of a group of plants and speciation of associated nematodes. In this model, there is a mutualistic association between flies of the genus *Fergusonina*, and *Fergusobia* species of nematodes, in which the nematode relies on the fly for dispersal, and the fly requires the nematode (at least in part) for gall induction (Figs 11.2(B), 11.8). *Fergusonina/Fergusobia* parasitize their hosts, and are apparently restricted to the *Myrtaceae*. Because it is less well known than the *Ficus*/fig wasp/nematode system, the individual components of the *Myrtaceae/Fergusonina/Fergusobia* association will be discussed here in some detail.

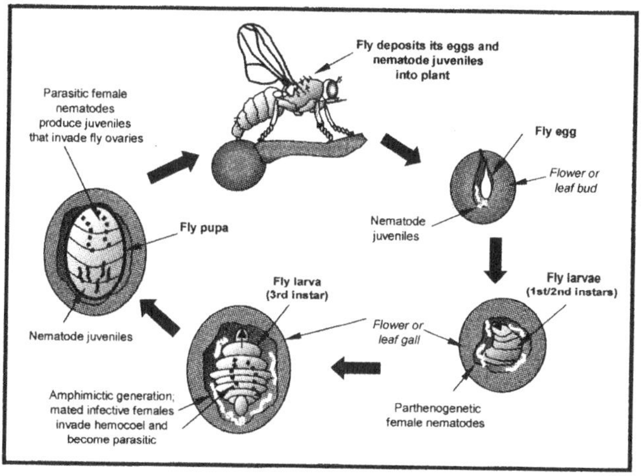

Figure 11.8 Generalized diagram of gall ontogeny in a myrtaceous host plant and the associated development of the *Fergusonina* fly and its mutualist nematode, *Fergusobia*.

The family *Myrtaceae* contains about 155 genera and over 4600 species. Fifty-five genera are endemic to Australia (Orlovich, et al., 1996). Most current workers accept that *Myrtaceae* has two subfamilies: the *Myrtoideae*, with fleshy fruits, and *Leptospermoideae*, with dry indehiscent fruits (Johnson and Briggs, 1981). From their phylogenetic analysis of the *Myrtaceae*, however, Johnson and Briggs (1984) argued that the recognition of the two subfamilies should be abandoned. *Syzigium*, a known host for *Fergusonina/Fergusobia*, is in the *Eugenia* alliance in the *Myrtoideae*. The other known hosts are found in the *Leptospermoideae*.

The distribution pattern of the *Myrtaceae* is clearly Gondwanic. It is probable that there was an early connection in ancestral *Myrtoideae* between Australia – Pacific – SE Asia region and South/Central America, inferred from the development of major and separate diversification in each region (Johnson and Briggs, 1981, 1984). One theory is that the *Leptospermoideae* evolved in east Gondwanaland and the *Myrtoideae* in the west, i. e., 110 mybp (Ladiges and Humphries, 1983). Australia was completely isolated 80 mybp. Another is that the origin of the *Myrtaceae* was in the South African part of west Gondwanaland (Johnson and Briggs, 1984), with much of the early diversification of the family occurring in east Gondwanaland, although the earliest reliable fossils of *Myrtaceae* appear to be from the Paleocene (60 – 53 mybp) (Lange, 1982). The presence in New Caledonia of the distinctive *Arillastrum*, which shares features with *Angophora*, suggests an early origin (80 – 60 mybp). Unfortunately, the fossil record for *Myrtaceae* and eucalypts is poor (Ladiges, 1997). Ambrose et al. (1979) suggested that fossil gumnuts in *Eucalyptus*-bearing sediments in central Australia came from the Miocene. Eucalypt-like fossils have been found from New Zealand (Pole, 1989, 1993), suggesting that eucalypts could have been widespread on Gondwanaland before 55 mybp, when New Zealand separated from Australia and Antarctica (Ladiges, 1997).

Given that most records of *Fergusonina/Fergusobia* galls come from *Eucalyptus* and *Melaleuca*, these groups will be considered in more detail here. Eucalypt taxonomy is regarded as difficult (Ladiges and Humphries, 1983). There are approximately 700 described species of *Eucalyptus* alone. Pryor and Johnson (1971) recognized eight subgenera of eucalypts based on a wide range of morphological characters, including development of the operculum, the inflorescence, seed and ovules, and genetic and ecological features. Below subgeneric level, they also recognized informal groupings of eucalypt species into sections and series. While they saw *Angophora* as a separate genus, they gave it the same taxonomic status as the other subgenera. *Angophora*, *Blakella*, and *Corymbia* were presented as more closely related to each other than to the other subgenera. A series of subsequent studies examined the cladistic relationships of *Angophora*, *Corymbia*, *Eucalyptus*, and the various eucalypt subgenera (Hill and Johnson, 1995, 1998; Ladiges and Humphries, 1983, 1986; Ladiges, et al., 1983, 1987, 1995; Pryor and

Johnson, 1981; Sale, et al., 1993), using both morphological and molecular data. These led to the development of phylogenies for the eucalypts, which could be compared with any phylogenetic trees developed for *Fergusonina* and *Fergusobia*. The integrity of subgenera, sections and series in *Eucalyptus* is strong (Ladiges, 1997).

Only four species of eucalypts do not occur on the Australian mainland - *E. deglupta*, *E. urophylla*, *E. wetarensis* and *E. orophila* (Ladiges, 1997; Pryor, et al., 1995). This distribution strongly suggests that Australia is the center of origin and primary diversification of *Eucalyptus*. *Eucalyptus* evolved in response to changing climate and poor soils (Pryor and Johnson, 1981). Speciation was promoted by the Kosciusko uplift 10 – 15 mybp, with the subsequent development of fertile basaltic soils, and the glacial and interglacial phases of the past 100,000 years or so (Ladiges and Humphries, 1983). Within the *Myrtaceae*, the *Eucalyptus* alliance is one of several with adaptations to dry conditions and fire (epicormic shoots and lignotubers) (Pryor and Johnston, 1981). It is likely that eucalypts developed adjacent to moist closed forests, and adapted to open forest conditions as Australia became drier (Pryor and Johnson, 1981). Subsequently, each line diversified to new niches as they developed. Many species are site-sensitive, and adapted to particular soils and topographies (Wardell-Johnson, et al., 1997). Others are distributed over wide geographic areas. For example, *E. camaldulensis* is the most widespread eucalypt consisting of fairly different but intergrading ecogeographic races (Pryor and Johnson, 1981). In general, there is a northern, summer rainfall group of eucalypts and a southern, winter rainfall group. A third group is highly endemic to southwestern Western Australia (Boland, et al., 1992). The *Angophora* and *Symphyomyrtus* alliances are less temperate than *Monocalyptus*. Morphologically, the eucalypts are now a very diverse group.

The open form of eucalypt flowers suggests a lack of adaptations to particular pollinators, and interfertility, hybridism, and species breakdowns are common (Pryor and Johnson, 1981). However, restrictions to interbreeding can occur within groups, e. g., differences in flowering period (Drake, 1980). With one reported exception (Ladiges, 1997), each of the subgenera is genetically isolated from all others (Griffin, et al., 1988; Pryor, 1959), but within subgenera, hybrids between species are frequent and generally fertile. Sympatric or parapatric species of interfertile groups tend to occupy different sites. Merging has occurred in many cases where the ranges of otherwise distinct species overlap. For example, the red gums, *E. tereticornis* and *E. blakelyi*, have an intermediate in much of the Hunter and Goulburn Valleys in New South Wales, and in the same area the grey box *E. moluccana* and white box *E. albens* have hybridized (Pryor and Johnson, 1981). In some cases, populations morphologically and probably physiologically intermediate between two species appear to have become isolated and stabilized by selection. Thus, hybridization is also a source of variation in eucalypts.

There are about 250 species of *Melaleuca* within Australia (Barlow, 1988), separated broadly into two species groups – the *leucadendra* group with broad leaves, and a second group with narrow leaves. The *leucadendra* group is distinctive in that it has all the ancestral states (broad leaves, indeterminate inflorescences, simple solitary fruits, entomophilous pollination), and may have been similar to the hypothetical ancestor of the genus (Barlow, 1988). It is tropical in distribution. *Melaleuca* is a Gondwanan genus, which originated in seasonally flooded habitats at the edges of tropical rainforests, and as scleromorphy evolved the genus radiated into nutrient-deficient, temperate, and semiarid habitats in the early to middle Tertiary, 30 – 60 mybp (Barlow, 1988; Beadle, 1981; Specht, 1981). *Melaleuca* seems to have differentiated in northern Australia, and become widely distributed there before the Miocene contact with Sundaland (30 mybp). There is evidence of recent east-west differentiation in northern Australia (Barlow, 1988). *Melaleuca* and *Eucalyptus* may have differentiated in parallel, on the edges of rainforests, one colonizing seasonally flooded habitats, and the other drier nutrient-deficient soils (Barlow, 1988; Pryor and Johnson, 1981). Most records of *Fergusonina/Fergusobia* galls on *Melaleuca* have come from species in the *leucadendra* group. Extensive sampling of narrow-leafed melaleucas from South Australia (Taylor and Davies, pers. obs.) has not provided any records, and there is only one confirmed record from New South Wales on *M. armillaris* from Sydney (Giblin-Davis, unpubl. data).

11.4.2 Galls and the Role of *Fergusonina* Flies

Fergusonina/Fergusobia galls are usually formed on young leaf or flower buds, where undifferentiated meristematic tissue is available for oviposition by the fly (Fig. 11.8), and are common in areas of regeneration, e.g., after fire (Davies, unpubl. data). The availability of undifferentiated meristematic tissue is probably the limiting factor for gall production. In *M. quinquenervia*, this means that the occurrence of many large galls in the winter (flowering and growing) months and small axial galls in the summer reflects the growing habits of the tea tree. The nematode is apparently responsible for gall induction (Currie, 1937; Giblin-Davis, et al., 2001a), and the fly for dispersal of the nematode, i.e., this is a mutualistic association (Currie, 1937). It is not clear whether the galls are induced by cecidogenic secretions produced by the nematodes alone, or whether secretions from the flies' accessory glands during oviposition also have a role (Giblin-Davis, et al., 2001b). Galls are formed before the fly eggs hatch (Taylor, et al., 1996), and nematodes can be found in galls in which the fly larvae have been killed. Thus, the fly larvae are unlikely to have a role in gall induction.

Fergusonina/Fergusobia galls are apparently restricted to the *Myrtaceae*, and have been recorded from 41 species of *Eucalyptus*, four species of *Corymbia*, two species of *Angophora*, eight species of *Melaleuca*, one species of *Callistemon* (closely related to *Melaleuca*), and *Syzygium cumini* (Colbran, 1964; Currie, 1937; Davies, et al., unpubl. data; Malloch, 1932; McLeod, et al., 1994; Siddiqi, 1986b, 1994). There are records from most regions of Australia, with some from India, the Philippines, and New Britain. There have been unconfirmed reports of *Fergusonina* galls on *Leptospermum*, but none from other myrtaceous genera such as *Lophostemon*, *Agonis*, *Calytrix*, *Micromyrtus*, *Darwinia*, *Kunzea*, *Acmena*, *Osbornia*, or *Baeckea*. There are no known records of *Fergusobia* or *Fergusonina* from Central or South America, but this may reflect lack of collecting in those areas.

Examination of the fossil record is unlikely to help to elucidate the evolution of the *Fergusonina/Fergusobia* gall interaction. *Eucalyptus* species cannot readily be identified from pollen or from leaves, and fossil fruits are limited. Galls in the Australian fossil record of *Myrtaceae* are unknown (Christophel, pers. comm.). There is a record of a drosophilid fly parasitized by a nematode from the Allantonematidae (a sphaerularioid family, to which *Fergusobia* also belongs) in Mid-Tertiary fossil amber from the Dominican Republic (Poinar, 1984). Thus, it is likely that an ancestral *Fergusobia* could have been present at about the time of development of the eucalypts.

Four basic gall forms are found, including shoot and inflorescence bud galls (terminal and axial; of varying forms from small and simple to large or complex), leaf galls (terminal and axial; flat, whole or part leaf; some callus-like), stem galls, and flower bud galls (Currie, 1937; Giblin-Davis, et al., 2001b; Taylor and Davies, pers. obs.). The selection pressures that resulted in these divergent morphologies are unknown, and it is not clear which is the "primitive" condition, i.e., the least specialized form of galls. The least differentiated galls appear to be the callus leaf galls, reported from *Angophora* (Colbran, 1964). Flower bud galls are probably the most specialized. Of the most commonly recorded gall forms, flower bud galls have been found from 16 species of eucalypts, flat leaf bud galls from nine species, terminal leaf bud from eight species, and "leafy" galls from six species (Currie, 1937; Davies and Taylor, unpubl. data). There is some apparent correlation of morphological characters of the nematodes with gall type (Davies, unpubl. data).

There is no evidence that a particular gall type is restricted to any one group of host *Myrtaceae*. For example, "moss" galls have been found associated with *Fergusonina/Fergusobia* on *Ang. costata* and *Ang. floribunda* (from New South Wales), *E. fasciculosa* (South Australia), and *M. nervosa* (northwestern Western Australia) (Davies and Taylor, unpubl. data). Five distinct gall forms, associated with at least three morphospecies of *Fergusonina* (Taylor, unpubl. data) and four morphospecies of *Fergusobia* (Davies, unpubl. data), have been recorded from *E. camaldulensis* in Australia. Three gall forms have been recorded from *E. microcarpa*, and two from *E. leucoxylon* (Taylor and Davies, unpubl. data) and *E. pauciflora* and *E. maculata* (Currie, 1937). In general, only one gall type is recorded for each fly species. In *M. quinquenervia*, however, axial and terminal gall forms vary from hairy to smooth depending upon whether the attacked bud is destined to be an inflorescence or a branch. Galls from inflorescence buds are typically smooth-textured while galls from leaf buds are hairy (Purcell, pers. comm.).

The life cycle of *Fergusonina* is shown in Fig. 11.8. Fly eggs hatch within the gall, and developing larvae feed on gall tissue. The fly has three larval stages, the third having a species-specific morphological structure called the dorsal shield, which appears to be correlated with gall form. The function of the dorsal shield is speculative, but it usually has a series of cuticular setae or projections that point anteriorly, and would dig into the plant material as the larva moved forward. The dorsal shield may be used to enlarge the chambers within the gall in which the flies and nematodes develop, as a scraper to form pellets of plant tissue that may be used as food, or to anchor the larva within the chamber. If so, it will have a role in determining the internal morphology of the gall. In general, the most elaborate dorsal shields occur in fly larvae associated with flower bud galls (Taylor, unpubl. data). The role of larvae in determining the final size of the gall is also unknown.

A number of morphological characters are likely to be useful in the taxonomy and systematics of *Fergusonina*. Ovipositor morphology and length is one such character that may be related to hardness and thickness of the tissue in which eggs are being deposited. The dorsal shield and cephalopharyngeal skeleton of third stage larvae may be related to plant tissue type or variation in modes of feeding. Male genitalia, particularly the shape and setation of the aedeagus, show interspecific variation (Taylor, unpubl. data; Tonnoir, 1937). To date, there are 21 described species of *Fergusonina*, 20 from Australian eucalypts (Currie, 1937; Malloch, 1924, 1925, 1932; Tonnoir, 1937) and one from *Syzygium* (Harris, 1982). Many new species from eucalypts await description (Taylor, unpubl. data). None are described from *Melaleuca*.

If nematode/fly/plant interactions are species specific, they may be regulated by semiochemicals. Eucalypts and melaleucas have large numbers of secondary compounds in their leaves, many of which are species specific (Hillis, 1966; Pryor, 1976). These are likely to have a role in the evolution of plant/insect/nematode interactions. Noble (1989) reviewed the ecological traits of the largest subgenera of *Eucalyptus*, *Symphyomyrtus*, and *Monocalyptus*, and found evidence for differences in herbivore and parasite damage to leaves of the two groups. *Symphyomyrtus* seems to be more damaged than *Monocalyptus* (Lowman and Heatwole, 1987), and to have more diverse assemblages of associated herbivores and pathogens (Burdon and Chilvers, 1974; Woinarski and Cullen, 1984). *Fergusonina/Fergusobia* galls have been collected from 19 series from 9 of 12 sections in sub-genus *Symphyomyrtus*, and from 7 series from 6 of 9 sections in sub-genus *Monocalyptus*. This suggests that *Fergusonina/Fergusobia* either speciated with the radiation of the eucalypts, or perhaps underwent considerable host-switching after speciation on one or the other subgenus.

11.4.3 Associated *Fergusobia* Nematodes

The generalized life cycle of *Fergusobia* is illustrated in Fig. 11.8. All female *Fergusonina* flies carry nematodes. Male flies do not carry nematodes, thus reducing chances for horizontal transmission. When the female fly deposits eggs in plant tissue she also deposits juveniles of *Fergusobia* (Currie, 1937). These nematodes feed in the gall and develop into parthenogenetic females. The feeding action of these nematodes apparently induces gall formation (Giblin-Davis, unpubl. data), although this needs verification. The females lay eggs that develop into male and female nematodes, which mate in the gall. Inseminated females are infective, and invade mature female fly larvae or puparia. Inside the fly, they develop into parasitic female nematodes that appear to undergo apolysis and ecdysis without the formation of a new cuticle. As a result, the stylet and cuticle are lost, the esophagus and intestine degenerate, and the epidermis becomes hypertrophied with the development of microvilli that expand the absorptive surface area of the nematode (Giblin-Davis, et al., 2001a). This stage then deposits eggs into the fly's hemolymph. The juvenile nematodes that hatch from these eggs move to the oviducts of the adult fly, and are deposited with her eggs. Thus, there are four distinct taxonomically important morphological forms of adult nematodes for each species of *Fergusobia*, namely the male, parthenogenetic female, infective female, and parasitic female.

Fergusobia was first described by Currie (1937), and its life cycle was further elucidated by Fisher and Nickle (1968). The genus was reviewed by Siddiqi (1986b). Since then, Siddiqi (1994) and Davies and Lloyd (1996) have described three new species bringing the total to six. However, this under-represents the potential diversity of the group. Davies (unpubl. data) has collected at least 17 new species of *Fergusobia*, from *Eucalyptus*, *Angophora*, and *Corymbia* from northern and eastern Australia, and at least six new species from *Melaleuca*. Morphological descriptions, measurements, and molecular studies have begun to confirm the species status and host specificity of these undescribed isolates.

Information on the distribution of *Fergusonina/Fergusobia* galls within Australia, and on host specificity of either the nematodes or the flies is limited. Some flies can attack more than one host plant species, e. g., *Fergusonina* (= *Fn.*) *brimblecombei*, *Fn. evansi*, and *Fn. greavesi*, but sibling species may occur. It is not known if specific fly/nematode combinations inducc specific gall forms or if gall forms depend on the part of the plant attacked, as with *M. quinquenervia*. More than one described species of *Fergusobia* has been found with one species of *Eucalyptus*, e. g., the nematodes *F. brevicauda* and *F. philippinensis* are found in *E. deglupta* in New Britain and the Philippines, respectively (Siddiqi, 1994), but no information on the associated flies is available. *Eucalyptus camaldulensis* has at least four fly/nematode associations, producing four gall forms (terminal leaf bud, two types of axial leaf bud, and flower bud). Furthermore, small morphological differences between nematodes have been found in terminal leaf bud galls from *E. camaldulensis* from widely different ranges (Kununurra, Western Australia, and Goolwa, South Australia), even though gall morphology is very similar. It is unclear if this is an example of clinal variation in one species of *Fergusobia*, or represents two closely related species.

Examination of gall, fly, and nematode material held in the Waite Institute Nematode Collection, USDA ABCL Brisbane collection, and the Duncan Swan Insect Collection (The University of Adelaide), including collections where nematodes can be directly associated with a particular fly species, suggests that each species of *Fergusobia* is associated with one species of *Fergusonina* (Taylor and Davies, unpubl. data), but not necessarily to one species of eucalypt or to one gall form. The nematode must have evolved mechanisms, which are probably host-specific, to escape the fly's immunological system during the generation in which it is an insect parasite (Davies and Lloyd, 1996). There is likely to be strong selection pressure for coevolution of nematodes and flies. However, given the relatively few morphological characters available for species determination, specificity of nematode/fly relations will have to be confirmed with molecular techniques.

Currie (1937) described one nematode species, *F. tumifaciens*, associated with the fly *Fn. carteri* from leaf galls on *E. stuartiana* (sic) (now probably

E. bridgesiana (D. Nicolle, pers. comm.)). He observed minor morphological differences between nematodes from a variety of gall forms on more than a dozen species of eucalypt. He suggested that "further work will show that the nematodes associated with the different flies will have differences in structure which entitle them to be considered as different species" (Currie, 1937). These differences included variation in the point of origin of the caudal alae and the size and position of the esophageal glands. Fisher and Nickle (1968) redescribed the nematode *F. curriei* (= *tumifaciens*, synonymized by Siddiqi (1986b)) from flower bud galls on *E. camaldulensis* associated with the fly *Fn. tillyardi*. *Fergusobia*, like the genera *Schistonchus* and *Parasitodiplogaster* (this chapter) and *Anguina* (Krall, 1991), may have a high degree of host specificity. If so, Currie (1937) and Fisher and Nickle (1968) may have described different species of nematodes, given that they were associated with different species of *Fergusonina*, from different gall forms on eucalypt species which are not closely related. Indeed, Currie refers to the margins of the caudal alae of males of his specimens of the nematode *F. tumifaciens* as being slightly crenate, a character not observed by Fisher and Nickle (1968) nor in specimens isolated from *E. camaldulensis* (Davies and Lloyd, 1996). Also, the esophageal glands of the parthenogenetic female of Currie's specimens of *F. tumifaciens* are shorter than that described by Fisher and Nickle (1968) (Davies and Lloyd, 1996). Molecular studies will provide independent data to help resolve some of this confusion.

Based upon preliminary work on 103 collections of *Fergusobia* (Davies, unpubl. data), the following morphological characters should be used in species diagnoses and for searching for evolutionary trends in *Fergusobia*. The shape of the head or labial disc appears to be an important species-level character (there is an elevated central conus in four species of *Fergusobia* from *Syzygium*, *E. deglupta*, *Corymbia*, and *Angophora*) (Davies, unpubl. data; Siddiqi, 1986b, 1994). The length of the esophageal gland relative to total body length is useful at the species level, and the apparent presence of an extra lobe in species of *Fergusobia* from *Melaleuca* may be a synapomorphy. The size of the excretory/secretory cell (this is very large in species of *Fergusobia* from the Series Tereticornes and one species of *Fergusobia* from *E. microcarpa*) may also be a synapomorphy. Stylet length and shape of knobs (relatively long in a species from *Angophora*), tail shape and length, size and shape of fixed parthenogenetic females, males and infective females ("a" ratio), size and shape (curved or angular) of spicules, length of caudal alae relative to body length (alae are relatively long in the *Fergusobia* species from *M. cajuputi*, *M. viridiflora*, and *C. ptychocarpa*), length and structure of gonads (may be straight or with one to two flexures), and presence or absence of caudal papillae can be useful for species separation and may be phylogenetically relevant when more species are elucidated.

There are few correlations between morphology of *Fergusobia* and gall type, plant host species, or fly host. Nematodes from *Corymbia* appear to have relatively small esophageal glands, and those from *Melaleuca* tend to be small nematodes, with large esophageal glands that seem to have a lobe not present in forms from eucalypts. This may be a physiological adaptation to deal with the particular terpenes and oils present in the *Melaleuca* leaves. If so, the morphology of the intestine of *Fergusobia*, like that of *Parasitodiplogaster*, may reflect the composition of the food ingested. All terminal leaf bud galls sampled were associated with parthenogenetic females that formed a tight C-shape when dead. Flat leaf galls tend to be associated with *Fergusobia* males that lie straight when killed.

11.4.4 Evolution of the *Fergusonina/Fergusobia* Mutualism

Given that *Fergusonina/Fergusobia* galls have been found on *Syzygium* (*Myrtaceae* subfamily *Myrtoideae*) as well as species from subfamily *Leptospermoideae*, it seems that an ancestral fly/nematode could have been associated with a very early, ancestral lineage of *Myrtaceae*. Alternatively, the fly/nematode association may have radiated with the *Leptospermoideae* in the Australian region and one or a few lineages switched hosts to sympatric *Myrtoideae*. Further collecting, both from Australia and from South and Central America and South Africa, would be needed to confirm a Gondwanan distribution of *Fergusonina/Fergusobia*, similar to that of their *Myrtaceae* hosts. If the *Fergusonina/Fergusobia* relationship developed in Australia, in parallel with eucalypt radiation, it would be expected that galls would only be found in Australia, Southeast Asia, and associated islands.

Is there any evidence that *Myrtaceae/Fergusonina/Fergusobia* species have coevolved? This is particularly important because the *Myrtaceae* seem to have developed new species readily in response to emergence of new niches, and it will be necessary to distinguish between cospeciation and speciation due to other factors, e. g., genetic drift after allopatry. Jolivet (1998) argued that coevolution between a gall insect species and a plant taxon is difficult to justify. However, the intimacy of the gall association could conceivably provide opportunities for each host (*Myrtaceae*) or symbiont combination (*Fergusonina/Fergusobia*) to have selective pressure on the associated partner in reciprocal ways. As mentioned above, specific coevolution may have been most important between the fly and the nematode during the development and maintenance of the mutualism.

It is likely that during radiation of the *Myrtaceae*, different populations of flies survived better on different populations of *Myrtaceae*, resulting in a radiation of host-specific plant/fly associations. Due to the association between the fly and the nematode, and the high improbability of the nematode dispersing on its own (without an insect vector), speciation of the nematode could have arisen by genetic drift from the ancestral species. The only way that nematodes from different individuals of any one species of host flies could come into contact would be if more than one fly oviposited into a bud, i.e., a gall had more than one "foundress". While we have no direct evidence on this point, we have noted that repeated oviposition into a bud by a single female on caged *M. quinquenervia* saplings can lead to a hypersensitive response by the plant and to death of the bud. In the field, very few oviposition scars (1 to 5) were found associated with young galls on *M. quinquenervia*, suggesting that individual galls were initiated by one or a small number of female flies. Nevertheless, it is possible that in some of the larger terminal leaf bud galls on *Eucalyptus* (i.e., *E. camaldulensis* and *E. intertexta*, where galls can be up to 50 mm in length and 30 mm in diameter in the former, and spheres with a diameter of 31 mm in the latter) that more than one female is responsible for gall formation. It is possible that, like some hymenopteran parasitoids (Quicke, 1997), individual flies "mark" an oviposition site to prevent over-exploitation of the resource. Questions associated with oviposition, e.g., the role of competition between females, the relation of gall size to the number of foundresses, would be amenable to experimental clarification. It may be that the plant supports continuation of single lines of nematodes via hypersensitive reactions that abort multi-foundress galls, i.e., it preserves the nematode genotype and restricts genetic recombinations which might have potential to evolve mechanisms to counter the plant defense reactions.

Thus, the *Fergusonina/Fergusobia* system might be analogous to a mitochondrial, matriarchal-based, system where symbionts are vertically transmitted through to the progeny of the next generation of female flies because only female flies carry nematodes and there may be no cross infection of plant galls. This could result in non-coalescing or non-intersecting maternal lineages. Gene flow outside of successful maternal lineages would be irrelevant to species cohesion. Speciation events would be the result of breaks in environmental homeostasis as happens in Model III allopatry (Wiley, 1981). Within lineages, recombination during the amphimictic generation prior to insect parasitism may function to reduce trends towards homozygosity whereas the parthenogenetic first generation in the plant would increase the trend.

While species specificity for (1) fly/plant or (2) fly/nematode would support coevolution of the associations, lack of host specificity (fly/plant) would not necessarily destroy the argument (Morrow, 1977). Specialist herbivore species can "overflow" onto less preferred hosts if food becomes limiting (Morrow, 1977), and this in itself may promote speciation. In both eucalypts and melaleucas, distributions of genetically similar species frequently overlap. Thus, it is probable that new species of fly could have developed through "overflow", and that (because of their isolation in separate chambers within galls) new species of nematode might have developed over time due to genetic drift.

Four species of fly, *Fn. tillyardi*, *Fn. greavesi*, *Fn. brimblecombei*, and *Fn. evansi*, have been collected from more than one host plant species (Currie, 1937; Taylor, unpubl. data; Tonnoir, 1937). *Fergusonina greavesi* has been recorded from species of *Eucalyptus* (*E. aromaphloia* and *E. polyanthemos*) that are placed in separate subgenera, but whose distribution overlaps in Victoria, Australia. Records of *Fn. tillyardi* are restricted to the Series Tereticornes, and the host species are widespread with frequently overlapping distributions. *Fergusonina evansi* has been found from two closely related series, both host species being widely distributed, and one (*E. leucoxylon*) has been described as possibly the most variable species of eucalypt. *Fergusonina brimblecombei* has been recorded from three series of the subgenus *Symphyomyrtus*, whose distributions overlap. Thus, the available evidence supports the suggestion that where flies have been recorded from more than one host species, they either are genetically similar and/or have overlapping distributions.

Cospeciation of plant/fly and fly/nematode would also be supported by matching evolutionary frameworks for the putative evolutionary partners. Work by Pryor and Johnson (1981) suggested that *Monocalyptus* was more primitive than *Angophora* and *Symphyomyrtus*, but most workers (e. g., Ladiges and Humphries, 1983; Sale, et al., 1993) put *Angophora* and *Corymbia* near the base of their eucalypt phylograms. If coevolution between hosts and *Fergusonina/Fergusobia* has occurred, the least derived forms of the flies and nematodes should be expected on the "ancestral" eucalypts and *Melaleucas* (the *leucadendra* group). Phylograms for the eucalypts will be compared with trees to be developed for both *Fergusobia* and *Fergusonina*.

Evidence tying *Fergusonina/Fergusobia* divergence to documented vicariance events in the history of associated plant hosts could be useful for understanding the timing of speciation events or the mechanisms for host specificity. For example, Marginson and Ladiges (1988) hypothesized an evolutionary sequence between stringybarks, *E. baxteri/E. arenacea/E. alpina*. Examination of the distribution pattern of the stringybarks suggests allopatric speciation with a vicariant event splitting a species ancestral to both *E. baxteri* and *E. arenacea* (Ladiges and Humphries, 1986). The event could have been the marine incursion in Australia from the late Eocene to the mid-Miocene. The Grampians Ranges, Victoria and areas above 200 m in the Mt Lofty, southern Flinders Ranges and on Fleurieu Peninsula, South Australia would have acted as dry refugia for plants during unfavourable periods (Boardman, 1986). Morphological and sequence comparisons of *Fergusonina/Fergusobia* from stringybarks from areas representing the disjunct distributions could be fruitful. Similarly, *E. macro-rhyncha* has a disjunct distribution, with discrete populations near the Grampians in Victoria and at Clare in South Australia, both isolated from a more widespread population along coastal New South Wales and Victoria (Boomsma, 1972; Nicolle, 1997) and would be worth study. The morphology of the two species of *Fergusobia* described by Siddiqi (1994), from *E. deglupta* from Papua New Guinea and the Philippines respectively, is so different that it suggests that they are probably associated with different fly species, and are not an example of genetic drift associated with geographic separation.

Because the nematode is at least partly responsible for gall induction, the *Fergusonina/Fergusobia* interaction is an example of mutualism. However, the nematode may affect fly fecundity. Competition may occur for resources between fly larvae and nematodes in the galls. Given the loss of the intestine after penetration of the third instar fly larva, the entomophilic parasitic female nematode appears to be "feeding" on the fly (larval, pupal, and adult) host (i. e., it is a true sphaerularioid nematode) with absorption of nutrients through a highly specialized cuticle or epidermis. Adult flies carry hundreds of second-stage juvenile nematodes, which appear to have a normal cuticle (through which nutrients cannot be absorbed). Thus, it is unlikely that the plant-infective juvenile nematodes feed (on the adult fly host) until they enter plant tissue, surviving on reserves derived from the fly host. As in the pollinator mutualism between fig wasps and *Ficus*, the balancing of the cost – benefit equation between the *Fergusonina* and *Fergusobia* mutualism creates the most possibilities for overexploitive adaptations to drive plausible sequential evolution or reciprocal coevolution.

It is difficult to decide which of the following possible associations is most likely to have evolved first: (1) plant/*Fergusobia*; (2) plant/*Fergusonina*; (3) *Fergusonina/Fergusobia*; or (4) plant/*Fergusonina/Fergusobia*. If the plant/*Fergusobia* association evolved first, the nematodes would have to have been able to survive on and in the aerial parts of plants, similar to contemporary *Anguina*, *Nothanguina*, or *Ditylenchus* (Tylenchida) or *Aphelenchoides* (Aphelenchida). This would require moist conditions during part of the life cycle or a wet environment (which could have occurred for an ancestral *Myrtaceae* on the edge of a rainforest or in swamps). Entry of nematodes into leaves could have occurred via the stomata, i. e., leaf galls would be the most primitive form of galls, or entry could have occurred into undifferentiated regions of developing seedlings or young plants, i. e., leaf bud or flat leaf galls might be the most primitive. Neither *Fergusobia*-only galls nor nematodes feeding on *Myrtaceae* leaves have been found. It is possible that the "ancestral" *Fergusobia* induced galls on roots. Siddiqi (1986b) pointed out that two species of *Fergusobia* were collected from soil, but suggested that these had emerged from fallen galls. Also, there are no known below-ground *Fergusobia/Fergusonina* associations, although a series of *Fergusonina* flies in the Australian National Insect Collection, Canberra (with dried plant material) emerged from bunched spherical galls at the interface between stem and roots from glasshouse cultures of cucalypt seedlings.

The plant/fly association could have evolved from ancestral agromyzids (leafminers) before evolution of the gall system. The nematode association could have developed similarly to the hypothetical *Ficus* system (Fig. 11.6) where the flies picked up nematode associates from other insects that invaded leafminer galleries developing into a successful mutualism that radiated with the *Myrtaceae*. Alternatively, some flies could have been carrying parasitic *Fergusobia* when they evolved an interaction with host plants. However, this might mean that *Fergusobia* evolved as a gall-inducer independently on many occasions, and had to develop new ways of avoiding the plant's defense mechanisms, unless the successful association evolved before the radiation of the *Myrtaceae*. Some means of regulating the numbers of nematodes invading and developing in the fly would have developed as the fly/nematode mutualism evolved. There are no known nematode associates of the Agromyzidae (Poinar, 1975) and all *Fergusonina* females that have been dissected and examined have been found associated with *Fergusobia*. If, however, it turns out that there are some *Fergusonina* species that are not associated with *Fergusobia* nematodes it could mean that: (1) *Fergusobia* is only associated with some *Fergusonina* species, i. e., the plant/fly association evolved first and different lineages became associated with *Fergusobia*; or (2) that nematodes have been secondarily lost from some species of flies.

It is easiest to imagine how a nematode/fly association might have evolved. There are several examples of phoresy apparently leading to the development of parasitic associations between nematodes and an insect, including *Schistonchus* (Giblin-Davis, 1996) and diplogasterids in figs (Poinar and Herre, 1991). If a plant-parasitic ancestor of *Fergusobia* had occurred in large numbers in plant tissue, some could have entered an agromyzid-like leaf-mining form of fly larvae. Once in, and able to avoid the fly's defense mechanisms, such nematodes would have a very high selective advantage over nematodes having to disperse by other means. However, the only nematode known to feed in leaves or stems of *Myrtaceae* today is *Fergusobia*, but there are many that feed on roots (McLeod, et al., 1994). Some flies could have been carrying parasitic Sphaerularioidea at the time of evolution of the plant/fly interaction. Siddiqi (1986b) speculated that the ancestors of the Hexatylina, to which *Fergusobia* belongs, might have been fungal feeders.

11.4.5 Preliminary Molecular Data

We collected nematodes and flies from a representative subset of *Myrtaceae* (*Melaleuca*, *Eucalyptus*, and *Corymbia*), including up to eight different geographic isolates from each plant host species from eastern Australia (Fig. 11.9). We tested for the existence of unique nematode species associated

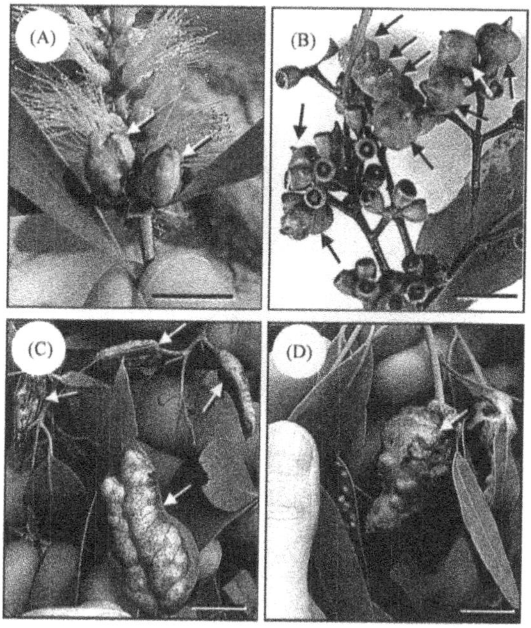

Figure 11.9 Different kinds of galls (arrows) induced by different species pairs of *Fergusonina/Fergusobia*. (A) Axial inflorescence bud galls on *Melaleuca quinquenervia*. (B) Flower bud galls on *Eucalyptus microcarpa*. (C) Flat leaf galls on *E. leucoxylon*. (D) Terminal leaf bud gall on *E. camaldulensis*. Bar = 1.4 cm.

with each plant host species using D2/D3 and for unique *Fergusonina* with the mitochondrial gene, cytochrome oxidase I. Although the data are still being processed, early analyses appear to confirm host-plant species specificity for nematodes and flies from the *M. leucadendra* group and several *Eucalyptus* species (Giblin-Davis et al., unpubl. data; Scheffer et al., unpubl. data).

11.5 Conclusions

We are studying biodiversity and cospeciation in the phylum Nemata using three nematode genera (*Schistonchus*, *Parasitodiplogaster*, and *Fergusobia*) that are obligatorily bound to their insect and plant hosts. We are asking the following questions about these model systems: (1) How many species are there? (2) Are they host specific? (3) Are the clades monophyletic? (4) Which species are the closest relatives and what do they look like and do? (5) Are phylogenies of nematode symbionts and insect and plant hosts congruent? (6) If they are congruent, can we estimate the timing of major divergences in the association? and (7) If they are not congruent, can host switching explain the pattern differences?

In all three nematode genera, it appears that there are large radiations of nematode species showing a high degree of host specificity (host switching appears to be conservative). There appear to be monophyletic nematode clades associated with monophyletic clades of insect and plant hosts suggesting congruence. The work is in its infancy and further generalizations must be postponed until more data are available.

References

Adams, B. J. 1998. Species concepts and the evolutionary paradigm in modern nematology. Journal of Nematology 30: 1 – 21.

Ambrose, G. J., R. A. Callen, R. B. Flint, and R. T. Lange. 1979. *Eucalyptus* fruits in stratigraphic context in Australia. Nature 290: 387 – 389.

Andrássy, I. 1984. Klasse Nematoda (Ordnungen Monhysterida, Desmoscolecida, Areolaimida, Chromodorida, Rhabditida). Berlin, Germany: Akademie-Verlag.

Arnheim, M., T. White, and W. E. Rainey. 1990. Application of PCR: Organismal and population biology. BioScience 40: 174 – 182.

Ashton, P. S. 1969. Speciation among tropical forest trees: Some deductions in the light of recent evidence. Biological Journal of the Linnaean Society of London 1: 155 – 196.

Baldwin, J. G., L. M. Frisse, J. T. Vida, C. D. Eddleman, and W. K. Thomas. 1997a. An evolutionary framework for the study of developmental evolution in a set of nematodes related to *Caenorhabditis elegans* (Nemata: Rhabditina). Molecular

Phylogenetics and Evolution 8: 249 – 259.

Baldwin, J. G., R. M. Giblin-Davis, C. D. Eddleman, D. S. Williams, J. T. Vida, and W. K. Thomas. 1997b. Buccal capsule of *Aduncospiculum halicti* (Nemata: Diplogasterina): an ultrastructural and molecular phylogenetic study. Canadian Journal of Zoology 75: 407 – 423.

Ballard, J. W. O., M. K. Thayer, A. F. Newton, Jr, and E. Grismer. 1998. Data sets, partitions and characters: philosophies and procedures for analyzing multiple data sets. Systematic Biology 47: 367 – 396.

Barlow, B. A. 1988. Pattern of differentiation in tropical species of *Melaleuca* L (Myrtaceae). Proceedings of the Ecological Society of Australia 15: 239 – 247.

Bawa, K. S., and J. E. Crisp. 1980. Wind-pollination in the understorey of a rain forest in Costa Rica. Journal of Ecology 68: 871 – 876.

Beadle, N. C. W. 1981. Origin of the Australian angiosperm flora. Pp. 409 – 426. In A. Keast, ed. Ecological Biogeography of Australia. The Hague: W. Junk.

Berg, C. C. 1989. Classification and distribution of *Ficus*. Experientia 45: 357 – 358.

Berg, C. C. 1990. Reproduction and evolution in *Ficus* (Moraceae): Traits connected with adequate rearing of pollinators. Memoirs of the New York Botanical Garden 55: 169 – 185.

Blaxter, M. L., P. De Ley, J. R. Barey, L. X. Liu, P. Schedleman, A. Wierstratete, J. R. Vanfleteren, L. Y. Mackey, M. Dorris, L. M. Frisse, J. T. Vida, and W. K. Thomas. 1998. A molecular evolutionary framework for the phylum Nematoda. Nature 392: 71 – 75.

Boardman, R. 1986. The history and evolution of South Australia's forests and woodlands. Pp. 16 – 31. In H. R. Wallace, ed. The Ecology of the Forests and Woodlands of South Australia. Adelaide, Australia: Government Printer.

Boland, D. J., M. I. H. Brooker, G. M. Chippendale, N. Hall, B. P. M. Hyland, R. D. Johnston, D. A. Kleinig, and J. D. Turner. 1992. Forest Trees of Australia. Victoria, Australia: CSIRO Publishing.

Boomsma, C. D. 1972. Native Trees of South Australia. South Australia: Woods and Forests Department.

Bromham, L., A. Rambaut, and P. H. Harvey. 1996. Determinants of rate variation in mammalian DNA sequence evolution. Journal of Molecular Evolution 43: 610 – 621.

Bronstein, J. L. 1992. Seed predators as mutualists: Ecology and evolution of the fig/pollinator interaction. Pp. 1 – 44. In E. Bernays, ed. Insect-plant Interactions, vol. 4. Boca Raton, FL: CRC Press.

Brooks, D. R. 1988. Macroevolutionary comparisons of host and parasite phylogenies. Annual Review of Ecology and Systematics 19: 235 – 259.

Brown, D. D., P. C. Wensink, and E. Jorday. 1972. A comparison of the ribosomal DNAs of *Xenopus laevis* and *Xenopus mulleri*: the evolution of tandem genes. Journal of Molecular Biology 63: 57 – 73.

Burdon, J. J., and G. A. Chilvers. 1974. Fungal and insect parasites contributing to niche differentiation in mixed species stands of eucalyp saplings. Australian Journal of Botany 22: 103 – 114.

Cabot, E. L., and A. T. Beckenbach. 1989. Simultaneous editing of multiple nucleic acid sequences with ESEE. Computational and Applied Biosciences 5: 233 – 234.

Center, B. J., R. M. Giblin-Davis, E. A. Herre, and G. C. Chung-Schickler. 1999. Histological comparisons of parasitism by *Schistonchus* spp. (Nemata: Aphelenchoididae) in Neotropical *Ficus* spp. Journal of Nematology 31: 393 – 406.

Chapela, I. H., S. A. Rehner, T. R. Schulta, and U. G. Mueller. 1994. Evolutionary history of the symbiosis between fungus-growing ants and their fungi. Science 266: 1691 – 1694.

Colbran, R. C. 1964. Studies of plant and soil nematodes 7. Queensland records of the order Tylenchida and the genera *Trichodorus* and *Xiphinema*. Queensland Journal of Agricultural Sciences 21: 77 – 123.

Collinson, M. E. 1989. The fossil record of the Moraceae. Evolution, systematics, and fossil history of the Hamamelidae. Pp. 319 – 339. In P. R. Crane, and S. Blackmore, eds. "Higher" Hamamelidae, vol. 2. Oxford, UK: Clarendon Press.

Compton, S. G. 1990. A collapse of host specificity in some South African fig wasps. South African Journal of Science 86: 39 – 40.

Corner, E. J. H. 1965. *Ficus* (Moraceae) and Hymenoptera (Chalcidoidea): Figs and their pollinators. Biological Journal of the Linnaean Society 25: 187 – 195.

Courtright, E. M., D. Wall, R. A. Virginia, L. M. Frisse, J. T. Vida, and W. K. Thomas. 2000. Nuclear and mitochondrial DNA sequence diversity in the Antarctic nematode *Scottnema lindsayae*. Journal of Nematology 32: 143 – 153.

Currie, G. A. 1937. Galls on *Eucalyptus* trees: A new type of association between flies and nematodes. Proceedings of the Linnaean Society of New South Wales 62: 147 – 174.

Davies, K. A., and J. Lloyd. 1996. Nematodes associated with Diptera in South Australia: A new species of *Fergusobia* Currie and a new record of *Syrphonema* Laumond and Lyon. Transactions of the Royal Society of South Australia 120: 13 – 20.

Davis, P. H., and V. H. Heywood. 1963. Principles of Angiosperm Taxonomy. Princeton, NJ: Van Nostrand.

DeCrappeo, N., and R. M. Giblin-Davis. 2001. *Schistonchus aureus* n. sp. and *S. laevigatus* n. sp. (Aphelenchoididae): Associates of native Floridian *Ficus* spp. and their *Pegoscapus* pollinators (Agaonidae). Journal of Nematology 33: 91 – 103.

Drake, D. W. 1980. Contrasting success of natural hybridization in two *Eucalyptus* species pairs. Australian Journal of Botany 28: 167 – 191.

Duncan, L. W., R. N. Inserra, W. K. Thomas, D. Dunn, I. Mustika, L. M. Frisse, M. L. Mendes, K. Morris, and D. T. Kaplan. 1999. Molecular and morphological analysis of isolates of *Pratylenchus coffeae* and closely related species. Nematropica 29: 61 – 80.

Felsenstein, J. 1985. Confidence limits on phylogenies: An approach using the bootstrap. Evolution 39: 783 – 791.

Felsenstein, J. 1993. PHYLIP (Phylogeny inference package), Version 3.5. University of Washington, Seattle, WA.

Ferris, V. R. 1983. Phylogeny, historical biogeography and the species concept in soil nematodes. Pp. 143 – 161. In A. R. Stone, H. M. Platt, and L. F. Khalil, eds. Concepts in Nematode Systematics. London: Academic Press.

Fisher, J. M., and W. R. Nickle. 1968. On the classification and life history of *Fergusobia curriei* (Sphaerulariidae: Nematoda). Proceedings of the Helminthological Society of Washington 35: 40 – 46.

Fitch, D. H. A. 1997. Evolution of the male tail development in rhabditid nematodes related to *Caenorhabditis elegans*. Systematic Biology 46: 145 – 179.

Fitch, D. H. A., and S. W. Emmons. 1995. Variable cell positions and cell contacts underlie morphological evolution of the rays in the male tails of nematodes related to *Caenorhabditis elegans*. Developmental Biology 170: 564 – 582.

Fitch, D. H. A., and W. K. Thomas. 1997. Evolution. Pp. 815 – 850. In D. Riddle, T. Blumenthal, B. Meyer, and J. Priess, eds. *C. elegans* II. New York: Cold Spring Harbor Laboratory Press.

Fitch, D. H. A., B. Bugaj-Gaweda, and S. W. Emmons. 1995. 18S ribosomal RNA gene phylogeny for some Rhabditidae related to *Caenorhabditis*. Molecular Biology and Evolution 12: 346 – 358.

Frost, D. R., and A. G. Kluge. 1994. A consideration of epistemology in systematic biology, with a special reference to species. Cladistics 10: 259 – 294.

Futuyma, D. J. 1983. Evolutionary interactions among herbivorous insects and plants. Pp. 207 – 231. In D. J. Futuyma, and M. Slatkin, eds. Coevolution. Sunderland, MA: Sinauer Associates.

Futuyma, D. J., and M. Slatkin. 1983. Introduction. Pp. 1 – 13. In D. J. Futuyma, and M. Slatkin, eds. Coevolution. Sunderland, MA: Sinauer Associates.

Gasparrini, G. 1864. Sulla maturazione e la qualita dei fichi dei contorni di Napoli. Atti Acad. Pontaniana 9: 99 – 118.

Gentry, A. H. 1992. Tropical forest biodiversity: Distribution patterns. Oikos 63: 19 – 28.

Giblin, R. M. 1986. Culture of nematode associates and parasites of insects. Pp. 408 – 413. In J. A. Veech, and D. W. Dickson, eds. Vistas on Nematology: A commemoration of the twenty-fifth anniversary of the society of nematologists. DeLeon Springs, FL: E. O. Painter Printing.

Giblin-Davis, R. M. 1996. Phoresy between diplogasterid, aphelenchid and tylenchid nematodes and insects. Nematropica 26: 205.

Giblin-Davis, R. M., B. J. Center, H. Nadel, J. H. Frank, and B. W. Ramirez. 1995. Nematodes associated with Fig wasps, *Pegoscapus* spp. (Agaonidae), and the syconia of native Floridian figs (*Ficus* spp.). Journal of Nematology 27: 1 – 14.

Giblin-Davis, R. M., K. A. Davies, D. S. Williams, and T. D. Center. 2001a. Cuticular changes in Fergusobiid nematodes associated with parasitism of Fergusoninid flies. Comparative Parasitology 68: 242 – 248.

Giblin-Davis, R. M., J. Makinson, B. J. Center, K. A. Davies, M. Purcell, G. S. Taylor, S. J. Scheffer, J. Goolsby, and T. D. Center. 2001b. *Fergusobia/ Fergusonina*-induced shoot bud gall development on *Melaleuca quinquenervia*. Journal of Nematology 33: 239 – 247.

Gordh, G. 1979. Superfamily Chacidoidea. Pp. 743 – 748. In K. Krombein, P. D. Hurd, and B. D. Burks, eds. Catalog of Hymenoptera in America North of Mexico, vol. 1. Washington, DC: Smithsonian Institution Press.

Gould, F. 1988. Genetics of pairwise and multispecies plant-herbivore coevolution. Pp. 13 – 55. In K. C. Spencer, ed. Chemical Mediation of Coevolution. New York: Academic Press.

Griffin, A. R., I. P. Burges, and L. Wolf. 1988. Patterns of natural and manipulated hybridization in the genus *Eucalyptus* L'herit.: A review. Australian Journal of Botany

36:41 -66.

Hafner, M. S., P. D. Sudman, F. X. Villablanca, T. S. Spradling, J. W. Demastes, and S. A. Nadler. 1994. Disparate rates of molecular evolution in cospeciating hosts and parasites. Science 265:1087 -1090.

Harris, K. M. 1982. First record of Fergusoninidae (Diptera: Schizophora) outside Australia: A new species of *Fergusonina* on *Syzygium* in India. Systematic Entomology 7:211 -216.

Herre, E. A. 1993. Population structure and the evolution of virulence in nematode parasites of fig wasps. Science 259:1442 -1445.

Herre, E. A. 1995. Factors affecting the evolution of virulence: Nematode parasites of fig wasps as a case study. Parasitology (Supplement) 111:179 -191.

Herre, E. A., C. A. Machado, E. Bermingham, S. S. McCafferty, J. D. Nason, D. Windsor, W. Van Houten, and K. Bachmann. 1996. Molecular phylogenies of figs and their pollinating wasps. Journal of Biogeography 23:521 -530.

Hill, K. D., and L. A. S. Johnson. 1995. Systematic studies in the eucalypts. 7. A revision of the bloodwoods, genus *Corymbia* (Myrtaceae). Telopea 6:185 -504.

Hill, K. D., and L. A. S. Johnson. 1998. Systematic studies in the eucalypts. 8. A review of the Eudesmoid eucalypts, *Eucalyptus* subgenus *Eudesmia* (Myrtaceae). Telopea 7:375 -414.

Hillis, D. M., and J. J. Bull. 1993. An empirical test of bootstrapping as a method for assessing confidence on phylogenetic analysis. Systematic Biology 42:182 -192.

Hillis, D. M., C. Morita, and B. K. Mable, eds. 1996. Molecular Systematics. Sunderland, MA: Sinauer Associates.

Hillis, W. E. 1966. Variation in polyphenol composition within species of *Eucalyptus* L'Herit. Phytochemistry 5:541 -546.

Hinkle, G., J. K. Wetterer, T. R. Schultz, and M. L. Sogin. 1994. Phylogeny of the attineant fungi based on analysis of small subunit ribosomal RNA gene sequences. Science 266:1695 -1697.

Hollar, L. J., and M. S. Springer. 1996. Old world fruitbat phylogeny: Evidence for convergent evolution and an endemic African clade. Proceedings of the National Academies of Science USA 94:5 716 -5721.

Huelsenbeck, J. P., J. J. Bull, and C. W. Cunningham. 1996. Combining data in phylogenetic analysis. Trends in Ecological Evolution 11:152 -158.

Huelsenbeck, J. P., B. Rannala, and Z. Yang. 1997. Statistical tests of host-parasite cospeciation. Evolution 51:410 -419.

Hunt, D. J. 1993. Aphelenchida, Longidoridae and Trichodoridae: Their Systematics and Bionomics. Cambridge, UK: CAB International Institute of Parasitology.

Janzen, D. H. 1979. How to be a fig. Annual Review of Ecology and Systematics 10:13 -51.

Johnson, L. A. S., and B. G. Briggs. 1981. Three old southern families - Myrtaceae, Proteaceae and Restionaceae. Pp. 429 -469. In A. Keast, ed. Ecological Biogeography of Australia. The Hague: W. Junk.

Johnson, L. A. S., and B. G. Briggs. 1984. Myrtales and Myrtaceae - a phylogenetic analysis. Annals of the Missouri Botanical Garden 71:700 -756.

Jolivet, P. 1998. Interrelationship between Insects and Plants. Boca Raton, FL: CRC

Press.
Jukes, T. H., and C. R. Cantor. 1969. Evolution of protein molecules. Pp. 21 – 132. In H. N. Munro, ed. Mammalian Protein Metabolism. New York: Academic Press.
Kampfer, S., C. Sturmbauer, and J. Ott. 1998. Phylogenetic analysis of rDNA sequences from adenophorean nematodes and implications for the Adenophorean-Secernentea controversy. Invertebrate Biology 117: 29 – 36.
Kishino, M., and M. Hasegawa. 1989. Evaluation of the maximum likelihood estimate of the evolutionary tree topologies from DNA sequence data, and the branching order in Hominoidea. Journal of Molecular Evolution 29: 170 – 179.
Kluge, A. G. 1989. A concern for evidence and a phylogenetic hypothesis of relationships among epicrates (Boide, Serpentes). Systematic Zoology 38: 7 – 25.
Kocher, T. D., W. K. Thomas, A. Meyer, S. V. Edwards, S. Pääbo, F. X. Villablanca, and A. C. Wilson. 1989. Dynamics of mitochondrial DNA evolution in animals: Amplification and sequencing with conserved primers. Proceedings of the National Academies of Science USA 86: 6196 – 6200.
Korenchenko, E. R. 1987. Experimental study of host specificity and species independence of nematodes of the genus *Parasitorhabditis*. Pp. 110 – 117. In M. D. Sonin, ed. Helminths of Insects. New Delhi, India: Amerind Publishing.
Krajewski, C., M. Blacket, L. Buckley, and M. Westerman. 1997. A multigene assessment of phylogenetic relationships within the dasyurid marsupial subfamily, Sminthopsinae. Molecular Phylogenetics and Evolution 8: 236 – 248.
Krall, E. L. 1991. Wheat and grass nematodes: *Anguina*, *Subanguina* and related genera. Pp. 721 – 760. In W. R. Nickle, ed. Manual of Agricultural Nematology. New York: Marcel Dekker.
Kumar, S., K. Tamaura, and M. Nei. 1993. MEGA: Molecular evolutionary genetic analysis, Version 1.01. The Pennsylvania State University, University Park, PA.
Kumari, R. V., and Y. N. Reddy. 1984. Studies on the association of a new nematode species *Schistonchus hispida* sp. n. (Aphelenchoidea, Nickle, 1971) and wasp. Proceedings on the Indian Academy of Parasitology 5: 21 – 25.
Ladiges, P. Y. 1997. Phylogenetic history and classification of eucalypts. Pp. 16 – 29. In J. E. Williams, and J. C. Z. Woinarski, eds. Eucalypt Ecology. Cambridge, UK: Cambridge University Press.
Ladiges, P. Y., and C. J. Humphries. 1983. A cladistic study of *Arillastrum*, *Angophora* and *Eucalyptus* (Myrtaceae). Botanical Journal of the Linnaean Society 87: 105 – 134.
Ladiges, P. Y., and C. J. Humphries. 1986. Relationships in the stringybarks, *Eucalyptus* L'Herit. informal subgenus *Monocalyptus* series *Capitellatae* and *Olsenianae*: Phylogenetic hypotheses, biogeography and classification. Australian Journal of Botany 34: 603 – 632.
Ladiges, P. Y., C. J. Humphries, and M. I. H. Brooker. 1983. Cladistic relationships and biogeographic patterns in the peppermint group of *Eucalyptus* (Informal Subseries Amygdalinae, Subgenus *Monocalyptus*) and the description of a new species, *E. willisii*. Australian Journal of Botany 31: 565 – 584.
Ladiges, P. Y., C. J. Humphries, and M. I. H. Brooker. 1987. Cladistic and biogeographic analysis of Western Australin species of *Eucalyptus* L'Herit., informal subgenus *Monocalyptus* Pryor and Johnson. Australian Journal of Botany 35: 251 – 281.

Ladiges, P. Y., F. Udovicic, and A. N. Drinnan. 1995. Eucalypt phylogeny – molecules and morphology. Australian Systematic Botany 8: 483 – 497.

Lambshead, P. J. D. 1993. Recent developments in marine benthic biodiversity. Oceanis 19: 5 – 24.

Lange, R. T. 1982. Australian tertiary vegetation: Evidence and interpretation. Pp. 44 – 89. In J. Smith, ed. A History of Australasian Vegetation. Sydney, Australia: McGraw Hill.

Lavergne, A., R. Douzery, T. Stichler, F. M. Catzeflis, and M. S. Springer. 1996. Interordinal mammalian relationships: Evidence for paenungulate monophyly is provided by complete mitochondrial 12S rRNA sequences. Molecular Phylogenetics and Evolution 6: 245 – 258.

Leignel, V., J. F. Humbert, and L. Elard. 1997. Study by ribosomal DNA ITS2 sequencing and RAPD analysis on the systematics of four *Metastrongylus* species (Nematoda: Metastrongyloidea). Journal of Parasitology 83: 608 – 611.

Lloyd, J., and K. A. Davies. 1997. Two new speices of *Schistonchus* (Tylenchida: Aphelenchoididae) associated with *Ficus macrophylla* from Australia. Fundamental and Applied Nematology 20: 79 – 86.

Lockhart, P. J., M. A. Steel, M. D. Hendy, and D. Penny. 1994. Recovering evolutionary trees under a more realistic model of sequence evolution. Molecular Biology and Evolution 11: 605 – 612.

Lowman, M. D., and H. Heatwole. 1987. The impact of defoliating insects on the growth of eucalypt saplings. Australian Journal of Ecology 12: 175 – 181.

Machado, C. A., E. A. Herre, S. McCafferty, and E. Bermingham. 1996. Molecular phylogenies of fig-pollinating and non-pollinating wasps and the implications for the origin and evolution of the fig-fig wasp mutualism. Journal of Biogeography 23: 531 – 542.

Malloch, J. R. 1924. Notes on Australian Diptera, No. iii. Proceedings of the Linnaean Society of New South Wales 49: 329 – 338.

Malloch, J. R. 1925. Notes on Australian Diptera, No. vi. Proceedings of the Linnaean Society of New South Wales 50: 80 – 97.

Malloch, J. R. 1932. Notes on Australian Diptera, No. xxxii. Proceedings of the Linnaean Society of New South Wales 57: 213 – 217.

Marginson, J. C., and P. Y. Ladiges. 1988. Variation in *Eucalyptus baxteri*. Australian Systematic Botany 1: 151 – 175.

Massey, C. L. 1974. Biology and Taxonomy of Nematode Parasites and Associates of Bark Beetles in the United States. Agriculture Handbook No. 446. Washington, DC: U. S. Government Printing Office.

Mayr, E. 1969. Prinicples of Systematic Zoology. New York: McGraw Hill.

McLeod, R., F. Reay, and J. Smyth. 1994. Plant Nematodes of Australia listed by Plant and by Genus. Sydney, Australia: New South Wales Agriculture.

Mitter, C., and D. R. Brooks. 1983. Phylogenetic aspects of coevolution. Pp. 65 – 98. In D. J. Futuyma, and M. Slatkin, eds. Coevolution. Sunderland, MA: Sinauer Associates.

Morrow, P. A. 1977. The significance of phytophagous insects in the *Eucalyptus* forests of Australia. Pp. 19 – 29. In W. J. Mattson, ed. The Role of Arthropods in Forest

Ecosystems. New York: Springer Verlag.

Nadler, S. A., and D. S. S. Hudspeth. 1998. Ribosomal DNA and phylogeny of the Ascaridoidea (Nemata: Secernentea): Implications for morphological evolution and classification. Journal of Molecular Evolution 10: 221 – 236.

Nelson, G. J., and N. I. Platnick. 1981. Systematics and Biogeography: Cladistics and Vicariance. New York: Columbia University Press.

Nicolle, D. 1997. Eucalypts of South Australia. South Australia: Lane Printing.

Noble, I. R. 1989. Ecological traits of the *Eucalyptus* L'Herit. Subgenera *Monocalyptus* and *Symphyomyrtus*. Australian Journal of Botany 37: 207 – 224.

NSF Workshop on Systematics and Inventories of Soil Nematodes. 1997. Tucson, Arizona. July, 1997. http://www.nrel.colostate.edu/soil/sisn.html

Orlovich, D. A., A. N. Drinnan, and P. Y. Ladiges. 1996. Floral development in the Metrisideros group (Myrtaceae) with special emphasis on the androecium. Telopea 6: 689 – 719.

Platt, H. M., K. M. Shaw, and P. J. D. Lambshead. 1984. Nematode species abundance patterns and their use in the detection of environmental perturbations. Hydrobiologia 118: 59 – 66.

Poinar, G. O., Jr. 1975. Entomogenous Nematodes. Leiden, Netherlands: E. J. Brill.

Poinar, G. O., Jr. 1979. *Parasitodiplogaster* sycophilon gen N., sp. N. (Diplogasteridae: Nematoda), a parasite of *Elisabethiella stuckenbergi* Grandi (Agaonidae: Hymenpotera) in Rhodesia. Proceedings of the Koninklijke Nederlandse Akademie van Wetenschappen, Series C, Biological and Medial Sciences 82: 375 – 381.

Poinar, G. O., Jr. 1984. Fossil evidence of nematode parasitism. Revue de Nematologie 7: 201 – 203.

Poinar, G. O., Jr., and E. A. Herre. 1991. Speciation and adaptive radiation in the fig wasp nematode *Parasitodiplogaster* (Diplogasteridae: Rhabditina) in Panama. Revue de Nematologie 14: 361 – 374.

Pole, M. S. 1989. Early Miocene flora from central Otago, New Zealand. Journal of the Royal Society of New Zealand 19: 121 – 125.

Pole, M. S. 1993. Early flora of the Manuherikia group, New Zealand 7. Myrtaceae, including *Eucalyptus*. Journal of the Royal Society of New Zealand 23: 313 – 328.

Prager, E. M., and A. C. Wilson. 1988. Ancient origin of lactalbumin from lysozyme analysis of SNA and amino acid sequences. Journal of Molecular Evolution 27: 326 – 335.

Price, P. W. 1980. Evolutionary Biology of Parasites. Princeton, NJ: Princeton University Press.

Price, P. W. 1996. Biological Evolution. New York: Saunders College Publishing.

Pryor, L. D. 1959. Evolution in Eucalyptus. Australian Journal of Science 22: 45 – 49.

Pryor, L. D. 1976. The Biology of Eucalypts. London, England: Edward Arnold.

Pryor, L. D., and L. A. S. Johnson. 1971. A Classification of the Eucalypts. Canberra, Australia: Australian National University Press.

Pryor, L. D., and L. A. S. Johnson. 1981. *Eucalyptus,* the universal Australian. Pp. 499 – 536. In A. Keast, ed. Ecological Biogeography of Australia. The Hague: W. Junk.

Pryor, L. D., E. R. Williams, and B. V. Gunn. 1995. A morphometric analysis of

Eucalyptus urophylla and related taxa with descriptions of two new species. Australian Systematic Botany 8: 57 – 70.

Quicke, D. L. J. 1997. Parasitic Wasps. London: Chapman and Hall.

Ramirez, B. W. 1970a. Host specificity of fig wasps (Agaonidae). Evolution 24: 681 – 691.

Ramirez, B. W. 1970b. Taxonomic and biological studies of neotropical fig wasps (Hymenoptera: Agaonidae). University of Kansas Scientific Bulletin 46: 1 – 44.

Ramirez, B. W. 1974. Specificity of the Agaonidae: the coevolution of *Ficus* and its pollinators. Ph. D. thesis, University of Kansas.

Reddy, Y. N., and P. N. Rao. 1986. *Schistonchus racemosa* sp. a nematode parasite of wasp (*Ceratosolen* sp.) associated with the fig, *Ficus racemosa* L. Indian Journal of Nematology 16: 135 – 137.

Richards, P. W. 1996. The Tropical Rainforest – an Ecological Study. 2 nd edition. Cambridge, UK: Cambridge University Press.

Rogstad, S. H. 1990. The biosystematics and evolution of the *Polyalthia hypoleuca* species complex (Annonaceae) of Malesia. II. Comparative distribution ecology. Journal of Tropical Ecology 6: 387 – 408.

Rühm, W. 1956. Die Nematoden der Ipiden. Parasitologische Schriftenreihe, Jena 6: 1 – 437.

Sale, M. M., B. M. Potts, A. K. West, and J. B. Reid. 1993. Relationships with *Eucalyptus* using chloroplast DNA. Australian Journal of Systematic Botany 6: 127 – 138.

Sanderson, M. J. 1989. Confidence limits on phylogenies: the bootstrap revisited. Cladistics 5: 113 – 129.

Sanderson, M. J. 1995. Objections to bootstrapping phylogenies: a critique. Systematic Biology 44: 299 – 320.

Siddiqi, M. R. 1986a. Tylenchida: Parasites of Plants and Insects. Farnham Royal, UK: Commonwealth Agricultural Bureaux

Siddiqi, M. R. 1986b. A review of the nematode genus *Fergusobia* Currie (Hexatylina) with descriptions of *F. jambophila* n. sp. and *F. magna*. Pp. 264 – 278. In G. Swarup, and D. R. Dasgupta, eds. Plant Parasitic Nematodes of India, Problems and Progress. New Delhi, India: Indian Agricultural Research Institute.

Siddiqi, M. R. 1994. *Fergusobia brevicauda* sp. n. and *F. philippinensis* sp. n. (Nematoda: Hexatylina) from *Eucalyptus deglupta*. Proceedings of the Second Afro-Asian Nematology Symposium. Pp. 96 – 100.

Sidow, A. R., and W. K. Thomas. 1994. A molecular evolutionary framework for eukaryotic model organisms. Current Biology 4: 596 – 603.

Specht, R. 1981. Ecophysiological principles determining the biogeography of major vegetation formations in Australia. Pp. 299 – 334. In A. Keast, ed. Ecological Biogeography of Australia. The Hague: W. Junk.

Springer, M. S., L. J. Hollar, and J. A. W. Kirsch. 1995. Phylogeny, molecules versus morphology, and rates of character evolution among fruit bats (Chiroptera: Megachiroptera). Australian Journal of Zoology 43: 557 – 582.

Stanhope, M. J., M. R. Smith, V. G. Waddell, C. A. Porter, M. S. Shivji, and M. Goodman. 1996. Mammalian evolution and the interphotoreceptor retinoid binding protein (IRBP) gene: Convincing evidence for several superordinal clades. Journal of Molecular Evolution 43: 83 – 92.

Swofford, D. L. 1998. PAUP*: Phylogenetic analysis using parsimony (* and other methods), Version 4.0. Sunderland, MA: Sinauer Associates.

Swofford, D. L., G. P. Olsen, P. J. Waddell, and D. M. Hillis. 1996. Phylogenetic inference. Pp. 405 – 514. In D. M. Hillis, C. Mortiz, and B. K. Mable, eds. Molecular Systematics. Sunderland, MA: Sinauer Associates.

Tamura, K., and N. Nei. 1993. Estimation of the number of nucleotide substitutions in the control region of mitochondrial DNA in humans and chimpanzees. Molecular Biology and Evolution 10: 512 – 526.

Taylor, G. S., A. D. Austin, and K. A. Davies. 1996. Biology of the gall-forming fly, *Fergusonina flavicornis* Malloch (Diptera: Fergusoninidae), and its associated Hymenoptera in South Australia, with a description of a new species of *Bracon* (Hymenoptera: Braconidae). Transactions of the Royal Society of South Australia 120: 131 – 146.

Templeton, A. R. 1983. Phylogenetic inference from restriction endonuclease cleavage site maps with particular reference to the evolution of humans and apes. Evolution 37: 221 – 224.

Thomas, W. K., and A. C. Wilson. 1991. Mode and tempo of molecular evolution in the nematode *Caenorhabditis*: cytochrome oxidase II and calmodulin sequences. Genetics 128: 269 – 279.

Thompson, J. N. 1989. Concepts of coevolution. Trends in Ecological Evolution 4: 179 – 183.

Tonnoir, A. L. 1937. Revision of the genus *Fergusonina* Mall. (Diptera, Agromyzidae). Proceedings of the Linnaean Society of New South Wales 62: 126 – 146.

Triantaphyllou, A. C. 1985. Cytogenetics, cytotaxonomy and phylogeny of root-knot nematodes. Pp. 113 – 121. In J. N. Sasser, and C. C. Carter, eds. An Advanced Treatise on *Meloidogyne*, Volume I: Biology and Control. Raleigh, NC: Department of Plant Pathology, NCSU and USAID.

Vanfleteren, J. R., Y. Van de Peer, M. L. Blaxter, S. A. R. Tweedie, C. Trotman, L. Lu, M. L. Van Hauwaert, and L. Moens. 1994. Molecular genealogy of some nematode taxa as based on cytochrome c and globin amino acid sequences. Molecular Phylogenetics and Evolution 3: 92 – 101.

Vovlas, N., R. N. Inserra, and N. Greco. 1992. *Schistonchus caprifici* parasitizing caprifig (*Ficus carica sylvestris*) florets and the relationship with its fig wasp (*Blastophaga psenes*) vector. Nematologica 38: 215 – 226.

Vovlas, N., A. Troccoli, S. Van Noort, and E. Van den Berg. 1998. *Schistonchus africanus* n. sp. (Aphelenchoididae) associated with *Ficus thonningii* and its Agaonidae pollinator wasp *Elizabethiella stuckenbergi*. Journal of Nematology 30: 404 – 410.

Wardell-Johnson, G. W., J. E. Williams, K. D. Hill, and R. Cumming. 1997. Evolutionary biogeography and contemporary distribution of eucalypts. Pp. 92 – 128. In J. E. Williams, ed. Eucalypt Ecology. Cambridge, UK: Cambridge University Press.

Weier, T. E., C. R. Stocking, and M. G. Barbour. 1974. Botany: An Introduction to Plant Biology. New York: John Wiley & Sons.

White, T. J., N. Arnheim, and H. A. Erlich. 1989. The polymerase chain reaction.

Trends in Genetics 5: 185 – 189.
Wiebes, J. T. 1979. Co-evolution of figs and their insect pollinators. Annual Review of Ecology and Systematics 10: 1 – 12.
Wiebes, J. T. 1982. The phylogeny of the Agaonidae (Hymenoptera, Chalcidoidea) and *Ficus* (Moraceae): Fig wasps and their figs, XV (Meso-American *Pegoscapus*). Proceedings of the Koninklijke Nederlandse Akademie van Wetenschappen 98: 167 – 183.
Wiley, E. O. 1981. Phylogenetics. New York: John Wiley & Sons.
Woinarski, J. C. Z., and J. M. Cullen. 1984. Distribution of invertebrates on foliage in forests of south-eastern Australia. Australian Journal of Ecology 9: 207 – 232.

Acknowledgements We thank Drs Rudi Scheffrahn and Nan-Yao Su for critical review of this chapter and A. Austin, T. Center, F. Driver, L. M. Frisse, J. Goolsby, E. A. Herre, J. Makinson, K. Morris, D. Nicolle, M. Purcell, S. Scheffer, and S. Van Noort for assistance, discussion, and suggestions that improved this chapter. This research was supported in part by a USDA T-STAR/CBAG grant in Tropical and Subtropical Agriculture CRSR-99-34135-8478, and USDA Specific Cooperative Agreement 58-6629-9-004 between the USDA-ARS Aquatic Weed Control Lab and the University of Florida. Florida Agricultural Experiment Station, Joural Series No. R-07325.

12 Cultivation of Nematodes

Paul De Ley and Manuel Mundo-Ocampo

12.1 Introduction

12.1.1 Purposes, Advantages and Drawbacks

The ability to culture organisms is always of great benefit to their scientific study, and sometimes simply essential. Cultures ensure that a steady supply of live material is available for teaching and research, eliminating the need for repeated sampling of inaccessible or temporally fluctuating habitats in search of living specimens. Cultures also provide all stages of development for observation, demonstration, and experiments, and allow for more standardized methods and for easier corroboration.

Despite the increasing availability of non-destructive tools and methods, the science of life often requires the death of the very creatures it aims to understand. This applies especially to biological investigations of fragile, minute, and short-lived creatures like nematodes. Live specimens usually provide the best starting point for a wide range of approaches, such as scanning or transmission electron microscopy (SEM/TEM), developmental studies, molecular analyses, etc.

With cultures, morphological and ecological variations can be characterized under controlled and reproducible conditions. Nematode reproduction, feeding, and locomotion can only be observed and recorded directly in the laboratory, and cultures often provide the first tangible evidence of the ecological and ethological properties of a certain nematode species. This is particularly important for studies on control of nematodes, or control of other organisms by nematodes, which require an array of experimental and observational procedures that rely on the living organism rather than on its preserved remains. And last but not least, as nematodes are invisible and unknown to most of us, the best illustration of their existence and abundance is the demonstration of a dense mass of writhing worms in a Petri dish or watch glass.

In all these investigations, it is not just an advantage to have live animals at your disposal, but also to have them in numbers as small or as large as the task may require. With more specimens available, a failed first attempt at ultrastructural, embryological, or biochemical preparation is only a hindrance and not a standstill. With multiple subpopulations in culture, it becomes possible to set up experiments using as many batches and replicates as statistical rigor or individual curiosity may demand. Just as importantly, cultures can be taken through the bottleneck of a single female or hermaphrodite, guaranteeing that all progeny of this parent animal is conspecific, and that subcultures derived from subsequent generations respond comparably to identical experimental conditions. For taxonomically confounding organisms like nematodes, species identity is often a major stumbling block, and single-female progeny often provides the most practical means of ensuring conspecificity of the organisms under study.

This being said, nematode cultures do have some drawbacks. Any organism kept under artificial conditions for successive generations, is likely to undergo selection pressures and bottlenecks that are quite different from its counterparts in nature. With short generation times and poorly understood physiology, nematodes in culture are quite susceptible to genetic drift and selective pressures, which may cause unforeseen changes to their biology. Unless direct comparisons with conspecific natural populations are done – and this is rarely feasible – we can never be sure that nematodes in culture have maintained all properties of their "wild" ancestors. In fact, there is substantial evidence that populations of organisms in culture can diverge dramatically from their natural counterparts (Goho and Bell, 2000; Notley-McRobb and Ferenci, 1999), especially over longer series of generations and in more artificial or extreme culture conditions.

Finally, the startup, maintenance, and exchange of nematode cultures often require a substantial investment of time and resources. Optimal environmental conditions and nutritional requirements are known for some species, but extensive and repeated experimentation may be necessary in order to culture others. Nematode cultures can be extremely easy to obtain and maintain in themselves, but they require regular attention as well as occasional improvisation and crisis management. Finally, it is often difficult to receive due academic credit and financial support for the time and costs involved, particularly so in modern scientific culture with its emphasis on short-term funding, high-tech equipment, and impact-based evaluations.

12.1.2 Outline of This Chapter

During the past century, an impressive multitude of culture systems has been described and applied to various groups of nematodes. Several classic review papers remain absolutely relevant today. Dougherty (1960) provided a first thorough overview of concepts and methods for culturing bacterial-feeding nematodes. Platzer (1977), Hooper (1986c), and Nicholas (1984) reviewed cultivation of a broader range of nematodes, focusing on methods and/or physio-ecological aspects. Moens and Vincx (1998) recently provided an overview of cultured species and culture methods for nematodes from marine and estuarine sediments.

In this chapter, our purpose is to present a practical and up-to-date manual on nematode culture by covering a range of basic cultivation techniques applicable or adaptable to most nematodes, and particularly by reviewing both recently developed and time-honored methods, including many that require little or no complex equipment. In order to be able to vouch for the feasibility and practicality of these techniques, we include many methods which we employ ourselves, or which are used by various colleagues and close collaborators. One major restriction applies to the range of nematodes covered: except for entomopathogenic nematodes, we will not discuss zooparasites. Their culture usually depends on the successful culture of their host, and since just about every kind of animal is parasitized by nematodes, this is well beyond the scope of this text and of our experience (see Platzer, 1977, for methods that allow partial completion of the life cycle without requiring a host).

Within these limits, we first attempt to summarize some of the scientific jargon commonly used in the context of nematode cultures, and to list some of the general equipment and principles used for culturing. We then present summaries and protocols of a range of culture methods employed for various groups of nematodes. For each of these groups, we also provide information on some "auxiliary methods" such as long-term storage of inocula, synchronization of hatching, control and removal of contaminants, etc.

Although many methods were first developed for one particular species or group of nematodes, they have been (or can easily be) adapted for use with other nematodes. To emphasize this point, we have organized subsequent sections with identical subheadings, and provide many cross-references between them. For those readers in particular who face troubleshooting problems, or who are looking for ideas to develop new methods, we recommend that they skip between subheadings of different sections in order to compare methods as currently applied in different groups. In some cases, we also provide pointers to methods that have not yet been fully developed, but which seem promising as possible alternative solutions. In this manner we hope to both provide beginning nematologists with a practical guideline to nematode cultivation, as well as presenting experienced "worm breeders" with an expanded repertoire that will allow them to culture a wider range of nematode taxa.

12.2 Terminology and General Principles

12.2.1 Some Important Terms, Definitions, and Parameters

A short enumeration of terms and components will clarify their usage in the subsequent text and simultaneously provide a brief overview of the many parameters that can influence the success of a culture. We will apply the term culture to any collection of organisms maintained in the laboratory under conditions that are designed to promote their reproduction and multiplication①. The entire assemblage of living and non-living components involved is referred to as a culture system, while the protocol and associated procedures for assembling and maintaining such a culture system are referred to as a culture method or a culture technique. If a successful combination of culture system and method result in the ability to maintain organisms indefinitely in the laboratory, then we will refer to the resulting culture as a continuous culture. Generally speaking, a culture system consists of three classes of components, all of which are of potential relevance to the applicable methods and their eventual success.

12.2.1.1 Biological Components of a Culture System

The following terms are used throughout this chapter to describe the biological effects and roles of organisms within a culture. The target organism or species of a culture system is the one intended to survive and reproduce within that system, for ultimate use in specific research activities. A prey or host organism/species is present in a culture system, and expected to survive and perhaps even reproduce, but only for the immediate purpose of providing food for the target species. All other organisms are considered contaminants, i. e., their presence in the culture system is unintentional. Adverse contaminants are those that inhibit survival and reproduction of the target organism, e. g., through such interactions as predation, competition, parasitism, pathogenicity, or

---

① This definition is looser than some previous proposals (e. g., Moens and Vincx, 1998) since it also applies to failed attempts where the organisms did not reproduce even once. However, our focus in this chapter lies on the methods and principles involved, rather than on particular instances of successful culture. Any given culture technique will work or not depending on a combination of *inter alia* the species, researchers, laboratory, and luck involved, and therefore counts as a technique for culturing organisms regardless of whether it works under a wide range of circumstances or not. Failure to obtain multiplication implies that the culture in question has failed, rather than being disqualified *a posteriori* as a true culture.

production of toxins. Neutral contaminants are those that do not measurably affect survival and reproduction of the target organism. Beneficial contaminants can also occur, however, and are those that promote survival and reproduction of the target species through any other role than that of prey or host (e. g., suppression of adverse contaminants, stabilization of chemical conditions, enhancement of the growth of a prey organism, etc.).

A number of terms are generally used to describe the number and known identity of co-occurring organisms in a culture system. A xenic or agnotobiotic culture is one in which multiple taxa are present, including at least one of unknown species identity. The opposite is a gnotobiotic system, i. e., one in which all species are known. If the target species is not the only one present in the system, then the culture is called synxenic. If exactly one other species is present besides the target, then the culture is called monoxenic (and the non-target species is usually a prey organism). If two, three, four, etc. species co-occur with the target species, then the culture system is respectively called dixenic, trixenic, quadrixenic, etc., or more generally referred to as polyxenic. Culture systems free from any other organisms than the target are called axenic.

12.2.1.2 Chemical Components of a Culture System

In analogy to the above, a number of terms are used to describe the roles of various chemical components in culture systems. A nutrient is used as food by a target and/or prey organism. An essential nutrient is a vitally required substance that must be present in minute quantities (e. g., sterols) because the target organism is unable to metabolize it from other nutrients and/or prey. When nutrients are provided in excess of what is consumed during optimal growth by target and/or prey, then the culture system is referred to as a climax or unlimited culture. When nutrients are purposely provided in restricted amounts, well below levels required for unrestricted growth and reproduction, then the system is called sparse or limited.

Nutrients are never the only chemicals in a culture system. For instance, a metabolite is a waste product of any organism present in the culture. A buffer is a compound or mixture of compounds that stabilizes chemical conditions in the system (usually pH). An inhibitor is a compound that affects the population dynamics of the target organism adversely without directly killing individuals (e. g., through interference effects on reproduction, mating, locomotion, feeding, etc.). A promotor is the opposite of an inhibitor, i. e., it increases growth and reproduction of the target without being in itself a nutrient (e. g., through toxicity to adverse contaminants). A toxin directly kills or weakens individual target organisms, even at low concentrations.

Various terms exist to describe available knowledge about the chemical composition of a given culture medium. A holidic medium is one in which all chemical constituents are known, usually including their initial concentrations. A meridic medium is one in which at least one chemical constituent is unknown, while an oligidic medium contains one or more complex ingredients of largely unknown chemical composition (e. g., liver extract, soil extract, etc.).

12.2.1.3 Physical Components of a Culture System

A third class of terms refers to the roles of various physical components that may occur in a culture system. The container encapsulates all other components and physically protects them from external disturbance, often with the aid of a sealant that prevents exchange from liquids and/or gases between the culture system and the exterior. Containers are often transparent to allow visual screening of the numbers and behavior of organisms in the culture. Cultures may need to be kept at a specific temperature in an incubator or refrigerator, but even if this is not required then it is still good practice to keep them in double containment, i. e., to store sets of containers inside clearly labeled plastic boxes with lids that close firmly. This will further reduce dehydration of the cultures and limit the spread of actively moving contaminants like mites.

The medium is the physical component containing most of the chemicals introduced and metabolized in the system, while the substrate is a more or less inert and usually solid structure that provides organisms with support, usually allowing locomotion of the target organism in at least two dimensions. Liquid cultures are systems in which the medium is liquid while the container provides the only substrate. For nematodes, liquid cultures are sometimes of limited use because they may inhibit egg-laying or mating in species incapable of active swimming. Gel cultures are systems in which a gelatinous substance is prepared with various added chemicals, so that the gel (typically agar) acts as both medium and substrate. These systems are usually very cheap and easy to handle and offer the added advantage that organisms can easily be observed inside the closed container. Wet gel cultures are identical to gel cultures, except for the intentional addition of a liquid layer on top of the set gel, so as to provide aquatic or moisture-loving target species with a more liquid environment than the gel itself (see 12.4.2.2(3)).

Filter and sponge cultures use porous substrates such as paper or glass fiber filters, or polyether polyurethane sponge; the substrate is usually partially submerged in liquid medium. Soil or sediment cultures use the natural substrate or a fraction thereof; they offer advantages in terms of being closer to natural conditions, but disadvantages in terms of ease of handling and screening, as well as the usually high degree of unknown chemical compounds in the system. Pot cultures are soil cultures in greenhouse pots, usually including a live vascular plant in the role of host organism.

Unlike biological and chemical components, the physical components and properties of a culture system are usually well-known or easy to find out. It should not be assumed that they are less relevant to cultivation as such, however. For nematode species that are difficult to culture, optimizing the temperature and physical properties of the substrate can be the key to success. This is particularly true for temperature and for presence and thickness of liquid layers on the substrate (see below and 12.4.2.2).

12.2.2 Three Basic Principles of Successful Nematode Cultivation

In order to successfully culture any organism, one must try to understand the basic aspects of its biology. Although some nematodes are notoriously easy to culture, others are much more difficult and will remain intractable until one or more unsuspected requirements or limiting factors are revealed. The more one knows about the natural conditions in which the target species occurs, the easier it is to imitate these conditions in the laboratory. When nothing is known about a certain nematode, attempts to culture it often provide the first clues about its ecology and behavior. In such cases it is essential to try out different culture methods and see which ones work best, or work at all. As can be expected, it is often easier to develop good culture conditions for an opportunistic species than for a specialized one, and the range of tolerance for diverse conditions can vary significantly, often even within a single genus.

An obvious first requirement for cultivation is to know what the target species feeds on in nature. For example, one of the main difficulties with cultivation of smaller Dorylaimida is the general lack of knowledge on their food sources. Chemical conditions are also important for many nematodes. Thus, although few quantitative studies exist (Khanna, et al., 1997; Lin, et al., 1997; Nicholas, 1984), it is safe to assume that different species have different ranges and optima for pH tolerance and osmotic pressure. In our own experience, nematodes with a more permeable cuticle or more indiscriminate feeding activity often appear to be affected more adversely by sudden changes in pH and concentrations of other ions. But even straightforward physical properties can be crucial. For example, many soil nematodes are easy to maintain in cultures at room temperature and quite a few have a temperature optimum above 25℃ (Vancoppenolle, et al., 1999), but for some reason *Rhabditophanes* sp. KR3021 cannot survive at temperatures above 20℃ (W. K. Thomas, pers. comm.). More generally, slender nematodes with thin cuticles are often unable to move efficiently in agar or minimal water films, and therefore require a basically liquid environment, while short and chubby nematodes are often trapped by the slightest accumulation of liquid, and require a nearly dry surface for efficient movement. Adjusting the culture system accordingly can make all the difference between success and failure.

A second principle of effective culture is to minimize the risk of losing established cultures. Just like demography, the maintenance of cultures depends not only on the rate of acquiring new ones but also on the rate of losing established ones. Accidents happen even in the best of circumstances, and in many cases cultures are kept in less-than-optimal places and conditions, where they run a substantial risk of contamination or destruction due to temperature fluctuations.

In order to lower the odds of culture loss, always spread the risk by dividing up each cultured strain over several subcultures. One way of doing this is to physically separate subcultures into series stored in different places, in different culture systems or even in different laboratories. Also, it is best to separate subcultures temporally by staggering the timing of fresh inoculations: unless an emergency prevents it, always inoculate a new set of subcultures from the oldest-running problem-free series. Thus, if you are culturing strains that require new inoculations every 2 months, then you should keep at least two series of subcultures and transfer these to fresh culture systems on alternate months. That way, if during transfer you happen to introduce adverse contaminants, or previously neutral contaminants are stimulated in ways that render them adverse to the target organism, then you will still have plenty of time to notice the problem and rescue the culture using inocula from the alternate series.

Finally, a third principle that is perhaps the most important in the long run, is that you should keep track of all documentation relating to your cultures. If you started a culture from your own samples and extractions, take note of the relevant collection data, your identification, and any remaining uncertainties about it. If you receive a culture from colleagues, always ask for all pertinent data. If you distribute a culture then you should also distribute its data of origin; otherwise it may return to you years later under a different species name and with an erroneous provenance!

In general, the most important facts to know are: locality and date of collection; the name of the original collector; who identified it and to what level of certainty; and the strain code (s) assigned to it. For free-living nematodes, a system of strain codes is now common practice, whereby each strain is given a code beginning with a few letters (two letters for *C. elegans* strains and three for other species) followed by a number. The letter code should be approved by collaborators of the *Caenorhabditis* Genetics Center (University of Minnesota, St. Paul; http://biosci.umn.edu/CGC/CGChomepage.htm) to avoid duplications, and refers to the scientist or laboratory where a given strain was first brought into culture, or where it first received a CGC-compliant strain code. If one strain turns out to have received multiple codes, then the CGC-compliant strain code that was applied first should be retained for all subsequent use.

12.2.3 Brief Outline of a Typical Cultivation Procedure

To start a culture, it is obviously necessary to collect and extract live specimens of the target species. For many opportunistic nematodes, it is sufficient to transfer 10 – 20 live animals from a soil/sediment extract into a standardized culture system, and then to inoculate the resulting populations regularly to fresh sets of the same culture system. However, as your culture collection expands you will be faced with increasing limitations of time and resources available, and you will inevitably have to deal with more specialized and/or less easily cultured species.

In that case, a more structured approach is appropriate (Fig. 12.1). After the initial extraction, it may be necessary to set up a transitional system that is closer to the target's original environment than the full-blown culture system(s) and method(s) that you intend to apply subsequently. Such a transitional system is often required to provide a few initial rounds of multiplication, simultaneously increasing the number of specimens available for the first inocula to subsequent cultures, and also providing the animals themselves with a chance of accommodating more gradually to the artificial conditions of your subsequent culture systems. We will refer to such a transitional system as a seed culture, its primary purpose being that of amplifying your chances of subsequent success at establishing truly continuous cultures. A seed culture is sometimes started by baiting rather than extracting, i.e., an attractant such as food or prey is introduced into the soil or sediment, in order to induce the nematodes to invade the bait and commence reproduction inside it. The bait can then be removed from the natural substrate and transferred to a container with artificial medium and/or substrate.

Once a seed culture has been established (Fig. 12.2), you may need to experiment with a range of culture methods in order to identify an approach that will hopefully allow you to culture the target species indefinitely. In order to minimize both the risk of losing the culture and the amounts of time involved, it is often advisable to maintain the target species in limiting conditions, so as to restrain growth and reproduction compared to optimal conditions. This reduces the risk of sudden population crashes and the necessity of frequent

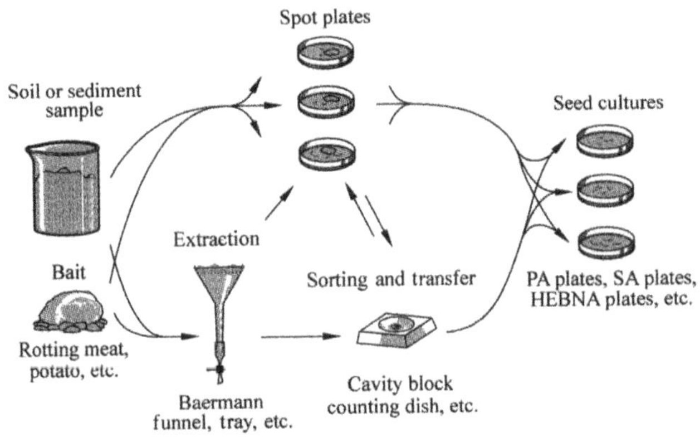

Figure 12.1 Basic steps of culture startup (see Section 12.2.3 in text), illustrated here for the specific case of bacterial-feeding nematodes. These can be obtained from a soil or sediment sample, or from various kinds of baits, either by spotting out the material directly onto plates, or via an extraction procedure suitable for recovery of live nematodes. In most cases, mixtures of species will be obtained, and these are transferred to an appropriate dish for sorting under the dissecting microscope. Of each recognizably different species, a number of healthy nematodes is then transferred to seed cultures, which, if successful, will provide the inocula for further culturing.

transfers to fresh cultures, while it provides you with more time for intervention in case of unforeseen problems. This type of culture is aimed primarily at maintenance of the target species, and will henceforth be called a stock culture.

For many research applications it is necessary to generate larger quantities of live nematodes than can easily be obtained from samples, seed cultures, or even stock cultures. In such cases, the simplest solution is to set up a climax culture system, in which the conditions for optimal growth and reproduction are met. Such cultures will here be referred to as harvest cultures, since they allow for the harvesting of maximal numbers of live specimens for subsequent research purposes.

Finally, it is often advisable to ensure that all nematodes in a culture are progeny from one parent animal, so as to exclude the possibility that the culture contains a mixture of morphologically similar species. Such cultures are typically started by inoculating single gravid females on separate fresh culture systems. Depending on the context, these single female cultures are best started: (1) from seed cultures if the nematodes in question prove to be relatively easy to obtain and maintain; (2) from stock cultures if the nematodes

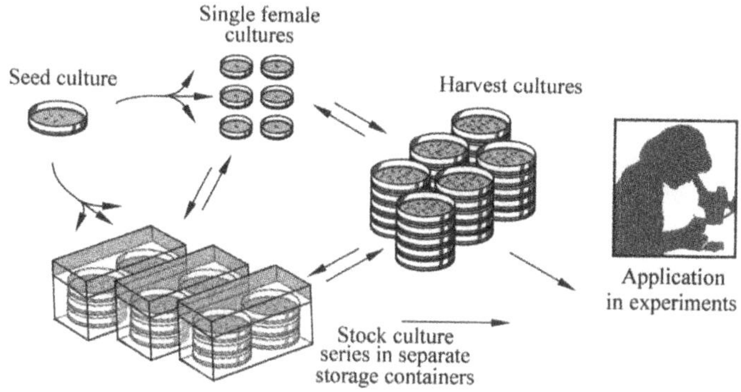

Figure 12.2 Specimen flow through the different intentionality types of cultures (see Section 12.2.3 in text), illustrated here for the specific case of Petri dish culture systems. First, one or more good seed cultures are chosen to provide inocula, which should not contain any species mixtures. If purity of the inocula is uncertain, single gravid females should be transferred to separate Petri dishes, and resulting progeny should be inspected for conspecificity (morphologically and/or in crossing experiments). If mixtures occur, then the single female progeny can be used as pure inocula of the different cryptic species involved. For the purpose of maintaining strains, keep them in limiting conditions (lower temperatures and limited nutrients) as stock cultures, preferably in two or three series stored separately and subcultured alternately over time (see text). Stock cultures may suffice for applications requiring few nematodes, but you will need to set up harvest cultures for experiments requiring greater numbers, by providing optimal conditions for growth and reproduction (see Fig. 12.3).

are harder to obtain and more sensitive, with individual females likely to suffer a high rate of attrition during transfer to separate systems; or (3) from harvest cultures if the research depends on particular selection steps such as e.g. mutant screening or virulence comparisons.

12.3 Standard Methods for *Caenorhabditis elegans* and Related Bacterial Feeders

Caenorhabditis elegans is by far the most widely cultured nematode species, being maintained in systems ranging from tiny Petri dishes to huge bioreactors (Gbewonyo, et al., 1994). It is a facultatively hermaphroditic bacterial feeder capable of very fast multiplication, specializing in a lifestyle of explosive short-term population growth alternating with survival of strongly limiting conditions and dispersal across less favorable microhabitats. Partly for these reasons, it has been chosen as a preferred model organism for nutritional, genetic, developmental, and neurological research (Brenner, 1974; Dougherty, 1949), and a wide range of methods were developed for its handling and culture in laboratory conditions. Many of these methods have become *de facto* standards for *C. elegans* research (Brenner, 1974; Lewis and Fleming, 1995; Stiernagle, 1999; Sulston and Hodgkin, 1988; Vanfleteren, pers. comm.), and most are also applicable with little or no modifications to a wide range of other bacterial-feeding nematodes, such as most other species of Rhabditidae, diplogasterids, panagrolaimids, and cephalobids that are capable of rapid reproduction and equipped with a relatively impermeable cuticle (Vancoppenolle, et al., 1999). Before elaborating on a more general range of culture systems and methods, we will first summarize basic *C. elegans* methods.

12.3.1 Collecting Live Material

Despite nearly 50 years of continuous culture of *C. elegans*, the ecology of this species has still not been investigated in any great detail (Sudhaus and Kiontke, 1996). Its behavior in culture clearly suggests that it is a specialist of micro-environments rich in bacteria, and that it primarily disperses to other such environments in the form of a non-feeding dauer juvenile. Sudhaus and Kiontke (1996) suggest a possible association with gastropods. *Caenorhabditis elegans* has been found in temperate regions around the world, and presumably only occurs densely in very localized microhabitats in natural soil and/or decomposing organic matter. One should not *a priori* expect success in isolating it with the various classical soil extraction techniques for nematodes, but it is possible: Abdul Kader and Côté (1996) isolated 93 populations in Quebec state using Baermann funnel extraction followed by hybridization tests with reference strains. Alternatively, it is easily baited from soils and compost using material such as a rotting potato or decomposing meat. And since it is so widely cultured by now, the easiest way of obtaining it is by simple request from a nearby institute or research facility, or from the *Caenorhabditis* Genetics Center, which keeps over 3500 strains of *C. elegans* in culture or frozen in liquid nitrogen, deriving both from natural sources and from mutational studies.

12.3.2 Basic Monoxenic Culture Methods

Caenorhabditis elegans and many of its relatives can be cultured under a wide range of conditions, partly because of the relatively impermeable nature of their cuticle. Thus, the methods given below can be modified in many ways and still allow successful culture of *C. elegans*, while conversely they can easily be applied with little or no modifications to many other bacterial-feeding species. Two basic restrictions apply to *C. elegans* wild-type strains: they only reproduce at temperatures of 10℃ – 26℃ (Abdul Kader and Côté, 1996), and they require sterols as an essential nutrient (as do the few other nematode species tested; Vanfleteren, 1980). Sterol can be administered in trace amounts from a stock solution of 5 mg · ml^{-1} cholesterol in ethanol. Protocols for monoxenizing or axenizing *C. elegans* are given below in 12.3.4.2.

12.3.2.1 Simple Handling and Transferring Methods

Caenorhabditis elegans is relatively sturdy with a resistant cuticle, and can therefore be transferred quite easily with various tools suitable for picking up individual nematodes. If the purpose is to establish and handle gnotobiotic cultures, then it is advisable to use picks that can be sterilized by flaming, such as a metal handle mounted with a steel 000 gauge insect pin with curved tip or with a platinum wire with flattened tip. The latter is a little more difficult to use when picking out small juveniles and/or single specimens from a dense culture, but the platinum wire lasts longer and its flattened tip is less likely to rupture the surface of the agar, thus reducing the likelihood that nematodes will burrow into the agar. Also, success at picking up nematodes with a platinum wire can be improved by dipping its tip into an *Escherichia coli* colony first, thus making it sticky.

While applying various auxiliary techniques (see 12.3.4), it is often necessary to transfer *C. elegans* to or through a liquid medium, e.g., for rinsing and/or centrifugation. In such cases, the nematodes do best in a relatively saline medium, presumably because they are adapted to micro-environments with relatively high concentrations of electrolytic compounds. The following two solutions are commonly used.

(1) M9 buffer

KH_2PO_4	3 g
Na_2HPO_4	6 g
NaCl	5 g
$MgSO_4$ 1M	1 ml

H_2O dist. add up to total volume of 1000 ml

(2) S buffer

NaCl	5.8 g
KH_2PO_4/K_2HPO_4 1M buffer at pH 6	50 ml
cholesterol stock solution	1 ml

H_2O dist. add up to total volume of 1000 ml

Both solutions can be autoclaved for sterile work.

12.3.2.2 Stock Cultures

In order to maintain stock cultures of *C. elegans*, it is usually better to cultivate the nematode in a nutrient-rich system with growth-restricted *E. coli* (see 12.3.4.1(1)) and to keep it at lower temperatures (15℃ – 18℃), rather than to restrict nutrients in the medium from the outset. In the absence of sufficient food, *C. elegans* will either die or transform into dauer stages, suspending all further development until fresh nutrients become available. Dauer juveniles greatly facilitate medium-term culture and storage because they can survive for extended periods (i.e., months), but in order to obtain a net increase of nematode numbers they should only be allowed to appear after several reproductive cycles.

(1) The most common culture system for *C. elegans* uses NGM (Nematode Growing Medium) agar (Brenner, 1974):

NaCl	3 g
agar	17 g
peptone	2.5 g
H_2O dist.	975 ml

After autoclaving for at least 20 minutes, the following solutions are added and mixed under sterile conditions:

cholesterol stock solution	1 ml
$CaCl_2$ 1M	1 ml
$MgSO_4$ 1M	1 ml
KH_2PO_4/K_2HPO_4 1M buffer at pH 6	25 ml

The NGM agar is then poured out into 5 cm or 9 cm Petri plates (use respectively about 4 ml or 12 ml of agar per plate) and allowed to set. The resulting gel will contain 1.7% (weight) polymerized agar, which is usually dense enough to prevent the nematodes from breaking through the surface and burrowing into the agar. If not, the agar can still be overlaid with a thin coating of 2% top agarose. Keep the plates at room temperature for 1 – 3 days to allow surface water to evaporate, and check for fungal or bacterial contaminants. Streak the plates with *E. coli* from liquid culture (see 12.3.4.1 (2)), or to avoid surface damage pipette the bacteria in liquid medium onto the plates, at about 25 μl per 5 cm plate or 100 μl per 9 cm plate. Allow the fresh plates to stand overnight and then transfer nematodes with a platinum wire or by transferring an agar block.

At 15℃ – 18℃, NGM plates can easily be kept for 6 to 8 weeks between transfers. After roughly 3 weeks, the nematodes will be crowded and will have exhausted the available food, leading to transition into dauer stage of all remaining hatchlings. Dauers will survive for extended periods at an inverted exponential rate, provided the culture system does not dehydrate and does not allow them to escape. Stock culture plates should therefore be sealed tightly with Parafilm M® (American National Can Co.) or a comparable sealing film. A good Parafilm M seal is produced by wrapping twice around the plate, which takes a film strip of about 4 inches or 10 cm long for a 5 cm plate, and 6 inches or 15 cm for a 9 cm plate.

Vancoppenolle et al. (1999) cultured a broad selection of species on NGM plates and recorded generation times at three different temperatures. Nearly all tested strains reproduced between 18℃ and 25℃, but 7 out of 12 tested species of Rhabditidae did not reproduce at 30℃, and of the five remaining species only one reproduced faster at 30℃ than at 25℃. In contrast, the single cephalobid and both panagrolaimid species tested all reproduced faster at 30℃, and there is anecdotal evidence to suggest that cephalobids and panagrolaimids generally reproduce fastest at temperatures of 30℃ or higher (Baujard, pers. comm.; Borgonie, pers. comm.).

(2) Another popular agar formulation is Nigon's medium, with the following composition (Nigon, 1949):

$MgSO_4 \cdot 7H_2O$	0.75 g
K_2HPO_4	0.75 g
NaCl	2.75 g
KNO_3	3.00 g
lecithin (dissolved in 25 ml ethanol)	1.00 g
bacteriological peptone	2.5 g
agar	15 g

H_2O dist. add up to total volume of 1000 ml

Autoclave for at least 20 minutes, adjust the pH if a value other than pH = 7.2 is desired, and pour plates.

(3) A simple but slightly more expensive alternative to NGM plates is provided by NA plates (Nutrient Agar), the main difference being that nutrient agar is available as a ready-made mix of agar and nutrients. The exact contents vary according to the supplier, but usually include beef extract and peptone. Some formulations include NaCl, but avoid these if the target organism does not support moderate to high salinity. NA plates are prepared by mixing and autoclaving the following:

Nutrient Agar	20 g
cholesterol stock solution	1 ml

H_2O dist. add up to total volume of 1000 ml

The pH of the autoclaved mixture varies between 6.5 and 7.5 depending on the supplier (adjust if necessary). NA plates are poured and handled exactly like

NGM plates (see 12.3.2.2(2) above).

12.3.2.3 Harvest Cultures

NGM or NA plates kept at 20℃ – 24℃ and seeded with wild-type *E. coli* will yield large numbers of nematodes in less than 2 weeks, when used in an exponential series of transfers (Fig. 12.3). Start with one plate containing large numbers of gravid females, and use the synchronization protocol described below (12.3.4.2) to produce a cohort of 100 – 1000 freshly hatched first-stage juveniles. Pipette these onto a new NGM or NA plate with *E. coli* and keep this plate 2 – 3 days at 20℃ – 24℃. As soon as the next generation

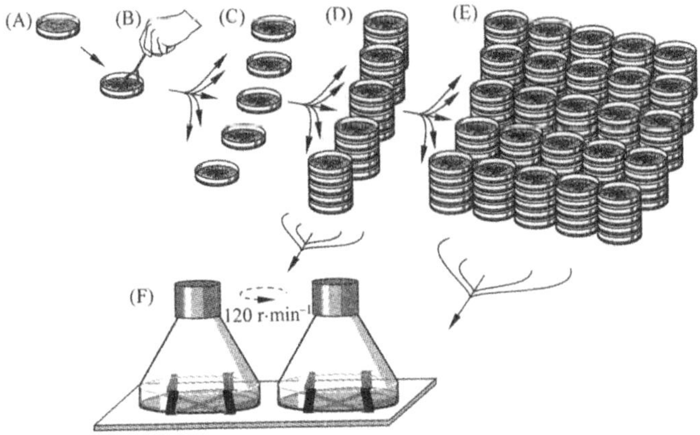

Figure 12.3 Harvest culture of *Caenorhabditis elegans* in an exponential series of Petri plates (see Section 12.3.2.3 in text). (A) Spread out a batch of 100 – 1000 synchronized juveniles on an NA or NGM plate with *E. coli*, and allow them to grow and mature for 3 days at 21℃ – 23℃. (B) Rinse the plates with 5 ml S buffer, using a Drigalski spatula to detach nematodes and eggs from the agar surface, and divide the resulting suspension equally over five new plates. (C) Leave the five plates for 3 days, then transfer the next generation of juveniles and eggs in the same way, resulting in 25 new plates. (D) Leave the 25 plates for another 3 days, and then transfer them all to one Fernbach bottle with 250 ml liquid medium (for medium yield) or divide them up once more over a total of 125 new plates (for maximum yield). (E) In the latter case, leave the 125 plates another 3 days, and then transfer all nematodes and eggs to four Fernbach bottles with 250 ml liquid medium each. (F) To allow most juveniles to reach maturity, leave the nematodes for 2 days in liquid culture in the baffled Fernbach bottles on a rotary shaker at 120 r· min^{-1}. Harvest the resulting nematodes on a sieve and wash them with sucrose (see text).

hatches, rinse the original plate with 5 ml S buffer (see 12.3.2.1(2)) by emptying a full 5 ml pipette of S buffer onto the plate and scraping its surface gently with a Drigalski spatula (glass rod or Pasteur pipette bent twice to produce a triangular end) to detach eggs and nematodes from the agar surface. Suck up into the 5 ml pipette all S buffer with eggs and nematodes, and distribute them over five new plates by releasing 1 ml onto each plate.

After another 2 – 3 days, repeat this procedure again with all five plates, dividing the harvest from each plate over five new plates, thus producing a batch of 25 subcultures. When harvested, this batch should yield in the order of 0.1 – 1 g of *C. elegans*. If more is required, wait another 2 – 3 days and then repeat the procedure of transfer to five plates again with all 25 plates, producing a batch of 125 culture plates. For maximum yield of biomass, you can finally transfer the nematodes from all 25 or 125 plates to 1.8 l Fernbach flasks with 250 ml S buffer, 0.25 ml cholesterol stock solution, and 3g thawed *E. coli* pellets (see 12.3.4.1(3)), and leave them capped but not sealed on a gently rotating shaker at 120 r· min^{-1} for another 1 – 2 days. This will allow all juveniles to reach maturity, but egg-laying will diminish in *C. elegans* (though not in some other nematodes) due to the liquid environment. Depending on the further use of the nematodes, it may be necessary to clean and concentrate them with a sucrose wash (see 12.3.4.5)

See Lewis and Fleming (1995) for methods for growing several hundred grams to several kilograms of *C. elegans*, using 20 l Carboy containers or 150 l fermentors.

12.3.3 Axenic Culture

For some biochemical and physiological studies, it is useful or necessary to culture *C. elegans* under more strictly controlled conditions than those provided by NGM or NA Petri plate systems. Axenic and/or holidic culture systems have been developed for such purposes, and allow efficient culture, but also require a greater investment of time, resources, and accuracy. In axenic conditions, *C. elegans* exhibits longer maturation times (Vanfleteren, 1980) and longer lifespans (Vanfleteren and De Vreese, 1995), and these methods are therefore better avoided if the scientific context does not require absolute control over the biological and chemical composition of the culture system. Protocols for monoxenizing or axenizing *C. elegans* are given below in 12.3.4.2.

12.3.3.1 Axenic Oligidic Culture

Vanfleteren (1980) provided an extensive review of various axenic media suitable for continuous culture of *C. elegans* and various other nematodes. The following paragraphs provide an updated protocol for the preparation of a complete but chemically poorly defined medium (Vanfleteren, pers. comm.).

(1) Prepare stock solution of 50 mg · ml^{-1} bovine hemoglobin in 0.1M KOH and autoclave 10 minutes. This will result in partially cleaved hemoglobin that dissolves more easily in the medium without loss of availability of the heme groups for the nematodes. Prepare basal medium consisting of 3% dry yeast extract and 3% soy peptone. Do not use yeast extract with high salt contents or peptone from other sources. Autoclave the basal medium for 20 minutes, let it cool down and then add hemoglobin supplement to a final concentration of 0.5 mg · ml^{-1}.

Transfer complete medium to a sterile container with flat bottom and tightly closing cap, making sure that the liquid is no thicker than 3 mm (without use of a shaker) or 5 mm (when using a shaker). Inoculate with axenized nematodes (see 12.3.4.2(1)) and store at an appropriate temperature (15℃ for stocks, 22℃ for harvest).

12.3.3.2 Axenic Holidic Culture

Various completely characterized media have been developed and tested for continuous culture of *C. elegans* (Lu and Goetsch, 1993; Vanfleteren, 1980). Early attempts at holidic media started with Dougherty et al. (1959) but failed without inclusion of complex extracts such as chicken embryo extract or liver protein fractions. Over the years, three essential or strongly limiting growth factors were found to be missing from the previous media: heme, sterol and a carbohydrate energy source. On this basis, Lu and Goetsch (1993) formulated a truly holidic medium that contained nine salts, 19 amino acids, five nucleotide precursors, 18 vitamins and vitamin-like growth factors, one sterol source, one heme source, and one carbohydrate. They also determined that five different carbohydrates elicited significantly different dose responses, D-glucose promoting growth most efficiently in terms of weight per volume. We refer to their paper for the detailed composition of the holidic medium.

12.3.4 Auxiliary Techniques

Apart from the cultivation itself, maintenance of *C. elegans* cultures usually also requires various other operations such as culturing of prey bacteria, removal of contaminants, permanent storage at ultralow temperatures, etc. Some of these are very straightforward, while others require considerable investment in additional equipment and consumables.

12.3.4.1 Culturing *Escherichia coli* Strains

One of the great conveniences of *C. elegans* is that it can be cultured continuously on *E. coli*, a prokaryote that is an even more popular model organism than *C. elegans* itself. A host of culture and handling protocols exist for *E. coli*, and numerous strains with interesting properties are known (Miller, 1992; Neihardt, 1996; WWW resources at http://web.bham.ac.uk/bcm4ght6/res.html). Many of these strains are freely available from microbiological institutes, and most can be cultured extremely easily in any biology laboratory.

(1) Prokaryote contaminants in culture systems often adversely affect the culture's target organism simply because they easily outgrow the target organism and quickly deplete vital resources such as oxygen and nutrients. The same is true of prokaryote prey organisms, and even a rapidly reproducing nematode like *C. elegans* can be overwhelmed by *E. coli*. For use in *C. elegans* culture, it is therefore advisable to use strains with metabolic and/or reproductive impairments, such as the leaky uracil-requiring strain OP50 (Brenner, 1974). Because of its restricted growth, plates with nematodes feeding on OP50 are often also cleaner and easier to screen than plates with wild-type *E. coli*.

Escherichia coli OP50 is cultured by itself in conditions that allow their optimal reproduction, while growing less prolifically on NGM plates. Although this limited growth of OP50 in the *C. elegans* culture system is beneficial to survival of stock cultures and facilitates screening of plates, it does have a downside in that it renders these cultures more vulnerable to take-over by contaminant prokaryotes. Even when it is not an absolute requirement that the *C. elegans* cultures remain gnotobiotic, it remains good practice to ensure maximum sterility of all tools and containers used for culture of *E. coli* and its inoculation on nematode culture plates.

(2) Petri plates can be used for stock cultures of *E. coli* by adding enriched agar media such as NA (see 12.3.2.2(3)) or LB (Luria-Bertani) agar. The latter is made up as follows:

Agar	15 g
Tryptone	10 g
Yeast extract	5 g
NaCl	10 g

H_2O dist. add up to total volume of 1000 ml

Add approximately 1 ml 1M NaOH to adjust the pH to 7.0, and autoclave for 20 minutes. After autoclaving, pouring and setting of the agar, *E. coli* from old plates or from liquid culture is streaked onto the agar, sealed and inoculated for several days at room temperature or overnight in a 37 ℃ incubator. The plates can then be stored for months at 4℃. For dense *E. coli* OP50 colonies, add 0.2 ml of an aqueous solution of 2 mg · ml^{-1} uracil before autoclaving.

(3) Liquid cultures of *E. coli* can be used both as stock and as harvest cultures for inoculation into a nematode culture system. Many liquid media allow continuous culture of *E. coli*, including for example LB medium:

Tryptone	10 g
Yeast extract	5 g
NaCl	10 g

H_2O dist. add up to total volume of 1000 ml

Add approximately 1 ml 1M NaOH to adjust the pH to 7.0, fill clean Erlenmeyer flasks with 1.5 cm or half an inch of this medium, cap the flasks and autoclave them. For dense *E. coli* OP50 cultures, add 0.2 ml uracil $2mg \cdot ml^{-1}$ aqueous solution before autoclaving. After cooling, inoculate each flask with *E. coli* from a clean stock plate or flask, and incubate overnight at room temperature or in a shaker at 37℃. Store the flasks at 4 ℃; enough bacteria should survive to provide inocula for about 3 months. For long-term storage, centrifuge liquid culture at 8000 r min^{-1} (~6000 g) for 10 min and remove supernatant, wash with S buffer, centrifuge and remove supernatant again, and finally add equal volume of S buffer and drip the suspension directly into a small amount of liquid nitrogen. The drops will freeze instantaneously into pellets and can then be transferred to cryovials for storage in an ultrafreezer or in liquid nitrogen.

(4) To minimize growth of bacterial contaminants, use minimal medium for *E. coli* OP50 by mixing the following sterile solutions (Lewis and Fleming 1995). To grow wild-type strains of *E. coli* in this manner, omit uracil:

M9 buffer (see 12.3.2.1(1))	10 ml
NH_4Cl 2M	0.1 ml
D-glucose 20% w/v	0.1 ml
uracil 2 $mg \cdot ml^{-1}$	0.02 ml

12.3.4.2 Synchronizing and Axenizing *C. elegans* Cultures

Because *C. elegans* has a high fecundity and significantly longer lifespan than generation time, the first two generations of progeny from single hermaphrodites will significantly overlap in time, and in less than a week it will be quite impossible to link the age of the resulting population to a particular generation. Also, *C. elegans* routinely digests only a fraction of the bacteria it ingests, and therefore cannot be surface-sterilized to remove all co-occurring prokaryotes.

A single method bypasses both potential problems, by briefly rinsing the nematodes in a caustic solution that kills all organisms present in the suspension, except for any eggs contained within gravid hermaphrodites. This simultaneously removes all contaminants and yields a cohort of newly hatched nematodes of nearly identical age. Such cohorts can be used both for transfer to axenic or synxenic culture systems, and for developmental, ecological, or biochemical studies requiring batches of nematodes of known age. This method does not always work well with relatives of *C. elegans*; in some species the female/hermaphrodite body cuticle does not break up in the caustic solution, and as a result the hatching juveniles often do not succeed in escaping from the maternal cuticle. In this case, the inclusion of sharp fragments (e. g., small pieces of razor blades) and briefly vortexing the suspension may improve fragmentation (Ehlers, pers. comm.).

(1) In order to obtain a small number of synchronized and/or axenized nematodes, first prepare the following caustic solution:

H_2O	5.5 ml
NaOCl 16%	4 ml (commercial bleach can be used if it is free of additives)
NaOH 10M	0.5 ml

Pipette 25 μl of the solution into a sterile dish and then transfer 10 – 20 gravid hermaphrodites into this drop using a platinum pick. Leave for 7 min, then rinse the surviving eggs with 5 ml sterile M9 or S buffer. Let the eggs settle and then draw off most of the buffer with a sterile pipette. If your aim is to start an axenic culture, then add 5 ml of the appropriate axenic medium at this point, seal the container and let the juveniles hatch and multiply over the following week(s). If instead your purpose is to obtain a synchronous cohort or a monoxenic culture, then add another 1 – 2 ml of sterile M9 or S buffer after the first rinse and removal of buffer. Leave the eggs for 6 hours at 17℃ – 20℃ in order to allow the first-stage juveniles to emerge without developing further. When all viable eggs have hatched, transfer the juveniles to fresh NGM plates with *E. coli* or to another container, as appropriate for your purpose.

(2) For subsequent applications requiring a larger inoculum of synchronized and/or axenized nematodes, take a Petri plate with numerous gravid hermaphrodites but no dauers (the cuticle of these will not break up, and may allow intestinal bacteria to survive the procedure). Rinse off the nematodes with a few ml of M9 or S buffer. Transfer the nematodes to a clean 15 ml centrifuge tube with screw cap and clear walls. Rinse the nematodes with 5 ml additional M9 or S buffer, spin them down for 3 min in a centrifuge at 3000 $r \cdot min^{-1}$ (~1600 g), and remove supernatant. Add 5 ml of freshly made caustic solution (see 12.3.4.2(1)), screw the cap tightly and gently shake the tube for 7 minutes, monitoring the decay of the nematodes to make sure eggs are released from gravid adults, but not yet dissolved. Spin down the eggs and debris for 3 min at 3000 $r \cdot min^{-1}$ and remove the supernatant. Immediately add 10 ml of M9 or S buffer, shake the tube to rinse the eggs,

spin another 3 min at 3000 r · min^{-1}, and remove supernatant. Repeat this rinse once or twice. Finally, add about 5 ml M9 or S buffer, and pipette the remaining eggs into a sterile Petri dish for regular observation.

In the next 6 hours, any surviving eggs will hatch, and the emerging first-stage juveniles will not grow significantly because of the absence of food. After 6 hours all viable juveniles have hatched, and this entire cohort of hatchlings can finally be transferred to fresh axenic medium or to NGM plates with *E. coli* (depending on the requirement of axenic or monoxenic culture), allowing them to proceed with postembryonic development and mature to a roughly synchronous batch of reproducing adults.

12.3.4.3 Removing Eukaryote Contaminants

Apart from prokaryotes, the other adverse contaminants most likely to occur in cultures of *C. elegans* are fungi and mites.

(1) Many mycelial fungi can be eliminated by transferring a block of agar from a contaminated plate to one end of a new 2% agar plate with at its other end an *E. coli* patch as bait for the nematodes. The nematodes will quickly move towards the *E. coli*, while the fungus will only grow minimally on the pure agar. After 6 – 12 hours, excise an agar block containing the *E. coli* patch with nematodes and repeat the baiting procedure. Finally, excise agar with *E. coli* and nematodes from the second baiting plate and transfer it to a new NGM or NA plate with *E. coli*.

(2) Some fungi, and especially yeast, are more difficult to remove from the system because they produce dispersal structures that can be carried phoretically on or inside the nematodes. In that case prepare NGM or NA plates with a fungostatic added to the agar after autoclaving (e. g., nystatin at a final concentration of 700 U· l^{-1}) and transfer nematodes to these plates, but do not use such plates for routine culture as they affect the behavior of the nematodes (Lewis and Fleming, 1995).

(3) Mites are omnipresent in most laboratories and easily invade Petri plates, even when these seem to be sealed tightly with parafilm. Depending on the species, they may feed on the agar, *E. coli*, fungal spores sticking to the sealant, and/or the nematodes inside a plate. A simple way to keep mites out of the cultures is to keep a beaker of paradichlorobenzene (moth crystals) in each incubator or plastic box that houses cultures. If one or more plates have been invaded by mites and it is essential that you rescue nematodes from these plates, then transfer agar blocks from these plates to new NGM or NA plates with *E. coli*, and store the new plates unsealed inside a closed plastic box, together with a small open vial of moth crystals. Leave the box for at least a week and then check whether the mites have disappeared. If so, seal the plates

and/or transfer nematodes to fresh plates for normal further storage.

12.3.4.4 Checking for Contaminants in Axenic Cultures

Liquid axenic medium should be clear, and nematodes usually live at least as long in unlimited axenic systems as in monoxenic ones. Cloudy medium, sudden death of the nematodes, or rapid decay of dead nematodes are all symptoms of contamination by prokaryotes. Cultures should also be screened under the dissection microscope to check for contaminant fungi and yeast. However, contaminants may even be present without obvious symptoms, e.g., because the initial stages of their growth are repressed by nematode feeding. Vanfleteren (1980) therefore recommended sterility checks as standard practice at the end of any experiment relying on axenic cultures.

(1) A simple verification of axenic cultures consists of extending their maintenance for 3 – 4 weeks longer after the finish of the desired experiments or other applications of these cultures. At the end of that period, they should be inspected under a dissection microscope for turbidity, eukaryotic contaminants, and nematode survival and appearance.

(2) A more thorough sterility screen requires that medium from the axenic cultures is transferred to different sterile microbiological media, and kept at two different temperatures (e.g., 20℃ and 37℃) for at least 2 weeks (Vanfleteren, 1980). NA plates and brain-heart infusion support growth of the most common prokaryotic contaminants (Dougherty, et al., 1959; Rothstein, 1974). Potato-dextrose agar plates can be used to detect fungal contaminants (see 12.7.1.1).

12.3.4.5 Washing and Concentrating Harvested Nematodes

Often, large amounts of nematodes collected from harvest cultures will be mixed with various kinds of detritus composed of, e.g., excess medium, substrate fragments, precipitated chemicals, bacterial mats, contaminants, etc. For many applications requiring a substantial amount of nematode biomass, removal of such detritus will greatly enhance the quality of subsequent results, and it is therefore advisable to wash and concentrate the harvested nematodes before proceeding with the application itself.

(1) A sucrose wash will efficiently remove most impurities from a batch of harvested nematodes (Vanfleteren, pers. comm.). However, it will also cause substantial physical and osmotic stress to the nematodes, and is not advisable for more sensitive species, as it may kill most or all individuals (in such a case, try $MgSO_4$ instead of sucrose solution). Even with the toughest species it is still essential to work quickly, in order to minimize damage to the harvested worms.

Transfer the harvested nematodes into a 15 ml centrifuge tube, e. g., if you start with a culture on one Petri plate, wash the worms off the plate with 3 ml of the appropriate buffer and pipette them into a 15 ml tube. Rinse with 5 ml buffer and cool the tube on ice for a few minutes. Spin the tube for 4 min at 3000 r min^{-1} and 4℃ in a cooled centrifuge. Remove supernatant and add fresh buffer to a total volume of 4 ml. Add 8 ml 60% sucrose solution and spin exactly 3 min at 3000 r min^{-1} and 4℃ in a cooled centrifuge. Pipette or aspirate the floating froth (which consists of cleaned worms) over to a new 15 ml tube. Rinse the walls of the old tube with buffer from a Pasteur pipette, to wash off any remaining worms sticking to the walls of that tube, and transfer the rinsing liquid to the new tube without mixing the buffer with the remaining sucrose in the tube. Add buffer to the new tube to a volume of 15 ml, cool on ice for a few minutes, and spin for 3 min at 3000 r · min^{-1} and 4℃ in a cooled centrifuge. Remove the supernatant and you have clean nematodes.

12.3.4.6 Long-Term Storage of Viable *C. Elegans* Strains

Inocula of *C. elegans* strains can be stored indefinitely in liquid nitrogen, provided they have been frozen rapidly under conditions that minimize the formation of intracellular ice crystals. The most widely used freezing protocol relies on the use of glycerin solution as an antifreeze. This solution most efficiently penetrates the thinner cuticle of early juvenile stages, and survival is therefore usually limited to first- and second-stage juveniles. Also, the glycerin itself is toxic at higher concentrations, and thawed inocula must be transferred quickly to conditions where the glycerin concentration drops significantly. Cephalobidae cannot be preserved in this manner, apparently because they do not absorb enough glycerin through their cuticle or mouth opening (see 12.4.3.2 for an alternative).

(1) Take one 9-cm or three 5-cm plates with numerous young juveniles, and pipette the nematodes to a 15 ml tube using 2.5 ml of M9 or S buffer. With a new 5 ml pipette, add 0.5 ml 80% glycerin solution, suck up and release the mixture twice using the same pipette, and then immediately divide it up over five cryotubes into 0.5 ml aliquots. Swiftly place the five cryotubes into a styrofoam box with tightly fitting tube compartments, or an isopropanol-containing cryobox (e. g., a Nalgene® Cryo 1℃ Freezing Container) at room temperature, and place the box inside a −80℃ ultrafreezer. The speed of this procedure is critical for subsequent revival because at room temperature the glycerin solution can kill the nematodes within 15 min after penetrating their tissues. The styrofoam box or cryobox should allow for a cooling rate of approximately 1℃ · min^{-1}, which will allow the nematode juveniles to stop their metabolic activity and become cryoprotected just in time to avoid death by

either ice crystallization or glycerin toxicity.

Leave the cooling box with cryovials overnight in the ultralow freezer. The next day, take out one cryotube for evaluation of revival. Thaw the cryotube in your hand, and as soon as its contents become partly molten drop the aliquot on a 9 cm 3-day-old 2% agar plate without *E. coli*. Leave the plate for 15 min. While the aliquot continues to melt, the glycerin solution will diffuse into the agar and thus cease to be toxic to the nematodes. After 15 – 30 min, the first nematodes should revive, and within an hour at least 1% (and preferably 20% – 50%) of the frozen nematodes should revive. Most of the *E. coli* frozen along with the nematodes will also revive and grow, using the remaining glycerin as a nutrient.

If the nematodes do not revive in reasonable numbers, then the viable freezing of the other four cryotubes has presumably also failed, and you should try the procedure again using larger numbers of nematodes and shorter handling times. On the other hand, if the control aliquot has a high rate of survival then the procedure was successful and the remaining four cryotubes can be transferred to liquid nitrogen for truly permanent storage. If a liquid nitrogen vat is not available to you, then you can just keep the cryotubes in the –80℃ ultrafreezer, but be aware of the certainty that each freezer must fail sooner or later and that you will need to store inocula in different freezers to spread the risk. Either way, the equipment needed for frozen storage is much more expensive than the basic infrastructure required for *C. elegans* culture.

12.4 Methods for Culturing a Wider Range of Bacterial-feeding Nematodes

Caenorhabditis elegans methods do not work well for many nematodes with more divergent and/or specialized requirements, which naturally warrant more appropriate culture systems. In this section, we provide an overview of various methods that can be tailored to suit the needs of less rapidly reproducing and/or less opportunistic target species. Some of these methods are extremely simple, while others require more laborious preparations and may involve a fair amount of tweaking.

12.4.1 Collecting Live Material

12.4.1.1 Handling Samples

Healthy nematodes are an essential requirement for cultivation attempts, and samples therefore need to be handled with due care prior to extraction. They obviously must not be fixed and should be protected from excessively high or low temperatures, as well as from desiccation, excessive humidity, or toxic trace elements leaching in from sample containers or extraction equipment. Depending on the origin, humidity, and nematode contents, it may be possible to store a sample for months at 4℃ and still recover significant numbers of active nematodes. Samples of naturally desiccated soil from rhizospheres in xeric habitats often contain surprisingly large numbers of anabiotic nematodes and are particularly easy to transport and store, but must be rehydrated for 8 – 16 hours prior to extraction in order to allow the nematodes to revive. Samples with high moisture contents on the other hand are best processed quickly after collection, in order to minimize asphyxiation and fungal growth.

12.4.1.2 Extraction

Nematodes can be extracted from soils and sediments with a wide range of devices and techniques (Bloemers and Hodda, 1995; Cagnolo, et al., 1999; Carta and Carta, 2000; Hooper, 1986a; Hunt and De Ley, 1996; Traunspurger, 2000; Yeates and Bongers, 1999). For some nematodes one must be aware of the particulars of their physical characteristics and modify standard procedures accordingly. Thus, Bunonematidae appear to be partly hydrostatic and will float on the surface of extracts, rather than sink to the bottom like most other nematodes. In this case one must actually be careful to sieve any excess extract liquid rather than discarding it as usual! Conversely, many nematodes with caudal glands and spinneret will adhere to the sides and bottom of the extraction equipment, and require vigorous rinsing and scrubbing in order to maximize their numbers in the final extract. Also, the most active species will rapidly exhaust their oxygen supply, and processing should be sped up (or oxygenation improved) according to the expected or observed activity of the target species. Baiting techniques are only appropriate for species capable of rapid invasion and multiplication, since any other attracted nematodes will be outcompeted exponentially by the progeny of the specialized colonizers that will inevitably turn up.

The larger the number of live nematodes recovered, the greater the likelihood that a continuous culture can be obtained, and it is therefore advisable to process large (sub)samples whenever possible, and also to use extraction equipment capable of quickly recovering numerous healthy nematodes. Elutriators and mist chambers (Hooper, 1986a) are well-suited to this task, while Baermann funnels require more time, are likely to lead to asphyxiation of very energetic species, and significantly constrain the sample volume that can be processed. Before using a funnel, it is therefore better to first concentrate nematodes from a large sample with the decanting and sieving technique (Cobb, 1918; Hooper, 1986a) and then to load the retained suspension on one or more funnels. Extraction trays are less likely to lead to asphyxiation and allow for a larger loading area than funnels, but the resulting extract usually requires further concentration, and nematodes with caudal glands are provided with larger surfaces for attachment. Centrifugation-flotation is undoubtedly the most strenuous method for live nematodes; it is only advisable if $MgSO_4$, Cecasol® 40C (Sobrep) or Ludox® AS (Du Pont) are available as separation liquids, since these are less toxic and less osmotically active than sucrose or Ludox® HS40 (Du Pont).

For marine nematodes, an unusual collection strategy was developed by Meyers and Hopper (1966), relying on the use of fungal mats as bait, not just for fungal-feeders but also for many other nematodes browsing on organisms trapped by or associated with the fungal hyphae (see 12.5.1.1(1)). These mats can then be used for spot plates (see 12.4.2.2(2)).

12.4.1.3 Sorting and Recognizing Nematodes in an Extract

After extraction, screen the resulting suspension of nematodes under a dissection microscope capable of magnifications up to 75 ×, and transfer recognizably different nematodes to separate seed cultures. The most common nematodes in any startup culture are ubiquitous opportunists like *Oscheius*, *Pristionchus*, *Panagrolaimus*, and *Acrobeloides* species. If these are not target species, nor prey for your target species, then they will behave like adverse contaminants, competing with the target species and most likely displacing it, unless sufficient specimens of the target are picked out in time and transferred to a clean culture.

It is possible to distinguish many common groups of nematodes at low power. Larger nematodes like dorylaims, mononchs, and enoplids can easily be recognized from the overall size and shape of body, stoma, pharynx, and tail. For smaller nematodes the behavior and appearance of living individuals often provide good clues to their identity. Thus, cephalobs, tylenchs, and some diplogasterids tend to be very slow moving with generally granular contents throughout all or most of the body. Healthy panagrolaimids, rhabditids, and other diplogasterids have a more clearly defined, dark intestine compared to the transparent neck and body cavity, and move actively and purposefully in very regular and sometimes rapid undulations. Monhysterids tend to oscillate frenetically for short periods and then rest for several minutes, either swimming freely in this manner, or forming a standing wave with the tail anchored to a solid surface. Plectids, teratocephalids, prismatolaimids, and chromadorids often move rapidly but jerkily with many reversals of direction, while their tail easily adheres to the substrate or entangles in detritus.

With a little practice, most of these groups are easy to distinguish even at final magnifications in the 50 – 100 × ranges. See 12.4.2.1(3) for appropriate tools for sorting and transferring a diversity of nematodes, including forms that are prone to dehydrate very quickly or easily sustain physical injury.

12.4.2 Basic Xenic or Synxenic Cultures

Many slower-growing nematodes can be maintained in agnotobiotic systems to which limited amounts of *E. coli* are occasionally added. After an initial phase of trial and error, it is often possible to culture such species with a limited repertoire of very simple methods. Nematodes from brackish or marine habitats usually require more complex modifications of these basic methods.

12.4.2.1 Handling

For the purposes of transferring or rinsing nematodes, the following buffers should be prepared in addition to M9 or S buffer (see 12.3.2.1(1) and 12.3.2.1(2)):

(1) P buffer: this is a 0.05M phosphate buffer devoid of other salts, particularly suited for handling freshwater nematodes with very permeable cuticle and hence sensitive to pH- and/or osmotic shocks. Measure the pH of the extract water and prepare P buffer accordingly (Table 12.1).

Table 12.1 Approximate amounts of KH_2PO_4 and K_2HPO_4 needed to prepare phosphate buffer solutions formulated to different pH (modified from Hale, 1965). For 1000 ml P buffer take 950 ml distilled water, add the volumes below, check pH, and adjust further if necessary.

Add this amount of KH_2PO_4 1M	and this amount of K_2HPO_4 1M	in order to obtain a final pH of
48.75 ml	1.25 ml	5.3
47.5 ml	2.5 ml	5.6
45 ml	5.0 ml	5.9
40 ml	10 ml	6.2
30 ml	20 ml	6.6
20 ml	30 ml	7.0
10 ml	40 ml	7.4
5.0 ml	45 ml	7.7
2.5 ml	47.5 ml	8.0

(2) Artificial seawater (ASW): ASW should be used with marine nematodes tolerating a salinity of approximately 35%. For nematodes from brackish habitats, dilute with distilled water to achieve the appropriate salinity (after Dietrich and Kalle in Moens and Vincx, 1998).

Solution A	
NaCl	239.0 g
$MgCl_2 \cdot 6H_2O$	108.3 g
$CaCl_2$ anhydrous	11.5 g
$SrCl_2 \cdot 6H_2O$	40 mg
KCl	6.82 g
KBr	0.99 mg

H_2O dist. add up to total volume of 8560 ml

Solution B	
$Na_2SO_4 \cdot 10H_2O$	90.6 g
$NaHCO_3$	200 mg
NaF	3 mg
H_3BO_3	27 mg

H_2O dist. add up to total volume of 1000 ml

Add solution B in a thin jet to solution A under constant stirring. Let the mixture stand for a day and pass it through a fine membrane filter (0.45 μm or less). Adjust pH with Tris-HCl to a final concentration of 5 mM.

(3) Unlike *C. elegans* and relatives, aquatic nematodes often desiccate in a matter of seconds when removed from a liquid environment. In such cases, nematodes may be killed or damaged by routine picking with a needle or eyelash, and it is faster and safer to use a modified Pasteur pipette (Hooper, 1986b) or micropipette tip (Fig. 12.4) in order to transfer each individual nematode with 2 – 5 μl of liquid. The latter tool requires less practice and dexterity than traditional picks, allowing very fast transfer of individual nematodes without exposing them to the air or to direct contact with the hard and/or sharp end of a pick.

12.4.2.2 Simple Methods for Varying Moisture and Surface Tension Effects in Petri Plates

The origin of the sample from which live nematodes are obtained usually provides basic clues as to the appropriate moisture level, pH, salinity, and temperature range. Depending on these parameters, try one or more of the following basic methods below as startup systems, and then expand your repertoire to determine the system that provides the best results in your laboratory. More details on appropriate choices for seed or stock cultures are given in Section 12.4.2.3.

(1) The simplest agar-based culture substrate for soil nematodes consists of plain agar with cholesterol (see 12.3.2) added to a final concentration of 5 $\mu l \cdot ml^{-1}$. Such PA plates should be made up to an agar thickness of 0.5% – 2% depending on the target nematode. Nematodes from moist soils or

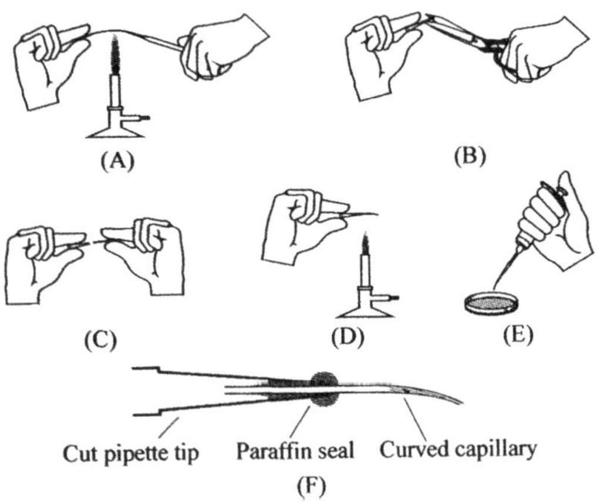

Figure 12. 4 Construction of glass-plastic compound micropipette tips (see Section 12. 4. 2. 1 in text). (A) Pulling steadily at both ends, heat the thin half of a long and fine Pasteur pipette over a flame. Smoothly detach the pipette's tip while stretching out the glass into a curved, hair-thin fiber. Break off the hair-thin part from the detached glass tip and inspect it at low power to check shape and size of the resulting opening. (B) Cut a plastic micropipette tip to obtain an opening slightly wider than the glass tip. (C) Slide the straight end of the glass tip into the cut plastic tip. (D) Seal the insertion point with a piece of paraffin molten above a flame. (E) Check the suction and expulsion action of the compound tip (if faulty, recover the glass tip and start again at B). For maximum sterility, autoclave all relevant components first, wear gloves during assembly, and immerse the assembled tip in ethanol between uses. (F) Diagram of assembled compound tip with a nematode in 2 – 3 μl of water. For best results, the curved capillary end should have a downward-slanting opening, and the straight end of the glass tip should be long enough to prevent spills into the plastic tip.

freshwater sediments may not survive long on the surface of the agar, and will do much better if they can dig into a freshly set soft agar (0. 5% – 1%). Smaller nematodes from xeric microhabitats are easily trapped by any remaining surface liquid, either from freshly set plates or from drops of condensation flowing down from the lid of an older sealed plate. Such nematodes will do much better on denser agar (1% – 2%) plates that have been left to stand unsealed for 3 days to allow evaporation of all surface moisture. Note also that thicker agar left to stand for some time will quickly absorb small amounts of liquid subsequently added to its surface, including liquid from bacterial culture or water deposited on the surface during transfer of nematodes with pipette or micropipette.

In order to prepare PA plates, autoclave agar and cholesterol at the desired strength(s), pour plates (4 ml PA for a 5 cm plate, 12 ml PA for a 9 cm plate) and swirl them to obtain an even agar surface. If you are not sure whether your target species has a short generation time and/or large food intake, then it is usually safer to err on the side of frugality rather than excess: add 1 drop (20 μl) of *E. coli* liquid culture to a 5 cm PA plate, or three drops (60 μl) to a 9 cm PA plate. If necessary, allow the plates to stand for 1 – 3 days to reduce or eliminate surface moisture. Transfer nematodes onto the plates, near the drop of bacterial culture but not inside it. Observe the transferred nematodes over the next hours and days to check whether they converge on the *E. coli* and can be seen to feed, mate, and lay eggs. If not, then you will need to modify the culture system more or less drastically, and look for clues as to other possible prey organisms or food types.

(2) A simple way to learn more about feeding behavior and food requirements of a target species, is to set up spot plates as seed cultures (Moens and Vincx, 1998). These are Petri plates with an appropriate agar (e.g., PA) in the center of which a depression is made. This depression is either filled with a small amount of soil or sediment as taken directly from a sample (Moens and Vincx, 1998), or with a concentrated extract including a mixture of nematodes and other micro-organisms (Esser, 1964), or else with pieces of fungal mat used to bait nematodes (Meyers and Hopper, 1966). Nematodes in the substrate or concentrate will emerge and start exploring the agar, while nematode eggs deposited in the fungal mat will hatch and release active juveniles. Separately sorted specimens of a possible target species can also be transferred onto the plate near the central well. If this well contains food items for these target nematodes, then they may converge on it and start feeding, thus revealing their preferred food source(s). You may also wish to add 1 – 3 drops of *E. coli* liquid culture to the spot plate, but no more since the culture liquid will stimulate rampant growth of contaminants introduced along with the sample substrate.

(3) Some freshwater nematodes are apparently truly aquatic, or at least they can be cultured more efficiently in PA plates with a liquid layer of P buffer on top than in PA plates with more or less dry surface. Such wet PA plates are basically prepared like ordinary PA plates, but after setting of the agar 1 ml of P buffer is added for a 5 cm plate, and 3 – 4 ml P buffer for a 9 cm plate. The plates are inoculated with 1 – 5 drops of *E. coli* liquid culture, nematodes are introduced, and the plates are sealed and kept at 15℃ – 18℃.

Wet PA plates offer the advantage that they are better protected against accidental desiccation and pH fluctuations. Using a pipette or micropipette, it is also extremely easy to inoculate series of other plates quickly from a wet PA plate, but on the other hand contaminants will invariably be carried over and cultures on such plates tend to suffer more during transportation by mail. Because of the excess liquid on the surface, asphyxiation must be avoided by adding only minimal amounts of *E. coli* and by keeping the plates at temperatures below 20℃.

(4) A few nematode species are extremely difficult to culture on agar, apparently because at the time of transferring to new plates a critical balance must be maintained between excess and lack of surface moisture. In such cases, semi-wet PA plates can provide an answer: these are wet PA plates of which the P buffer is poured off just before inoculation with nematodes. The remaining surface film is apparently of just the right thickness to prevent the nematodes from either being pinned down by surface tension or asphyxiating in excess liquid. A nematode that seems to prefer such conditions, in our experience, is *Myolaimus heterurus*.

It is also possible to further fine-tune the thickness of the surface film. Wilsonematids like *Tylocephalus auriculatus* present a special case, in that they are highly active and require a moist substrate that is nevertheless free of any excess liquid. The best results for transfer of these nematodes are obtained by preparing semi-wet plates, transferring a block of agar with nematodes from an old plate to the new one (active individuals retreat too fast from the needle's touch to allow picking), and leaving the new plate on a running flowbench for 1 – 2 hours with partly open lid while visually inspecting regularly, until the nematodes move efficiently over the fresh agar without being trapped in remaining liquid between the old agar block and the new substrate.

(5) Addition of soil extract to a mixture of agar and known chemicals produces an oligidic or a meridic medium that includes traces of a wide range of elements and compounds, and may therefore yield successes in cases where holidic media appear to lack some unknown essential substance for the target organism. Soil extracts are prepared to avoid inclusion of large soil particles in the agar, typically by boiling an amount of soil or sediment, decanting the liquid and passing it through paper and millipore filters (Arpin, 1976; Lee, et al., 1970; Maertens, 1975; Moens and Vincx, 1998). Such extracts can be stored for weeks in a refrigerator, or for months in a freezer. After mixing and autoclaving with plain agar (a cholesterol supplement is usually not required), the resulting SXA plates will allow faster growth of contaminants than PA plates and should therefore be sealed with parafilm when they are stored for some time before being used.

However, an even simpler approach is possible and has been applied by us with good results to a range of species. Addition of 1% (mass) of soil to an agar mixture before autoclaving will produce soil agar rather than soil extract agar. With a little care, freshly autoclaved soil agar can be poured almost as cleanly and easily as any other agar medium, provided at least the soil in question is fairly sandy and does not contain significant amounts of plant debris, so that during pouring all soil particles will remain at the bottom of the autoclaved bottle. Unlike soil extract agar, soil agar does not require a separate preparatory procedure, but it must be autoclaved for longer (at least 25 min) to make sure that all cysts and fungal spores in the soil are truly dead. The resulting SA plates are otherwise exchangeable with SXA plates, and should also be sealed for storage.

(6) For nematodes from brackish or marine sediments, seed cultures on agar should always use soft agar (0.5% –0.7%) to allow burrowing. The agar can be molten in one of the following two media (modified from Moens and Vincx, 1998) prepared using ASW (see 12.4.2.1(2)) diluted to the appropriate salinity for nematodes from brackish habitats.

Holidic Erdschreiber medium:

$NaNO_3$	100.0 mg
NH_4Cl	0.5 mg
K_2HPO_4	10.0 mg
$Na_2SiO_3 \cdot 9H_2O$	10.0 mg
thiamine	1.0 μg
cobalamine	1.0 μg
ASW	add up to total volume of 1000 ml

Holidic Killian medium: Add 2 ml of solution A and 1 ml of solution B to 997 ml ASW.

Solution A	
$NaNO_3$	200 mg
KNO_3	200 mg
NH_4NO_3	100 mg
H_2O dist.	add up to total volume of 100 ml

Solution B	
NaH_2PO_4	400 mg
$CaCl_2$	400 mg
$FeCl_3$	200 mg
conc. HCl	0.2 ml
H_2O dist.	add up to total volume of 80 ml

Both media were originally formulated to include soil extract① but Moens and

① For 1 liter Killian medium include 20 ml soil extract in the final mixture, for 1 liter Erdschreiber medium include 10 ml soil extract.

Vincx (1998) report success with media from which this was omitted. Autoclave 0.5% – 0.7% agar in simplified Erdschreiber or Killian medium with 5 $\mu g \cdot ml^{-1}$ final concentration cholesterol. Pour in Petri plates and inoculate with nematodes and/or substrate spots. The resulting HEA plates (holidic Erdschreiber agar) or HKA plates (holidic Killian agar) will support moderate growth of bacteria and good growth of various microalgae, allowing culture of both bacterial- and algal-feeding nematodes.

(7) For all systems described above (12.4.2.2(1) – 12.4.2.2(6)), higher production of both bacteria and nematodes can usually be obtained by mixing a small amount of nutrient agar with pure agar. The proportion of the pure bacto-agar to the nutrient agar can be varied between e.g., 10:1 and 4:1 to determine the best compromise between growth of prey, contaminant, and target organisms. Our own observations and those of Moens and Vincx (1998) show that the resulting BNA plates or relevant variations (including e.g., holidic Erdschreiber medium, holidic Killian medium, soil, or soil extract) work well for a range of marine and soil nematodes.

12.4.2.3 Agar-based Seed, Stock, and Harvest Cultures

For marine bacterial-feeding nematodes, culture systems should probably always be based on media with ASW, and seed cultures are best started using spot plates with HEA and HKA. For stock and harvest cultures, successfully reproducing nematodes can then be transferred to pure BNA plates (strength 0.6% – 1%, proportion of the two agar types 4: 1 for the latter and 10: 1 in the former two systems), or to BNA plates prepared with either holidic Erdschreiber medium or holidic Killian medium. Such plates support good growth of several marine or estuarine Rhabditida and Monhysterida (Moens and Vincx, 1998). Success rates with other groups are low, however, and replication of previously successful systems appears to be erratic (e.g., see comments by Moens and Vincx, 1998 on the Vlasblom medium of Vranken, et al., 1984).

Dry SA plates, SA spot plates, and wet SA plates (all at 1% strength) provide a sufficient range of seed culture systems to obtain initial success with culture of many bacterial-feeding soil or freshwater nematodes. During trials with one or more of these culture systems, the target species will hopefully adapt and reproduce sufficiently to allow experimentation with a wider range of systems, and eventual selection of the preferred type of stock or harvest culture system for that particular species. Ideally, the stock culture system will require a minimal investment of time and effort, yet produce sufficient numbers of nematodes to indefinitely allow transfer to new sets of the same system.

In our own experience, soil and freshwater nematodes incapable of forming dauers will not survive for very long on agar plates that include supplements (BNA, NA, Nigon's medium, and NGM), as inactive nematodes will die, shrivel and blacken within a matter of days, and eventually decompose completely. Such types of plates are therefore unsuitable for stock cultures of sensitive and/or slow-growing nematodes, although in some cases they can still be used effectively for harvest cultures, usually of species that do not require wet plates. By comparison, PA plates and SA plates are much better suited for stock cultures of such nematodes, because they often allow the last active generation to become quiescent without dying, and to persist for an extended time without visible deterioration of the body structures. In many cases, old PA or SA plates at 15℃ – 18℃ can be stored for several more months after cessation of all nematode activity, and still produce good inocula for fresh plates. Even when few intact nematodes are left, adding a single drop of *E. coli* liquid culture to such a quiescent plate may suffice to reactivate the remaining viable nematodes within a few hours, inducing them to converge on the fresh bacteria, and easily allowing their subsequent transfer as an inoculum for a fresh plate.

12.4.2.4 Some Other Systems

(1) The vinegar eelworm *Turbatrix aceti* can be cultured in 50% vinegar and only requires subculturing every 3 – 6 months, which can simply be done by pipetting 25 – 50 ml of the old vinegar into a new flask (Hooper, 1986c). This nematode is therefore very easy to maintain with minimal effort in most laboratory settings.

(2) Asparagine-mannitol agar was recommended by Overgaard Nielsen (1949) for bacterial-feeders such as Plectidae, which are not adapted to very saprobic environments. A detailed description of the method can be found in Hooper (1986c).

12.4.3 Auxiliary Techniques

Applicability to slower-growing nematodes of the auxiliary techniques for *C. elegans* (see 12.3.4) depends greatly on the reproduction rate, generation time, and cuticular properties of these nematodes.

12.4.3.1 Removing Contaminants

The hypochlorite axenization protocol for *C. elegans* (12.3.4.1) can also be used efficiently for some slower-growing bacterial-feeders, provided the eggs inside gravid females are not too permeable or few in number. Nematodes that can be cultured on dry plates are fairly easy to separate from fungal contaminants, but those that require wet plates are more difficult to rescue.

(1) One of the few workable methods for physically separating aquatic nematodes from some of their contaminants is to put pieces of agar with nematodes (and contaminants) on top of three layers of tissue paper in a dish with P buffer. After 6 to 12 hours, collect the nematodes that have succeeded to crawl through the paper, rinse them in P buffer and transfer them to another dish with tissue paper in P buffer. Repeat the procedure, rinse the nematodes collected from the bottom of this second dish, and finally inoculate them on fresh plates.

(2) Contamination of microalgae in cultures of freshwater or marine nematodes can be reduced or removed by keeping the cultures in the dark.

(3) Bacterial contaminants can sometimes be removed by rinsing sensitive nematodes or nematode eggs in solutions of antibiotics like streptomycin sulfate (final concentration 0.1% –0.5%), penicillin (final conc. 0.1%), or novobiocin (final conc. 0.1%). More aggressive chemicals (see 12.6.4.3) are only appropriate if the nematodes are sufficiently resistant, and in that case it is worth trying the *C. elegans* protocol first (12.3.4.1) since that will also kill intraintestinal contaminants.

12.4.3.2 Medium-term and Long-term Storage of Viable Strains of Bacterial-feeding Nematodes

The glycerin protocol for freezing *C. elegans* (12.3.4.6) can be adapted to work with many other nematodes, but it does not work well with species that absorb the glycerin too quickly or too slowly. In the former case (e.g., Plectidae), they will die almost immediately upon mixing with the glycerin solution, and in the latter case (e.g., Cephalobidae), the cryoprotectant does not enter the body in sufficient amounts and hence cannot prevent damage due to ice crystal formation. At least three different approaches can be tried to circumvent such problems: extended preincubation in dilute glycerin, different antifreeze solutions, or gradual desiccation. All three work reasonably to very well with various strains of entomopathogenic nematodes (see 12.6.4.4), and they are undoubtedly easily adapted for non-pathogenic bacterial feeders.

(1) In certain groups of nematodes (e.g., Cephalobidae) most species are naturally capable of surviving slow desiccation, and this allows for a simple, albeit erratic, procedure for cryoprotection. PA plates with such nematodes can be allowed to desiccate gradually, e.g., by making a few perforations in the seal of the plate and leaving it at 20℃ –25℃ for 2 –3 weeks. If desiccation is sufficiently gradual, the nematodes will assemble on the most slowly desiccating part of the gel, where they become anhydrobiotic by bunching together and curling up in characteristic shriveled little spirals. Slivers of the desiccated agar with desiccated nematodes can then be transferred to cryotubes

and permanently stored frozen in an ultrafreezer at −70℃ to −80℃. For unknown reasons, subsequent transfer to liquid nitrogen does not appear to work, but dry nematodes kept at −80℃ can be stored and rehydrated to yield good inocula for fresh cultures.

(2) Precise methods for desiccating various nematodes under strictly controlled conditions (see also 12.6.4.4(1)) were developed by Simons (1973) and adapted by Demeure et al. (1979). These methods rely on transferring live nematodes to a millipore filter, which is then placed in a relative humidity (RH) chamber where the RH is gradually reduced from 100% to 97.7%. After 2 to 4 weeks in such a chamber, about 20% −30% of the nematodes on the millipore filter are capable of revival. In our own experience, on gradually dried agar plates recovery of the nematodes sometimes exceeds 50%, but the crude desiccation procedure makes success highly erratic and it is therefore likely that the RH chamber represents a better tool for drying nematodes in preparation for freezing.

(3) Desiccated nematodes not only allow long-term storage in an ultrafreezer, but also represent a good medium-term storage option at room temperature. If stored well-sealed and at constant temperature, they can safely be kept for 4 to 8 months, and then revived through rehydration.

(4) Trehalose plays a role in cryo-and anhydroprotection in many organisms, and has also been detected in various nematodes (see review by Womersley, et al., 1998). Its precise function in nematodes is not entirely clear, however: on the one hand, research on entomopathogenic nematodes has shown that they increase trehalose synthesis at low temperatures (Qiu and Bedding, 1999), but on the other hand, dauers with higher trehalose contents are incapable of entering true cryptobiosis (Womersley, 1990). The picture is further complicated by evidence that different nematodes employ different strategies of entering cryptobiosis (Womersley, et al., 1998) and thus probably rely on different biochemical processes. From anecdotal evidence (see comments below on *Prionchulus*), we suspect that in at least some cases slow desiccation and cooling do stimulate similar protective biochemical pathways improving subsequent recovery from both desiccation and freezing. For bacterial feeders with low or zero recovery from simpler procedures, it may therefore be possible to develop more efficient cryopreservation protocols by including an initial incubation step at low temperature, perhaps in combination with extended exposure to dilute glycerin and/or gradual desiccation prior to freezing.

12.5 Methods for Culturing Omnivores and Predators

In very general terms, diplogasterid and seinurid predators/omnivores appear to be easier to culture than their trophic counterparts among enoplids, dorylaims, and mononchs. Many of the former can be cultured using modifications of standard systems for *C. elegans* and relatives, often in the absence of eukaryote prey, and we will not discuss them further here. Adenophorean predators, however, tend to have much longer generation times, often lack obvious adaptations for surviving harsh conditions, and seemingly depend strongly on the availability of prey nematodes. An obligate predator will by definition not thrive in the absence of prey, but neither will it do well if the prey multiplies exponentially and saturates the culture medium with its waste products. Arpin (1976) emphasized that various kinds of enriched agar media were quite unsuitable for culture of two Mononchida, apparently because the slightest increase in osmotic pressure interfered with their survival.

Nevertheless, the ecological sensitivity of some of these predators may have been significantly overestimated. Many species probably feed on bacteria as well (Arpin and Kilbertus, 1981), and some can actually be kept in continuous culture in the absence of eukaryote prey (Arpin, 1976). Also, circumstantial evidence suggests that certain species may be capable of surviving desiccation, and even freezing. Thus, desert and dune sands often contain a remarkably high proportion of dorylaims (De Ley, 1992), some of which appear to be dispersed by wind (Baujard and Martiny, 1994; Krnjaic and Krnjaic, 1970). In desiccation experiments with soil *in situ*, Arpin (1969) repeatedly found persistence of dorylaims in soil with less than 20% water contents. Furthermore, mosses are subject to frequent desiccation but can nevertheless support high densities of dorylaims or mononchs. On several occasions, we have obtained large numbers of *Prionchulus muscorum* directly from rehydrated dry moss, and we found that most of these specimens survived immediate freezing to −80℃ in extract suspension without further additives (Vierstraete and De Ley, unpubl. data). Perhaps these mononchs had elevated anhydroprotectant levels as a result of dehydration, and were thereby coincidentally also protected against freezing (see 12.4.3.2(4) and 12.6.4.4(2)).

12.5.1 Collecting Live Material

The collection, handling, and extraction methods mentioned above (12.4.1) for live bacterial-feeding nematodes will also efficiently yield predatory and omnivorous nematodes, although baiting techniques are less rewarding if one is looking for chemically sensitive species. One exception applies to marine predators, however, and is explained here more fully.

12.5.1.1 Extraction Through Baiting

(1) Meyers and Hopper (1966) cultured the marine fungi *Dendryphiella arenaria* and *Halosphaeria mediosetigera* in flasks containing 0.1% bacto-yeast extract, sterile natural sea water, and 4.2 cm discs of cellulose or glass paper filter. After 8 days on a reciprocating shaker at 25℃, the fungal mats were harvested and washed, and transferred to wooden stakes with one or more plastic attachments (Fig. 12.5) capable of containing the mats between vertical layers of 52 × 52 mesh Saran (Dow Chemical Company, Midland, Michigan) screening. These stakes were then driven into the sediment until the lowermost fungal mat was level with the sediment surface. After various baiting times, the attachments were removed and the mats transferred back to flasks with sterile sea water for incubation. A range of nematodes was isolated in this way, including the predator *Metoncholaimus* sp.

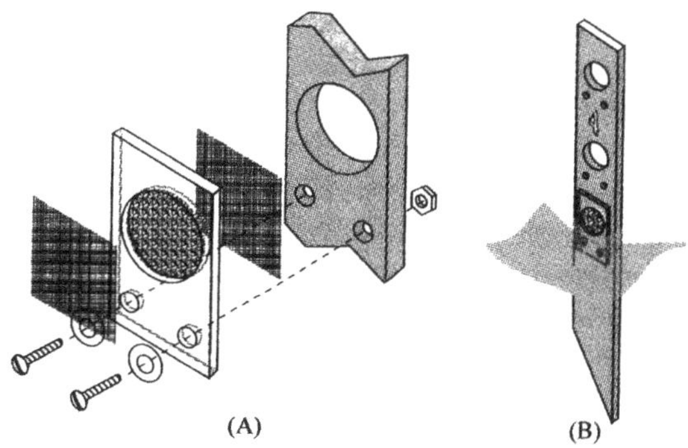

Figure 12.5 Fungal mat mounts used by Meyers and Hopper (1966) to bait estuarine and marine nematodes (see Section 12.5.1.1 in text). (A) Fungal mat on a cellulose or glass fiber filter is mounted between two layers of 52 × 52 mesh Saran screening, and one or more mounts are screwed to a wooden stake with matching openings. (B) The stake is driven into sediment until the bottom mount is level with the surface, and then left for at least 1 day before removal and transfer to the lab.

12.5.2 Culture Systems Adapted for Predators or Omnivores

Major differences occur in the rate at which a given predatory species will prey on other nematode species (Bilgrami, 1993; Khan, et al., 1995; Moens, et al., 2000). Prey nematodes in later stages of development and/or with thicker cuticle are attacked less frequently and survive a greater proportion of these attacks. For maximum effect, it is therefore advisable to try out various small and slow-moving bacterial-feeding nematodes as prey. Choose a small prey species, or selectively harvest first-stage juvenile prey by, e. g., sieving, centrifuging or synchronizing.

12.5.2.1 Gel-based Culturc Systems

(1) Nelmes (1974) and Maertens (1975) cultured *Prionchulus punctatus* (Mononchida) on a modified enchytraeid culture medium. Maertens (1975) first prepared humus extract by boiling 200 g forest humus for 90 min in 2000 ml distilled water and then filtering it. The extract was kept at −18℃ as frozen stock, and it was used to prepare SXA plates containing 0.6% −1.4% agar and 10% humus extract. He used various bacterial-feeding nematode species as prey organisms: some that burrow rapidly into the agar (*Caenorhabditis briggsae* and *Eucephalobus striatus*) were inoculated to the plates 4 days before introduction of *P. punctatus*, while others that remain more on the agar surface (*Panagrellus redivivus* and *Turbatrix aceti*) were only added as weekly food supply after inoculation with *P. punctatus*.

(2) Arpin (1976) developed a method for culturing Mononchida on agar containing soil extract, without any intentionally added prey organism. He obtained soil extract by boiling 200 g of soil in 500 ml water for 30 min, filtering the suspension through cloth, and then diluting the filtrate with water to a volume of 500 ml. The soil extract was stored at 15℃, and then a few drops were added to 3 day old 0.5% agar plates (without cholesterol). Another 3 or 4 days later, mononchs were transferred to the new plates, which were then stored in the dark at 18℃. In this way, Arpin (1976) recorded generation times of about 5 months for *P. muscorum* and *Clarkus papillatus*. He emphasized that additions of sterols, RNA, or amino acids to the agar invariably interfered with the survival of the nematodes. He also observed that small soil particles were ingested preferentially by the nematodes, and that juveniles died within 2 weeks when such particles were absent from the culture system. It is not entirely clear from his paper or from later comments (Arpin and Kilbertus, 1981) whether this culture system was free from live bacteria or not.

(3) A very straightforward culture system for predatory dorylaims and

diplogasterids consists of 1% PA plates with 5 – 10 mg of milk powder spread on the surface as nutrient for bacterial growth, and inoculated with *Rhabditis* sp. as nematode prey (Bilgrami and Jairajpuri, 1988; Khan, et al., 1995).

(4) Hopper and Meyers (1966) extracted an assemblage of marine nematodes (*Acanthonchus cobbi* and *Monhystera* spp.) from fungal mats incubated *in vitro* for 28 days (see 12.5.1.1(1)), and then cultured these for another 48 days on 0.5% SWA with added cells of the yeast *Kluyveromyces aestuarii*. They did not determine the exact food source of *A. cobbi* (Cyatholaimidae), but the observations of Moens and Vincx (1997) on various Chromadoridae with similar buccal morphology suggest that they could be in their terminology "epigrowth feeders", i.e., preying on unicellular eukaryotes as well as on bacteria.

(5) Moens and Vincx (1998) devised an interesting technique for maintaining the slowly reproducing facultative predator *Oncholaimus oxyuris* for several generations, without frequently having to transfer this marine/estuarine nematode to fresh plates. They placed concentric partitions, made by removing the bottom from a 5 cm and a 9 cm Petri plate (Fig. 12.6), in a 14 cm Petri dish and filled the resulting compartments evenly with Killian medium BNA (see 12.4.2.2(6)). The central compartment was first inoculated with *Diplolaimelloides meyli* and associated unknown bacteria, and after 8 days 20 *O. oxyuris* were added (together with some relatively benign protozoan contaminants).

When the agar in this compartment began to liquefy as a result of bacterial and nematode activity, the 5 cm partition was removed, the *O. oxyuris* were moved to the surrounding compartment, and the central patch with deteriorated agar was removed to avoid excessive growth by the contaminants. This was repeated subsequently with the 9 cm ring, and in this manner three generations of *O. oxyuris* were hatched and raised in the laboratory. Adults of the last generation were transferred to a new plate with concentric partitions, but failed to reproduce further.

Variations of such partitioned plate cultures could easily be applied to other nematode species (not only marine but also freshwater or terrestrial) that are difficult or impossible to maintain in continuous culture, due to their long generation time and/or apparent sensitivity to the drastic changes resulting from transfer from old to new cultures. Other shapes and sizes of containers could be used and might allow longer series of colonization steps of adjacent compartments.

12.5.2.2 Soil-based Culture Systems

(1) Even if more artificial culture systems fail, it is still possible to cultivate

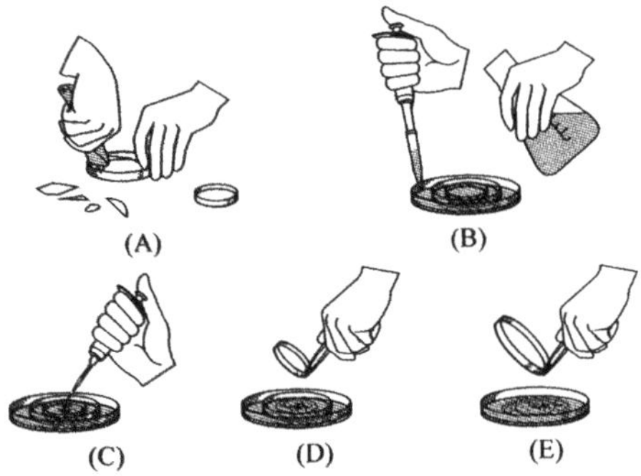

Figure 12.6 Partitioned plate cultures (see Section 12. 5. 2. 1(5) in text) for maintaining several generations of marine predatory nematodes (Moens and Vincx, 1998). (A) Cut the bottoms out of a 5 cm and a 9 cm Petri dish. (B) Place the 5 cm and 9 cm rims inside a 14 cm Petri dish, and fill the resulting compartments equally with plain BNA or holidic Erdschreiber BNA. (C) Transfer sediment particles and prey nematodes to the central compartment, leave for 8 days and then introduce the predatory nematodes. (D) When the agar in the central compartment starts to liquefy, remove the 5 cm rim. (E) When the agar in the next compartment starts to liquefy, remove the 9 cm rim. To restrain growth of contaminants, it may be necessary in steps (D) and (E) to manually move nematodes to the next section and remove the liquefied agar in the center.

"difficult" target predatory species successfully in soil-based systems. Steiner and Heinly (1922) cultured *Mononchus papillatus* in sterilized soil in glass tubes and fed them daily with various nematode prey species. In an updated setting with a more methodical approach, it should be possible to adapt one or more Baujard incubators (see 12. 8. 2. 2(6)) to the purpose of determining optimal soil, moisture, temperature, and prey combinations for the culture in soil of a wide range of predatory nematodes, if at least sufficient specimens of the target species are initially available.

12.5.2.3 Liquid Culture Systems

(1) To our knowledge, no permanent liquid culture system has been developed as yet for Mononchidae, but Arpin (1976) noted that females of *Prionchulus muscorum* and *Clarkus papillatus* could be kept alive for 2 months (at 25℃) to 2 years (at 6℃) in water washed through soil, and that on average some 50 viable eggs were laid per female at temperatures of 12℃ – 14℃. However, juveniles hatched from these eggs did not survive for more than 3 months.

12.6 Methods for Culturing Entomopathogenic Nematodes

The biology of entomopathogenic nematodes (EPNs) of the genera *Steinernema* and *Heterorhabditis* is intimately linked to that of their bacterial symbionts and insect hosts, in a triangle of interactions that is both scientifically fascinating and directly relevant to plant protection. With regards to their cultivation, EPNs are similar to *C. elegans* and relatives in that they are bacterial-feeders with fairly impermeable cuticle, short generation time, high reproductive rate, and capability to develop dauer juveniles. On the other hand, their unusual association with insects and bacteria implies that a very specific methodology is required for their cultivation. As a result, a range of methods are applied to EPNs, some of which are directly borrowed from or applicable to *C. elegans* and relatives, while others are much more specific to their unique ecological properties.

For use in biocontrol experiments, it must be kept in mind that the choice of cultivation method can significantly affect the infectivity of the resulting infective juveniles (Grewal, et al., 1999; Stuart and Gaugler, 1996; Yang, et al., 1997). The temperature of cultivation will also influence infectivity, and it is possible to select for populations that are more infective at low temperatures than preceding generations (Schirocki *and* Hague, 1997). Other properties influenced by culture conditions include tolerance of freezing (Surrey, 1996) and reproductive mode (Strauch, et al., 1994).

Due to space limitations, we will not deal with Mermithidae in this chapter, as these nematodes require protocols that are individually attuned to the different requirements of each species. For recent information on culture systems for mermithids, we refer the reader to the papers of Giblin and Platzer (1987), Gordon et al. (1990), Luhring and Mullens (1997), Peng et al. (1998), Santamarina Mijares and Bellini (2000), and the review of Petersen (1985).

12.6.1 Sources of Live Material

12.6.1.1 Baiting

Dutky (1959) developed cultivation methods relying on the use of the wax moth *Galleria mellonella*[①] as host for EPNs, and Bedding and Akhurst (1975) used wax moth larvae as bait for EPNs in a very simple detection and collection

① Culture methods for wax moth are given in 12.6.4.1.

technique. Samples of preferably sandy soil are placed in jars with five to ten *Galleria* larvae at the bottom, which will attract and be infected by EPN juveniles. After six days at 24℃, the larvae are recovered and dissected or transferred to White traps (White, 1927; see 12. 6. 2. 1). Note that this method also attracts other nematode species (particularly some rhabditids and diplogasterids) which are presumably not entomopathogenic in their own right, but can participate opportunistically in the invasion of the insect, and unless manually removed also in the invasion of subsequent cultures!

12.6.1.2 Exchange of Strains

A wide selection of species and strains are cultured in various laboratories around the world. If you are not specifically interested in obtaining species from your local surroundings and prefer to avoid the complications of identifying strains from soil samples yourself, a simple solution is to request reference strains from an EPN laboratory. However, you should be aware that the patents and other commercial aspects of biopesticide development might intervene and result in your request being turned down.

12.6.1.3 Shopping for worms

The commercial application of EPNs has resulted in the production and distribution of several brand products that can now be bought in many garden centers. General information, including lists of commercial suppliers in the USA, can be found at http://www.nysaes.cornell.edu/ent/biocontrol/pathogens/nematodes.html and at http://ianrwww.unl.edu/ianr/plntpath/nematode/wormepns.htm. Websites of European suppliers of EPN can be found at http://www.e-nema.de, http://www.biocontrol.ch, http://www.bionema.se, http://www.koppert.nl, and http://www.microbio.dial.pipex.com.

12.6.2 Xenic and Monoxenic Cultures

Depending on the ultimate purpose and the available facilities, many culture methods can be used for EPNs. In xenic or monoxenic systems, these nematodes must be cultured in association with their own natural symbionts (*Photorhabdus luminescens* or *Xenorhabdus* spp.). It is sometimes possible to use a single strain of these entomopathogenic bacteria as common food source for multiple species of EPNs, but the procedure is impractical because it affects infectivity (Grewal, et al., 1997) and because care must be taken to avoid the transition of the bacteria from primary to secondary form (Gerritsen and Smits, 1997). Also, some bacterial strains are actually toxic to other nematode species than the one with which they are naturally associated (Han and Ehlers,

1999; Hu, et al., 1999). *Escherichia coli* is not an appropriate food source, because it delays growth by at least a week in *Steinernema* species (Ehlers, et al., 1990). Boemare et al. (1997) provide an up-to-date review of the biological aspects of the symbiotic relationship between the nematodes and the bacteria.

12.6.2.1 White Traps

White traps are simple culture systems for maintaining EPNs on *Galleria* wax moth larvae and for harvesting the infective juveniles from these (Dutky, 1959; White, 1927). This method is quite simple in itself but does require maintenance of wax moth cultures (12.6.4.1).

(1) A White trap consists of an overturned 5 cm Petri dish lid placed inside a 9 cm Petri dish and covered with circular filter paper. The 9 cm dish is filled with water or S buffer, and one or more infected and dead wax moth larvae are placed in the center of the paper. As increasing numbers of nematodes are produced inside the larvae, numerous infective juveniles and some adults will crawl out of the host's body and collect in the liquid around the rims of the filter paper. They can then be transferred to another dish with wet filter paper and allowed to infect new wax moth larvae, or they can be injected into the larvae with a syringe.

(2) A simple modification of the White trap consists of using a PA plate rather than an overturned 5 cm dish with filter paper inside a larger dish (Wouts, 1991). Infected wax moth larvae are placed in a ring on the PA plate, and then the infective juveniles will emerge and accumulate along the rim and in the center of the agar, ready for transfer to new larvae. Like dauer stages of other nematodes, infective EPN juveniles are more or less hydrophobic and can easily climb out of a Petri dish. PA plates should therefore be sealed tightly in between inoculation and transfer.

12.6.2.2 Monoxenic Culture

Monoxenic culture ensures that other bacteria will not compete with the bacterial symbiont for nutrients, and that their metabolites will not hamper or kill the nematodes. Protocols for monoxenizing EPNs are given below in 12.6.4.3.

(1) Vast quantities of EPNs can be produced in Erlenmeyer flasks filled with polyether polyurethane sponge, autoclaved after addition of various tissue homogenates or broth. The final yield depends not only on the temperature, amount of medium, and elapsed time, but also on inoculum size (Han, et al., 1993; Wang and Bedding, 1998). Bedding (1981) mixed 60% or 70% ground pigs' kidneys with 20% or 10% melted beef fat and 20% tap water,

while Wouts (1981) developed a slightly less malodorous medium by combining 13% corn oil with 18% soy flour, 1.1% nutrient broth, 0.04% yeast extract, and 67.5% water. The Erlenmeyer is autoclaved and inoculated with EPNs, and subsequently left at room temperature. Over time, massive numbers of infective juveniles are produced, which can be stored for months in the flask (Wouts, 1981). This type of system can therefore both be used as a stock and harvest culture, and is also easily adapted for cultivation of *C. elegans* and relatives (see e.g., culture of *Panagrellus redivivus* by Taylor, et al., 1957). A variation of the same approach consists of using autoclavable polyethylene bags (Bedding, 1984).

(2) EPNs can also be cultured monoxenically on agar plates. One of the simplest variants consists of using lipid agar to pour LA plates (Wouts, 1981):

Bacto nutrient broth	16g
Agar	12g
Corn oil	5g

H_2O dist. add up to total volume of 1000 ml

After autoclaving, pouring, and setting, the LA plates are inoculated with the appropriate bacterial symbiont from liquid culture (12.6.4.2(1)) and the next day with axenized EPN juveniles. Other suitable types of agar can be found in Lunau et al. (1993).

(3) Monoxenic liquid media are also capable of generating very high densities of EPNs, and can be applied at industrial scale in bioreactor system (Ehlers, et al., 2000; Friedman, 1990), but many of these media were developed by commercial companies and are therefore shrouded in secrecy or protected by patents. Males of *Heterorhabditis* spp. cannot attach to females in a liquid medium (Stoessel and Ehlers, 1992), and liquid culture is therefore strongly biased towards hermaphroditic reproduction (Strauch, et al., 1994). Endotokia matricida or "matricidal hatching" is common in this culture system, especially so with lower bacterial densities or reduced nutrient concentrations (Johnigk and Ehlers, 1999). After entering the dauer stage, *Heterorhabditis* juveniles require a bacterial signal inducing their recovery before they will resume development (Strauch and Ehlers, 1998), and fluctuations in the strength of this signal can cause drastic differences in the yield of liquid cultures (Ehlers, et al., 1998).

12.6.2.3 Other Culture Systems

A range of other insect hosts can be used for culture of EPNs in laboratory conditions. Examples include the domestic cricket *Acheta domesticus* (Grewal, et al., 1999) and the Japanese beetle *Popillia japonica* (Stuart and Gaugler, 1996). It must be noted that some soil-inhabiting insects are very poor hosts, on the other hand (Peters, et al., 1997).

12.6.3 Axenic Culture

As with *C. elegans*, few scientific purposes absolutely require axenicity of the cultures used, and in that case it is usually advisable to avoid the sustained effort of maintaining axenic cultures. Nevertheless, they may be necessary for nutritional studies, investigations of symbiosis and pathogenicity, or for the efficient establishment of multiple series of strictly monoxenic cultures. Axenization protocols are listed in 12.6.4.3.

12.6.3.1 Axenic culture of *Steinernema*

Several systems were developed for growing different quantities of nematodes in more or less strictly defined media.

(1) The following simple method was used by Glaser (1940): 1 cm^3 cubes of raw sterile liver or kidney were added to NA plates inoculated with axenized juveniles and incubated at 25℃.

(2) Lunau et al. (1993) grew *Steinernema carpocapsae* DD-136 axenically in a medium consisting of 40 $g \cdot l^{-1}$ soy peptone, 30 $g \cdot l^{-1}$ yeast extract, 0.2 $g \cdot l^{-1}$ cholesterol, and 0.1% horse hemoglobin. A 200 ml medium in liter flasks kept at room temperature and shaken at 80 $r \cdot min^{-1}$ yielded some 19,000 nematodes ml^{-1} after 16 days.

(3) Attempts to culture *Heterorhabditis* axenically have not yet yielded any published media or protocols that work. However, it is clear from the experiments of Han and Ehlers (1998) and Ehlers et al. (1998) that bacteria-free dauer juveniles of this genus can be produced if nematodes are grown in monoxenic cultures with non-specific *Photorhabdus* spp. (see 12.6.4.3(4)). Furthermore, Strauch and Ehlers (1998) obtained recovery of dauers by adding a filtrate from liquid culture of their bacterial symbiont. Such filtrates were obtained by centrifuging 2-day-old bacterial liquid culture and filtering the supernatant with a 0.2 μm filter. However, development after recovery did not proceed further under axenic conditions.

12.6.4 Auxiliary Techniques

12.6.4.1 Culturing wax moth

The greater wax moth *Galleria mellonella* can be cultured indefinitely on an artificial diet consisting of the following (Wang and Bedding, 1996):

Farex Blended cereal	300 g
glycerin	140 g
honey	140 g

yeast	20 g
H_2O	24 ml

This mixture is applied to the bottom of large glass or plastic pots or containers with ventilation holes or screen, wax moth larvae are introduced and left to develop at 30℃. Wang and Bedding (1996) used fifth instar larvae of about 15 g for infection with *S. carpocapsae* or *H. bacteriophora*, and followed population dynamics of the nematodes inside the host/cadaver.

12.6.4.2 Culturing Entomopathogenic Bacteria

(1) Strain-specific *Xenorhabdus* or *Photorhabdus* spp. are best isolated from infected insects (Ehlers, pers. comm.). At the moment of death, usually about 30 h after nematode invasion, the immune system of the infected insect has often eliminated most contaminating bacteria. In most cases, therefore, streaking a drop of hemolymph onto NBTA agar (Akhurst, 1980) will provide a pure culture of the symbiotic bacterium. If contaminants are still present, their colonies can easily be distinguished from nematode symbiont colonies (see 12.6.4.2(3)). Pure cultures can then be transferred to yeast salt or YS broth (Dye, 1968) for production of larger amounts of bacteria:

Yeast extract	5 g
NaCl	5 g
$NH_4H_2PO_4$	0.5 g
K_2HPO_4	0.5 g
$MgSO_4 \cdot 7H_2O$	0.2 g

H_2O dist. add up to total volume of 1000 ml

The bacterial culture can be stored in 15% glycerol at −30℃ (Lunau, et al., 1993).

(2) If too few dauers are available to guarantee successful infection of an insect host, bacterial symbionts can instead be isolated from the infective juveniles. First, surface-sterilize the juveniles (see 12.6.4.3(1)). Next, inoculate them on a sterile medium appropriate for monoxenic culture (12.6.2.2), supplemented with bacterial filtrate of a pre-established culture of symbionts from a related nematode strain (to induce recovery: see 12.6.3.1(3)). Finally, after the juveniles have recovered, streak the released bacteria onto NBTA plates.

(3) In order to ensure that the bacterial associates are cultured in their primary form (the secondary form usually yields much fewer nematodes), they can be screened by growing them on NBTA plates (Akhurst, 1980) or MacConkey agar plates (Lunau, et al., 1993). Primary form colonies will be blue-green in color on the former medium, or red on the latter.

12.6.4.3 Monoxenizing and Axenizing/Synchronizing Protocols

(1) Removing all, or all but one, co-occurring organisms from an EPN culture poses problems that are quite similar to those of axenizing *C. elegans* (see 12.3.4.2). However, a critical difference is that EPN dauer juveniles retain monospecific inocula of their bacterial symbiont in an intestinal pouch. This simplifies monoxenization compared to *C. elegans* because it allows the use of these dauer juveniles for surface sterilization and transfer to sterile media.

Lunau et al. (1993) tested several surface sterilants on *S. feltiae* and obtained the best results with $HgCl_2$ administered as follows. Approximately 50 infective juveniles were rinsed in sterile water and centrifuged for 5 min at 5000 $r \cdot min^{-1}$. The pellet was subsequently washed for 10 min in 2 ml 0.1% $HgCl_2$ on a shaker, and then rinsed twice in sterile Ringer's solution. After this procedure, the infective juveniles can be inoculated on monoxenic growth medium where they will release their symbionts and start feeding. $HgCl_2$ is highly toxic and should be handled with due care.

(2) Axenizing can be done according to the *C. elegans* axenization protocol (see 12.3.4.2(1)) but does not always work in *Steinernema* due to the more resistant cuticle of the adults (Vanfleteren, pers. comm.). In a modified version of this approach, Lunau et al. (1993) first digested egg-bearing adults for 15 min in 0.4 M NaOH, during which time the nematodes were repeatedly passed through a Pasteur pipette, and then removed debris and supernatant before leaving the eggs for another 10 min in fresh 0.4 M NaOH. The eggs were finally rinsed twice in sterile water and then transferred to LA, YS, or TSB (Tripcase Soy Broth) plates to check for bacterial growth during the next 4 – 6 days. Best results were obtained with gravid adults that did not yet contain endotokous juveniles, and with eggs that had developed no further than the "pretzel" stage.

(3) A simple procedure was established by Lunau et al. (1993), which effectively bypasses problems with cuticular strength of the adults in *Steinernema*: egg-bearing females can be left in 0.1% merthiolate for 2 hours, during which time they will deposit eggs. The females are then removed, while the eggs are given another 2 hours in fresh merthiolate 0.1% before being rinsed and transferred to plates. This procedure did not work well with *Heterorhabditis* because the hermaphrodites retained most of their eggs.

(4) Some strains of *Photorhabdus luminescens* allow growth and reproduction of other *Heterorhabditis* species than their natural associate, but are not retained in the intestinal pouch of the infective juveniles of these other species (Han and Ehlers, 1998). A prudently chosen "recombination" of *Heterorhabditis* and *Photorhabdus* strains therefore allows culture systems in

which the dauers can subsequently be axenized by surface sterilization. Han and Ehlers (1998) performed this sterilization by incubating dauers for 6 hours in 0.5% streptomycin sulfate. The procedure can also be applied to *Steinernema* spp. using non-specific *Xenorhabdus* spp. (Ehlers, pers. comm.). The monoxenic cultures should always be started from axenic nematode eggs to avoid the presence of the natural symbiont.

12.6.4.4 Long-term Storage of Viable EPN Strains

Extensive research has been done in recent years on cryopreservation of viable inocula of EPNs. Although the *C. elegans* protocol (12.3.4.6) sometimes works well with little or no modification, individual strains can exhibit markedly different responses and recovery rates, and it is therefore often advisable to include an extended preincubation step (Popiel and Vasquez, 1991) and to optimize protocols for individual strains and species (Nugent, et al., 1996). Various aspects of freezing tolerance in EPNs were investigated by Wharton and Surrey (1994) and Brown and Gaugler (1996, 1998).

(1) Popiel et al. (1988) patented a cryopreservation protocol with an initial desiccation step consisting of 3 days, exposure to 97% RH. Nugent et al. (1996) compared the efficiency of 4 days, evaporative desiccation at 95% RH with preincubation in various concentrations of glycerin for various amounts of time (see 12.6.4.4(2)). They obtained 57% recovery after desiccation and freezing for the particular strain tested (*Heterorhabditis* sp. HI82), but even better percentages resulted from some of the tested variations of glycerin preincubation.

(2) Popiel and Vasquez (1991) developed a two-stage incubation procedure using glycerin and methanol for freezing dauer juveniles of *H. bacteriophora* and *S. carpocapsae*. Curran et al. (1992) and Nugent et al. (1996) modified this method, the latter authors comparing results across different *Heterorhabditis* strains and species as obtained with different cryoprotectants and different duration of the two (or even three) incubation stages. The following approach can be distilled from these studies (and especially from Nugent, et al., 1996) as an outline of a first-attempt protocol.

Harvest dauer juveniles, wash and concentrate them using centrifugation or sedimentation. Preincubate the juveniles for 6 days in 15% glycerin at 23℃. Remove the excess cryoprotectant by vacuum pumping through a Whatman® No. 1 or No. 42 filter paper and rinse the nematodes under suction with 2 ml of 70% methanol at 0℃. Transfer the nematodes on the filter to a 3 cm Petri dish with 3 ml of 0℃ 70% methanol, stir to resuspend the juveniles and leave the dish on ice for 7 min. Transfer them to 1.5 ml microcentrifuge tubes and spin for 2 min at 3000 – 4000 r· min^{-1}. Reduce the methanol volume to 60 μl by pipetting, then transfer the nematodes to Whatman No. 1 filter paper strips and place them in 2 ml round bottom cryovials. Cap and plunge immediately into liquid nitrogen. When subsequently thawing one or more inocula, immediately add 2 ml water at room temperature, vortex, pour into a Petri dish and leave for 6 – 24 hours.

If this protocol does not yield satisfactory results, or if you want to optimize it and get maximal recovery, try both longer preincubation with lower concentrations of glycerin (e. g., 8 or 10 days in 11% – 13% glycerin) as well as shorter preincubation at higher concentrations (e. g., 1 – 4 days in 17% – 22% glycerin). Extended exposure to low temperatures leads to increased synthesis of trehalose (Qiu and Bedding, 1999), mimicking the effect of glycerin preincubation (Popiel and Vasquez, 1991). In cases of poor recovery, it may therefore help to preincubate the nematodes at 5℃ – 8℃ rather than at room temperature. Instead of freezing the juveniles on filter paper, they can also be frozen in suspension or on glass slivers, but this usually leads to lower recovery rates, presumably because of slower thawing (Nugent, et al., 1996; Popiel and Vasquez, 1991).

12.7 Cultivation of Fungal- or Yeast-feeding Nematodes

12.7.1 Xenic and Monoxenic Cultures

Monoxenic cultures of fungal-feeding nematodes are relatively easy to establish and will maintain themselves for long periods of time without much intervention. Temperature regulation allows you to modify the rate of reproduction as is most convenient (Estey, 1985).

12.7.1.1 Nematodes Feeding on Terrestrial Fungi

(1) A range of species and genera can be grown on agar plates with *Botrytis cinerea* or *Rhizoctonia solani*, but some experimenting may be required to determine the preferences of the target species with regards to strain and age of the fungus (Hooper, 1986c). A very common system for culturing these fungi and nematodes is potato-dextrose agar, which can be used both for seed and stock cultures. PDA plates are routinely made at 1% – 2% and kept at 20℃ – 28℃ after inoculation with fungi and nematodes. Established fungus-nematode cultures can often be stored for some time at temperatures as low as 4℃. Coiled individuals indicate depletion of the medium. At that point, small pieces of agar containing both organisms should be transferred to plates with fresh PDA medium.

Visibility in an old culture is usually poor and it may be difficult to find nematodes for picking out with a needle. It is usually much easier to rinse the nematodes off the surface of the agar, or off the lid of the Petri dish, using a Pasteur pipette with sterile P buffer or water. Nematodes that have buried into the agar can be extracted by placing agar pieces on tissue paper in a small dish with water or P buffer.

(2) *Agaricus hortensis* (the common mushroom) can be used to grow and harvest larger quantities of some fungal-feeding species (Evans, 1970; Hooper, 1986c). Autoclave 60 ml of wheat grains and 90 ml water for 1 hour at 130℃ in an Erlenmeyer. Leave for 3 days. Inoculate with commercial spawn and agitate from time to time. Leave at 20℃ for 2 months. Inoculate with large numbers of nematodes (2000 individuals in 2 ml water). Leave at 20℃ for 2 – 3 months before harvesting. The latter can be done by rinsing with water, or by transferring the hyphenated grains to an extraction dish or funnel.

12.7.1.2 Nematodes Feeding on Intravascular Fungi

The pine wood nematode, *Bursaphelenchus xylophilus*, is a mycophagous insect associate severely detrimental to pine, causing enormous losses to the lumber industry in Japan and in other countries around the world (Jikumaru and Togashi, 2000). It feeds both on living pine cells and on several fungi developing in the dying or dead wood (Fukushige, 1991).

(1) Cultures of *B. xylophilus* have been successfully maintained on several fungi grown on agar media or grains, e. g., *Botrytis cinerea* growing on autoclaved barley grains (Maehara and Futai, 1996). Fukushige (1991) tested several species of fungi inoculated alone or in combinations into pine shoot segments, and concluded that *Ceratocystis* spp. growing on PDA plates are suitable for nematode population increase. However, other tested fungal species did not support feeding and reproduction of the nematodes. The type and age of the fungi cultures also affected nematode population development.

(2) Warren et al. (1995) have applied other approaches, such as inoculation of Jack pine (*Pinus banksiana*) wood logs with the blue stain fungus *Ophiostoma minus*, followed by inoculation with *Bursaphelenchus xylophilus* and then introduction of the beetle *Monochamus carolinensis*. This procedure resulted in an increased population density of *B. xylophilus* in bolts infested with the blue stain fungus. Furthermore, beetles recovered from logs infected with the blue stain fungus contained a significantly greater average number of *B. xylophilus* specimens than beetles recovered from logs without the fungus. This combined culture system for both the nematode and its insect vector is well suited for study under laboratory conditions of the specific interactions between nematode and insect.

Subsequent studies by Maehara and Futai (1996) confirmed that pine wood (in this case *Pinus densiflora*) infested with the blue stain fungus *Ophiostoma minus* promotes higher propagation of *B. xylophilus*. Furthermore, when the beetle *Monochamus alternatus* coexisted with the fungus and the nematode, larger populations of dauer juveniles were encountered and higher numbers were transferred to the beetles. Negative results were observed when other fungi such as *Trichoderma* colonized the dead wood. This indicated again that species of fungi present in the pine wood play an important role in the dispersal of nematodes by the insect vector.

(3) Survival and dispersion of *B. xylophilus* may occur when infested wood chips are imported from nematode infested areas. Tomminen et al. (1991) placed eastern white pine wood chips in plastic bags, inoculated them with *B. xylophilus* and kept the bags sealed at various temperatures for up to 130 days. They observed that temperature and host tree moisture both were important factors in the development and dispersal of *B. xylophilus*. Temperatures of 40℃ did not support the nematodes in this culture system, but this could have been due to depletion of oxygen rather than intrinsic physiological limitations. Temperatures of 30℃ favored development to adults, while temperatures as low as 3℃ allowed high rates of survival of propagative and dauer larvae.

12.7.1.3 Nematodes Feeding on Marine Fungi

(1) Meyers et al. (1964) cultured *Aphelenchoides marina* in fungal mats, similar to those described above in 12.5.1.1(1).

12.7.2 Axenic Culture

Aphelenchus avenae, *Aphelenchoides rutgersi*, and *B. xylophilus* have been cultured successfully in axenic media (Bolla and Jordan, 1982; Buecher, et al., 1970; Hansen, et al., 1970; Moens, et al., 1996; Myers, 1992; Myers, et al., 1971; Thirugnanam, 1976) suggesting that these nematodes do not require contact with food cells to initiate pharyngeal activity (unlike at least some plant parasites: see 12.8.3). We refer to the original papers for detailed methods; the media used were similar to those applied for *C. elegans* (see 12.3.2.2). It is interesting to note that chick embryo extract has often been used as a growth promotor in axenic culture of fungal-feeders, and not in axenic culture of bacterial-feeding nematodes or EPNs (see Table II in Vanfleteren, 1980).

12.7.3 Auxiliary Methods

Surface sterilization techniques can be applied similar to 12.4.3.1 and 12.6.4.3. Cultures of *R. salina* and *B. cinerea* can be maintained on PDA agar and stored at 4℃. Depending on the preferences of the target species, the fungi should be allowed to grow for 1 or more weeks before inoculation of nematodes. Riga and Webster (1991) modified the *C. elegans* protocol for cryopreservation in dilute glycerin and applied it successfully to *B. xylophilus*.

12.8 Methods for the Cultivation of Plant-parasitic Nematodes

A wide range of nematode species can be maintained in plant pots in the greenhouse or lath house, and some plant-parasitic nematodes (PPNs) can also be cultured using more artificial *in vitro* systems (review of Platzer, 1977). Depending on the host-parasite relationships, several approaches should be considered for establishing plant-parasitic nematode cultures. As obligate parasites, they must feed on host cells for development and reproduction. So far, no successful continuous axenic cultures of plant-parasitic nematodes have been developed, and the cultivation of these nematodes therefore depends directly on the nature of the host cells and their possible modes of maintenance in a culture system.

12.8.1 Collecting Live Material

12.8.1.1 From Field Populations

To establish seed cultures, soil samples containing the desired nematode species must be collected from rhizospheres in natural sites or agricultural fields. It is obviously of crucial importance to determine as accurately as possible to which plant(s) the sampled rhizospheres belong, and it may be necessary to collect these plants, or their fruits or seeds, for maintenance in the greenhouse. In the laboratory, a portion of the collected soil should then be processed and extracted to determine the nematode diversity and population density. If present in sufficient numbers, the target species can be picked out from the extract and transferred directly to seed culture systems. If present in low numbers, attempts can be made to inoculate pots with a probable host plant, in the hope that populations of the target species will eventually build up to useful numbers.

12.8.1.2 From Culture Collections

Perhaps even more so than for any other group of nematodes, numerous plant-parasitic species are already being cultured in laboratories around the world. Depending on your needs, it may be much more convenient to obtain an established culture from colleagues than to have to start one up yourself, all the more so because those colleagues will be able to advise you extensively on the appropriate culture conditions. Nevertheless, care must be taken to comply with quarantine regulations and to avoid turning your own laboratory into a bridgehead for invading plant pests.

12.8.1.3 Quarantine and Clean Handling Procedures

If you wish to culture economically important plant parasites that are not yet omnipresent in your area, then you should make a maximum effort to contain those cultures and avoid accidental release of the nematodes into the environment. This requires a number of facilities (Baldwin, et al, 1999), such as:

(1) Hot water available at all sinks in your laboratory. Rinsing all equipment thoroughly with hot water (preferably above 65℃, or even higher when dealing with species from xeric habitats) in a plugged sink will kill nematodes that would otherwise be flushed down the drain alive and well.

(2) A collection tank receiving water from the extraction sink(s). This water must be treated chemically or thermally to kill all organisms released during extraction.

(3) A (big) autoclave for sterilizing the remainders of discarded soil samples, extracted soil and plant material.

(4) Restricted access to quarantine cultures, i.e., by keeping doors to laboratories, cupboards, and containers locked and clearly labeled as quarantine areas.

12.8.2 Xenic and Monoxenic Soil-based Culture Systems

Ectoparasitic plant parasites do not penetrate the host roots but feed externally, moving from feeding site to feeding site. Only the stylet is inserted into one or several layers of the root tissue. Most ectoparasites have the ability to feed on multiple host-plant species, i.e., they are polyphagous. Also, they interact directly with other micro-organisms in the soil environment and are strongly influenced by soil factors such as temperature, pH, moisture content, host requirements, etc., which may play an important role in the establishment of certain species (Koenning and Barker, 1985). In contrast, nematodes living and feeding inside the root are more specialized in terms of host specificity, feeding mechanisms, and complexity of the host-parasite relationship. In some of the most important groups of such endoparasites, the infective stage penetrates the host tissue, establishes a feeding site, and permanently (in the case of females) or temporarily (in the case of males) becomes sedentary (Mundo-Ocampo, 1983).

While inside the root, endoparasites are better protected against fluctuations in soil conditions than ectoparasites, and it has been suggested that *in vitro* cultures of the latter are harder to maintain because they are more dependent on biotic and physico-chemical conditions (Dasgupta, et al., 1970). This may explain why for instance *Belonolaimus longicaudatus* has proven quite difficult to culture under *in vitro* conditions (Huang and Becker, 1997). Soil texture may affect the development of some ectoparasitic nematodes in different ways (Mashela, et al., 1991). We have observed that mixing an original soil sample with an artificially created soil often results in loss of the target species, apparently due to dilution or alteration of some unknown soil factor. However, in other instances, such a mixture seems not to affect the increase of target nematode populations.

12.8.2.1 Plots and Microplots

(1) Small plots or concrete containers in the open air can provide a convenient intermediate between collecting samples in the field versus maintaining cultures *in vitro*. Such systems are obviously much more exposed to the local weather and other uncontrollable influences, and will evidently work better with local nematode species than with exotic ones. However, they also provide the nematodes with a large quantity of more or less natural soil, a broad range of organisms, and a presumably stable network of well-balanced ecological interactions between these. As such, they are generally agnotobiotic.

The partitions of adjacent plots or walls of neighboring containers should be raised above the soil surface, and must reach sufficient depth underground, in order to minimize cross-contamination between adjacent nematode populations. The appropriate depth depends very much on the rhizosphere dimensions of the host plants, and the vertical movements of the target nematodes, but we would recommend that partitions or walls should reach at least 1.5 m depth. Also, at the bottom edge of the partitions each plot is preferably provided with a grille or strong mesh, as well as a layer of gravel, both acting as barriers against passage of roots, rodents, nematodes, etc.

After initial establishment of host plants and target nematodes, much less maintenance and supervision is required than in greenhouse or laboratory cultures. Nevertheless, the plots should be weeded, protected to some degree against carry-over of other nematode species, and replanted with host plants when necessary. They should also be screened at least seasonally to monitor fluctuations in the density of the target species, and to verify that it is indeed permanently established. Such screening must consist of some form of soil extraction and is therefore in itself more time-consuming than for most *in vitro* systems, but the extract can often serve other purposes as well, e.g., in preparation of teaching and demonstration material.

12.8.2.2 Cultures in Pots and Greenhouses

(1) Compared to the open field or microplots, greenhouses often provide more controlled conditions that are also more advantageous for optimal growth of both host and parasite. If the host plant can be maintained in the greenhouse, then pot cultures represent a simple and inexpensive way of maintaining live plant-parasitic nematodes for considerable periods of time (Fig. 12.7). Irrigation of plants growing under these conditions is very critical, and may require the installation of an automatic irrigation system, consisting of microsprinklers or emitters that dispense water at selected time intervals according to the specific needs of the plants.

As a relatively natural system, pot cultures can be used both as seed and stock cultures, and in the latter case they should be inspected routinely to determine the population fluctuations. Water schedules have to be designed according to water quality and quantity requirements, which have to be determined for each combination of plant and nematode species. Some species require high levels of moisture (i. e., *Juncus/Ekphymatodera, Carex/Verutus*), while others are sensitive to salts or to other minerals present in the water. As a rule, avoid drastic fluctuations since these may affect nematodes and plants to a lethal degree.

The pots should be weeded regularly, since weeds may adversely affect both the host plant and its parasites. The greenhouse floor, if not completely covered with concrete or another hardened surface, should also be kept weed-free, as the presence of weeds will allow the uncontrolled establishment and buildup of nematodes (both those that have escaped from pots as well as those that may enter the greenhouse environment through soil or accidental introduction) which may subsequently contaminate pots. Great care should also be taken to avoid cross-contamination between pots (unless nursery conditions are to be replicated for experimental purposes; see, e. g., Ko, et al., 1997), especially during automated or manual watering, e. g., by putting all planted pots on benches with surfaces from which any spilled water drains away immediately.

A few clean host plants in fresh pots should be kept at hand, ready for inoculation with nematodes from old cultures in the event of the latter's decline. Individual pot cultures may last for many years, but unknown factors or accidents (e. g., failure of the temperature regulation of the greenhouse, omissions in the watering schedule, inappropriate use of pesticides inside the greenhouse) can never be excluded, and losses of individual cultures or whole strains will inevitably occur. If permanent survival of a strain is critical to your work, then it is as always best to use several culture systems and to spread the

Figure 12.7 Pot cultures of plant parasitic nematodes on various hosts and in various settings (see Section 12.8.2.2 in text). (A) Plastic pots with cabbage plants sustaining *Heterodera cruciferae* in the lath house. (B) Resin pots with *Ficus* sp. sustaining *Heterodera fici* (on left) and California black walnut sustaining *Pratylenchus vulnus* and *Xiphinema americanum* (on right). (C) Emptied concrete temperature tanks in the greenhouse, showing the suspended platform on which pots are placed, the water supply (middle left side), the thermostat (front), and the water circulation pump (upper left corner). (D) Corn plants in compressed paper pots, inoculated with *Paratrichodorus christiei*. (A) and (B) are not kept in a greenhouse but in a lath house, i.e., a structure covered on all sides with thin slats rather than glass, allowing free air circulation and natural rainfall, while providing more shade and more natural temperature fluctuations than a greenhouse. On the other hand, environmental conditions are impossible to control in this structure, and risks of contamination are higher.

risks over several culture series.

(2) Temperature fluctuations within and between pots in an experimental series should be kept at a minimum. This can be achieved by submerging them partly in a large thermo-regulated water bath or temperature tank (Fig. 12.7). Water is maintained in constant circulation by a water pump to homogenize the temperature across the tank. Watertight sealed containers made of a conductive material such as plastic, polystyrene, or stoneware can be used to grow the plants and to establish nematode cultures. To avoid contamination across pots, plants must evidently be watered from above, and every precaution must be taken to prevent nematodes from entering the water in the tank itself.

(3) Leaf parasites can be maintained by inoculation of nematodes in suspension onto the leaves of suitable hosts grown in pots. We have, e.g., kept *Aphelenchoides besseyi* in this manner on the bird's-nest fern *Asplenium nidus*. A similar technique can be applied with individual leaves: Sanwal (1959) cultured *A. ritzemabosi* on separate chrysanthemum leaves placed on wet sand and kept in high air humidity.

(4) The pot soil itself can be sterilized (e.g., with steam or methyl bromide) prior to planting of the host and inoculation with selected micro-organisms and nematodes, allowing for gnotobiotic micro- or mesocosm studies of ecological interactions in simplified soil ecosystems (see, e.g., Ingham, et al., 1985). Depending on the degree of containment required, ordinary plant pots can be replaced by containers that effectively seal off the soil from external influences and contaminants, and sterile diluted medium can be used for watering the host plant. If only a moderate degree of containment is required, regular tap water can be used after filtration on filters and/or sieves appropriate for removal of, e.g., bacteria, fungi, nematodes, and other organisms.

(5) Greenhouses are also well-suited for maintenance and containment of cultures of quarantine organisms (Baldwin, et al., 1999), provided access is strictly controlled and quarantine procedures are rigorously adhered to. At UC-Riverside, we have maintained a number of nematode cultures under standard and quarantine greenhouse conditions (Table 12.2).

(6) Baujard (1995) developed a miniaturized system of pots (actually PVC tubes) inserted through a grid of circular openings in an otherwise closed incubator, which was equipped with a thermostat and two light bulbs to control air temperature around the PVC tubes. Combined with a controlled water supply, they were used to culture a wide range of plant-parasitic nematodes in sandy soil with various host plants, in conditions imitating as closely as possible those observed in the field in various arid and semiarid habitats in Senegal. The incubators were also used for an extensive series of experiments on the ecological requirements and preferences of various species, each incubator

Table 12.2 List of plant-parasitic nematode cultures previously or currently maintained as part of the UC-Riverside nematode collection (some cultures are no longer available).

Nematode species	Host	Greenhouse
Sedentary endoparasites and semi-endoparasites:		
Cactodera cacti	Christmas cactus (*Schlumbergera bridgesii*) Cactus (*Opuntia* spp.)	Standard
Punctodera chalcoensis	Maize (*Zea mays*)	Quarantine
Globodera spp.	Nightshade (*Solanum dulcamara*)	Quarantine
Ekphymatodera thomasoni	Rush (*Juncus effusus*)	Standard
Verutus californicus	Sedge (*Carex* sp.)	Standard
Meloidodera charis	Mesquite (*Prosopis* sp.)	Standard
Meloidodera floridensis	Pine (*Pinus* sp.)	Standard
Heterodera schachtii	Sugarbeet (*Beta vulgaris*)	Standard
Heterodera cruciferae	Cabbage (*Brassica oleracea*)	Standard
Heterodera betulae	Birch (*Betula alva*)	Standard
Heterodera fici	Rubber plant (*Ficus* sp.)	Standard
Meloidogyne spp.	Tomato (*Lycopersicon sculentum*)	Standard
Migratory endoparasites:		
Pratylenchus vulnus	Walnut (*Juglans liindsi*)	Standard
Nacobbus aberrans	Tomato (*Licopersicum esculentum*)	Quarantine
Ectoparasites:		
Tylenchorhynchus maximus	Bermuda Grass (*Cynodon dactylon*)	Standard
Scutellonema brachyurum	*Oleander* sp.	Standard
Helicotylenchus spp.	Bermuda Grass (*Cynodon dactylon*) Banana (*Musa* sp.)	Standard
Paratrichodorus sp.	Sudan grass (*Sorghum bicolor*)	Standard
Belonolaimus longicaudatus	Bermuda Grass (*Cynodon dactylon*)	Quarantine
Tylenchorhynchus sp.	Unidentified grass	Standard

easily accommodating multiple replicate series for parallel studies on different nematodes (Baujard and Martiny, 1995).

(7) Another approach to increased control consists of using soilless substrates: combinations of various mineral, synthetic, and organic substances can be used in pot cultures of some migratory endoparasitic nematodes. Wang et al. (1997) measured population development of *Radopholus similis* in *Anthurium andreanum* grown in pots with various combinations of pine bark compost, peat, cinder, perlite, and rockwool. They found that soil-less substrate combinations with organic components resulted in increased damage

to the host plants. Amsing (1990) reported that movements of nutrient solution quickly disseminated *Pratylenchus vulnus* among roses grown on rockwool in an ebb/flow cultivation system. Such studies suggest that one could efficiently use soil-less pot cultures to produce large numbers of these nematodes, as a culture system that is less artificial than *in vitro* culture, but more easily controlled and characterized than ordinary pot cultures.

12.8.3 *In vitro* Culture Systems

During the last few decades, *in vitro* culturing of plant-parasitic nematodes on plants or plant tissues has become a common technique. Various procedures for obtaining and maintaining gnotobiotic cultures of endoparasitic and ectoparasitic nematodes are available in the literature (Eisenback and Zunke, 1998; Huettel, 1990; Orion and Bar-Eyal, 1995; Sijmons, et al., 1991; Zuckerman, 1971; Zuckerman, et al., 1985). We will provide the most practical techniques available and focus on some of the more recent information. It should be noted that no axenic culture technique has so far been developed for Tylenchina. Part of the problem appears to be that the economically important species do not initiate feeding behavior until they detect contact with plant roots or plant cell walls (Borgonie, pers. comm.).

12.8.3.1 *In vitro* Aseptic Dual Cultures with Entire Plants

(1) Certain plants can be fitted inside an *in vitro* system in their entirety. Monoxenic cultures of several plant-parasitic nematodes have been successfully established on the cruciferous plant, *Arabidopsis thaliana*. Because of its small genome size and other agronomic features, this plant is an effective model host for studies of plant/nematode interactions. Cultures of various endo- and ectoparasites have been established on *A. thaliana* in order to study life cycle and plant resistance (Orion and Bar-Eyal, 1995; Sijmons, et al., 1991).

(2) Soilless plastic pouches for seed germination (manufactured by Vaughan's Inc., Downers Grove, IL) have been used widely to obtain pure cultures of plant-parasitic nematodes (Den Ouden, 1960; Klinkenberg, 1963) (Fig. 12.8). The bags offer several advantages: they minimize space usage, are handy for direct observations of host-parasite interactions, and the obtained nematode specimens or plant tissues are ideal for ultrastructural observations (TEM/SEM) because materials are free of soil particles and other solid contaminants, thus avoiding optical obstruction of important structures under SEM, and preventing knife damage when sectioning specimens for TEM observations.

Surface-sterilized seeds are germinated in the pouches. Once the roots have developed, surface-sterilized second-stage juveniles are inoculated in the

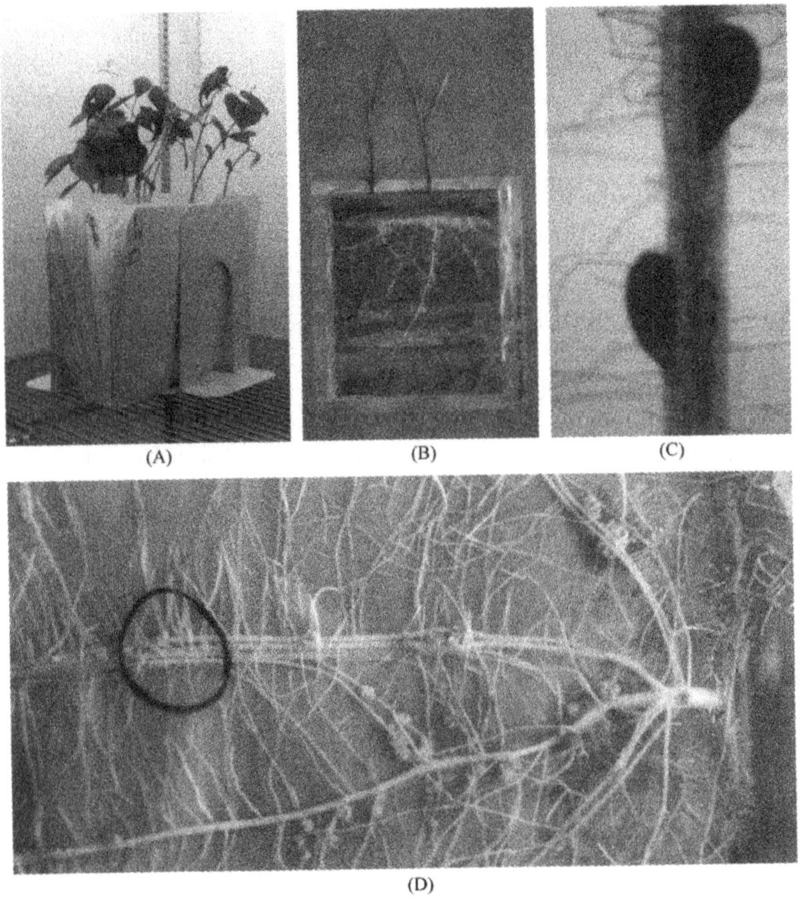

Figure 12.8 Soil-less plastic pouches for seed germination (manufactured by Vaughan's Inc., Downers Grove, IL) and close-ups of plant-parasitic nematodes in different developmental stages (Section 12.8.3.1(2) in text). (A) Several cowpea plants maintained under controlled light and temperature conditions. A growth chamber is calibrated to simulate natural conditions for plant and nematode development. (B) A plastic pouch including two "teocintle" seedlings (*Zea mexicana*). (C) A close up of teocintle roots showing different developmental stages of *Punctodera chalcoensis*. (D) A close up of tomato plant roots showing galls and mature females (circle) of *Nacobbus aberrans*, as observed directly through the lining of a plastic pouch.

pouches, which are maintained in a growth chamber at light and temperature settings as appropriate for the plant and nematode species. The plants in the pouches need to be systematically watered with a nutrient solution (Hoagland and Arnon, 1950), and care must be taken to avoid contamination by bacteria or fungi. Several sedentary endoparasites, including *Meloidogyne* spp., *Punctodera chalcoensis*, *Globodera* spp., and *Nacobbus aberrans*, have been successfully cultured in seed-pack growth pouches (Mundo-Ocampo, 1993; Souza, 1999). So far, attempts at culturing the ectoparasite *Belonolaimus longicaudatus* on corn roots with this technique have not provided positive results (Becker, pers. comm).

12.8.3.2 *In vitro* Aseptic Dual Cultures with Excised Plant Tissues

(1) Excised roots of various plant species can be established under axenic conditions in Petri plates or flasks with media containing nutrients and other chemical compounds necessary to sustain plant development. Agar is the most commonly used substrate, in combination with different media. Other gelling agents have been tested (e. g., Gellan gum by Eyre and Caswell, 1991; Pluronic F127 by Ko and Van Gundy, 1988 and Ko, et al., 1996). The choice of a medium is not only important with respect to essential elements provided for plant development, but also affects the nematodes' infection rate (Huang and Becker, 1997; Ko and Van Gundy, 1988; Sijmons, et al., 1991). This type of system allows manipulation of many components of the culture environment that are more difficult to control under field conditions or in pots.

According to host specificity, nematodes are surface-sterilized with different types of disinfectants (see 12.4.3.1(1), 12.4.3.1(3), 12.6.4.3(1), and 12.6.4.3(3)), including antibiotics and other compounds. This step is very critical since excessive doses may be toxic, while insufficient dosage will not allow efficient action, resulting respectively in poor nematode survival and in contamination of the system (Bavaresco and Walker, 1994; Huang and Becker, 1997; Zuckerman, 1971).

If the tissue/nematode combination is compatible, then the nematodes will be attracted to the tissues and the feeding process is initiated. As specimens feed and develop, observations can be made under the microscope. New computerized equipment can be used to record image sequences pertaining to life cycles as well as developmental and morphological aspects (Eisenback and Zunke, 1998; Wyss, 1978).

Two examples of protocols are given below. The first (modified from Cordero, et al., 1991) is used to culture *Cactodera cacti* on Christmas cactus (*Schlumbergera* sp.). Wash Christmas cactus cladodes with a commercial detergent. Disinfect the plant tissue by immersion in a 10% solution of commercial bleach for four minutes. Rinse the plant tissue in sterile water. Transfer the stems to Petri dishes containing Gambor's B-5 medium (GIBCO, Grand Island, New York) plus White's organics at pH 6.5. Maintain the Petri dishes at 25℃ under a 16 hour illumination cycle (3000 Lux) for about a month to induce the root development. Disinfect second-stage juveniles of *C. cacti* with four rinses of sterile water and 10 min incubation in a saturated rifampicin solution. Young roots are inoculated with about 25 juveniles per plate. Plates with infected roots are incubated and maintained at 25℃ in a growth chamber under an illumination cycle of 16 hours and 300 Lux (higher light levels may create unnecessary risks of overheating).

The second example protocol (modified from Cordero and Baldwin, 1990) allows culture of *Heterodera schachtii* on sugarbeet (*Beta vulgaris*). Seeds of sugarbeet (we use cv. USH 11, Holly hybrid large) are surface-sterilized by immersion in 10% commercial bleach for 10 min. Rinse seeds with sterile water and plate them out on 1% water agar. After germination of seeds in the dark, transfer them to Petri dishes containing Gambor's B-5 medium plus White's organics at 6.5 pH, with 0.25% Gelrite® (a gelling agent). Young roots are inoculated with about 25 second-stage juveniles of *Heterodera schachtii* disinfected as above, and plates with infected roots are also incubated and maintained at 25℃ in the growth chamber under a 16 hour/300 Lux cycle.

This type of *in vitro* cultivation has been applied to a range of plant-parasitic nematodes, including several ectoparasites. Recent examples include *Xiphinema index* cultured on grape roots growing on a rich nutrient medium. Low salt concentrations in the medium were found to have a beneficial effect on nematode development and behavior, presumably resembling the natural soil conditions in which these nematodes occur (Bavaresco and Walker, 1994). *In vitro* cultures of *Belonolaimus longicaudatus* have been established successfully on excised corn roots. Observations under these conditions have provided information on the life cycle and other aspects of the mode of reproduction of this economically important nematode (Huang and Becker, 1997, 1999).

Many semi-endoparasites and endoparasites can be cultured on excised tissues. Cultures of *Rotylenchulus reniformis* have been obtained on tomato root explants growing on a modified White's nutrient medium with Gellan gum and iron sulfate chelated with disodium ethylenediaminetetraacetate (EDTA). This technique provided an adequate clear support for nematode feeding and development and yielded viable eggs and vermiform stages. However, quantitative data are not available and the effect of EDTA on nematode infectivity was not determined (Eyre and Caswell, 1991). Meyer et al. (1997) have compared yields of monoxenic cultures of *Heterodera glycines* on excised root tips of 24 soybean cultivars, and found up to seven-fold differences in numbers of first-generation females produced on different cultivars. *In vitro* banana shoot cultures provide a convenient means of cultivating populations of the burrowing nematode *Radopholus similis*: large numbers of specimens can be obtained by this method (Mateille, 1990). Lesion nematodes (*Pratylenchus* spp.) have been successfully cultivated in root cultures of several plants including corn, tomato, and soybean (Huettel and Rebois, 1985; Rebois et al., 1984).

(2) Several migratory endoparasitic nematodes can be cultured and maintained in sealed Petri plates with carrot discs. This is particularly common practice for cultivation of burrowing nematodes *Radopholus* spp. and lesion nematodes *Pratylenchus* spp. (Huettel, 1985; Moody, et al., 1973; O'Bannon and Taylor, 1968).

(3) Huettel (1990) provided an updated review of nematode cultivation on plant callus tissue. Carrot callus tissue allows efficient culture of migratory endoparasites such as *Pratylenchus brachyurus*, *P. agilis*, *P. scribneri*, and *Radopholus similis*. The carrot callus procedure yields high numbers of nematodes and reduces the need for additional sterilization during continuous maintenance of the cultures (Reise, et al., 1987). Alfalfa callus tissue can also be used for at least some *Pratylenchus* and *Radopholus* species (Castrol and Ferraz, 1990; Riedel, 1985). De Waele et al. (1991) succeeded in culturing *Ditylenchus destructor* on potato callus tissue, while Plowright and Akehurst (1992) obtained reproduction of the "ufra nematode" *D. angustus* on wheat callus tissue, at rates that could exceed reproduction on rice seedlings, albeit with greater variability.

The protocol established by Reise et al. (1987) for culture of migratory endoparasites on carrot callus consists of the following steps. Under sterile conditions, callus tissue is dissected out from 4-week-old surface-sterilized carrot discs. Callus tissue is transferred to Petri plates with Gamborg's B5 Medium plus 0.1 mg · l^{-1} 2,4-dichlorophenoxy-acetic acid, 12.5 mg · l^{-1} gentamycin sulfate and 1.5% agar. The tissue is maintained at 28℃ for 6–8 weeks, with subsequent subculturing if additional replicates are needed. Before inoculation, nematodes are surface-sterilized, e.g., by repeated rinsing with chlorhexidine, streptomycin sulfate, and mercuric chloride (Huettel, 1990). About 100 nematodes are then transferred to plates containing 1.5 g of callus tissue each and kept at 28℃. Nematodes reproduce and remain viable for at least 4 months.

12.8.4 Auxiliary Techniques

12.8.4.1 Axenizing

Surface sterilization techniques similar to 12.4.3.1 and 12.6.4.3 work even better with most tylenchs than with many bacterial-feeders, because the latter may continue sucking up medium regardless of the presence of toxins, thus effectively poisoning themselves. An extensive review of axenizing methods for a range of plant parasites can be found in Platzer (1977), and we will just mention one recent variant here.

(1) Heungens et al. (1996) developed an elegant modification of the *C. elegans* protocols (12.3.4.2) for use with cysts. Cysts are placed in a modified syringe with a 30 μm mesh nylon sieve at its end, and this syringe is then filled and emptied in successive washes, NaOCl and more washes, in similar fashion to the *C. elegans* protocol but without requiring centrifugation or transfer between tubes.

12.8.4.2 Long-term Storage of Viable Inocula

(1) As in other groups, species of plant parasites with naturally occurring survival stages are often easier to preserve in frozen form during that stage. A very simple and efficient method of storing cysts of *Globodera rostochiensis* and *G. pallida* at −80℃ was described by Folkertsma et al. (1997), and consisted of air-drying the cysts, followed by storage at 4℃, and finally transfer to −80℃ in containers with punctured lids. Ultrafreezing is not required, however: Ireholm (1996) compared various ways of storing *Heterodera avenae* cysts at temperatures between −18℃ and +20℃, and obtained reasonable hatching rates (albeit lower than those recorded by Folkertsma, et al., 1997) from freeze-dried cysts stored up to 5 years at −18℃.

(2) Similar to EPNs (see 12.6.4.4(2)), plant-parasitic nematodes can be cryopreserved with greater success after an extended preincubation phase. Galway and Curran (1995) adapted the methods of Popiel and Vasquez (1991) and Curran et al. (1992) for use with *Pratylenchus* spp. After 4 or 5 days preincubation in 14% −17% glycerin, they obtained recovery rates above 70% for *P. thornei* stored in liquid nitrogen. Applying the procedure to three other species yielded 51% −67% recovery.

(3) Bridge and Ham (1985) modified a cryopreservation protocol originally designed for helminths and applied it to *Meloidogyne graminicola* juveniles. In a first step, infective juveniles harvested from pot cultures were first incubated in 10% aqueous ethanediol (ethylene glycol) at 37℃ for 15 min, then transferred to an ice bath, topped up with equal volume 70% ethanediol to a

final concentration of 40%, and incubated for a further 10 - 45 min. The suspension was then pipetted onto glass slivers and these were dropped into liquid nitrogen. Thawing was performed by dropping the frozen slivers into 30 ml of water at 30℃. The highest recovery obtained with this method was 25%, after 30 min incubation in 40% ethanediol at 0℃, but infectivity was highest in recovered juveniles obtained after 45 min incubation in 40% ethanediol.

Triantaphyllou and McCabe (1989) adapted this approach further, and applied it to seven *Meloidogyne* species as well as three *Heterodera* species. They recommend a first preincubation step of 1 - 3 hours in 10% ethanediol at room temperature, followed by 40 - 60 min in 40% solution at 0℃, and then freezing on chromatography paper rather then glass slivers. A 50% - 90% recovery was obtained in this way with juveniles, but eggs hardly ever recovered, even after pretreatment of the egg shell with NaOCl to increase permeability. Extending further on this work, van der Beek et al. (1996) optimized it for use of microcentrifuge tubes with three other *Meloidogyne* species and noted that, although recovery declined with first step duration between 3 and 4.5 hours, for unclear reasons it increased again when the first step lasted over 4.5 hours. Standardized duration of respectively 120 min and 45 min incubation were then applied to other species and populations, and yielded significant differences in recovery. Confirming earlier observations, they also noted a significant decline in recovery when the nematodes were taken out of the nitrogen and exposed to room temperature before re-submerging, or when they were not immediately immersed in tepid water after removal from the nitrogen.

12.9 Conclusions

As illustrated in this chapter, a wide range of methods has been established for rearing nematodes under more or less controlled conditions. Some of these are easily applied or adapted to a wide range of species, while others are only appropriate for a few or even a single species. Likewise, some methods are fast, robust, or require only minimal equipment and facilities, while others are more time-consuming, disaster-prone, or technically complicated. Perhaps most important of all, much depends on the ingenuity, persistence, and good fortune of the researchers involved.

There is ample scope for combining steps of different approaches, and the possibilities for developing new protocols are still wide open. New technology and novel approaches will undoubtedly play an important role in simplifying and optimizing culture methods, and we expect that many more nematode species will be brought into culture in the years to come. However, it remains an open question whether or not these species will mostly belong to nematode families that are already well represented in culture collections.

An interesting comparison can be drawn between nematology, microbiology, and mycology, three areas of biology that resemble one another in terms of the practical limits and problems encountered. Intriguingly, it appears that only a tiny fraction of the total number of bacterial species can be maintained in culture with established techniques (e. g., Kozdroj and van Elsas, 2001; Torsvik, et al., 1998; Ward, et al., 1990). The diversity of non-culturable strains is less well documented in the case of fungi, but first indications suggest that it may be significant (Viaud, et al., 2000).

In nematodes, at least a few representatives have been cultured successfully from all major groups of soil nematodes, but certain taxa do present much greater difficulties than others, and some species have as yet resisted all attempts at transfer to monoxenic and/or *in vitro* systems. No doubt many a research project has run aground, or drastically changed course, because of unexpected difficulties with rearing its target nematodes. In a vicious circle, such recalcitrant organisms often cannot be cultured because we still have a much too limited understanding of the crucial environmental and biotic requirements for nematode development and reproduction; and yet we cannot easily gain such knowledge without experimenting on cultured strains! The difficulties are much greater still for marine nematodes, which remain as understudied with respect to cultivation requirements as they are in so many other scientific aspects.

Nevertheless, we assume that most nematodes can at least be cultured in sediment or soil, and that such cultures can provide a good starting point for developing more artificial systems. As we learn more and more from in-depth studies of model species like *C. elegans*, the number of potentially answerable comparative questions increases with the expansion of our repertoire of other culturable nematodes. We therefore hope that our review will both stimulate the development of novel rearing techniques for previously non-culturable species, as well as contributing to the increasingly widespread cultivation of the great diversity of more amenable nematodes.

References

Abdul Kader, N., and M. G. Côté. 1996. Isolement, identification et caractérization de souches québécoises du nématode *Caenorhabditis elegans*. Fundamental and Applied Nematology 19: 381 –389.

Akhurst, R. J. 1980. Morphological and functional dimorphism in *Xenorhabdus* spp., bacteria symbiotically associated with the insect pathogenic nematodes *Neoaplectana* and *Heterorhabditis*. Journal of General Microbiology 121: 303 – 309.

Amsing, J. J. 1990. Spread and development of *Pratylenchus vulnus* populations: ebb and flow ideal for attacks by root nematodes. Vakblad voor de Bloemisterij 45: 34 – 37.

Arpin, P. 1969. Etude préliminaire d'un facteur écologique important pour les nématodes: L'humidité actuelle du sol. Revue d'Ecologie et de Biologie du Sol 6: 429 – 435.

Arpin, P. 1976. Cultivation medium for Mononchidae (Nematoda). Revue d'Ecologie et de Biologie du Sol 13: 629 – 634.

Arpin, P., and G. Kilbertus. 1981. Ultrastructure du contenu digestif et de l'épithelium intestinal chez quelques nématodes prédateurs (Mononchida) et bactériophages. Revue de Nématologie 4: 131 – 143.

Baldwin, J. G., M. Monson, and E. Paine. 1999. University of California (Riverside) quarantine and isolation facility for soil and water nematodes. Pp. 166 – 181. In R. P. Kahn, and S. B. Mathur, eds. Containment Facilities and Safeguards. St. Paul, MN: APS Press.

Baujard, P. 1995. Laboratory methods used for the study of the ecology and pathogenicity of Tylenchida, Longidoridae and Trichodoridae from rainy and semiarid tropics of West-Africa. Fundamental and Applied Nematology 18: 63 – 66.

Baujard, P., and B. Martiny. 1994. Transport of nematodes by wind in the peanut cropping area of Senegal, West Africa. Fundamental and Applied Nematology 17: 543 – 550.

Baujard, P., and B. Martiny. 1995. Ecology and pathogenicity of the Hoplolaimidae (Nemata) from the sahelian zone of West-Africa. 7. *Helicotylenchus dihystera* (Cobb, 1893) Sher, 1961 and comparison with *Helicotylenchus multicinctus* (Cobb, 1893) Golden, 1956. Fundamental and Applied Nematology 18: 503 – 511.

Bavaresco, L., and M. A. Walker, 1994. Techniques for successfully establishing *Xiphinema index* in dual culture with grape. American Journal Viticulture 45: 273 – 277.

Bedding, R. A. 1981. Low cost *in vitro* mass production of *Neoaplectana* and *Heterorhabditis* species (Nematoda) for field control of insect pests. Nematologica 27: 109 – 114.

Bedding, R. A. 1984. Large scale production, storage and transport of the insect-parasitic nematodes *Neoaplectana* spp. and *Heterorhabditis* spp. Annals of Applied Biology 104: 117 – 120.

Bedding, R. A., and R. J. Akhurst. 1975. A simple technique for the detection of insect parasitic nematodes in soil. Nematologica 21: 109 – 110.

Bilgrami, A. L. 1993. Analysis of the predation of *Aporcelaimellus nivalis* on prey nematodes from different prey trophic categories. Nematologica 39: 356 – 365.

Bilgrami, A. L., and M. S. Jairajpuri. 1988. Attraction of *Mononchoides longicaudatus* and *M. floridensis* (Nematoda: Diplogasterida) towards prey and factors influencing attraction. Revue de Nématologie 11: 195 – 202.

Bloemers, G. F., and M. Hodda. 1995. A method for extracting nematodes from a tropical forest soil. Pedobiologia 39: 331 – 343.

Boemare, N., A. Givaudan, M. Brehelin, and C. Laumond. 1997. Symbiosis and pathogenicity of nematode-bacterium complexes. Symbiosis 22: 21 – 45.

Bolla, R. I., and W. Jordan. 1982. Cultivation of the pine wilt nematode,

Bursaphelenchus xylophilus, in axenic culture media. Journal of Nematology 14: 377 – 381.

Brenner, S. 1974. The genetics of *Caenorhabditis elegans*. Genetics 77: 71 – 94.

Bridge, J., and P. J. Ham. 1985. A technique for the cryopreservation of viable juveniles of *Meloidogyne graminicola*. Nematologica 31: 185 – 189.

Brown, I. M., and R. Gaugler. 1996. Cold tolerance of steinernematid and heterorhabditid nematodes. Journal of Thermal Biology 21: 115 – 121.

Brown, I. M., and R. Gaugler. 1998. Survival of steinernematid nematodes exposed to freezing. Journal of Thermal Biology 23: 75 – 80.

Buecher, E. J., E. L. Hansen, and R. F. Myers. 1970. Continuous axenic culture of *Aphelenchoides* sp. Journal of Nematology 2: 189 – 190.

Cagnolo, S. R., M. M. A. de Doucet, M. E. Doucet, and J. A. Di Rienzo. 1999. A method for the extraction of aquatic Mermithidae (Nematoda) from riverbed samples. Nematology 1: 757 – 758.

Carta, L. K., and D. G. Carta. 2000. Nematode specific gravity profiles and applications to flotation extraction and taxonomy. Nematology 2: 201 – 210.

Castrol, M. E. A., and S. Ferraz. 1990. Multiplication of *Pratylenchus brachyurus*, *P. zeae*, *Radopholus similis* and *Tylenchorhynchus* sp. in monoxenic culture on alfalfa callus tissues. Nematologia Brasileira 14: 103 – 120.

Cobb, N. A. 1918. Estimating the nema population of soil. Agriculture Techniques Circular 1, Bureau of Plant Industry, US Department of Agriculture.

Cordero, A. D., and J. G. Baldwin. 1990. Effect of age on body wall cuticle morphology of *Heterodera schachtii* Schmidt females. Journal of Nematology 22: 356 – 361.

Cordero, A. D., J. G. Baldwin, and M. Mundo-Ocampo. 1991. Fine structure of the posterior cone of females of *Cactodera cacti* Filip'ev & Schuurmans Stekhoven (Nemata: Heteroderidae). Revue de Nématologie 14: 455 – 465.

Curran, J., C. Gilbert, and K. Butler. 1992. Routine cryopreservation of isolates of *Steinernema* and *Heterorhabditis* spp. Journal of Nematology 24: 269 – 270.

Dasgupta, D. R., S. Nand, and A. R. Seshadri. 1970. Culturing, embryology, and life cycle history studies on the lance nematode *Hoplolaimus indicus*. Nematologica 16: 235 – 248.

De Ley, P. 1992. The nematode community of a marginal soil at Cambérène, Senegal, with special attention to functional morphology and niche partitioning in the family Cephalobidae. Mededelingen van de Koninklijke Academie voor Wetenschappen, Letteren en Schone Kunsten van België – Klasse der Wetenschappen 53 (1991): 107 – 153.

Demeure, Y., D. W., Freckman, and S. D. Van Gundy. 1979. *In vitro* response of four species of nematodes to desiccation and discussion of this and related phenomena. Revue de Nématologie 2: 203 – 210.

Den Ouden, H. 1960. An improved method for making thin agar layers in polyethylene bags. Nematologica 5: 255 – 259.

De Waele, D., R. Wilken, and J. M. Lindeque. 1991. Response of potato cultivars to *Ditylenchus destructor* isolated from groundnut. Revue de Nématologie 14: 123 – 126.

Dougherty, E. C. 1949. The role of free-living nematodes in genetic research. Pp. 562 – 564. In Proceedings of the VIII International Congress of Genetics, Hereditas

Supplementary Volume.

Dougherty, E. C. 1960. Cultivation of Aschelminths, especially Rhabditid nematodes. Pp. 297 – 318. In J. N. Sasser, and W. R. Jenkins, eds. Nematology. Chapel Hill, NC: University of North Carolina Press.

Dougherty, E. C., E. L. Hansen, W. L. Nicholas, J. A. Mollett, and E. Yarwood. 1959. Axenic cultivation of *Caenorhabditis briggsae* (Nematoda: Rhabditidae) with unsupplemented and supplemented chemically defined media. Annals of the New York Academy of Sciences 77: 176 – 217.

Dutky, S. R. 1959. Insect microbiology. Pp. 175 – 200. In W. W. Umbreit, ed. Advances in Applied Microbiology, Vol. 1. New York: Academic Press.

Dye, D. W. 1968. A taxonomic study of the genus *Erwinia*: I. The "amylovora" group. New Zealand Journal of Science 11: 590 – 607.

Ehlers, R.-U., S. Stoessel, and U. Wyss. 1990. The influence of phase variants of *Xenorhabdus* spp. and *Escherichia coli* (Enterobacteriaceae) on the propagation of entomopathogenic nematodes of the genera *Steinernema* and *Heterorhabditis*. Revue de Nématologie 13: 417 – 424.

Ehlers, R.-U., S. Lunau, K. Krasomil-Osterfeld, and K. H. Osterfeld. 1998. Liquid culture of the entomopathogenic nematode-bacterium-complex *Heterorhabditis megidis/ Photorhabdus luminescens*. BioControl 43: 77 – 86.

Ehlers, R.-U., I. Niemann, S. Hollmer, O. Strauch, D. Jende, M. Shanmugasundaram, U. K. Mehta, S. K. Easwaramoorthy, and A. Burnell. 2000. Mass production potential of the bacto-helminthic biocontrol complex *Heterorhabditis indica – Photorhabdus luminescens*. Biocontrol Science and Technology 10: 607 – 616.

Eisenback, J. D., and U. Zunke, 1998. Extraction, culturing and microscopy. Pp. 140 – 155. In S. B. Sharma, ed. The Cyst Nematodes. Dordrecht: Kluwer Academic Publishers.

Esser, R. P. 1964. Nematode interactions in plates of non-sterile water agar. Proceedings of the Soil Crop Science Society of Florida 23 (1963): 121 – 138.

Estey, R. H. 1985. Culturing mycophagous nematodes. Pp. 149 – 152. In B. M. Zuckerman, W. F. Mai, and M. B. Harrison, eds. Plant Nematology Laboratory Manual. Amherst, MA: University of Massachusetts Agricultural Experiment Station Amherst.

Evans, A. A. F. 1970. Mass culture of mycophagous nematodes. Journal of Nematology 2: 99 – 100.

Eyre, J. M., and E. P. Caswell, 1991. Sterile culture of *Rotylenchulus reniformis* on tomato root with Gellan Gum as a supporting medium. Journal of Nematology 23: 229 – 231.

Folkertsma, R. T., J. Helder, F. J. Gommers, and J. Bakker. 1997. Storage of potato cyst nematodes at – 80 degrees C. Fundamental and Applied Nematology 20: 299 – 302.

Friedman, M. J. 1990. Commercial production and development. Pp. 153 – 172. In R. Gaugler, and H. K. Kaya, eds. Entomopathogenic Nematodes in Biological Control. Boca Raton, FL: CRC Press.

Fukushige, H. 1991. Propagation of *Bursaphelenchus xylophilus* (Nematoda: Aphelenchoididae) on fungi growing in pine shoots segments. Applied Entomology and

Zoology 26: 371 – 376.

Galway, N. J., and J. Curran. 1995. Cryopreservation of *Pratylenchus* spp. Journal of Nematology 27: 483 – 485.

Gbewonyo, K., S. P. Rohrer, L. Lister, B. Burgess, D. Cully, and B. Buckland. 1994. Large-scale cultivation of the free-living nematode *Caenorhabditis-elegans*. Biotechnology 12: 51 – 54.

Gerritsen, L. J. M., and P. H. Smits. 1997. The influence of *Photorhabdus luminescens* strains and form variants on the reproduction and bacterial retention of *Heterorhabditis megidis*. Fundamental and Applied Nematology 20: 317 – 322.

Giblin, R. M., and E. G. Platzer. 1987. Evaluation of culture media for axenic growth of *Romanomermis culicivorax*. Revue de Nématologie 10: 67 – 74.

Glaser, R. W. 1940. The bacteria-free culture of a nematode parasite. Proceedings of the Society of Experimental Biology and Medicine 43: 512 – 514.

Goho, S., and G. Bell. 2000. The ecology and genetics of fitness in *Chlamydomonas*. IX. The rate of accumulation of variation of fitness under selection. Evolution 54: 416 – 424.

Gordon, R., M. Cornect, T. L. Young, and K. T. Kean. 1990. Empirical and physiological assessment of *in vitro* growth in the mermithid nematode *Romanomermis culicivorax*. Canadian Journal of Zoology 68: 511 – 516.

Grewal, P. S., M. Matsuura, and V. Converse. 1997. Mechanisms of specificity of association between the nematode *Steinernema scapterisci* and its symbiotic bacterium. Parasitology 114: 483 – 488.

Grewal, P. S., V. Converse, and R. Georgis. 1999. Influence of production and bioassay methods on infectivity of two ambush foragers (Nematoda: Steinernematidae). Journal of Invertebrate Pathology 73: 40 – 44.

Hale, L. J. 1965. Biological Laboratory Data. London: Methuen.

Han, R. C., and R.-U. Ehlers. 1998. Cultivation of axenic *Heterorhabditis* juveniles and their response to non-specific *Photorhabdus luminescens* food signals. Nematologica 44: 425 – 435.

Han, R. C., and R.-U. Ehlers. 1999. Trans-specific nematicidal activity of *Photorhabdus luminescens*. Nematology 1: 687 – 693.

Han, R. C., L. Cao, and X. Liu. 1993. Effects of inoculum size, temperature and time on *in vitro* production of *Steinernema carpocapsae* Agriotos. Nematologica 39: 366 – 375.

Hansen, E. L., E. J. Buecher, and A. A. F. Evans. 1970. Axenic culture of *Aphelenchus avenae*. Nematologica 16: 328 – 329.

Heungens, K., D. Mungniéry, M. Van Montagu, G. Gheysen, and A. Niebel. 1996. A method to obtain disinfected *Globodera* infective juveniles directly from cysts. Fundamental and Applied Nematology 19: 91 – 93.

Hoagland, D. R., and D. I. Arnon. 1950. The water-culture method for growing plants without soil. Berkeley, CA: Agricultural Experiment Station, College of Agriculture.

Hooper, D. J. 1986a. Extraction of free-living stages from soil. Pp. 5 – 30. In J. F. Southey, ed. Laboratory Methods for Work with Plant and Soil Nematodes, 6th edn. Ministry of Agriculture, Fisheries and Food reference book 402. London: Her Majesty's Stationery Office.

Hooper, D. J. 1986b. Handling, fixing, staining and mounting nematodes. Pp. 59 – 80. In J. F. Southey, ed. Laboratory Methods for Work with Plant and Soil Nematodes, 6th edn. Ministry of Agriculture, Fisheries and Food reference book 402. London: Her Majesty's Stationery Office.

Hooper, D. J. 1986c. Culturing nematodes and related experimental techniques. Pp. 133 – 157. In J. F. Southey, ed. Laboratory Methods for Work with Plant and Soil Nematodes, 6 th edn. Ministry of Agriculture, Fisheries and Food reference book 402. London: Her Majesty's Stationery Office.

Hopper, B. E., and S. P. Meyers. 1966. Aspects of the life cycle of marine nematodes. Helgoländer wissenschaftliche Meeresuntersuchungen 13: 444 – 449.

Hu, K. J., J. X. Li, and J. M. Webster. 1999. Nematicidal metabolites produced by *Photorhabdus luminescens* (Enterobacteriaceae), bacterial symbiont of entomopathogenic nematodes. Nematology 1: 457 – 469.

Huang, X., and J. O. Becker. 1997. *In vitro* culture and feeding behavior of *Belonolaimus longicaudatus* on excised *Zea mays* roots. Journal of Nematology 29: 411 – 415.

Huang, X., and J. O. Becker. 1999. Life cycle and mating behavior of *Belonolaimus longicaudatus* in gnotobiotic culture. Journal of Nematology 31: 70 – 74.

Huettel, R. N. 1985. Carrot disc culture. Pp. 153 – 154. In B. M. Zuckerman, W. F. Mai, and M. B. Harrison, eds. Plant Nematology Laboratory Manual. Amherst, MA: University of Massachusetts Agricultural Experiment Station Amherst.

Huettel, R. N. 1990. Monoxenic culturing of plant-parasitic nematodes using carrot discs, callus tissue and root-explants. Pp. 163 – 172. In B. M. Zuckerman, W. F. Mai, and L. R. Krusberg, eds. Plant Nematology Laboratory Manual, 2nd edn. Amherst, MA: University of Massachusetts Agricultural Experiment Station Amherst.

Huettel, R. N., and R. V. Rebois. 1985. Culturing plant-parasitic nematodes using root explants. Pp. 155 – 158. In B. M. Zuckerman, W. F. Mai, and M. B. Harrison, eds. Plant Nematology Laboratory Manual. Amherst, MA: University of Massachusetts Agricultural Experiment Station Amherst.

Hunt, D. J., and P. De Ley. 1996. Nematodes in soils. Pp. 227 – 240. In G . S. Hall, ed. Methods for the Examination of Organismal Diversity in Soils and Sediments. Wallingford, UK: CAB International.

Ingham, R. E., J. A. Trofymow, E. R. Ingham, and D. C. Coleman. 1985. Interactions of bacteria, fungi, and their nematode grazers; effects on nutrient cycling and plant growth. Ecological Monographs 55: 119 – 140.

Ireholm, A. 1996. Long-term storage of *Heterodera avenae* cysts. Fundamental and Applied Nematology 19: 357 – 361.

Jikumaru, S., and K. Togashi. 2000. Temperature effects on the transmission of *Bursaphelenchus xylophilus* (Nemata: Aphelenchoididae) by *Monochamus alternatus* (Coleoptera: Cerambycidae). Journal of Nematology 32: 110 – 116.

Johnigk, S.-A., and R.-U. Ehlers. 1999. Endotokia matricida in hermaphrodites of *Heterorhabditis* spp. and the effect of food supply. Nematology 1: 717 – 726.

Khan, Z., A. L. Bilgrami, and M. S. Jairajpuri. 1995. A comparative study on the predation by *Allodorylaimus americanus* and *Discolaimus silvicolus* (Nematoda: Dorylaimida) on different species of plant-parasitic nematodes. Fundamental and Applied Nematology 18: 99 – 108.

Khanna, N., C. P. Cressmann, III, C. P. Tatara, and P. L. Willams. 1997. Tolerance of the nematode *Caenorhabditis elegans* to pH, salinity, and hardness in aquatic media. Archives of Environmental Contamination and Toxicology 32: 110 – 114.

Klinkenberg, C. H. 1963. Observations on the feeding habits of *Rotylenchulus reniformis*, *Pratylenchus crenatus*, *P. penetrans*, *Tylenchorhynchus dubius*, and *Hemicycliophora similis*. Nematologica 9: 502 – 506.

Ko, M. P., and S. D. Van Gundy. 1988. An alternative gelling agent for culture and studies of nematodes, bacteria, fungi and plant tissue. Journal of Nematology 20: 478 – 485.

Ko, M. P., D. P. Schmitt, and B. S. Sipes. 1996. Axenizing and culturing endomigratory plant-parasitic nematodes using Pluronic F127, including its effects on population dynamics of *Pratylenchus penetrans*. Journal of Nematology 28: 115 – 123.

Ko, M. P., D. P. Schmitt, and M. Saxby. 1997. Effects of container bases on the spread of *Meloidogyne incognita* in a Hawaiian ornamental nursery. Plant Disease 81: 607 – 613.

Koenning, S. R., and K. R. Barker. 1985. Gnotobiotic techniques for plant-parasitic nematodes. Pp. 49 – 64. In K. R. Barker, C. C. Carter, and J. N. Sasser, eds. An Advanced Treatise on *Meloidogyne* Volume II: Methodology. Raleigh, NC: North Carolina State University Graphics.

Kozdroj, J., and J. D. van Elsas. 2001. Structural diversity of microorganisms in chemically perturbed soil assessed by molecular and cytochemical approaches. Journal of Microbiological Methods 43: 197 – 212.

Krnjaic, D., and S. Krnjaic. 1970. Dispersion of nematodes by wind. Bolletino del Laboratorio di Entomologia Agraria "Filippo Silvestri" di Portici 30: 66 – 70.

Lee, J. J., J. H. Tietjen, R. J. Stone, W. A. Muller, J. Rullman, and M. McEnery. 1970. The cultivation and physiological ecology of members of salt marsh epiphytic communities. Helgoländer wissenschaftlicher Meeresuntersuchungen 20: 136 – 156.

Lewis, J. A., and J. T. Fleming. 1995. Basic culture methods. Pp. 3 – 29. In H. F. Epstein, and D. C. Shakes, eds. *Caenorhabditis elegans*: Modern Biological Analysis of an Organism. San Diego, CA: Academic Press.

Lin, M., X. Liu, W. Ling, and Z. Fan. 1997. Observations on the overwintering of *Ditylenchus destructor* in sweet potato and its tolerance against pH and salt concentration. Jiangsu Journal of Agricultural Sciences 13: 36 – 39.

Lu, N. C., and K. M. Goetsch. 1993. Carbohydrate requirement of *Caenorhabditis elegans* and the final development of a chemically defined medium. Nematologica 39: 303 – 311.

Luhring, K. A., and B. A. Mullens. 1997. Improved rearing methods for *Heleidomermis magnapapula* (Nematoda: Mermithidae), a larval parasite of *Culicoides variipennis sonorensis* (Diptera: Ceratopogonidae). Journal of Medical Entomology 34: 704 – 709.

Lunau, S., S. Stoessel, A. J. Schmidt-Peisker, and R. U. Ehlers. 1993. Establishment of monoxenic inocula for scaling up *in vitro* cultures of the entomopathogenic nematodes *Steinernema* spp. and *Heterorhabditis* spp. Nematologica 39: 385 – 399.

Maehara, N., and K. Futai. 1996. Factors affecting both the numbers of the pinewood nematode, *Bursaphelenchus xylophilus* (Nematoda: Aphelenchoididae), carried by the Japanese pine sawyer, *Monochamus alternatus* (Coleoptera: Cerambycidae), and the nematode's life history. Applied Entomology and Zoology 31: 443 – 452.

Maertens, D. 1975. Observations on the life cycle of *Prionchulus punctatus* (Cobb, 1917) (Nematoda) and some culture conditions. Biologisch Jaarboek Dodonaea 43: 197 – 218.

Mashela, P., R. McSorley, L. W. Duncan, and R. A. Dunn. 1991. Correlation of *Belonolaimus longicaudatus*, *Hoplolaimus galateus* and soil texture with yield of alyceclover (*Alysicarpus* spp.). Nematropica 21: 177 – 184.

Mateille, T. 1990. Monoxenic culture of banana-parasitic nematodes on *Musa acuminata* cv. Poyo shoots. Journal of Nematology 22: 608 – 611.

Meyer, S. L. F., D. J. Chitwood, and P. Crowley. 1997. Influence of soybean cultivar on reproduction of *Heterodera glycines* in monoxenic culture. Journal of Nematology 29: 389 – 394.

Meyers, S. P., and B. E. Hopper. 1966. Attraction of the marine nematode, *Metoncholaimus* sp., to fungal substrates. Bulletin of Marine Science 16: 142 – 150.

Meyers, S. P., W. A. Feder, and K. M. Tsue. 1964. Studies of the relationships among nematodes and filamentous fungi in the marine environment. Developments in Industrial Microbiology 5: 354 – 364.

Miller, J. H. 1992. A Short Course in Bacterial Genetics: A Laboratory Manual and Handbook for *Escherichia Coli* and Related Bacteria. Cold Spring Harbor: Cold Spring Harbor Laboratory Press.

Moens, T., and M. Vincx. 1997. Observations on the feeding ecology of estuarine nematodes. Journal of the Marine Biology Association of the United Kingdom 77: 211 – 227.

Moens, T., and M. Vincx. 1998. On the cultivation of free-living marine and estuarine nematodes. Helgoländer Meeresuntersuchungen 52: 115 – 139.

Moens, T., A. Coomans, and D. De Waele. 1996. Effect of temperature on the *in vitro* reproduction of *Aphelenchoides rutgersi*. Fundamental and Applied Nematology 19: 47 – 52.

Moens, T., P. M. J. Herman, L. Verbeeck, M. Steyaert, and M. Vincx. 2000. Laboratory experiments on predation rates and prey selectivity in two predacious estuarine nematode species. Marine Ecology Progress Series 205: 185 – 193.

Moody, H., B. F. Lownsbery, and J. M. Ahmed. 1973. Culture of the root-lesion nematode *Pratylenchus vulnus* on carrot disks. Journal of Nematology 5: 225 – 226.

Mundo-Ocampo, M. 1983. Host parasite relationships of the Heteroderidae (Tylenchida – Heteroderoidea) with emphasis in host responses. Ph. D. Thesis. University of California, Riverside, CA.

Mundo-Ocampo, M. 1993. A practical method for culturing *Globodera* and *Punctodera*. Nematropica 23: 128 (Abstr.).

Myers, R. F. 1992. Modification of the oligidic medium for axenic culture of *Aphelenchoides rutgersi*. Journal of Nematology 24: 275 – 280.

Myers, R. F., E. J. Buecher, and E. L. Hansen. 1971. Oligidic medium for axenic culture of *Aphelenchoides* sp. Nematologica 3: 197 – 198.

Neihardt, F. C. 1996. *Escherichia coli* and *Salmonella*: Cellular and Molecular Biology, 2nd edn. Washington, DC: American Society of Microbiology.

Nelmes, A. J. 1974. Evaluation of the feeding behavior of *Prionchulus punctatus* (Cobb), a nematode predator. Journal of Animal Ecology 43: 553 – 565.

Nicholas, W. L. 1984. The Biology of Free-living Nematodes, 2 nd edn. Oxford: Oxford University Press.

Nigon, V. 1949. Les modalités de la reproduction et le déterminisme du sexe chez quelques nématodes libres. Annales des Sciences Naturelles – Biologie et Zoologie Animale 11: 1 – 132.

Notley-McRobb, L., and T. Ferenci. 1999. Adaptive mgl-regulatory mutations and genetic diversity evolving in glucose-limited *Escherichia coli* populations. Environmental Microbiology 1: 33 – 43.

Nugent, M. J., S. A. O'Leary, and A. M. Burnell. 1996. Optimised procedures for the cryopreservation of different species of *Heterorhabditis*. Fundamental and Applied Nematology 19: 1 – 6.

O'Bannon, J. H., and A. L. Taylor. 1968. Migratory endoparasitic nematodes reared on carrot discs. Phytopathology 58: 385.

Orion, D., and M. Bar-Eyal. 1995. Observations on the life cycle and behavior of *Helicotylenchus multicinctus* in monoxenic culture on *Arabidopsis thaliana*. Nematropica 25: 67 – 70.

Overgaard Nielsen, C. 1949. The Soil Inhabiting Nematodes. Studies on the Soil Microfauna-Natura Jutlandica 2. Aarhus: Naturhistorisk Museum.

Peng, Y., J. Song, G. Tian, Q. Xue, F. Ge, J. Yang, and Q. Shi. 1998. Field evaluations of *Romanomermis yunanensis* (Nematoda: Mermithidae) for control of Culicinae mosquitoes in China. Fundamental and Applied Nematology 21: 227 – 232.

Peters, A., D. H. Gouge, R.-U. Ehlers, and N. G. M. Hague. 1997. Avoidance of encapsulation by *Heterorhabditis* spp. infecting larvae of *Tipula oleracea*. Journal of Invertebrate Pathology 70: 161 – 164.

Petersen, J. J. 1985. Nematodes as biological-control agents. 1. Mermithidae. Advances in Parasitology 24: 307 – 344.

Platzer, E. G. 1977. Culture media for nematodes. Pp. 29 – 59. In M. Rechcigl, Jr, ed. CRC Handbook Series in Nutrition and Food – Section G: Diets, Culture Media, Food Supplements – Volume II: Food Habits of, and Diets for Invertebrates and Vertebrates – Zoo Diets. Cleveland, OH: CRC Press.

Plowright, R. A., and T. E. Akehurst. 1992. Monoxenic culture of the Ufra nematode *Ditylenchus angustus*. Fundamental and Applied Nematology 15: 327 – 330.

Popiel, I., and E. M. Vasquez. 1991. Cryopreservation of *Steinernema carpocapsae* and *Heterorhabditis bacteriophora*. Journal of Nematology 23: 432 – 437.

Popiel, I., K. D. Holtemann, I. Glazer, and C. Womersley. 1988. Commercial storage and shipment of entomogenous nematodes. World Intellectual Property Organisation, International Patent No. WO8801134A1.

Qiu, L. H., and R. Bedding. 1999. Low temperature induced cryoprotectant synthesis by the infective juveniles of *Steinernema carpocapsae*: Biological significance and mechanisms involved. Cryo-Letters 20: 393 – 404.

Rebois, R. V., and R. N. Huettel. 1986. Population dynamics, root penetration, and feeding behavior of *Pratylenchus agilis* in monoxenic root cultures of corn, tomato and soybean. Journal of Nematology 18: 392 – 397.

Rebois, R. V., R. N. Huettel, and A. Lauritis. 1984. A comparison of media for the monoxenic culture of corn cyst, soybean cyst, reniform, root knot, and lesion nematodes on root explants. First International Congress of Nematology Aug. 5 – 10, 1984. Guelph, Ontario, Canada, p. 85 (Abstr.).

Reise, R. W., R. N. Huettel, and R. M. Sayre. 1987. Carrot callus-tissue for culture of endoparasitic nematodes. Journal of Nematology 19: 387 – 389.

Riedel, R. M. 1985. Establishing *Pratylenchus* spp. in monoxenic culture on alfalfa (*Medicago sativa*) callus tissue. Pp. 159 – 162. In B. M. Zuckerman, W. F. Mai, and M. B. Harrison, eds. Plant Nematology Laboratory Manual. Amherst, MA: University of Massachusetts Agricultural Experiment Station Amherst.

Riga, E., and J. M. Webster. 1991. Cryopreservation of the pinewood nematode, *Bursaphelenchus* spp. Journal of Nematology 23: 438 – 440.

Rothstein, M. 1974. Practical methods for axenic culture of free-living nematodes *Turbatrix aceti* and *Caenorhabditis briggsae*. Comparative Biochemistry and Physiology B 49: 669 – 678.

Santamarina Mijares, A., and A. C. Bellini. 2000. Produccion masiva de *Romanomermis iyengari* (Nematoda: Mermithidae) y su aplicacion en criaderos de anofelinos en Boa Vista (Roraima), Brasil. Pan-American Journal of Public Health 7: 155 – 161.

Sanwal, K. C. 1959. A simple method for rearing pure populations of the foliar nematode, *Aphelenchoides ritzemabosi*, in the laboratory. Canadian Journal of Zoology 37: 707 – 711.

Schirocki, A. G., and N. G. M. Hague. 1997. The effect of selective culture of *Steinernema feltiae* at low temperature on establishment, pathogenicity, reproduction and size of infective juveniles. Nematologica 43: 481 – 489.

Sijmons, P. C., F. M. W. Grundler, N. Von Mende, P. R. Burrows, and U. Wyss. 1991. *Arabidopsis thaliana* as a new model host for plant-parasitic nematodes. The Plant Journal 1: 245 – 254.

Simons, W. 1973. Nematode survival in relation to soil moisture. Mededelingen van de Landouwhogeschool Wageningen 73: 1 – 85.

Souza, M. R. 1999. Esophageal gland development linked to the life cycle of *Nacobbus aberrans* (Nemata: Pratylenchidae). Ph. D. Dissertation University of California. Riverside, CA.

Steiner, G., and H. Heinly. 1922. The possibility of control of *Heterodera radicicola* and other plant-injurious nemas by means of predatory nemas, especially by *Mononchus papillatus* Bastian. Journal of the Washington Academy of Sciences 12: 367 – 386.

Stiernagle, T. 1999. Maintenance of *C. elegans*. Pp. 51 – 67. In I. A. Hope, ed. *C. elegans* – A Practical Approach. The practical approach series 213. Oxford: Oxford University Press.

Stoessel, S., and R.-U. Ehlers. 1992. Mating of *Heterorhabditis* spp. in submerged and emerged *in vitro* cultures. Nematologica 38: 437.

Strauch, O., and R.-U. Ehlers. 1998. Food signal production of *Photorhabdus luminescens* inducing the recovery of entomopathogenic nematodes *Heterorhabditis* spp. in liquid culture. Applied Microbiology and Biotechnology 50: 369 – 374.

Strauch, O., S. Stoessel, and R.-U. Ehlers. 1994. Culture conditions define automictic or amphimictic reproduction in entomopathogenic rhabditid nematodes of the genus *Heterorhabditis*. Fundamental and Applied Nematology 17: 575 – 582.

Stuart, R. J., and R. Gaugler. 1996. Genetic adaptation and founder effect in laboratory populations of the entomopathogenic nematode *Steinernema glaseri*. Canadian Journal of

Zoology – Revue Canadienne de Zoologie 74: 164 – 170.

Sudhaus, W., and K. Kiontke. 1996. Phylogeny of *Rhabditis* subgenus *Caenorhabditis* (Rhabditidae, Nematoda). Journal of Zoological Systematics and Evolutionary Research 34: 217 – 233.

Sulston, J., and J. Hodgkin. 1988. Methods. Pp. 587 – 606. In W. B. Wood, ed. The Nematode *Caenorhabditis elegans*. New York: Cold Spring Harbor Laboratory Press.

Surrey, M. R. 1996. The effect of rearing method and cool temperature acclimation on the cold tolerance of *Heterorhabditis zealandica* infective juveniles (Nematoda: Heterorhabditidae). Cryo-letters 17: 313 – 320.

Taylor, A. L., J. Feldemesser, and W. A. Feder. 1957. A new technique for preliminary screening of nematocides. Plant Disease Reporter 41: 527 – 530.

Thirugnanam, M. 1976. *Aphelenchoides-rutgersi* growth requirements for heme and purine nucleotides in crude chick-embryo extract. Experimental Parasitology 40: 149 – 157.

Tomminen, J., S. Halik, and D. R. Bergdahl. 1991. Incubation-temperature and time effects on life stages of *Bursaphelenchus xylophilus* in wood chips. Journal of Nematology 23: 477 – 484.

Torsvik, V., F. L. Daae, R. A. Sandaa, and L. Ovreas. 1998. Novel techniques for analyzing microbial diversity in natural and perturbed environments. Journal of Biotechnology 64: 53 – 62.

Traunspurger, W. 2000. The biology and ecology of lotic nematodes. Freshwater Biology 44: 29 – 45.

Triantaphyllou, A. C., and E. McCabe. 1989. Efficient preservation of root-knot and cyst nematodes in liquid nitrogen. Journal of Nematology 21: 423 – 426.

van der Beek, H. J. G., W. B. J. Veldhuis, C. Zijlstra, and C. H. van Silfhout. 1996. Preservation of *Meloidogyne hapla* and *M. chitwoodi* in liquid nitrogen: Differences in response between populations. Fundamental and Applied Nematology 19: 227 – 234.

Vancoppenolle, B., G. Borgonie, and A. Coomans. 1999. Generation times of some free-living nematodes cultured at three temperatures. Nematology 1: 15 – 18.

Vanfleteren, J. 1980. Nematodes as nutritional models. Pp. 47 – 79. In B. M. Zuckerman, ed. Nematodes as Biological Models, Vol. 3. New York: Academic Press.

Vanfleteren, J. R., and A. De Vreese. 1995. The gerontogenes *age-1* and *daf-2* determine metabolic-rate potential in aging *Caenorhabditis elegans*. FASEB Journal 9: 1355 – 1361.

Viaud, M., A. Pasquier, and Y. Brygoo. 2000. Diversity of soil fungi studied by PCR-RFLP of ITS. Mycological Research 104: 1027 – 1032.

Vranken, G., D. Van Brussel, R. Vanderhaeghen, and C. Heip. 1984. Research on the development of a standardized ecotoxicological test on marine nematodes I. Culturing conditions and criteria for two monhysterids, *Monhystera disjuncta* and *Monhystera microphthalma*. Pp. 159 – 184. In G. Persoone, E. Jaspers, and C. Claus, eds. Exotoxicological Testing for the Marine Environment, Vol. 2. Ghent: State University of Ghent.

Wang, J. X., and R. A. Bedding. 1996. Population development of *Heterorhabditis bacteriophora* and *Steinernema carpocapsae* in the larvae of *Galleria mellonella*. Fundamental and Applied Nematology 19: 363 – 367.

Wang, J. X., and R. A. Bedding. 1998. Population dynamics of *Heterorhabditis bacteriophora* and *Steinernema carpocapsae in vitro* monoxenic solid culture.

Fundamental and Applied Nematology 21: 165 – 171.
Wang, K. H., B. S. Sipes, and A. R. Kuehnle. 1997. Effect of soilless media on the growth of *Anthurium andraeanum* infected by *Radopholus similis*. Nematropica 27: 77 – 84.
Ward, D. M., R. Weller, and M. M. Bateson. 1990. 16s ribosomal-RNA sequences reveal numerous uncultured microorganisms in a natural community. Nature 345: 63 – 65.
Warren, J. E., O. R. Edwards, and M. J. Linit. 1995. Influence of bluestain fungi on laboratory rearing of pinewood nematode infested beetles. Fundamental and Applied Nematology 18: 89 – 92.
Wharton, D. A., and M. R. Surrey. 1994. Cold tolerance mechanisms of the infective larvae of the insect parasitic nematode, *Heterorhabditis zealandica* Poinar. Cryo-Letters 15: 353 – 360.
White, G. F. 1927. A method for obtaining infective nematode larvae from cultures. Science 66: 302 – 303.
Womersley, C. Z. 1990. Dehydration survival and anhydrobiotic potential of entomopathogenic nematodes. Pp. 117 – 130. In R. Gaugler, and H. K. Kaya, eds. Entomopathogenic Nematodes in Biological Control. Boca Raton, FL: CRC Press.
Womersley, C. Z., D. A. Wharton, and L. M. Higa. 1998. Survival biology. Pp. 271 – 302. In R. N. Perry, and D. J. Wright, eds. The Physiology and Biochemistry of Free-living and Plant-parasitic Nematodes. Wallingford, UK: CAB International.
Wouts, W. M. 1981. Mass production of the entomogenous nematode *Heliorhabditis heliothidis* (Nematoda: Heterorhabditidae) on artificial media. Journal of Nematology 13: 467 – 469.
Wouts, W. M. 1991. *Steinernema (Neoaplectana)* and *Heterorhabditis* species. Pp. 855 – 897. In W. R. Nickle, ed. Manual of Agricultural Nematology. New York: Marcel Dekker.
Wyss, U. 1978. Root and cell response to feeding by *Xiphinema index*. Nematologica 24: 159 – 166.
Yang, H. W., H. Jian, S. G. Zhang, and G. Y. Zhang. 1997. Quality of the entomopathogenic nematode *Steinernema carpocapsae* produced on different media. Biological Control 10: 193 – 198.
Yeates, G. W., and T. Bongers. 1999. Nematode diversity in agroecosystems. Agriculture Ecosystems and Environment 74: 113 – 135.
Zuckerman, B. M. 1971. Gnotobiology. Pp. 159 – 184. In B. M. Zuckerman, W. F. Mai, and R. A. Rohde, eds. Plant-parasitic Nematodes. Vol. 2. New York: Academy Press.
Zuckerman, B. M., W. F. Mai, and M. B. Harrison. 1985. Plant Nematology Laboratory Manual. Amherst, MA: University of Massachusetts Agricultural Experiment Station Amherst.

Acknowledgements We are greatly indebted to Gaetan Borgonie, Ralf-Udo Ehlers, Robert McSorley, Tom Moens, and Jacques Vanfleteren for critical reading and for many invaluable suggestions.

Index

Index

Printed and bound by CPI Group (UK) Ltd, Croydon, CR0 4YY

16/09/2026

14963431-0007